# Pesticide residues in food 2006

## Evaluations
## Part I – Residues

FAO
PLANT
PRODUCTION
AND PROTECTION
PAPER

**189/1**

**Sponsored jointly by FAO and WHO**

Joint meeting of the
FAO Panel of Experts on Pesticide Residues
in Food and the Environment
and the
WHO Core Assessment Group
Rome, Italy, 3–12 October 2006

WORLD HEALTH ORGANIZATION
FOOD AND AGRICULTURE ORGANIZATION OF THE UNITED NATIONS
Rome, 2007

Monographs containing summaries or residue data and toxicological
data considered at the 2006 JMPR, together with recommendations, are available upon
request from FAO or WHO under the title:

Pesticide residues in food 2006
Evaluations
Part I: Residues
FAO Plant Production and Protection Paper
and
Part II: Toxicology
WHO

This report contains the collective views of two international groups of
experts and does not necessarily represent the decisions or the stated
policy of the Food and Agriculture Organization of the United Nations or
of the World Health Organization.

INTERNATIONAL PROGRAMME ON CHEMICAL SAFETY

The preparatory work for the toxicological evaluation of pesticide residues carried out
by the WHO Expert Group on Pesticide Residues for consideration by the FAO/WHO Joint
meeting on Pesticide Residues in Food and the Environment is actively supported by the
International Programme on Chemical Safety (IPCS).
IPCS is a joint venture of the United Nations Environment Progamme, the International
Labour Organization and the World Health Organization. One of the main objectives
of IPCS is to carry out and disseminate evaluations of the effects of chemicals on human
health and the quality of the environment.

ISBN 978-92-5-105722-3

# CONTENTS

pages

Participants.................................................................................................................. v

Abbreviations............................................................................................................... ix

Use of JMPR reports and evaluations by registration authorities ...............................xiii

Introduction............................................................................................................... xv

## Monographs

### Volume 1

Acephate (095) ................................................................................................................ 1

Aldicarb (117) ................................................................................................................ 7

Aminopyralid (220) [2/]................................................................................................... 9

Bifenazate (219) [2/] ........................................................................................................ 85

Boscalid (221) [2]............................................................................................................ 159

Chlorpyrifos (017) .......................................................................................................... 301

Diazinon (022) ................................................................................................................ 305

Dimethoate (027) ............................................................................................................ 309

Disulfoton (074)............................................................................................................... 313

Endosulfan (032) [1/] ...................................................................................................... 331

Fenamiphos (085) ............................................................................................................ 503

### Volume 2

Fenpropathrin (185) ....................................................................................................... 523

Fludioxonil (211) ............................................................................................................. 535

Imidacloprid (206) .......................................................................................................... 547

Methoxyfenozide (209)................................................................................................... 551

Pirimicarb (101) [1/] ....................................................................................................... 555

Propamocarb (148) [1/]................................................................................................... 719

Propargite (113) .............................................................................................................. 803

Propiconazole (160)......................................................................................................... 817

Pyraclostrobin (210)........................................................................................................ 821

Quinoxyfen (222) [2/] ..................................................................................................... 867

Thiabendazole (065) ........................................................................................................ 975

Thiacloprid (223)[2/]........................................................................................................ 983

Annex I, Previous FAO and WHO documents ............................................................. 1149

[1/] Evaluated for the Periodic Review Programme of the Codex Committee on Pesticide Residues.
[2/] New compound.

# LIST OF PARTICIPANTS

## 2006 JOINT FAO/WHO MEETING ON PESTICIDE RESIDUES
## ROME, 3–12 OCTOBER 2006

### FAO Members

Dr Ursula Banasiak, Federal Institute for Risk Assessment, Thielallee 88-92, D-14195 Berlin, Germany
Tel.: (49 30) 8412 3337; Fax: (49 30) 8412 3008; E-mail: u.banasiak@bfr.bund.de

Professor Eloisa Dutra Caldas, University of Brasilia, College of Health Sciences, Pharmaceutical Sciences Department, Campus Universitário Darci Ribeiro, 70919-970 Brasília/DF, Brazil
Tel.: (55 61) 3307 3671; Fax: (55 61) 3273 0105; E-mail: eloisa@unb.br

Mr Stephen Funk, Health Effects Division (7509P), United States Environmental Protection Agency, 1200 Pennsylvania Avenue NW, Washington DC 20460, USA *(FAO Chairman)*
Tel.: (1 703) 305 5430; Fax: (1 703) 305 0871; E-mail: funk.steve@epa.gov

Mr Denis J. Hamilton, Principal Scientific Officer, Biosecurity, Department of Primary Industries and Fisheries, PO Box 46, Brisbane, QLD 4001, Australia
Tel.: (61 7) 3239 3409; Fax: (61 7) 3211 3293; E-mail: denis.hamilton@dpi.qld.gov.au

Mr David Lunn, Programme Manager (Residues–Plants), Export Standards Group, New Zealand Food Safety Authority, PO Box 2835, Wellington, New Zealand
Tel.: (644) 463 2654; Fax: (644) 463 2675; E-mail: dave.lunn@nzfsa.govt.nz

Dr Bernadette C. Ossendorp, Centre for Substances and Integrated Risk Assessment, National Institute of Public Health and the Environment (RIVM), Antonie van Leeuwenhoeklaan 9, PO Box 1, 3720 BA Bilthoven, the Netherlands
Tel.: (31 30) 274 3970; Fax: (31 30) 274 4475; E-mail: bernadette.ossendorp@rivm.nl

Dr Yukiko Yamada, Director, Food Safety and Consumer Policy Division, Food Safety and Consumer Affairs Bureau, Ministry of Agriculture, Forestry and Fisheries, 1-2-1 Kasumigaseki, Tokyo 100-8950, Japan
Tel.: (81) 3 3591 4963; Fax: (81) 3 3597 0329; E-mail:yukiko_yamada@nm.maff.go.jp

### WHO Members

Professor Alan R. Boobis, Experimental Medicine & Toxicology, Division of Medicine, Faculty of Medicine, Imperial College London, Hammersmith Campus, Ducane Road, London W12 0NN, England *(WHO Chairman)*
Tel.: (44 20) 8383 3221; Fax: (44 20) 8383 2066; E-mail: a.boobis@imperial.ac.uk

Dr Les Davies, Chemical Review, Australian Pesticides & Veterinary Medicines Authority, PO Box E240, Kingston ACT 2604, Australia
Tel.: (61 2) 6272 3898; Fax: (61 2) 6272 3218; E-mail: les.davies@apvma.gov.au

Dr Vicki L. Dellarco, Office of Pesticide Programs (7509P), United States Environmental Protection Agency, Health Effects Division; 1200 Pennsylvania Avenue NW, Washington, DC 20460, USA *(WHO Rapporteur)*
Tel.: (1 703) 305 1803; Fax: (1 703) 305 5147; E-mail: dellarco.vicki@epa.gov

Dr Helen Hakansson, Institute of Environmental Medicine, Karolinska Institutet, Unit of Environmental Health Risk Assessment, Box 210, Nobels väg 13, S-171 77 Stockholm, Sweden
Tel.: (46 8) 5248 7527; Fax: (46 8) 3438 49, E-mail: Helen.Hakansson@ki.se

Dr Angelo Moretto, Department of Occupational Medicine and Public Health, University of Milan, ICPS Ospedale Sacco, Via Grassi 74, 20157 Milan, Italy
Tel.: (39 02) 3568661; Fax: (39 02) 38203163; E-mail: angelo_moretto@fastwebnet.it

Professor David Ray, MRC Applied Neuroscience Group, Biomedical Sciences, University of Nottingham, Queens Medical Centre, Nottingham NG7 2UH, England
Tel.: (44 115) 82 30138; Fax: (44 115) 823 0142; E-mail: David.Ray@nottingham.ac.uk

Dr Roland Solecki, Safety of Substances and Preparations, Coordination and Overall Assessment, Federal Institute for Risk Assessment, Thielallee 88-92, D-14195 Berlin, Germany
Tel.: (49 188) 8412 3232; Fax: (49 188) 8412 3260; E-mail: Roland.Solecki@bfr.bund.de

Dr Maria Tasheva, National Center of Public Health Protection (NCPHP), 15 Iv. Ev. Geshov boulevard, 1431 Sofia, Bulgaria
Tel.: (3592) 954 11 97; Fax: mtasheva@aster.net

**Secretariat**

Mr Árpád Ambrus, Consultant, Homezo utca 41, H-1221, Budapest, Hungary *(FAO Temporary Adviser)*
Tel.: (36 20) 2096785; E-mail: aa22632003@yahoo.co.uk

Dr Raj Bhula, Pesticide Residues Manager, Australian Pesticides and Veterinary Medicines Authority, PO Box E240, ACT 2604, Australia *(FAO Temporary Adviser)*
Tel.: (61 2) 6271 6551; Fax: (61 2) 6272 3551; E-mail: raj.bhula@apvma.gov.au

Dr Hong Chen, International Program Manager, USDA/IR-4 Headquarters, 500 College Road East, Suite 201W, Princeton, NJ 08540, USA *(FAO Temporary Adviser)*
Tel.: (1 732) 932 9575 ext 4627; Fax: (1 609) 514 2612; E-mail: hchen@AESOP.Rutgers.edu

Mr Bernard Declercq, 13 impasse du court Riage, 91360 Epinay sur Orge, France *(FAO Temporary Adviser)*
Tel.: (33 1) 64488369; Fax: (33 1) 64488369; E-mail: bernard-declercq@wanadoo.fr

Dr Ian C Dewhurst, Pesticides Safety Directorate, Mallard House, Kings Pool, 3 Peasholme Green, York YO1 7PX, England *(WHO Temporary Adviser)*
Tel.: (44 1904) 455 890; Fax: (44 1904) 455 711; E-mail: ian.dewhurst@psd.defra.gsi.gov.uk

Dr Ronald D. Eichner, 13 Cruikshank Street, Wanniassa ACT 2903, Australia *(FAO Editor)*
Tel.: (61 2) 629 62118; Fax (61 2) 629 46564; E-mail: eichners@tpg.com.au

Dr D. Kanungo, Additional DG, Directorate General of Health Services, Ministry of Health and Family Welfare, West, Block No. 1, R.K. Puram, New Delhi, India *(WHO Temporary Adviser)*
Tel.: (91 11) 261 01 268; Fax: (91 11) 261 89 307; E-mail: dkanungo@nic.in

Dr Sandhya Kulshrestha, Secretary, Central Insecticides Board and Registration Committee, Directorate of Plant Protection, Quarantine & Storage, Department of Agriculture and Cooperation, Ministry of Agriculture, NH IV, Faridabad-121001, India *(WHO Temporary Adviser)*
Tel.: (91 129) 241 3002; Fax: (91 129) 241 21 25/(91 11) 2 338 4182; E-mail: skulsh57@yahoo.co.in

Dr Yibing He, Pesticide Residue Division, Institute for the Control of Agrochemicals, Ministry of Agriculture, Building 22, Maizidian Street, Chaoyang District, Beijing 100026, China *(FAO Temporary Adviser)*
Tel.: (86 10) 641 94253; Fax: (86 10) 641 94252; E-mail: heyibing@agri.gov.cn

Dr Chaido Lentza-Rizos, National Agricultural Research Foundation, 1 Sofocli Venizelou Street, 141 23 Lycovrissi, Athens, Greece *(FAO Temporary Adviser)*
Tel.: (30) 210 2819 728; Fax: (30) 210 2818 735; E-mail: rizos_ch@otenet.gr

Dr Manfred Lützow, Feldhofweg 38, CH-5432 Neuenhof, Switzerland *(FAO Consultant)*
Tel.: (41 56) 406 2357; Fax: (41 56) 406 2359; E-mail: Manfred.luetzow@luetzow.ch

Dr Katerina Mastovska, Eastern Regional Research Center (ERRC), Agricultural Research Service (ARS), United States Department of Agriculture, 600 E. Mermaid Lane, Wyndmoor, PA 19038, USA *(FAO Temporary Adviser)*
Tel.: (1 215) 233 6645; Fax: (1 215) 233 6642; E-mail: katerina.mastovska@ars.usda.gov

Dr Heidi Mattock, 21bis rue du Mont Ouest, 38230 Tignieu-Jameyzieu, France *(WHO Editor)*
Tel.: (33 478) 32 0758; E-mail: heidimattock@yahoo.com

Dr Douglas B. McGregor, Toxicity Evaluation Consultants, 38 Shore Road, Aberdour KY3 0TU, Scotland *(WHO Temporary Adviser)*
Tel.: (44 1383) 860901; Fax: (44 1383) 860901; E-mail: mcgregortec@btinternet.com

Dr Utz Mueller, Principal Toxicologist, Risk Assessment – Chemical Safety, Food Standards Australia New Zealand, PO Box 7186, Canberra, BC ACT 2610, Australia *(WHO Temporary Adviser)*
Tel.: (61 2) 6271 2631; Fax: (61 2) 6271 2278; E-mail utz.mueller@foodstandards.gov.au

Dr Rudolf Pfeil, Toxicology of Pesticides, Federal Institute for Risk Assessment, Thielallee 88-92, D-14195 Berlin, Germany *(WHO Temporary Adviser)*
Tel.: (49 30) 8412 3828; Fax: (49 30) 8412 3260; E-mail: rudolf.pfeil@bfr.bund.de

Mr Derek Renshaw, Food Standards Agency, Aviation House 125 Kingsway, London WC2B 6NH, England *(WHO Temporary Adviser)*
Tel.:(4420)72768505; Fax: (4420) 72768513; E-mail:derek.renshaw@foodstandards.gsi.gov.uk

Mr Christian Sieke, Federal Institute for Risk Assessment, Thielallee 88-92, D-14195 Berlin, Germany *(FAO Temporary Adviser)*
Tel.: (49 1) 88 8412 3336; Fax: (49 1) 88 8412 3008; E-mail: c.sieke@bfr.bund.de

Dr Atsuya Takagi, Division of Toxicology, Chief of the Third Section, National Institute of Health Sciences, 1-18-1 Kamiyoga, Setagaya-ku, Tokyo 158-8501, Japan *(WHO Temporary Adviser)*
Tel.: (81 3) 3700 9646; Fax: (81 3) 3707 6950; E-mail: takagi@nihs.go.jp

Dr Prakashchandra V. Shah, US Environmental Protection Agency (EPA), Mail Stop: 7509C, 1200 Pennsylvania Ave., NW, Washington DC 20460, USA *(WHO Temporary Adviser)*
Tel.: (1 703) 308 1846; Fax: (1 703) 305 5147; E-mail: Shah.PV@epa.gov

Dr Angelika Tritscher, WHO Joint Secretary, International Programme on Chemical Safety, World
    Health Organization, 1211 Geneva 27, Switzerland *(WHO Joint Secretary)*
    Tel.: (41 22) 791 3569; Fax: (41 22) 791 4848; E-mail: tritschera@who.int

Dr Gero Vaagt, FAO Joint Secretary, FAO Plant Protection Service (AGPP), Viale delle Terme di
    Caracolla, 00153 Rome, Italy *(FAO Joint Secretary)*
    Tel.: (office): (39 06) 5705 5757; Phone (mobile): (39 348) 083 1999; Fax: (39 06) 5705
    6347/3224; E-mail: Gero.Vaagt@fao.org

Dr Christiane Vleminckx, Toxicology Division, Scientific Institute of Public Health, FPS – Public
    Health, Food Chain Safety and Environment, Rue Juliette Wytsman, 14, B-1050 Brussels,
    Belgium *(WHO Temporary Adviser)*
    Tel.: (32 2) 642 5351; Fax: (32 2) 642 5224; E-mail: c.vleminckx@iph.fgov.be

Dr Gerrit Wolterink, Centre for Substances & Integrated Risk Assessment, National Institute of Public
    Health and the Environment (RIVM), Antonie van Leeuwenhoeklaan 9, PO Box 1, 3720 BA
    Bilthoven, the Netherlands *(WHO Temporary Adviser)*
    Tel.: (31 30) 274 4531; Fax: (31 30) 274 4475; E-mail: Gerrit.Wolterink@rivm.nl

Dr Jürg Zarn, Swiss Federal Office of Public Health, Food Toxicology Section, Stauffacherstrasse
    101, CH-8004 Zurich, Switzerland *(WHO Temporary Adviser)*
    Tel.: (41 43) 322 21 93; Fax: (41 43) 322 21 99; E-mail: Juerg.Zarn@bag.admin.ch

# ABBREVIATIONS

(Well-known abbreviations in general use are not included. Specific abbreviations for pesticide degradation products, etc., may be used in the monographs and these are either identified where first used or in a table within the monograph. Two-letter codes for pesticide formulations are given in the Manual on development and use of FAO and WHO specifications for pesticides, 1st Ed., FAO Plant Production and Protection Paper 173, FAO, Rome, 2002.)

| | |
|---|---|
| AChE | anti-acetylcholinesterase |
| ACN | acetonitrile |
| ADI | acceptable daily intake |
| AFID | alkali flame-ionization detection or detector (equivalent to TSD, forerunner of NPD) |
| ai | active ingredient |
| AR | Applied radioactivity |
| ARfD | acute reference dose |
| AUC | area under the curve for concentration–time |
| BBCH | Biologische Bundesanstalt, Bundessortenamt and Chemical industry. |
| $BMDL_{10}$ | benchmark-dose lower 95% confidence level |
| bw | body weight |
| CA | Chemical Abstracts |
| CAC | Codex Alimentarius Commission |
| CAS | Chemical Abstracts Services |
| CCN | Codex classification number (for compounds or commodities) |
| CCPR | Codex Committee on Pesticide Residues |
| CCRVDF | Codex Committee on Residue of Veterinary Drugs in Food |
| CEC | cation exchange capacity |
| CI | chemical ionization |
| CV | coefficient of variation (RSD) |
| CXL | Codex Maximum Residue Limit (Codex MRL).  See MRL |
| d | days |
| DAT | days after (last) treatment |
| DCM | dichloromethane |
| DFG | Deutsche Forschungsgemeinschaft |
| $DT_{50}$ | time for 50% decomposition (i.e. half-life) |
| $DT_{90}$ | time for 90% decomposition |
| 2D-TLC | two dimensional thin layer chromatography |
| dw | dry weight |
| ECD | electron capture detection or detector |
| EI | electron-impact (ionization), now more usually electron ionization |
| EPA | Environmental Protection Agency (usually US EPA) |
| eq | residue expressed as ai equivalent |
| $F_1$ | first filial generation |
| $F_2$ | second filial generation |
| FAO | Food and Agriculture Organization of the United Nations |
| FID | flame-ionization detection or detector |
| FPD | flame-photometric detection or detector |
| GAP | good agricultural practice(s) |

| | |
|---|---|
| GC | gas chromatography; the detector system used is usually also abbreviated as a suffix |
| GEMS/Food | Global Environment Monitoring System–Food Contamination Monitoring and Assessment Programme |
| GLP | good laboratory practice (i.e. the defined system, not in the general sense) |
| GPC | gel-permeation chromatography |
| GSH | glutathione |
| | |
| HPLC | high-performance liquid chromatography |
| HPLC-MS | high-performance liquid chromatography – mass spectrometry |
| HPLC-UV | high-performance liquid chromatography with UV absorption detection |
| hr | hour |
| HR | highest residue in the edible portion of a commodity found in trials used to estimate a maximum residue level in the commodity |
| HR-P | highest residue in a processed commodity calculated by multiplying the HR of the raw commodity by the corresponding processing factor |
| | |
| IARC | International Agency for Research on Cancer |
| IEDI | international estimated daily intake |
| IESTI | international estimate of short-term dietary intake |
| IPCS | International Programme on Chemical Safety |
| IR | infrared spectroscopy |
| ITD | ion-trap detector or detection |
| IUPAC | International Union of Pure and Applied Chemistry |
| | |
| JECFA | Joint Expert Committee on Food Additives |
| JMPR | Joint Meeting on Pesticide Residues |
| JMPS | Joint FAO/WHO Meeting on Pesticide Specifications |
| | |
| LC | liquid chromatography |
| LC-MS | liquid chromatography – mass spectrometry |
| $LC_{50}$ | median lethal concentration |
| $LD_{50}$ | median lethal dose |
| LOAEL | lowest-observed-adverse-effect level |
| LOAEC | lowest-observed-adverse-effect concentration |
| LOD | limit of detection |
| LOQ | limit of quantification |
| LSC | liquid scintillation counting or counter |
| | |
| M | molar = mole/L |
| MID | multiple ion detection (mass spectrometric) |
| MRL | Maximum Residue Limit.  MRLs include draft MRLs and Codex MRLs (CXLs).  The MRLs recommended by the JMPR on the basis of its estimates of maximum residue levels enter the Codex procedure as draft MRLs.  They become Codex MRLs when they have passed through the procedure and have been adopted by the Codex Alimentarius Commission. |
| MS | mass spectrometry or mass spectrometric detector (suffix to GC- or LC-) |
| MSD | mass-selective detection or detector |
| MS/MS | tandem mass spectrometry |
| | |
| NOAEL | no-observed-adverse-effect level |
| NMR | nuclear magnetic resonance |
| NPD | nitrogen/phosphorus detector |
| | |
| OECD | Organization for Economic Co-operation and Development |
| om | amount of organic matter in soil |

| PES | post extracted solids |
|---|---|
| PF | processing factor |
| PHI | pre-harvest interval |
| ppm | parts per million (used only with reference to the concentration of a pesticide in a diet, in all other contexts the terms mg/kg or mg/l are used) |
| $P_{ow}$ | octanol–water partition coefficient |
| RAC | raw agricultural commodity |
| r.d. | relative density (formerly called specific gravity) |
| RfD | reference dose (usually in phrase "acute RfD") |
| RSD | precision under repeatability conditions (measurements within one day or one run) expressed as relative standard deviation (= coefficient of variation) |
| SD | standard deviation |
| SPE | solid-phase extraction (may also describe a post-extraction clean-up process) |
| STMR | supervised trials median residue |
| STMR-P | supervised trials median residue in a processed commodity calculated by multiplying the STMR of the raw commodity by the corresponding processing factor |
| t | tonne (metric ton) |
| TAR | total applied (or administered) radioactivity |
| TLC | thin-layer chromatography |
| TRR | total radioactive residue |
| TMDI | theoretical maximum daily intake |
| TSD | thermionic specific detection or detector (equivalent to AFID, forerunners of NPD) |
| USDA | US Department of Agriculture |
| US FDA | US Food and Drug Administration |
| UV | ultraviolet (radiation) |
| W | the previous recommendation is withdrawn, or withdrawal of the existing Codex or draft MRL is recommended |
| WHO | World Health Organization |

## USE OF JMPR REPORTS AND EVALUATIONS BY REGISTRATION AUTHORITIES

Most of the summaries and evaluations contained in this report are based on unpublished proprietary data submitted for use by JMPR in making its assessments. A registration authority should not grant a registration on the basis of an evaluation unless it has first received authorization for such use from the owner of the data submitted for the JMPR review or has received the data on which the summaries are based, either from the owner of the data or from a second party that has obtained permission from the owner of the data for this purpose.

# INTRODUCTION

The Report of the Joint Meeting of the FAO Panel of Experts on Pesticide Residues in Food and the Environment and the WHO Core Assessment Group (JMPR), held in Rome, 3-12 October 2006, contains a summary of the evaluations of residues in foods of the various pesticides considered, as well as information on the general principles followed by the Meeting (JMPR, 2006). The present document contains summaries of the residues data considered, together with the recommendations made.

The Evaluations are issued in two parts:
Part I: Residues (by FAO);
Part II: Toxicology (by WHO).

For those interested in both aspects of pesticide evaluation, both parts and the Report containing summaries of residues and toxicological considerations are available.

Some of the compounds considered at the Meeting were previously evaluated and reported on in earlier publications. In general, only new information is summarized in the relevant monographs but reference is made to previously published evaluations, which should also be consulted. In the case of older compounds which are re-evaluated as part of the periodic review programme of the CCPR, a review of all available data, including data which may have previously been submitted, is carried out. Compounds evaluated for the first time and those evaluated in the CCPR periodic review programme are identified in the Table of Contents.

Summaries of recommended MRLs, STMR and HR levels and assessments of dietary intake, are published as Annexes 1, 3 and 4 in the Report, and reference is made to this report.

The name of the compound appearing as the title of each monograph is followed by its Codex Classification Number in parentheses.

References to previous Reports and Evaluations of Joint Meetings are listed in Annex I.

## Acknowledgements

The monographs in these Evaluations were prepared by the following participants in the 2006 JMPR, for the FAO Panel of Experts on Pesticide Residues in Food and the Environment:

Dr. A. Ambrus, Dr U. Banasiak, Dr R. Bhula, Dr H. Chen, Mr B. Declercq, Dr E. Dutra Caldas, Dr S. Funk, Mr D. J. Hamilton, Dr C. Lentza-Rizos, Mr D. Lunn, Dr K. Mastovska, Dr B. C. Ossendorp, Dr Y. Yamada and Dr Yibing He.

**Note**. Any comment on pesticide residues in food and their evaluation should be addressed to the:

Plant Protection Service
Plant Production and Protection Division
Food and Agricultural Organization
Viale delle Terme di Caracalla
00100 Rome, Italy

## Reference

JMPR, 2006. Pesticide residues in Food – 2006. Report of the Joint Meeting of the FAO Panel of Experts on Pesticide Residues in Food and the Environment and the WHO Core Assessment Group on Pesticide Residues, Rome, Italy, 3-12 October 2006. WHO and FAO, Rome, 2006.

# ACEPHATE (095)

*First draft was prepared by Arpad Ambrus, Hungary*

## EXPLANATION

Acephate was last evaluated by the JMPR in 2003 and 2005 when an ADI of 0-0.03 mg/kg bw/day and an ARfD of 0.1 mg/kg bw/day were established, and a number of maximum residue levels were estimated.

The cranberry industry performed a number of supervised trials within the Interregional Research Project No. 4 to provide data for the establishment of US tolerances for acephate residues in cranberry. The relevant labels and reports of supervised trials were submitted for evaluation to the 2006 JMPR.

## RESIDUE ANALYSIS

### *Analytical methods*

The cranberry samples were analyzed according to analytical methods RM-12A-5 and RM 12A-6a for determining residues of acephate and methamidophos by gas chromatography/thermionic detection. The methods are based on ethyl acetate extraction, GPC or Silicagel cleanup and detection by GC with NPD. The stated limits of quantitation (LOQs) for both methods were 0.02 mg/kg for acephate and 0.01 mg/kg for methamidophos. The average of concurrent recoveries of acephate from mature berries fortified at 0.25 mg/kg was 84.8% with a coefficient of variation of 8.9% (n=14). For methamidophos, the average of concurrent recoveries from mature berries fortified at 0.1 mg/kg was 80.6% with a coefficient of variation of 6.3% (n=15).

### *Stability of pesticide residues in stored analytical samples*

Exact storage intervals were not reported for each sample in these studies. Samples were stored frozen from harvest to analysis for approximately 16 months or less.

Storage stability tests were reported by the 2003 JMPR for several crops and indicated that residues were stable for a number of months; e.g., beans and lettuce (15 months), peas (14 months) and bell peppers (13 months) (FAO, 2004).

## USE PATTERN

In the USA, Orthene 75S and 75WG, containing 75% active substance, can be applied to protect cranberry as shown in Table 1.

Table 1. Registered uses of acephate.

| Country | Method | Product | No | Water L/ha | Rate kg ai/ha | PHI day |
|---------|--------|---------|----|------------|---------------|---------|
| Canada | Broadcast | 75S | 2 | 225-1650 | 0.56 | |
| USA | Broadcast foliar | 75S | 1 | Min 7.6[1] | 1.12 | 90 |
| USA | Broadcast or chemigation | 75WG | 2[2] | Min 140 | 1.68 | 60 |

[1] For aerial application

[2] The minimum interval between applications is 10 days.

## RESIDUES RESULTING FROM SUPERVISED TRIALS

During the 1976 to 1981 growing seasons, at 17 established trial locations, 64 tests were conducted on cranberries in three geographical regions of the USA.

For each test, single or multiple broadcast foliar applications of acephate were made using the 75% SC/S formulation (Daussin 2006). Single application rates were 0.56 to 3.36 kg ai/ha resulting in total application rates of 1.12 to 13.44 kg ai/ha/season. Ground equipment was used at all trial locations except four trials where applications were made aerially. Spray volumes ranged from 7.6-38 l/ha for the air applications and from 379 to 1136 l/ha for the ground applications. No use of adjuvants was indicated. Cranberry fruit samples were taken between 6 or 131 days following the last application.

In addition, berries from three trial locations were processed into cranberry cocktail. The procedure for preparing cocktails was not reported. The residues detected are summarized in Table 2.

Table 2. Residue data from field trials with acephate.

| Trial ID | Location (City, State) | Trial Start Year | Variety | No. of Appl | Single Appl. Rate, kg ai/ha | DAT[1] | Residues (mg/kg) Fruit | | | Cocktail |
|---|---|---|---|---|---|---|---|---|---|---|
| | | | | | | | A[2] | MA[3] | Ave[4] | Total |
| T-3534 | East Wareham, MA | 1976 | Early Black | 1 | 1.68 | 83 | 0.18 0.18 | ND ND | 0.18 (< 0.01) | |
| T-3538 | E. Wareham, MA | 1976 | Early Black | 1 | 2.52 | 38 | 0.68 0.58 | 0.02 0.02 | 0.63 (0.02) | |
| T-3539 [c] | E. Wareham, MA | 1976 | Howes | 1 | 1.68 | 28 | 0.77 0.80 | 0.03 0.03 | 0.78 0.83 | |
| | | | | | | 131 | 0.04 0.03 | ND ND | 0.04 (< 0.01) | |
| | | | | 1 | 2.52 | 28 | 1.4 1.6 | 0.05 0.05 | 1.5 (0.05) | |
| | | | | | | 131 | 0.05 0.04 | ND ND | 0.04 (< 0.01) | |
| | | | | 2 | 3.36 | 119 | 0.06 0.07 | ND ND | 0.06 (< 0.01) | |
| T-3665 | Long Beach, WA | 1976 | McFarlin | 1 | 2.52 | 30 | 1.2 1.05 | 0.06 0.05 | 1.1 (0.06) | |
| | | | | 1 | 3.36 | 30 | 1.6 1.1 | 0.07 0.06 | 1.4 (0.06) | |
| T-4033 | Whitesbog, NJ | 1977 | Early Black | 2 | 1.12 | 7 | 0.23 0.24 | 0.01 0.01 | 0.24 (0.01) | |
| | | | | | | 14 | 0.24 0.25 | 0.01 0.01 | 0.24 (0.01) | |
| | | | | | | 21 | 0.25 0.26 | 0.01 0.01 | 0.26 (0.01) | |
| | | | | 2 | 1.68 | 7 | 0.30 0.46 | 0.01 0.02 | 0.38 (0.02) | |
| | | | | | | 14 | 0.24 0.51 | ND 0.02 | 0.36 (0.02) | |
| | | | | | | 21 | 0.47 0.49 | 0.02 0.02 | 0.48 (0.02) | |
| T-4034 | Whitesbog, NJ | 1977 | Early Black | 2 | 1.12 | 7 | 1.2 1.2 | 0.04 0.04 | 1.2 (0.04) | |
| | | | | | | 14 | 1.4 1.4 | 0.06 0.06 | 1.4 (0.06) | |
| | | | | | | 21 | 1.1 1.2 | 0.05 0.05 | 1.2 (0.05) | |

| Trial ID | Location (City, State) | Trial Start Year | Variety | No. of Appl | Single Appl. Rate, kg ai/ha | DAT[1] | Residues (mg/kg) | | | Cocktail |
|---|---|---|---|---|---|---|---|---|---|---|
| | | | | | | | Fruit | | | |
| | | | | | | | A[2] | MA[3] | Ave[4] | Total |
| T-4034 (cont.) | Whitesbog, NJ | 1977 | Early Black | 2 | 1.68 | 7 | 2.2 2.4 | 0.07 0.08 | 2.4 (0.08) | |
| | | | | | | 14 | 2.0 2.3 | 0.07 0.07 | 2.2 (0.07) | |
| | | | | | | 21 | 1.9 1.9 | 0.07 0.06 | 1.9 (0.06) | |
| T-4035 | East Wareham, MA | 1977 | Bergman | 3 | 1.68 | 7 | 0.83 0.49 | 0.05 0.04 | 0.62 (0.04) | |
| | | | | | | 14 | 0.41 0.40 | 0.03 0.03 | 0.40 (0.03) | |
| | | | | | | 22 | 0.49 0.48 | 0.04 0.04 | 0.48 (0.04) | |
| T-4036 [c] | E. Wareham, MA | 1977 | Early Black | 4 | 1.68 | 6 | 4.5 6.3 | 0.17 0.22 | 5.4 (0.2) | |
| | | | | | | 14 | 2.5 3.2 | 0.11 0.13 | 2.8 (0.12) | |
| | | | | | | 21 | 1.9 1.9 | 0.08 0.08 | 1.9 (0.08) | |
| T-4037 [c] | E. Wareham, MA | 1977 | Early Black | 4 | 1.68 | 6 | 7.1 6.9 | 0.23 0.20 | 7.0 (0.22) | |
| | | | | | | 14 | 6.2 4.9 | 0.14 0.13 | 5.6 (0.14) | |
| | | | | | | 21 | 5.6 8.3 | 0.12 0.15 | 7.0 (0.14) | |
| T-4038 | E. Wareham, MA | 1977 | Early Black | 3 | 1.68 | 7 | 0.51 0.54 | 0.04 0.04 | 0.52 (0.04) | |
| | | | | | | 14 | 0.49 0.55 | 0.05 0.05 | 0.53 (0.05) | |
| | | | | | | 22 | 0.26 0.24 | 0.03 0.02 | 0.25 (0.02) | |
| T-4046 [c] | Whitesbog, NJ | 1977 | Early Black | 2 | 1.12 | 7 | 0.09 | ND | 0.09, (< 0.01) | |
| | | | | | | 14 | 0.49 0.47 | 0.02 0.01 | 0.48 (0.02) | |
| | | | | | | 21 | 0.30 0.29 | ND ND | 0.30 (< 0.01) | |
| | | | | | 2 | 1.68 | 7 | 0.27 | ND | 0.27 (< 0.01) | |
| | | | | | | 14 | 0.20 0.32 | 0.01 ND | 0.26 (0.01) | |
| | | | | | | 21 | 0.22 0.15 | ND ND | 0.18 (< 0.01) | |
| T-4047 | Long Beach, WA | 1977 | McFarlin | 3 | 1.12 | 21 | 0.30 0.27 | 0.03 0.03 | 0.28 (0.03) | |
| | | | | 3 | 1.68 | 21 | 0.39 0.39 | 0.06 0.06 | 0.39 (0.06) | |
| T-5006 | E. Wareham, MA | 1979 | Early Black | 3 | 0.84 | 21 | 0.41 | 0.01 | 0.42[5] | 0.15 |
| | | | | 3 | 1.68 | 21 | 0.71 | 0.03 | 0.74[5] | 0.26 |
| | | | | 3 | 3.36 | 21 | 2.4 | 0.06 | 2.5[5] | 0.71 |
| | | | | 3 | 0.84 | 14 | 1.06 | 0.04 | 1.1[5] | |
| | | | | 4 | 1.68 | 14 | 1.1 | 0.04 | 1.1[5] | 0.26 |
| | | | | 4 | 3.36 | 14 | 1.8 | 0.06 | 1.9[5] | 0.62 |

| Trial ID | Location (City, State) | Trial Start Year | Variety | No. of Appl | Single Appl. Rate, kg ai/ha | DAT[1] | Residues (mg/kg) | | | Cocktail |
|---|---|---|---|---|---|---|---|---|---|---|
| | | | | | | | Fruit | | | |
| | | | | | | | A[2] | MA[3] | Ave[4] | Total |
| T-5055 | E. Wareham, MA | 1980 | Early Black | 2 | 0.56 | 88 | ND<br>ND | ND<br>ND | < 0.01<br>(< 0.01) | |
| | | | | 2 | 1.12 | 88 | 0.07<br>0.05 | ND<br>ND | 0.06<br>(< 0.01) | |
| | | | | 2 | 2.24 | 88 | 0.13<br>0.11 | ND<br>ND | 0.12<br>(< 0.01) | |
| | | | | 4 | 0.56 | 46 | 0.32<br>0.31 | 0.02<br>0.02 | 0.32<br>(0.02) | |
| | | | | 4 | 1.12 | 46 | 1.2<br>0.82 | 0.05<br>0.04 | 1.1[5] | 0.13 |
| | | | | 4 | 2.24 | 46 | 2.6<br>2.8 | 0.1<br>0.11 | 2.7<br>(0.10) | |
| | | | | 4 | 0.56 | 28 | 1.4<br>0.66 | 0.04<br>0.06 | 1.03<br>(0.05) | |
| | | | | 5 | 1.12 | 28 | 1.2<br>0.49 | 0.05<br>0.02 | 1.4<br>(0.04) | |
| | | | | 5 | 2.24 | 28 | 2.8<br>3.3 | 0.10<br>0.12 | 2.2[5] | 0.59 |
| T-5056 | Chatsworth, NJ | 1980 | Early Black | 3 | 1.68 | 30 | 1.3<br>1.4<br>1.1 | 0.05<br>0.05<br>0.05 | 1.4<br>1.2<br>(0.05) | |
| T-5155 | Warrens, WI | 1980 | Not Specified | 3 | 1.12 | 34 | 0.33<br>0.35<br>0.29 | 0.03<br>0.03<br>0.02 | 0.32<br>(0.03) | |
| | | | | 2 | 1.12 | 45 | 0.22<br>0.18<br>0.20 | 0.02<br>0.01<br>0.02 | 0.20<br>(0.02) | |
| | | | | | | 66 | 0.11<br>0.13<br>0.12 | ND<br>ND<br>ND | 0.12<br>(< 0.01) | |
| T-5456 | Coos County, OR | 1981 | Stevens | 5 | 1.12 | 30 | 0.45<br>0.42 | 0.03<br>0.03 | 0.44<br>(0.03) | |
| | | | | | | 50 | 0.42<br>0.47 | 0.03<br>0.03 | 0.44<br>(0.03) | |
| | | | | 5 | 2.24 | 30 | 0.80<br>0.81 | 0.06<br>0.05 | 0.80<br>(0.06) | |
| | | | | | | 30 | 0.80<br>0.80 | 0.06<br>0.06 | 0.80<br>(0.06) | |

1. DAT: days after last application; 2. A: acephate; 3. MA: methamidophos; 4. Ave: average acephate residues in replicate samples with average methamidophos residues in brackets; 5. Fruits used for preparing cocktail.

## APPRAISAL

Acephate was last evaluated by the JMPR in 2005 and 2003 when an ADI of 0-0.03 mg/kg bw/day, an ARfD of 0.1 mg/kg bw/day were established and a number of maximum residue levels were estimated. The residue for compliance with MRLs was defined as acephate, but for dietary intake estimation it was decided that methamidophos residues should also be taken into consideration.

Results of supervised trials carried out on cranberry according the US registered uses were submitted for evaluation.

### *Results of supervised residue trials*

The compound can be applied either as broadcast treatment at a rate of 1.12 kg ai/ha with a 90 day PHI, or as two chemigation (applied in irrigation) treatments at a rate of 1.68 kg ai/ha with a PHI of 60 days.

During the 1976 to 1981 growing seasons 39 field trials were conducted on cranberries in three geographical regions of USA. For each test, single or multiple broadcast foliar applications of acephate were made with individual application rates of 0.56–3.36 kg ai/ha resulting in total application rates of 1.12 to 13.4 kg ai/ha/season. Cranberry fruit samples were taken between 6 and 131 days following the last application.

Exact storage intervals were not reported for each sample in these studies. Samples were stored frozen from harvest to analysis for up to 16 months. Storage stability tests for various intervals were reported by the 2003 JMPR for several crops. The results suggest that the decrease in residues during storage was not significant.

The cranberry samples were extracted with ethyl acetate, cleaned up on GPC column and detected with NPD. The stated limits of quantification were 0.02 mg/kg for acephate and 0.01 mg/kg for methamidophos. Average of concurrent recoveries of acephate and methamidophos from mature berries fortified at 0.25 mg/kg and 0.1 mg/kg were 84.8% and 80.6%, respectively.

Three trials were performed at approximate US GAP. The average acephate residues in replicate samples were 0.06, 0.18 and 0.2 mg/kg (methamidophos residue was not detectable).

The Meeting estimated a maximum residue level of 0.5, HR of 0.2 and STMR of 0.18.

The Methamidophos residues in cranberry fruits should be below 0.01 mg/kg.

In seven trials, performed with 3 to 5 applications of 0.84–3.36 kg ai/ha at each timing, cranberry cocktail (cranberry juice) was prepared from the fruits harvested between 14–46 days after the last application. The total residues (sum of acephate and methamidophos) in fruit and cocktail (juice) were:

| | Total residues mg/kg | | | | | | |
|---|---|---|---|---|---|---|---|
| Fruit | 0.425 | 0.745 | 1.15 | 1.15 | 1.15 | 1.95 | 2.25 |
| Cocktail | 0.15 | 0.26 | 0.26 | 0.13 | 0.62 | 0.59 | 0.71 |
| Fp | 0.35 | 0.35 | 0.23 | 0.11 | 0.32 | 0.26 | 0.28 |
| Estimated processing factor | | | median: | 0.28 | average: | 0.27 | |

## RECOMMENDATION

On the basis of the data from supervised trials, the Meeting concluded that the residue levels listed below are suitable for establishing maximum residue limits and for dietary intake assessment.

Summary of recommendations for MRLs, STMRs and HRs for acephate

| CCN | Commodity | MRL, mg/kg | | STMR or STMR-P, mg/kg | HR or HR/P mg/kg |
|---|---|---|---|---|---|
| | | New | Previous | | |
| FB 0265 | Cranberry | 0.5 | | 0.18 | 0.2 |

## DIETARY RISK ASSESSMENT

### Long-term intake

The GEMS/Food regional diets specify the following long-term cranberry consumption (g/day/person) for various cluster diets: A (0.1); D (0.3); F (0.6); M (2.5). The consumption of cranberry in other regions is nil.

The highest IEDI in the 13 GEMS/Food regional diets, based on estimated STMR, was 0.03% of the maximum ADI (0.03 mg/kg bw).

The Meeting concluded that the long-term intake of residues of acephate from use on cranberry will not practically increase the intake of residues from other uses considered earlier by the JMPR.

### *Short-term intake*

The GEMS/Food regional diet specifies the large portion sizes of cranberry of 3.53 g/kg bw for adults and 6.78 g/kg bw for children (both are from the USA).

The IESTIs of acephate calculated on the basis of the large portion size and the estimated HR of 0.2 mg/kg are 0.71% and 1.4% of the ARfD for adults and children, respectively.

The Meeting concluded that the short-term intake of residues resulting from the use of acephate on cranberry that have been considered by the JMPR is unlikely to present a public health concern.

## REFERENCES

**Author, Date, Title, Institute, Report Reference, Document No.**

Daussin, S. 2006. JMPR Format of Petition Proposing a Tolerance for Acephate Use in Cranberry Production. Interregional Project No. 4. MRID 0089

Elliott, E.J. and J.B. Leary.1978. Residue analysis of acephate and methamidophs in crops, soil, water and milk, Chevron Chemical Company RM-12A-5

FAO 2004. Pesticide residues in food, Evaluations 2003, Plant Production and Protection Paper Plant Production and Protection Paper No. 177

Slagowski, J.L and J.B. Leary. 1980. Determination of acephate and methamidophs in crops, Chevron Chemical Company RM-12A-6a

# ALDICARB

*First draft prepared by Eloisa Dutra Caldas, University of Brasilia, Brasilia, Brazil*

## EXPLANATION

Aldicarb residues were last evaluated by the JMPR in 2001 and 2002. In 2001 residues in individual units of banana and potato and a processing study on potato were evaluated. The Meeting recommended a maximum residue level of 0.2 mg/kg for banana and confirmed a previous recommendation of 0.5 mg/kg for potato.

The IESTI calculated for banana, potato and microwaved potato exceed the ARfD for aldicarb (0.003 mg/kg bw) for both the general population (excluding women of child-bearing age) (140–160% above the ARfD) and children (330–560% above the ARfD). A variability factor of 5 was used in the calculation of the intake for banana. In 2002, based on additional residue data provided from individual banana fingers and composite samples, the Meeting recommended the application of a variability factor of 3 for the calculation of the IESTI of aldicarb in banana. The refined IESTI represented 40% ARfD for the general population and 110% ARfD for children.

For potato, a highest residue from individual unit data of 1.2 mg/kg was used in the first term of the equation for case 2a, with no variability factor (see Chapter 3 of 2003 JMPR Report). The HR from a composite sample (0.45 mg/kg) was used in the second part of the equation.

At its 35th, 36th, 37th and 38th Sessions, the CCPR returned the draft MRL for banana and potato to Step 6 due to acute intake concerns. The Committee requested that the JMPR consider using alternative GAPs to estimate lower MRLs (see General Considerations 2.3). The present Meeting received GAP information for potato from the government of The Netherlands.

The residue definition of aldicarb is the sum of aldicarb, aldicarb sulfone and aldicarb sulfoxide, expressed as aldicarb.

## *Results of supervised residue trials*

### *Banana*

Based on twenty four trials conducted in France (Guadalupe and Martinique) and Côte d'Ivoire submitted to the 2001 JMPR with bagged and unbagged banana according to GAP, the Meeting recommended a HR of 0.10 mg/kg in banana pulp and a maximum residue level of 0.2 for aldicarb in banana. No additional residues trials or GAP information were provided.

The Meeting concluded that none of the residue data relating to available GAP suggests a lower maximum residue level to replace the current proposal of 0.2 mg/kg for aldicarb in banana.

### *Potato*

Forty five trials conducted in Europe and USA with aldicarb in potatoes according to GAP were evaluated by the 2001 JMPR. Twenty trials were conducted according to GAP of the Netherlands (furrow application at 12.8 g/100m, corresponding to 1.7 kg ai/ha or broadcast application of 3.36 kg ai/ha), giving a highest residue of 0.36 mg/kg. Nine trials conducted according to GAP in Greece, Italy and Spain (furrow at 2.5 kg ai/ha) gave a highest residue of 0.45 mg/kg. In Europe, the PHI is 90 days. Sixteen trials conducted in the USA (GAP of 3.36 kg ai/ha with a PHI of 150 days, positive displacement) gave a highest residue of 0.20 mg/kg.

New GAP information in the Netherlands indicates a furrow application of 0.75 kg ai/ha and broadcast application of 3 kg ai/ha with no specified PHI. Ten furrow trials (at 13 g/100m) previously evaluated against the Netherlands GAP do not match the new GAP.

The residues derived from the 35 trials, according to the current GAP, in ranked order were: < 0.02, < 0.03 (3), 0.02 (2), 0.03 (6), 0.04 (6), 0.05, 0.06 (4), 0.09, 0.10, 0.11, 0.12, 0.13, 0.14, 0.18, 0.20 (2), 0.27 (2), 0.36 and 0.45 mg/kg.

The Meeting concluded that none of the residue data relating to available GAP suggests a lower maximum residue level to replace the current proposal of 0.5 mg/kg for aldicarb in potato.

# AMINOPYRALID (220)

*First draft was prepared by Dr Raj Bhula, Australian Pesticides and Veterinary Medicines Authority, Canberra, Australia.*

## EXPLANATION

Aminopyralid is a new herbicide that is used for the control of broadleaf weeds in pastures and cereal crops. It was advanced to the 2006 JMPR schedule of new compounds by the 37th session of the CCPR. The manufacturer has submitted studies on metabolism in plants and animals, analytical methods, storage stability, environmental fate and degradation, supervised field trials, processing and farm animal (livestock) feeding.

## IDENTITY

| | |
|---|---|
| ISO common name | Aminopyralid |
| Synonyms | DE-750; XDE-750; XR-750; X660750 |
| IUPAC name | 4-amino-3,6-dichloropyridine-2-carboxylic acid |
| Chemical Abstracts name | 2-pyridinecarboxylic acid, 4-amino-3,6-dichloro |
| CAS Number | 150114-71-9 |
| Molecular formula | $C_6H_4Cl_2N_2O_2$ |
| Molecular weight | 207.026 |

Structural formula

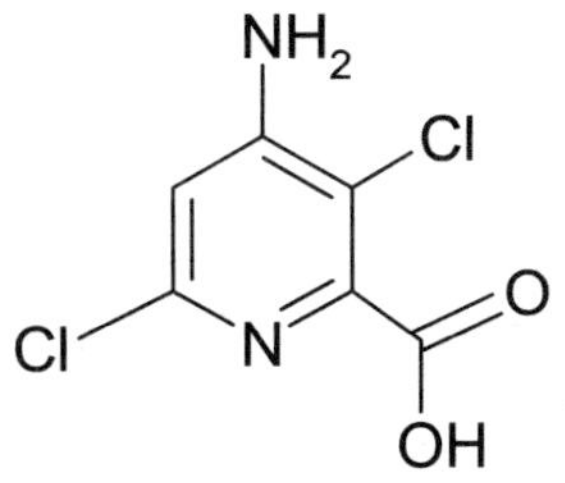

## PHYSICAL AND CHEMICAL PROPERTIES

Pure active ingredient

Purity: 99.5% wt/wt ± 0.1%

| Physico-chemical properties | | Ref. |
|---|---|---|
| Appearance | off-white powder (at 22.7 °C) | FAPC 013087 |
| Odour | odourless | FAPC 013087 |
| Melting Point | 163.5 °C | FAPC 023263 |
| Density | 1.72 ± 0.065 at 20 °C | FAPC 013087 |
| pH | 2.31 (1% w/w aq solution at 23.4 °C) | FAPC 013087 |
| Vapour pressure | $9.52 \times 10^{-9}$ at 20 °C | DECO GL-AL MD-2001-000921 |
| | $2.59 \times 10^{-8}$ Pa at 25 °C | DECO GL-AL MD-2001-000921 |
| Henry's Law constant | $5.18 \times 10^{-10}$ Pa m$^3$/mol unbuffered at 18 °C | NAFST540 |
| | $9.30 \times 10^{-12}$ Pa m$^3$/mol at pH 5 and 20 °C | |
| | $9.61 \times 10^{-12}$ Pa m$^3$/mol at pH 7 and 20 °C | |
| | $9.71 \times 10^{-12}$ Pa m$^3$/mol at pH 9 and 20 °C | |
| | | |
| Solubility in water | 2.48 g/L unbuffered at 18 °C (pH 2.35) | FOR01015 |

| Physico-chemical properties | | Ref. |
|---|---|---|
| | 212 g/L in pH 5 buffer at 20 °C | |
| | 205 g/L in pH 7 buffer at 20 °C | |
| | 203 g/L in pH 9 buffer at 20 °C | |
| | | |
| Dissociation constant in water | average pKa 2.56 ± 0.03 (pH range 2 to 3.4; n=9) | 01-822-AG; GHF-P-2332 |
| | | |
| Solubility in organic solvents | acetone         29,195 µg/mL | GHE-P-9228 |
| | 1,2-dichloroethane 189 µg/mL | |
| | ethyl acetate      3,939 µg/mL | |
| | Heptane      insoluble (< 10 µg/mL) | |
| | methanol      52,191 µg/mL | |
| | n-octanol     4,548 µg/mL | |
| | xylene      43 µg/mL | |
| | | |
| Octanol/water partition coefficient | log $K_{ow}$ = 0.201 ± 0.049 in unbuffered water (mean $K_{ow}$ = 1.59 ± 0.18 (n=6)) | FOR01009 |
| | | |
| Hydrolysis in sterile buffer at 20 °C and 50 °C | Hydrolytically stable at pH 5, 7 and 9 at 20 °C (31 days) and at 50 °C (5 days). | 020067 |
| | | |
| Photolysis in water at 20 °C | Half-life of 0.6 days at pH 5; $DT_{90}$ 2 days at pH 5* | 020066 |

*The half-life of photolysis of $^{14}$C-aminopyralid in sterile pH 5 buffer solution is 0.6 days, with a $DT_{90}$ of 2 days. Samples were continuously irradiated at 20°C for the equivalent of 38 days summer sunlight at 40° N latitude. Photolysis products identified were oxamic acid, malonamic acid and $CO_2$ (≈ 28% of applied radiocarbon).

## METABOLISM AND ENVIRONMENTAL FATE

### Animal metabolism

The meeting received metabolism studies for aminopyralid administered to lactating goats and laying hens. In both studies, aminopyralid (or XDE-750) was labelled in the 2- and 6-positions of the pyridine ring, as shown below.

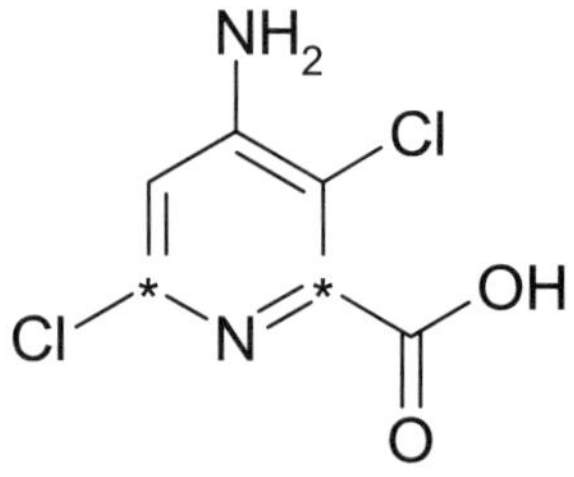

### Lactating goat

A single lactating goat was orally dosed for 6 days with $^{14}$C aminopyralid at 13.96 ppm in the feed (0.26 mg/kg bw/day) (MacPherson and Gedik 2003). Mean food consumption over 6 days was 1.38 kg dry matter; the mean dose was 19.23 mg/goat. The dose was administered immediately after the morning milking and prior to the morning feeding.

Urine and faeces were collected prior to dose administration and at 24 hour intervals thereafter for the duration of the dosing period. The animal cage was washed at the end of the excreta collection period.

Milk was collected immediately prior to dosing (am) and in the afternoon (pm) each day; the final milk sample was collected immediately prior to sacrifice. Less than 24 hours after the final dose, the goats were slaughtered and samples of liver, kidney, omental fat, perirenal fat, triceps muscle, semi-membranous muscle and longissimus dorsi muscle were collected for analysis. The two fats (omental and perirenal) were composited in equal amounts to form one fat sample; similarly the muscle samples were composited into one muscle sample. Liver and kidney samples were processed on the day of collection and placed in freezer storage (−20° C); muscle and fat samples were placed in the freezer pending analysis. All samples were analysed within 245 days of collection.

Total radioactivity was measured in all samples of milk, urine, faeces, cage wash and tissues using scintillation counting and combustion analysis. Total radioactive residues (TRR) as a percentage of the administered dose is shown below.

| Urine | Faeces | Cage wash | Milk | Liver | Kidneys | **Total** |
|-------|--------|-----------|------|-------|---------|-----------|
| 46.03% | 46.46% | 2.88% | 0.05% | 0.01% | 0.01% | **95.6%** |

Elimination of the radioactivity via urine and faeces was similar, with each accounting for approximately 46% of the total administered dose; total eliminated 92%. Approximately 9% and 8% of the administered dose was excreted daily via faeces and urine, respectively, during 48 to 144 hours of the study.

The radioactivity in milk was found to plateau within 24 to 48 hours after dosing had commenced, i.e., by dose 2 to 3. The total radioactivity in milk was about 0.05% of the total administered dose, with concentrations ranging from 0.003 to 0.008 mg/kg parent equivalents[1].

TRR in tissues were 0.008 mg/kg equivalents in liver, 0.071 mg/kg equivalents in kidneys, 0.001 mg/kg equivalents in composite fat and non-detectable in composite muscle.

Extraction of milk samples with MeOH recovered 72.6% (0.004 mg/kg equivalents) and 72.4% TRR (0.006 mg/kg equivalents) from the 32 and 128 hour samples, respectively.

The MeOH extracts were partitioned with ether:hexane (1:1 v/v) followed by hexane to remove lipids which led to 70.7% and 70.9% TRR in the MeOH, respectively (processed extracts) for the 32 hour and 128 hour samples. Pepsin hydrolysis of the remaining non-extracted radioactivity in the post-extraction solids (PES) released another 8.5 and 4.4% TRR, respectively (see Table 1). The milk extracts were not analysed further by HPLC due to very low levels of radioactivity being present.

Extraction of liver and kidney samples with MeOH recovered 75.6% and 95.5% TRR, respectively. Partitioning of the MeOH extracts from liver with hexane gave an overall recovery of 57.8% TRR (0.004 mg/kg equivalents). Approximately 24% TRR remained in the PES. The liver extracts were not analysed with HPLC due to the low levels of radioactivity present.

The hexane extracts from kidney (initial extracts) were further processed by drying and reconstitution in 0.1% formic acid in $H_2O:CH_3CN$ giving an overall recovery of 79.9% TRR. One radiolabelled component was identified by HPLC analysis of the processed kidney extracts; this was parent aminopyralid present at 0.057 mg/kg.

The distribution of radioactivity following various extraction steps outlined above is shown in Table 1.

Table 1. Distribution of radioactivity in extracts of milk and tissues from a lactating goat.

| TRR and extracts | % TRR (mg/kg equivalents) in extracted milk and tissues | | | |
|---|---|---|---|---|
| | Milk (32 hrs) | Milk (128 hrs) | Liver | Kidney |
| TRR (mg/kg equiv)① | 0.006 | 0.008 | 0.008 | 0.071 |
| Initial extracts② | 72.6 (0.004) | 72.4 (0.006) | 75.6 (0.006) | 95.5 (0.067) |
| Processed extracts③ | 70.7 (0.004) | 70.9 (0.006) | 57.8 (0.004) | 79.9 (0.057) |
| Pepsin hydrolysates④ | 8.5 (0.001) | 4.4 (< 0.001) | – | – |
| PES⑤ | 10 (0.001) | 12.9 (0.001) | 24.4 (0.002) | 4.5 (0.003) |

① From radioanalysis data. ② Extracted with MeOH (3×), (1× for milk). ③Combined initial extracts were partitioned with hexane (milk partitioned with ether:hexane 1:1), then concentrated and centrifuged. Kidney initial extracts were dried and reconstituted in 0.1% formic acid in $H_2O:CH_3CN$ (95:5 v/v). ④The milk pepsin hydrolysates were not processed further due to low levels of radioactivity. ⑤Post extracted solids after extraction and/or enzyme hydrolysis.

---

[1] LOD for milk was reported as 0.00007 mg/kg parent equivalents.

Although urine is not normally considered important compared to edible tissues and milk, TRR in urine was investigated further. Centrifugation and concentration of urine samples collected at 24 and 120 hours resulted in 100% and 94.8% of the TRR being recovered without further clean-up; this represented 3.10% and 13.72% of the total administered dose, respectively. HPLC analysis of the concentrated urine samples at 24 and 120 hours revealed that the major component of the radioactivity was aminopyralid (96.8% and 90.9% of the 24 and 120 hr samples, respectively). An unknown component was present at 3.2%TRR in the 24 hour urine sample and two unknown components were present at 1.4% and 2.5% of the TRR in the 120 hour sample. Similarly, in faeces collected at 24 and 120 hours, parent aminopyralid comprised 94.4% and 95.6% of the TRR, respectively. Therefore > 94% the total amount of recovered radioactivity that was eliminated via urine and faeces was composed of unchanged aminopyralid.

In summary aminopyralid was readily excreted from the goat, with residues of 0.06 mg/kg aminopyralid present in kidneys only.

*Laying hens*

Ten laying hens of 45 weeks age were orally dosed for 7 days with $^{14}$C aminopyralid at the equivalent of 10.5 ppm in the feed (daily dose of 1.7 mg/bird; mean bodyweight 1.7 kg and mean feed consumption 161 g/day) (Magnussen 2004). Another 10 birds were used as control animals.

The hens received capsules containing the dose each morning following collection of eggs and excreta. Eggs were collected twice a day (morning and evening) and prepared for analysis; eggs from the evening collection were refrigerated overnight (4°C) and pooled with those collected the following morning, to give a composite sample for each day. Both control and test groups were sampled in an identical fashion. Eggs were cracked open, the shells discarded and the composite samples weighed, homogenised (hand-whisk) and stored frozen prior to analysis.

Prior to each dose, excreta were collected from the previous 24-hour period from each pen with the excreta pooled for the treatment group. All samples were stored frozen prior to analysis for TRR.

Approximately 24 hours after administration of the final dose, hens were slaughtered and samples of liver, abdominal fat, muscle (light and dark meat) and skin with subcutaneous fat were collected from each bird. Each sample type was pooled and stored frozen prior to analysis. Samples of eggs, excreta and tissues were assayed for radioactivity using LSC and/or combustion. Assays of total radioactivity were completed within two weeks of sample collection.

TRR in excreta collected daily ranged from 55% to 87% of the nominal daily dose over 1 to 6 days; approximately 55% on days 1 and 2 and > 80% day 3 onwards. This indicates that a large percentage of the dose was excreted daily. TRR in eggs over 7 days and in tissues on day 7 is shown in Table 2.

Table 2. TRR in eggs and hen tissues dosed for 7 days at 10 ppm in the diet.

| Sample | Sampling Interval (days) | mg/kg aminopyralid equivalents[1] |
|--------|--------------------------|-----------------------------------|
| Eggs | 1 | <LOD (< 0.0012) |
|      | 2 | 0.0023 |
|      | 3 | 0.0029 |
|      | 4 | 0.003 |
|      | 5 | 0.0036 |
|      | 6 | 0.0036 |
|      | 7 | 0.004 |
| Skin/fat | 7 | 0.0029 |
| Muscle | 7 | ND (0.0018) |
| Fat | 7 | ND (0.0017) |
| Liver | 7 | 0.0024 |

[1] based on sample mean dpm/g/666000 µg/mg activity of test material

With the exception of egg samples at day 1, all eggs contained low levels of radioactivity that gradually increased to a plateau level within 5 to 7 days of the study. Levels on day 2 were 0.0023 mg/kg equivalents, which had not quite doubled to 0.004 mg/kg equivalents by day 7. TRR in all eggs collected over the study period accounted for < 0.01% of the total administered dose.

TRR in tissues were << 0.01% of the total administered dose. TRR in muscle and fat were comparable, corresponding to 0.0018 and 0.0017 mg/kg aminopyralid equivalents, respectively and were reported as non-detectable. TRR in skin (with fat) and liver were detectable and were 0.0029 and 0.0024 mg/kg aminopyralid equivalents, respectively. The levels of radioactivity in eggs and tissues were low and were not further characterised.

Residues in day 7 excreta were however further characterised and identified. Extraction and analysis occurred approximately 49 days after slaughter. Residues were readily extracted into $CH_3CN:H_2O$ with < 96% TRR being recovered. HPLC analysis of the extracts indicated that 93% of the extracted radioactivity was due to aminopyralid, while another 3% was attributed to a component that was more polar than the parent compound.

In summary, aminopyralid is readily excreted by hens following oral dosing for 7 days at 10 ppm in the feed. TRR in eggs on days 6 and 7 of the study were 0.004 mg/kg aminopyralid equivalents. TRR in muscle and fat were non-detectable (< 0.002 mg/kg aminopyralid equivalents), while TRR in liver and skin/fat were 0.002 – 0.003 mg/kg aminopyralid equivalents.

aminopyralid

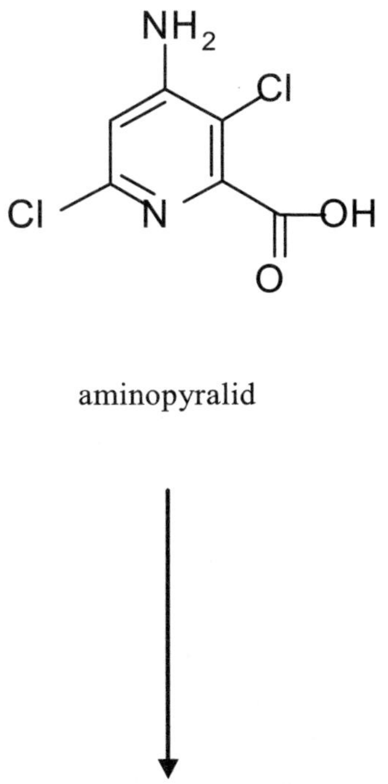

Goats: Less than 3 – 4% converted to unknown components.
Hens: Less than 10% converted to aqueous soluble residues that in part can be converted back to aminopyralid using base or acid hydrolysis.

Figure 1. Proposed metabolic pathway for goats and hens.

## *Plant metabolism*

The meeting received metabolism studies for aminopyralid applied to wheat and grass/pastures. In both studies, aminopyralid (or XDE-750) was labelled in the 2- and 6-positions of the pyridine ring, as shown below.

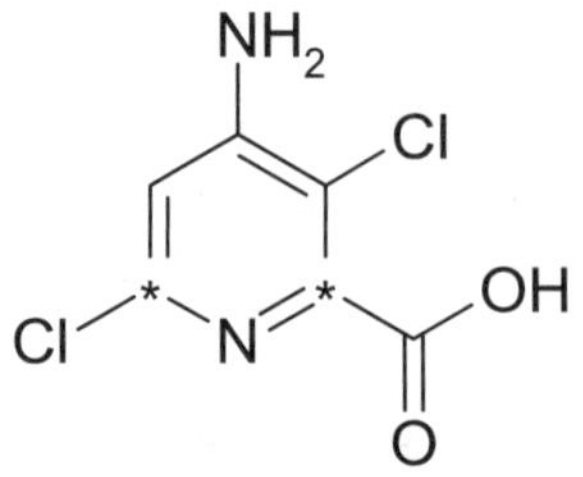

## *Wheat*

$^{14}C$ aminopyralid was applied to spring wheat at rates of 40 and 80 g ai/ha (2× and 4× nominal application rate) (Graper *et al.* 2003). The wheat was at stages BBCH 26–28 (6 to 8 tillers detectable) when treated and grown outdoors. Samples of plant material (forage) were taken at 0 and 14 days after treatment (DAT); hay was sampled at 35 DAT and grain and straw at 86 DAT.

Harvested samples were analysed by combustion; TRR in the various samples are shown below:

Table 3. TRR in various wheat samples.

| Sample | TRR aminopyralid XDE-750 equivalents (mg/kg) | |
|---|---|---|
| | 40 g ai/ha | 80 g ai/ha |
| Forage  0 DAT | 2.022 | 4.121 |
| Forage  14 DAT | 0.418 | 0.874 |
| Hay      35 DAT | 0.284 | 0.691 |
| Straw   86 DAT | 0.281 | 0.623 |
| Grain   86 DAT | 0.039 | 0.084 |

LOD and LOQ from combustion data are 0.001 and 0.003 mg/kg equivalents, respectively.

Homogenised samples were initially extracted with 70:30 $CH_3CN:H_2O$. Extraction occurred within 15 to 42 days after sample collection. All samples, except for the 0 DAT forage, were subjected to a cold extraction followed by reflux in the same 70:30 $CH_3CN:H_2O$ mixture. The initial $CH_3CN/H_2O$ extraction and reflux of grain and straw samples was followed by partitioning of extractable radioactivity ($CH_3CN/CH_2Cl_2$) prior to HPLC analysis. Extraction and recovery of $^{14}C$ in $CH_3CN/H_2O$ is shown in Table 4.

All extractable radioactivity was analysed using HPLC and the results are shown in Tables 5 and 6. HPLC analyses occurred within 38 to 93 days after sample collection. Non-extracted radioactivity (extracted tissue) increased from < 2% TRR in 0 DAT forage to 25% TRR in 86 DAT grain. For all samples, the initial extractions recovered 75% to 98% TRR.

Table 4. Distribution of $^{14}C$ in $CH_3CN/H_2O$ extracted wheat samples as %TRR and mg/kg equivalents.

| Sample TRR (mg/kg equivalents) | %TRR and (mg/kg aminopyralid equivalents) | | | | | |
|---|---|---|---|---|---|---|
| | $CH_3CN/H_2O$ extract | Reflux filtrate | Total $^{14}C$ extracted | Organic phase | Aqueous phase | Extracted tissue |
| Forage 0 DAT 2× (2.022) | 98.3% (1.988) | – – | 98.3% (1.988) | – – | – – | 1.7% (0.034) |
| Forage 0 DAT 4× 4.121 | 98.7% 4.069 | – | 98.7% 4.069 | – | – | 1.3% 0.052 |

| Sample TRR (mg/kg equivalents) | %TRR and (mg/kg aminopyralid equivalents) | | | | | |
|---|---|---|---|---|---|---|
| | CH$_3$CN/H$_2$O extract | Reflux filtrate | Total $^{14}$C extracted | Organic phase | Aqueous phase | Extracted tissue |
| Forage 14 DAT 2× | 91% | 2.3% | 93.3% | – | – | 6.7% |
| 0.418 | 0.38 | 0.01 | 0.390 | – | – | 0.028 |
| Forage 14 DAT 4× | 90.7% | 1.8% | 92.5% | – | – | 7.5% |
| 0.874 | 0.793 | 0.016 | 0.809 | – | – | 0.066 |
| Hay 35 DAT 2× | 88.8% | 2.1% | 90.9 | – | – | 9.1% |
| 0.284 | 0.252 | 0.006 | 0.258 | – | – | 0.026 |
| Hay 35 DAT 4× | 88.4% | 2.3% | 90.6% | – | – | 9.4% |
| 0.691 | 0.610 | 0.016 | 0.626 | – | – | 0.065 |
| Straw 86 DAT 2× | 64.2% | 16.6% | 80.8% | 1.8% | 79.1% | 19.2% |
| 0.281 | 0.18 | 0.047 | 0.227 | 0.005 | 0.222 | 0.054 |
| Straw 86 DAT 4× | 78.6% | 4.4% | 83% | 1.7% | 81.3% | 17% |
| 0.623 | 0.490 | 0.027 | 0.517 | 0.011 | 0.506 | 0.106 |
| Grain 86 DAT 2× | 63.7% | 16% | 79.8% | 11.4% | 68.4% | 20.2% |
| 0.039 | 0.025 | 0.006 | 0.031 | 0.004 | 0.026 | 0.008 |
| Grain 86 DAT 4× | 60.1% | 14.9% | 75% | 13.4% | 61.6% | 25% |
| 0.084 | 0.051 | 0.013 | 0.063 | 0.011 | 0.052 | 0.021 |

The wheat forage and hay samples were analysed with HPLC after initial extraction. The levels of aminopyralid and other metabolites at various HPLC retention times (regions) are shown in Table 5.

Table 5: HPLC Characterisation of radioactivity in CH$_3$CN/H$_2$O extracts of wheat forage and hay samples.

| Sample (mg/kg parent equivalents extracted) | %TRR and (mg/kg aminopyralid equivalents) | | | | |
|---|---|---|---|---|---|
| | aminopyralid (Region 3) | Region 1 | Region 5 | Region 6 | Total of XDE-750 + regions 1 & 5 |
| Forage 0 DAT 2× | 87.2% | 5.3% | 3.5% | 1.8% | 96% |
| (1.988) | 1.762 | 0.108 | 0.071 | 0.037 | 1.941 |
| Forage 0 DAT 4× | 89.9% | 4.6% | 2.1% | 2.2% | 96.6% |
| (4.069) | 3.703 | 0.19 | 0.087 | 0.089 | 3.980 |
| Forage 14 DAT 2× | 40.6% | 34.7% | 17.9% | – | 93.2% |
| (0.390) | 0.170 | 0.145 | 0.075 | – | 0.39 |
| Forage 14 DAT 4× | 37.9% | 36% | 18.6% | – | 92.5% |
| (0.809) | 0.332 | 0.314 | 0.163 | – | 0.809 |
| Hay 35 DAT 2× | 11.6% | 55.1% | 23.7% | – | 90.4% |
| (0.258) | 0.033 | 0.156 | 0.067 | – | 0.256 |
| Hay 35 DAT 4× | 12.7% | 50% | 25.7% | – | 88.4% |
| (0.626) | 0.088 | 0.345 | 0.178 | – | 0.611 |

The radioactivity that was attributed to parent aminopyralid (based on co-chromatography with reference standard) was present in Region 3. The levels of parent compound ranged from almost 90% extracted TRR in day 0 forage samples to 11% extracted TRR in day 35 hay samples. As parent aminopyralid decreases, the levels of radioactivity present in Regions 1 and 5 increase and form the majority of the extracted TRR. The radioactivity in Regions 1 and 5 comprised a glucose conjugate of aminopyralid and a glucose conjugate of hydroxylated aminopyralid, which were identified in hay samples following base and acid hydrolysis.

The levels of aminopyralid and other metabolites in straw and grain at various HPLC retention times (regions) are shown in Table 6.

Table 6. HPLC Characterisation of radioactivity in aqueous phase and organic partitioned extracts of wheat straw and grain samples.

| Sample | %TRR and (mg/kg aminopyralid XDE-750 equivalents) | | | | |
|---|---|---|---|---|---|
| | aminopyralid (Region 3) | Region 1 | Region 2 | Region 5 | Total of aminopyralid + regions 1 & 5 |
| Straw      86 DAT 2× Aqueous phase (0.227) | 7.9% (0.022) | 42% (0.118) | 6.1% (0.017) | 23.1% (0.065) | 73% (0.205) |
| Straw      86 DAT 4× Aqueous phase | 11.3% (0.071) | 32.9% (0.205) | – – | 37% (0.230) | |
| Organic phase | 0.1% (< 0.001) | < 0.1% (< 0.001) | – – | 0.3% (0.002) | |
| Total (0.517) | 11.4% (0.071) | 33% (0.205) | – – | 37.3% (0.232) | 81.7% (0.508) |
| Grain      86 DAT 2× Aqueous phase (0.031) | 17.7% (0.007) | 11.6% (0.004) | – – | 39% (0.015) | 68.3% (0.026) |
| Grain      86 DAT 4× Aqueous phase | 15.9% (0.013) | 12.3% (0.01) | – – | 33.3% (0.028) | |
| Organic phase | 0.3% (< 0.001) | <LOQ <LOQ | <LOQ <LOQ | 6.9% (0.006) | |
| Total (0.063) | 16.2% (0.014) | 12.3% (0.01) | <LOQ <LOQ | 40.2% (0.034) | 68.7% (0.058) |

Following partitioning of the initial extracts with $CH_2Cl_2:CH_3CN$, the radioactivity is predominantly soluble in the aqueous phase. Acid/base treatment of the aqueous phase and the extracted tissues resulted in further release of radioactivity that was identified as aminopyralid. The analysis of straw and grain (86 DAT 4×) is shown in Table 7.

Table 7. HPLC Characterisation of radioactivity in wheat straw and grain samples following acid/base treatment of aqueous phase and extracted tissues.

| Sample | %TRR and (mg/kg aminopyralid  equivalents) | | | | |
|---|---|---|---|---|---|
| | XDE-750 | Region 1 | Region 2 | Region 5 | Region 6 |
| Straw      86 DAT 4× | | | | | |
| Aqueous phase | 71.9% (0.448) | 6.1% (0.038) | – – | – – | 3.3% (0.02) |
| Extracted tissue | 6.6% (0.041) | 1.5% (0.009) | | 0.7% (0.004) | 0.9% (0.005) |
| Total | 78.5% (0.489) | 7.6% (0.047) | | 0.7% (0.004) | 4.2% (0.025) |
| Grain      86 DAT 4× | | | | | |
| Aqueous phase | 48.6% (0.041) | 1.8% (0.002) | 1.1% (0.001) | 0.5% (< 0.001) | 5.4% (0.005) |
| Extracted tissue | 11% (0.009) | <LOQ (<LOQ) | | <LOQ (<LOQ) | 1.6% (0.001) |
| Total | 59.6% (0.05) | 1.8% (0.002) | 1.1% (0.001) | 0.5% (< 0.001) | 7% (0.006) |

LOQ = 0.1%TRR.

Of the total extractable [14]C as reported in Table 4, aminopyralid comprised 78.5% of the TRR in wheat straw either as free, conjugated or bound forms. Approximately 12 – 13 % remained in other HPLC regions, and up to 3.5% TRR remained in unextracted tissue following acid/base treatment.

In grain, aminopyralid comprised 60% of the TRR (86 DAT 4×) again as free, conjugated or bound forms. Approximately 10% remained in other HPLC regions, and 0.7% TRR remained in unextracted tissue following acid/base treatment.

The data for the straw and grain demonstrate that acid/base treatment led to another HPLC region (region 6) that formed < 10% of the total extractable radioactivity for both sample types. This region is thought to be due to a minor component of region 5.

Acid/base stability of aminopyralid was demonstrated by subjecting $^{14}$C aminopyralid to acid/base treatments used in the metabolism studies. Approximately 97% of the radioactivity from the parent compound was recovered in HPLC region 3 as aminopyralid.

In relation to storage stability, repeated analyses were conducted for samples that had been stored frozen for up to 277 days (– 20 °C). The results of repeated analyses were similar to the initial analyses, indicating that the extracts were stable under conditions of frozen storage.

In summary, aminopyralid when applied to wheat is predominantly present as parent compound either in free, conjugated or bound forms. Acid/base treatment leads to good recovery of the applied material as parent compound.

*Grass/Pastures*

Three types of grasses that comprise pastures (perennial ryegrass, big bluestem and *Panicum maximum*) were treated with a single spray application of $^{14}$C aminopyralid at a rate of 360 g ai/ha (Magnussen and Balcher 2004). $^{14}$C aminopyralid was labelled in the 2- and 6-positions of the pyridine ring. The grasses were germinated and established in large tubs in a greenhouse. The containers were then moved outdoors and the grasses were cut to give a uniform stand (height 12 to 14 cm) in each container, prior to treatment.

Samples of grass were collected for analysis at 0 time (2 to 4 hours after application) and at 7, 14, 21 and 42 days after treatment (DAT). All samples were rinsed and then kept in frozen storage (– 20 °C) until further analysis. For the 42 DAT samples, half of the sample was weighed and then allowed to air dry over 1 to 2 days to produce a hay sample. After drying, the sample was reweighed and kept in frozen storage pending analysis. All samples were analysed within 2 months of collection, with the exception of the storage stability samples which were analysed within 660 days after collection.

TRR in all samples was determined using combustion analysis and LSC of liquid samples or extracts. Extracted material was analysed using HPLC and LC/MS. TRR in the unrinsed samples of the three grass species is shown in Table 8.

Table 8. TRR in unrinsed grass samples at varying times after application.

| Sampling Interval | | TRR (mg/kg aminopyralid equivalents) | | |
|---|---|---|---|---|
| | | Ryegrass | Big Bluestem | *Panicum maximum* |
| 0 time | | 48.84 | 25.84 | 17.96 |
| 7 DAT | | 36.91 | 21.03 | 18.87 |
| 14 DAT | | 22.90 | 8.29 | 10.08 |
| 21 DAT | | 20.85 | 9.03 | 10.03 |
| 42 DAT | grass | 6.57 | 5.61 | 4.81 |
| 42 DAT | hay | 23.34 | 12.64 | 19.13 |

The results indicate that there is a decline in TRR with time, with 0 day values being highest in ryegrass compared to the other two grass species. The differences reported for the 42 DAT grass and hay samples are due to loss of moisture and the dry matter content.

Extraction of radioactive residues from each of the grasses involved rinsing of the samples sequentially with water then MeOH, and HPLC analysis of the water and MeOH extracts. The rinsed grass was homogenised with $CH_3CN:H_2O$ (70:30), filtered and the resulting extract analysed by HPLC. Acid and base hydrolysis of the spent grass extracts resulted in aqueous and organic extracts that were further characterised using HPLC and LC/MS. Data from the extracts of each pasture type are shown in Table 9.

Table 9. Characterisation of TRR in various extracts of grasses by HPLC analysis.

| Sampling Interval | % TRR (mg/kg parent equivalents) and % recovered residue | aminopyralid (% recovered) | C-1[a] | C-2[b] | C-3[c] |
|---|---|---|---|---|---|
| Ryegrass | | | | | |
| 0 time | *48.84* | | | | |
|   H$_2$O/MeOH | 40.1 (82.1) | 80.3 | 1.4 | 0.3 | – |
|   CH$_3$CN/H$_2$O | 8.30 (17) | 16.7 | < 0.18% | 0.2 | – |
|   spent grass | 0.44 (0.9) | – | – | – | – |
|   Totals | | 97 | 1.4 | 0.5 | – |
| 7 DAT | *36.91* | | | | |
|   CH$_3$CN/H$_2$O | 30.23 (81.9) | 42 | 4.7 | 2.3 | 32.9 |
|   CH$_3$CN/acid | 6.27 (17) | 5.8 | 3.8 | 1 | 6.4 |
|   spent grass | 0.41 (1.1) | – | – | – | – |
|   Totals | | 47.8 | 8.5 | 3.3 | 39.3 |
| 14 DAT | *22.90* | | | | |
|   CH$_3$CN/H$_2$O | 13.26 (57.9) | 23.8 | 3.6 | 0.9 | 29.5 |
|   CH$_3$CN/acid | 8.89 (38.8) | 10.9 | 4.6 | 4 | 19.3 |
|   spent grass | 0.76 (3.3) | – | – | – | – |
|   Totals | | 34.7 | 8.2 | 4.9 | 48.8 |
| 21 DAT | *20.85* | | | | |
|   CH$_3$CN/H$_2$O | 11.57 (55.5) | 17.1 | 3 | 0.6 | 34.7 |
|   CH$_3$CN/acid | 8.32 (39.9) | 8.2 | 3.8 | 4 | 23.3 |
|   spent grass | 0.76 (4.6) | – | – | – | – |
|   Totals | | 25.3 | 6.8 | 4.6 | 58 |
| 42 DAT  grass | *6.57* | | | | |
|   CH$_3$CN/H$_2$O | 3.52 (53.6) | 13.8 | 3.4 | 1.2 | 35.2 |
|   CH$_3$CN/acid | 2.76 (42) | 8.1 | 4.1 | 5.5 | 24.3 |
|   spent grass | 0.29 (4.4) | – | – | – | – |
|   Totals | | 21.9 | 7.5 | 6.7 | 59.5 |
| 42 DAT  hay | *23.34* | | | | |
|   CH$_3$CN/H$_2$O | 9.67 (41.5) | 12.6 | 2 | ND | 26.9 |
|   CH$_3$CN/acid | 11.27 (48.3) | 11.2 | 4.2 | 3.9 | 29.1 |
|   spent grass | 2.38 (10.2) | – | – | – | – |
|   Totals | | 23.8 | 6.2 | 3.9 | 56 |
| Big Bluestem | | | | | |
| 0 time | *25.84* | | | | |
|   H$_2$O/MeOH | 22.58 (87.4) | 85.4 | 1.8 | 0.1 | – |
|   CH$_3$CN/H$_2$O | 3.10 (12) | 11.7 | ND | 0.3 | – |
|   spent grass | 0.16 (0.6) | – | – | – | – |
|   Totals | | 97.1 | 1.8 | 0.4 | – |
| 7 DAT | *21.03* | | | | |
|   CH$_3$CN/H$_2$O | 18.02 (85.7) | 61.3 | 6.2 | 3.5 | 14.6 |
|   CH$_3$CN/acid | 2.71 (12.9) | 5.3 | 3.2 | < 0.39 | 4.2 |
|   spent grass | 0.27 (1.3) | – | – | – | – |
|   Totals | | 66.6 | 9.4 | 3.5 | 18.8 |
| 14 DAT | *8.29* | | | | |
|   CH$_3$CN/H$_2$O | 6.83 (82.4) | 38.1 | 7.4 | 5.2 | 31.5 |
|   CH$_3$CN/acid | 1.3 (15.7) | 6.8 | 2.8 | ND | 6.1 |
|   spent grass | 0.16 (1.9) | – | – | – | – |
|   Totals | | 44.9 | 10.2 | 5.2 | 37.6 |
| 21 DAT | *9.03* | | | | |
|   CH$_3$CN/H$_2$O | 6.75 (74.7) | 28.2 | 8.5 | 3.3 | 34.8 |
|   CH$_3$CN/acid | 2.06 (22.8) | 9.8 | 2.5 | 1.7 | 8.8 |
|   spent grass | 0.23 (2.5) | – | – | – | – |
|   Totals | | 38 | 11 | 5 | 43.6 |
| 42 DAT  grass | *5.61* | | | | |
|   CH$_3$CN/H$_2$O | 3.8 (67.7) | 18.3 | 8.9 | 2.3 | 37.9 |
|   CH$_3$CN/acid | 1.65 (29.5) | 10.3 | 2.3 | 3.7 | 12.2 |
|   spent grass | 0.16 (2.8) | – | – | – | – |
|   Totals | | 28.6 | 11.2 | 6 | 50.1 |

| Sampling Interval | % TRR (mg/kg parent equivalents) and % recovered residue | aminopyralid (% recovered) | C-1[a] | C-2[b] | C-3[c] |
|---|---|---|---|---|---|
| 42 DAT    hay | *12.64* | | | | |
| CH$_3$CN/H$_2$O | 7.46 (59) | 18.8 | 6.3 | 1.2 | 32.8 |
| CH$_3$CN/acid | 4.61 (36.5) | 13 | 2.6 | 2.1 | 18.8 |
| spent grass | 0.57 (4.5) | – | – | – | – |
| Totals | | 31.8 | 8.9 | 3.3 | 51.6 |
| *Panicum maximum* | | | | | |
| 0 time | *17.96* | | | | |
| H$_2$O/MeOH | 16.07 (89.5) | 86.2 | 0.1 | < 0.14 | 2.4 |
| CH$_3$CN/H$_2$O | 1.86 (10.3) | 9.9 | ND | < 0.14 | 0.2 |
| spent grass | 0.04 (0.2) | – | – | – | – |
| Totals | | 96.1 | 0.1 | < 0.14 | 2.6 |
| 7 DAT | *18.87* | | | | |
| CH$_3$CN/H$_2$O | 17.08 (90.5) | 64.3 | 5.5 | 1.6 | 19 |
| CH$_3$CN/acid | 1.59 (8.4) | 3.4 | 1.9 | < 0.35 | 2.7 |
| spent grass | 0.21 (1.1) | – | – | – | – |
| Totals | | 67.7 | 7.4 | 1.6 | 21.7 |
| 14 DAT | *10.08* | | | | |
| CH$_3$CN/H$_2$O | 8.10 (80.4) | 27 | 3.6 | 2.2 | 47.6 |
| CH$_3$CN/acid | 1.72 (17.1) | 5.9 | 1.8 | 1.8 | 7.6 |
| spent grass | 0.24 (2.4) | – | – | – | – |
| Totals | | 32.9 | 5.4 | 4 | 55.2 |
| 21 DAT | *10.03* | | | | |
| CH$_3$CN/H$_2$O | 7.29 (72.9) | 25.9 | 4 | 1.4 | 41.4 |
| CH$_3$CN/acid | 2.37(23.6) | 8 | 2 | 2.8 | 10.8 |
| spent grass | 0.37 (3.7) | – | – | – | – |
| Totals | | 33.9 | 6 | 4.2 | 52.2 |
| 42 DAT    grass | *4.81* | | | | |
| CH$_3$CN/H$_2$O | 2.69 (56) | 19.9 | 3.9 | 2.1 | 30 |
| CH$_3$CN/acid | 1.92 (39.9) | 10.7 | 2.5 | 5.6 | 21.1 |
| spent grass | 0.2 (4.1) | – | – | – | – |
| Totals | | 30.6 | 6.4 | 7.7 | 51.1 |
| 42 DAT    hay | *19.13* | | | | |
| CH$_3$CN/H$_2$O | 10.43 (54.5) | 24 | 3.7 | 0.8 | 26 |
| CH$_3$CN/acid | 7.86 (41.1) | 10.9 | 2.6 | 5.2 | 22.4 |
| spent grass | 0.84 (4.4) | – | – | – | – |
| Totals | | 34.9 | 6.3 | 6 | 48.4 |

[a] The C-1 Region represented those residues eluting just after the column void volume (4-7 minutes);

[b] The C-2 Region represented those residues eluting just after XDE-750 (12-15 minutes);

[c] The C-3 Region represented those residues that were even less polar than the C-2 residues (17-21 minutes). The retention time for XDE-750 in this system was approximately 9 minutes.

The data show that during the first 2 weeks after application, aminopyralid forms a large proportion of the radioactivity, however it is slowly metabolised and transformed to other components denoted C-1, C-2 and C-3. Metabolite C-3 forms a large proportion of the isolated radioactivity (> 50% TRR) from day 7 onwards. It is easily extracted into the solvent mixtures that also extract aminopyralid and radiochromatograms showed that the extracts elute as single peaks.

To further characterise components C-1 to C-3, the CH$_3$CN/H$_2$O and CH$_3$CN/acid extracts from the 21 DAT grass samples were subjected to acid (1 or 3 N HCl) and/or base (1 N NaOH) hydrolysis, followed by organic solvent partitioning. Base hydrolysis was found to be more efficient than acid hydrolysis and 75% to 90% of the TRR in the C-1 and C-3 components from the CH$_3$CN/H$_2$O and CH$_3$CN/acid fractions was released following base hydrolysis. HPLC and LC/MS analysis of the organosoluble residues following base hydrolysis showed that 87–96% of the released radioactivity in the three grasses was eluted as aminopyralid.

The stability of selected samples of ryegrass and *Panicum maximum* after 650 days of frozen storage was investigated and details are given in Table 10.

Table 10. Stability of extracts from grass samples following frozen storage.

| Sample | Storage Interval | Extracts | %TRR (mg/kg parent equivalents) and % recovered residue | %TRR | | | |
|---|---|---|---|---|---|---|---|
| | | | | aminopyralid | C-1 | C-2 | C-3 |
| Rye grass 14 DAT | 0 days | $CH_3CN/H_2O$ | 57.9 | 23.8 | 3.6 | 0.9 | 29.5 |
| | | $CH_3CN/acid$ | 38.8 | 10.9 | 4.6 | 4.0 | 19.3 |
| | | Spent grass | 3.3 | – | – | – | – |
| | 665 days | $CH_3CN/H_2O$ | 54.2 | 23.1 | 2.8 | – | 26.7 |
| | | $CH_3CN/acid$ | 40.1 | 10.2 | 4.5 | 2.4 | 22.9 |
| | | Extracted grass | 5.7 | – | – | – | – |
| Pan. max 14 DAT | 0 days | $CH_3CN/H_2O$ | 80.4 | 27 | 3.6 | 2.2 | 47.6 |
| | | $CH_3CN/acid$ | 17.1 | 5.9 | 1.8 | 1.8 | 7.6 |
| | | Spent grass | 2.4 | – | – | – | – |
| | 671 days | $CH_3CN/H_2O$ | 71.8 | 25.5 | 3.5 | 2 | 40.8 |
| | | $CH_3CN/acid$ | 22.9 | NA | NA | NA | NA |
| | | Extracted grass | 5.4 | – | – | – | – |
| Pan. max 21 DAT | 0 days | $CH_3CN/H_2O$ | 72.7 | 25.9 | 4 | 1.4 | 41.4 |
| | | $CH_3CN/acid$ | 23.6 | 8 | 2 | 2.8 | 10.8 |
| | | Spent grass | 3.7 | – | – | – | – |
| | 698 days | $CH_3CN/H_2O$ | 62.1 | 25.7 | 3.2 | – | 33.2 |
| | | $CH_3CN/acid$ | 33 | NA | NA | NA | NA |
| | | Extracted grass | 4.8 | – | – | – | – |

NA = not analysed.

The data show that although there are small differences in the extractability of the TRR in the $CH_3CN/H_2O$ and $CH_3CN/acid$ fractions, the overall amount of aminopyralid and other components recovered was the same. There were no significant issues in relation to storage stability of aminopyralid residues.

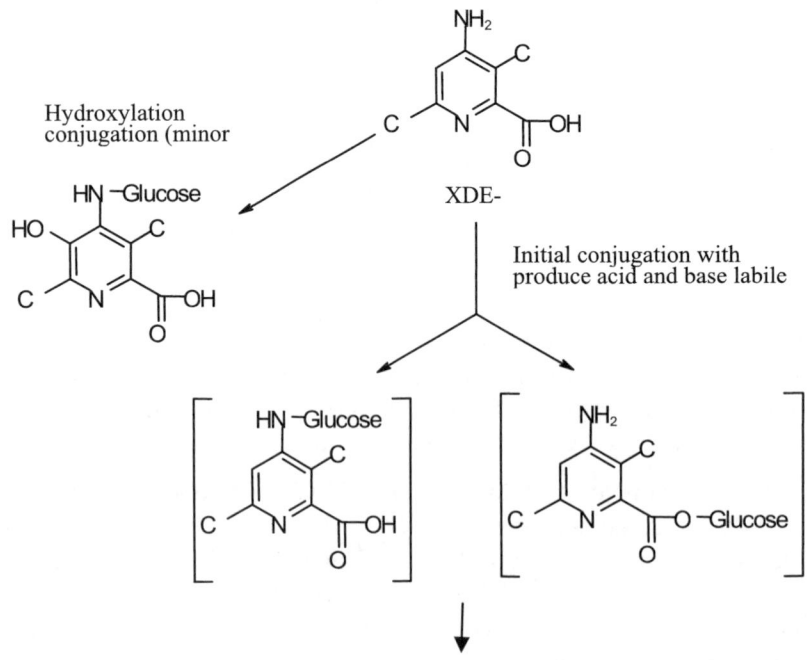

Figure 2: Proposed metabolic pathway for wheat and pasture grasses.

## Environmental fate in soil

The meeting received information on the environmental fate of aminopyralid in soil, including studies on photodegradation on soil, aerobic soil degradation and crop rotation studies (confined and field). The studies are reported in the following sections. $^{14}$C aminopyralid was labelled in the 2- and 6-positions of the pyridine ring in all soil studies.

### Photodegradation on soil

The photodegradation of aminopyralid was investigated in a German silt loam soil (Parabraun Erde) at 25 °C and 75% of ⅓ bar moisture using a Xe lamp as a light source (Rutherford 2004). Samples were irradiated at the equivalent of 28 days summer at 40 °N latitude.

The $^{14}$C aminopyralid was applied to samples of soil in flasks at a concentration of 5.2 mg ai/kg; samples were intermittently irradiated in a 12-hour irradiation/12-hour dark cycle to simulate regular sunlight. Treated samples were connected to traps for collection of $CO_2$ and other volatiles. Control samples were stored in the dark in an incubator at 25 °C. Samples were taken at 0, 2, 7, 14, 26, 35 and 44 days after treatment for the determination of parent compound and possible transformation products. A chemical actinometer was used to measure the amount of light that the samples had been subjected to; 44 DAT was equivalent to 28 days of irradiation in summer sun at 40 °N latitude.

Samples were extracted with acetone: 1 N HCl (90:10) and $^{14}$C residues were analysed using HPLC. Parent aminopyralid accounted for < 90% of the radioactivity present in any HPLC determination of organic extracts. The mass balance in the dark and irradiated samples was $100 \pm 4\%$ and $94 \pm 7\%$, respectively. After 44 days, 92% of the applied $^{14}$C remained as aminopyralid in the dark samples; no transformation had occurred in the dark samples.

In the irradiated samples, the concentration of aminopyralid decreased from 104% at day 0 to 69% at day 44; up to 3% of the applied $^{14}$C was present in the traps as $CO_2$ and acid gases. Extracted $^{14}$C residues decreased from 105% of the applied amount at day 0 to 74% and 93% of the applied amount in the irradiated and dark samples, respectively. In irradiated samples, non-extracted $^{14}$C (as determined by combustion analysis) increased from 0.6% at day 0 to 8% at day 44. In the dark samples, the corresponding values were 0.6% at day 0 and 3.8% at day 44.

Aminopyralid had degraded into non-extracted $^{14}$C and volatile components. The half-life ($DT_{50}$) for photodegradation of aminopyralid was 61 days and $DT_{90}$ was 203 days.

### Aerobic soil degradation

The rate of aerobic soil degradation of $^{14}$C aminopyralid in four European soils was investigated (Yoder and Smith 2003). The characteristics of the four soils are shown in Table 11.

Table 11. Characteristics of European soils used in a soil degradation study.

| Parameter | Soils and characteristics | | | |
|---|---|---|---|---|
| Location, Country | Thessaloniki, Greece | Cuckney, UK | Charentilly, France | Parabraun Erde, Germany |
| Texture | Clay loam | Sand | Light clay | Loam |
| Sand (%) | 41 | 91 | 42 | 60 |
| Silt (%) | 36 | 6 | 32 | 26 |
| Clay(%) | 23 | 3 | 26 | 14 |
| pH | 7.7 | 5.6 | 5.8 | 7.7 |
| Organic matter (%) | 2.5 | 2.4 | 1.9 | 2 |
| Soil biomass (µg/g) Initial | 216.1 | 111.4 | 54.7 | 40.6 |
| Final | 94.1 | 39.5 | 50.8 | 72 |
| CEC① (meq/100 g) | 14.4 | 6.8 | 15.1 | 10 |
| MHC② (%) | 86.9 | 43.7 | 68.4 | 56 |
| Bulk density (g/cm$^3$) | 0.88 | 1.28 | 1.08 | 1.18 |

①CEC is cation exchange capacity. ②MHC is moisture holding capacity.

Samples of each soil type were prepared 7 days before treatment to achieve 40% moisture holding capacity (MHC) and were incubated in the dark at 20 °C. Additional samples of the Parabraun Erde loam were used to investigate the effect of temperature on aerobic soil degradation by incubating in the dark at 10° and 30 °C for 8 days. Additional samples of Parabraun Erde soil were also sterilised before treatment to differentiate between microbial and aerobic degradation of aminopyralid. The samples (as prepared above at 20 °C) were irradiated with γ rays for 350 minutes prior to treatment with aminopyralid.

The $^{14}C$ aminopyralid was applied at concentrations of 0.14–0.17 mg ai/kg soil to simulate actual application rates of 110–120 g ai/ha. The treated samples were connected to NaOH traps (0.2 M) to collect $CO_2$ and to maintain constant relative humidity. All samples were incubated in the dark for up to 4 months after treatment. Duplicate samples of each soil type were taken at 0, 4, 7, 14, 28, 61, 92 and 123 days after treatment. At each sampling time, aliquots of the trapping solution were also taken and analysed by LSC. All soil samples were extracted several times using acetone: 1 N HCl 90:10; all extracts were combined and assayed for $^{14}C$ using LSC and HPLC. The extracted soil was also combusted to determine the amount of non-extracted $^{14}C$. The data is presented in Table 15, calculated $DT_{50}$ and $DT_{90}$ values are shown in Table 12.

Table 12. $DT_{50}$ and $DT_{90}$ values determined for European soils from an aerobic degradation study.

| Soil | $DT_{50}$ (days) | $DT_{90}$ (days) |
|---|---|---|
| Thessaloniki (clay loam) | 18 | 58 |
| Cuckney (sand) | 143 | 476 |
| Charentilly (light clay) | 22 | 72 |
| Parabraun Erde (loam) | 84 | 280 |
| Parabraun Erde (10° C) | 409 | 1359 |
| Parabraun Erde (30° C) | 115 | 382 |
| Parabraun Erde (sterile) | NA | NA |

NA = degradation was not observed.

The degradation half-life of aminopyralid ranged from 18 days in a clay loam to 143 days in a sandy soil.

In the Parabraun Erde soils, degradation at 10 °C was significantly slower than at 20 °C, however at 30 °C there was no increase in degradation or temperature-based dependence observed. No degradation was found in the sterile soil, indicating that abiotic degradation had not occurred.

The relative percentage of $^{14}C$ detected as $CO_2$ at the end of the study ranged from 30 to 70% of the applied radioactivity for samples incubated at 20 °C. Non-extracted radioactivity (NER) accounted for up to 23% of the applied $^{14}C$. $^{14}CO_2$ was higher in both clay-based soils, with 65% and 70% in the Thessaloniki clay and Charentilly light clay, respectively.

In the HPLC analysis of the soil extracts, only aminopyralid was present; no other component was present in the chromatograms.

In summary, aminopyralid degrades to $CO_2$ and non-extracted radioactivity or residues under ambient aerobic conditions.

In another study conducted in the US, the degradation of $^{14}C$ aminopyralid was investigated on five soils (Yoder and Smith 2002). Soils were selected from sites that were representative of typical pasture growing regions in the US and Canada. The characteristics of the 5 soils are shown in Table 13.

Table 13. Characteristics of US soils used in a soil degradation study.

| Parameter | Soils and characteristics | | | | |
|---|---|---|---|---|---|
| Location, Country | Kansas, USA | Manitoba, Canada | N Dakota, USA #1 | N. Dakota, USA # 2 | Texas, USA |
| Texture | Silt loam | Loam | Sandy loam | Clay loam | Clay |
| Soil series | Holdrege | Regent | Manning | Barnes | Houston Black |
| Sand (%) | 21 | 45 | 74 | 34 | 17 |
| Silt (%) | 60 | 40 | 16 | 34 | 32 |
| Clay(%) | 19 | 15 | 10 | 32 | 51 |
| pH | 4.6 | 7.5 | 7.3 | 4.8 | 7.5 |
| Organic matter (%) | NA | NA | 2.4 | 7.1 | 5.9 |
| Soil biomass ($\mu$g/g) | | | | | |
| Initial | 81.5 | 313.5 | NA | NA | 672.8 |
| Final | 61.3 | 220.4 | 105.3 | 93.6 | 759.9 |
| CEC (meq/100 g) | 14.5 | 25.6 | 15.5 | 23.5 | 45 |
| MHC (%)* | 25.9 | 30.3 | 14.3 | 33.1 | 40.8 |
| Bulk density (g/cm$^3$) | 1.06 | 1.03 | 1.24 | 0.96 | 0.99 |

* at ⅓ bar moisture

Samples of each soil were prepared 7 days before treatment to achieve 75% of ⅓ bar moisture content and incubated at 25 °C in the dark. The $^{14}$C aminopyralid was applied at soil concentrations equivalent to 5, 20, 60 and 120 g ai/ha. The Houston Black clay soil was treated at 60 and 120 g ai/ha rates only; the remaining four soils were treated at all rates. Treated samples were connected to NaOH traps (0.2 M) to collect $CO_2$ and to maintain constant relative humidity. All samples were incubated in the dark up to 12 months after treatment and samples were taken at intervals of 0, 1, 4, 8, 14, 22 days and 1, 2, 3, 4, 6, 9 and 12 months after treatment. At each sampling time, the trapping solution was assayed by LSC. As in the European study described above, all soil samples were extracted several times using acetone: 1 N HCl 90:10; all extracts were combined and assayed for $^{14}$C using LSC and HPLC. The extracted soil was also combusted to determine the amount of non-extracted $^{14}$C.

Calculated half-lives (DT$_{50}$) and DT$_{90}$ values are shown below.

Table 14. DT$_{50}$ and DT$_{90}$ values determined for US soils from an aerobic degradation study.

| Soil series | Application Rate (g ai/ha) | DT$_{50}$ (days) | DT$_{90}$ (days) |
|---|---|---|---|
| Holdrege (Kansas USA) | 5 | 60 | 199 |
| | 20 | 48 | 161 |
| | 60 | 59 | 195 |
| | 120 | 46 | 153 |
| Regent (Manitoba, Canada) | 5 | 25 | 83 |
| | 60 | 49 | 163 |
| | 120 | 34 | 113 |
| Manning (Nth Dakota, USA) | 5 | 14 | 48 |
| | 60 | 21 | 70 |
| | 120 | 14 | 45 |
| Barnes (Nth Dakota, USA) | 5 | ND | ND |
| | 20 | 266 | 885 |
| | 60 | 341 | 1134 |
| | 120 | 343 | 1141 |
| Houston Black (Texas, USA) | 60 | 5 | 16 |
| | 120 | 5 | 16 |

ND = degradation was not observed.

The half-lives ranged from 5 days in the Houston Black soil to 343 days in the Barnes soil; little difference was observed in the rate of degradation in the range of concentrations tested for each soil. Again, as observed in the European soil study, the clay based soil resulted in the shortest degradation half-life (DT$_{50}$).

The relative percentage of $^{14}$C detected as $CO_2$ at the end of the study ranged from 30 to 70% of the applied radioactivity for all samples. Non-extracted radioactivity (NER) accounted for 2 % to

25% of the applied [14]C at the end of the incubation period, with the highest levels present in the Houston Black clay. In the HPLC analysis of the soil extracts, only aminopyralid was present, no other component was present in the chromatograms.

In summary, aminopyralid degrades to $CO_2$ and non-extracted radioactivity or residues under aerobic conditions.

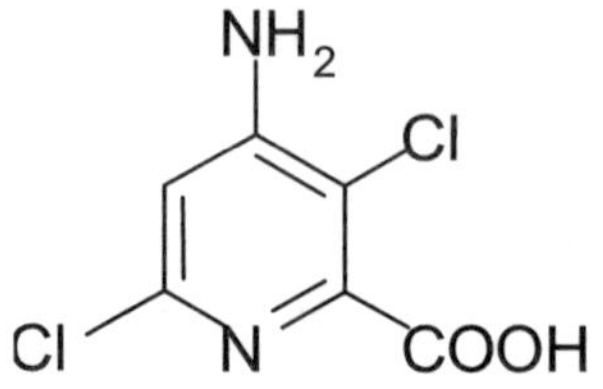


minor (< 5% AR soil metabolites)

incorporation into soil NER and/or formation of $CO_2$

Figure 3: Proposed degradation of aminopyralid in soil.

Table 15. Material balance of radioactivity for each of four European soil types.

| DAT | %$^{14}$C of applied aminopyralid | | | | | | | | | | | | | | | |
|---|---|---|---|---|---|---|---|---|---|---|---|---|---|---|---|---|
| | Thessaloniki clay loam (20 °C) | | | | Cuckney sand (20 °C) | | | | Charentilly light clay (20 °C) | | | | Parabraun Erde loam (20 °C) | | | |
| | $CO_2$ | Extract | NER | Total | $CO_2$ | Extract | NER | Total | $CO_2$ | Extract | NER | Total | $CO_2$ | Extract | NER | Total |
| 0 | NA | 94.1 | 0.9 | 95 | NA | 92.9 | 1.1 | 94 | NA | 92 | 1.9 | 93.9 | NA | 94.3 | 2.5 | 96.8 |
| 0 | NA | 97 | 0.9 | 97.8 | NA | 91.3 | 0.9 | 92.1 | NA | 91.7 | 2.5 | 94.2 | NA | 91.3 | 2.3 | 93.7 |
| 4 | 2.1 | 91.8 | 3 | 96.9 | 1.2 | 94 | 1.2 | 96.4 | 3.3 | 91.5 | 2 | 97.1 | 1.3 | 94.1 | 2.3 | 97.7 |
| 4 | 2.2 | 92.9 | 3.1 | 98.2 | 1.2 | 94 | 1 | 96.3 | 3.3 | 92.5 | 2 | 97.9 | 1.3 | 95 | 1.9 | 98.2 |
| 7 | 3.9 | 88.1 | 4.3 | 96.4 | 1.8 | 92.3 | 1.6 | 95.6 | 5.6 | 85. | 2.3 | 93.5 | 2.2 | 90.2 | 2.6 | 95 |
| 7 | 3.7 | 88.9 | 4.1 | 96.7 | 1.2 | 90.8 | 1 | 93 | 5.1 | 87 | 2.3 | 94.3 | 2.3 | 90.7 | 2.2 | 95.2 |
| 14 | 6.2 | 81.5 | 5.5 | 93.3 | 3.2 | 88.4 | 1.4 | 93 | 10.4 | 79.3 | 2.8 | 92.5 | 2.2 | 86.1 | 2. | 90.8 |
| 14 | 9.1 | 78.4 | 6.9 | 94.3 | 3.3 | 87.6 | 1.2 | 92 | 10.9 | 77.5 | 3.4 | 91.8 | 4.3 | 83.5 | 2.9 | 90.7 |
| 28 | 22.4 | 55.6 | 13.1 | 91.1 | 5.9 | 84.1 | 1.8 | 91.8 | 23.6 | 57.6 | 13.2 | 94.4 | 9.5 | 78 | 3.2 | 90.7 |
| 28 | 18.2 | 62.6 | 11.5 | 92.2 | 6 | 84.4 | 1.5 | 91.9 | 23.5 | 56 | 14 | 93.5 | 9.7 | 78.1 | 3.7 | 91.5 |
| 61 | 56.7 | 8.4 | 22.4 | 87.5 | 14.4 | 72.6 | 3.5 | 90.5 | 55.2 | 14.6 | 20.8 | 90.6 | 22.7 | 58.8 | 5.6 | 87.1 |
| 61 | 55.7 | 9.1 | 22.7 | 87.6 | 13.3 | 73.3 | 2.5 | 89.1 | 53 | 17.4 | 20.6 | 91.1 | 23 | 59.1 | 5.4 | 87.6 |
| 92 | 68.6 | 1.5 | 21.5 | 91.6 | 24.1 | 60.2 | 11.2 | 95.5 | 69.4 | 4.2 | 17.7 | 91.3 | 35.4 | 44.8 | 13.3 | 93.5 |
| 92 | 67.9 | 1.5 | 21.7 | 91.1 | | | | | 69.2 | 4.2 | 15.4 | 88.8 | 33.7 | 46.9 | 14.2 | 94.8 |
| 123 | 64.8 | 2.4 | 19.6 | 86.8 | 28.4 | 52.7 | 8 | 89.1 | | | | | 41.7 | 33.7 | 14.3 | 89.7 |
| 123 | | | | | 28.8 | 50.4 | 9.2 | 88.3 | | | | | 40.9 | 34.4 | 15.2 | 90.5 |

| DAT | %$^{14}$C of applied aminopyralid | | | | | | | | | | | |
|---|---|---|---|---|---|---|---|---|---|---|---|---|
| | Parabraun Erde loam (10 °C) | | | | Parabraun Erde loam (30 °C) | | | | Parabraun Erde loam (sterile) | | | |
| | $CO_2$ | Extract | NER | Total | $CO_2$ | Extract | NER | Total | $CO_2$ | Extract | NER | Total |
| 0 | NA | 98.4 | 1.4 | 99.8 | NA | 96.7 | 1 | 97.7 | NA | 98.2 | 0.4 | 98.6 |
| 0 | NA | 96.2 | 1.1 | 97.3 | NA | 98 | 1.4 | 99.4 | NA | 98.3 | 0.5 | 98.9 |
| 7 | 0.7 | 98.2 | 1.5 | 100.4 | 4 | 88.5 | 2.3 | 94.8 | 0.1 | 97.1 | 0.7 | 97.9 |
| 7 | 0.7 | 94.7 | 1.3 | 96.7 | 4 | 88.7 | 2.4 | 95 | 0.1 | 93.9 | 0.8 | 94.8 |
| 14 | 1.4 | 91.3 | 1.7 | 94.4 | 8.1 | 81.6 | 3.3 | 92.9 | 0.7 | 95.1 | 1 | 96.8 |
| 14 | 1.4 | 91.8 | 1.7 | 95 | 6.1 | 83.7 | 3.3 | 93.1 | 0.7 | 94.3 | 1.1 | 96.1 |
| 28 | 2.1 | 91.2 | 2 | 95.3 | 13.3 | 73 | 4.6 | 91 | 0.8 | 99.4 | 2 | 102.3 |
| 28 | 2.7 | 90.4 | 2 | 95.1 | 13.1 | 73.5 | 4.4 | 90.9 | 0.8 | 97 | 1.3 | 99.1 |

| DAT | Thessaloniki clay loam (20 °C) | | | | Cuckney sand (20 °C) | | | | Charentilly light clay (20 °C) | | | | Parabraun Erde loam (20 °C) | | | |
|---|---|---|---|---|---|---|---|---|---|---|---|---|---|---|---|---|
| | $CO_2$ | Extract | NER | Total | $CO_2$ | Extract | NER | Total | $CO_2$ | Extract | NER | Total | $CO_2$ | Extract | NER | Total |
| 61 | 7.1 | 84.5 | 5.5 | 97 | 27.7 | 59.5 | 9.9 | 97.1 | 1.6 | 98.5 | 1.6 | 101.6 | | | | |
| 61 | 6.4 | 86.4 | 6.3 | 99 | 26.5 | 58.9 | 11.4 | 96.8 | 1.3 | 96.1 | 1.5 | 98.9 | | | | |
| 92 | 8.7 | 82.3 | 6.5 | 97.6 | 32.8 | 51.4 | 9.5 | 93.6 | 2.1 | 100.5 | 2.1 | 104.7 | | | | |
| 92 | 8.3 | 82.4 | 5.3 | 95.9 | 32.2 | 51.5 | 9.3 | 93 | 2.5 | 95.3 | 2 | 99.8 | | | | |
| 123 | 10.2 | 78.2 | 7.1 | 95.5 | 37 | 44.8 | 12.7 | 94.6 | 3.3 | 100.2 | 2.3 | 105.8 | | | | |
| 123 | 10.6 | 78.3 | 7 | 96 | 35.4 | 46.3 | 11.3 | 93 | 3 | 100.7 | 2.2 | 106 | | | | |

*Field Dissipation*

The Meeting received data from field dissipation studies conducted in Europe, US and Canada.

The dissipation of aminopyralid in soil under field conditions in Europe was investigated (Unsworth *et al.* 2003). Sites in the U.K., Germany, Northern and Southern France that represented typical pasture growing areas were selected for the study. Three soil types were included in the study; clay loam in the U.K., sandy loam in Germany and Northern France and clay in Southern France. A commercial formulation containing both aminopyralid and triclopyr was used in the study, however only aminopyralid residues were determined.

Aminopyralid was applied at a rate of 60 g acid equivalents/ha (g ae/ha) using a boom sprayer. Each trial plot was divided into 4 sub-plots. A single application was made to bare soil and samples of soil were collected from 0 days to 12 months after application. At each sampling point, five core samples (30 cm depth) were taken from each of the four sub-plots and bulked to give a single sample of 20 cores. All samples were stored frozen (< –18°C) prior to analysis. Each core was cut into 10 cm "horizons" and replicate horizons from each sample were bulked for extraction. Although samples were taken up to 12 months after application, only samples up to 5 months were analysed, as $DT_{90}$ had been reached within this time. Additional samples were taken for biomass determination immediately prior to application and 6 and 12 months after application.

Residues of aminopyralid were determined using method GRM 02.34 which has a reported LOQ of 1.5 µg/kg. The method itself, including extractions, is described in the analytical methods section of this report. Fortified recoveries were conducted over the concentration range of 1.5–150 µg/kg and ranged 70 to 95% with a mean of 84% (n = 11) for extraction using 1 N HCl and 68 to 101% with a mean of 82% (n = 18) for extraction using 9 N HCl.

Soil characteristics and residues in the different soils are given in Tables 16 and 17.

Table 16. Characteristics of European soils used in a field dissipation study.

| Parameter | Soils and characteristics | | | |
|---|---|---|---|---|
| Location, Country | Derbyshire, U.K. | Dollern, Germany | Challons le Verger, Nth France | Sourges, Sth France |
| Texture | Clay loam | Sandy loam | Sandy loam | Clay |
| Sand (%) | 44 | 71 | 76 | 18 |
| Silt (%) | 32 | 19 | 15 | 35 |
| Clay(%) | 24 | 10 | 9 | 47 |
| pH | 6.6 | 6.2 | 7.5 | 8 |
| Organic carbon (%) | 1.5 | 3.6 | 0.9 | 3 |
| Soil biomass[1] (mg/100 g soil) | | | | |
| Initial | 43.94 (2.3) | 33.82 (0.99) | 28.87 (2.62) | 34.38 (0.9) |
| 6 months | 29.04 (1.56) | 28.42 (0.84) | 24.44 (2.22) | 56.29 (1.54) |
| 12 months | 38.06 (2.05) | 23.68 (0.76) | 28.69 (2.47) | 21.30 (0.68) |

| Parameter | Soils and characteristics | | | |
|---|---|---|---|---|
| Location, Country | Derbyshire, U.K. | Dollern, Germany | Challons le Verger, Nth France | Sourges, Sth France |
| Texture | Clay loam | Sandy loam | Sandy loam | Clay |
| CEC (meq/100 g) | 10.8 | 11.2 | 10.5 | 38.7 |
| MHC (%)②* | 13.7 | 5 | 6.6 | 20.3 |
| Bulk density (g/cm³) | 1.2 | 1.3 | 1.1 | 0.9 |

① Also expressed as % organic matter carbon in ().                    ② at 15 bar.

Table 17. Aminopyralid residues at varying soil depths and days after application.

| Trial Location | DAA | Aminopyralid (µg/kg dry wgt) | | |
|---|---|---|---|---|
| | | 0 – 10 cm | 10 – 20 cm | 0 – 20 cm |
| UK | Pre-application | ND | ND | ND |
| | 0 | 43.98 | ND | 22.06 |
| | 3 | 52.49 | ND | 26.32 |
| | 7 | 53.64 | ND | 26.89 |
| | 14 | 40.23 | ND | 20.19 |
| | 28 | 30.57 | (0.43) | 15.50 |
| | 61 | 13.41 | (0.44) | 6.92 |
| | 119 | 1.95 | ND | (1.05) |
| Germany | Pre-application | ND | ND | ND |
| | 0 | 50.43 | (1.28) | 25.86 |
| | 3 | 45.83 | ND | 22.99 |
| | 7 | 29.72 | 2.84 | 16.28 |
| | 14 | 18.50 | 14.18 | 16,34 |
| | 28 | 12.53 | 12.94 | 12.74 |
| | 55 | 8.77 | 8.68 | 8.72 |
| | 158 | (0.88) | (0.73) | (0.80) |
| Nth France | Pre-application | ND | ND | ND |
| | 0 | 34.58 | (0.71) | 17.65 |
| | 3 | 21.15 | (1.22) | 11.19 |
| | 7 | 15.37 | ND | 7.76 |
| | 14 | 19.88 | ND | 10.02 |
| | 28 | 15.74 | (0.58) | 8.16 |
| | 59 | 4.54 | (0.36) | 2.45 |
| | 127 | 2.17 | 2.56 | 2.37 |
| Sth France | Pre-application | ND | ND | ND |
| | 0 | 43.94 | 3.67 | 23.80 |
| | 3 | 23.29 | 2.79 | 13.04 |
| | 7 | 14.88 | 6.46 | 10.67 |
| | 14 | 11.83 | 4.86 | 8.34 |
| | 28 | 4.60 | (1.02) | 2.81 |
| | 61 | (0.77) | ND | (0.46) |
| | 125 | (0.35) | ND | (0.25) |

ND = not detected (< 0.3 µg/kg). For the values at 0 – 20 cm, ND was < 0.15 µg/kg. Residues values in brackets are > 0.3 µg/kg but < 0.15 µg/kg.

The calculated $DT_{50}$ and $DT_{90}$ values were as follows:

| | UK | Germany | Nth France | Sth France |
|---|---|---|---|---|
| $DT_{50}$ (days) | 35 | 31 | 26 | 8 |
| $DT_{90}$ (days) | 116 | 105 | 87 | 26 |

The data indicate that $t_{1/2}$ or $DT_{50}$ values are highly soil dependant and ranged from 8 days in a clay soil to 31 days in a sandy loam to 35 days in a clay loam.

In the field dissipation studies conducted in the US and Canada in 2002, aminopyralid was applied to test sites at a rate of 150 g ae/ha (Roberts and Schelle, 2004). Three test sites were selected in the US and five in Canada. The soil characteristics were determined for core samples taken down to 90cm depth, however only measurements for the 0–15 cm and 15–30 cm soil samples are shown in Table 19, for ease of comparison with the European study. Where values for specific parameters differ for the two core samples, the values for the 15–30 cm samples are given in parentheses.

Aminopyralid was applied using a tractor mounted spray rig and broadcast boom. At each site there were three replicated test plots, each consisting of five sub-plots. At each sampling interval (0 – 180 days in the US and 0–450 days in Canada), the five sub-plots were sampled from each replicate to produce a total of 15 core samples for analysis. The 15 core samples were cut and/or combined to produce three composite samples of 15 cm depth segments. Composite samples stored at − 20°C prior to analysis in accordance with method GRM 02.34. The data are shown in Tables 20 and 21 and $DT_{50}$ and $DT_{90}$ values are reported in Table 18.

The $DT_{50}$ or $t_{1/2}$ values from the field studies range from 8 to 54 days, with the majority of values approximating 30 to 35 days (Table 18). The $DT_{90}$ values are not as consistent, ranging from 26 days to 430 days. These values may be compared to the values derived from the radiolabelled aerobic soil degradation study where the $DT_{50}$ values ranged from 18 to 143 days in the European study and 5 to 343 days in the US study. The temperatures at the sites in the US and Canada ranged 18–26 C and 10–18 °C, respectively. Temperatures were not given for the European sites.

Table 18. $DT_{50}$ ($t_{1/2}$) and $DT_{90}$ values from all field dissipation studies.

| Trial Site | $DT_{50}$ (days) | $DT_{90}$ (days) | Study Reference |
|---|---|---|---|
| UK | 35 | 116 | GHE-P-10573; Unsworth et al. 2003 |
| Germany | 31 | 105 | |
| Nth France | 26 | 87 | |
| Sth France | 8 | 26 | |
| CA, USA | 26 | 85 | 020032; Roberts & Schelle 2004 |
| MS, USA | 34 | 114 | |
| MT, US | 35 | 220 | |
| NB, Canada | 21 | 60 | 020031, Roberts & Schelle, 2004 |
| ON, Canada | 31 | 260 | |
| MB, Canada | 30 | 110 | |
| SK, Canada | 9 | 40 | |
| AB, Canada | 54 | 430 | |

Table 19. Soils characteristics of test sites for dissipation studies conducted in US and Canada.

| Parameter | US and Canada Soils and Characteristics | | | | | | | |
|---|---|---|---|---|---|---|---|---|
| Location, Country | Fresno County, CA, USA | Washington County, MS, USA | Chouteau County, MT, USA | Kings County, NB, Canada | Brant County, ON, Canada | Whitewater County, MB, Canada | Corman Park County, SK, Canada | Lacombe County AB, Canada |
| Texture | Sandy loam | Silt loam | Sandy loam | Sandy loam | Loam | Loam/Clay loam | Loam | Clay loam |
| Sand (%) | 52 (54) | 29 | 77 (81) | 60 (66) | 35 | 35 (33) | 41 | 31 (27) |
| Silt (%) | 30 | 60 (56) | 14 (10) | 28 (26) | 46 (40) | 42 (38) | 38 (36) | 38 (40) |
| Clay(%) | 18 (16) | 11 (15) | 9 | 12 (8) | 19 (25) | 23 (29) | 21 (23) | 31 (33) |
| pH | 7.2 | 6 (6.5) | 8.2 | 5.4 (5.7) | 7.2 (7.7) | 7.5 (8.1) | 5.9 (6.9) | 5.3 (6) |
| Organic matter (%) | 1.3 (0.5) | 1.1 (0.8) | 1.3 (0.9) | 7.1 (6) | 3 (2.1) | 5.3 (2) | 4.6 (2.5) | 13.2 (8.1) |
| Soil biomass (µg/g) | 121.2 (14.1) | 52.1 (7.1) | 156 (81) | 292 (229) | 158 (70) | 236 (117) | 114 (119) | 459 (142) |
| CEC[1] (meq/100 g) | 16.4 (17.4) | 10.3 (13.7) | 14.1 (12.2) | 17.5 (15.4) | 14.1 (14.6) | 28.1 (26.7) | 20 | 25.5 (29.1) |
| MHC[2] (%) 1/3 bar | 19.2 (16) | 19.3 (23.7) | 12.8 (9.8) | 39.2 (30.3) | 24.4 | 29.2 (26.7) | 25.3 (21.9) | 40 (38) |
| 15 bar | 9.6 (8.7) | 7.2 (9.8) | 7.6 (6) | 21.3 (18.8) | 12 (11) | 21.6 (18.4) | 18 (15.6) | 30.2 (27.8) |
| Bulk density (g/cm$^3$) | 1.15 (1.17) | 1.18 | 1.24 (1.29) | 1.03 (0.96) | 1.17 | 1.07 | 1.12 (1.07) | 0.91 (0.98) |

[1] CEC is cation exchange capacity. [2] MHC is moisture holding capacity.

Table 20. Aminopyralid residues in soils in US and Canada.

| Fresno County, CA, USA | | | Washington County, MS, USA | | | Chouteau County, MT, USA | | | Kings County, NB, Canada | | |
|---|---|---|---|---|---|---|---|---|---|---|---|
| DAT | Residues µg/kg soil | % AR | DAT | Residues µg/kg soil | % AR | DAT | Residues µg/kg soil | %AR | DAT | Residues µg/kg soil | %AR |
| 0① | 940, 65 | 100, 91 | 0② | 1232, 152 | 100, 99 | 0③ | 1353, 68 | 100, 83 | 0④ | 976, 56.7 | 100, 81 |
| 9 | 86.20 | 97 | 8 | 47.4 | 59 | 7 | 31.5 | 59.2 | 6 | 60.8 | 81 |
| 15 | 55.1 | 64 | 15 | 92.6 | 105 | 14 | 38.2 | 64 | 15 | 42.1 | 58 |
| 22 | 53.3 | 59 | 29 | 35.8 | 46 | 28 | 24.8 | 51 | 29 | 30.4 | 43 |
| 65 | 10.7 | 11 | 57 | 20.1 | 27 | 56 | 12.6 | 26 | 62 | 7.9 | 8.4 |
| 91 | 0.78 | 1 | 93 | 8 | 11 | 98 | 7.9 | 14 | 90 | 4.6 | 5.9 |
| 126 | 0.83 | 1.7 | 122 | 12.9 | 17 | 129 | 8.4 | 15 | 121 | 2.1 | 2.7 |
| 182 | 0.52 | 0.6 | 183 | 1.2 | 1.6 | 372 | ND | 0 | | | |

① 15 and 45 minutes after treatment, soil pans and 15 cm cores, respectively.

② 5 and 35 minutes after treatment, soil pans and 15 cm cores, respectively.

③ 6 and 50 minutes after treatment, soil pans and 15 cm cores, respectively.

④ 10 and 35 minutes after application, soil pans and 15 cm cores, respectively.

Table 21. Aminopyralid residues in soils in US and Canada.

| Brant County, ON, Canada | | | Whitewater County, MB, Canada | | | Corman Park County, SK, Canada | | | Lacombe County, AB, Canada | | |
|---|---|---|---|---|---|---|---|---|---|---|---|
| DAT | Residues µg/kg soil | % AR | DAT | Residues µg/kg soil | % AR | DAT | Residues µg/kg soil | %AR | DAT | Residues µg/kg soil | %AR |
| 0⑤ | 1914, 58 | 100, 67 | 0⑥ | 2629, 98 | 100, 84 | 0⑦ | 9706, 95 | 100, 85 | 0⑦ | 2084, 101 | 100, 54 |
| 9 | 61.8 | 98 | 7 | 80.8 | 86 | 7 | 27.3 | 30 | 7 | 106 | 60 |
| 16 | 45.5 | 66 | 14 | 94.5 | 103 | 14 | 22.8 | 25 | 14 | 83.7 | 54 |
| 32 | 30.7 | 36 | 28 | 75.7 | 78 | 30 | 10.9 | 12 | 28 | 96.3 | 60 |
| 62 | 17.8 | 30 | 63 | 36.9 | 37 | 63 | 2.1 | 2.4 | 61 | 80 | 48 |
| 97 | 8.9 | 14 | 92 | 15.6 | 16 | 88 | 4.7 | 3.7 | 91 | 62.6 | 37 |
| 130 | 14.8 | 18 | 121 | 5.2 | 6 | 123 | 1.6 | 1.9 | 123 | 39.3 | 24 |
| 368 | 2.5 | 2.7 | | | | | | | 370 | 39.6 | 22 |
| | | | | | | | | | 455 | 10.8 | 6 |

⑤ 11 and 60 minutes after treatment for soil pans and 15 cm cores, respectively.

⑥ 10 and 60 minutes after treatment for soil pans and 15 cm cores, respectively.

⑦ 1 hour after treatment for soil pans and 15 cm cores.

## Environmental fate in water-sediment systems

The meeting received studies on the anaerobic aquatic metabolism of aminopyralid and degradation in a sediment-water system. Due to the nature of the use of the compound, these studies are not considered necessary for the meeting, as indicated in the *Revised Data Requirements for Studies of Environmental Fate* (JMPR 2003, section 2.11, p. 12).

## Crop Rotation Studies

A confined crop rotation study was conducted in the US in a sandy loam soil (Magnussen 2004). [14]C aminopyralid was applied to the soil at a rate of 10 g ai/ha, equivalent to the application rate for wheat. The characteristics of the soil were as follows: 71% sand; 20% silt; 9% clay; pH 6.1; CEC 9.5 meq/100 g; 2.3% OM; 10% MHC at ⅓ bar; 1.2 g cm$^{-3}$ density. The [14]C aminopyralid was applied to each test plot (1.38 mg/plot) and allowed to remain fallow for 90 or 120 days outdoors. Plots were tilled to a depth of 7–8 cm immediately prior to sowing with lettuce, turnips or sorghum.

Samples of immature lettuce were collected from each of the plots at 46–50 days after planting and mature samples were collected at harvest at 59–61 days after planting. Immature turnip plants were collected at 39–42 days after planting, while mature turnips were sampled at 73–82 days after planting. For sorghum, "early forage" samples were taken at 29–34 days after planting and "late forage" samples were taken at 80–97 days after planting. At harvest (110–127 days after planting) samples of grain and stover (fodder) were collected for analysis. Soil samples were collected at the time of application, at the time of planting and at the time that each sample was collected. At all other sampling times, core samples were collected at random from each plot. All samples (except turnip roots) were homogenised and stored frozen (-20 °C) prior to combustion analysis. The turnip roots were washed in water prior to frozen storage; the water rinsates were retained for LSC analysis.

TRR in representative crop and soil samples were determined by combustion before extraction of the remainder of the samples. Early sorghum forage, sorghum stover and mature turnip tops were the only ones that required extraction and HPLC analysis of radioactivity. Homogenised samples were extracted using CH3CN:H2O (70:30) and partitioned against hexane. Following acidification (pH 2), the samples were partitioned using CH3CN:CH2Cl2. All fractions were assayed using LSC and the CH3CN:CH2Cl2 and aqueous fractions were analysed using HPLC. Values of TRR in each of the samples are shown in Table 22.

Table 22. TRR in various follow crops at 90 and 120 DAT in a crop rotation study.

| Sample | 90 day plots | | 120 day plots | |
|---|---|---|---|---|
| | Days after planting | TRR (mg/kg parent equivalents)① | Days after planting | TRR (mg/kg parent equivalents)① |
| Lettuce | | | | |
| immature | 50 | 0.002 | 45 | 0.001 |
| mature | 61 | < 0.002 | 58 | < 0.001 |
| Turnip | | | | |
| immature | 39 | 0.007 | 42 | 0.007 |
| mature tops | 79 | 0.004 | 72 | 0.01 |
| mature roots | 79 | < 0.001 | 72 | < 0.001 |
| Sorghum | | | | |
| early forage② | 29 | 0.027 | 33 | 0.017 |
| late forage③ | 97 | 0.003 | 80 | 0.003 |
| stover | 127 | 0.027 | 110 | 0.003 |
| grain | 127 | 0.006 | 110 | 0.003 |

① mg/kg aminopyralid equivalents. ②18″ high crop. ③ Equivalent to soft to hard dough stage. LOD = 0.0005 mg/kg aminopyralid equivalents; LOQ = 0.002 – 0.003 mg/kg aminopyralid equivalents.

The data show that TRR above the reported LOQs are present in immature turnips, turnip tops, sorghum forage and sorghum stover harvested from both the 90 day and 120 day plots. Very little radioactivity is present in mature crop parts, such as lettuce, turnips or grain. The distribution of the radioactivity in extracted samples is shown in Table 23.

Table 23. Distribution and characterisation of radioactivity in extracts of follow crops.

| Sample Extracts | Total % TRR | % TRR | | | |
|---|---|---|---|---|---|
| | | aminopyralid | C-1 | C-2 | NER |
| Sorghum forage (90 days (early)) | | | | | |
| CH3CN/CH2Cl2 | 49 | 39.6 | – | 9.4 | – |
| Extracted tissue | 39.9 | 4.6 | 20.5 | 13.7 | |
| Spent tissue① | 11.1 | – | – | – | 11.1 |
| Total % TRR | | 44.2 | 20.5 | 23.1 | 11.1 |
| (mg/kg) | | (0.012) | (0.006) | (0.006) | (0.003) |
| Sorghum forage (120 days (early)) | | | | | |
| CH3CN/CH2Cl2 | 38.4 | 23.7 | – | 14.2 | – |
| Extracted tissue | 45.8 | 3.2 | 17 | 24 | – |
| Spent tissue① | 15.8 | – | – | – | 15.8 |
| Total % TRR | | 26.9 | 17 | 38.2 | 15.8 |
| (mg/kg) | | (0.005) | (0.003) | (0.006) | (0.003) |
| Sorghum stover (90 days) | | | | | |
| CH3CN/CH2Cl2 of CH3CN/H2O | 15.3 | 5.4 | 1.4 | 8.5 | – |
| Extracted CH3CN/H2O | 50.2 | 7.2 | 15.9 | 27.2 | – |
| CH3CN/CH2Cl2 of reflux | 7.7 | 5.5 | – | 2.2 | – |
| Reflux spent aqueous | 16.7 | NA | NA | NA | – |

| Sample Extracts | Total % TRR | % TRR | | | |
|---|---|---|---|---|---|
| | | aminopyralid | C-1 | C-2 | NER |
| Spent tissue① | 11.1 | – | – | – | 11.1 |
| Total % TRR (mg/kg) | | 18.1 (0.005) | 17.3 (0.005) | 37.9 (0.01) | 11.1 (0.003) |
| Turnip tops (120 days) CH$_3$CN/CH$_2$Cl$_2$ (1$^{st}$ CH$_3$CN/H$_2$O) | 15.5 | 4.4 | – | 11.1 | – |
| Extracted aqueous | 75 | 12.6 | 5.4 | 56.8 | – |
| 2$^{nd}$ CH$_3$CN/H$_2$O | 7.2 | NA | NA | NA | – |
| Spent tissue① | 2.3 | – | – | – | 2.3 |
| Total % TRR (mg/kg) | | 17.2 (0.002) | 5.4 (0.001) | 67.9 (0.007) | 2.3 (0.001) |

① Spent tissue = PES or post extraction solids. C-1 and C-2 are metabolite components, NER: non-extractable residue

A large proportion of the radioactivity was extracted, with TRR in spent tissues (or post extraction solids) ranging from 2% to 16% of the TRR in the selected samples. The data in Table 23 indicate that the residue profile in the extracted fractions is similar to that found in the wheat metabolism study. The majority of the extracted radioactivity is composed of aminopyralid and two metabolite components C-1 and C-2. C-1 is more polar than aminopyralid, while C-2 is less polar than aminopyralid. As the crop matures, less aminopyralid is present, with a large proportion of the radioactivity being associated with another component, C-2. Results from base hydrolysis of the spent aqueous fractions show that there is significant conversion of both C-1 and C-2 metabolites to aminopyralid. The results are very similar to those seen in the wheat and grass metabolism studies, where base hydrolysis of aqueous and organosoluble fractions followed by purification resulted in the identification of free and bound forms of aminopyralid.

In relation to soil, the amounts of radioactivity remaining at the time of planting and at various times up until harvest are shown in Table 24. At the time of planting, approximately 35 to 60% of the applied radioactivity was present in the plots. Of the remaining TRR in the soil of both plots at planting, approximately 16% to 32% was extracted as aminopyralid and another 15 to 25% comprised non-extracted residues. Additional components in the HPLC profile, being either more or less polar than aminopyralid, were present at levels of 1–3% of the TRR.

Table 24. TRR in soil at various times after application.

| Sample | DAT | %TRR in Soil | % of Applied Radioactivity in Soil (90 days) | | | |
|---|---|---|---|---|---|---|
| | | | parent | Spent AQ | NER① | Other② |
| Lettuce | | | | | | |
| 0 DAT | 0 | 88 | 80.3 | 3.6 | 0.7 | 3.2 |
| Planting | 90 | 51.6 | 24.5 | 2 | 22.7 | 1.6 |
| Harvest | 151 | 31.4 | 13.5 | 0.9 | 15.9 | 1 |
| Turnips | | | | | | |
| 0 DAT | 0 | 87.5 | 79.9 | 3.6 | 0.7 | 3.2 |
| Planting | 90 | 40.5 | 19.6 | 2.5 | 16.7 | < MQL③ |
| Immature | 129 | 37.3 | 16.4 | 2.1 | 17.1 | 1.2 |
| Harvest | 179 | 16.1 | 5.4 | 0.1 | 9.8 | 0.7 |
| Sorghum | | | | | | |
| 0 DAT | 0 | 106 | 91.3 | 4.1 | 0.8 | 3.6 |
| Planting | 90 | 61.6 | 32.2 | 1.6 | 22.9 | 4.4 |
| Early forage | 119 | 21.1 | NA | NA | NA | NA |
| Late forage | 187 | 22.7 | 2.8 | 0.6 | 17.1 | 2.2 |
| Harvest | 217 | 20.9 | 3.7 | 2.2 | 13.2 | 1.5 |
| Sample | DAT | %TRR in Soil | % of Applied Radioactivity in Soil (120 days) | | | |
| | | | parent | Spent AQ | NER① | Other② |
| Lettuce | | | | | | |
| 0 DAT | 0 | 87.5 | 79.1 | 0.3 | 5.7 | 2.2 |
| Planting | 120 | 57.2 | 31.8 | 1.6 | 20.2 | 3.5 |
| Harvest | 179 | 38.5 | 14.5 | 3.6 | 17.8 | 2.2 |
| Turnips | | | | | | |
| 0 DAT | 0 | 87.9 | 79.5 | 0.3 | 5.8 | 2.2 |
| Planting | 120 | 37 | 18.2 | 1 | 15.8 | 2 |
| Harvest | 193 | 32.9 | 9.4 | 4.5 | 18 | 0.9 |

| Sample | DAT | %TRR in Soil | % of Applied Radioactivity in Soil (90 days) | | | |
|--------|-----|-------------|---------|----------|-------|--------|
|        |     |             | parent | Spent AQ | NER① | Other② |
| Sorghum | | | | | | |
| 0 DAT | 0 | 93.3 | 85.2 | 3.8 | 0.8 | 3.4 |
| Planting | 120 | 36.4 | 16.3 | 1.8 | 17.3 | < MQL③ |
| Early forage | 154 | 38 | 9 | 1.5 | 25.7 | 1.3 |
| Late forage | 200 | 34.4 | 5.5 | 2.7 | 23.4 | 2.8 |
| Harvest | 230 | 23.9 | 3.6 | 1.9 | 17.2 | 1 |

① NER = Non-extractable residues ② Other = HPLC regions other than parent aminopyralid composed of 1 to 2 other components. ③ MQL = minimum quantifiable limit equivalent to 1% of the total applied radioactivity.

At harvest, aminopyralid had decreased to levels ranging 3.6–14.5% of the applied radioactivity for both plots.

Further extraction and workup of the non-extracted residues in soil using strong acid resulted in the release of another 15% of the TRR. In all cases, the additional $^{14}$C released was not aminopyralid *per se*, but an aqueous soluble component.

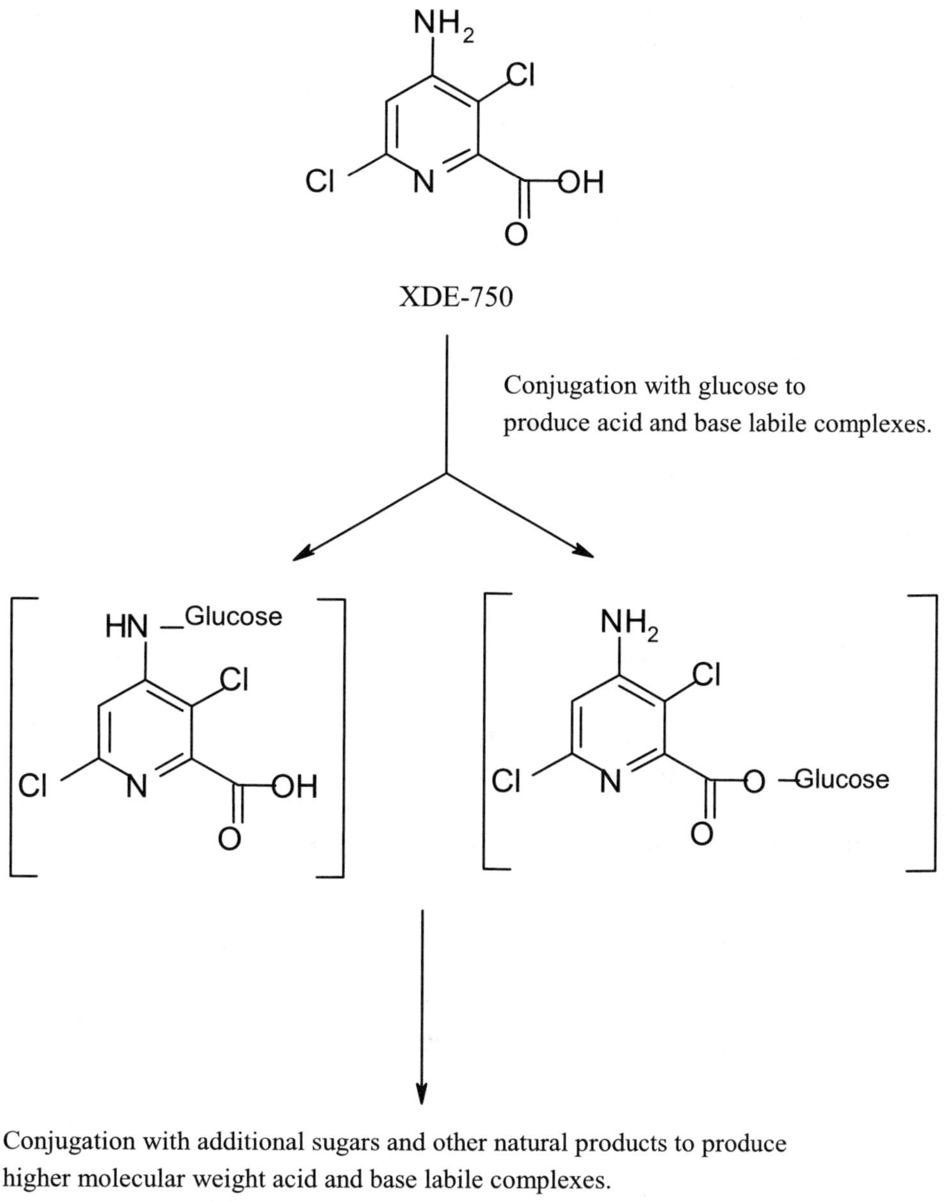

Figure 4: Proposed metabolic pathway of aminopyralid in rotational crops.

## RESIDUE ANALYSIS

The Meeting received details of analytical methods for the determination of residues of aminopyralid in agricultural commodities (namely pasture, cereal grains, cereal forage and straw), bovine tissues and milk, soil, water, air and human blood and urine. Only relevant studies are summarised in this evaluation.

### *Analytical methods*

#### *Crop Matrices*

Method GRM 02.31 (Olberding, Arnold and Hastings, 2004) was developed for the quantitative determination of aminopyralid residues in barley, sorghum, wheat and grass pasture. Residues are determined using LC/MS/MS. The validated LOQ for all matrices is reported as 0.01 mg/kg.

Residues of aminopyralid are extracted from the sample matrices by homogenisation with 0.1 N NaOH. An aliquot of the NaOH solution is acidified using HCl and the solution is heated at 80 °C for approximately 90 minutes. Following hydrolysis, the sample is purified using a mixed-mode polymeric anion exchange solid-phase extraction (SPE) column. After elution from the SPE column with ethyl acetate/trifluoroacetic acid (99:1), an internal standard is added ($^{13}C_2^{15}N$-aminopyralid) and the eluate is evaporated to dryness. The remaining residue is dissolved in $CH_3CN$/pyridine/1-BuOH solution (22:2:1) and then derivatized with butyl chloroformate to form the 1-butyl esters (1-BE) of both the analyte and the internal standard. Following derivatization, the mixture is diluted to volume with $MeOH/H_2O/CH_3COOH$ (50:49.9:0.1) and analysed using liquid chromatography with positive-ion electrospray tandem mass spectrometry (LC/MS/MS). The MS/MS ion transitions are $(M + H)^+$ at $m/z$ 263 and fragment ions (daughter ions) at $m/z$ 134 and $m/z$ 161 for aminopyralid-BE. For the internal standard, $^{13}C_2^{15}N$-aminopyralid 1-BE, the ion transitions are $(M + H)^+$ +2 at $m/z$ 268 and fragment ions at $m/z$ 139 and $m/z$ 166. HPLC operating conditions include use of a Diazem 3000 $C_{18}$ column (4.6 mm × 100mm, 3 µm particle size) operating at 35 °C; mobile phase A: $MeOH/CH_3COOH$ (99.9:0.1) and mobile phase B: $H_2O/CH_3COOH$ (99.9: 0.1).

Isotopic overlap between the analyte and the internal standard was determined by analysing standard solutions of each compound individually, i.e. determining peak areas for each derivatised compound. The analyte → ISTD crossover factor was determined as 0.00532 and the ISTD → analyte crossover factor was 0.00266.

Calculated LOQs ranged from 0.004 to 0.006 mg/kg, which support the method LOQ of 0.01 mg/kg. The calculated LOD was < 0.002 mg/kg. In actual samples, results are reported as ND for residues that were > LOD and < validated LOQ if ≥ LOD.

In the method validation component of the study, untreated control samples of barley, sorghum and wheat grain; barley, sorghum and wheat forage; barley straw, sorghum stover and wheat straw; grass forage and hay were fortified with aminopyralid at concentrations ranging 0.01 – 0.5 mg/kg for grain, 0.01–5 mg/kg for forage and straw and 0.01–20 mg/kg for grass forage and hay. Recoveries are summarised in Table 26.

Table 25. Recovery of aminopyralid from barley, sorghum, wheat matrices and grass forage and hay.

| Sample matrix | Fortification level (mg/kg) | Number of samples (n) | Recovery Range (%) | Mean Recovery (%) |
|---|---|---|---|---|
| Cereal grains (barley, sorghum, wheat) | 0.01 | 10 | 92 – 112 | 102 |
| | 0.025 | 2 | 107, 109 | 108 |
| | 0.05 | 5 | 96 – 111 | 104 |
| | 0.1 | 2 | 109, 110 | 109 |
| | 0.25 | 3 | 95, 101, 103 | 100 |
| | 0.5 | 5 | 97 – 107 | 103 |

| Sample matrix | Fortification level (mg/kg) | Number of samples (n) | Recovery Range (%) | Mean Recovery (%) |
|---|---|---|---|---|
| Cereal forages (barley, sorghum, wheat) | 0.01 | 12 | 88 – 105 | 97 |
| | 0.05 | 6 | 86 – 104 | 97 |
| | 0.25 | 3 | 93, 97, 98 | 96 |
| | 1 | 3 | 101, 106, 106 | 104 |
| | 5 | 6 | 90 – 106 | 99 |
| Cereal straw/stover (barley, sorghum, wheat) | 0.01 | 12 | 89 – 108 | 97 |
| | 0.05 | 6 | 95 – 104 | 100 |
| | 0.25 | 3 | 93, 94, 94 | 94 |
| | 1 | 3 | 94, 97, 97 | 96 |
| | 5 | 6 | 92 – 97 | 95 |
| Grass forage | 0.01 | 8 | 87 – 101 | 92 |
| | 0.05 | 4 | 94 – 98 | 96 |
| | 0.25 | 4 | 99 – 103 | 101 |
| | 1 | 4 | 95 – 101 | 98 |
| | 5 | 3 | 95, 95, 97 | 96 |
| | 20 | 6 | 93 – 103 | 100 |
| Grass hay | 0.01 | 8 | 89 – 105 | 98 |
| | 0.05 | 4 | 95 – 105 | 100 |
| | 0.25 | 4 | 92 – 101 | 97 |
| | 1 | 4 | 93 – 100 | 98 |
| | 5 | 4 | 92 – 108 | 99 |
| | 20 | 6 | 98 – 102 | 100 |

Independent laboratory validation (Reed, 2005) confirmed the LOQ of 0.01 mg/kg in fortified wheat grain and grass forage samples, using LC/MS/MS. Recoveries in wheat grain at concentrations of 0.01 and 0.1 mg/kg ranged 100–116% and 93–118%, respectively. The mean recovery in wheat grain was $110 \pm 8.2\%$ (n = 9). Recoveries in grass forage at concentrations of 0.01 and 60 mg/kg ranged 94–105% and 111–120%, respectively. The mean recovery in grass forage was $108 \pm 9.5\%$ (n = 10).

A GC/MS method for the determination of aminopyralid, fluroxypyr and 2,4-D residues in pastures was provided (Pinheiro and De Vito, 2003). The method is a modified version of GRM 02.31 and has a validated LOQ of 1 mg/kg. Modifications of method GRM 02.31 include acidification with $H_2SO_4$ in place of HCl and ethyl acetate partitioning prior to derivatisation to form the 1-butyl ester. Recoveries were validated over a range of concentrations (1 – 100 mg/kg) and ranged 78 – 115% with a mean recovery of $95 \pm 7\%$. The reported LOD was 0.2 mg/kg in pasture samples.

An assessment of European multi-residue methods for the determination of aminopyralid was provided to the meeting (Class, 2003). It was concluded that the German DFG S19 or the Netherlands multi-residue methods are not appropriate for the determination of aminopyralid in plant materials, foodstuffs of animal origin or soil without major adaptations. The lack of basic and/or acidic extraction procedures in the multi-residue methods preclude their use for aminopyralid. It was recommended that for enforcement purposes, methods GRM 02.31, GRM 02.34 and GRM 03.18 be used until a LC/MS/MS based multi-residue method with extraction and clean-up steps targeting carboxylic acid analytes was made available.

*Animal matrices*

Method GRM 03.18 was developed for the determination of aminopyralid residues in bovine tissues (muscle, fat, liver and kidney) and milk (Rutherford and Hastings, 2003). The validated limit of quantitation is 0.01 mg/kg.

Aminopyralid residues are extracted from tissue samples by homogenising and shaking with $MeOH/NaHCO_3$ solution (20:1, v:w). An aliquot is diluted with $H_2O$ and purified using a SPE plate (Waters Oasis MAX solid-phase extraction plate). The SPE plate is washed with $CH_3CN$ and a $MeOH/CH_3COOH$ solution (97:3) and eluted with an ethyl acetate/trifluoroacetic acid solution (99:1).

The internal standard ($^{13}C_2$$^{15}N$-aminopyralid) is added to the eluate, which is evaporated to dryness and the remaining residues are re-dissolved in $CH_3CN$/pyridine/BuOH (22:2:1) coupling reagent. The samples and standards are derivatized at room temperature with butyl chloroformate to form the 1-butyl esters of the analyte and internal standard. Following derivatization, the samples are diluted with $MeOH/H_2O/CH_3COOH$ (50:50:0.1) mobile phase. The final solution is analysed by liquid chromatography with positive-ion electrospray ionisation (ESI) tandem mass spectrometry (LC/MS/MS). All instrumental conditions are identical to those reported for method GRM 02.31.

The method was validated over the concentration range of 0.01–2.5 mg/kg in kidney and 0.01–1 mg/kg for all other tissues and milk. The recoveries in various tissues and milk are shown in Table 27.

Table 26. Recovery of aminopyralid from animal matrices.

| Sample | Fortification level (mg/kg) | Number of samples (n) | Recovery Range (%) | Mean Recovery (%) |
|---|---|---|---|---|
| Bovine fat | 0.01 | 5 | 85 – 99 | 92 |
| | 0.1 | 2 | 94, 97 | 96 |
| | 1 | 5 | 88 – 97 | 93 |
| Bovine muscle | 0.01 | 5 | 81 – 96 | 86 |
| | 0.1 | 2 | 91, 92 | 92 |
| | 1 | 5 | 75 – 85 | 79 |
| Bovine kidney | 0.01 | 5 | 67 – 86 | 79 |
| | 0.5 | 2 | 83, 89 | 86 |
| | 2.5 | 5 | 72 – 95 | 82 |
| Bovine liver | 0.01 | 5 | 77 – 88 | 83 |
| | 0.1 | 2 | 89, 93 | 91 |
| | 1 | 5 | 77 – 82 | 79 |
| Whole milk | 0.01 | 5 | 78 – 87 | 81 |
| | 0.1 | 2 | 86, 91 | 89 |
| | 1 | 5 | 76 – 83 | 79 |
| Milk cream | 0.01 | 2 | 71, 91 | 81 |
| | 0.1 | 1 | 88 | |
| | 1 | 2 | 64, 71 | 68 |
| Skim milk | 0.01 | 2 | 77, 82 | 79 |
| | 0.1 | 1 | 99 | |
| | 1 | 2 | 71, 75 | 73 |

Recoveries were conducted in whole milk, milk cream and skim milk. The mean recovery in all milk fractions was 80 ± 10% (n = 22) over the concentration range 0.01 – 1 mg/kg (whole milk 82%, cream 77% and skim milk 81%).

In an independent laboratory validation of method GRM 03.18 (Reed 2004), the LOQ of 0.01 mg/kg was confirmed for aminopyralid in bovine milk and kidneys using LC/MS/MS. Recoveries in bovine milk at concentrations of 0.01 and 0.1 mg/kg ranged 72 – 97% and 84 – 90%, respectively. The mean recovery in bovine milk was 83 ± 8% (n = 10). Recoveries in bovine kidney at concentrations of 0.01 and 0.5 mg/kg ranged 72 – 97% and 93 – 97%, respectively. The mean recovery in bovine kidney was 87 ± 11% (n = 10).

*Soil*

Residues of aminopyralid in soil are determined using method GRM 02.34 (Lindsey and Hastings, 2004). Aminopyralid is extracted from soil by shaking with a $CH_3CN$/1 N HCl solution (90:10). The sample is then centrifuged and the extract is decanted. A second extraction is performed by adding the $CH_3CN$/HCl solution to the soil and mechanically shaking for 30 minutes. The sample is centrifuged and the second extract is combined with the first extract. The extract is evaporated to dryness and re-dissolved in 1N HCl. An aliquot of the extract is purified on an SPE plate which is then washed with $H_2O$/MeOH (95:5) and eluted with $CH_3CN$. Internal standard ($^{13}C_2$$^{15}N$-aminopyralid) is added to the eluate and evaporated to dryness and taken up in $CH_3CN$/pyridine/BuOH solution (22:2:1) prior to derivatisation. The remaining residue is derivatised using butyl chloroformate and diluted with $MeOH/H_2O/CH_3COOH$ mobile phase (50:50:0.1). The purified extract is analysed using HPLC with ESI LC/MS/MS. Residue identity is confirmed by matching retention time in conjunction with monitoring the MS/MS ion transitions of aminopyralid butyl ester at $m/z$ 263/134 and $^{13}C_2{}^{15}N$-aminopyralid butyl ester at $m/z$ 268/139.

The method was validated with recoveries being conducted in four soil types at concentrations ranging 0.0015–0.1 mg/kg; the validated LOQ was 0.0015 mg/kg. A summary of the soil recoveries is given in Table 28. The overall mean recovery at fortifications ranging 0.0015 – 0.1 mg/kg was 88 ± 6%.

Table 27. Recoveries of aminopyralid in soils samples.

| Matrix | Number of samples | Fortification Level (mg/kg) | Recovery Range (%) | Mean Recovery (%) |
|---|---|---|---|---|
| Sandy soil | 10 | 0.0015 – 0.1 | 80 – 102 | 92 ± |
| Silt | 10 | 0.0015 – 0.1 | 86 – 94 | 90 ± |
| Loam | 10 | 0.0015 – 0.1 | 81 – 91 | 87 ± |
| Clay | 10 | 0.0015 – 0.1 | 81 – 91 | 85 ± |
| All soils | 40 | 0.0015 – 0.1 | 80 – 102 | 88 ± 6 |

*Extraction efficiency*

To determine the extraction efficiency of the procedures in method GRM 02.31, samples taken in the $^{14}C$ plant metabolism studies were extracted and hydrolysed in the manner described above ('cold' method). The 42 DAT samples of ryegrass, big bluestem and *Panicum maximum* from the grass metabolism study (Magnussen and Balcher 2004) and samples of the wheat forage (14 DAT), wheat hay (35 DAT) and wheat grain and straw (86 DAT) from the spring wheat metabolism study (Graper *et al* 2003) were extracted, hydrolysed and prepared for quantitation according to Method GRM 02.31. Following extraction and preparation, the remaining residues were analysed using capillary gas chromatography with negative-ion chemical ionisation mass spectrometry (GC/NCI-MS). The results are shown in Table 30. For grass samples the extraction efficiency ranged from 88 – 114%; for 14 DAT wheat forage and 35 wheat DAT hay the extraction efficiency was 101% and 72%, respectively. For 86 DAT straw, the extraction efficiency was 87%. For the 86 DAT grain, the extraction efficiency was 101% when calculated based on total radioactivity but 170% when calculated on the basis of extractable $^{14}C$ labelled aminopyralid.

Table 28. Summary of extraction efficiency of aminopyralid from grass and wheat matrices.

| Sample Matrix | Radioactivity (mg/kg) | | Aminopyralid (mg/kg) | | Extraction Efficiency② |
|---|---|---|---|---|---|
| | Total | Extractable | $^{14}C$ | 'Cold'① | |
| **42 DAT grass** | | | | | |
| Ryegrass | 6.57③ | 6.28④ | 5.84⑤ | 6.65 | 114% |
| Big bluestem | 5.61③ | 5.45④ | 5.04⑤ | 4.45 | 88% |
| Panicum maximum | 4.81③ | 4.61④ | 4.24⑤ | 4.85 | 114% |
| **Wheat** | | | | | |
| Forage (14 DAT) | 0.418⑥ | 0.390⑥ | – | 0.393 | 101% |
| Hay (35 DAT) | 0.691⑥ | 0.611⑥ | – | 0.440 | 72% |
| Grain (86 DAT) | 0.084⑥ | 0.058⑥ | 0.050⑦ | 0.085 | 170% |
| Straw (86 DAT) | 0.623⑥ | 0.508⑥ | 0.489⑦ | 0.424 | 87% |

① Values obtained using sample preparation procedures in Method GRM 02.31.

② Extraction efficiency = 'cold' results/ $^{14}C$ results × 100.

③ Values from Table 9 (%TRR from unrinsed samples).

④ Values from Table 9 (data from extracts of each species).

⑤ Total $^{14}C$ as aminopyralid +C-1+C-3 Table 9.

⑥ Values from Table 4 (total $^{14}C$ extractable).

⑦ Values from Table 7 (total aminopyralid equivalents).

The extraction efficiency of method GRM 02.34 was determined by extracting soil samples treated in accordance with the test conditions of the European aerobic soil degradation study (Yoder and Smith, 2003). Samples of soil were treated with $^{14}$C-aminopyralid at the equivalent of 120 g ai/ha (0.016 mg/kg). Samples were extracted at 9, 21, 57 and 79 days after dosing using the extraction procedures described in method GRM 02.34. Replicate aliquots of each extract were counted by LSC to determine $^{14}$C aminopyralid. Each extract was then analysed using HPLC with ESI LC/MS/MS for qualitative and quantitative analysis of aminopyralid. The resulting levels were compared to verify extraction efficiency and are shown in Table 31.

Table 29. Summary of extraction efficiency of aminopyralid residues in soil samples.

| Soil sample | % recovery with LSC and HPLC | | | | % recovery by LC/MS/MS | | | |
|---|---|---|---|---|---|---|---|---|
|  | Day 9 | Day 21 | Day 57 | Day 79 | Day 9 | Day 21 | Day 57 | Day 79 |
| Sand | 82 | 75 | 63 | 56 | 79 | 69 | 59 | 52 |
| Clay | 68 | 68 | 26 | 10 | 64 | 60 | 23 | 7 |
| Loam | 72 | 63 | 52 | 43 | 72 | 60 | 49 | 38 |

The results show that the recoveries using the "cold" method are acceptable and demonstrate that method GRM 02.34 is applicable for the determination of aminopyralid residues in soils.

### *Stability of pesticide residues in stored analytical samples*

The storage stability of aminopyralid in pasture grass and hay and wheat grain and straw was investigated (Lindsay, 2004). Samples of hay, forage (grass), wheat grain and wheat straw were fortified with aminopyralid at a concentration of 0.1 mg/kg and placed in frozen storage at -20 °C for up to 488 days (grass) and 469 days (wheat). The conditions were consistent with the storage of actual field samples.

Samples were analysed at 0, 28, 130, 187 and 488 days after fortification for the grass samples and at 0, 113, 168, 273 and 469 days after fortification for the wheat samples (n = 3 at each time point). Concurrent (procedural) recoveries were conducted at each of the sampling intervals by freshly fortifying control samples (n = 2) with aminopyralid at a concentration of 0.1 mg/kg. Residues were then extracted and determined in all samples in accordance with method GRM 02.31 which has an LOQ of 0.01 mg/kg. The data are shown in Table 32.

Table 30. Storage stability of aminopyralid residues in wheat and grass matrices.

| Matrix | Storage interval (days) | Fortification concentration (mg/kg) | Residue found Mean (n = 3) (mg/kg) | Procedural recovery (%) |
|---|---|---|---|---|
| Grass | 0 | 0.1 | 0.0882 | 94 |
| Hay | 28 | | 0.0911 | 92 |
|  | 130 | | 0.0712 | 76 |
|  | 187 | | 0.0787 | 82 |
|  | 489 | | 0.0865 | 88 |
| Forage | 0 | 0.1 | 0.0858 | 93 |
|  | 28 | | 0.0911 | 95 |
|  | 130 | | 0.0757 | 75 |
|  | 187 | | 0.0835 | 84 |
|  | 489 | | 0.0884 | 87 |
| Wheat | 0 | 0.1 | 0.0848 | 86 |
| Grain | 113 | | 0.0866 | 90 |
|  | 168 | | 0.0892 | 95 |
|  | 273 | | 0.0977 | 103 |
|  | 469 | | 0.0910 | 89 |
| Straw | 0 | 0.1 | 0.0839 | 83 |
|  | 113 | | 0.0840 | 84 |
|  | 168 | | 0.0912 | 91 |
|  | 273 | | 0.0924 | 101 |
|  | 469 | | 0.0869 | 85 |

The data indicate that aminopyralid residues are stable under conditions of frozen storage for up to 489 days in grass hay and forage, and up to 469 days in wheat grain and straw.

The storage stability of aminopyralid residues in sandy loam soil was investigated (Lindsay 2004). The characteristics of the soil were as follows: bulk density 1.03 g/cm$^3$; organic matter 7.1%; CEC 17.5 meq/100g; $H_2O$ content at ⅓ bar 39.2% and at 15 bar 21.3%; 60% sand, 28% silt and 12% clay; pH 5.4.

Samples of soil were fortified with aminopyralid at 0.015 mg/kg and were stored in both polyethylene and tin containers at -20°C for up to 497 days. The conditions were consistent with the storage of actual field samples. Samples were analysed at 0, 41, 133, 194, 460 and 497 days after fortification. Concurrent (procedural) recoveries were conducted at each interval by freshly fortifying soil samples (n = 2) with aminopyralid at a concentration of 0.0015 mg/kg. Residues were extracted and determined in accordance with method GRM 02.34, as described above, which has an LOQ of 0.015 mg/kg. The concentration of aminopyralid remaining from 0 to 497 days, ranged from 0.0091 – 0.0151 mg/kg (uncorrected) or 61% – 101% of the fortified concentration.

## USE PATTERNS

Table 31. Registered uses of aminopyralid. Expressed as acid equivalents/ha (ae/ha) or acid equivalents/100L (ae/100L).

| Crop/Situation | Country | Application | | | | No. | PHI (days) | | PSI/ |
|---|---|---|---|---|---|---|---|---|---|
| | | Form. | Type | Rate (g ae/ha) | Timing | | Harvest | Graze/ Cut | ESI① |
| Pastures | Australia | EO | Ground foliar/spot spray | 150 – 210② | | 1 | Nil | Nil | 3 days |
| Pastures maintenance | Brazil | SL | Ground & aerial | 40 – 100③ | | 1 | Not determined | Not determined | |
| renewal | | EO | | 40 – 80④ | | 1 | | | |
| Pastures | Canada | SL | Ground & aerial | 60 – 120⑤ | | | Nil | Nil | ⟿ |
| Pastures | Colombia | SL | Ground | 10 – 50 g ae/100L⑥ | Post-emergence | 1 | | | |
| Pastures | Mexico | SL | Ground foliar | 20 – 120 | | | 7 days | 7 days | |
| | Mexico | SC | Ground foliar | 10 – 40 g ae/100L | | | 7 days | 7 days | |
| Pastures | New Zealand | SC | Ground foliar/broadcast | 60 | | | Nil | Nil | 4 days⑦ |
| | | | Spot spray | 6 – 9 g ae/100L | | | | | |
| Pastures | UK | EO | Ground foliar | 60⑧ | | 1 | | 7 days | |
| Pastures | US | SL | Ground/aerial | 53 – 123⑨ | | | 0⑩ or 7❶days | 0 or 7 days | ⟿ |
| Pastures | Venezuela | SL | Ground foliar | 120 – 160❶ or 30 – 40 g ae/100L | | | | | |
| Wheat | Argentina | WG | Ground foliar | 3.75 – 5 | From 3$^{rd}$ leaf to end of tillering | NS | Nil | | |
| Wheat, Barley Oats, Triticale | Australia | EO | Ground foliar | 5 – 7.5❷ | 3-leaf to 1$^{st}$ node (Z13 to Z31) | 1 | Nil | 7 days | 3 days |
| Wheat | Canada | | Ground foliar | 10 | 2 to 6 leaf stage | | 50 days | Nil | ⟿ |

| Crop/Situation | Country | Application | | | | No. | PHI (days) | | PSI/ |
| | | Form. | Type | Rate (g ae/ha) | Timing | | Harvest | Graze/ Cut | ESI① |
|---|---|---|---|---|---|---|---|---|---|
| Wheat | Mexico | SL | Ground foliar | 30 g ai/ha | Not stated | | 7 days | | |
| Wheat | US | | Ground/aerial | 7.6 – 10❸ | 3-leaf to early jointing (Z30) | 1 | 50 days | 14 days (8.75 label low rate Nil) | ↶ |
| | | | Spot spray | 38 g ai/100L | | | | | |

① PSI/ESI = Pre-slaughter Interval or Export Slaughter Interval. ② with up to 98 g ai/ha fluroxypyr. ③ with up to 800 g ai/ha 2,4-D or 200 g ai/ha fluroxypyr. ④ with up to 640 g ai/ha 2,4-D or 160 g ai/ha fluroxypyr. ⑤ with up to 1440 g ae/ha 2,4-D. ⑥ with up to 400 g ai/100L 2,4-D. ⑦ Slaughter Interval (domestic and export). ⑧ with 200 g ai/ha fluroxypyr. ⑨ with 982 g ae/ha 2,4-D. ⑩ 0 days on Milestone label. ❶ 7 days on Forefront label. ❶ with 1280 g ae/ha 2,4-D. ❷ with up to 105 g ai/ha fluroxypyr. ❸ with 143 g ai/ha fluroxypyr. ↶3 days removal to clean feed before introducing to another sensitive broadleaf cropping area. (crop safety statement).

## RESIDUES RESULTING FROM SUPERVISED TRIALS

The Meeting received information on supervised field trials for aminopyralid uses on cereals and pastures. Tables are listed below:

| | |
|---|---|
| Barley grain | Table 37 |
| Oat grain | Table 38 |
| Wheat grain | Table 39 |
| Barley straw (Europe) | Table 40 |
| Barley straw (Australia) | Table 41 |
| Barley forage | Table 42 |
| Oat straw | Table 43 |
| Oat forage | Table 44 |
| Wheat straw (Europe and USA) | Table 45 |
| Wheat straw (Australia) | Table 46 |
| Wheat forage (Europe and USA) | Table 47 |
| Wheat forage (Australia) | Table 48 |
| Wheat hay (USA) | Table 49 |
| Pastures and hay (Europe and USA) | Table 50 |
| Pastures and hay (Australia and New Zealand) | Table 51 |

All trials submitted to the meeting were conducted in accordance with the OECD principles of Good Laboratory Practice (GLP). All trials were well documented in both the analytical and field phases of the reports. Analytical reports included method validation with procedural recoveries from fortification at concentrations similar to those occurring in samples from supervised field trials. Dates of analyses and/or duration of residue sample storage were provided. Intervals of freezer storage between sampling and analysis were recorded for all trials and were covered by the conditions of the freezer storage stability studies.

All trials included analyses of control samples, however no control data are recorded in the tables except in instances where residues in the control samples exceeded the validated LOQ. Residue data reported in the following tables are not adjusted for analytical recoveries.

In most trials, field samples from an unreplicated plot were taken at each sampling interval and were analysed. For the purposes of the evaluation, where replicate samples were analysed separately, the mean of the two results is reported as the residue from the plot in following tables.

When residues were not detected and were reported as ND in the studies, they are indicated as being below the LOQ (e.g. < 0.01 mg/kg). Residues, application rates and spray concentrations have generally been rounded to two significant figures or, for residues near the LOQ, to one significant figure. Residue values from trials that reflect the maximum GAP have been used for the estimation of maximum residue levels. Residue trials conducted at exaggerated or higher than label application rates are also included in the estimation of maximum residue levels, where the resulting residues are less than the LOQ. The values that are included in the MRL estimation are underlined in the tables below.

Formulations used in the residue trials include soluble granules (SG), water dispersible granules (WG), soluble concentrates (SL), water-in-oil emulsions (EO), micro-emulsions (ME) containing salt forms of aminopyralid. Some were combination products with other herbicides such as fluroxypyr, 2,4-D and triclopyr. The types of formulations used in various trials are tabulated below.

Table 32. Formulations used in supervised trials for cereals and pastures.

| Formulation Code/ Designation | g ae/kg or g ae/L | Form of active | Trials |
|---|---|---|---|
| GF-1118; SG 75% | 750 g ae/kg | Potassium salt | Spain, Hungary, Poland, Italy |
| GF-1118; WG 750 | 750 g ae/kg | Not stated | Argentina |
| GF-871; SL 240 | 240 g ae/L | TIPA salt[①] | Australia, USA |
| GF-982[②]; EO | 10 g ae/L | TIPA salt[①] | Australia, USA, Canada |
| GF-389; SL 247 | 247 g ae/L | Potassium salt | USA |
| GF-1004; SL 40 | 40 g ae/L | TIPA salt[①③] | Brazil |
| GF-843; EO 40 | 40 g ae/L | Potassium salt[④] | Brazil |
| GF-839; ME 30 #1 | 30 g ae/L | Potassium salt[⑤] | UK, Germany, Spain, France, Italy |
| GF-819, ME 30 #2 | 30 g ae/L | Potassium salt[⑥] | UK, Germany, France, Spain |

① Triisopropanolammonium salt. ② Combination with 140 g ae/L fluroxypyr (1-methylheptyl ester or meptyl ester).
③ Combination with 320 g ae/L 2,4-D. ④ Combination with 80 g ae/L fluroxypyr (meptyl ester).
⑤ Combination with 100 g ae/L fluroxypyr (as meptyl ester) ⑥ Combination with 240 g ae/L triclopyr (as butyl

Application rates and/or spray concentrations are expressed as g acid equivalents/ha (g ae/ha) or g acid equivalents/100l (g ae/100l). In all cases residues were determined and expressed as aminopyralid. Methods used to determine residues in cereal matrices were GRM 02.31 and GRM 03.25 (Argentina) and GRM 02.31 for grass pasture and hay matrices.

*Cereal grains*

Residue data were provided for barley, oats and wheat from trials conducted in Argentina, Australia, Canada, Europe and USA.

Table 33. Aminopyralid residues in barley grain.

| BARLEY GRAIN | Application | | | | | PHI | Residues | Ref./Study No. |
|---|---|---|---|---|---|---|---|---|
| Country, Year (variety) | Form. | Rate (g ae/ha) | Water (L/ha) | No. | Timing | (days) | (mg/kg) | |
| Torredalomar, Spain, 2003 (Astoria) | SG 75% | 10 | 302 | 1 | BBCH 30 | 47 | 0.04 | GHE-P-10576① CEMS-2063 |
| Zaragoza, Spain, 2004 (Unia) | SG 75% | 10 | 298 | 1 | BBCH 30 | 58 | 0.08 | GHE-P-10690② CEMS-2246 |
| SA, Australia, 2002 (Barque) | SL 240 | 10 | 99 | 1 | Z15 (5-leaf) | 103 | <u>0.07</u> | 020060-03③ |
| | | 15 | 90 | 1 | | 103 | 0.10 | |
| | | 20 | 90 | 1 | | 103 | 0.12 | |
| NSW, Australia, 2002 (Grimnett) | SL 240 | 10 | 96 | 1 | 5 – 6 leaf | 134 | <u>0.06</u> | 020060-06③ |
| | | 15 | 96 | 1 | | 134 | < 0.01 | |
| | | 20 | 100 | 1 | | 134 | 0.10 | |
| NSW, Australia, 2003 (Tantangara) | SL 240 | 7.5 | 95 | 1 | 1st node | 80 | <u>0.07</u> | 030072-02④ |
| | | 15 | 95 | 1 | | 80 | 0.16 | |
| | EO | 7.5 | 95 | 1 | | 80 | <u>0.06</u> | |
| | | 15 | 95 | 1 | | 80 | 0.11 | |

| BARLEY GRAIN | Application | | | | | PHI | Residues | Ref./Study No. |
|---|---|---|---|---|---|---|---|---|
| Country, Year (variety) | Form. | Rate (g ae/ha) | Water (L/ha) | No. | Timing | (days) | (mg/kg) | |
| QLD, Australia, 2003 (Cullum) | SL 240 | 7.5 | 106 | 1 | 1 – 3 tiller | 88 | <u>0.04</u> | 030072-06④ |
| | | 15 | 105 | 1 | | 88 | 0.04 | |
| | EO | 7.5 | 106 | 1 | 1 – 3 tiller | | <u>0.03</u> | |
| | | 15 | 107 | 1 | | | 0.05 | |

① Recoveries were 72% and 75% at 0.01 and 0.1 mg/kg, respectively. ② Recoveries were 71% and 101% at 0.01 and 0.1 mg/kg, respectively. ③Recoveries were 76% and 99% at 0.01 and 0.05 mg/kg, respectively. ④ Recoveries ranged 89 – 101% over concentrations 0.01 – 1 mg/kg.

Table 34. Aminopyralid residues in oat grain.

| OAT GRAIN | Application | | | | | PHI | Residues | Ref./Study No. |
|---|---|---|---|---|---|---|---|---|
| Country, Year (variety) | Form. | Rate (g ae/ha) | Water (L/ha) | No. | Timing | (days) | (mg/kg) | |
| NSW, Australia, 2002 (Graza) | SL 240 | 10 | 100 | 1 | 5 – 6 leaf | 134 | <u>0.02</u> | 020060-07① |
| | | 15 | | | | 134 | 0.04 | |
| | | 20 | | | | 134 | 0.04 | |
| SA, Australia, 2003 (Euro) | SL 240 | 7.5 | 100 | 1 | 5 leaf 1 tiller | 112 | <u>0.03</u> | 030071-02② |
| | | 15 | | | | 112 | 0.06 | |
| | EO | 7.5 | 100 | 1 | 5 leaf 1 tiller | 112 | <u>0.03</u> | |
| | | 15 | | | | 112 | 0.06 | |
| NSW, Australia, 2003 (Graza 68) | SL 240 | 7.5 | 95 | 1 | 1st node | 80 | <u>≤ 0.01</u> | 030072-03③ |
| | | 15 | 95 | | | 80 | 0.03 | |
| | EO | 7.5 | 95 | | 1st node | 80 | <u>0.01</u> | |
| | | 15 | 95 | | | 80 | 0.03 | |
| SA, Australia, 2003 (Euro) | SL 240 | 7.5 | 100 | 1 | 6 leaf 2 tillers | 106 | <u>0.03</u> | 030072-08③ |
| | | 15 | 100 | | | 106 | 0.05 | |
| | EO | 7.5 | 100 | | 6 leaf 2 tillers | 106 | <u>0.03</u> | |
| | | 15 | 100 | | | 106 | 0.07 | |

①Recovery 116% at 0.01 mg/kg. ② Recoveries were 94 and 97% at 1 mg/kg. ③ Recoveries ranged 89 – 101% over concentrations 0.01 – 1 mg/kg.

Table 35. Aminopyralid residues in wheat grain.

| WHEAT GRAIN | Application | | | | | PHI | Residues | Ref./Study No. |
|---|---|---|---|---|---|---|---|---|
| Country, Year (variety) | Form. | Rate (g ae/ha) | Water (L/ha) | No. | Timing | (days) | (mg/kg) | |
| Szolnok, Hungary, 2003 (MV Emma) | SG 75% | 9.5 | 210 | 1 | BBCH 30 | 72 | < 0.01 | GHE-P-10577① CEMS-2064A |
| Rzechino, Poland, 2003 (Candos) | SG 75% | 10.5 | 208 | 1 | BBCH 30 | 84 | 0.01 | GHE-P-10577① CEMS-2064B |
| Bologna, Italy, 2003 (Soisson) | SG 75% | 10 | 299 | 1 | BBCH 30 | 69 | 0.01 | GHE-P-10575② CEMS-2059A |
| Bologna, Italy, 2003 (Mieti) | SG 75% | 10 | 298 | 1 | BBCH 31-32 | 73 | < 0.01 | CHE-P-10575② CEMS-2059B |
| Zaragoza, Spain, 2003 (Scalibur) | SG 75% | 10 | 300 | 1 | BBCH 30 | 49 | < 0.01 | CHE-P-10575② CEMS-2059C |
| Zaragoza, Spain, 2004 (Kilopondio) | SG 75% | 10 | 299 | 1 | BBCH 30 | 70 | < 0.01 | GHE-P-10689③ CEMS-2245D |

| WHEAT GRAIN | Application | | | | | PHI | Residues | Ref./Study No. |
|---|---|---|---|---|---|---|---|---|
| Country, Year (variety) | Form. | Rate (g ae/ha) | Water (L/ha) | No. | Timing | (days) | (mg/kg) | |
| Mainar, Spain, 2004 (Amaroc) | SG 75% | 10 | 298 | 1 | BBCH 30 | 80 | < 0.01 | GHE-P-10689[3] CEMS-2245E |
| Villareal, Spain, 2004 (Amilcar) | SG 75% | 10 | 302 | 1 | BBCH 30 | 80 | < 0.01 | GHE-P-10689[3] CEMS-2245F |
| Lower Saxony, Germany 2004 (Drifter) | SG 75% | 11 | 330 | 1 | BBCH 30 | 118 | 0.01 | GHE-P-10691[4] CEMS-2247A |
| Kolobrzeski, Poland 2004 (Ritmo) | SG 75% | 11 | 327 | 1 | BBCH 30 | 116 | < 0.01 | GHE-P-10691[4] CEMS-2247B |
| Argentina, 2003 (Delfin) | WG 750 | 7.5 | 200 | 1 | Z 22/23 | 97 | ≤ 0.01 | 030070[5] |
| | | 15 | 200 | 1 | | 97 | < 0.01 | GHB-P 1096 |
| Argentina, 2003 (Don Enrique (klein)) | WG 750 | 7.5 | 200 | 1 | 2 tillers | 95 | ≤ 0.01 | 030070[5] |
| | | 15 | 200 | 1 | | 95 | < 0.01 | GHB-P 1096 |
| Argentina, 2003 (Prointa granar) | WG 750 | 7.5 | 200 | 1 | 1 tiller | 111 | ≤ 0.01 | 030070[5] |
| | | 15 | 200 | 1 | | 111 | < 0.01 | GHB-P 1096 |
| Argentina, 2004 (Buck sureno) | WG 750 | 7.5 | 200 | 1 | Z21 | 111 | ≤ 0.01 | 030070.01[5] |
| | | 15 | 200 | 1 | | 111 | < 0.01 | GHB-P 1109 |
| Argentina, 2004 (Klein scorpion) | WG 750 | 7.5 | 200 | 1 | BBCH 14 | 121 | ≤ 0.01 | 030070.01[5] |
| | | 15 | 200 | 1 | | 121 | 0.02 | GHB-P 1109 |
| Argentina, 2004 (Granar) | WG 750 | 7.5 | 200 | 1 | 1-2 tillers | 86 | ≤ 0.01 | 030070.01[5] |
| | | 15 | 200 | 1 | | 86 | < 0.01 | GHB-P 1109 |
| WA, Australia, 2002 (Calingiri) | SL 240 | 10 | 101 | 1 | 5-6 leaf stage | 98 | ≤ 0.01 | 020060-01[6] |
| | | 15 | 102 | 1 | | 98 | < 0.01 | |
| | | 20 | 101 | 1 | | 98 | < 0.01 | |
| SA, Australia, 2002 (Frame) | SL 240 | 10 | 99 | 1 | Z15 | 103 | ≤ 0.01 | 020060-02[6] |
| | | 15 | 90 | 1 | | 103 | < 0.01 | |
| | | 20 | 90 | 1 | | 103 | < 0.01 | |
| NSW, Australia, 2002 (Babbler) | SL 240 | 10 | 100 | 1 | 5 – 6 leaf | 134 | ≤ 0.01 | 020060-05[6] |
| | | 15 | 96 | 1 | | 134 | < 0.01 | |
| | | 20 | 100 | 1 | | 134 | < 0.01 | |
| QLD, Australia, 2002 (Petrel) | SL 240 | 10 | 102 | 1 | 4 – 6 leaf | 88 | 0.07 | 020060-08[6] |
| | | 15 | 99 | 1 | | 88 | 0.07 | |
| | | 20 | 101 | 1 | | 88 | 0.11 | |
| NSW, Australia, 2003 (Wollaroi) | SL 240 | 7.5 | 101 | 1 | 6-leaf | 80 | 0.01 | 030071-01[7] |
| | | 15 | 101 | 1 | | 80 | 0.02 | |
| | EO | 7.5 | 101 | 1 | 6-leaf | 80 | ≤ 0.01 | |
| | | 15 | 101 | 1 | | 80 | 0.01 | |
| NSW, Australia, 2003 (Mulgara) | SL 240 | 7.5 | 95 | 1 | 1st node | 80 | 0.01 | 030072-01[8] |
| | | 15 | 105 | 1 | | 80 | 0.02 | |
| | EO | | 95 | 1 | 1st node | 80 | ≤ 0.01 | |
| | | | 95 | 1 | | 80 | 0.01 | |
| QLD, Australia, 2003 (Strzelecki) | SL 240 | 7.5 | 103 | 1 | 5 leaf 1 tiller | 99 | 0.02 | 030072-04[8] |
| | | 15 | 102 | 1 | | 99 | 0.04 | |
| | EO | 7.5 | 102 | 1 | 5 leaf 1 tiller | 99 | 0.01 | |
| | | 15 | 100 | 1 | | 99 | 0.02 | |
| QLD, Australia, 2003 (Baxter) | SL 240 | 7.5 | 103 | 1 | 3 – 6 tiller | 88 | 0.03 | 030072-05[8] |
| | | 15 | 102 | 1 | | 88 | 0.06 | |
| | EO | 7.5 | 99 | 1 | 3 – 6 tiller | 88 | 0.02 | |
| | | 15 | 102 | 1 | | 88 | < 0.01 | |
| SA, Australia, 2003 (Yitpi) | SL 240 | 7.5 | 96 | 1 | 1st tiller | 107 | ≤ 0.01 | 030072-07[8] |
| | | 15 | 100 | 1 | | 107 | < 0.01 | |
| | EO | 7.5 | 100 | 1 | 1st tiller | 107 | ≤ 0.01 | |
| | | 15 | 100 | 1 | | 107 | < 0.01 | |
| VA, USA, 2003 (Coker9835) | EO | 10 | 135 | 1 | BBCH 32 | 73 | 0.025 | 030042[9] |
| AR, USA, 2003 (Natchez) | EO | 10 | 116 | 1 | BBCH 41-45 | 50 | 0.023 | |

| WHEAT GRAIN | Application | | | | | PHI | Residues | Ref./Study No. |
|---|---|---|---|---|---|---|---|---|
| Country, Year (variety) | Form. | Rate (g ae/ha) | Water (L/ha) | No. | Timing | (days) | (mg/kg) | |
| IN, USA, 2003 (Bravo) | EO | 10 | 188 | 1 | BBCH 31 | 80 | ≤ 0.01 | |
| MN, USA, 2003 (Oxen) | EO | 10 | 150 | 1 | BBCH 37 | 72 | < 0.01 | |
| SD, USA, 2003 (Marshall) | EO | 10 | 152 | 1 | BBCH 37 | 72 | < 0.01 | |
| ND, USA, 2003 (Oxen) | EO | 10 | 151 | 1 | BBCH 37 | 72 | < 0.01 | |
| NE, USA, 2003 (VNS HRW) | EO | 10 | 187 | 1 | BBCH 31 | 61 | 0.025 | |
| OK, USA, 2003 (Coker9663) | EO | 10 | 119 | 1 | BBCH 33 | 69 | 0.013 | |
| | EO | 50 | 119 | 1 | BBCH 33 | 69 | 0.055 | |
| SD, USA, 2003 (Marshall) | EO | 10 | 152 | 1 | BBCH 37 | 72 | < 0.01 | |
| | SL 240 | 9.7 | 152 | 1 | BBCH 37 | 72 | < 0.01 | |
| | SL 247 | 10.4 | 152 | 1 | BBCH 38 | 64 | 0.012 | |
| ND, USA, 2003 (Oxen) | EO | 10 | 152 | 1 | BBCH 37 | 72 | < 0.01 | |
| | SL 240 | 9.7 | 152 | 1 | BBCH 37 | 72 | < 0.01 | |
| | SL 247 | 10.4 | 152 | 1 | BBCH 38 | 65 | 0.01 | |
| ND, USA, 2003 (Alsen) | EO | 10.4 | 147 | 1 | BBCH 37 | 50 | 0.013 | |
| | SL 240 | 10.5 | 146 | 1 | BBCH 37 | 50 | 0.014 | |
| | SL 247 | 10.5 | 146 | 1 | BBCH 37 | 50 | 0.014 | |
| NE. USA, 2003 (Forge HRS) | EO | 10 | 187 | 1 | BBCH 31 | 56 | 0.022 | |
| | SL 240 | 10 | 188 | 1 | BBCH 31 | 56 | 0.013 | |
| | SL 247 | 10 | 187 | 1 | BBCH 31 | 56 | 0.023 | |
| SD, USA, 2003 (Walworth) | EO | 10 | 140 | 1 | BBCH 30-31 | 64 | 0.011 | |
| | SL 240 | 10 | 140 | 1 | BBCH 30-31 | 64 | 0.01 | |
| | SL 247 | 10 | 140 | 1 | BBCH 30-31 | 64 | 0.01 | |
| TX#1, USA, 2003 (TAM 105) | EO | 10 | 143 | 1 | BBCH 39 | 57 | ≤ 0.01 | |
| TX#2, USA, 2003 (TAM 105) | EO | 10 | 143 | 1 | BBCH 37 | 69 | ≤ 0.01 | |
| NM, USA, 2003 (Jagger) | EO | 10 | 141 | 1 | BBCH 37 | 62 | ≤ 0.01 | |
| TX#4, USA, 2003 (TAM 200) | EO | 10 | 177 | 1 | BBCH 33 | 67 | 0.011 | |
| TX#5, USA, 2003 (Jagger) | EO | 10 | 174 | 1 | BBCH 37 | 67 | ≤ 0.01 | |
| KS, USA, 2003 (Ike) | EO | 10 | 110 | 1 | BBCH 30 | 80 | ≤ 0.01 | |
| WA, USA, 2003 (Declo) | EO | 10 | 141 | 1 | BBCH 39 | 57 | 0.021 | |
| SK, Canada, 2003 (ACEatonia) | EO | 10 | 107 | 1 | BBCH 30 | 55 | 0.01 | |
| | SL 240 | 10 | 109 | 1 | BBCH 30 | 54 | ≤ 0.01 | |
| | SL 247 | 9.4 | 107 | 1 | BBCH 30 | 54 | 0.011 | |
| SK, Canada, 2003 (ACEatonia) | EO | 10 | 109 | 1 | BBCH 37 | 49 | ≤ 0.01 | |
| | SL 240 | 10 | 111 | 1 | BBCH 37 | 49 | 0.013 | |
| | SL 247 | 10 | 113 | 1 | BBCH 37 | 49 | 0.013 | |

①Recoveries 79 and 123% @ 0.01 and 0.1 mg/kg. ②Recoveries 84 and 119% @ 0.01 and 0.1 mg/kg. ③Recoveries were 71% and 89% at 0.01 and 0.1 mg/kg, respectively. ④Recoveries were 71% and 101% at 0.01 and 0.1 mg/kg. ⑤Recoveries ranged 71 – 102% at 0.01 mg/kg; 71 – 79% at 0.1 mg/kg and 71 – 107% at 1 mg/kg. ⑥ Recoveries ranged 88 – 98% over concentrations 0.01 – 0.4 mg/kg (n = 3). ⑦ Recoveries were 94 and 97% at 1 mg/kg. ⑧ Recoveries ranged 89 – 101% over concentrations 0.01 – 1 mg/kg. ⑨ Recoveries ranged 80 – 103% over concentrations over concentration 0.05 and 0.5 mg/kg (n = 18).

### Livestock feeds

Residue data were provided for forage and straw from barley, oats, wheat and grass pastures from trials conducted in Argentina, Australia, Canada, Europe and USA.

Table 36. Aminopyralid residues in barley straw from trials in Europe

| BARLEY STRAW | Application | | | | | PHI | Residues | Ref./Study No. |
|---|---|---|---|---|---|---|---|---|
| Country, Year (variety) | Form. | Rate (g ae/ha) | Water (L/ha) | No. | Timing | (days) | (mg/kg) as received | |
| Torredalomar Spain, 2003 (Astoria) | SG 75% | 10 | 302 | 1 | BBCH 30 | 47 | < 0.01 | GHE-P-10576① CEMS-2063 |
| Zaragoza, Spain, 2004 (Unia) | SG 75% | 10 | 298 | 1 | BBCH 30 | 58 | 0.06 | GHE-P-10690② CEMS-2246 |

①Recoveries were 80 and 102% at concentrations of 0.01 and 0.1 mg/kg, respectively. ② Recoveries were 68% and 79% at concentrations of 0.01 and 0.1 mg/kg, respectively.

Table 37. Aminopyralid residues in barley straw from trials in Australia.

| BARLEY STRAW | Application | | | | | PHI | Residues | Ref./Study No. |
|---|---|---|---|---|---|---|---|---|
| Country, Year (variety) | Form. | Rate (g ae/ha) | Water (L/ha) | No. | Timing | (days) | (mg/kg) dry weight | |
| SA, Australia, 2002 (Barque) | SL 240 | 10 | 99 | 1 | Z15 (5-leaf) | 103 | 0.03 | 020060-03①② |
| | | 15 | 90 | 1 | | 103 | 0.07 | |
| | | 20 | 90 | 1 | | 103 | 0.07 | |
| NSW, Australia, 2002 (Grimnett) | SL 240 | 10 | 96 | 1 | 5 – 6 leaf | 134 | 0.03 | 020060-06①③ |
| | | 15 | 96 | 1 | | 134 | 0.04 | |
| | | 20 | 100 | 1 | | 134 | 0.05 | |
| NSW, Australia, 2003 (Tantangara) | SL 240 | 7.5 | 95 | 1 | 1st node | 80 | 0.08 | 030072-02④⑤ |
| | | 15 | 95 | 1 | | 80 | 0.18 | |
| | EO | 7.5 | 95 | 1 | | 80 | 0.07 | |
| | | 15 | 95 | 1 | | 80 | 0.16 | |
| QLD, Australia, 2003 (Cullum) | SL 240 | 7.5 | 106 | 1 | 1 – 3 tiller | 88 | 0.03 | 030072-06④⑥ |
| | | 15 | 105 | 1 | | 88 | 0.04 | |
| | EO | 7.5 | 106 | 1 | 1 – 3 tiller | | 0.04 | |
| | | 15 | 107 | 1 | | | 0.06 | |

①Recoveries 99 and 107% @ 0.4 and 2 mg/kg. ②Moisture contents ranged 10.7 – 12.4%. ③Moisture contents ranged 25.5 – 41.3%. ④Recoveries ranged 78 – 100% over concentrations 0.01 – 1 mg/kg (n = 7). ⑤Moisture contents ranged 7.6 – 8.3%. ⑥Moisture contents ranged 55 – 68.6%.

Table 38. Aminopyralid residues in barley forage.

| BARLEY FORAGE | Application | | | | | PHI | Residues | Ref./Study No. |
|---|---|---|---|---|---|---|---|---|
| Country, Year (variety) | Form. | Rate (g ae/ha) | Water (L/ha) | No. | Timing | (days) | (mg/kg) dry weight | |
| SA, Australia, 2002 (Barque) | SL 240 | 10 | 99 | 1 | Z15 (5-leaf) | 0 | 3.89 | 020060-03①② |
| | | | | | | 1 | 2.60 | |
| | | | | | | 3 | 1.32 | |
| | | | | | | 7 | 0.54 | |
| | | | | | | 14 | 0.26 | |
| | | | | | | 28 | 0.13 | |
| | | 15 | 90 | 1 | Z15 (5-leaf) | 0 | 5.18 | |
| | | | | | | 1 | 4.19 | |
| | | | | | | 3 | 2.15 | |
| | | | | | | 7 | 1.04 | |
| | | | | | | 14 | 0.47 | |
| | | | | | | 28 | 0.21 | |

| BARLEY FORAGE | Application | | | | | PHI | Residues | Ref./Study No. |
|---|---|---|---|---|---|---|---|---|
| Country, Year (variety) | Form. | Rate (g ae/ha) | Water (L/ha) | No. | Timing | (days) | (mg/kg) dry weight | |
| | | 20 | 90 | 1 | Z15 (5-leaf) | 0 | 8.96 | |
| | | | | | | 1 | 5.23 | |
| | | | | | | 3 | 3.74 | |
| | | | | | | 7 | 1.12 | |
| | | | | | | 14 | 0.47 | |
| | | | | | | 28 | 0.28 | |
| NSW, Australia, 2002 (Grimnett) | SL 240 | 10 | 96 | 1 | 5 – 6 leaf | 0 | 1.99 | 020060-06②③ |
| | | | | | | 1 | 3.33 | |
| | | | | | | 3 | 1.29 | |
| | | | | | | 7 | 0.71 | |
| | | | | | | 14 | 0.30 | |
| | | | | | | 28 | 0.27 | |
| | | 15 | 96 | 1 | 5 – 6 leaf | 0 | 3.20 | |
| | | | | | | 1 | 3.09 | |
| | | | | | | 3 | 2.06 | |
| | | | | | | 7 | 1.12 | |
| | | | | | | 14 | 0.41 | |
| | | | | | | 28 | 0.34 | |
| | | 20 | 100 | 1 | 5 – 6 leaf | 0 | 5.10 | |
| | | | | | | 1 | 3.80 | |
| | | | | | | 3 | 2.71 | |
| | | | | | | 7 | 1.51 | |
| | | | | | | 14 | 0.44 | |
| | | | | | | 28 | 0.40 | |

①Moisture contents ranged 84.4 to 87.9%. ②Recoveries ranged 86 – 108 % over concentrations 0.01 – 5 mg/kg (n = 7). ③ Moisture contents ranged 76.3 – 86.3%.

Table 39. Aminopyralid residues in oat straw.

| OAT STRAW | Application | | | | | PHI | Residues | Ref./Study No. |
|---|---|---|---|---|---|---|---|---|
| Country, Year (variety) | Form. | Rate (g ae/ha) | Water (L/ha) | No. | Timing | (days) | (mg/kg) dry weight | |
| NSW, Australia, 2002 (Graza) | SL 240 | 10 | 100 | 1 | 5 – 6 leaf | 134 | 0.11 | 020060-07①② |
| | | 15 | | | | 134 | 0.11 | |
| | | 20 | | | | 134 | 0.17 | |
| SA, Australia, 2003 (Euro) | SL 240 | 7.5 | 100 | 1 | 5 leaf 1 tiller | 112 | 0.02 | 030071-02③④ |
| | | 15 | | | | 112 | 0.05 | |
| | EO | 7.5 | 100 | 1 | 5 leaf 1 tiller | 112 | 0.03 | |
| | | 15 | | | | 112 | 0.10 | |
| NSW, Australia, 2003 (Graza 68) | SL 240 | 7.5 | 95 | 1 | 1st node | 80 | 0.05 | 030072-03④⑤ |
| | | 15 | 95 | | | 80 | 0.13 | |
| | EO | 7.5 | 95 | | | 80 | 0.04 | |
| | | 15 | 95 | | | 80 | 0.08 | |
| SA, Australia, 2003 (Euro) | SL 240 | 7.5 | 100 | 1 | 6 leaf 2 tillers | 106 | 0.04 | 030072-08④⑥ |
| | | 15 | 100 | | | 106 | 0.07 | |
| | EO | 7.5 | 100 | | 6 leaf 2 tillers | 106 | 0.04 | |
| | | 15 | 100 | | | 106 | 0.09 | |

①Recoveries 73 and 77% @ 0.4 and 2 mg/kg. ② Moisture contents ranged 25.2 – 46.7%. ③ Moisture contents ranged 12.9 – 19.5%. ④ Recoveries ranged 78 – 100% over concentrations 0.01 – 1 mg/kg (n = 7). ⑤ Moisture contents ranged 42.8 – 56.2%. ⑥ Moisture contents ranged 23.9 – 33.5%.

Table 40. Aminopyralid residues in oat forage.

| OAT FORAGE | Application | | | | | PHI | Residues | Ref./Study No. |
|---|---|---|---|---|---|---|---|---|
| Country, Year (variety) | Form. | Rate (g ae/ha) | Water (L/ha) | No. | Timing | (days) | (mg/kg) dry weight | |
| NSW, Australia, 2002 (Graza) | SL 240 | 10 | 100 | 1 | 5 – 6 leaf | 0<br>1<br>3<br>7<br>14<br>28 | 1.19<br>1.46<br>0.90<br>0.40<br>0.15<br>0.34 | 020060-03①② |
| | | 15 | 100 | 1 | 5 – 6 leaf | 0<br>1<br>3<br>7<br>14<br>28 | 1.72<br>1.81<br>1.19<br>0.59<br>0.18<br>0.26 | |
| | | 20 | 100 | 1 | 5 – 6 leaf | 0<br>1<br>3<br>7<br>14<br>28 | 3.46<br>2.77<br>1.57<br>0.85<br>0.27<br>0.52 | |
| SA, Australia, 2003 (Euro) | SL 240 | 7.5 | 100 | 1 | 5 leaf 1 tiller | 0<br>1<br>3<br>7<br>14<br>28 | 1.74<br>1.27<br>0.81<br>0.34<br>0.29<br>0.18 | 030071-02③④ |
| | | 15 | 100 | 1 | 5 leaf 1 tiller | 0<br>1<br>3<br>7<br>14<br>28 | 3.87<br>2.95<br>1.53<br>0.66<br>0.54<br>0.30 | |
| | EO | 7.5 | 100 | 1 | 5 leaf 1 tiller | 0<br>1<br>3<br>7<br>14<br>28 | 2.93<br>2.04<br>1.05<br>0.79<br>0.31<br>0.23 | |
| | | 15 | 100 | 1 | 5 leaf 1 tiller | 0<br>1<br>3<br>7<br>14<br>28 | 5.77<br>3.25<br>2.03<br>1.43<br>0.79<br>0.59 | |

① Moisture contents ranged 81.8 – 87.2%. ②Recoveries ranged 87 – 99% over concentrations 0.01 – 5 mg/kg.
③ Recoveries ranged 85 – 124% over concentrations 0.01 – 1 mg/kg (n = 15). ④ Moisture contents ranged 82.7 – 87.2%.

Table 41. Aminopyralid residues in wheat straw from Europe and USA.

| WHEAT STRAW | Application | | | | | PHI | Residues | Ref./Study No. |
|---|---|---|---|---|---|---|---|---|
| Country, Year (variety) | Form. | Rate (g ae/ha) | Water (L/ha) | No. | Timing | (days) | (mg/kg) as received | |
| Szolnok, Hungary, 2003 (MV Emma) | SG 75% | 9.5 | 210 | 1 | BBCH 30 | 72 | 0.04 | GHE-P-10577①<br><br>CEMS-2064A |
| Rzechino, Poland, 2003 (Candos) | SG 75% | 10.5 | 208 | 1 | BBCH 30 | 84 | 0.03 | GHE-P-10577①<br><br>CEMS-2064B |

| WHEAT STRAW | Application | | | | | PHI | Residues | Ref./Study No. |
|---|---|---|---|---|---|---|---|---|
| Country, Year (variety) | Form. | Rate (g ae/ha) | Water (L/ha) | No. | Timing | (days) | (mg/kg) as received | |
| Bologna, Italy, 2003 (Soisson) | SG 75% | 10 | 299 | 1 | BBCH 30 | 69 | 0.09 | GHE-P-10575② CEMS-2059A |
| Bologna, Italy, 2003 (Mieti) | SG 75% | 10 | 298 | 1 | BBCH 31-32 | 73 | 0.07 | CHE-P-10575② CEMS-2059B |
| Zaragoza, Spain, 2003 (Scalibur) | SG 75% | 10 | 300 | 1 | BBCH 30 | 49 | 0.13 | CHE-P-10575② CEMS-2059C |
| Zaragoza, Spain, 2004 (Kilopondio) | SG 75% | 10 | 299 | 1 | BBCH 30 | 70 | 0.16 | GHE-P-10689③ CEMS-2245D |
| Mainar, Spain, 2004 (Amaroc) | SG 75% | 10 | 298 | 1 | BBCH 30 | 80 | 0.03 | GHE-P-10689③ CEMS-2245E |
| Villareal, Spain, 2004 (Amilcar) | SG 75% | 10 | 302 | 1 | BBCH 30 | 80 | 0.04 | GHE-P-10689③ CEMS-2245F |
| Lower Saxony, Germany 2004 (Drifter) | SG 75% | 11 | 330 | 1 | BBCH 30 | 118 | 0.08 | GHE-P-10691④ CEMS-2247A |
| Kolobrzeski, Poland 2004 (Ritmo) | SG 75% | 11 | 327 | 1 | BBCH 30 | 116 | 0.06 | GHE-P-10691④ CEMS-2247B |
| VA, USA, 2003 (Coker9835) | EO | 10 | 135 | 1 | BBCH 32 | 73 | 0.06 | 030042⑤ |
| AR, USA, 2003 (Natchez) | EO | 10 | 116 | 1 | BBCH 41-45 | 50 | 0.04 | |
| IN, USA, 2003 (Bravo) | EO | 10 | 188 | 1 | BBCH 31 | 80 | 0.03 | |
| MN, USA, 2003 (Oxen) | EO | 10 | 150 | 1 | BBCH 37 | 72 | < 0.01 | |
| SD, USA, 2003 (Marshall) | EO | 10 | 152 | 1 | BBCH 37 | 72 | < 0.01 | |
| ND, USA, 2003 (Oxen) | EO | 10 | 151 | 1 | BBCH 37 | 72 | < 0.01 | |
| NE, USA, 2003 (VNS HRW) | EO | 10 | 187 | 1 | BBCH 31 | 61 | 0.13 | |
| OK, USA, 2003 (Coker9663) | EO | 10 | 119 | 1 | BBCH 33 | 69 | 0.06 | |
| SD, USA, 2003 (Marshall) | EO SL 240 SL 247 | 10 9.7 10.4 | 152 152 152 | 1 1 1 | BBCH 37 BBCH 37 BBCH 38 | 72 72 64 | < 0.01 0.02 0.06 | |
| ND, USA, 2003 (Oxen) | EO SL 240 SL 247 | 10 9.7 10.4 | 152 152 152 | 1 1 1 | BBCH 37 BBCH 37 BBCH 38 | 72 72 65 | 0.01 0.02 0.04 | |
| ND, USA, 2003 (Alsen) | EO SL 240 SL 247 | 10.4 10.5 10.5 | 147 146 146 | 1 1 1 | BBCH 37 BBCH 37 BBCH 37 | 50 50 50 | 0.13 0.07 0.07 | |
| NE. USA, 2003 (Forge HRS) | EO SL 240 SL 247 | 10 10 10 | 187 188 187 | 1 1 1 | BBCH 31 BBCH 31 BBCH 31 | 56 56 56 | 0.1 0.06 0.08 | |
| SD, USA, 2003 (Walworth) | EO SL 240 SL 247 | 10 10 10 | 140 140 140 | 1 1 1 | BBCH 30-31 BBCH 30-31 BBCH 30-31 | 64 64 64 | 0.1 0.07 0.06 | |
| TX#1, USA, 2003 (TAM 105) | EO | 10 | 143 | 1 | BBCH 39 | 57 | 0.04 | |
| TX#2, USA, 2003 (TAM 105) | EO | 10 | 143 | 1 | BBCH 37 | 69 | 0.02 | |
| NM, USA, 2003 (Jagger) | EO | 10 | 141 | 1 | BBCH 37 | 62 | 0.12 | |
| TX#4, USA, 2003 (TAM 200) | EO | 10 | 177 | 1 | BBCH 33 | 67 | 0.04 | |
| TX#5, USA, 2003 (Jagger) | EO | 10 | 174 | 1 | BBCH 37 | 67 | 0.04 | |

| WHEAT STRAW | Application | | | | | PHI | Residues | Ref./Study No. |
|---|---|---|---|---|---|---|---|---|
| Country, Year (variety) | Form. | Rate (g ae/ha) | Water (L/ha) | No. | Timing | (days) | (mg/kg) as received | |
| KS, USA, 2003 (Ike) | EO | 10 | 110 | 1 | BBCH 30 | 80 | 0.04 | |
| WA, USA, 2003 (Declo) | EO | 10 | 141 | 1 | BBCH 39 | 57 | 0.04 | |
| SK, Canada, 2003 (ACEatonia) | EO | 10 | 107 | 1 | BBCH 30 | 55 | 0.07 | |
| | SL 240 | 10 | 109 | 1 | BBCH 30 | 54 | 0.07 | |
| | SL 247 | 9.4 | 107 | 1 | BBCH 30 | 54 | 0.14 | |
| SK, Canada, 2003 (ACEatonia) | EO | 10 | 109 | 1 | BBCH 37 | 49 | 0.07 | |
| | SL 240 | 10 | 111 | 1 | BBCH 37 | 49 | 0.05 | |
| | SL 247 | 10 | 113 | 1 | BBCH 37 | 49 | 0.07 | |

①Recoveries 84 and 71% @ 0.01 and 0.1 mg/kg. ②Recoveries 87 and 88% @ 0.01 and 0.1 mg/kg. ③Recoveries were 75% and 89% at 0.01 and 0.1 mg/kg, respectively. ④ Recoveries were 68% and 79% at 0.01 and 0.1 mg/kg, respectively. ⑤Recoveries ranged 72 – 93% over concentrations 0.05, 0.5 and 5 mg/kg (n = 17).

Table 42, Aminopyralid residues in wheat straw from Australia.

| WHEAT STRAW | Application | | | | | PHI | Residues | Ref./Study No. |
|---|---|---|---|---|---|---|---|---|
| Country, Year (variety) | Form. | Rate (g ae/ha) | Water (L/ha) | No. | Timing | (days) | (mg/kg) dry weight | |
| WA, Australia, 2002 (Calingiri) | SL 240 | 10 | 100 | 1 | 5-6 leaf stage | 98 | 0.10 | 020060-01③④ |
| | | 15 | | | | | 0.18 | |
| | | 20 | | | | | 0.20 | |
| SA, Australia, 2002 (Frame) | SL 240 | 10 | 99 | 1 | Z15 | 103 | 0.07 | 020060-02③⑤ |
| | | 15 | 90 | 1 | | 103 | 0.1 | |
| | | 20 | 90 | 1 | | 103 | 0.1 | |
| NSW, Australia, 2002 (Babbler) | SL 240 | 10 | 100 | 1 | 5 – 6 leaf | 134 | 0.04 | 020060-05③⑥ |
| | | 15 | 96 | 1 | | 134 | 0.05 | |
| | | 20 | 100 | 1 | | 134 | 0.06 | |
| QLD, Australia, 2002 (Petrel) | SL 240 | 10 | 102 | 1 | 4 – 6 leaf | 88 | 0.07 | 020060-08③⑦ |
| | | 15 | 99 | 1 | | 88 | 0.10 | |
| | | 20 | 101 | 1 | | 88 | 0.14 | |
| NSW, Australia, 2003 (Wollaroi) | SL 240 | 7.5 | 101 | 1 | 6-leaf | 80 | 0.06 | 030071-01⑧⑨ |
| | | 15 | 101 | 1 | | 80 | 0.14 | |
| | EO | 7.5 | 101 | 1 | 6-leaf | 80 | 0.08 | |
| | | 15 | 101 | 1 | | 80 | 0.18 | |
| NSW, Australia, 2003 (Mulgara) | SL 240 | 7.5 | 95 | 1 | 1st node | 80 | 0.09 | 030072-01⑨⑩ |
| | | 15 | 105 | 1 | | 80 | 0.2 | |
| | EO | | 95 | 1 | 1st node | 80 | 0.12 | |
| | | | 95 | 1 | | 80 | 0.24 | |
| QLD, Australia, 2003 (Strzelecki) | SL 240 | 7.5 | 103 | 1 | 5 leaf 1 tiller | 99 | 0.02 | 030072-04⑨❶ |
| | | 15 | 102 | 1 | | 99 | 0.12 | |
| | EO | 7.5 | 102 | 1 | 5 leaf 1 tiller | 99 | 0.02 | |
| | | 15 | 100 | 1 | | 99 | 0.03 | |
| QLD, Australia, 2003 (Baxter) | SL 240 | 7.5 | 103 | 1 | 3 – 6 tiller | 88 | 0.04 | 030072-05⑨❷ |
| | | 15 | 102 | 1 | | 88 | 0.1 | |
| | EO | 7.5 | 99 | | 3 – 6 tiller | 88 | 0.06 | |
| | | 15 | 102 | | | 88 | 0.07 | |
| SA, Australia, 2003 (Yitpi) | SL 240 | 7.5 | 96 | 1 | 1st tiller | 107 | 0.05 | 030072-07⑨❸ |
| | | 15 | 100 | 1 | | 107 | 0.2 | |
| | EO | 7.5 | 100 | 1 | 1st tiller | 107 | 0.13 | |
| | | 15 | 100 | 1 | | 107 | 0.23 | |

③ Recoveries ranged 85 – 94% over concentrations 0.05 – 1 mg/kg (n = 3). ④ Moisture contents ranged 22.2 – 26%. ⑤ Moisture contents ranged 9.9 – 12%. ⑥ Moisture contents ranged 13.7 – 18.3%. ⑦ Moisture contents ranged 11.2 – 36.3 %. ⑧ Moisture contents ranged 7.1 – 8.8% for the SL formulation and 8.2 – 9.8% for the EO formulation. ⑨ Recoveries ranged 78 – 100% over concentrations 0.01 – 1 mg/kg (n = 7). ⑩ Moisture contents ranged 7.2 – 9.3%. ❶ Moisture contents ranged 24.1 – 33.4%. ❷ Moisture contents ranged 24.1 – 55%. ❸ Moisture contents ranged 8.1 – 9.6%.

Table 43. Aminopyralid residues in wheat forage from Europe and USA.

| WHEAT FORAGE Country, Year (variety) | Application | | | | | PHI (days) | Residues (mg/kg) as received | Ref./Study No. |
|---|---|---|---|---|---|---|---|---|
| | Form. | Rate (g ae/ha) | Water (L/ha) | No. | Timing | | | |
| Bologna, Italy, 2003 (Soisson) | SG 75% | 10 | 299 | 1 | BBCH 30 | 0<br>14<br>28<br>64 | 0.4<br>0.06<br>0.05<br>0.05 | GHE-P-10575① CEMS-2059A |
| Zaragoza, Spain, 2004 (Kilopondio) | SG 75% | 10 | 299 | 1 | BBCH 30 | 0<br>14<br>28<br>38 | 0.35<br>0.03<br>0.05<br>0.06 | GHE-P-10689② CEMS-2245D |
| VA, USA, 2003 (Coker9835) | EO | 10 | 135 | 1 | BBCH 32 | 0<br>7<br>14<br>21<br>28 | 0.79<br>0.26<br>0.19<br>0.16<br>0.11 | 030042③ |
| AR, USA, 2003 (Natchez) | EO | 10 | 116 | 1 | BBCH 41-45 | 0<br>7 | 0.30<br>0.07 | |
| IN, USA, 2003 (Bravo) | EO | 10 | 188 | 1 | BBCH 31 | 0<br>7 | 0.52<br>0.13 | |
| MN, USA, 2003 (Oxen) | EO | 10 | 150 | 1 | BBCH 37 | 0<br>7 | 0.19<br>0.17 | |
| SD, USA, 2003 (Marshall) | EO | 10 | 152 | 1 | BBCH 37 | 0<br>7 | 0.29<br>0.19 | |
| ND, USA, 2003 (Oxen) | EO | 10 | 151 | 1 | BBCH 37 | 0<br>7 | 0.32<br>0.16 | |
| NE, USA, 2003 (VNS HRW) | EO | 10 | 187 | 1 | BBCH 31 | 0<br>7 | 0.54<br>0.12 | |
| OK, USA, 2003 (Coker9663) | EO | 10 | 119 | 1 | BBCH 33 | 0<br>7<br>14<br>21<br>28 | 0.57<br>0.12<br>0.10<br>0.10<br>0.12 | |
| SD, USA, 2003 (Marshall) | EO | 10 | 152 | 1 | BBCH 37 | 0<br>7<br>14<br>21<br>28 | 0.31<br>0.17<br>0.07<br>0.04<br>0.03 | |
| | SL 240 | 9.7 | 152 | 1 | BBCH 37 | 0<br>7<br>14<br>21<br>28 | 0.25<br>0.05<br>0.11<br>0.15<br>0.08 | |
| | SL 247 | 10.4 | 152 | 1 | BBCH 38 | 0<br>7<br>14<br>21<br>28 | 0.21<br>0.13<br>0.16<br>0.28<br>0.11 | |
| ND, USA, 2003 (Oxen) | EO | 10 | 152 | 1 | BBCH 37 | 0<br>7 | 0.30<br>0.16 | |
| | SL 240 | 9.7 | 152 | 1 | BBCH 37 | 0<br>7 | 0.16<br>0.16 | |
| | SL 247 | 10.4 | 152 | 1 | BBCH 38 | 0<br>7 | 0.21<br>0.09 | |
| ND, USA, 2003 (Alsen) | EO | 10.4 | 147 | 1 | BBCH 37 | 0<br>7 | 0.41<br>0.12 | |
| | SL 240 | 10.5 | 146 | 1 | BBCH 37 | 0<br>7 | 0.43<br>0.06 | |
| | SL 247 | 10.5 | 146 | 1 | BBCH 37 | 0<br>7 | 0.38<br>0.06 | |

| WHEAT FORAGE | Application | | | | | PHI | Residues | Ref./Study No. |
|---|---|---|---|---|---|---|---|---|
| Country, Year (variety) | Form. | Rate (g ae/ha) | Water (L/ha) | No. | Timing | (days) | (mg/kg) as received | |
| NE. USA, 2003 (Forge HRS) | EO | 10 | 187 | 1 | BBCH 31 | 0<br>7 | 0.49<br>0.11 | |
| | SL 240 | 10 | 188 | 1 | BBCH 31 | 0<br>7 | 0.37<br>0.05 | |
| | SL 247 | 10 | 187 | 1 | BBCH 31 | 0<br>7 | 0.63<br>0.08 | |
| SD, USA, 2003<br><br>(Walworth) | EO | 10 | 140 | 1 | BBCH 30-31 | 0<br>7 | 0.53<br>0.13 | |
| | SL 240 | 10 | 140 | 1 | BBCH 30-31 | 0<br>7 | 0.41<br>0.05 | |
| | SL 247 | 10 | 140 | 1 | BBCH 30-31 | 0<br>7 | 0.40<br>0.05 | |
| TX#1, USA, 2003 (TAM 105) | EO | 10 | 143 | 1 | BBCH 39 | 0<br>7 | 0.37<br>0.12 | |
| TX#2, USA, 2003 (TAM 105) | EO | 10 | 143 | 1 | BBCH 37 | 0<br>7 | 0.26<br>0.03 | |
| NM, USA, 2003 (Jagger) | EO | 10 | 141 | 1 | BBCH 37 | 0<br>7 | 0.45<br>0.11 | |
| TX#4, USA, 2003 (TAM 200) | EO | 10 | 177 | 1 | BBCH 33 | 0<br>7 | 0.63<br>0.1 | |
| TX#5, USA, 2003 (Jagger) | EO | 10 | 174 | 1 | BBCH 37 | 0<br>7 | 0.36<br>0.06 | |
| KS, USA, 2003 (Ike) | EO | 10 | 110 | 1 | BBCH 30 | 0<br>7 | 0.67<br>0.05 | |
| WA, USA, 2003 (Declo) | EO | 10 | 141 | 1 | BBCH 39 | 0<br>7 | 0.16<br>0.05 | |
| SK, Canada, 2003<br><br>(ACEatonia) | EO | 10 | 107 | 1 | BBCH 30 | 0<br>7 | 0.11<br>0.09 | |
| | SL 240 | 10 | 109 | 1 | BBCH 30 | 0<br>7 | 0.72<br>0.11 | |
| | SL 247 | 9.4 | 107 | 1 | BBCH 30 | 0<br>7 | 0.85<br>0.22 | |
| SK, Canada, 2003<br><br>(ACEatonia) | EO | 10 | 109 | 1 | BBCH 37 | 0<br>7 | 0.49<br>0.05 | |
| | SL 240 | 10 | 111 | 1 | BBCH 37 | 0<br>7 | 0.53<br>0.05 | |
| | SL 247 | 10 | 113 | 1 | BBCH 37 | 0<br>7 | 0.42<br>0.04 | |

① Recoveries 100 and 119% @ 0.01 and 0.1 mg/kg. ②Recoveries were 70% and 86% at 0.01 and 0.1 mg/kg, respectively. ③Recoveries ranged 75 – 105% over concentrations 0.05, 0.5 and 5 mg/kg (n = 34).

Table 44. Aminopyralid residues in wheat forage from Australia.

| WHEAT FORAGE | Application | | | | | PHI | Residues | Ref./Study No. |
|---|---|---|---|---|---|---|---|---|
| Country, Year (variety) | Form. | Rate (g ae/ha) | Water (L/ha) | No. | Timing | (days) | (mg/kg) dry weight | |
| WA, Australia, 2002 (Calingiri) | SL 240 | 10 | 100 | 1 | 5-6 leaf stage | 0<br>1<br>3<br>7<br>14<br>28 | 3.81<br>2.64<br>1.21<br>0.71<br>0.42<br>0.37 | 020060-01①② |
| | | 15 | 100 | 1 | 5-6 leaf stage | 0<br>1<br>3<br>7<br>14<br>28 | 5.26<br>4.80<br>1.35<br>1.33<br>0.79<br>0.62 | |

| WHEAT FORAGE | Application | | | | | PHI | Residues | Ref./Study No. |
|---|---|---|---|---|---|---|---|---|
| Country, Year (variety) | Form. | Rate (g ae/ha) | Water (L/ha) | No. | Timing | (days) | (mg/kg) dry weight | |
| | | 20 | 100 | 1 | 5-6 leaf stage | 0<br>1<br>3<br>7<br>14<br>28 | 7.86<br>5.77<br>2.84<br>1.40<br>0.96<br>0.70 | |
| SA, Australia, 2002 (Frame) | SL 240 | 10 | 99 | 1 | Z15 (5-leaf) | 0<br>1<br>3<br>7<br>14<br>28 | 3.81<br>2.82<br>ND *c 1.35*<br><u>0.48</u><br>0.32<br>0.11 | 020060-02①③ |
| | | 15 | 90 | 1 | Z15 (5-leaf) | 0<br>1<br>3<br>7<br>14<br>28 | 4.33<br>4.41<br>1.01<br>0.50<br>0.37<br>0.11 | |
| | | 20 | 90 | 1 | Z15 (5-leaf) | 0<br>1<br>3<br>7<br>14<br>28 | 7.43<br>4.89<br>2.47<br>0.83<br>0.47<br>0.20 | |
| NSW, Australia, 2002 (Babbler) | SL 240 | 10 | 100 | 1 | 5 – 6 leaf | 0<br>1<br>3<br>7<br>14<br>28 | 2.59<br>2.90<br>2.11<br><u>1.02</u><br>0.33<br>0.44 | 020060-05①④ |
| | | 15 | 96 | 1 | 5 – 6 leaf | 0<br>1<br>3<br>7<br>14<br>28 | 4.53<br>3.38<br>2.44<br>1.09<br>0.38<br>0.53 | |
| | | 20 | 100 | 1 | 5 – 6 leaf | 0<br>1<br>3<br>7<br>14<br>28 | 7.18<br>4.56<br>3.72<br>1.57<br>0.36<br>0.68 | |
| QLD, Australia, 2002 (Petrel) | SL 240 | 10 | 102 | 1 | 4 – 6 leaf | 0<br>1<br>3<br>7<br>14<br>28 | 4.50<br>3.02<br>1.42<br><u>0.77</u><br>0.49<br>0.34 | 020060-08①⑤ |
| | | 15 | 99 | 1 | 4 – 6 leaf | 0<br>1<br>3<br>7<br>14<br>28 | 9.57<br>5.68<br>1.93<br>1.27<br>0.78<br>0.43 | |
| | | 20 | 101 | 1 | 4 – 6 leaf | 0<br>1<br>3<br>7<br>14<br>28 | 13.3<br>10<br>3.22<br>1.88<br>1.08<br>0.55 | |

| WHEAT FORAGE Country, Year (variety) | Application | | | | | PHI (days) | Residues (mg/kg) dry weight | Ref./Study No. |
|---|---|---|---|---|---|---|---|---|
| | Form. | Rate (g ae/ha) | Water (L/ha) | No. | Timing | | | |
| NSW, Australia, 2003 (Wollaroi) | SL 240 | 7.5 | 101 | 1 | 6-leaf | 0 1 3 6 14 28 | 1.58 1.26 0.95 0.16 0.08 0.14 | 030071-01⑥⑦ |
| | | 15 | 101 | 1 | 6-leaf | 0 1 3 6 14 28 | 2.38 2.99 2.29 0.45 0.31 0.26 | |
| | EO | 7.5 | 101 | 1 | 6-leaf | 0 1 3 6 14 28 | 1.63 0.89 0.72 0.45 0.35 0.22 | |
| | | 15 | 101 | 1 | 6-leaf | 0 1 3 6 14 28 | 2.66 1.44 0.88 0.99 0.56 0.33 | |

①Recoveries ranged 87 – 106% over concentrations 0.01 – 5 mg/kg (n = 17). ②Moisture contents ranged 82.2–86.6%. ③Moisture contents ranged 75.1 – 85.6%. ④Moisture contents ranged 77.1 – 81.4%. ⑤Moisture contents ranged 73.5 – 88.5%. ⑥Moisture contents ranged 70.6 – 84.1%. ⑦ Recoveries ranged 85 – 124% over concentrations 0.01 – 1 mg/kg (n = 15).

Table 45. Aminopyralid residues in wheat hay from USA; wheat hay was allowed to dry for 2 to 7 days to obtain proper moisture for hay, prior to freezing for analysis.

| WHEAT HAY Country, Year (variety) | Application | | | | | PHI (days) | Residues (mg/kg) as received | Ref./Study No. |
|---|---|---|---|---|---|---|---|---|
| | Form. | Rate (g ae/ha) | Water (L/ha) | No. | Timing | | | |
| VA, USA, 2003 (Coker9835) | EO | 10 | 135 | 1 | BBCH 32 | 0 7 14 21 28 | 1.99 0.99 0.61 0.45 0.41 | 030042① |
| AR, USA, 2003 (Natchez) | EO | 10 | 116 | 1 | BBCH 41-45 | 0 7 | 0.76 0.28 | |
| IN, USA, 2003 (Bravo) | EO | 10 | 188 | 1 | BBCH 31 | 0 7 | 1.31 0.33 | |
| MN, USA, 2003 (Oxen) | EO | 10 | 150 | 1 | BBCH 37 | 0 7 | 0.45 0.39 | |
| SD, USA, 2003 (Marshall) | EO | 10 | 152 | 1 | BBCH 37 | 0 7 | 0.69 0.56 | |
| ND, USA, 2003 (Oxen) | EO | 10 | 151 | 1 | BBCH 37 | 0 7 | 0.69 0.46 | |
| NE, USA, 2003 (VNS HRW) | EO | 10 | 187 | 1 | BBCH 31 | 0 7 | 0.98 0.23 | |
| OK, USA, 2003 (Coker9663) | EO | 10 | 119 | 1 | BBCH 33 | 0 7 14 21 28 | 1.08 0.29 0.34 0.24 0.22 | |

| WHEAT HAY | Application | | | | | PHI | Residues | Ref./Study No. |
|---|---|---|---|---|---|---|---|---|
| Country, Year (variety) | Form. | Rate (g ae/ha) | Water (L/ha) | No. | Timing | (days) | (mg/kg) as received | |
| SD, USA, 2003 (Marshall) | EO | 10 | 152 | 1 | BBCH 37 | 0<br>7<br>14<br>21<br>28 | 0.86<br>0.45 c 0.017<br>0.25<br>0.044<br>0.044 | |
| | SL 240 | 9.7 | 152 | 1 | BBCH 37 | 0<br>7<br>14<br>21<br>28 | 0.37<br>0.24 c 0.017<br>0.26<br>0.18<br>0.06 | |
| | SL 247 | 10.4 | 152 | 1 | BBCH 38 | 0<br>7<br>14<br>21<br>28 | 0.35<br>0.27 c 0.017<br>0.21<br>0.24<br>0.06 | |
| ND, USA, 2003 (Oxen) | EO | 10 | 152 | 1 | BBCH 37 | 0<br>7 | 0.76<br>0.32 | |
| | SL 240 | 9.7 | 152 | 1 | BBCH 37 | 0<br>7 | 0.34<br>0.36 | |
| | SL 247 | 10.4 | 152 | 1 | BBCH 38 | 0<br>7 | 0.54<br>< 0.01 | |
| ND, USA, 2003 (Alsen) | EO | 10.4 | 147 | 1 | BBCH 37 | 0<br>7 | 1.67<br>0.28 | |
| | SL 240 | 10.5 | 146 | 1 | BBCH 37 | 0<br>7 | 1.32<br>0.19 | |
| | SL 247 | 10.5 | 146 | 1 | BBCH 37 | 0<br>7 | 0.87<br>0.16 | |
| NE. USA, 2003 (Forge HRS) | EO | 10 | 187 | 1 | BBCH 31 | 0<br>7 | 1.23<br>0.37 | |
| | SL 240 | 10 | 188 | 1 | BBCH 31 | 0<br>7 | 0.71<br>0.17 | |
| | SL 247 | 10 | 187 | 1 | BBCH 31 | 0<br>7 | 1.24<br>0.23 | |
| SD, USA, 2003 (Walworth) | EO | 10 | 140 | 1 | BBCH 30-31 | 0<br>7 | 1.88<br>0.44 | |
| | SL 240 | 10 | 140 | 1 | BBCH 30-31 | 0<br>7 | 1.38<br>0.19 | |
| | SL 247 | 10 | 140 | 1 | BBCH 30-31 | 0<br>7 | 1.33<br>0.16 | |
| TX#1, USA, 2003 (TAM 105) | EO | 10 | 143 | 1 | BBCH 39 | 0<br>7 | 0.98<br>0.33 | |
| TX#2, USA, 2003 (TAM 105) | EO | 10 | 143 | 1 | BBCH 37 | 0<br>7 | 0.83<br>0.06 | |
| NM, USA, 2003 (Jagger) | EO | 10 | 141 | 1 | BBCH 37 | 0<br>7 | 1.02<br>0.32 | |
| TX#4, USA, 2003 (TAM 200) | EO | 10 | 177 | 1 | BBCH 33 | 0<br>7 | 0.54<br>0.19 | |
| TX#5, USA, 2003 (Jagger) | EO | 10 | 174 | 1 | BBCH 37 | 0<br>7 | 0.43<br>0.14 | |
| KS, USA, 2003 (Ike) | EO | 10 | 110 | 1 | BBCH 30 | 0<br>7 | 1.46<br>0.12 | |
| WA, USA, 2003 (Declo) | EO | 10 | 141 | 1 | BBCH 39 | 0<br>7 | 0.38<br>0.18 | |
| SK, Canada, 2003 (ACEatonia) | EO | 10 | 107 | 1 | BBCH 30 | 0<br>7 | 0.37<br>0.13 | |
| | SL 240 | 10 | 109 | 1 | BBCH 30 | 0<br>7 | 2.32<br>0.62 | |
| | SL 247 | 9.4 | 107 | 1 | BBCH 30 | 0<br>7 | 2.36<br>0.64 | |

| WHEAT HAY | Application | | | | | PHI | Residues | Ref./Study No. |
|---|---|---|---|---|---|---|---|---|
| Country, Year (variety) | Form. | Rate (g ae/ha) | Water (L/ha) | No. | Timing | (days) | (mg/kg) as received | |
| SK, Canada, 2003 | EO | 10 | 109 | 1 | BBCH 37 | 0 7 | <u>1.28</u> 0.13 | |
| (ACEatonia) | SL 240 | 10 | 111 | 1 | BBCH 37 | 0 7 | <u>1.48</u> 0.08 | |
| | SL 247 | 10 | 113 | 1 | BBCH 37 | 0 7 | <u>1.46</u> 0.12 | |

① Recoveries ranged 70 – 107% over concentrations 0.05, 0.5 and 5 mg/kg (n = 35).

## *Pastures*

Hay samples in the European trials were cut at the PHI indicated and left to dry for 3 to 5 days prior to analysis. In the US and Canadian trials, hay samples were dried for 2 to 5 days after cutting at the PHI. All results are expressed on an as received basis. Samples for silage were cut at the PHI indicated in the tables below and left to silage prior to collection for analysis at 115 days after application.

Table 46. Aminopyralid residues in pastures and hay from trials conducted in Europe and USA.

| PASTURES/HAY Country, Year (variety) | Application | | | | Sample | PHI (days) | Residues (mg/kg) as received | Ref./Study No. |
|---|---|---|---|---|---|---|---|---|
| | Form. | Rate (g ae/ha) | Water (L/ha) | No. | | | | |
| Mogi Mirim, Brazil, 2002 (*Brachiaria decumbens*) | SL 40 | 120 | 250 | 1 | Grass | 0 7 14 21 28 | <u>1.9, 1</u> 2.1, 1.7 1.7, 2.7 1.6, 2.8 2.1, 1.2 | 020131①② |
| | | 240 | 250 | 1 | Grass | 0 7 14 21 28 | 4.9, 5.6 6.5, 6 8.6, 6.4 8, 6.4 7.2, 6.5 | |
| Cedral, Brazil, 2002 (*Brachiaria brizantha*) | SL 40 | 120 240 | 250 | 1 1 | Grass | 0 0 | <u>11.2, 12</u> 15.5, 15.6 | |
| Londrina, Brazil, 2002 (*panicum maximum*) | SL 40 | 120 240 | 200 | 1 1 | Grass | 0 0 | 12.2, 13.5 11.6, 11.8 | |
| Mogi Mirim, Brazil, 2002 (*Brachiaria decumbens*) | EO 40 | 100 | 250 | 1 | Grass | 0 7 14 21 28 | <u>2.1, 0.6</u> 2.8, 2.8 1.8, 2.6 2.3, 2.6 1.7, 3 | 020132①③ |
| | | 200 | 250 | 1 | Grass | 0 7 14 21 28 | 0.2, 1 4.2, 1.4 4.3, 4.3 8.2, 2.5 3.4, 3.5 | |
| Cedral, Brazil, 2002 (*Brachiaria brizantha*) | EO 40 | 100 200 | 250 | 1 1 | Grass | 0 0 | <u>8.9, 3.8</u> 12.7, 26.6 | |
| Londrina, Brazil, 2002 (*panicum maximum*) | EO 40 | 100 200 | 200 | 1 1 | Grass | 0 0 | <u>2.4, 4.8</u> 20.2, 16.3 | |
| Derbyshire, UK, 2002 (Italian and periennial Ryegrass ley) | ME 30 #1 | 60 | 301 | 1 | Grass | 0 | 4.22 | GHE-P-10448 295/155/1 |
| Lamstedt, Germany, 2002 (*Lolium perenne*) | ME 30 #1 | 65 | 327 | 1 | Grass | 0 | 4 | 295/155/2 |
| Derbyshire, UK, 2002 (Italian ryegrass) | ME 30 #2 | 60 | 299 | 1 | Grass Hay | 0 3 | 1.7 1.18 | GHE-P-10449④ 295/153/9 |

| PASTURES/HAY Country, Year (variety) | Application | | | | Sample | PHI (days) | Residues (mg/kg) as received | Ref./Study No. |
|---|---|---|---|---|---|---|---|---|
| | Form. | Rate (g ae/ha) | Water (L/ha) | No. | | | | |
| Lamstedt, Germany, 2002 (*Lolium perenne*) | ME 30 #2 | 63 | 313 | 1 | Grass<br>Hay | 0<br>3 | 1.62<br>1.42 | 295/153/2 |
| Emilia Romagna, Italy 2002 (*Lolium multiflorum, Avene spp*) | ME 30 #2 | 60 | 302 | 1 | Hay | 3 | 4 | 295/153/3 |
| Noilhan, France, 2002 (*Dactylis spp*) | ME 30 #2 | 62 | 310 | 1 | Hay | 3 | 8 | 295/153/4 |
| Derbyshire, UK, 2002 (Long term ryegrass ley) | ME 30 #2 | 60 | 299 | 1 | Grass | 0<br>1<br>3<br>7<br>14<br>21 | 3.15<br>2.56<br>3.2<br>1.53<br>1.46<br>1.93 | 295/153/5 |
| Lower Saxony, Germany 2002 (*Lolium spp*) | ME 30 #2 | 60 | 299 | 1 | Grass | 0<br>1<br>4<br>7<br>14<br>21 | 2.95<br>1.97<br>1.31<br>1.81<br>0.91<br>0.71 | 295/153/6 |
| Tarn-et-Garonne, France, 2002 (*Vicia spp Trifolium spp.*) | ME 30 #2 | 61 | 303 | 1 | Grass | 0<br>1<br>3<br>7<br>14<br>21 | 3.92<br>2.15<br>0.96<br>0.84<br>0.55<br>0.81 | 295/153/7 |
| Montardon, France, 2002 (*Festuca arundinacea*) | ME 30 #2 | 61 | 255 | 1 | Grass | 0<br>1<br>3<br>7<br>15<br>21 | 4.25<br>3.04<br>2.65<br>2.18<br>1.38<br>1.10 | 295/153/8 |
| Neidersachsen, Germany, 2003 (*Lolium*) | ME 30 #1 | 66 | 330 | 1 | Grass<br>Hay | 0<br>3 | 13.65<br>8.5 | GHE-P-10578⑤<br>CEMS-2065A |
| Sheffield, UK, 2003 (Perennial ryegrass) | ME 30 #1 | 58 | 289 | 1 | Grass<br>Hay | 0<br>3 | 9.47<br>6.59 | CEMS-2065B |
| Rabastens, France, 2003 (*Festuca, Lolium spp*) | ME 30 #1 | 60 | 301 | 1 | Grass<br>Hay | 0<br>3 | 4.55<br>1.68 | CEMS-2065C |
| Pastriz, Spain, 2003 (Fuego) | ME 30 #1 | 60 | 302 | 1 | Grass<br>Hay | 0<br>3 | 7.06<br>8.85 | CEMS-2065D |
| Chesterfield, UK, 2003 (Ryegrass, fescue and pasture mixture) | ME 30 #1 | 60 | 300 | 1 | Grass<br><br><br><br><br><br>Silage | 0<br>1<br>3<br>7<br>14<br>21<br>3 | 2<br>1.83<br>1.69<br>1.03<br>0.63<br>0.41<br>1.45 | GHE-P-10579<br>CEMS-2066A⑥ |
| Maine-et-Loire, France, 2003 (*Festuca spp*) | ME 30 #1 | 60 | 302 | 1 | Grass | 0<br>1<br>3<br>7<br>14<br>21 | 8.86<br>5<br>3.21<br>2.97<br>0.79<br>0.4 | CEMS-2066B |
| Maine-et-Loire, France, 2003 (*Dactyle spp*) | ME 30 #1 | 58 | 292 | 1 | Grass<br>Hay | 0<br>0 | 5.49<br>13.08 | CEMS-2066C |
| Niedersachsen, Germany, 2003 (*Lolium perenne*) | ME 30 #1 | 63 | 210 | 1 | Grass<br><br>Hay | 0<br><br>3 | 3.22<br><br>2.02 | CEMS-2066D |

| PASTURES/HAY Country, Year (variety) | Application | | | | Sample | PHI (days) | Residues (mg/kg) as received | Ref./Study No. |
|---|---|---|---|---|---|---|---|---|
| | Form. | Rate (g ae/ha) | Water (L/ha) | No. | | | | |
| Montardon, France, 2003 (*Festuca spp.*) | ME 30 #1 | 61 | 305 | 1 | Grass | 0 | 6.48 | GHE-P-10580⑦ |
| | | | | | | 1 | 3.56 | CEMS-2067A |
| | | | | | | 3 | 3.75 | |
| | | | | | | 7 | 1.96 | |
| | | | | | | 14 | 2.11 | |
| | | | | | | 21 | 1.37 | |
| Olaeta-Aramaio, Spain, 2003 (Ryegrass Barsilo) | ME 30 #1 | 61 | 306 | 1 | Grass | 0 | 2.34 | CEMS-2067B |
| | | | | | | 1 | 2.4 | |
| | | | | | | 3 | 1.91 | |
| | | | | | | 7 | 1.32 | |
| | | | | | | 14 | 1.07 | |
| | | | | | | 21 | 0.95 | |
| Tournan, France, 2003 (Italian ryegrass) | ME 30 #1 | 55 | 273 | 1 | Grass | 0 | 1.65 | CEMS-2067C |
| | | | | | Hay | 3 | 3.37 | |
| Bologna, Italy, 2003 (*Festuca spp, Dacus spp, Digitaria spp*) | ME 30 #1 | 60 | 302 | 1 | Grass | 0 | 7.24 | CEMS-2067D |
| | | | | | Hay | 3 | 5.79 | |
| AB, Canada, 2002 (Brome fescue mix) | SL 240 | 118 | 148 | 1 | Grass | 0 | 13.7 | 020018①⑧ |
| | | | | | | 3 | 5.2 | |
| | | | | | | 7 | 4.1 | |
| | | | | | | 14 | 1.9 | |
| | | | | | | 21 | 2 | |
| | | | | | | 28 | 2.2 | |
| | | | | | Hay | 0 | 15.9 | |
| | | | | | | 14 | 3.8 | |
| | | | | | | 21 | 4.2 | |
| | | | | | | 28 | 4.6 | |
| | SL 240* | 118 | 149 | 1 | Grass | 0 | 13.2 | * +2,4-D tank mix |
| | | | | | | 3 | 6.9 | |
| | | | | | | 7 | 5.9 | |
| | | | | | | 14 | 2.3 | |
| | | | | | | 21 | 1.6 | |
| | | | | | | 28 | 2.1 | |
| | | | | | Hay | 0 | 25.1 | |
| | | | | | | 14 | 4.7 | |
| | | | | | | 21 | 3.3 | |
| | | | | | | 28 | 5.5 | |
| AB, Canada, 2002 (Fescue Timothy mix) | SL 240 | 116 | 97 | 1 | Grass | 0 | 14.6 | |
| | | | | | | 3 | 6.7 | |
| | | | | | | 7 | 3.8 | |
| | | | | | | 14 | 2.3 | |
| | | | | | | 22 | 2.6 | |
| | | | | | | 28 | 1.9 | |
| | | | | | Hay | 0 | 16.8 | |
| | | | | | | 14 | 3.2 | |
| | | | | | | 22 | 4.5 | |
| | | | | | | 28 | 3.7 | |
| | S 240* | 125 | 104 | 1 | Grass | 0 | 13.8 | * +2,4-D tank mix |
| | | | | | | 3 | 6.3 | |
| | | | | | | 7 | 4.8 | |
| | | | | | | 14 | 1.7 | |
| | | | | | | 22 | 2.1 | |
| | | | | | | 28 | 1.9 | |
| | | | | | Hay | 0 | 21.6 | |
| | | | | | | 14 | 3 | |
| | | | | | | 22 | 3.3 | |
| | | | | | | 28 | 3.4 | |
| SK, Canada, 2002 (Brome grass mix) | SL 240 | 119 | 150 | 1 | Grass | 0 | 10.7 | |
| | | | | | | 7 | 1.5 | |
| | | | | | | 14 | 1.4 | |
| | | | | | Hay | 0 | 26.2 | |
| | | | | | | 14 | 2.9 | |

| PASTURES/HAY Country, Year (variety) | Application | | | | Sample | PHI (days) | Residues (mg/kg) as received | Ref./Study No. |
|---|---|---|---|---|---|---|---|---|
| | Form. | Rate (g ae/ha) | Water (L/ha) | No. | | | | |
| | | | | | | 22 | 2.8 | |
| AB, Canada, 2002 (Fescue Brome mix) | SL 240 | 119 | 149 | 1 | Grass | 0 | 12.2 | |
| | | | | | | 7 | 4.4 | |
| | | | | | | 14 | 2.6 | |
| | | | | | Hay | 0 | 22.2 | |
| | | | | | | 14 | 5.1 | |
| | | | | | | 21 | 5.1 | |
| AB, Canada, 2002 (Orchard Fescue Timothy mix) | SL 240 | 121 | 101 | 1 | Grass | 0 | 9.1 | |
| | | | | | | 7 | 4.6 | |
| | | | | | | 13 | 2.9 | |
| | | | | | Hay | 0 | 15.1 | |
| | | | | | | 13 | 3.9 | |
| | | | | | | 21 | 5.3 | |
| MB, Canada, 2002 (Brome grass) | SL 240 | 121 | 101 | 1 | Grass | 0 | 12.7 | |
| | | | | | | 7 | 3.5 | |
| | | | | | | 14 | 2.3 | |
| | | | | | Hay | 0 | 55 | |
| | | | | | | 14 | 6.6 | |
| | | | | | | 20 | 4.2 | |
| SK, Canada, 2003 (Native grasses) | SL 240 | 120 | 150 | 1 | Grass | 0 | 12.8 | |
| | | | | | | 8 | 3.8 | |
| | | | | | | 14 | 3.7 | |
| | | | | | Hay | 0 | 30 | |
| | | | | | | 14 | 9.2 | |
| | | | | | | 21 | 7.8 | |
| VA, USA, 2002 (Fescue/orchard grass) | SL 240 | 120 | 140 | 1 | Grass | 0 | 4.8 | |
| | | | | | | 3 | 4.6 | |
| | | | | | | 7 | 3.6 | |
| | | | | | | 14 | 2.8 | |
| | | | | | | 21 | 1.8 | |
| | | | | | | 28 | 2.3 | |
| | | | | | Hay | 0 | 17.9 | |
| | | | | | | 14 | 8.6 | |
| | | | | | | 21 | 7.8 | |
| | | | | | | 28 | 6.4 | |
| | SL 240* | 120 | 140 | 1 | Grass | 0 | 8.7 | |
| | | | | | | 3 | 6.4 | |
| | | | | | | 7 | 4.5 | |
| | | | | | | 14 | 5.8 | |
| | | | | | | 21 | 3.4 | |
| | | | | | | 28 | 3.5 | |
| | | | | | Hay | 0 | 26.3 | |
| | | | | | | 14 | 11 | |
| | | | | | | 21 | 16.6 | |
| | | | | | | 28 | 4.8 | |
| WA, USA, 2002 (Fescue) | SL 240 | 120 | 140 | 1 | Grass | 0 | 4.8 | |
| | | | | | | 3 | 1.1 | |
| | | | | | | 7 | 1.2 | |
| | | | | | | 14 | 1.1 | |
| | | | | | | 21 | 1.1 | |
| | | | | | | 28 | 1 | |
| | | | | | Hay | 0 | 10.6 | |
| | | | | | | 14 | 3.1 | |
| | | | | | | 21 | 3.2 | |
| | | | | | | 28 | 3 | |
| | SL 240* | 120 | 140 | 1 | Grass | 0 | 6.1 | |
| | | | | | | 3 | 1.3 | |
| | | | | | | 7 | 1.8 | |
| | | | | | | 14 | 1.6 | |
| | | | | | | 21 | 1.2 | |
| | | | | | | 28 | 1 | |
| | | | | | Hay | 0 | 12.4 | |
| | | | | | | 14 | 3.8 | |

| PASTURES/HAY Country, Year (variety) | Application | | | | Sample | PHI (days) | Residues (mg/kg) as received | Ref./Study No. |
|---|---|---|---|---|---|---|---|---|
| | Form. | Rate (g ae/ha) | Water (L/ha) | No. | | | | |
| | | | | | | 21 | 4.1 | |
| | | | | | | 28 | 3 | |
| GA, USA, 2002 (Bermuda grass) | SL 240 | 120 | 164 | 1 | Grass | 0 | 12.5 | |
| | | | | | | 7 | 5.9 | |
| | | | | | | 14 | 2 | |
| | | | | | Hay | 0 | 25.8 | |
| | | | | | | 14 | 8.8 | |
| | | | | | | 21 | 5.6 | |
| ID, USA, 2002 (Mixture) | SL 240 | 120 | 163 | 1 | Grass | 0 | 6.9 | |
| | | | | | | 7 | 0.7 | |
| | | | | | | 14 | 0.9 | |
| | | | | | Hay | 0 | 18.2 | |
| | | | | | | 14 | 2.4 | |
| | | | | | | 20 | 2.3 | |
| MS, USA, 2002 (Bermuda grass) | SL 240 | 125 | 148 | 1 | Grass | 0 | 10 | |
| | | | | | | 7 | 2.2 | |
| | | | | | | 14 | 3 | |
| | | | | | Hay | 0 | 18.6 | |
| | | | | | | 14 | 7.1 | |
| | | | | | | 21 | 3.6 | |
| MT-1, USA, 2002 (Crested wheat) | SL 240 | 122 | 190 | 1 | Grass | 0 | 11.1 | |
| | | | | | | 7 | 7.1 | |
| | | | | | | 14 | 2.4 | |
| | | | | | Hay | 0 | 15.9 | |
| | | | | | | 14 | 5 | |
| | | | | | | 21 | 4.4 | |
| MT-2, USA, 2002 (Orchard grass) | SL 240 | 121 | 188 | 1 | Grass | 0 | 9.4 | |
| | | | | | | 6 | 1.5 | |
| | | | | | | 13 | 1.3 | |
| | | | | | Hay | 0 | 15.7 | |
| | | | | | | 13 | 2.2 | |
| | | | | | | 20 | 1.3 | |
| ND, USA, 2002 (Brome grass) | SL 240 | 120 | 151 | 1 | Grass | 0 | 11.2 | |
| | | | | | | 7 | 4.4 | |
| | | | | | | 14 | 1.4 | |
| | | | | | Hay | 0 | 18.1 | |
| | | | | | | 14 | 3.4 | |
| | | | | | | 22 | 1 | |
| NY, USA, 2002 (Orchard grass) | SL 240 | 115 | 181 | 1 | Grass | 0 | 4.6 | |
| | | | | | | 7 | 3 | |
| | | | | | | 14 | 2.8 | |
| | | | | | Hay | 0 | 10.9 | |
| | | | | | | 14 | 10 | |
| | | | | | | 21 | 7.8 | |
| OH, USA, 2002 (Fescue/orchard grass) | SL 240 | 123 | 175 | 1 | Grass | 0 | 5 | |
| | | | | | | 7 | 2. | |
| | | | | | | 15 | 2.1 | |
| | | | | | Hay | 0 | 16 | |
| | | | | | | 15 | 10.4 | |
| | | | | | | 21 | 5.6 | |
| PA, USA, 2002 (Tall fescue) | SL 240 | 125 | 157 | 1 | Grass | 0 | 10.1 | |
| | | | | | | 7 | 2 | |
| | | | | | | 14 | 2.8 | |
| | | | | | Hay | 0 | 32.2 | |
| | | | | | | 14 | 5.9 | |
| | | | | | | 21 | 5.9 | |
| TX, USA, 2002 (Tall fescue) | SL 240 | 121 | 171 | 1 | Grass | 0 | 16 | |
| | | | | | | 7 | 1.5 | |
| | | | | | | 14 | 2 | |
| | | | | | Hay | 0 | 18.2 | |
| | | | | | | 14 | 2.1 | |
| | | | | | | 21 | 3.9 | |

| PASTURES/HAY Country, Year (variety) | Application | | | | Sample | PHI (days) | Residues (mg/kg) as received | Ref./Study No. |
|---|---|---|---|---|---|---|---|---|
| | Form. | Rate (g ae/ha) | Water (L/ha) | No. | | | | |
| WI, USA, 2002 (Reed canary grass) | SL 240 | 121 | 176 | 1 | Grass | 0 | 6.5 | |
| | | | | | | 7 | 3.7 | |
| | | | | | | 14 | 2 | |
| | | | | | Hay | 0 | 20.3 | |
| | | | | | | 14 | 10 | |
| | | | | | | 21 | 5.9 | |

① Joint oil or adjuvant added to sprays. ②Recoveries ranged 78 – 108% over concentrations 1, 50, 100 and 200 mg/kg. ③Recoveries ranged 78 – 107% over concentrations 1, 50 and 100 mg/kg. ④ Recoveries ranged 84 – 122% over concentrations 0.01, 1 and 5 mg/kg in pastures and 78% and 87% at 0.01 mg/kg and 5 mg/kg in hay, respectively. ⑤ Recoveries were 77 and 112% at 0.01 and 0.1 mg/kg in grass and 72% and 93% at 0.01 and 0.1 mg/kg in hay. ⑥ Recoveries were 84% and 70% at 0.01 mg/kg and 93% and 113% at 0.1 mg/kg in grass, 75% at 0.01 mg/kg and 88% at 0.1 mg/kg in hay and 95% at 0.01 mg/kg and 90% at 0.1 mg/kg in silage. ⑦ Recoveries were 84% and 94% at 0.01 mg/kg and 0.1 mg/kg in grass respectively; 105% and 93% at 0.01 mg/kg and 0.1 mg/kg in hay, respectively. ⑧Mean recoveries ranged 79 – 92% over concentrations 0.01, 0.1, 20, 30 and 60 mg/kg in forage; mean recoveries ranged 84 – 100% over concentrations 0.01, 0.02, 0.1, 1, 20, 30 and 60 mg/kg.

Table 47. Aminopyralid residues in pastures from trials conducted in Australia and New Zealand.

| PASTURES/HAY Country, Year (variety) | Application | | | | Sample | PHI (days) | Residues (mg/kg) dry weight | Ref./Study No. |
|---|---|---|---|---|---|---|---|---|
| | Form. | Rate (g ae/ha) | Water (L/ha) | No. | | | | |
| NSW, Australia, 2004 (Native pasture) | SL 240 | 270 | 245 | 1 | Forage | 0 | 37.1 | 030103-01①② |
| | | | | | | 3 | 18.2 | |
| | | | | | | 7 | 14.5 | |
| | | | | | | 14 | 6.1 | |
| | | | | | | 30 | 6.9 | |
| | | | | | | 56 | 2.6 | |
| | | 540 | 253 | 1 | Forage | 0 | 86.5 | |
| | | | | | | 3 | 30 | |
| | | | | | | 7 | 28.2 | |
| | | | | | | 14 | 12.5 | |
| | | | | | | 30 | 12.8 | |
| | | | | | | 56 | 9.6 | |
| NSW, Australia, 2004 (Native pasture) | SL 240 | 270 | 253 | 1 | Forage | 0 | 52.5 | 030103-02①③ |
| | | | | | | 3 | 8.4 | |
| | | | | | | 7 | 7.9 | |
| | | | | | | 14 | 4.3 | |
| | | | | | | 30 | 3.4 | |
| | | | | | | 56 | 3.2 | |
| | | 540 | 253 | 1 | Forage | 0 | 76.7 | |
| | | | | | | 3 | 12.4 | |
| | | | | | | 7 | 15.4 | |
| | | | | | | 14 | 9.9 | |
| | | | | | | 30 | 9.3 | |
| | | | | | | 56 | 7.3 | |
| VIC, Australia, 2004 (Kangaroo grass) | SL 240 | 270 | 238 | 1 | Forage | 0 | 103 | 030103-03①④ |
| | | | | | | 3 | 90.6 | |
| | | | | | | 7 | 13.8 | |
| | | | | | | 14 | 13.3 | |
| | | | | | | 28 | 11.5 | |
| | | | | | | 56 | 1.2 | |
| | | 540 | 241 | 1 | Forage | 0 | 212 | |
| | | | | | | 3 | 26 | |
| | | | | | | 7 | 28.7 | |
| | | | | | | 14 | 25.7 | |
| | | | | | | 28 | 25.1 | |
| | | | | | | 56 | 2.5 | |

| PASTURES/HAY | Application | | | | Sample | PHI | Residues | Ref./Study No. |
|---|---|---|---|---|---|---|---|---|
| Country, Year (variety) | Form. | Rate (g ae/ha) | Water (L/ha) | No. | | (days) | (mg/kg) dry weight | |
| QLD, Australia, 2004 (Kikuyu grass) | SL 240 | 210 | 240 | 1 | Forage | 0<br>3<br>7<br>14<br>28<br>56 | 9<br>5.3<br>4.4<br>1.4<br>1.3<br>0.85 | 030102-01❷❸ |
| | | 420 | 253 | 1 | Forage | 0<br>3<br>7<br>14<br>28<br>56 | 19<br>5.3<br>7.7<br>6.4<br>2.3<br>2.4 | |
| | EO | 210 | 243 | 1 | Forage | 0<br>3<br>7<br>14<br>28<br>56 | 0.7<br>4.6<br>4.6<br>3.3<br>2.1<br>0.7 | |
| QLD, Australia, 2004 (Qld blue grass) | SL 240 | 210 | 241 | 1 | Forage | 0<br>3<br>7<br>14<br>28<br>56 | 19<br>5.5<br>8.3<br>6.6<br>3.1<br>2.4 | 030102-02❷❹ |
| | | 420 | 253 | 1 | Forage | 0<br>3<br>7<br>14<br>28<br>56 | 36<br>11<br>9.1<br>2.6<br>4.8<br>2.2 | |
| | EO | 210 | 255 | 1 | Forage | 0<br>3<br>7<br>14<br>28<br>56 | 12<br>5.3<br>4.4<br>5.5<br>2.3<br>2.2 | |
| QLD, Australia, 2004 (Native pasture) | SL 240 | 210 | 249 | 1 | Forage | 0<br>3<br>7<br>14<br>28<br>56 | 12<br>1.5<br>1.5<br>0.61<br>0.44<br>0.46 | 030102-03❷❺ |
| | | 420 | 251 | 1 | Forage | 0<br>3<br>7<br>14<br>28<br>56 | 21<br>2.8<br>2.6<br>1.8<br>1.1<br>0.9 | |
| | EO | 210 | 260 | 1 | Forage | 0<br>3<br>7<br>14<br>28<br>56 | 12<br>3.3<br>3.3<br>1.7<br>1.3<br>1.6 | |
| Leeston, NZ, 2003 (Perennial ryegrass & clover) (5 cm height) | SL 240 | 60 | 192 | 1 | Forage | 0<br>1<br>3<br>7<br>14<br>28 | 38.8<br>22.4<br>11.3<br>9.6<br>10.9<br>7.4 | 030028-01⑤⑥ |

| PASTURES/HAY Country, Year (variety) | Application Form. | Application Rate (g ae/ha) | Application Water (L/ha) | Application No. | Sample | PHI (days) | Residues (mg/kg) dry weight | Ref./Study No. |
|---|---|---|---|---|---|---|---|---|
| | | 120 | 198 | 1 | Forage | 0 | 96.6 | |
| | | | | | | 1 | 56.9 | |
| | | | | | | 3 | 28.8 | |
| | | | | | | 7 | 30.1 | |
| | | | | | | 14 | 25.5 | |
| | | | | | | 28 | 15.7 | |
| | | 300 | 390 | 1 | Forage | 0 | 245 | |
| | | | | | | 1 | 150 | |
| | | | | | | 3 | 40 | |
| | | | | | | 7 | 53 | |
| | | | | | | 14 | 64 | |
| | | | | | | 28 | 41 | |
| | | 600 | 381 | 1 | Forage | 0 | 482 | |
| | | | | | | 1 | 294 | |
| | | | | | | 3 | 139 | |
| | | | | | | 7 | 125 | |
| | | | | | | 14 | 4.6 | |
| | | | | | | 28 | 77 | |
| Leeston, NZ, 2003 (Perennial ryegrass & clover) (15 cm height) | SL 240 | 60 | 182 | 1 | Forage | 0 | 11.3 | 030028-02⑤⑦ |
| | | | | | | 1 | 21.9 | |
| | | | | | | 3 | 15 | |
| | | | | | | 7 | 11.7 | |
| | | | | | | 14 | 12.8 | |
| | | | | | | 28 | 6.5 | |
| | | 120 | 191 | 1 | Forage | 0 | – | |
| | | | | | | 1 | 47 | |
| | | | | | | 3 | 3.1 | |
| | | | | | | 7 | 28 | |
| | | | | | | 14 | 24 | |
| | | | | | | 28 | 17 | |
| | | 300 | 371 | 1 | Forage | 0 | 221 | |
| | | | | | | 1 | 24 | |
| | | | | | | 3 | 75 | |
| | | | | | | 7 | 80 | |
| | | | | | | 14 | 51 | |
| | | | | | | 28 | 42 | |
| | | 600 | 375 | 1 | Forage | 0 | 430 | |
| | | | | | | 1 | 268 | |
| | | | | | | 3 | 156 | |
| | | | | | | 7 | 139 | |
| | | | | | | 14 | 84 | |
| | | | | | | 28 | 75 | |
| Oakura, NZ, 2003 (Ryegrass and white clover) | SL 240 | 60 | 156 | 1 | Forage | 0 | 48 | 030028-03⑤⑧ |
| | | | | | | 1 | 29 | |
| | | | | | | 3 | 11 | |
| | | | | | | 7 | 10 | |
| | | | | | | 14 | 7 | |
| | | | | | | 28 | 5 | |
| | | 120 | 162 | 1 | Forage | 0 | 75 | |
| | | | | | | 1 | 60 | |
| | | | | | | 3 | 20 | |
| | | | | | | 7 | 19 | |
| | | | | | | 14 | 17 | |
| | | | | | | 28 | 11 | |
| | | 300 | 299 | 1 | Forage | 0 | 238 | |
| | | | | | | 1 | 111 | |
| | | | | | | 3 | 51 | |
| | | | | | | 7 | 55 | |
| | | | | | | 14 | 51 | |
| | | | | | | 28 | 27 | |

| PASTURES/HAY Country, Year (variety) | Application | | | Sample | PHI (days) | Residues (mg/kg) dry weight | Ref./Study No. |
|---|---|---|---|---|---|---|---|
| | Form. | Rate (g ae/ha) | Water (L/ha) | No. | | | |
| | | 600 | 299 | 1 | Forage | 0 | 402 | |
| | | | | | | 1 | 244 | |
| | | | | | | 3 | 71 | |
| | | | | | | 7 | 87 | |
| | | | | | | 14 | 89 | |
| | | | | | | 28 | 7 | |
| Okato, NZ, 2003 (Ryegrass) | SL 240 | 60 | 157 | 1 | Forage | 0 | 42.4 | 030028-04⑤⑨ |
| | | | | | | 1 | 1.5 | |
| | | | | | | 3 | 13.2 | |
| | | | | | | 7 | 9.5 | |
| | | | | | | 14 | 8.7 | |
| | | | | | | 28 | 2.3 | |
| | | 120 | 154 | 1 | Forage | 0 | 77 | |
| | | | | | | 1 | 54 | |
| | | | | | | 3 | 20 | |
| | | | | | | 7 | 56 | |
| | | | | | | 14 | 13 | |
| | | | | | | 28 | 8 | |
| | | 300 | 309 | 1 | Forage | 0 | 141 | |
| | | | | | | 1 | 87 | |
| | | | | | | 3 | 76 | |
| | | | | | | 7 | 46 | |
| | | | | | | 14 | 41 | |
| | | | | | | 28 | 23 | |
| | | 600 | 299 | 1 | Forage | 0 | 5.6 | |
| | | | | | | 1 | 324 | |
| | | | | | | 3 | 125 | |
| | | | | | | 7 | 82 | |
| | | | | | | 14 | 71 | |
| | | | | | | 28 | 45 | |
| Dunsandel, Canterbury, NZ, 2002 (Browntop, white clover | SL 240 | 60 | 218 | 1 | Forage | 0 | 6.2 | 020049-01⑩❶ |
| | | | | | | 7 | 2.1 | |
| | | | | | | 14 | 1.4 | |
| | | | | | | 28 | 1.7 | |
| | | 120 | 204 | 1 | Forage | 0 | 11 | |
| | | | | | | 7 | 3.3 | |
| | | | | | | 14 | 3.3 | |
| | | | | | | 28 | 8.9 | |
| | | 240 | 204 | | Forage | 0 | 24 | |
| | | | | | | 7 | 7.3 | |
| | | | | | | 14 | 7 | |
| | | | | | | 28 | 4.5 | |
| | | 480 | 189 | | Forage | 0 | 37 | |
| | | | | | | 7 | 14.5 | |
| | | | | | | 14 | 15 | |
| | | | | | | 28 | 12 | |
| Leeston, NZ, 2002 (Perennial ryegrass and white clover) | SL 240 | 60 | 204 | 1 | Forage | 0 | 7.3 | 020049-02⑩❶ |
| | | | | | | 7 | 1.5 | |
| | | | | | | 14 | 0.8 | |
| | | | | | | 28 | < 0.01 | |
| | | 120 | 218 | 1 | Forage | 0 | 13 | |
| | | | | | | 7 | 2.9 | |
| | | | | | | 14 | 2 | |
| | | | | | | 28 | 1.2 | |
| | | 240 | 204 | 1 | Forage | 0 | 31 | |
| | | | | | | 7 | 5.3 | |
| | | | | | | 14 | 4.1 | |
| | | | | | | 28 | 1.7 | |

| PASTURES/HAY | Application | | | Sample | PHI | Residues | Ref./Study No. |
|---|---|---|---|---|---|---|---|
| Country, Year (variety) | Form. | Rate (g ae/ha) | Water (L/ha) | No. | | (days) | (mg/kg) dry weight | |
| | | 480 | 204 | 1 | Forage | 0 7 14 28 | 64 14 12 4.7 | |

①Recoveries ranged 77 – 137% over concentrations 0.01, 1 and 10 mg/kg. ②Moisture contents ranged 41 – 63%. ③Moisture contents ranged 33 – 62%. ④Moisture contents ranged 8 – 28%. ⑤Recoveries ranged 40 – 166% at 0.01 mg/kg; recoveries ranged 67 – 103% over concentrations 0.1, 1, 10 and 100 mg/kg. ⑥Moisture contents ranged 80 – 87%. ⑦Moisture contents ranged 82 – 89%. ⑧Moisture contents ranged 85 – 90%. ⑨Moisture contents ranged 83 – 90%. ⑩Recoveries ranged 79 - 129% at concentrations of 1, 10 and 100 mg/kg. ❶Moisture contents ranged 71 – 77%. ❶Moisture contents ranged 78 – 83%. ❷ Recoveries ranged 73 -97% over concentrations of 1 and 10 mg/kg. ❸Moisture contents ranged 70 – 83%. ❹ Moisture contents ranged 45 –70%. ❺Moisture contents ranged 41 – 75%.


## FATE OF RESIDUES IN STORAGE AND PROCESSING

### In processing

As part of a supervised field trial study (Roberts, Schelle and Knuteson, 2004), wheat crops at a site in Oklahoma were treated at 50 g ae/ha (equivalent to 5× the US GAP) at crop stage BBCH 33, and bulk grain was collected at 69 days after application for processing into various wheat fractions (Figure 5).

Method GRM 02.31 was used to determine aminopyralid residues in whole grain and wheat fractions. A summary of the results is shown below.

Table 48. Aminopyralid residues in processed wheat fractions.

| Application rate | Sample/Fraction | Aminopyralid residues (mg/kg) | Processing Factor (PF) |
|---|---|---|---|
| 50 g ae/ha | Whole (bulked) wheat | 0.055 | – |
| | Bran | 0.133  c 0.016 | 2.4 |
| | Flour | <LOQ | 0.2 |
| | Shorts① | 0.067 | 1.2 |
| | Middlings② | 0.032 | 0.58 |
| | Germ | 0.02 | 0.36 |
| | Aspirated grain fractions | 0.338  c 0.03 | 6.1 |

① Low grade mill product containing principally germ and fine bran particles.
② Also known as semolina.

Recoveries in various wheat fractions are shown below.

Table 49. Recoveries in various wheat fractions.

| Sample matrix | % Recovery at Fortification Levels | |
|---|---|---|
| | 0.05 mg/kg | 2.5 mg/kg |
| Wheat grain | 96, 102 | |
| Aspirated grain fractions | 104 | 102 |
| Bran | 109 | 102 |
| Middlings | 115 | 104 |
| Shorts | 127 | 109 |
| Flour | 112 | 109 |
| Germ | 97 | 104 |

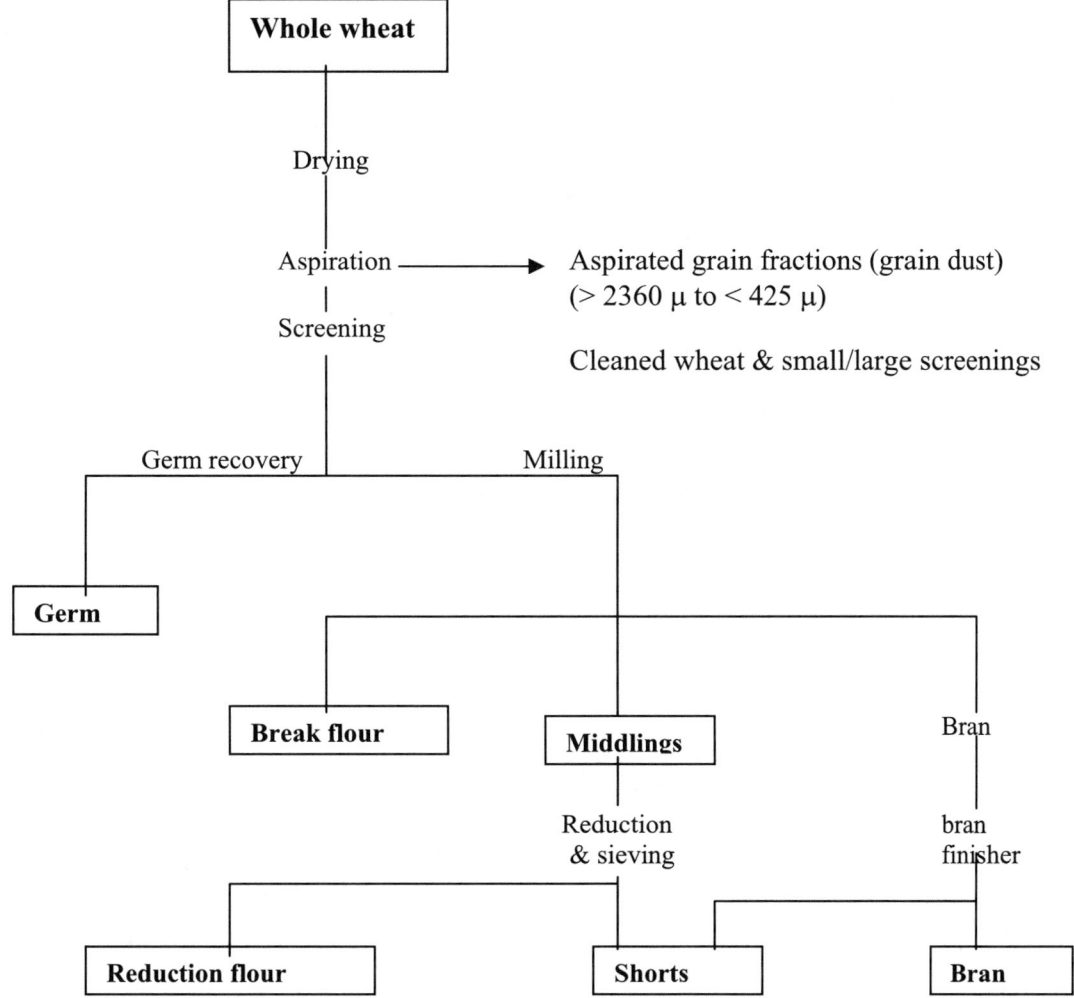

Figure 5. Schematic of wheat grain processing as described by Roberts, Schelle and Knuteson, 2004.

## *Residues in Animal Commodities*

### *Farm Animal (Livestock) Feeding Studies*

#### *Lactating Cow Feeding Study*

The meeting received a lactating dairy cow feeding study, which provided information on likely residues resulting in animal tissues and milk from residues in the animal diet (Rosser, Rutherford and McFarlane, 2004).

In this study, Holstein dairy cows were divided into a control group I and four treatment groups (T-I–T-IV). The four treatment groups corresponded to nominal doses of 0.5×, 1×, 3× and 10×, where 1× was 64.5 ppm in the feed. Summaries of bodyweight ranges, feed intakes and actual doses are shown in Table 54. There were four animals in the control group (C-1–C-4) and three animals per dose group for T-I to T-III (0.5× to 10×). In the T-IV (or 10×) group, an additional six animals were included to allow for a withdrawal or depuration phase following the 28 days dosing period; two animals included for slaughter at 3, 7 and 14 days after dosing had ceased. Bodyweights in the control group ranged 502 – 590 kg at the commencement of the study and 501 – 600 kg on day 28 of the study.

Table 50. Summary of dosing regime, animal bodyweights and feed consumption.

| Treatment Group | Animal No. | Mean mg aminopyralid /day | Mean Feed consumption① (kg) | Dose in the feed (ppm)② | Bodyweight (kg)③ | Dose (mg/kg bw)④ |
|---|---|---|---|---|---|---|
| T-I-1 | 2 | 466.3 | 17.2 | 27.1 | 522 | 0.89 |
| T-I-2 | 5 | 567.5 | 17.1 | 33.2 | 608 | 0.93 |
| T-I-3 | 14 | 821.5 | 22.2 | 37 | 565 | 1.46 |
| *Mean* | | *618.4* | *18.8* | *32.8* | *565* | *1.10* |
| T-II-1 | 20 | 1333.6 | 22.8 | 58.5 | 499 | 2.67 |
| T-II-2 | 11 | 1577.1 | 23.3 | 67.7 | 656 | 2.40 |
| T-II-3 | 17 | 1312.7 | 19.4 | 67.7 | 550 | 2.38 |
| *Mean* | | *1407.8* | *21.8* | *64.5* | *569* | *2.48* |
| T-III-1 | 26 | 3482.8 | 19.5 | 178.6 | 553 | 6.29 |
| T-III-2 | 7 | 3334.4 | 19.3 | 172.8 | 592 | 5.63 |
| T-III-3 | 15 | 4365.4 | 22.8 | 191.5 | 602 | 7.25 |
| *Mean* | | *3727.7* | *20.5* | *181.5* | *583* | *6.40* |
| T-IV-1 | 24 | 14028.9 | 21.6 | 649.5 | 513 | 27.34 |
| T-IV-2 | 10 | 15007.1 | 21.6 | 694.8 | 570 | 26.35 |
| T-IV-3 | 23 | 14550.3 | 20.8 | 699.5 | 612 | 23.78 |
| T-IV-4 | 13 | 11747.3 | 18.4 | 638.4 | 547 | 21.49 |
| T-IV-5 | 16 | 12991.8 | 21.1 | 615.7 | 714 | 18.19 |
| T-IV-6 | 21 | 12725.8 | 19.0 | 669.8 | 565 | 22.52 |
| T-IV-7 | 1 | 14870.6 | 23.2 | 641.0 | 586 | 25.36 |
| T-IV-8 | 22 | 14737.6 | 23.2 | 635.2 | 579 | 25.45 |
| T-IV-9 | 12 | 12227.2 | 21.7 | 563.5 | 595 | 20.54 |
| *Mean* | | *13654.1* | *21.2* | *644.7* | *587* | *23.27* |

① Mean daily feed consumption on a dry weight basis for the 4 weeks during dosing.

② Dose of aminopyralid expressed as mg aminopyralid per kg feed dry weight consumed.

③ Mean of bodyweights recorded on days 8, 15, 22 and 28 of the study.

④ Dose of aminopyralid expressed as mg aminopyralid per kg bodyweight.

The actual mean dose levels were 32.8, 64.5, 181.5 and 644.7 ppm in the feed.

Milk from each cow was collected at morning and afternoon milkings (AM and PM) and both milkings were pooled to form a single whole milk sample for each day for each animal. Samples of whole milk were collected for analysis the day before dosing began and during the first 7 days of the study, then every third day until the end of the dosing period (study days 10, 13, 16, 19, 22, 25 and 28). In addition, samples of whole milk were collected from each cow in the depuration phase of the study from the time that dosing has ceased until slaughter. On days 13 and 28 of the study, additional samples of pooled milk were collected and separated into cream and skim milk from 3 cows in the control group, three cows in the 1× group and three cows in 10× group that were not used in the depuration phase of the study.

Cows in the control group and groups T-I, T-II and T-III and three animals in group T-IV were slaughtered within 24 hours after receiving the final dose (day 29). The remaining animals in group T-IV were slaughtered on days 32, 36 and 43 or 3, 7 and 14 days after dosing has ceased. Samples of liver, kidney, muscle (equal portions flank, loin and leg composited) and fat (equal portions perirenal, abdominal and subcutaneous composited) were collected from each animal. All samples (tissues and milk) were stored frozen at -17°C prior to analysis. All samples were analysed within 1 month of sampling (4 to 26 days); control samples were analysed after 17 to 33 days after storage.

Residues in liver, kidney, muscle, fat, whole milk, skim milk and cream were determined using method GRM 03.18. Residue levels in milk and tissues and concurrent recoveries are shown in Tables 55, 56 and 57.

Table 51. Recovery of aminopyralid in milk, skim milk, cream and edible tissues.

| Matrix | | Fortification level (mg/kg) | Number of samples (n) | Recovery Range (%) | Mean Recovery (%) |
|---|---|---|---|---|---|
| Milk | | 0.01 | 34 | 74 – 116 | 89 ± 10 |
| | | 0.1 | 15 | 83 – 103 | 93 ± 6 |
| | | 1 | 8 | 77 – 92 | 86 ± 5 |
| | Mean | 0.01 – 1 | 57 | 74 – 116 | 90 ± 9 |
| Skim milk | | 0.01 | 4 | 80 – 91 | 85 ± 5 |
| | | 0.1 | 2 | 77, 89 | 83 ± 8 |
| | Mean | 0.01 – 0.1 | 6 | 77 – 91 | 84 ± 6 |
| Cream | | 0.01 | 4 | 80 – 90 | 85 ± 5 |
| | | 0.1 | 2 | 82, 92 | 87 ± 7 |
| | Mean | 0.01 – 0.1 | 6 | 80 – 92 | 86 ± 5 |
| Fat | | 0.01 | 8 | 70 – 93 | 80 ± 7 |
| | | 0.1 | 3 | 85 – 91 | 88 ± 3 |
| | | 1 | 2 | 91, 92 | 91 ± 1 |
| | Mean | 0.01 – 1 | 13 | 70 – 93 | 84 ± 8 |
| Muscle | | 0.01 | 6 | 68 – 83 | 78 ± 5 |
| | | 0.1 | 3 | 86 – 89 | 87  1 |
| | | 1 | 2 | 82, 92 | 87 ± 7 |
| | Mean | 0.01 – 1 | 11 | 68 – 92 | 82 ± 7 |
| Liver | | 0.01 | 4 | 73 – 89 | 81 ± 7 |
| | | 0.1 | 1 | 87 | 87 |
| | | 1 | 2 | 93, 94 | 94 ± 1 |
| | Mean | 0.01 – 1 | 7 | 73 – 94 | 85 ± 8 |
| Kidney | | 0.01 | 6 | 81 – 102 | 92 ± 8 |
| | | 0.1 | 1 | 83 | 83 |
| | | 1 | 2 | 89 | 89 |
| | | 3 | 2 | 92, 99 | 96 ± 5 |
| | Mean | 0.01 – 3 | 11 | 81 – 102 | 92 ± 7 |

Table 52: Aminopyralid residues in skim milk and cream for dose levels T-II and T-IV.

| Study Day | Skim milk | | Cream | |
|---|---|---|---|---|
| | T-II (64.5 ppm) | T-IV (644.7 ppm) | T-II (64.5 ppm) | T-IV (644.7 ppm) |
| 13 | < 0.01 (2), 0.015 | 0.058, 0.054, 0.068 | < 0.01 (2), 0.012 | 0.052, 0.046, 0.059 |
| 28 | < 0.01 (3) | 0.043, 0.048, 0.074 | < 0.01 (3) | 0.037, 0.043, 0.065 |

Following dosing at 32.8 ppm in the feed, aminopyralid residues in milk were < 0.01 mg/kg over the 28 days of dosing. Residues reached plateau within 2 to 3 days of dosing. Residues in milk ranged < 0.01–0.024 mg/kg and 0.011–0.028 mg/kg following dosing at 64.5 and 181.5 ppm, respectively. Aminopyralid residues ranged 0.023–0.127 mg/kg following dosing at 644.7 ppm. There was little difference between the residues found in skim milk and cream indicating that aminopyralid is not fat soluble.

Residues had declined to < 0.01 mg/kg within 2 days of withdrawal from dosing at the highest level of 644.7 ppm.

Table 53. Aminopyralid residues in tissues at the end of the dosing period.

| Sample/Tissue | Aminopyralid residues in tissues at day 29 at each dose level (mg/kg) | | | |
|---|---|---|---|---|
| | T-I (32.8 ppm) | T-II (64.5 ppm) | T-III (181.5 ppm) | T-IV (644.7 ppm) |
| Muscle | < 0.01 (3) | < 0.01 (3) | 0.01, 0.017, 0.046 | 0.012, 0.021, 0.029 |
| Fat | < 0.01 (2), 0.011 | < 0.01 (2), 0.013 | 0.016, 0.073, 0.095 | 0.025, 0.039, 0.042 |
| Liver | < 0.01 (3) | < 0.01 (2), 0.014 | 0.026, 0.033, 0.054 | 0.059, 0.064, 0.116 |
| Kidney | 0.043, 0.052, 0.102 | 0.099, 0.139, 0.202 | 0.456, 0.507, 1.537 | 0.902, 1.242, 2.549 |

Highest residues were found in kidney at all dose levels; the magnitude of residues were muscle < fat < liver < kidney. The depuration data show that following withdrawal from dosing for 3 days, residues in all tissues, including kidney, decline to < 0.01 mg/kg.

Table 54. Aminopyralid residues in tissues following depuration up to 14 days after dosing.

| Sample/Tissue | Aminopyralid residues in tissues following depuration (mg/kg) | |
|---|---|---|
| | Days of study | T-IV (644.7 ppm) |
| Muscle | 29 (0) | 0.012, 0.021, 0.029 |
| | 32 (+3) | < 0.01, < 0.01 |
| | 36 (+7) | < 0.01, < 0.01 |
| | 43 (+14) | < 0.01, < 0.01 |
| Fat | 29 (0) | 0.025, 0.039, 0.042 |
| | 32 (+3) | < 0.01, 0.01 |
| | 36 (+7) | < 0.01, < 0.01 |
| | 43 (+14) | < 0.01, < 0.01 |
| Liver | 29 (0) | 0.059, 0.064, 0.116 |
| | 32 (+3) | < 0.01, < 0.01 |
| | 36 (+7) | ND, ND |
| | 43 (+14) | ND, ND |
| Kidney | 29 (0) | 0.902, 1.242, 2.549 |
| | 32 (+3) | < 0.01, < 0.01 |
| | 36 (+7) | ND, < 0.01 |
| | 43 (+14) | < 0.01, < 0.01 |

## RESIDUES IN FOOD IN COMMERCE OR AT CONSUMPTION

No information was available on residues monitoring data for aminopyralid as it is a new compound.

## NATIONAL MAXIMUM RESIDUE LIMITS

The meeting was aware that the following national MRLs had been established.

Table 55. National MRLs for aminopyralid.

| Country | Commodity | MRL (mg/kg) |
|---|---|---|
| Argentina | Wheat, grain | 0.04 |
| Australia | Cereal grains | 0.1 |
| | Edible offal (mammalian)[except kidney] | 0.02 |
| | Eggs | *0.01 |
| | Kidney (mammalian0 | 0.3 |
| | Meat (mammalian) | *0.01 |
| | Milks | *0.01 |
| | Poultry meat | *0.01 |
| | Poultry, edible offal of | *0.01 |
| | Wheat bran, unprocessed | 0.3 |
| | Forage of cereal grains (green) | 3 |
| | Mixed pastures (leguminous/grasses) | 200 |
| | Straw and fodder of cereal grains (dry) | 0.2 |
| USA | Grass, forage | 25 |
| | Grass, hay | 50 |
| | Wheat, forage | 2.0 |
| | Wheat, hay | 4.0 |
| | Wheat, grain | 0.04 |
| | Wheat, straw | 0.25 |
| | Wheat, bran | 0.1 |
| | Wheat, aspirated grain fractions | 0.2 |
| | Milk | 0.03 |
| | Cattle, meat | 0.02 |
| | Goat, meat | 0.02 |
| | Horse, meat | 0.02 |

| Country | Commodity | MRL (mg/kg) |
|---|---|---|
| | Sheep, meat | 0.02 |
| | Cattle, fat | 0.02 |
| | Goat, fat | 0.02 |
| | Horse, fat, | 0.02 |
| | Sheep, fat | 0.02 |
| | Cattle, meat by-products, except kidney | 0.02 |
| | Goat, meat by-products, except kidney | 0.02 |
| | Horse, meat by-products, except kidney | 0.02 |
| | Sheep, meat by-products, except kidney | 0.02 |
| | Cattle, kidney | 0.3 |
| | Goat, kidney | 0.3 |
| | Horse, kidney | 0.3 |
| | Sheep, kidney | 0.3 |
| Canada | Hay or fodder (dry) of grasses | 65 |
| | Wheat grain | 0.05 |
| | Wheat straw and fodder, dry | 0.5 |
| | Wheat bran, processed | 0.1 |
| | Wheat flour | 0.01 |
| | Wheat germ | 0.02 |
| | Milk of cattle, goat, and sheep | 0.02 |
| | Milk, cream of cattle, goats, and sheep | 0.02 |
| | Meat from mammals other than marine mammals | 0.05 |
| | Fat from mammals other than marine mammals | 0.05 |
| | Liver from mammals other than marine mammals | 0.05 |
| | Kidney from mammals other than marine mammals | 1.0 |
| Japan | Wheat, grain | 0.04 |
| Colombia (has adopted US-EPA MRLs): | Almond hulls | 25 |
| | Grass, forage | 25 |
| | Grass, hay | 50 |
| | Wheat, forage | 2.0 |
| | Wheat, hay | 4.0 |
| | Wheat, grain | 0.04 |
| | Wheat, straw | 0.25 |
| | Wheat, bran | 0.1 |
| | Wheat, aspirated grain fractions | 0.2 |
| | Milk | 0.03 |
| | Cattle, meat | 0.02 |
| | Goat, meat | 0.02 |
| | Horse, meat | 0.02 |
| | Sheep, meat | 0.02 |
| | Cattle, fat | 0.02 |
| | Goat, fat | 0.02 |
| | Horse, fat, | 0.02 |
| | Sheep, fat | 0.02 |
| | Cattle, meat by-products, except kidney | 0.02 |
| | Goat, meat by-products, except kidney | 0.02 |
| | Horse, meat by-products, except kidney | 0.02 |
| | Sheep, meat by-products, except kidney | 0.02 |
| | Cattle, kidney | 0.3 |
| | Goat, kidney | 0.3 |
| | Horse, kidney | 0.3 |
| | Sheep, kidney | 0.3 |
| United Kingdom (tMRLs) | Milk | 0.01* |
| | Meat, muscle | 0.01* |
| | Meat, fat | 0.01* |
| | Meat, liver | 0.01* |
| | Meat, kidney | 0.1 |

* MRL set at or about LOQ.

## APPRAISAL

Aminopyralid is a new herbicide that is used for the control of broadleaf weeds in pastures and cereal crops. It was advanced to the 2006 JMPR schedule of new compounds by the 37th session of the CCPR. The manufacturer submitted studies on metabolism in plants and animals, analytical methods, storage stability, environmental fate and degradation, supervised field trials, processing and farm animal (livestock) feeding.

### Chemical name and structure:

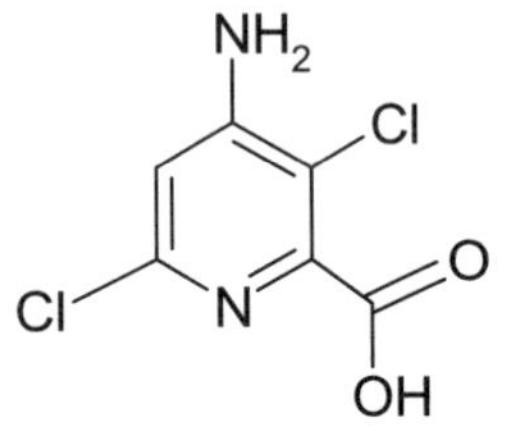

4-amino-3, 6-dichloropyridine-2-carboxylic acid

### Animal metabolism

The meeting received metabolism studies in lactating goats and laying hens. In both studies, aminopyralid was labelled in the 2- and 6-positions of the pyridine ring, as shown below.

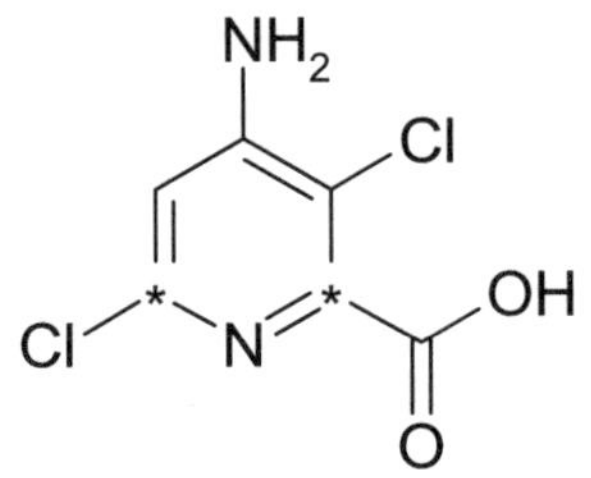

A single <u>lactating goat</u> was orally dosed for 6 days with [$^{14}$C] aminopyralid at the equivalent of 14 ppm in the feed (0.26 mg/kg bw/day). Total radioactivity was measured in samples of milk, urine, faeces, cage wash and tissues. A large proportion of the administered dose was eliminated via urine and faeces, with each accounting for approximately 46% of the total administered dose; the total eliminated was 92% of the administered dose. Approximately 3% of the administered dose was present in cage wash and < 0.1% was present in milk and tissues.

The radioactivity in milk was found to plateau within 24 to 48 hours after dosing had commenced. The total radioactivity in milk was approximately 0.05% of the total administered dose, with concentrations ranging 0.003–0.008 mg/kg aminopyralid equivalents.

TRR in tissues were 0.008 mg/kg equivalents in liver, 0.071 mg/kg equivalents in kidneys, 0.001 mg/kg equivalents in composite fat and non-detectable in composite muscle.

Methanol extraction of milk followed by partitioning with ether and hexane resulted in recovery of 71% TRR. Methanol extraction of liver and kidney samples followed by partitioning with hexane resulted in release of 58% and 80% TRR, respectively. Further characterisation was not conducted in milk or liver due to very low levels of radioactivity being present.

One radiolabelled component representing approximately 80% $^{14}$C in kidney was identified by HPLC analysis of kidney extracts; this was parent aminopyralid present at 0.057 mg/kg. In summary, aminopyralid was rapidly excreted from the goat, with detectable residues of 0.06 mg/kg present in kidneys only.

Ten <u>laying hens</u> of 45 weeks age were orally dosed for 7 days with $^{14}C$ aminopyralid at the equivalent of 10.5 ppm in the feed. TRR in excreta collected daily ranged from 55% to 87% of the nominal daily dose over 1 to 6 days of the study. All eggs contained low levels of radioactivity that gradually increased to a plateau level of 0.004 mg/kg aminopyralid equivalents within 5 to 7 days of the study. TRR in all eggs collected over the study period accounted for < 0.01% of the total administered dose.

TRR in tissues were < 0.01% of the total administered dose. TRR in muscle and fat were comparable, corresponding to 0.0018 and 0.0017 mg/kg aminopyralid equivalents, respectively. TRR in skin (with fat) and liver were 0.0029 and 0.0024 mg/kg aminopyralid equivalents, respectively. The levels of radioactivity in eggs and tissues were low and were not further characterised.

Residues in excreta were readily extracted into $CH_3CN:H_2O$ with approximately 96% TRR being recovered. HPLC analysis of the excreta extracts indicated that 93% of the radioactivity was composed of unchanged aminopyralid.

In summary, aminopyralid is readily excreted by hens following oral dosing for 7 days at 10 ppm in the feed. TRR in eggs on days 6 and 7 of the study were 0.004 mg/kg aminopyralid equivalents. TRR in muscle and fat were non-detectable (< 0.002 mg/kg aminopyralid equivalents), while TRR in liver and skin/fat were 0.002–0.003 mg/kg aminopyralid equivalents.

The Meeting concluded that aminopyralid is readily eliminated by lactating goats and laying hens following oral dosing, and that it does not undergo any significant metabolism. Most of the eliminated radioactivity was recovered as unchanged aminopyralid.

### *Plant metabolism*

Metabolism studies for aminopyralid applying to wheat and grass/pastures were submitted to the Meeting. In both studies, aminopyralid was labelled in the 2- and 6-positions of the pyridine ring, as shown above.

$^{14}C$ aminopyralid was applied to <u>spring wheat</u> at rates of 40 and 80 g ai/ha. The wheat was at growth stages BBCH 26 – 28 (6 to 8 tillers detectable) at the time of application. Samples of plant material (forage) were taken at 0 and 14 days after application; hay was sampled at 35 days after application and grain and straw at 86 days after application. Homogenised samples were extracted with $CH_3CN:H_2O$ followed by partitioning with $CH_3CN/CH_2Cl_2$ prior to HPLC analysis. In forage and hay 88–96% of the extracted radioactivity was composed of free and/or conjugated forms of aminopyralid. Similarly in straw and grain 89 and 69%, respectively, of the extracted radioactivity was composed of free and conjugated aminopyralid. Acid/base treatment of the extracted solids and aqueous phase extracts released another 4% and 7% of the TRR in straw and grain, respectively.

Three types of <u>pasture grasses</u> (perennial ryegrass, big bluestem and *Panicum maximum*) were treated with a single application of $^{14}C$ aminopyralid at a rate of 360 g ai/ha. Samples of grass were collected for analysis at 0, 7, 14, 21 and 42 days after application.

During the first 7 days after application, aminopyralid comprised approximately 49–97% of the TRR. After 7 days, the amount of extracted aminopyralid decreased and large proportions of other metabolites formed, ranging 18–60% of the TRR at 7 to 42 days after application in grass forage. Acid/base hydrolysis of the metabolites released radioactivity that was subsequently identified by HPLC analysis as aminopyralid. The metabolites were glucose conjugates of aminopyralid that were easily released by base and/or acid hydrolysis. Overall, 87–96% of the TRR in grass forage and hay was identified as aminopyralid and conjugates of aminopyralid.

In summary, the two plant metabolism studies demonstrated that aminopyralid forms the major proportion of the radioactivity when applied to wheat and pasture grasses. With time, the plant converts aminopyralid to glucose conjugates that are easily released by base and/or acid hydrolysis to parent compound.

## Environmental fate

The meeting received information on the environmental fate of aminopyralid in soil, including studies on photodegradation on soil, aerobic soil degradation and crop rotation studies (confined and field). $^{14}C$ aminopyralid was labelled in the 2- and 6-positions of the pyridine ring in all soil studies.

In a photodegradation study on soil, aminopyralid degraded into non-extracted $^{14}C$ and $CO_2$. The half-life ($DT_{50}$) for photodegradation of aminopyralid was 61 days and $DT_{90}$ was 203 days.

Aerobic soil degradation of $^{14}C$ aminopyralid was investigated in a range of European and US soils. As found in the soil photodegradation study, aminopyralid degrades to $CO_2$ and non-extracted residues under aerobic conditions; no other residues were identified. Reported $DT_{50}$ values ranged from 18 – 343 days and $DT_{90}$ values ranged 45 – 1141 days.

Data were reported from field dissipation studies that were conducted in Europe, US and Canada. The reported $DT_{50}$ values from the field studies ranged from 8 to 54 days with the majority of values approximating 30 to 35 days. $DT_{90}$ values ranged 26 to 430 days.

Crop rotation studies were conducted in a leafy vegetable (lettuce), a root crop (turnip) and a cereal crop (sorghum). The crops were grown in soil treated with $^{14}C$ aminopyralid either 90 or 120 days prior to sowing. The results from the different crops were similar, with radioactivity ranging 0.002 – 0.007 mg/kg equivalents in lettuce, turnip roots and tops and sorghum grain and 'late forage'. The highest levels of radioactivity were present in 'early forage' of sorghum at 0.03 mg/kg equivalents from sorghum sown at 90 days after application. Extraction of the radioactivity led to results similar to those in the wheat and grass metabolism studies, i.e. the radioactivity was composed of aminopyralid and conjugates that were easily released by base hydrolysis of extracted phases and non-extracted solids.

## Methods of analysis

The Meeting received details of analytical methods for the determination of residues of aminopyralid in agricultural commodities (namely grass pastures, cereal grains, cereal forage and straw), bovine tissues and milk and soil.

Methods for the quantitative determination of aminopyralid residues in barley, sorghum, wheat and grass pasture were provided to the Meeting (method GRM 02.31). Residues are determined using LC/MS/MS. The validated LOQ for all matrices is reported as 0.01 mg/kg. Residues of aminopyralid and its conjugates are extracted from the sample matrices by homogenisation with mild base, followed by acidification and purification on an SPE column. Aminopyralid in the purified extract is derivatized to its butyl ester form and quantified using LC/MS/MS.

In the method validation component of the study, untreated control samples of barley, sorghum and wheat grain; barley, sorghum and wheat forage; barley straw, sorghum stover and wheat straw; pasture grass forage and hay were fortified with aminopyralid at concentrations ranging 0.01–0.5 mg/kg for grain, 0.01–5 mg/kg for forage and straw and 0.01–20 mg/kg for grass forage and hay. Mean recoveries in all samples ranged 92–109% over all concentrations tested. Independent laboratory validation confirmed the LOQ of 0.01 mg/kg, and mean recoveries ranged 93–120% over all matrices tested.

The extraction efficiency of aminopyralid was determined by extracting wheat grain, forage, straw, hay and pasture grass samples from the metabolism studies in accordance with method GRM 02.31. The extraction efficiencies of the pasture grass samples ranged 88 – 114% and of the wheat matrices ranged 72–101% when calculated based on the total radioactivity.

A GC/MS method for the determination of aminopyralid, fluroxypyr and 2,4-D residues in pastures was provided, which was a modified version of the method described above and has a validated LOQ of 1 mg/kg. Recoveries were validated over a range of concentrations (1–100 mg/kg) and ranged 78–115% with a mean recovery of 95%. The reported LOD was 0.2 mg/kg in grass pasture samples.

Aminopyralid residues in animal tissues and milk are also determined by LC/MS/MS with a validated limit of quantitation of 0.01 mg/kg (method GRM 03.18). Residues are extracted with MeOH/NaHCO₃ solution, purified using a SPE plate and derivatized to form the 1-butyl ester. The method was validated over the concentration range of 0.01–2.5 mg/kg in kidney and 0.01–1 mg/kg for all other tissues and milk. The mean recoveries in bovine fat, muscle, liver and kidney ranged 79–96% and in whole milk, skim milk and cream ranged 73–89%. In an independent laboratory validation, the LOQ of 0.01 mg/kg was confirmed for aminopyralid in bovine milk and kidneys using LC/MS/MS. The mean recoveries in bovine milk and kidney were 83% and 87%, respectively.

Methods were provided for the determination of aminopyralid residues in soil (Method GRM 02.34). The soil method was validated with recoveries being conducted in four soil types at concentrations ranging 0.0015–0.1 mg/kg with a mean recovery of 88%; the validated LOQ was 0.0015 mg/kg.

### Stability of pesticide residues in stored analytical samples

The storage stability of aminopyralid in pasture grass and hay and wheat grain and straw was investigated. Samples of hay, forage (grass), wheat grain and wheat straw were fortified with aminopyralid at a concentration of 0.1 mg/kg and placed in frozen storage at –20 °C. Aminopyralid residues are stable under conditions of frozen storage for up to 489 days in grass hay and forage, and up to 469 days in wheat grain and straw, with 86% and 88% aminopyralid remaining in fortified grass hay and forage, respectively and 91% and 87% remaining in wheat grain and straw, respectively

Storage stability in animal tissues and milk was not conducted as samples from the metabolism and livestock feeding study were analysed within 3 months of sample collection.

### Residue definition

The results of the plant metabolism studies on wheat and pasture grasses indicate that aminopyralid is not significantly metabolised and is transformed to glucose conjugates. Greater than 90% extracted TRR is identified as aminopyralid and conjugates with no metabolites formed to any extent.

In goats and hens, administered aminopyralid is readily eliminated following oral dosing, and it does not undergo any significant metabolism. Greater than 93% of the eliminated radioactivity was recovered as unchanged aminopyralid. Detected radioactivity in goat kidney was identified as aminopyralid.

Analytical methods for plant and animal matrices and soil determine aminopyralid and any conjugates that are hydrolysed by acid/base as the butyl ester derivative of aminopyralid.

On the basis of the metabolism in plants and animals and the analytical methodology submitted, the Meeting recommended a residue definition for aminopyralid for plants and animals.

Definition of the residue (for compliance with the MRL for all commodities and for estimation of dietary intake for plant and animal commodities): aminopyralid and its conjugates that can be hydrolysed, expressed as aminopyralid.

The residue is not fat-soluble.

### Results of supervised trials on crops

In all supervised field trials, commercial formulations containing either the potassium salt or triisopropanolammonium (TIPA) salt were used. In several trials both salt forms were used and in addition many formulations included combinations of the aminopyralid salts with other herbicides such as 2,4-D, fluroxypyr and triclopyr. The aminopyralid TIPA salt dissociates rapidly in water to the aminopyralid acid at pH values greater than 2.56.

Supervised trials for the foliar application of aminopyralid on cereals, namely barley, oats and wheat, and pasture grasses were provided to the Meeting.

*Cereal grains*

### Barley

Data for barley was received from trials conducted in Australia. The registered use in Australia allows a single application between 3-leaf to $1^{st}$ node (BBCH 13-31) at a rate of 5–7.5 g ai/ha with a nil PHI for harvest and a non-grazing interval of 7 days after application.

The residues in barley grain from four trials in Australia are in rank order: 0.03, 0.04, 0.06, 0.06, and 0.07 (2) mg/kg.

Data for barley was received from two trials conducted in Spain, however there is no GAP for barley in Spain.

The Meeting considered that there were insufficient trials to estimate a maximum residue level for barley.

### Oats

Data for oats were received from trials conducted in Australia where two different formulations were used. The registered use in Australia allows a single application between 3-leaf to $1^{st}$ node (BBCH 13-31) at a rate of 5–7.5 g ai/ha with a nil PHI and grazing at 7 days after application.

The residues in oat grain from four trials in Australia are in rank order: < 0.01, 0.01, 0.02 and 0.03 (4) mg/kg.

The Meeting considered that there were insufficient trials to estimate a maximum residue level for oats.

### Wheat

Trials on wheat were conducted in Argentina, Australia, Canada, Hungary, Italy, Poland, Spain and the USA. GAP in Argentina is a single application at rates of 3.75–5 g ai/ha from $3^{rd}$ leaf to end of tillering, with a nil PHI for harvest. Residues in wheat grain from trials in Argentina were: < 0.01 (6) mg/kg.

GAP for wheat in Australia is the same as that for barley and oats; a single application between 3-leaf to $1^{st}$ node (BBCH 13-31) at a rate of 5–7.5 g ai/ha with a nil PHI and grazing at 7 days after application. Residues in wheat grain from the Australian trials are in rank order were: < 0.01 (7), 0.01 (3), 0.02 (2), 0.03 and 0.07 mg/kg.

In the US, GAP allows a single application from 3-leaf to early jointing (BBCH 13 to 30–31) at rates of 7.6–10 g ai/ha, with a PHI of 50 days for harvest and 14 days for grazing/cutting. GAP in Canada is a single application from 2 to 6-leaf stage at a rate of 10 g ai/ha with a PHI of 50 days for harvest and no restrictions for grazing. The Meeting considered that as the GAP in both Canada and USA is specified as an application timing as well as PHI, the stage of crop growth at which the spray is applied is the important determinant compared to the PHI in relation to the final residues in grain. Therefore, trials where the application timings were within the specified GAP for Canada and USA were considered relevant, even though the actual PHIs may have been longer than 50 days.

Residues in wheat grain from the Canadian and US trials are in rank order: < 0.01 (8), 0.01 (14) and 0.02 (5) mg/kg.

Although there are no registered uses of aminopyralid on wheat in Europe, the Meeting considered that the data from the European trials may be included in the estimation of the maximum residue level for wheat as the application rates and application timings are similar to those registered uses in Argentina, Australia, Canada and the USA. The Meeting made reference to the outcomes of the OECD/FAO Zoning project and considered that cropping practices and herbicide use in broadacre crops such as wheat are similar in many regions and therefore the European trials may be considered as being supportive of GAP in other regions, in this case specifically Argentina, Australia, Canada and USA. The residues from the European trials are in rank order: < 0.01 (7) and 0.01 (3) mg/kg.

The Meeting agreed that the data sets for wheat from trials in Argentina, Australia, Canada, Europe and the USA could be combined, therefore the residues in wheat grain in ranked order were: < 0.01 (23), 0.01 (10), 0.011 (3), 0.12, 0.013, (4), 0.014(2), 0.02 (2), 0.021, 0.022, 0.023, 0.025(2), 0.03 and 0.07 mg/kg.

The Meeting also agreed that the data set for wheat could support the data sets for barley and oats, thereby allowing maximum residue levels to be estimated for barley and oats. In addition, registered uses in Australia apply to the minor crop triticale, therefore the combined data set for barley, oats and wheat may also be applied to triticale by crop extrapolation.

Residues for the combined data sets for barley, oats and wheat in ranked order were: < 0.01 (24), 0.01 (11), 0.011 (3), 0.012, 0.013 (4), 0.014 (2), 0.02 (3), 0.021, 0.022, 0.025, 0.025, 0.023, 0.03 (6), 0.04, 0.06 (2), and 0.07 (3)

The Meeting recommended a maximum residue level of 0.1 mg/kg for aminopyralid in barley, oats, triticale and wheat, with an STMR of 0.01 mg/kg.

### Livestock feed commodities

In trials conducted in Australia and New Zealand, residues in livestock feed commodities were reported on a 'dry weight' basis with associated moisture contents reported for samples at harvest and therefore require no correction for the estimation of MRLs. However the data from trials conducted in all other regions are reported on an 'as received' basis and require correction for moisture content by using default factors that are presented in the FAO Manual 2002. The default factors used are indicated and the 'as received' figures are adjusted for moisture content in the relevant sections below.

### Cereal Straw

The registered use of aminopyralid on barley and oats in Australia allows a single application between 3-leaf to 1st node (BBCH 13-31) at a rate of 5–7.5 g ai/ha with a nil PHI and grazing at 7 days after application.

· The data from the Australian trials are reported on a dry weight basis and therefore require no correction for moisture content. Residues in barley straw from trials that correspond to Australian GAP in ranked order were: 0.03 (3), 0.04, 0.07 and 0.08 mg/kg.

Residues in oat straw from trials that correspond to Australian GAP in ranked order were: 0.02, 0.03, 0.04 (3), 0.05 and 0.11 mg/kg.

Data for barley were received from two trials in Spain; these trials corresponded to Australian GAP. Residues when corrected for moisture content (88%) were 0.01 and 0.07 mg/kg.

### Wheat straw

The registered use of aminopyralid on wheat in Australia allows a single application between 3-leaf to 1st node (BBCH 13–31) at a rate of 5–7.5 g ai/ha with a nil PHI and grazing at 7 days after application.

The data from the Australian trials are reported on a 'dry weight' basis and therefore require no correction for moisture content. Residues in wheat straw from trials that correspond to Australian GAP in ranked order were: 0.02 (2), 0.04 (2), 0.05, 0.06 (2), 0.07 (2), 0.08, 0.09, 0.1, 0.12 and 0.13 mg/kg.

In the US, GAP allows a single application from 3-leaf to early jointing (BBCH 13-30) at rates of 7.6–10 g ai/ha, with a PHI of 50 days for harvest and 14 days for grazing/cutting. GAP in Canada is a single application from 2 to 6-leaf stage at a rate of 10 g ai/ha with a PHI of 50 days for harvest and a nil PHI for grazing. The Meeting considered that as the GAP in both Canada and USA is specified as application timing as well as PHI, the stage of crop growth at which the spray is applied is the important determinant when compared to the PHI in relation to the final residues in grain. Therefore, trials where the application timings were within the specified GAP for Canada and USA were considered relevant, even though the actual PHIs may have been longer than 50 days. Residues in wheat straw on an 'as received' basis from the Canadian and US trials in ranked order were: 0.02, 0.03, 0.04 (6), 0.05, 0.06 (4), 0.07 (7), 0.08, 0.1 (2), 0.12, 0.13 (2) and 0.14 mg/kg. When corrected for moisture content (12%), residues in wheat straw in ranked order were: 0.02, 0.03, 0.04 (6), 0.06, 0.07 (4), 0.08 (7), 0.09, 0.11 (3), 0.15 (2) and 0.16 mg/kg.

As with the wheat grain, there are no registered uses of aminopyralid on wheat straw in Europe, however the data from the European trials may be included in the estimation of the maximum residue level for wheat as the application rates and application timings are similar to those for registered uses in Argentina, Australia, Canada and USA. In addition, cropping practices and herbicide use in broadacre crops such as wheat are similar in many regions and therefore the European trials may be considered as being supportive of GAP in Argentina, Australia, Canada and USA. Residues in wheat straw from the European trials were reported on an 'as received' basis and in ranked order were: 0.03 (2), 0.04 (2), 0.06, 0.07, 0.08, 0.09, 0.13 and 0.16 mg/kg. When corrected for moisture content (12%), residues in wheat straw in ranked order were: 0.03 (2), 0.04 (2), 0.07, 0.08, 0.09, 0.1, 0.15 and 0.21 mg/kg.

The Meeting agreed that the data sets for wheat straw from trials in Argentina, Australia Canada, Europe and the US were from a single population and could be combined, therefore the residues in ranked order were: 0.02 (3), 0.03 (3), 0.04 (10), 0.05, 0.06 (3), 0.07 (7), 0.08 (9), 0.09 (3), 0.1 (5), 0.12, 0.13, 0.15 (3), 0.16 and 0.21 mg/kg.

The Meeting also agreed that the data set for wheat straw could support the data sets for barley and oat straw. In addition, registered uses in Australia apply to the minor crop triticale; therefore the combined data set for barley, oats and wheat may also be applied to triticale straw by crop extrapolation.

Residues from the combined data sets for barley, oat and wheat straw in rank order were: 0.01, 0.02 (4), 0.03 (7), 0.04 (14), 0.05 (2), 0.06 (3), 0.07 (9), 0.08 (10), 0.09 (3), 0.1 (6), 0.12, 0.13, 0.15 (3), 0.16 and 0.21 mg/kg. The Meeting recommended a maximum residue level of 0.3 mg/kg for aminopyralid in straw of barley, oats, triticale and wheat, with a highest residue of 0.21 mg/kg and an STMR of 0.07 mg/kg.

*Cereal forage*

GAP in Australia for barley and oats allows grazing of forage at 7 days after an application between 3-leaf to 1st node (BBCH 13–31) at a rate of 5–7.5 g ai/ha. Residues in barley forage (as reported on a dry weight basis) from trials that correspond to GAP were 0.54 and 0.71 mg/kg. Residues in oat forage were 0.34, 0.4 and 0.79 mg/kg.

The Meeting considered that the trials for barley and oat forage could be combined with the data set for wheat forage for the purposes of estimating the livestock dietary burden.

GAP in Australia for wheat allows application between 3-leaf to 1st node (BBCH 13–31) at a rate of 5–7.5 g ai/ha, with grazing of forage at 7 days after application. Residues in wheat forage (as reported on a dry weight basis) from trials that corresponded to GAP were 0.16, 0.45, 0.48, 0.71, 0.77, 1.02 mg/kg.

In the US, GAP allows a single application from 3-leaf to early jointing (BBCH 30-31) at rates of 7.6–10 g ai/ha, with an interval of 14 days after application for grazing or cutting. Residues in wheat forage that correspond to US GAP (as reported on an 'as received' basis) in ranked order were: 0.07, 0.1 (2), 0.16 and 0.19 mg/kg. When corrected for moisture content (75%), residues in wheat forage were: 0.28, 0.4 (2), 0.64 and 0.76 mg/kg.

GAP in Canada is a single application from 2 to 6-leaf stage at a rate of 10 g ai/ha with no restrictions on grazing. Residues in wheat forage (on an 'as received' basis) from trials that corresponded to Canadian GAP were: 0.11, 0.42, 0.49, 0.53, 0.72 and 0.85 mg/kg. An additional 25 US trials corresponded to GAP in Canada and residues in ranked order were: 0.16 (2), 0.19, 0.21,

0.26, 0.29, 0.3 (2), 0.32, 0.36, 0.37 (2), 0.38, 0.4, 0.41 (2), 0.42, 0.45, 0.49, 0.52, 0.53, 0.54, 0.63 (2) and 0.67 mg/kg. When corrected for moisture content (75%), residues were: 0.64 (2), 0.76, 0.84, 1.04, 1.16, 1.2 (2), 1.28, 1.44, 1.48 (2), 1.52, 1.6, 1.64 (2), 1.68, 1.8, 1.96, 2.08, 2.12, 2.16, 2.52 (2), and 2.68 mg/kg.

The Meeting considered that the trials corresponding to GAP in Australia, Canada and the US were from the same population and decided to combine the data for the purposes of estimating the livestock dietary burden. The residues in barley, oat and wheat forage in ranked order were: 0.16, 0.28, 0.34, 0.4 (2), 0.4, 0.45, 0.48, 0.54, 0.64 (3), 0.71, 0.71, 0.76 (2), 0.77, 0.79, 0.84, <u>1.02</u>, <u>1.04</u>, 1.16, 1.2 (2), 1.28, 1.44, 1.48 (2), 1.52, 1.6, 1.64 (2), 1.68, 1.8, 1.96, 2.08, 2.12, 2.16, 2.52 (2) and 2.68 mg/kg. The highest residue is 2.7 mg/kg and the STMR is 1.03 mg/kg.

*Wheat hay*

Residues in wheat hay were reported from trials conducted in the US and Canada. The GAP for wheat in the US is single application from 3-leaf to early jointing (BBCH 13–30) at rates of 7.6–10 g ai/ha, with a PHI of 50 days for harvest and 14 days after application for grazing or cutting. The GAP for wheat in Canada is a single application from 2 to 6-leaf stage at a rate of 10 g ai/ha with a PHI of 50 days for harvest and no restrictions for grazing.

Residues in wheat hay from US trials that correspond to US GAP (as reported on an 'as received' basis) were: 0.21, 0.25, 0.26, 0.34 and 0.61 mg/kg. When corrected for moisture content (88%), residues on a dry weight basis were: 0.24, 0.28, 0.29, 0.39 and 0.69 mg/kg.

Residues from Canadian trials that correspond to Canadian GAP were: 0.37, 1.28, 1.46, 1.48, 2.32 and 2.36 mg/kg. When corrected for moisture content (12%), residues on a dry weight basis were: 0.42, 1.45, 1.66, 1.68, 2.64, 2.68 mg/kg.

An additional 25 US trials corresponded to Canadian GAP and residues on an 'as received' basis were: 0.34, 0.38, 0.43, 0.45, 0.54 (2), 0.69 (2), 0.71, 0.76 (2), 0.83, 0.87, 0.98 (2), 1.02, 1.23, 1.24, 1.31, 1.32, 1.33, 1.37, 1.46, 1.67 and 1.88 mg/kg. When corrected for moisture content (12%), residues on a dry weight basis were: 0.38, 0.43, 0.49, 0.51, 0.61 (2), 0.78, 0.81, 0.86 (2), 0.94, 0.99, 1.1 (2), 1.16, 1.39, 1.41, 1.49, 1.5 (2), 1.56, 1.66, 1.89, 2.14 mg/kg.

The Meeting agreed to combine the data sets from Canada and the USA and residues in wheat hay on a dry weight basis were: 0.24, 0.28, 0.29, 0.38, 0.39, 0.42, 0.43, 0.49, 0.51, 0.61 (2), 0.69, 0.78, 0.81, 0.86 (2), 0.94, <u>0.99</u>, <u>1.1</u> (2), 1.16, 1.39, 1.41, 1.45, 1.49, 1.5 (2), 1.56, 1.66 (2), 1.68, 1.89, 2.14, 2.64 and 2.68 mg/kg. The Meeting recommended a maximum residue level of 3 mg/kg in wheat hay (n = 36), with a highest residue of 2.7 mg/kg and an STMR of 1 mg/kg.

The Meeting decided that wheat forage (when expressed on a dry weight basis) is similar to wheat hay for the purposes of estimating the livestock dietary burden. Therefore only one of the two commodities is considered necessary for inclusion in the livestock dietary burden tables. The highest residue in cereal forage/hay is 2.7 mg/kg and the STMR is 1 mg/kg.

*Grass Pastures: forage and hay*

The meeting received data for grass pastures (forage) and hay from trials conducted in Australia, Brazil, Canada, France, Germany, Italy, New Zealand, Spain, the UK and USA.

GAP in Brazil allows a single application at rates of 40–100 g ae/ha with no specified interval to harvest or for grazing. Forage residue data, calculated as the average of replicate measurements that correspond to Brazilian GAP expressed on an 'as received' basis were: 1.3, 1.4, 3.6, 6.3 and 11.6 mg/kg. When corrected for moisture content (75%) residues were: 5.2, 5.6, 14.4, and 46.4 mg/kg.

Trials from Europe were evaluated against UK GAP, which allows a single application at 60 g ae/ha with a 7 days interval for grazing/cutting. Forage data from trials in France, Germany, Italy, Spain and the UK that correspond to UK GAP expressed on an as received basis were: 0.8, 1, 1.3, 1.5, 1.8, 2.0, 2.2, and 3 mg/kg. Correcting for moisture content (75%), residues in forage on a dry weight basis were: 3.2, 4, 5.2, 6, 7.2, 7.6, 8.8 and 12 mg/kg. The sampling intervals for hay did not correspond to the PHI of 7 days.

GAP in Canada allows application at 60–120 g ae/ha with no restrictions for grazing/cutting. Residues in forage as expressed on an as received basis were: 9.1, 10.7, 12.2, 12.7, 12.8, 13.2, 13.7, 13.8 and 14.6 mg/kg. Correcting for moisture content (75%), residues on a dry weight basis were: 36.4, 42.8, 48.8, 50.8, 52.1, 52.8, 54.8, 55.2 and 58.4 mg/kg. Residues in hay that correspond to GAP are on an as received basis were: 15.1, 15.9, 16.8, 21.6, 22.2, 25.1, 26.2, 30, and 55 mg/kg. When corrected for moisture content (12%), residues were: 17.1, 18.1, 19.1, 24.5, 25.2, 29.8, 34.1 and 62.5 mg/kg.

GAP in the US allows application at 50–120 g ae/ha with no restrictions for grazing/cutting of forage and hay. Residues in forage as expressed on an as received basis were: 4.6, 4.8 (2), 5, 6.1, 6.5, 6.9, 8.7, 9.4, 10, 10.1, 11.1, 11.2, 12.5, and 16 mg/kg. Correcting for moisture content (75%), residues on a dry weight basis were: 18.4, 19.2 (2), 20, 24.4, 26, 27.6, 34.8, 37.6, 40 (2), 44 92), 50 and 64 mg/kg.

Residues in hay on an 'as received' basis were: 10.6, 10.9, 12.4, 15.7, 15.9, 16, 17.9, 18.1, 18.2 (2), 18.6, 20.3, 25.8, 26.3 and 32.2 mg/kg. Correcting for moisture content (12%), residues on a dry weight basis were: 12, 12.4, 14.1, 17.8, 18.1, 18.2, 20.3, 20.5, 20.7, 21.1, 23.1, 29.3, 29.9 and 36.6 mg/kg.

GAP in New Zealand is application at 60 g ae/ha with a nil PHI for gazing/cutting. Forage data from trials that correspond to GAP expressed on a dry weight basis were: 6.2, 7.3, 11.3, 38.8, 42.4, and 48 mg/kg.

GAP in Australia is application at rates ranging 150–210 g ae/ha with a nil PHI for grazing/cutting. Forage data from trials that correspond to GAP were (as expressed on a 'dry weight' basis): 9, 12 (3), 19, 37.1, 52.5, and 103 mg/kg.

The Meeting agreed that for the purposes of estimating livestock dietary burden, the GAP from Australia led to the highest residues in grass pasture. However the data from the US and Canadian trials also fall within the spread of the values from the limited Australian trials. The Meeting therefore agreed to combine the data sets from Australia, Canada and the US; residues in forage on a dry weight basis were: 18.4, 19.2 (2), 20, 24.4, 26, 27.6, 34.8, 36.4, 37.1, 37.6, 40 (2), 42.8, 44 (2), 48.8, 50, 50.8, 52.1, 52.5, 52.8, 54.8, 55.2, 58.4, 64 and 103 mg/kg. The Meeting recommended a highest residue of 103 mg/kg for the purposes of estimating the livestock dietary burden, with an STMR of 41 mg/kg.

Residues in hay from trials conducted in Canada and the US can also be combined on the basis of application rate and nil PHI. Residues are in rank order and on a dry weight basis: 12, 12.4, 14.1, 17.1, 17.8, 18.1 (2), 18.2, 19.1, 20.3, 20.5, 20.7, 21.2, 23.1, 24.5, 25.2, 29.3, 29.8, 29.9, 34.1, 36.6 and 62.5 mg/kg. The Meeting recommended a maximum residue level of 70 mg/kg for grass hay, with a highest residue of 63 mg/kg and an STMR of 21 mg/kg for the purposes of estimating livestock dietary burden.

## Fate of residues during processing

A processing study for wheat was provided to the Meeting. A summary of the processing factors and the resulting STMR-P values is provided.

| Raw Agricultural Commodity | | | | Processed Commodity | | | |
|---|---|---|---|---|---|---|---|
| Commodity | MRL (mg/kg) | STMR (mg/kg) | HR (mg/kg) | Commodity | PF | MRL (mg/kg) | STMR-P (mg/kg) |
| Wheat | 0.1 | 0.01 | 0.07 | Wheat bran | 2.4 | 0.3 | 0.024 |
| | | | | Flour | 0.2 | – | 0.002 |
| | | | | Germ | 0.36 | – | 0.0036 |
| | | | | Aspirated grain fraction | 6.1 | – | 0.06 |

Data from the processing study indicate that there is concentration of aminopyralid residues in wheat bran and aspirated grain fractions, with processing factors of 2.4 and 6.1, respectively. The Meeting agreed to recommend a maximum residue level of 0.3 mg/kg for wheat bran. An HR-P of 0.17 mg/kg is estimated for wheat bran (wheat milled by-products) and 0.43 mg/kg is estimated for aspirated grain fractions for inclusion in the livestock dietary burden. The corresponding STMR-P values for livestock burden are 0.024 mg/kg for wheat bran (wheat milled by-products) and 0.06 mg/kg for aspirated grain fractions.

### Farm animal feeding studies

Groups of lactating dairy cows received the equivalent of 0, 32.8, 64.5, 181.5 and 644.7 ppm in the feed for 28 days. Following the dosing period, there was an additional depuration phase of 14 days, with slaughter intervals of 3, 7 and 14 days after withdrawal from dosing.

Residues were determined in liver, kidney, muscle, fat, whole milk, skim milk and cream.

Residues in whole milk following dosing at 32.8 ppm in the feed were < 0.01 mg/kg over the 28 days period. Residues reached plateau within 2 to 3 days of dosing. Residues in milk ranged < 0.01–0.024 mg/kg and 0.011–0.028 mg/kg following dosing at 64.5 and 181.5 ppm, respectively. Aminopyralid residues ranged 0.023–0.127 mg/kg following dosing at 644.7 ppm. Residues had declined to < 0.01 mg/kg within 2 days of withdrawal from dosing at the highest level of 644.7 ppm.

The highest aminopyralid residues in tissues following dosing at 32.8 ppm level were: muscle < 0.01 mg/kg, fat 0.01 mg/kg, liver < 0.01 mg/kg, and kidney 0.1 mg/kg. Following dosing at 64.5 ppm, aminopyralid residues were < 0.01 mg/kg in muscle, 0.01 mg/kg in fat and liver and 0.2 mg/kg in kidney.

The highest aminopyralid residues in tissues following dosing at 181.5 ppm level were 0.05 mg/kg in muscle and liver, 0.09 mg/kg in fat, and 1.5 mg/kg in kidney. The highest aminopyralid residues in tissues following dosing at 644.7 ppm level were 0.03 mg/kg in muscle, 0.04 mg/kg in fat, 0.06 mg/kg in liver, and 2.5 mg/kg in kidney.

As there is no hen or poultry feeding study, the hen metabolism study is used to recommend appropriate maximum residue levels in hen tissues and eggs. The dose level in the hen metabolism study was 10.5 ppm in the feed and hens were dosed daily for 7 days. TRR in muscle, skin/fat, fat, liver and eggs were < 0.01 mg/kg which is the limit of quantitation in the method used to determine aminopyralid residues in animal tissues and milk. Although the method was not validated for eggs, the Meeting considered that as aminopyralid is not fat-soluble and that the method had been validated for bovine tissues and milk, it would also be applicable to eggs.

### Farm animal dietary burden

The Meeting estimated the dietary burden of aminopyralid residues in livestock (farm animals) on the basis of the livestock diets listed in the FAO Manual 2002.

The maximum dietary burden calculations include the highest residues (HR) and STMR-P values which are used for the estimation of maximum residue levels in animal commodities such as milk, eggs, meat and offal. The STMR dietary burden calculations for livestock allow an estimate of the median residues in milk, eggs, meat and offal that can be used in the chronic dietary assessments and there STMR and STMR-P values for feeds are used.

The percentage dry matter is taken as 100% when highest residues and STMR values are already expressed on a dry weight basis.

*Estimated maximum dietary burden of farm animals*

| Commodity | Group | Residue (mg/kg) | HR/STMR | Diet content (%) | | | Residue Contribution (mg/kg) | | |
|---|---|---|---|---|---|---|---|---|---|
| | | | | Beef cattle | Dairy cattle | Poultry | Beef cattle | Dairy cattle | Poultry |
| Cereal forage | AF | 2.7 | HR | 100 | 60 | 10 | 2.7 | 1.62 | 0.27 |
| Cereal straw | AS | 0.21 | HR | 80 | 20 | 10 | 0.17 | 0.04 | 0.02 |
| Grass forage | AF | 103 | HR | **100** | **100** | **10** | *103* | *103* | *10.3* |
| Grass hay | AS | 63 | HR | 100 | 60 | 10 | 63 | 37.8 | 6.3 |
| Cereal grain | GC | 0.01 | HR | 80 | 40 | **40** | 0.06 | 0.03 | ***0.03*** |
| Wheat milled by-products | CF | 0.02 | STMR-P | 40 | 40 | **50** | 0.07 | 0.07 | ***0.08*** |
| AGF* | – | 0.06 | STMR-P | 5 | – | | 0.003 | | |
| **Total** | | | | **100** | **100** | **100** | **103** | **103** | **10.4** |

* Aspirated grain fractions

The calculated highest dietary burdens for beef cattle, dairy cattle and poultry are 103, 103, and 10.4 ppm, respectively.

*Estimated STMR dietary burden of farm animals*

| Commodity | Group | Residue (mg/kg) | STMR/ STMR-P | Diet content (%) | | | Residue Contribution (mg/kg) | | |
|---|---|---|---|---|---|---|---|---|---|
| | | | | Beef cattle | Dairy cattle | Poultry | Beef cattle | Dairy cattle | Poultry |
| Cereal forage | AF | 1 | STMR | 100 | 60 | 10 | 1 | 0.6 | 0.1 |
| Cereal straw | AS | 0.07 | STMR | 80 | 20 | 10 | 0.06 | 0.01 | 0.01 |
| Grass forage | AF | 37 | STMR | **100** | **100** | **10** | *37* | *37* | *3.7* |
| Grass hay | AS | 21 | STMR | 100 | 60 | 10 | 21 | 12.6 | 2.1 |
| Cereal grain | GC | 0.01 | STMR | 80 | 40 | **40** | < 0.01 | < 0.01 | ***< 0.01*** |
| Wheat milled by-products | CF | 0.02 | STMR-P | 40 | 40 | **50** | 0.01 | 0.01 | ***0.01*** |
| AGF* | – | 0.06 | STMR-P | 5 | – | – | 0.003 | – | – |
| **Total** | | | | **100** | **100** | **100** | **37** | **37** | **3.7** |

* Aspirated grain fractions

The STMR dietary burdens for beef cattle, dairy cattle and poultry are 37, 37 and 3.7 mg/kg, respectively.

### Animal commodity maximum residue levels

The livestock dietary burdens used for the estimation of the maximum residue levels for animal commodities were 103 ppm for beef and dairy cattle and 10.4 ppm for poultry. The livestock dietary burdens used for the STMR estimation for dietary risk assessment were 37 ppm for beef and dairy cattle and 3.7 ppm for poultry.

For poultry, the maximum dietary burden of 10.3 ppm is very similar to the dose level of 10.5 ppm in the hen metabolism study. Residues in hen tissues and eggs were < 0.01 mg/kg following dosing for 7 consecutive days. On the basis of the hen metabolism study, the Meeting recommended maximum residue levels of *0.01 mg/kg poultry meat, poultry offal and eggs. These values of *0.01 mg/kg were also used in the dietary risk assessment.

For cattle, the maximum dietary burden of 103 ppm is between the dose levels of 64.5 ppm and 181.5 ppm. The residues in kidney are the highest of all tissues and range 0.45–2.5 mg/kg between these two feed levels. On the basis of interpolation between the two feed levels, the Meeting recommended a maximum residue level of 1 mg/kg for mammalian kidney. For liver and other offal, the Meeting recommended a maximum residue level of 0.05 mg/kg. For milk, the meeting recommended a maximum residue level of 0.02 mg/kg. For meat, the Meeting recommended a maximum residue level of 0.1 mg/kg on the basis of higher levels being present in fat compared to muscle at the same dose levels.

| Dietary burden (ppm) | | | | Residues (mg/kg) | | | | | |
|---|---|---|---|---|---|---|---|---|---|
| Feed level [ppm] | Milk | | Meat (fat) | | Liver | | Kidney | | |
| | Highest | Mean | Highest | Mean | Highest | Mean | Highest | Mean | |
| **MRL estimate  (beef and dairy cattle)** | | | | | | | | | |
| (103 ppm) | | 0.016 | 0.026, (0.054) | | 0.031 | | 0.87 | | |
| [64.5/181.5] | [0.024, 0.028] | | [< 0.01, 0.046] | | [0.014, 0.054] | | [0.202, 1.537] | | |
| **STMR estimate  (beef and cattle)** | | | | | | | | | |
| (37 ppm) | | 0.01 | | 0.01 | | 0.01 | | 0.1 | |
| [32.8 ppm] | | | | | | | | | |

For the purposes of the dietary risk assessment, the STMR estimate for the livestock dietary burden is 37 ppm for beef and dairy cattle and 3.7 ppm for poultry. Based on the poultry metabolism study, the STMR values for poultry and eggs are 0.01 mg/kg. In the lactating cattle study, the feed level of 32.8 is close to the STMR level of 37 ppm. The STMR values for milk, meat (fat) and liver are 0.01 mg/kg and 0.1 mg/kg for kidney.

## Conclusions

On the basis of the data from supervised trials and farm animal feeding studies, the Meeting provisionally estimated maximum residue levels, but these are not recommended for use as maximum residue limits (MRLs) because of the deferred toxicological evaluation.

## DIETARY RISK ASSESSMENT

### Long-term intake and short-term intake

As the Meeting received an incomplete toxicology data submission, the dietary risk assessment for aminopyralid could not be finalised.

## REFERENCES

**Author, Date, Title, Institute, Report Reference, Document No.**

Cathie, C., 2001. Determination of Dissociation Constant of XR750 Using UV-Visible Spectrophotometry. GHF-P-2332; 01-822-AG

Class, T. 2004. The Development and Validation of a Method for the Determination of Aminopyralid in Air. PTRL Europe Study P 722 G. 031145

Cook, W.L., 2003. Aqueous Photolysis of XDE-750 in pH 5 Buffer Under Xenon Light. 020066

Cook. W.L. 2003. Hydrolysis of XDE-750 at pH 5, 7, and 9. 020067.Covance study No. 295/155

Cowles, J. 2002. XDE-750 Residues in Grass Dominant New Zealand Pasture after One Application of GF-871 in the 2002 Season. 020049

Cowles, J. 2003. Aminopyralid Residues in Australian Winter Cereals after One Application of GF-871 in the 2002 Season. 020060

Cowles, J. 2004. Aminopyralid and Fluroxypyr Residues in Australian Winter Cereal Forage, Grain and Straw after One Application of GF-871 or GF-982 in the 2003 Season. 030071

Cowles, J. 2004. Aminopyralid Residues in New Zealand Pasture Grass after One Application of GF-871 Herbicide in the 2003 Season. 030028

Cowles, J. 2004. Aminopyralid and Fluroxypyr Residues in Australian Winter Cereal Grain and Straw at Harvest after One Application of GF-871 or GF-982 in the 2003 Season. 030072

Cowles, J. 2004. Residues of Aminopyralid (XDE-750) in Grass Pasture in Southern Australia after the Application of GF-871 in the 2003 C 2004 Season. 030103

Cowles, J., 2004. Residues of Aminopyralid (XDE-750) and Fluroxypyr in Grass Pasture in Northern Australia after the Application of GF0871, Starane herbicide and GF-982 in the 2003 C 2004. Season. 030102

Dusek, D. 2003. Generation of Aspirated Grain Fraction and Wheat Processing. 030042

Ghaoui, L., 2001. Determination of the Purity and Identity of TSN102417, XDE-750 Pure Active Ingredient. FAPC 003276

Ghaoui, L., 2002. Calculation of Henry's Law Constant for XDE-750 From Unbuffered and pH 5, 7 and 9 Buffered Water NAFST540

Graper, L.K., Smith, K.P. and Hilla, S., 2003. A Nature of the Residue Study with 14C-labelled XDE-750 Applied to Spring Wheat. 020022

Griffin, K.A. 2001. Vapor Pressure of XR-750 by Knudsen-Effusion Weight Loss Method. DECO GL-AL MD-2001-000921

Hastings, M.J. 2003. Determination of Residues of Aminopyralid in Water by Liquid Chromatography with Tandem Mass Spectrometry Detection. GRM 01.32. 011159

Hastings, M.J. 2003. Method Validation Report for the Determination of Residues of Aminopyralid in Water by Liquid Chromatography with Tandem Mass Spectrometry Detection Using Dow AgroSciences Method GRM 01.32. 011159

Kalvan, H.C. 2005. Residues of XDE-750 in Wheat After Application of GF-1118 Herbicide e Argentina 2004   2005. GHB-P 1109. 030070.01

Kalvan, H.C. 2005. Residues of XDE-750 in Wheat After Application of GF-1118 Herbicide K Argentina, 2003. GHB-P 1096 030070

Lindsay, D.A. 2004. Frozen Storage Stability of XDE-750 in Range Land and Pasture Grass and Hay and Wheat Straw and Wheat Grain. 030004.01

Lindsay, D.A. 2004. Frozen Storage Stability of XDE-750 in Soil. 030002.01

Lindsey, A.E. and Hastings, M.J. 2004. Determination of Residues of Aminopyralid in Soil by Liquid Chromatography with Tandem Mass Spectrometry Detection. GRM 02.34. 021295

Lindsey, A.E. and Hastings, M.J. 2004. Method Validation Report for the Determination of Residues of Aminopyralid in Soil by Liquid Chromatography with Tandem Mass Spectrometry Detection Using Dow AgroSciences Method GRM 02.34. 021295

MacPherson, D. and Gedik, L., 2003, The Distribution and Metabolism of [14C]-XDE-750 in the Lactating Goat. Inveresk Number 201893

Madsen, S. 2003. Determination of the n-Octanol/Water Partition Coefficient (Shake Flask method) of XDE-750 Technical. FOR01009Magnussen, J.D. 2004. A Confined Rotational Crop Study with 14C XDE-750. 030008

Magnussen, J.D. and Balcer, J.L. 2004. 14C XDE 750 Grass Nature of the Residue Study. 010071

Magnussen, J.D. 2004. 14C XDE-750 Poultry Nature of the Residue Study. 030009

McCormick, R.W., Schelle, G.E. and Dolder, S.C., 2004. Magnitude of Residue of XDE-750 and 2,4-D in Rangeland and Pasture Grasses. 020018

Mollica, J. and West, S. 2003. Method Validation for the Analysis of XDE-750 in Human Blood and Urine. 031005

Nelson, R.M. 2002. Determination of the Water Solubility of XDE-750. FOR01015

Olberding, E.L. and Hastings, M.J. 2004. Validation Report for Method GRM 02.31 n Determination of Residues of Aminopyralid in Agricultural Commodities by Liquid Chromatography with Tandem Mass Spectrometry Detection. 021310

Olberding, E.L., Arnold, B.H. and Hastings, M.J. 2004. Determination of Residues of Aminopyralid in Agricultural Commodities by Liquid Chromatography with Tandem Mass Spectrometry Detection. GRM 02.31. 021310

Pinheiro, A.C. 2003. Residuos de Aminopyralid (XDE-750) E 2.4-D em Pastagem Após Aplicação de GF-1004 Herbicida A Brasil 2002   2003. 020131

Pinheiro, A.C. 2003. Residuos de Aminopyralid (XDE-750) E Fluroxypyr em Pastagem Após Aplicação de GF-843 Herbicida P Brasil 2002 B 2003. 020132

Pinheiro, A.C. and De Vito, R. 2003. Determination of Residues of Aminopyralid (XDE-750), 2,4-D and Fluroxypyr in Pasture by Gas Chromatography with Mass Spectrometry Detection. GRM 03.25

Rawle, N.W. 2005. Residues of XDE-750 in Winter Wheat at Intervals and at Harvest Following a Single Application of GF-1118, Spain E 2004. GHE-P-10689

Rawle, N.W., 2003. Residues of XDE-750 in Winter Barley at Harvest Following a Single Application of GF-118, Spain R 2003. GHE-P-10576

Rawle, N.W., 2003. Residues of XDE-750 in Winter Wheat at Harvest Following a Single Application of GF-118, Hungary and Poland R 2003. GHE-P-10577

Rawle, N.W., 2003. Residues of XDE-750 in Winter Wheat at Intervals and at Harvest Following a Single Application of GF-118, Italy and Spain F 2003 GHE-P-10575

Rawle, N.W., 2004. Residues of XDE-750 in Pasture at Harvest Following a Single Application of GF-839, Germany, UK, Southern France and Spain R 2003. GHE-P-10578

Rawle, N.W., 2004. Residues of XDE-750 in Pasture at Intervals and at Harvest Following a Single Application of GF-839, Southern France, Spain and Italy , 2003. GHE-P-10580

Rawle, N.W., 2004. Residues of XDE-750 in Pasture at Intervals and at Harvest Following a Single Application of GF-839, UK, Northern France and Germany i 2003. GHE-P-10579

Rawle, N.W., 2005. Residues of XDE-750 in Winter Barley at Harvest Following a Single Application of GF-1118, Spain R 2004. GHE-P-10690

Rawle, N.W., 2005. Residues of XDE-750 in Winter Wheat at Harvest Following a Single Application of GF-1118, Germany and Poland 2 2004. GHE-P-10691

Reed, R.L. 2004. Independent Laboratory Validation of Dow AgroSciences LLC Method GRM 03.18 R Determination of Residues of Aminopyralid in Bovine Tisues by Liquid Chromatography with Tandem Mass Spectrometry. 030098

Reed, R.L. 2005. Independent Laboratory Validation of Dow AgroSciences LLC Method GRM 02.31 R Determination of Residues of Aminopyralid in Agricultural Commodities by Liquid Chromatography with Tandem Mass Spectrometry Detection. 020157

Robaugh, D. and Randolph, R. 2003. Determination of Residues of XDE-750 in Whole Human Blood and Human Urine by Liquid Chromatography with Tandem Mass Spectrometry Detection. Pyx-1419, Rev.0. 031005

Roberts, D.W. and Schelle, G.E. 2004, Terrestrial Field Dissipation of XDE-750 in Canada. 020031
Roberts, D.W. and Schelle, G.E. 2004. Terrestrial Field Dissipation of XDE-750 in the USA. 020032

Roberts, D.W., Schelle, G.E. and Knuteson, J.A. 2004. Magnitude of Residues of XDE-750 in Wheat Agricultural Commodities. 030042

Rosser, S.W., Rutherford, L.A. and McFarlane, J.H. 2004. Magnitude of XDE-750 Residues in Bovine Tissues and Milk from a 28-Day Feeding Study. 030061

Roulin, S., 2001, Determination of the Organic Solvent Solubility of XR-750. GHE-P-9228/01003/DA

Russell, M.W., 2001. Determination of Color, Physical State, Odor, pH and Density of XDE-750, Pure Active Ingredient. FAPC 013273

Russell, M.W., 2003 (amended). Determination of Color, Physical State, Odor, pH and Density of XDE-750, TGAI-Tox. FAPC 013087

Russell, M.W., 2003. Determination of Melting, Boiling and Decomposition Points of DE-750 Pure Active Ingredient FAPC 023263

Rutherford, L.A. and Hastings, M.J. 2003. Determination of Residues of Aminopyralid in Bovine Tissues by Liquid Chromatography with Tandem Mass Spectrometry. GRM 03.18. 021327

Rutherford, L.A. and Hastings, M.J. , 2003. Method Validation Report for the Determination of Aminopyralid in Bovine Tissues by Liquid Chromatography with Tandem Mass Spectrometry Using Dow AgroSciences LLC Method. 021327

Rutherford, L.A., 2004. Photodegradation of XDE-750 on Soil. 020080

Unsworth, C., 2003. A One year Field Study to Determine the Residues of XDE-750 in Pasture Following a Single Application of GF-839, Europe 2002 2 2003. GHE-P-10449

Unsworth, C., Oxspring, S., Balluff, M. and Rosser, S. 2003. A One year Field Study to Determine the Residues of XDE-750 in Pasture Following a Single Application of GF-839, Northern Europe 2002 R 2003. GHE-P-10448 Covance study No. 295/153

Unsworth, C., Scrimshaw, O., Balluff, M., Lagrasse, S., Morgan, A.J. and Schelle, G., 2003. A One Year Field Study to Determine the Dissipation of XDE-750 Through Soil Following A Single Application of GF-819, Europe 2002 u 2003. GHE-P-10573

Yoder, R.N. and Smith, K.P., 2002. Aerobic Soil Degradation of XDE-750 in Five North American Soils. 010091

Yoder, R.N. and Smith, K.P., 2003. Aerobic Soil Degradation of XDE-750 in Four European Soils. 020054

## CROSS REFERENCES

| Author | Document Code | Year |
| --- | --- | --- |
| Cathie, C. | GHF-P-2332; 01-822-AG | 2001 |
| Class, T. | 031145 | 2004 |
| Cook, W.L. | 020066 | 2003 |
| Cook. W.L. | 020067 | 2003 |
| Cowles, J. | 020049 | 2002 |
| Cowles, J. | 020060 | 2003 |
| Cowles, J. | 030028 | 2004 |
| Cowles, J. | 030071 | 2004 |
| Cowles, J. | 030072 | 2004 |
| Cowles, J. | 030102 | 2004 |
| Cowles, J. | 030103 | 2004 |
| Dusek, D. | 030042 | 2003 |
| Ghaoui, L. | NAFST540 | 2002 |
| Ghaoui, L. | FAPC 003276 | 2001 |
| Graper, L.K., Smith, K.P. and Hilla, S. | 020022 | 2003 |
| Griffin, K.A. | DECO GL-AL MD-2001-000921 | 2001 |
| Hastings, M.J. | 011159 | 2003 |
| Hastings, M.J. | 011159 | 2003 |
| Kalvan, H.C. | 030070 | 2005 |
| Kalvan, H.C. | 030070.01 | 2005 |
| Lindsay, D.A. | 030002.01 | 2004 |
| Lindsay, D.A. | 030004.01 | 2004 |
| Lindsey, A.E. and Hastings, M.J. | 021295 | 2004 |
| Lindsey, A.E. and Hastings, M.J. | 021295 | 2004 |
| MacPherson, D. and Gedik, L. | Inveresk Number 201893 | 2003 |
| Madsen, S. | FOR01009 | 2003 |
| Magnussen, J.D. | 030008 | 2004 |
| Magnussen, J.D. | 030009 | 2004 |
| Magnussen, J.D. and Balcer, J.L. | 010071 | 2004 |
| McCormick, R.W., Schelle, G.E. and Dolder, S.C. | 020018 | 2004 |
| Mollica, J. and West, S. | 031005 | 2003 |
| Nelson, R.M. | FOR01015 | 2002 |
| Olberding, E.L. and Hastings, M.J. | 021310 | 2004 |
| Olberding, E.L., Arnold, B.H. and Hastings, M.J. | 021310 | 2004 |
| Pinheiro, A.C. | 020131 | 2003 |
| Pinheiro, A.C. | 020132 | 2003 |
| Pinheiro, A.C. and De Vito, R. | GRM 03.25 | 2003 |
| Rawle, N.W. | GHE-P-10575 | 2003 |
| Rawle, N.W. | GHE-P-10576 | 2003 |
| Rawle, N.W. | GHE-P-10577 | 2003 |
| Rawle, N.W. | GHE-P-10578 | 2004 |
| Rawle, N.W. | GHE-P-10579 | 2004 |
| Rawle, N.W. | GHE-P-10580 | 2004 |
| Rawle, N.W. | GHE-P-10689 | 2005 |
| Rawle, N.W. | GHE-P-10690 | 2005 |
| Rawle, N.W. | GHE-P-10691 | 2005 |
| Reed, R.L. | 020157 | 2005 |
| Reed, R.L. | 030098 | 2004 |
| Robaugh, D. and Randolph, R. | 031005 | 2003 |
| Roberts, D.W. and Schelle, G.E. | 020031 | 2004 |
| Roberts, D.W. and Schelle, G.E. | 020032 | 2004 |
| Roberts, D.W., Schelle, G.E. and Knuteson, J.A. | 030042 | 2004 |
| Rosser, S.W., Rutherford, L.A. and McFarlane, J.H. | 030061 | 2004 |
| Roulin, S. | GHE-P-9228/01003/DA | 2001 |
| Russell, M.W. | FAPC 023263 | 2003 |

| | | |
|---|---|---|
| Russell, M.W. | FAPC 013273 | 2001 |
| Russell, M.W. | FAPC 013087 | 2003 (amended) |
| Rutherford, L.A. | 020080 | 2004 |
| Rutherford, L.A. and Hastings, M.J. | 021327 | 2003 |
| Rutherford, L.A. and Hastings, M.J. | 021327 | 2003 |
| Unsworth, C. | GHE-P-10449 Covance study No. 295/153 | 2003 |
| Unsworth, C., Oxspring, S., Balluff, M. and Rosser, S. | GHE-P-10448 Covance study No. 295/155 | 2003 |
| Unsworth, C., Scrimshaw, O., Balluff, M., Lagrasse, S., Morgan, A.J. and Schelle, G. | GHE-P-10573 | 2003 |
| Yoder, R.N. and Smith, K.P. | 010091 | 2002 |
| Yoder, R.N. and Smith, K.P. | 020054 | 2003 |

## BIFENAZATE (219)

*First draft was prepared by Denis Hamilton, Biosecurity, Queensland Department of Primary Industries and Fisheries, Australia*

## IDENTITY

| | |
|---|---|
| ISO common name | Bifenazate |
| Synonyms: | D2341 |
| IUPAC name | Isopropyl 2-(4-methoxybiphenyl-3-yl)hydrazinoformate |
| Chemical Abstracts name | 1-methylethyl 2-(4-methoxy[1,1'-biphenyl]-3-yl) hydrazinecarboxylate |

| | |
|---|---|
| CAS Number | 149877-41-8 |
| CIPAC Number | 736 |
| Molecular formula | $C_{17}H_{20}N_2O_3$ |
| Molecular mass | 300.4 |
| Structural formula | |

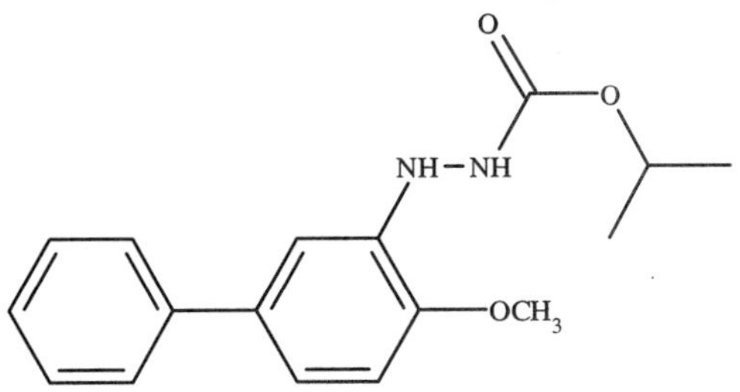

## PHYSICAL AND CHEMICAL PROPERTIES

Pure active ingredient:

| Property | Result | Ref |
|---|---|---|
| Colour (purity 99.7%) | White | Friedlander, 1998, GRL FR-11291 |
| Odour (purity 99.7%) | Slight odour characteristic of aromatic compounds | Friedland, 1998, GRL FR-11293-01 |
| Appearance (purity 99.7%) | Crystalline solid consisting of small needles of less than 1 mm in length | Friedlander, 1998, GRL FR-11292-01 |
| Melting point (purity 98.1%) | 121.5–123 °C | Dunn, 2003, GRL-12057 |
| Relative density (purity 99.7%) | 1.19 g/cm$^3$ | Stevenson, 1998, GRL-FR-11296 |
| Vapour pressure (purity 98.1%, light brown powder) | $3.8 \times 10^{-7}$ Pa at 25 °C[1] | Tremain, 2003, 666/079 |
| Solubility in water (purity 98.1%, pale brown powder) | 1.52 mg/L water at 20 °C<br>1.66 mg/L aqueous pH 5 buffer at 20 °C | Riggs, 2004, GRL-11907 |

---

[1] Vapour pressure balance method with measurements over a temperature range of 100-113 °C and extrapolation to 25 °C.

| Property | Result | Ref |
|---|---|---|
| Solubility in organic solvents at 20 °C (purity 99.7%) | ethyl acetate    102 g/L<br>acetonitrile     96 g/L<br>methanol 45 g/L<br>toluene         24.7 g/L<br>hexane        0.23 g/L<br>n-octanol     8.9 g/L | Riggs, 1998, GRL-FR-11295 |
| Octanol/water partition coefficient (purity 98.1%, pale brown powder) | Log $P_{OW}$ = 3.5 at 38 °C | Riggs, 2003, GRL-12061 |
| Hydrolysis rate (purity ≥ 98%) | Conditions: 25 °C, <1% acetonitrile in sterile buffers in the dark, approx 1 mg/L.<br>pH          $DT_{50}$<br>4           9.1 days<br>5           5.4 days<br>7           20 hours<br>9           1.6 hours | Shah, 1997, 6337-95-0006-EF-001 |
| Dissociation constant in water | No evidence of dissociation in solutions where bifenazate is stable and could be tested, i.e. between pH 6.5 and 2. | Riggs, 2004, GRL-12062 |

*Hydrolysis of bifenazate (Shah, 1997, 6337-95-0006-EF-001)*

Bifenazate, at approximately 1 mg/L, was hydrolysed in sterile aqueous buffers in the dark. The buffers contained less than 1% acetonitrile as a co-solvent. The first step was an oxidation reaction producing bifenazate-diazene, which then hydrolysed to produce methoxy- and hydroxy- biphenyls. Bifenazate-diazene reached a maximum 21.3% (pH 4, day 14), 27.3% (pH 5, day 10), 58.5% (pH 7, 27 hours) and 23.5% (pH 9, 1.8 hours) of the starting material.

The data allowed estimates for hydrolysis $DT_{50}$s of bifenazate-diazene: 58, 50, 18 and 0.28 hours at pH 4, 5, 7 and 9 respectively.

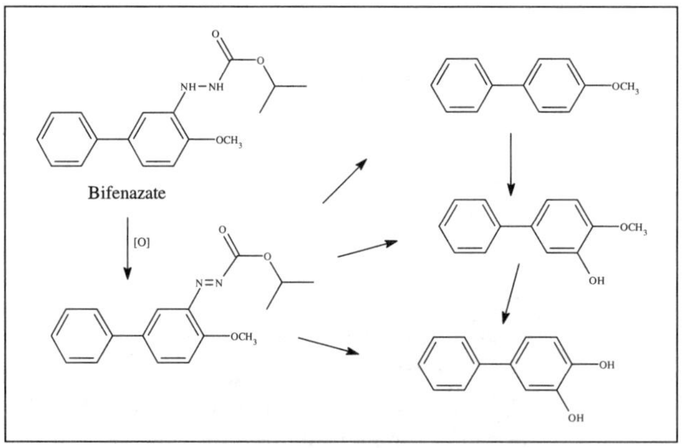

Figure 1. Proposed hydrolysis pathway for bifenazate (Shah, 1997, 6337-95-0006-EF-001).

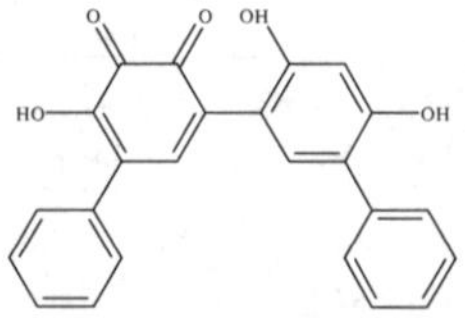

Findak, DiFrancesco and Shah (1999, 6337-95-0006-EF-001-001) subsequently identified a product of hydrolysis at pH 5 and 7. The compound accounted for a maximum of approximately 14% and 24% of the [14]C during hydrolysis at pH 5 (day 30) and pH 7 (day 5).

*Photolysis of bifenazate (Shah, 1997, 6337-95-0006-EF-001)*

Bifenazate ([14]C labelled), at approximately 1 mg/L, was photolysed in a sterile aqueous pH 5 acetate buffer in simulated sunlight, with 12 hours light and 12 hours dark for 6 days.

The photolysis $DT_{50}$ for parent bifenazate was 0.72 days and for bifenazate + bifenazate-diazene it was 3.3 days. The dark solution $DT_{50}$ for parent bifenazate was 1.8 days and for bifenazate

+ bifenazate-diazene the $DT_{50}$ was 4.4 days. Three other products were identified in the photolysis solution: 3,4-dihydroxybiphenyl and 3-hydroxy-4-methoxybiphenyl accounted for 16% and 32% of the initial $^{14}C$ at the end of the 6 days, while 3-methoxybiphenyl reached a maximum of 13% of the $^{14}C$ after 54 hours of photolysis. Polar products and carbon dioxide were also produced.

*Photolysis of bifenazate (Lewis, 2001, 217/27)*

Bifenazate, at approximately 1 mg/L, was photolysed in a sterile aqueous pH 5 buffer in simulated sunlight filtered to remove wavelengths below 290 nm, with 12 hours light and 12 hours dark for 30 days. The intensity of the light was adjusted so as to approximate the same number of summer day's sunlight in UK or Florida. The $DT_{50}$ for parent bifenazate was 0.9 days (10.8 hours). The initial half-life calculated from the 0–2 days data for bifenazate was 0.8 days (9 hours) and for combined bifenazate + bifenazate-diazene was 1.5 days (18 hours). The identified products of photolysis are shown in Figure 2. Polar products and carbon dioxide were also produced.

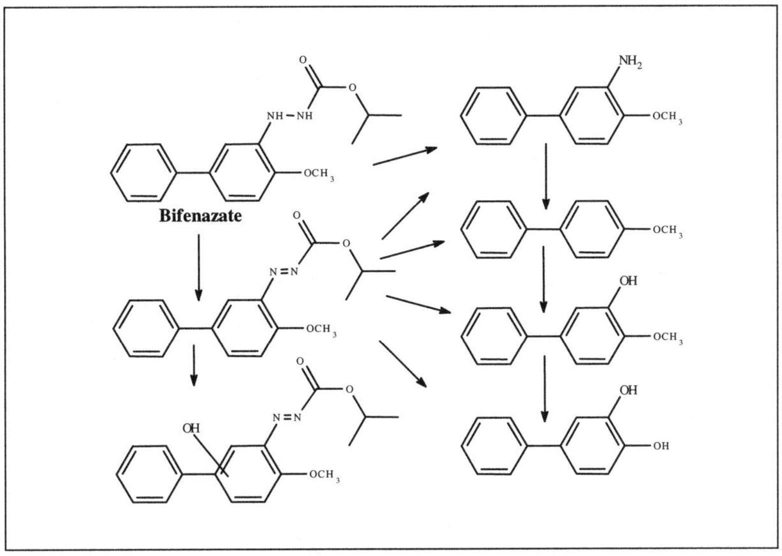

Figure 2. Proposed photolysis pathway for bifenazate (Lewis, 2001, 217/27)

Technical material:

| Property | Result | | Ref |
|---|---|---|---|
| Solubility in organic solvents at 20 °C | ethyl acetate | 11.3 g/100 mL | Riggs, 1997, GRL-FR-10806 |
| | acetonitrile | 11.1 g/100 mL | |
| | methanol 5.07 g/100 mL | | |
| | toluene | 2.62 g/100 mL | |
| | hexane | 0.0232 g/100 mL | |
| | n-octanol | 0.954 g/100 mL | |

## FORMULATIONS

Bifenazate is available as commercial WP, SC and WG formulations.

| Code | Description | Concentration | Examples |
|---|---|---|---|
| WP | wettable powder | 500 g/kg | Acramite 50WP, Acramite 50WS (WP in water soluble bags) |
| SC | suspension concentrate | 480, 240 or 200 g/L | Acramite 480 SC, Acramite 4L, Floramite 240 SC, Mitokohne 20 SC |

| Code | Description | Concentration | Examples |
|------|-------------|---------------|----------|
| WG | water dispersible granules | 750 g/kg | Acramite 75 WG |

## METABOLISM AND ENVIRONMENTAL FATE

Animal and plant metabolism and environmental fate studies used bifenazate $^{14}$C labelled in the substituted phenyl ring.

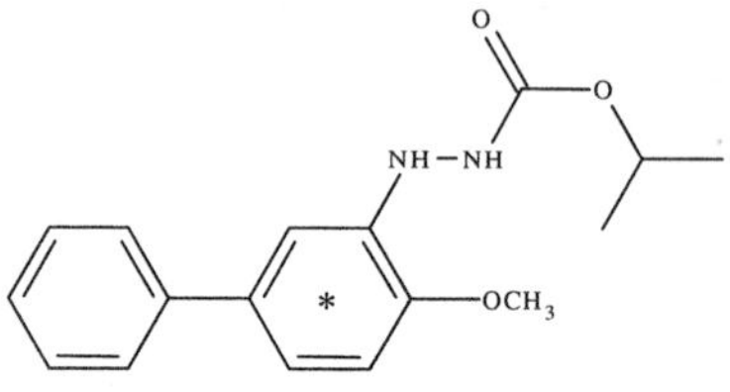

bifenazate $^{14}$C labelled in the
substituted phenyl ring

Structures, names and codes for metabolites are summarised below.

| | |
|---|---|
| Bifenazate-diazene<br>  diazenecarboxylic acid, 2-(4-methoxy-[1,1'-biphenyl]-, 1-methylethyl ester<br>  CAS number: 149878-40-0<br>  Code:          D3598 | 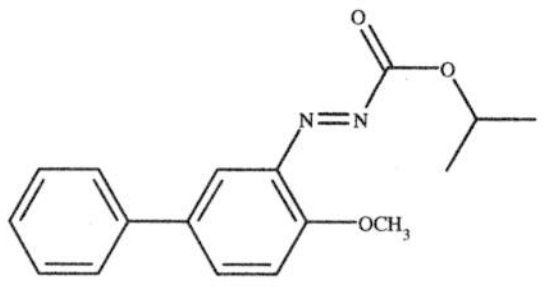 |
| Bifenazate-diazene oxide<br>  diazenecarboxylic acid, 2-(4-methoxy-[1,1-biphenyl]-3-yl)-, 1-methylethyl ester 2-oxide<br>  Code:          D4642 | 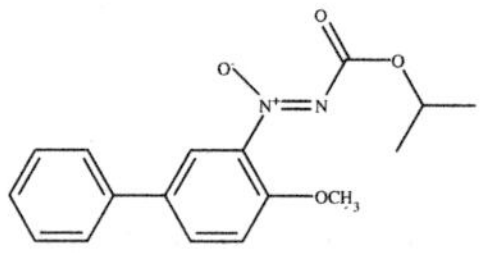 |
| 3-hydroxy-4-methoxybiphenyl<br>  CAS number: 37055-80-4<br>  Code:          D9963 | 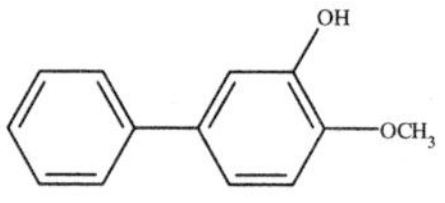 |
| Bifenazate-carbamate<br>  carbamic acid, (4-methoxy-[1,1'-biphenyl]-3-yl)-, 1-methylethyl ester<br>  Code:          D6887 | 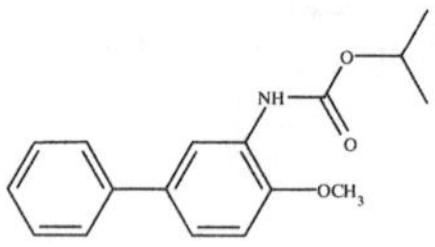 |
| 4-hydroxy-4'-methoxybiphenyl<br>  CAS number: 16881-71-3<br>  Code:          D9477 | 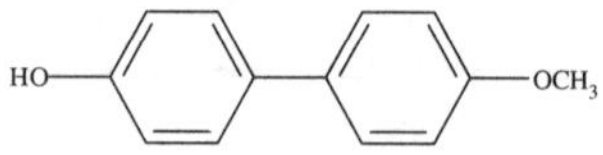 |
| Bifenazate glucuronide | 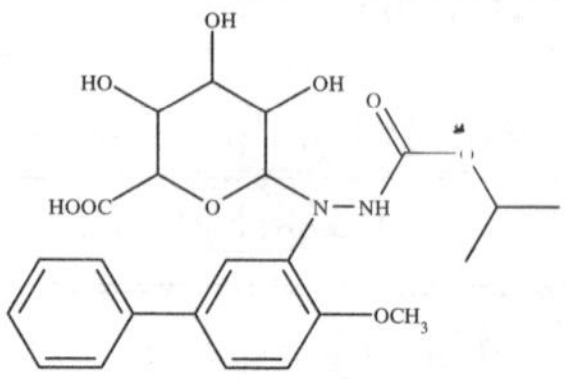 |

| 4-hydroxybiphenyl glucuronide | 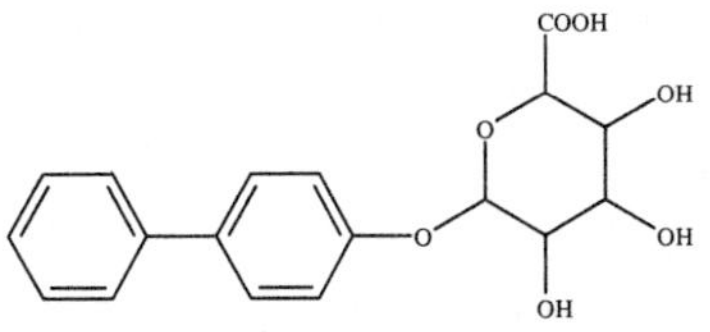 |
|---|---|
| 4-hydroxybiphenyl<br>  CAS number: 92-69-3<br>  Code:        A1530 | 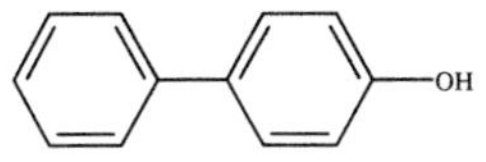 |
| 4,4'-dihydroxybiphenyl<br>  CAS number: 92-88-6<br>  Code:        D9569 | 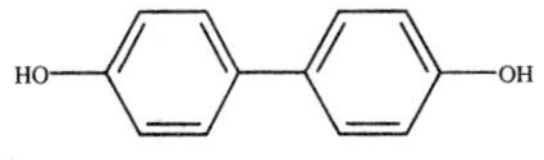 |
| 4-methoxybiphenyl<br>  CAS number: 613-37-6<br>  Code:        D1989 | 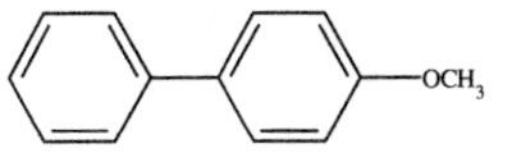 |
| 4-hydroxybiphenyl sulphate | 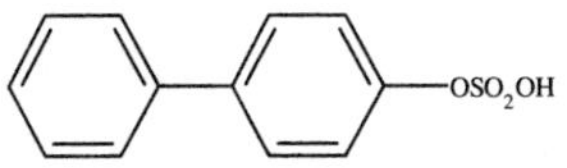 |
| 4-hydroxy bifenazate | 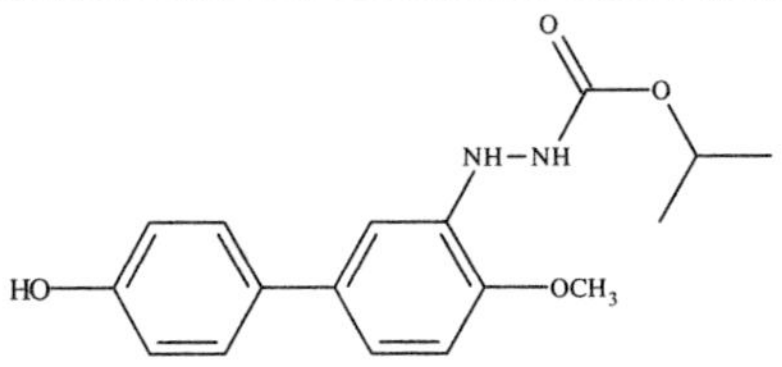 |
| 4-hydroxy bifenazate-diazene | 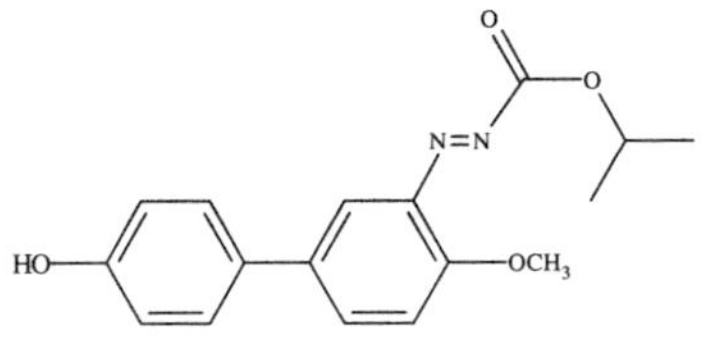 |

### Animal metabolism

The Meeting received animal metabolism studies with bifenazate in rats, lactating goats and laying hens.

Bifenazate is readily converted to bifenazate-diazene by mild oxidation. Primary metabolites are readily produced by removal of the side chain and by hydroxylation of the biphenyl rings. Glucuronide and sulphate conjugates are also produced.

### Rats

McClanahan (1998, 95236) studied the distribution, metabolism and excretion of bifenazate following a single oral dose (10 or 1000 mg/kg bw) of [$^{14}$C]bifenazate in rats. Parent bifenazate and the following metabolites were identified in excreta: bifenazate glucuronide, bifenazate-diazene, 4-hydroxy bifenazate, 4-hydroxy bifenazate-diazene, 4-hydroxybiphenyl and its sulphate conjugate, 4,4'-dihydroxybiphenyl and its glucuronate and sulphate conjugates, 4-methoxybiphenyl and 4-hydroxy-4'-methoxybiphenyl and its conjugates.

### Lactating goats

A lactating goat weighing 31 kg (day 1 and day 4) was dosed orally once daily for 4 consecutive days by gelatin capsule with 21 mg/animal/day of [$^{14}$C]bifenazate, equivalent to 10 ppm in the feed (McClanahan and Bayus, 1999, 96-0064) for a 2.10 kg/day feed consumption. Milk was collected twice daily; a day's sample began in the afternoon after dosing and ended with the morning milking preceding the next dose. Milk production averaged 1.78 kg/day during the dosing period. The animal was slaughtered approximately 8 hours after the final dose for tissue collection. Recovery of administered $^{14}$C was 68%. Samples were stored in a freezer below -10 °C for the following intervals before the initial extraction for metabolite identification (milk 1 day; kidney 27 days; muscle, fat and liver 32 days). See Table 14 and Table 15 for freezer storage stability of tissues and milk of lactating goats.

The majority of the administered $^{14}$C was present in the excreta and stanchion wash (19.5% in urine, 46.5% in faeces). Milk accounted for 0.22% and tissues and blood 2.0% of the administered $^{14}$C. The distribution of the radiolabel and identified metabolites in goat milk and tissues are summarised in Table 1. Residues of $^{14}$C were higher in liver (1.77 mg/kg) than in other tissues.

In day 4 goat milk, 4-hydroxybiphenyl sulphate was the major identified residue at 0.019 mg/kg (41% of TRR). Bifenazate + bifenazate-diazene at 0.004 mg/kg comprised 9% of the TRR in milk. In muscle, the highest identified residue was 4-hydroxybiphenyl at 0.002 mg/kg (13% of TRR). Bifenazate + bifenazate-diazene comprised approximately 3–9% of the TRR in muscle. In fat, bifenazate was the major component, accounting for 53-58% of the TRR. Bifenazate + bifenazate-diazene comprised approximately 58–67% of the TRR in fat (0.070-0.072 mg/kg). The residue levels and patterns in omental and perirenal fat were quite similar.

In goat liver, only 10% of the TRR was extractable. The major identified metabolite was 4-hydroxybiphenyl glucuronide at 0.017 mg/kg (0.93% of TRR). Bifenazate + bifenazate-diazene comprised 0.98% of the TRR. The extracted liver was subjected to hydrolysis and extraction procedures to investigate the nature of the 87% non-extractable residues. The evidence suggested that some of the unextractable TRR was covalently bound to liver protein.

In goat kidney, 4-hydroxybiphenyl glucuronide and sulphate were the major identified components at 0.023 mg/kg, equivalent to approximately 14% of the TRR. Bifenazate + bifenazate-diazene comprised less than approx 2% of the TRR. Approximately 47% of the TRR was unextractable. The kidney was subjected to the same treatment as previously described for the liver, again suggesting that some of the unextractable $^{14}$C was bound to protein.

Table 1. Distribution of $^{14}$C residue and identified metabolites in tissues and milk of a lactating goat dosed orally for 4 days with 21 mg/animal/day of [$^{14}$C] bifenazate, equivalent to 10 ppm in the feed (McClanahan and Bayus, 1999, 96-0064).

| Component | Concentration, mg/kg, expressed as parent | | | | | | | |
|---|---|---|---|---|---|---|---|---|
| | Loin muscle | Leg muscle | Omental fat | Peri-renal fat | Liver | Kidney | Milk, day 3 | Milk, day 4 |
| Total $^{14}$C residue (TRR) | 0.013 | 0.014 | 0.10 | 0.13 | 1.77 | 0.26 | 0.032 | 0.047 |
| Extracted residue | 0.008 | 0.008 | 0.090 | 0.096 | 0.175 | 0.13 | 0.031 | 0.044 |
| Unextractable | 0.005 | 0.005 | 0.021 | 0.013 | 1.54 | 0.12 | 0.002 | 0.002 |
| Bifenazate | 0.001 | nd | 0.061 | 0.066 | 0.011 | 0.003 | | 0.0003 |
| Bifenazate glucuronide | | | | | 0.005 | | | |
| Conjugates, includes 4-hydroxybiphenyl sulphate and glucuronide and bifenazate glucuronide | 0.002 | 0.003 | 0.004 | 0.005 | | | | |
| 4-hydroxybiphenyl | 0.002 | 0.002 | 0.006 | 0.007 | 0.012 | 0.009 | | 0.0008 |
| 4-hydroxybiphenyl glucuronide | | | | | 0.017 | 0.004 | | |
| 4-hydroxybiphenyl sulphate | | | | | 0.005 | 0.029 | 0.012 | 0.019 |
| 4-methoxybiphenyl | | | 0.003 | 0.004 | 0.006 | 0.005 1/ | | 0.0017 |
| Bifenazate-diazene | | | 0.009 | 0.006 | 0.006 | | | 0.0039 |

nd: not detected. Detection limit approx 0.0001 mg/kg.

1/ In kidney, combined 4-methoxybiphenyl + bifenazate-diazene residue = 0.005 mg/kg.

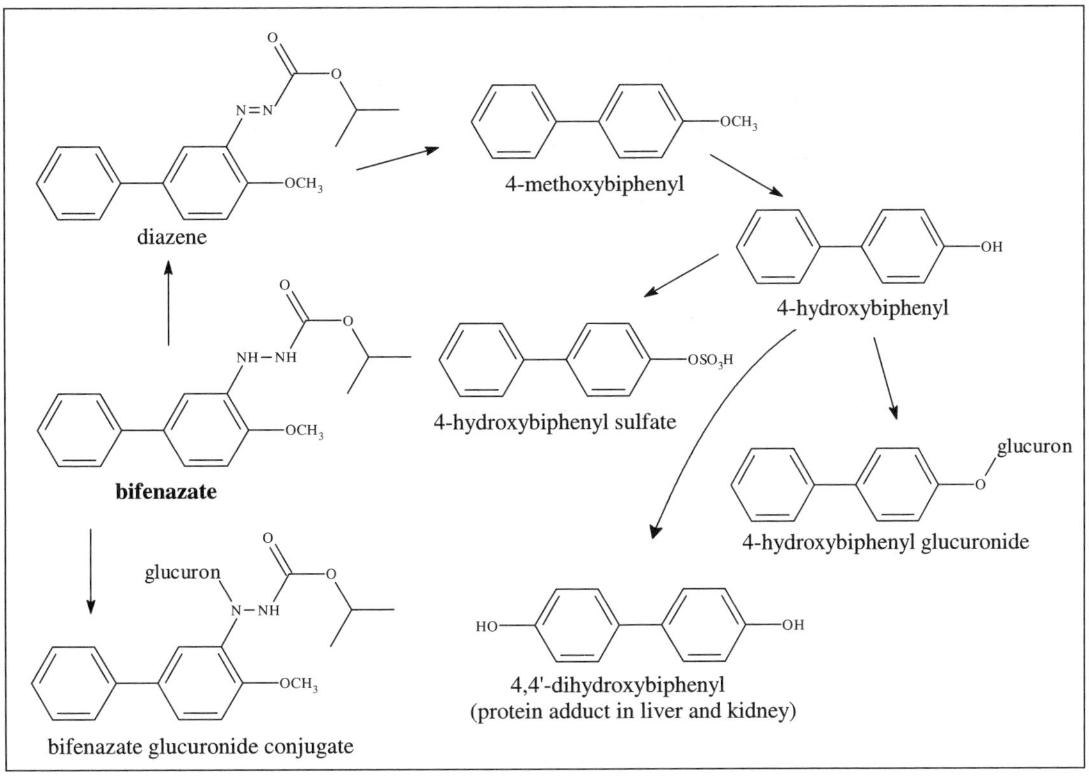

Figure 3. Proposed bifenazate metabolic pathway in lactating goats (McClanahan and Bayus, 1999, 96-0064).

*Laying hens*

A group of white leghorn laying hens (10 birds), approximately 25 weeks old, mean body weight 1.60 kg at study initiation and completion were dosed orally once daily via gelatin capsule for 4 consecutive days with 1.3 mg/bird/day of [$^{14}$C]bifenazate, equivalent to 10 ppm in the feed (McClanahan *et al.*, 1999, 96-0265) for a 127 g/day feed consumption. Eggs were collected twice daily. The birds were slaughtered approximately 8.5 hours after the final dose for tissue collection (breast and thigh muscle, skin and fat and liver). Recovery of administered $^{14}$C was approximately 85%. Samples were stored in a freezer below -10 °C for 64 days before the initial extraction for metabolite identification. See Table 16 for freezer storage stability testing of residues in tissues and eggs of laying hens.

Most of the administered $^{14}$C was present in excreta and cages (83.6%), with 1.38% in the tissues and 0.01% in the eggs. Residues were not detectable (< 0.005 mg/kg) in breast muscle and egg white. The highest $^{14}$C residue was present in liver at 0.61 mg/kg, of which 0.19 mg/kg was extractable. The major identified residues in liver, skin + fat and egg yolk were hydroxybiphenyl, bifenazate-diazene and bifenazate, respectively. The distribution of residues is summarised in Table 2.

Table 2. Distribution of $^{14}$C residue and identified metabolites in tissues and eggs of laying hens dosed orally for 4 days with 1.3 mg/bird/day of [$^{14}$C]bifenazate, equivalent to 10 ppm in the feed (McClanahan *et al.*, 1999, 96-0265).

| Component | | Concentration, mg/kg, expressed as parent | | | | | |
|---|---|---|---|---|---|---|---|
| | | Skin + fat | Liver | Thigh muscle | Breast muscle | Egg white, day 4 | Egg yolk, day 4 |
| Total $^{14}$C residue (TRR) | | 0.048 | 0.61 | 0.006 | < 0.005 | < 0.005 | 0.025 |
| Extracted residue | | 0.025 | 0.19 | 0.002 | | | 0.012 |
| Unextractable | | 0.015 | 0.37 | 0.003 | | | 0.007 |
| Bifenazate | | 0.001 | 0.002 | < 0.0005 | | | 0.005 |
| Conjugates, mostly bifenazate glucuronide | | 0.001 | 0.017 | < 0.0005 | | | < 0.0005 |
| 4-hydroxybiphenyl 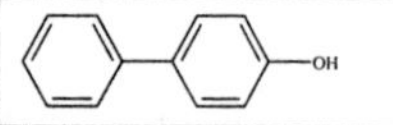 | | 0.001 | 0.013 | < 0.0005 | | | 0.001 |
| 4-methoxybiphenyl 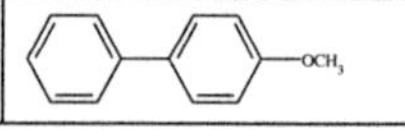 | | 0.005 | 0.001 | < 0.0005 | | | 0.001 |
| Bifenazate-diazene | | 0.008 | 0.002 | < 0.0005 | | | 0.001 |

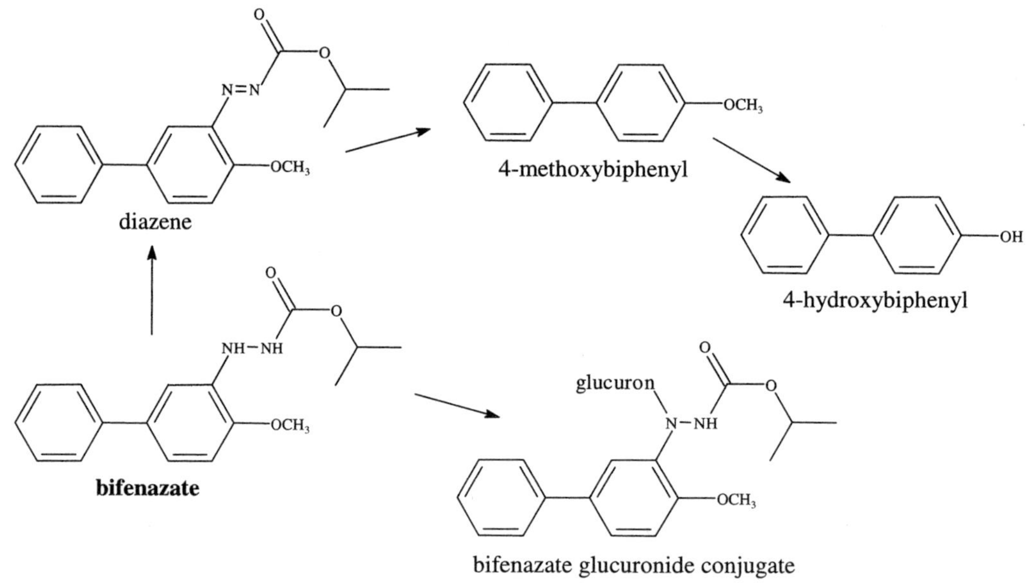

Figure 4. Proposed bifenazate metabolic pathway in laying hens (McClanahan *et al.*, 1999, 96-0265).

## Plant metabolism

The Meeting received plant metabolism studies with bifenazate on oranges, apples, grapes, radish and cotton.

In plants, most of the resultant residue from the use of bifenazate was a surface residue. Parent bifenazate was the major component of the residue at shorter intervals and the major identified component at longer intervals after treatment. Bifenazate-diazene was usually also present, but at much lower levels than parent bifenazate. Very little of the residue translocated to the roots from treated radish foliage.

## Oranges

In an orange metabolism study in California USA, Panthani and Hatzenbeler (1998, 6381-95-0028-EF-001) foliar sprayed Valencia oranges once with [$^{14}$C]bifenazate formulated as a WP (wettable powder) at 0.42 and 2.2 kg ai/ha and harvested mature fruit 43, 184, 274 and 442 days later for analysis. The mature fruits were maintained at refrigeration temperature (approx 5°C) for shipment to the laboratory and were processed immediately upon receipt from the field.

Surface residues in the oranges harvested 43 days after treatment, constituting 78% and 81% of the total residues for whole fruit, were removed with an acetonitrile rinse. Parent bifenazate was the major component of the surface residue (Table 3). Bifenazate-diazene was also identified in the surface rinse together with minor metabolites bifenazate-diazene oxide, 4-methoxybiphenyl and 3-hydroxy-4-methoxybiphenyl. The rinsed oranges were peeled and the peeled fruits were homogenized and centrifuged to produce pulp and juice. The TRR distribution was mostly into the peel (20% and 18% of the total residues for whole fruit) with very minor amounts into the pulp (0.9% and 0.9%) and juice (1.2% and 0.8%).

Total residue levels (TRR) in the oranges declined substantially for the fruit harvested at 184, 274 and 442 days after treatment (Table 3). The percentage of the residue on the fruit surface also declined (8% and 12% of the total residues for whole fruit at day 442). Bifenazate and bifenazate-diazene were identifiable components of the residue even at the longer intervals after treatment. At the longer intervals, polar materials constituted higher percentages of the residue. Most of the unextractable residue in peel was released by acid hydrolysis, but none of the individual components represented a residue exceeding 0.005 mg/kg.

In summary, for oranges most of the residues were surface residues, the majority of which was parent bifenazate. Very little of the residue appeared in orange pulp or juice.

Table 3. Residues in and on oranges treated with WP formulated [14C]bifenazate, at the equivalent of 0.42 and 2.2 kg ai/ha and harvested 43, 184, 274 and 442 days later (Panthani and Hatzenbeler, 1998, 6381-95-0028-EF-001).

| Samples and metabolites | Concentration, mg/kg of [14]C expressed as parent | | | |
|---|---|---|---|---|
| | Treatment 0.42 kg ai/ha | | Treatment 2.2 kg ai/ha | |
| Whole fruit, 43 days, TRR | 0.353 | | 1.47 | |
| Unextractable TRR | | 0.024 | | 0.082 |
| Extractable TRR | | 0.330 | | 1.38 |
| bifenazate | | 0.266 | | 1.16 |
| bifenazate-diazene | | 0.026 | | 0.088 |
| bifenazate-diazene oxide | | 0.001 | | 0.004 |
| 4-methoxybiphenyl | | 0.001 | | 0.003 |
| 3-hydroxy-4-methoxybiphenyl | | 0.002 | | 0.012 |
| Surface wash, 43 days, TRR | 0.275 | | 1.18 | |
| bifenazate | | 0.259 | | 1.12 |
| bifenazate-diazene | | 0.016 | | 0.060 |
| bifenazate-diazene oxide | | < 0.001 | | < 0.001 |
| 4-methoxybiphenyl | | < 0.001 | | < 0.001 |
| 3-hydroxy-4-methoxybiphenyl | | < 0.001 | | < 0.001 |
| Peel, 43 days, TRR | 0.070 | | 0.259 | |
| Unextractable of peel, 43 days, TRR | | 0.022 | | 0.075 |
| Extractable of peel, 43 days, TRR | | 0.048 | | 0.185 |
| bifenazate | | 0.004 | | 0.037 |
| bifenazate-diazene | | 0.010 | | 0.028 |
| bifenazate-diazene oxide | | 0.001 | | 0.004 |
| 4-methoxybiphenyl | | 0.001 | | 0.003 |
| 3-hydroxy-4-methoxybiphenyl | | 0.002 | | 0.012 |

| Samples and metabolites | Concentration, mg/kg of $^{14}$C expressed as parent | |
|---|---|---|
| | Treatment 0.42 kg ai/ha | Treatment 2.2 kg ai/ha |
| Pulp, 43 days, TRR | 0.003 | 0.013 |
| Unextractable of pulp, 43 days, TRR | 0.001 | 0.004 |
| Extractable of pulp, 43 days, TRR | 0.003 | 0.009 |
| bifenazate | 0.001 | 0.001 |
| bifenazate-diazene | < 0.001 | < 0.001 |
| bifenazate-diazene oxide | < 0.001 | < 0.001 |
| 4-methoxybiphenyl | < 0.001 | < 0.001 |
| 3-hydroxy-4-methoxybiphenyl | < 0.001 | < 0.001 |
| Juice, 43 days, TRR | 0.005 | 0.012 |
| bifenazate | 0.003 | 0.001 |
| bifenazate-diazene | < 0.001 | < 0.001 |
| bifenazate-diazene oxide | < 0.001 | < 0.001 |
| 4-methoxybiphenyl | < 0.001 | < 0.001 |
| 3-hydroxy-4-methoxybiphenyl | < 0.001 | < 0.001 |
| Whole fruit, 184 days, TRR | 0.096 | |
| bifenazate | 0.035 | |
| bifenazate-diazene | 0.010 | |
| Surface wash, 184 days, TRR | 0.042 | |
| Peel, 184 days, TRR | 0.050 | |
| Pulp, 184 days, TRR | 0.002 | |
| Juice, 184 days, TRR | 0.003 | |
| Whole fruit, 274 days, TRR | 0.095 | 0.081 |
| bifenazate | 0.015 | |
| bifenazate-diazene | 0.006 | |
| Surface wash, 274 days, TRR | 0.021 | 0.020 |
| Peel, 274 days, TRR | 0.068 | 0.045 |
| Pulp, 274 days, TRR | 0.004 | 0.010 |
| Juice, 274 days, TRR | 0.003 | 0.005 |
| Whole fruit, 442 days, TRR | 0.013 | 0.032 |
| bifenazate | 0.001 | |
| bifenazate-diazene | 0.001 | |
| Surface wash, 442 days, TRR | 0.001 | 0.004 |
| Peel, 442 days, TRR | 0.009 | 0.020 |
| Pulp, 442 days, TRR | 0.002 | 0.005 |
| Juice, 442 days, TRR | 0.001 | 0.003 |

## Apples

In an apple metabolism study in California USA, Panthani and Hatzenbeler (1998, 6850-96-0101-EF-001) foliar sprayed Granny Smith apples once with [$^{14}$C]bifenazate formulated as a WP (wettable powder) at 0.42 and 2.2 kg ai/ha and harvested the fruit at maturity 101 days later for analysis. The mature fruits were processed immediately after harvest.

Surface residues, constituting 55% and 66% of the total residues for whole apples, were removed with an acetonitrile rinse. Parent bifenazate was the major component of the surface residue (Table 4). Bifenazate-diazene was also identified in the surface rinse together with minor metabolites bifenazate-diazene oxide and 4-methoxybiphenyl. The rinsed apples were homogenized and centrifuged to produce pomace and juice. The TRR distribution was mostly into the pomace (35% and 26% of the total residues for whole apples) with lesser amounts into the juice (10% and 9%).

Very little of the residue was identified in pomace and juice components. The juice contained mainly polar fractions. None of the extractable components from the pomace or juice, apart from bifenazate and bifenazate-diazene, corresponded to available metabolite standards. Acid and base hydrolysis did not release any identifiable components.

In summary, for apples most of the residues were surface residues, the majority of which was parent bifenazate. Residues penetrating the fruit were mostly metabolized to polar compounds.

Table 4. Residues in and on apples treated with WP formulated [$^{14}$C]bifenazate, at the equivalent of 0.42 and 2.2 kg ai/ha and harvested 101 days later (Panthani and Hatzenbeler, 1998, 6850-96-0101-EF-001).

| Samples and metabolites | Concentration, mg/kg of $^{14}$C expressed as parent | |
| --- | --- | --- |
| | Treatment 0.42 kg ai/ha | Treatment 2.2 kg ai/ha |
| Whole fruit TRR | 0.088 | 0.37 |
| Unextractable TRR | 0.015 | 0.053 |
| Extractable TRR | 0.071 | 0.32 |
| bifenazate | 0.030 | 0.18 |
| bifenazate-diazene | 0.005 | 0.017 |
| bifenazate-diazene oxide | 0.001 | 0.003 |
| 4-methoxybiphenyl | < 0.001 | 0.001 |
| bifenazate-carbamate 1/ | | 0.001 |
| Surface wash TRR | 0.049 | 0.244 |
| bifenazate | 0.029 | 0.17 |
| bifenazate-diazene | 0.004 | 0.015 |
| bifenazate-diazene oxide | 0.001 | 0.003 |
| 4-methoxybiphenyl | < 0.001 | 0.001 |
| bifenazate-carbamate | | < 0.001 |
| Pomace TRR | 0.031 | 0.096 |
| Unextractable of pomace TRR | 0.015 | 0.053 |
| Extractable of pomace TRR | 0.014 | 0.038 |
| bifenazate | 0.001 | 0.001 |
| bifenazate-diazene | 0.001 | 0.001 |
| bifenazate-diazene oxide | < 0.001 | < 0.001 |
| 4-methoxybiphenyl | < 0.001 | < 0.001 |
| bifenazate-carbamate | | < 0.001 |
| Juice TRR | 0.009 | 0.033 |
| bifenazate | < 0.001 | < 0.001 |
| bifenazate-diazene | < 0.001 | < 0.001 |
| bifenazate-diazene oxide | < 0.001 | < 0.001 |
| 4-methoxybiphenyl | < 0.001 | < 0.001 |
| bifenazate-carbamate | | < 0.001 |

1/ Bifenazate-carbamate was a very minor component, detected in only the high-rate treatment extractable TRR from the whole fruit. The sponsor expressed the opinion that it may have been a small impurity in the labelled bifenazate.

*Grapes*

In a grape metabolism study in California USA, McManus and DeMatteo (2001, 2000-097) foliar sprayed Thompson seedless grapes with [$^{14}$C]bifenazate formulated as a WP (wettable powder) at 0.56 and 1.1 kg ai/ha and harvested the crop at maturity 30 days later for analysis. Foliage and fruit were also taken on day 0 after the single treatment.

Harvested samples were stored at temperatures below -10°C until processing within 30 days of harvest.

Grapes were thawed to room temperature and washed with methanol. The rinsed grapes were mixed with dry ice and homogenized to a fine powder. After the dry ice sublimed, the ground samples were centrifuged to produce juice and pomace. Pomace was rinsed with methanol, which was added to the juice. Almost the entire residue was in the surface rinsings (Table 5). HPLC analysis of the surface rinsings revealed only two peaks corresponding to bifenazate and bifenazate-diazene in the chromatogram and together accounting for 98% and 95% of TRR for the 0.56 and 1.1 kg ai/ha treatments respectively. Only trace amounts of $^{14}$C were released from the pomace by acid, alkali and enzymic hydrolysis. HPLC analysis of grape juice revealed the presence of bifenazate and bifenazate-diazene and other trace metabolites.

The residues on grapes were mostly surface residues consisting essentially of bifenazate and bifenazate-diazene.

Table 5. Residues in and on grapes treated with WP formulated [$^{14}$C]bifenazate, at the equivalent of 0.56 and 1.1 kg ai/ha (McManus and DeMatteo, 2001, 2000-097).

| Samples | Concentration, mg/kg of $^{14}$C expressed as parent | |
|---|---|---|
| | Treatment 0.56 kg ai/ha | Treatment 1.1 kg ai/ha |
| Unwashed grapes TRR | 3.5 | 12.0 |
| Surface wash TRR | 3.4 | 11.7 |
| Aqueous juice TRR | 0.075 | 0.11 |
| Pomace TRR | 0.10 | 0.24 |

*Radish*

In a plant metabolism study in USA, Charlton and Tecle (2002, 2001-147) sprayed the foliage of radish plants (variety French Breakfast) with [$^{14}$C]bifenazate formulated as a WS (water-dispersible powder) at 1.1 and 2.2 kg ai/ha and harvested the crop 7 days later for analysis. Harvested samples were stored in freezers or refrigerators and sample processing and analysis began within 32 days of harvest. Most of the $^{14}$C remained on the foliage, with very little reaching the roots (Table 6). The nature of the residue in the roots was not investigated further because the levels were very low.

More than half of the residue associated with the foliage was removed with a surface wash of 50% acetonitrile in water. Parent bifenazate was the major component of the surface residue.

Washed and unwashed tops were extracted by homogenization with acetonitrile+methanol+water (1+1+1, v+v+v) to examine the nature of the residue by HPLC techniques. Bifenazate and bifenazate-diazene were the major identified components of the extractable residues. Another metabolite that appeared in small quantities in the extracts was identified as a ring-hydroxylated diazene. The position of the hydroxyl group on the biphenyl ring was not determined. Another minor metabolite was also observed, but its structure was not determined.

Table 6. Residues in and on radish plants treated with WS formulated [$^{14}$C]bifenazate, at the equivalent of 1.1 and 2.2 kg ai/ha (Charlton and Tecle, 2002, 2001-147).

| Samples and metabolites | Concentration, mg/kg of $^{14}$C expressed as parent | |
|---|---|---|
| | Treatment 1.1 kg ai/ha | Treatment 2.2 kg ai/ha |
| Unwashed tops TRR | 13 | 21 |
| Unextractable of unwashed tops TRR | 0.74 | 1.7 |
| Extract of unwashed tops TRR | 14 | 20 |
| bifenazate | 1.7 | 14 |

| Samples and metabolites | Concentration, mg/kg of $^{14}$C expressed as parent | |
|---|---|---|
| | Treatment 1.1 kg ai/ha | Treatment 2.2 kg ai/ha |
| bifenazate-diazene | 5.7 | 1.0 |
| ring-hydroxylated bifenazate-diazene 1/ | 1.2 | 1.3 |
| Surface wash TRR | 7.9 | 17 |
| bifenazate | 7.3 | 15 |
| bifenazate-diazene | 0.2 | 1.3 |
| ring-hydroxylated bifenazate-diazene | nd | nd |
| Washed tops TRR | 5.7 | 10 |
| Unextractable of washed tops TRR | 0.39 | 1.4 |
| Extract of washed tops TRR | 4.8 | 7.7 |
| bifenazate | 0.9 | 4.0 |
| bifenazate-diazene | 1.7 | 0.8 |
| ring-hydroxylated bifenazate-diazene | 0.4 | 0.9 |
| Roots TRR | 0.0023 | 0.0043 |

nd: not detected.

ring-hydroxylated bifenazate-diazene

1/   $C_{17}H_{18}N_2O_4$

## Cotton

In a cotton metabolism study in California USA, Panthani and Hatzenbeler (2000, 7137-97-0024-EF-001) foliar sprayed cotton plants (variety Maxxa), at late bloom to early boll set, once with [$^{14}$C]bifenazate formulated as a WP (wettable powder) at 0.56 and 2.2 kg ai/ha and harvested fuzzy seed, lint and gin trash 112 days later for analysis.

For each plant, cotton seed was removed by hand from the mature open bolls. Each plant was then stripped of all leaves, petioles, calyx and unopened immature bolls to represent the gin trash. The cotton seed was mechanically ginned to produce the lint and fuzzy seed. The gin trash was dried overnight under heat lamps and then portions were ground to a fine powder in a coffee grinder and the samples were placed in a freezer. Fuzzy seed and gin trash were shipped to the laboratory 3 and 10 days after harvest respectively. All samples were stored in a freezer below -5°C and were extracted and analysed within 30 days of harvest.

Bifenazate was extensively metabolised in cotton seed (Table 7). The majority of the hexane extractable residues were shown to be $^{14}$C incorporated into triglycerides, while the aqueous acetonitrile extracts contained mainly polar compounds. Bifenazate, bifenazate-diazene, bifenazate-diazene oxide and 4-methoxybiphenyl were present at trace levels (each < 0.001 mg/kg). Acid and base catalysed hydrolyses did not release any recognized compound from possible conjugation.

A high proportion (77–82%) of the gin trash residue was extractable, with bifenazate approximately 50% of the extractable residue and bifenazate-diazene, bifenazate-diazene oxide and 4-methoxybiphenyl identified as minor residue components. Enzyme, acid and base catalysed hydrolysis of the unextractable material suggested that the $^{14}$C was incorporated and covalently bound into natural compounds.

In summary, the residues in cotton seed from the use of bifenazate were extensively metabolized, with identified primary metabolites at very low levels. Parent bifenazate was the major identified component of the residue in cotton trash.

Table 7. Residues in cotton seed and gin trash from cotton plants treated with WP formulated [$^{14}$C]bifenazate, at the equivalent of 0.56 and 2.2 kg ai/ha and harvested 112 days later (Panthani and Hatzenbeler, 2000, 7137-97-0024-EF-001).

| Samples and metabolites | Concentration, mg/kg of $^{14}$C expressed as parent | |
|---|---|---|
| | Treatment 0.56 kg ai/ha | Treatment 2.2 kg ai/ha |
| Cotton seed, TRR | 0.075 | 0.125 |
| Unextractable of cotton seed, TRR | 0.049 | 0.079 |
| Extractable of cotton seed, TRR | 0.026 | 0.046 |
| bifenazate | < 0.001 | < 0.001 |
| bifenazate-diazene | < 0.001 | < 0.001 |
| bifenazate-diazene oxide/4-methoxybiphenyl | < 0.001 | < 0.001 |
| Gin trash, TRR | 0.410 | 0.838 |
| Unextractable of gin trash, TRR | 0.150 | 0.288 |
| Extractable of gin trash, TRR | 0.317 | 0.685 |
| bifenazate | 0.154 | 0.338 |
| bifenazate-diazene | 0.018 | 0.051 |
| bifenazate-diazene oxide/4-methoxybiphenyl | 0.006 | |
| 4-methoxybiphenyl | | 0.009 |
| bifenazate-diazene oxide | | 0.004 |

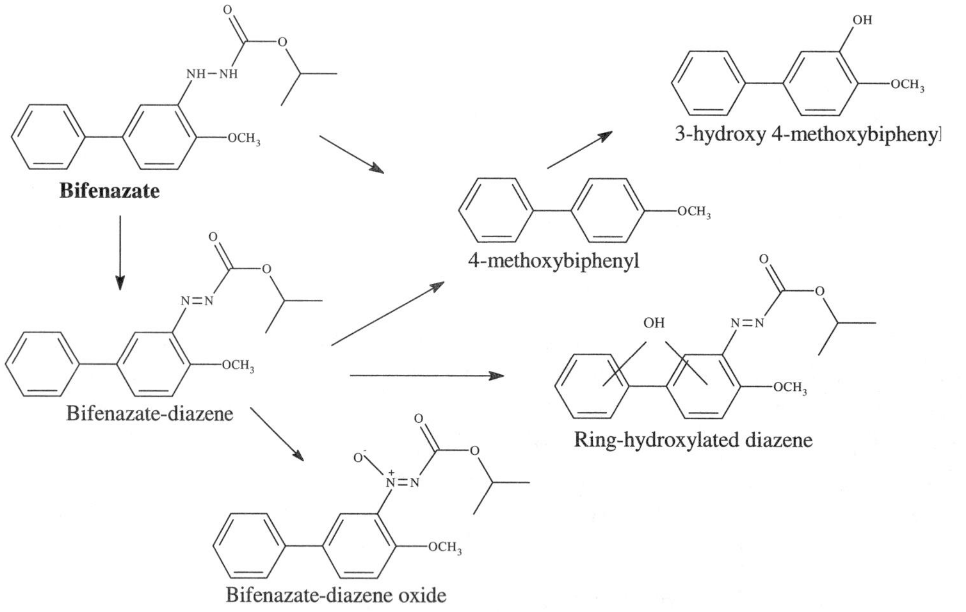

Figure 5. Proposed bifenazate metabolic pathway in crops.

## *Environmental fate in soil*

The Meeting received information on crop rotational studies for bifenazate. Information on soil metabolism and field dissipation was not required because no bifenazate uses as seed treatments or on root crops, if there are such uses, were provided for evaluation.

## *Crop rotation studies*

Information on the fate of radiolabelled bifenazate in a confined crop rotational study was made available to the meeting.

In a confined rotational crop study in USA (Findak, 2000, 6507-95-0124-EF-001) soil (Ohio loamy sand, 1.1% organic matter, 77% sand, 18% silt, 5.2% clay, pH 6.5) was treated directly with $^{14}C$ labelled bifenazate at a rate equivalent to 0.56 kg ai/ha and allowed to age under greenhouse conditions prior to sowing of the rotational crops. Crops of carrots, lettuce and wheat were sown into the treated soil in pots at intervals of 30, 125 and 360 days after treatment.

Immature lettuce plants were sampled at the 4-5 leaf stage. Immature carrot plants were sampled when carrots were approximately 6 mm in diameter. Wheat forage samples were taken approximately 5 weeks after sowing. The remainder of the crops were grown to maturity, subsequently harvested and analysed for $^{14}C$ (TRR) content (Table 8). A parallel treatment at the exaggerated rate of 5.6 kg ai/ha was used to assist in the identification of potential metabolites. Precautions were taken to prevent $^{14}CO_2$ from being released into the greenhouse from the treated soils and crops.

Samples were further examined by extraction (acetonitrile and acetonitrile/water). Where extractable residues exceeded 0.01 mg/kg they were examined by HPLC analysis but no parent compound or reference metabolite was observed. Low levels of unidentifiable components were observed as broad unresolved areas in the chromatograms.

The unextractable residual solids from the wheat straw and fractions from the wheat forage were subjected to acid, base and enzyme hydrolysis, but no parent bifenazate or recognizable metabolite was released.

Table 8. Confined rotational crop studies with $^{14}C$ labelled bifenazate. Soil was treated with $^{14}C$ bifenazate at a rate equivalent to 0.56 kg ai/ha (Findak, 2000, 6507-95-0124-EF-001).

| Application country, year, ref. | Rotational crop (variety) | TSI 1/ days | THI 2/ days | Sample | TRR as bifenazate mg/kg | Extractable, % of TRR | Residues, mg/kg |
|---|---|---|---|---|---|---|---|
| Bare soil, USA, 1996, 6507-95-0124-EF-001 | | | | | | | |
| | carrot (Nantes Coreless) | 30 | 91 | roots | 0.033 | 36% | 3/ |
| | | 30 | 148 | roots | 0.007 | na | na |
| | lettuce (Grand Rapids) | 30 | 81 | leaves | 0.015 | 33% | 3/ |
| | | 30 | 163 | leaves | 0.014 | 29% | 3/ |
| | wheat (Clark) | 30 | 65 | forage | 0.038 | 53% | 3/ |
| | | 30 | 192 | straw | 0.12 | 38% | 3/ |
| | | 30 | 192 | chaff | 0.031 | 13% | na |
| | | 30 | 192 | grain | 0.016 | 25% | na |
| | carrot (Nantes Coreless) | 125 | 205 | roots | 0.010 | 40% | na |
| | | 125 | 252 | roots | 0.006 | na | na |
| | lettuce (Grand Rapids) | 125 | 169 | leaves | 0.013 | 23% | na |
| | | 125 | 203 | leaves | 0.005 | na | na |
| | wheat (Clark) | 125 | 169 | forage | 0.020 | 25% | 3/ |
| | | 125 | 309 | straw | 0.051 | 24% | 3/ |
| | | 125 | 309 | chaff | 0.025 | 16% | na |
| | | 125 | 309 | grain | 0.019 | 16% | na |

| Application country, year, ref. | Rotational crop (variety) | TSI 1/ days | THI 2/ days | Sample | TRR as bifenazate mg/kg | Extractable, % of TRR | Residues, mg/kg |
|---|---|---|---|---|---|---|---|
| | wheat (Clark) | 360 | 391 | forage | 0.018 | 17% | na |
| | | 360 | 498 | straw | 0.033 | 29% | na |
| | | 360 | 498 | chaff | 0.015 | 53% | na |
| | | 360 | 498 | grain | 0.011 | 55% | na |

na: not analysed.

1/ TSI: interval between treatment on soil and sowing of rotation crop, days.

2/ THI: interval between treatment on soil and harvest of rotation crop (or sampling of soil), days.

3/ HPLC analysis did not detect (LOQ 0.01 mg/kg) bifenazate or identifiable metabolites in any raw agricultural commodity.

## METHODS OF RESIDUE ANALYSIS

### *Analytical methods*

The Meeting received descriptions and validation data for analytical methods for residues of bifenazate in raw agricultural commodities, processed commodities, feed commodities, animal tissues, milk and eggs.

Because bifenazate and bifenazate-diazene are readily interconverted by mild oxidation and reduction conditions, the measured residue includes both compounds. The analytical methods use a mild reduction with ascorbic acid to convert the bifenazate-diazene residue to bifenazate before the measurement step. Residues are typically extracted with acetonitrile and water acidified with acetic acid. After a partition clean-up and reduction with ascorbic acid, the residue is analysed by HPLC with coulometer detection.

Jablonski (1998, 6998-97-0237-CR-001) developed an HPLC method with coulometric detection for analysis of the combined residues of bifenazate and bifenazate-diazene in apples and citrus. The oxidative coulometric detection system is quite selective. Substituted hydrazines such as bifenazate are oxidised at 200 mV, but most sample matrix components are not.

Jablonski (1998, 6998-98-0051-CR-001) analysed apples and oranges from [$^{14}$C]bifenazate crop metabolism studies by the HPLC-coulometer method (6998-97-0237-CR-001) and by an HPLC radiometric method. The sample extraction procedure for the HPLC radiometric analysis was similar to the extraction procedure of method 6998-97-0237-CR-001. The HPLC-coulometer results were approximately 60% of those from the radiometric method (Table 9).

Table 9. Comparison of analyses for bifenazate and bifenazate-diazene on samples from $^{14}$C crop metabolism studies by an HPLC-coulometer method (6998-97-0237-CR-001) and an HPLC radiometric method (Jablonski, 1998, 6998-98-0051-CR-001).

| Sample | Residues of bifenazate + bifenazate-diazene (mg/kg) by method 6998-97-0237-CR-001 | Residues of bifenazate + bifenazate-diazene (mg/kg) by HPLC radiometric method |
|---|---|---|
| Apple | 0.107 | 0.186 |
| Apple | 0.117 | 0.170 |
| Orange | 0.178 | 0.342 |
| Orange | 0.196 | 0.327 |

*Apples and citrus* (Jablonski, 1998, 6998-97-0237-CR-001)

Analyte:     bifenazate and bifenazate-diazene     HPLC-coulometer    Method 6998-97-0237-CR-001

LOQ:     0.01 mg/kg

Description     Residues are extracted twice from homogenized matrix with acetonitrile + acetic acid. An aliquot of filtered extract is partitioned with aqueous sodium sulphate and dichloromethane. The organic phase, which contains the residues, is dried and evaporated to near dryness. The residue is taken up in HPLC mobile phase (5% acetonitrile and 95% sodium acetate buffer 50 mM pH 4) containing ascorbic acid. The ascorbic acid reduces bifenazate-diazene to bifenazate. The oxidative coulometric detection system (150-200 mV) after the reversed phase HPLC measures the combined residue as bifenazate. Without the ascorbic acid, bifenazate may readily oxidise to bifenazate-diazene during the analysis. The coulometric detector provides a high degree of selectivity with little background interference observed. If the oil from some citrus samples causes interferences, a hexane wash should be introduced at an early stage of the cleanup.

*Fruit and fruit matrices: peaches, plums, grapes, grape juice, raisins, prunes* (Wiedmann, 1999, RP-98018)

Analyte:     bifenazate and bifenazate-diazene     HPLC-coulometer    Method 7543-98-0072-CR-002

LOQ:     0.01 mg/kg

Description     Residues are extracted twice from chopped matrix by blending with acetonitrile + acetic acid. Grape juice is mixed with the extraction solvent. An aliquot of filtered extract is partitioned with aqueous sodium sulphate and dichloromethane. The organic phase, which contains the residues, is dried and evaporated to near dryness. The residue is taken up in HPLC mobile phase (5% acetonitrile and 95% sodium acetate buffer 50 mM pH 4) containing ascorbic acid. The ascorbic acid reduces bifenazate-diazene to bifenazate. The oxidative coulometric detection system (200 mV) after the reversed phase HPLC measures the combined residue as bifenazate.

*Peaches, raisins and almonds* (Wood, 2003, RP-02009)

Analyte:     bifenazate and bifenazate-diazene     LC-MS-MS     Method NCL ME 245

LOQ:     0.01 mg/kg

Description     Homogenized matrix is extracted with 0.25% acetic acid in 70:30 acetonitrile:water. A portion of the extract is mixed with sodium chloride solution and the residues are partitioned into dichloromethane. The dichloromethane extract is evaporated to a small volume, then mixed with reducing solution (0.25% ascorbic acid in water:acetonitrile) and incubated in a water bath at 50 °C for 1 hour. A portion of the extract is then filtered through a syringe-tip filter and analysed by LC-MS-MS. The $[M+H]^+$ ion is used as the precursor ion for bifenazate. Transitions 301.1/198.1 (for quantification) and 301.1/170.1 are observed. For almonds, an extra step with a hexane wash is introduced after the extraction to remove the almond oil. The ascorbic acid reduction converts bifenazate-diazene residues to bifenazate so the method measures bifenazate and bifenazate-diazene as bifenazate.

The method is suitable as a confirmatory method for bifenazate residues.

*Bovine liver, kidney, milk and fat* (Wood, 2003, 2003-016)

Analyte:     bifenazate and bifenazate-diazene     LC-MS-MS     Method NCL ME 259

LOQ:     0.01 mg/kg

Description     Homogenized bovine liver, kidney or milk is extracted with 0.25% acetic acid in 70:30 acetonitrile:water. A portion of the extract is diluted with water and subjected to a solid-phase-extraction column cleanup and a mild reduction with ascorbic acid ready for analysis by LC-MS-MS. The $[M+H]^+$ ion is used as the precursor ion for bifenazate. Transitions 301.1/198.1 (for quantification) and 301.1/170.1 are observed.

Homogenized bovine fat is extracted with acetonitrile and excess lipid is removed by centrifugation, cooling and filtration. The extract is subjected to the ascorbic acid reduction and the remainder of the procedure follows that of the other tissues.

The method is suitable as a confirmatory method for bifenazate residues.

*Bovine tissues and milk* (Jablonski, 1999, 7473)

Analyte:            bifenazate and bifenazate-diazene            HPLC-coulometer    Method 7473-98-0115-CR-001
LOQ:               0.01 mg/kg
Description        Extraction, cleanup and ascorbic acid reduction procedures are similar to those described in Method NCL ME 259. The measurement step relies on HPLC with oxidative coulometric detection.

Recovery data from the internal and independent laboratory validation (ILV) testing are summarised in Table 10.

Table 10. Analytical recoveries for spiked bifenazate and diazene in various substrates. Diazene means bifenazate-diazene.

| Commodity | Spiked analyte | Spike conc, mg/kg | n | Mean recov% | Range recov% | Method | Ref |
|---|---|---|---|---|---|---|---|
| almond | bifenazate | 0.01-1.0 | 9 | 91 | 83-100 | NCL ME 245 | RP-02009 |
| almond | diazene | 0.01-1.0 | 9 | 99 | 92-112 | NCL ME 245 | RP-02009 |
| almond hull | bifenazate | 0.01-10 | 9 | 83 | 72-97% | 6998-97-0237-CR-001 | GRL-11866 |
| almond hull | diazene | 0.01-10 | 9 | 92 | 80-118% | 6998-97-0237-CR-001 | GRL-11866 |
| almond kernel | bifenazate | 0.01-0.5 | 9 | 92 | 87-95% | 6998-97-0237-CR-001 | GRL-11866 |
| almond kernel | diazene | 0.01-0.5 | 9 | 83 | 76-91% | 6998-97-0237-CR-001 | GRL-11866 |
| apple | bifenazate | 0.01 | 5 | 81 | 75-88 | 6998-97-0237-CR-001 | 6998-98-0051-CR-001 |
| apple | bifenazate | 0.01-0.10 | 14 | 91 | 71-114% | 6998-97-0237-CR-001 | 6998-97-0237-CR-001 |
| apple | bifenazate | 0.01-0.10 | 6 | 82 | 75-89% [2] | 6998-97-0237-CR-001 | 99214    ILV |
| apple | bifenazate | 0.1 | 5 | 100 | 96-103 | 6998-97-0237-CR-001 | 6998-98-0051-CR-001 |
| apple | diazene | 0.01 | 5 | 79 | 74-82 | 6998-97-0237-CR-001 | 6998-98-0051-CR-001 |
| apple | diazene | 0.01-0.10 | 14 | 81 | 73-102% | 6998-97-0237-CR-001 | 6998-97-0237-CR-001 |
| apple | diazene | 0.01-0.10 | 6 | 87 | 82-91% | 6998-97-0237-CR-001 | 99214    ILV |
| apple | diazene | 0.1 | 5 | 91 | 83-96 | 6998-97-0237-CR-001 | 6998-98-0051-CR-001 |
| apricot | bifenazate | 0.01-0.1 | 9 | 84 | 75-92 | 6998-97-0237-CR-001 | GRL-11929 |
| apricot | diazene | 0.01-0.1 | 9 | 85 | 77-93 | 6998-97-0237-CR-001 | GRL-11929 |
| bovine fat | bifenazate | 0.01 | 5 | 77 | 69-84 | 7473-98-0115-CR-001 | 7473 |
| bovine fat | bifenazate | 0.1 | 5 | 94 | 86-103 | 7473-98-0115-CR-001 | 7473 |
| bovine fat | diazene | 0.01 | 5 | 74 | 69-85 | 7473-98-0115-CR-001 | 7473 |
| bovine fat | diazene | 0.2 | 5 | 93 | 90-99 | 7473-98-0115-CR-001 | 7473 |
| bovine fat | bifenazate | 0.01, 0.1 | 10 | 103 | 97-110 | NCL ME 259 | 2003-016 |
| bovine fat | diazene | 0.01, 0.1 | 10 | 92 | 76-106 | NCL ME 259 | 2003-016 |
| bovine kidney | bifenazate | 0.01 | 5 | 93 | 83-99 | 7473-98-0115-CR-001 | 7473 |
| bovine kidney | bifenazate | 0.1 | 5 | 105 | 101-107 | 7473-98-0115-CR-001 | 7473 |
| bovine kidney | diazene | 0.01 | 5 | 77 | 70-81 | 7473-98-0115-CR-001 | 7473 |
| bovine kidney | diazene | 0.1 | 5 | 83 | 78-87 | 7473-98-0115-CR-001 | 7473 |
| bovine kidney | bifenazate | 0.01, 0.1 | 10 | 90 | 79-102 | NCL ME 259 | 2003-016 |
| bovine kidney | diazene | 0.01, 0.1 | 10 | 85 | 80-88 | NCL ME 259 | 2003-016 |
| bovine liver | bifenazate | 0.01 | 5 | 79 | 78-80 | 7473-98-0115-CR-001 | 7473 |
| bovine liver | bifenazate | 0.1 | 5 | 96 | 91-100 | 7473-98-0115-CR-001 | 7473 |
| bovine liver | diazene | 0.01 | 5 | 69 | 56-75 | 7473-98-0115-CR-001 | 7473 |
| bovine liver | diazene | 0.1 | 5 | 83 | 78-86 | 7473-98-0115-CR-001 | 7473 |
| bovine liver | bifenazate | 0.01, 0.1 | 10 | 106 | 100-110 | NCL ME 259 | 2003-016 |
| bovine liver | diazene | 0.01, 0.1 | 10 | 84 | 79-90 | NCL ME 259 | 2003-016 |
| bovine milk | bifenazate | 0.01, 0.1 | 10 | 95 | 87-103 | 7473-98-0115-CR-001 | 7473 |
| bovine milk | diazene | 0.01 | 5 | 77 | 69-89 | 7473-98-0115-CR-001 | 7473 |
| bovine milk | diazene | 0.1 | 5 | 99 | 90-106 | 7473-98-0115-CR-001 | 7473 |
| bovine milk | bifenazate | 0.01, 0.1 | 10 | 96 | 93-99 | NCL ME 259 | 2003-016 |
| bovine milk | diazene | 0.01, 0.1 | 10 | 77 | 72-99 | NCL ME 259 | 2003-016 |
| bovine muscle | bifenazate | 0.01 | 5 | 80 | 73-83 | 7473-98-0115-CR-001 | 7473 |
| bovine muscle | bifenazate | 0.1 | 5 | 116 | 102-121 | 7473-98-0115-CR-001 | 7473 |
| bovine muscle | diazene | 0.01 | 5 | 75 | 67-83 | 7473-98-0115-CR-001 | 7473 |
| bovine muscle | diazene | 0.1 | 5 | 104 | 99-109 | 7473-98-0115-CR-001 | 7473 |
| cantaloupe | bifenazate | 0.01-0.5 | 9 | 98 | 82-118% | 6998-97-0237-CR-001 | GRL-11662 |
| cantaloupe | diazene | 0.01-0.5 | 9 | 83 | 70-97% | 6998-97-0237-CR-001 | GRL-11662 |

---

[2] A recovery of 131% for a 0.01 mg/kg spike was reported as an outlier.

| Commodity | Spiked analyte | Spike conc, mg/kg | n | Mean recov% | Range recov% | Method | Ref |
|---|---|---|---|---|---|---|---|
| cucumber | bifenazate | 0.01 | 2 | 112 | 111, 113 | 6998-97-0237-CR-001 | GRL-11670 |
| cucumber | bifenazate | 0.1, 1.0 | 6 | 76 | 71-79 | 6998-97-0237-CR-001 | GRL-11670 |
| cucumber | diazene | 0.01 | 3 | 77 | 69-87 | 6998-97-0237-CR-001 | GRL-11670 |
| cucumber | diazene | 0.1, 1.0 | 6 | 97 | 83-117 | 6998-97-0237-CR-001 | GRL-11670 |
| grape juice | bifenazate | 0.01-1.0 | 16 | 95 | 82-105 | 7543-98-0072-CR-002 | RP-98018 |
| grape juice | diazene | 0.01-1.0 | 16 | 89 | 79-93 | 7543-98-0072-CR-002 | RP-98018 |
| grapes | bifenazate | 0.01-1.0 | 16 | 82 | 72-99 | 7543-98-0072-CR-002 | RP-98018 |
| grapes | diazene | 0.01-1.0 | 16 | 74 | 66-84 | 7543-98-0072-CR-002 | RP-98018 |
| orange | bifenazate | 0.01 | 5 | 85 | 76-94 | 6998-97-0237-CR-001 | 6998-98-0051-CR-001 |
| orange | bifenazate | 0.01-0.50 | 15 | 87 | 76-100% | 6998-97-0237-CR-001 | 6998-97-0237-CR-001 |
| orange | bifenazate | 0.1 | 5 | 91 | 90-2 | 6998-97-0237-CR-001 | 6998-98-0051-CR-001 |
| orange | diazene | 0.01 | 5 | 96 | 93-105 | 6998-97-0237-CR-001 | 6998-98-0051-CR-001 |
| orange | diazene | 0.01-0.50 | 15 | 87 | 70-110% | 6998-97-0237-CR-001 | 6998-97-0237-CR-001 |
| orange | diazene | 0.1 | 5 | 80 | 75-84 | 6998-97-0237-CR-001 | 6998-98-0051-CR-001 |
| peach | bifenazate | 0.01-0.1 | 9 | 95 | 87-102 | 6998-97-0237-CR-001 | GRL-11930 |
| peach | diazene | 0.01-0.1 | 9 | 92 | 80-103 | 6998-97-0237-CR-001 | GRL-11930 |
| peach | bifenazate | 0.01-1.0 | 16 | 79 | 64-85 | 7543-98-0072-CR-002 | RP-98018 |
| peach | diazene | 0.01-1.0 | 16 | 76 | 70-81 | 7543-98-0072-CR-002 | RP-98018 |
| peach | bifenazate | 0.01-1.0 | 9 | 101 | 94-106 | NCL ME 245 | RP-02009 |
| peach | diazene | 0.01-1.0 | 9 | 101 | 96-109 | NCL ME 245 | RP-02009 |
| pecans | bifenazate | 0.01-0.5 | 9 | 88 | 70-118 | 6998-97-0237-CR-001 | GRL-11868 |
| pecans | diazene | 0.01-0.5 | 9 | 87 | 77-108 | 6998-97-0237-CR-001 | GRL-11868 |
| peppers | bifenazate | 0.01-0.5 | 9 | 88 | 70-120% | 6998-97-0237-CR-001 | GRL-11668 |
| peppers | diazene | 0.01-0.5 | 9 | 79 | 70-91% | 6998-97-0237-CR-001 | GRL-11668 |
| plum | bifenazate | 0.01-0.1 | 9 | 91 | 84-100 | 6998-97-0237-CR-001 | GRL-11928 |
| plum | diazene | 0.01-0.1 | 9 | 83 | 71-90 | 6998-97-0237-CR-001 | GRL-11928 |
| plum | bifenazate | 0.01-1.0 | 16 | 83 | 72-101 | 7543-98-0072-CR-002 | RP-98018 |
| plum | diazene | 0.01-1.0 | 16 | 81 | 73-89 | 7543-98-0072-CR-002 | RP-98018 |
| prunes | bifenazate | 0.01-1.0 | 16 | 81 | 71-91 | 7543-98-0072-CR-002 | RP-98018 |
| prunes | diazene | 0.01-1.0 | 24 | 71 | 60-83 | 7543-98-0072-CR-002 | RP-98018 |
| raisins | bifenazate | 0.01-1.0 | 16 | 80 | 71-96 | 7543-98-0072-CR-002 | RP-98018 |
| raisins | diazene | 0.01-1.0 | 16 | 74 | 63-86 | 7543-98-0072-CR-002 | RP-98018 |
| raisins | bifenazate | 0.01-1.0 | 9 | 101 | 91-114 | NCL ME 245 | RP-02009 |
| raisins | diazene | 0.01-1.0 | 9 | 100 | 94-108 | NCL ME 245 | RP-02009 |
| strawberry | bifenazate | 0.01-1.0 | 9 | 91 | 80-99% | 6998-97-0237-CR-001 | GRL-11940 |
| strawberry | diazene | 0.01-1.0 | 9 | 87 | 84-90% | 6998-97-0237-CR-001 | GRL-11940 |
| summer squash | bifenazate | 0.01-1.0 | 9 | 87 | 72-115% | 6998-97-0237-CR-001 | GRL-11664 |
| summer squash | diazene | 0.01-1.0 | 9 | 96 | 78-122% | 6998-97-0237-CR-001 | GRL-11664 |
| tomato | bifenazate | 0.01-0.5 | 9 | 100 | 88-120 | 6998-97-0237-CR-001 | GRL-11666 |
| tomato | diazene | 0.01-0.5 | 9 | 94 | 75-114 | 6998-97-0237-CR-001 | GRL-11666 |
| tomato paste | bifenazate | 0.01-0.5 | 8 | 82 | 76-87 | 6998-97-0237-CR-001 | GRL-11666 |
| tomato paste | diazene | 0.01-0.5 | 9 | 100 | 84-119 | 6998-97-0237-CR-001 | GRL-11666 |
| tomato puree | bifenazate | 0.01-0.5 | 9 | 87 | 81-95 | 6998-97-0237-CR-001 | GRL-11666 |
| tomato puree | diazene | 0.01-0.5 | 9 | 84 | 70-94 | 6998-97-0237-CR-001 | GRL-11666 |

Hackert Anderson and Koch (1999, 45552) examined the applicability of multiresidue methods to the residue analysis of bifenazate and bifenazate-diazene. Because both compounds exhibit fluorescence, testing began with an HPLC-fluorescence method. Bifenazate was not sufficiently stable in methanol and bifenazate-diazene was poorly separated from two decomposition peaks, so the method was not successful. In a GLC multiresidue method, acceptable chromatography was achieved on a DB-1 type column with NPD although some conversion of bifenazate to bifenazate-diazene was observed during chromatography. In the method without Florisil cleanup, recoveries of 24–43% were achieved with spiking an apple sample at 2 mg/kg. Interferences were too much to observe recoveries at 0.1 mg/kg. When a Florisil column cleanup was introduced, recoveries of both compounds were poor (2–22%). Partial conversion of bifenazate to bifenazate-diazene occurred during the cleanup.

In summary, none of the tested multiresidue methods was suitable for the analysis of bifenazate and bifenazate-diazene.

*Extraction efficiency of analytical methods*

Tissue and milk samples from a goat dosed orally for 4 consecutive days, with [$^{14}$C]bifenazate at the equivalent of 20 ppm in the feed as in a goat metabolism study, were used for radiovalidation of an analytical enforcement method for bifenazate in animal commodities (Gupta and Cassidy, 2005, 2005-013).

For the enforcement method, samples of milk, fat or liver were extracted by the procedures described for Method 7473-98-0115-CR-001 or Method NCL ME 259. For analysis of 4-hydroxybiphenyl and its sulphate conjugate, a portion of the extract was hydrolysed with hydrochloric acid for 2 hours at 60°C to convert the sulphate conjugate to free 4-hydroxybiphenyl before HPLC analysis.

The results for the enforcement method and the radiolabel measurement showed good agreement for residue analysis of bifenazate, bifenazate-diazene and 4-hydroxybiphenyl in milk, fat and liver (Table 11).

Table 11. Comparison of radiolabel measurement and enforcement analytical method for residues of bifenazate and metabolites in milk and tissues of a goat dosed orally with [$^{14}$C]bifenazate (Gupta and Cassidy, 2005, 2005-013). All residues are expressed as parent bifenazate.

| Sample | Analyte | Determined from radiolabel detection, mg/kg | Enforcement method, mg/kg |
|---|---|---|---|
| Day 3 milk | 4-hydroxybiphenyl + sulphate conjugate | 0.0015 | < 0.0025 |
| Day 4 milk | 4-hydroxybiphenyl + sulphate conjugate | 0.0022 | 0.004 |
| Fat | bifenazate + bifenazate-diazene | 0.043 | 0.045 |
| Liver | 4-hydroxybiphenyl + sulphate conjugate | 0.0062 | 0.006 |
| Liver | bifenazate + bifenazate-diazene | 0.0082 | < 0.01 |

**Stability of residues in stored analytical samples**

The Meeting received information on the stability of residues of bifenazate residues in apples, apricots, cantaloupe, cherries, cotton seed, cotton seed hulls, cotton seed meal, cotton seed refined oil, egg yolk, fat, gin trash, grape juice, grapes, kidney, liver, milk, mint, muscle, oranges, peaches, peppers, plums, potatoes, poultry liver, poultry muscle, poultry skin + fat, prunes, tomato, tomato paste and tomato puree.

Bifenazate residues (measured as bifenazate + bifenazate-diazene) are not particularly stable in some substrates. Stability is improved where the commodity is stored unchopped and in processed commodities presumably where enzymes are denatured. Bifenazate residues are stable in fat and milk, but are particularly unstable in kidney.

Storage stability data are recorded in the tables unadjusted for concurrent procedural recoveries. If the concurrent procedural recoveries were outside of the 70–120% range the data from that sampling occasion were not taken into account.

Buckrell (2001, GRL-FR-11667) fortified aliquots (approximately 20 g) of homogenised tomato matrix in glass jars with bifenazate for freezer storage stability testing at a temperature below -18°C. After each storage interval, an aged aliquot and a freshly fortified aliquot acting as a procedural recovery were analysed for bifenazate + bifenazate-diazene (Table 12). Analytical results were reported as bifenazate + bifenazate-diazene (as bifenazate).

Black (2002, GRL-FR-11853) fortified aliquots (approximately 20 g) of homogenised cherry matrix in plastic bags with bifenazate for freezer storage stability testing. After each storage interval, aged aliquots and freshly fortified aliquots acting as procedural recoveries were analysed (Table 12). Analytical results were reported as bifenazate + bifenazate-diazene (as bifenazate).

Benstead (2001, GRL-FR-11669) fortified aliquots (approximately 20 g) of homogenised peppers matrix in glass jars with bifenazate for freezer storage stability testing. After each storage interval, an aged aliquot and a freshly fortified aliquot acting as a procedural recovery were analysed for bifenazate + bifenazate-diazene (Table 12). Analytical results were reported as bifenazate + bifenazate-diazene (as bifenazate).

Benstead (2001, GRL-FR-11663) fortified aliquots (approximately 20 g) of homogenised cantaloupe matrix in glass jars with bifenazate for freezer storage stability testing. After each storage interval, an aged aliquot and a freshly fortified aliquot acting as a procedural recovery were analysed for bifenazate + bifenazate-diazene (Table 12). Analytical results were reported as bifenazate + bifenazate-diazene (as bifenazate).

Black (2002, GRL-FR-11911) fortified aliquots (approximately 20 g) of homogenised mint tops matrix in plastic bags with bifenazate for freezer storage stability. After each storage interval, aged aliquots and freshly fortified aliquots acting as procedural recoveries were analysed (Table 12). Analytical results were reported as bifenazate + bifenazate-diazene (as bifenazate).

Black (2003, GRL-11936) fortified aliquots (approximately 20 g) of homogenised potato tuber matrix in plastic bags with bifenazate for freezer storage stability testing. After each storage interval, aged aliquots and freshly fortified aliquots acting as procedural recoveries were analysed (Table 12). Analytical results were reported as bifenazate + bifenazate-diazene (as bifenazate). Because early results were questionable (low recoveries and residues not detected in stored samples), reserve samples were also analysed and two additional samples were run as a second trial.

Black (2004, GRL-12140) fortified aliquots (approximately 20 g) of chopped potato tuber in plastic bags with bifenazate for freezer storage stability testing. After each storage interval, aged aliquots and freshly fortified aliquots acting as procedural recoveries were analysed (Table 12). Analytical results were reported as bifenazate + bifenazate-diazene (as bifenazate). Because early results were questionable (low recoveries and residues not detected in stored samples), reserve samples were also analysed and two additional samples were run as a second trial.

Black (2005, GRL-12171) fortified aliquots (approximately 20 g) of sectioned plum tissue in plastic bags with bifenazate for freezer storage stability testing. After each storage interval, aged aliquots and freshly fortified aliquots acting as procedural recoveries were analysed (Table 12). Analytical results were reported as bifenazate + bifenazate-diazene (as bifenazate).

Black (2005, GRL-12172) fortified aliquots (approximately 20 g) of tomato paste and tomato puree in plastic bags with bifenazate for freezer storage stability testing. After each storage interval, aged aliquots and freshly fortified aliquots acting as procedural recoveries were analysed (Table 12). Analytical results were reported as bifenazate + bifenazate-diazene (as bifenazate).

Wiedmann and Korpalski (1999, RP-98019) fortified aliquots (approximately 20 g) of homogenates of peaches, grapes, apples, oranges, grape juice and prunes in glass jars with bifenazate and bifenazate-diazene for freezer storage stability testing. They also fortified some sample types on the intact surface (whole grapes and sections of apples and peaches) to determine the stability when stored as whole fruit. After each storage interval, aged aliquots and freshly fortified aliquots acting as procedural recoveries were analysed (Table 12). Analytical results were reported as bifenazate + bifenazate-diazene (as bifenazate). Homogenized tissues appeared to cause faster degradation of the residues than whole fruit. Residues were reasonably stable in processed commodities such as prunes and grape juice.

Table 12. Freezer storage stability data for bifenazate and metabolites spiked into matrices of apples, apricots, cantaloupe, cherries, cotton seed, cotton seed hulls, cotton seed meal, cotton seed refined oil, gin trash, grape juice, grapes, mint, oranges, peaches, peppers, plums, potatoes, prunes, tomato, tomato paste and tomato puree.

| Storage interval | Procedural recov % | Bifenazate + bifenazate-diazene, mg/kg | Storage interval | Procedural recov % | Bifenazate + bifenazate-diazene, mg/kg |
|---|---|---|---|---|---|
| Tomato, homogenized matrix fortified with bifenazate at 0.1 mg/kg (Buckrell, 2001, GRL-FR-11667) storage temp below below -18°C. | | | Cherries, homogenized matrix, fortified with bifenazate at 0.1 mg/kg (Black, 2002, GRL-FR-11853), storage temp below below -20°C. | | |
| 1 month | 80% | 0.067 | 1 month | 105% 113% | 0.080 0.077 0.094 |
| 3 months | 101% | 0.075 | 3 months | 92% 90% | 0.046 0.044 0.054 |
| 6 months | 96% | 0.080 | 6 months | 93% 81% | 0.038 0.046 0.038 |
| | | | | | 30% decline in 2.6 months. |
| Peppers, homogenized matrix fortified with bifenazate at 0.1 mg/kg (Benstead, 2001, GRL-FR-11669), storage temp below below -18°C. | | | Cantaloupe, homogenized matrix fortified with bifenazate at 0.1 mg/kg (Benstead, 2001, GRL-FR-11663), storage temp below below -18°C. (30% decline in | | |
| 1 month | 84% | 0.083 | 1 month | 83% | 0.074 |
| 3 months | 95% | 0.076 | 3 months | 90% | 0.063 |
| 6 months | 96% | 0.077 | 6 months | 95% | 0.047 |
| | | | | | 30% decline in 3.9 months |
| Mint tops, homogenized matrix fortified with bifenazate at 0.1 mg/kg (Black, 2002, GRL-FR-11911), storage temp between -17.7°C and -26.5°C. | | | Potatoes, chopped tuber fortified with bifenazate at 0.1 mg/kg (Black, 2003, GRL-11936), storage temp between -22.1°C and -27.1°C. | | |
| 28 days | 90% 79% | 0.095 0.068 0.074 | 0 | 38% | |
| 102 days | 81% 88% | 0.11 0.095 0.098 | 1 month | 112% 101% | < 0.005 (3) |
| | | | Trial 2 | 93% 100 % | < 0.005 (3) |
| | | | 1.5 months | 45% 97% | < 0.005 (3) |
| | | | 3 months | 82% 30% | < 0.01 < 0.005 (2) |
| Potatoes, chopped tuber fortified with bifenazate at 0.1 mg/kg (Black, 2004, GRL-12140), storage temp between -23.4°C and -27.0°C. | | | Plums, sectioned tissue fortified with bifenazate at 0.1 mg/kg (Black, 2005, GRL-12171), storage temp between -22.4°C and -26.8°C. | | |
| 0 | 79% 70% 80% | | 0 | 78% 81% 76% | |
| 1 week | 103% 104% | 0.014 0.018 0.019 | 1 week | 73% 72% 78% | 0.078 0.074 |
| 2 weeks | 107% 89% | 0.011 0.012 0.013 | 2 weeks | 76% 75% 61% | 0.073 0.071 |
| 4 weeks | 89% 87% | 0.012 0.017 0.020 | 4 weeks | 70% 71% 77% | 0.070 0.074 |
| Tomato paste, fortified with bifenazate at 1 mg/kg (Black, 2005, GRL-12172), storage temp between -18.7°C and -27.2°C. | | | Tomato puree, fortified with bifenazate at 1 mg/kg (Black, 2005, GRL-12172), storage temp between -18.7°C and -27.2°C. | | |
| 0 | 80% 80% 78% | | 0 | 83% 85% 84% | |
| 1 week | 83% 83% | 0.85 0.81 0.80 | 1 week | 100% 92% | 0.89 1.00 0.90 |
| 2 weeks | 87% 82% | 0.80 0.76 0.81 | 2 weeks | 95% 90% | 0.99 0.97 0.83 |
| 4 weeks | 84% 87% | 0.75 0.73 0.81 | 4 weeks | 87% 89% | 0.82 0.80 0.85 |
| Cottonseed matrix, fortified at 0.1 mg/kg bifenazate (Wiedmann, 2000, 10495-1) | | | Cottonseed matrix, fortified at 0.1 mg/kg bifenazate-diazene (Wiedmann, 2000, 10495-1) | | |
| 0 | 78% 100% 98% | | 0 | 93% 85% 90% | |
| 21 days | 99% | 0.038 0.038 | 21 days | 88% | 0.038 0.046 |
| 56 days | 99% | 0.059 0.067 | 56 days | 83% | 0.040 0.043 |
| Gin trash, fortified at 0.1 mg/kg bifenazate (Wiedmann, 2000, 10495-1) | | | Gin trash, fortified at 0.1 mg/kg bifenazate-diazene (Wiedmann, 2000, 10495-1) | | |
| 0 | 76% 76% 83% | | 0 | 73% 75% 76% | |
| 44 days | 79% | 0.051 0.055 | 44 days | 84% | 0.049 0.041 |
| Cotton seed hulls, fortified at 0.1 mg/kg bifenazate (Wiedmann, 2000, 10495-1) | | | Cotton seed hulls, fortified at 0.1 mg/kg bifenazate-diazene (Wiedmann, 2000, 10495-1) | | |
| 0 | 89% 87% 77% | | 0 | 84% 83% 83% | |
| 52 days | 98% | 0.070 0.063 | 52 days | 93% | 0.067 0.064 |

| Storage interval | Procedural recov % | Bifenazate + bifenazate-diazene, mg/kg | Storage interval | Procedural recov % | Bifenazate + bifenazate-diazene, mg/kg |
|---|---|---|---|---|---|
| Cotton seed meal, fortified at 0.1 mg/kg bifenazate (Wiedmann, 2000, 10495-1) | | | Cotton seed meal, fortified at 0.1 mg/kg bifenazate-diazene (Wiedmann, 2000, 10495-1) | | |
| 0 | 95% 94% 99% | | 0 | 88% 91% 88% | |
| 43 days | 96% | 0.058 0.055 | 43 days | 92% | 0.071 0.080 |
| Cotton seed refined oil, fortified at 1 mg/kg bifenazate (Wiedmann, 2000, 10495-1) | | | Cotton seed refined oil, fortified at 1 mg/kg bifenazate-diazene (Wiedmann, 2000, 10495-1) | | |
| 0 | 55% 66% 70% | | 0 | 73% 77% 79% | |
| 28 days | 63% 1/ | 0.74 0.77 | 28 days | 85% | 0.72 0.75 |
| Homogenized apricots fortified at 0.1 mg/kg bifenazate (Wesley, 2002, GRL11934) stored at -20.3 °C to -28.9 °C | | | | | |
| 0 | 76% | | | | |
| 1 month | 92% 77% | 0.066 0.076 0.070 | | | |
| 1.5 months | 84% 76% | 0.065 0.059 0.061 | | | |
| 2 months | 78% 72% | 0.052 0.065 | | | |
| Homogenized apples, fortified at 0.1 mg/kg bifenazate (Wiedmann and Korpalski, 1999, RP-98019) stored at freezer temperature -24 °C to -20 °C. | | | Homogenized apples, fortified at 0.1 mg/kg bifenazate-diazene (Wiedmann and Korpalski, 1999, RP-98019) stored at freezer temperature -24 °C to -20 °C. | | |
| 0 | 94% 92% 97% 109% | | 0 | 95% 96% 95% 97% | |
| 7 days | 91% 95% | 0.078 0.080 | 7 days | 94% 91% | 0.086 0.081 |
| 14 days | 72% 69% | 0.056 0.057 | 14 days | 70% 78% | 0.065 0.063 |
| 21 days | 68% 68% 1/ | 0.046 0.041 | 21 days | 63% 67% 1/ | 0.048 0.048 |
| 29 days | 73% 77% | 0.053 0.058 | 29 days | 81% 81% | 0.065 0.063 |
| 42 days | 67% 67% 1/ | 0.052 0.052 | 42 days | 68% 72% | 0.050 0.048 |
| 70 days | 91% 88% | 0.057 0.055 | 70 days | 86% 79% | 0.051 0.056 |
| 107 days | 87% 88% | 0.042 0.042 | 107 days | 77% 81% | 0.041 0.044 |
| 182 days | 89% 91% | 0.038 0.038 | 182 days | 79% 71% | 0.037 0.041 |
| | | 30% decline in 106 days | | | 30% decline in 97 days |
| Apples, fortified on the skin surface at 0.1 mg/kg bifenazate (Wiedmann and Korpalski, 1999, RP-98019) stored at freezer temperature -24 °C to -20 °C. | | | Apples, fortified on the skin surface at 0.1 mg/kg bifenazate-diazene (Wiedmann and Korpalski, 1999, RP-98019) stored at freezer temperature -24 °C to -20 °C. | | |
| 0 | 88% 93% 88% 90% | | 0 | 77% 76% 73% 76% | |
| 14 days | 81% 92% | 0.088 0.092 | 14 days | 81% 83% | 0.075 0.090 |
| 28 days | 94% 88% | 0.078 0.090 | 28 days | 79% 79% | 0.080 0.069 |
| 56 days | 88% 84% | 0.098 0.086 | 56 days | 89% 90% | 0.070 0.068 |
| 126 days | 92% 87% | 0.093 0.096 | 126 days | 90% 83% | 0.073 0.051 |
| 224 days | 94% 92% | 0.098 0.088 | 224 days | 86% 80% | 0.072 0.082 |
| Homogenized grapes (stems removed), fortified at 0.1 mg/kg bifenazate (Wiedmann and Korpalski, 1999, RP-98019) stored at freezer temperature -24 °C to -20 °C. | | | Homogenized grapes (stems removed), fortified at 0.1 mg/kg bifenazate-diazene (Wiedmann and Korpalski, 1999, RP-98019) stored at freezer temperature -24 °C to -20 °C. | | |
| 0 | 92% 90% 95% 99% | | 0 | 88% 89% 86% 86% | |
| 7 days | 90% 92% | 0.066 0.062 | 7 days | 89% 89% | 0.069 0.073 |
| 14 days | 81% 79% | 0.051 0.049 | 14 days | 73% 74% | 0.050 0.049 |
| 21 days | 76% 79% | 0.045 0.035 | 21 days | 74% 75% | 0.039 0.039 |
| 29 days | 81% 90% | 0.046 0.031 | 29 days | 82% 78% | 0.045 0.044 |
| 42 days | 79% 72% | 0.044 0.031 | 42 days | 77% 74% | 0.031 0.037 |
| 70 days | 87% 83% | 0.023 0.019 | 70 days | 85% 87% | 0.020 0.024 |
| | | 30% decline in 23 days | | | 30% decline in 22 days |
| Grapes (stems removed), fortified on the surface at 0.1 mg/kg bifenazate (Wiedmann and Korpalski, 1999, RP-98019) stored at freezer temperature -24 °C to -20 °C. | | | Grapes (stems removed), fortified on the surface at 0.1 mg/kg bifenazate-diazene (Wiedmann and Korpalski, 1999, RP-98019) stored at freezer temperature -24 °C to -20 °C. | | |
| 0 | 71% 81% 100% 95% | | 0 | 71% 76% 86% 89% | |
| 14 days | 107% 98% | 0.088 0.088 | 14 days | 93% 92% | 0.086 0.086 |
| 28 days | 94% 89% | 0.083 0.081 | 28 days | 84% 90% | 0.074 0.074 |
| 56 days | 85% 87% | 0.079 0.070 | 56 days | 86% 90% | 0.079 0.068 |
| 126 days | 96% 94% | 0.081 0.083 | 126 days | 94% 90% | 0.073 0.078 |
| 224 days | 84% 97% | 0.073 0.076 | 224 days | 82% 84% | 0.066 0.067 |

| Storage interval | Procedural recov % | Bifenazate + bifenazate-diazene, mg/kg | Storage interval | Procedural recov % | Bifenazate + bifenazate-diazene, mg/kg |
|---|---|---|---|---|---|
| Homogenized peaches (seeds removed), fortified at 0.1 mg/kg bifenazate (Wiedmann and Korpalski, 1999, RP-98019) stored at freezer temperature -24 °C to -20 °C. | | | Homogenized peaches (seeds removed), fortified at 0.1 mg/kg bifenazate-diazene (Wiedmann and Korpalski, 1999, RP-98019) stored at freezer temperature -24 °C to -20 °C. | | |
| 0 | 89% 91% 91% 88% | | 0 | 88% 92% 90% 91% | |
| 7 days | 81% 84% | 0.068 0.067 | 7 days | 81% 79% | 0.070 0.070 |
| 14 days | 81% 80% | 0.058 0.030 | 14 days | 96% 86% | 0.061 0.067 |
| 21 days | 74% 79% | 0.048 0.053 | 21 days | 76% 67% | 0.049 0.054 |
| 28 days | 71% 71% | 0.049 0.052 | 28 days | 68% 72% | 0.057 0.059 |
| 42 days | 91% 84% | 0.062 0.056 | 42 days | 92% 89% | 0.063 0.074 |
| 70 days | 84% 81% | 0.047 0.055 | 70 days | 86% 81% | 0.055 0.057 |
| 105 days | 83% 85% | 0.041 | 105 days | 75% 82% | 0.036 0.033 |
| 182 days | 87% 84% | 0.033 0.033 | 182 days | 77% 80% | 0.035 0.037 |
| | | 30% decline in 126 days | | | 30% decline in 92 days |
| Peaches, fortified on the skin surface at 0.1 mg/kg bifenazate (Wiedmann and Korpalski, 1999, RP-98019) stored at freezer temperature -24 °C to -20 °C. | | | Peaches, fortified on the skin surface at 0.1 mg/kg bifenazate-diazene (Wiedmann and Korpalski, 1999, RP-98019) stored at freezer temperature -24 °C to -20 °C. | | |
| 0 | 81% 86% 97% 93% | | 0 | 67% 66% 82% 86% | |
| 14 days | 98% 92% | 0.075 0.084 | 14 days | 80% 89% | 0.066 0.072 |
| 28 days | 83% 78% | 0.058 0.059 | 28 days | 83% 74% | 0.050 0.042 |
| 56 days | 91% 95% | 0.060 0.077 | 56 days | 79% 90% | 0.052 0.044 |
| 126 days | 101% 88% | 0.049 0.062 | 126 days | 89% 82% | 0.035 0.040 |
| 223 days | 98% 94% | 0.063 0.067 | 223 days | 77% 79% | 0.044 0.056 |
| Homogenized whole oranges, fortified at 0.1 mg/kg bifenazate (Wiedmann and Korpalski, 1999, RP-98019) stored at freezer temperature -24 °C to -20 °C. | | | Homogenized whole oranges, fortified at 0.1 mg/kg bifenazate-diazene (Wiedmann and Korpalski, 1999, RP-98019) stored at freezer temperature -24 °C to -20 °C. | | |
| 0 | 87% 84% 81% 89% | | 0 | 86% 82% 82% 85% | |
| 7 days | 79% 83% | 0.063 0.063 | 7 days | 78% 79% | 0.068 0.071 |
| 14 days | 89% 88% | 0.056 0.066 | 14 days | 79% 74% | 0.061 0.062 |
| 28 days | 92% 87% | 0.070 0.059 | 28 days | 81% 83% | 0.060 0.076 |
| 40 days | 91% 92% | 0.064 0.062 | 40 days | 72% 69% | 0.069 0.065 |
| 75 days | 95% 94% | 0.066 0.068 | 75 days | 87% 80% | 0.069 0.072 |
| 105 days | 82% 87% | 0.055 0.050 | 105 days | 72% 71% | 0.058 0.066 |
| 186 days | 95% 96% | 0.052 0.051 | 186 days | 81% 83% | 0.065 0.068 |
| Homogenized grape juice, fortified at 0.1 mg/kg bifenazate (Wiedmann and Korpalski, 1999, RP-98019) stored at freezer temperature -24 °C to -20 °C. | | | Homogenized grape juice, fortified at 0.1 mg/kg bifenazate-diazene (Wiedmann and Korpalski, 1999, RP-98019) stored at freezer temperature -24 °C to -20 °C. | | |
| 0 | 95% 95% 99% 98% | | 0 | 89% 89% 92% 93% | |
| 7 days | 99% 99% | 0.089 0.105 | 7 days | 87% 84% | 0.085 0.087 |
| 14 days | 98% 95% | 0.096 0.098 | 14 days | 88% 88% | 0.084 0.083 |
| 28 days | 94% 94% | 0.096 0.101 | 28 days | 86% 89% | 0.085 0.082 |
| 40 days | 91% 90% | 0.091 0.089 | 40 days | 81% 80% | 0.081 0.079 |
| 75 days | 97% 102% | 0.103 0.096 | 75 days | 90% 91% | 0.089 0.084 |
| 107 days | 96% 94% | 0.095 0.095 | 107 days | 82% 77% | 0.087 0.086 |
| 186 days | 99% 97% | 0.104 | 186 days | 83% 84% | 0.085 0.087 |
| Homogenized prunes, fortified at 0.1 mg/kg bifenazate (Wiedmann and Korpalski, 1999, RP-98019) stored at freezer temperature -24 °C to -20 °C. | | | Homogenized prunes, fortified at 0.1 mg/kg bifenazate-diazene (Wiedmann and Korpalski, 1999, RP-98019) stored at freezer temperature -24 °C to -20 °C. | | |
| 0 | 73% 72% 76% 72% | | 0 | 72% 70% 66% 67% | |
| 7 days | 76% 80% | 0.075 0.069 | 7 days | 71% 73% | 0.070 0.068 |
| 14 days | 75% 73% | 0.082 0.073 | 14 days | 77% 81% | 0.063 0.063 |
| 28 days | 83% 80% | 0.073 0.070 | 28 days | 78% 70% | 0.064 0.061 |
| 42 days | 80% 85% | 0.072 0.073 | 42 days | 72% 79% | 0.057 0.062 |
| 70 days | 86% 87% | 0.067 0.066 | 70 days | 75% 71% | 0.056 0.052 |
| 105 days | 84% 84% | 0.071 0.068 | 105 days | 79% 71% | 0.056 0.057 |
| 182 days | 86% 94% | 0.080 0.078 | 182 days | 70% 66% 1/ | 0.055 0.053 |

Jablonski (1999, 7475) tested the freezer storage stability of bifenazate, bifenazate-diazene and metabolite 4-hydroxybiphenyl fortified in bovine milk and tissues at 0.20 mg/kg (Table 13). It should be noted that the tests for 2 days storage were done in a follow-up experiment after substantial losses had been found after 14 days storage.

The compounds were stable in milk for the interval tested (202 days). The compounds were also reasonably stable in fat. In muscle, bifenazate and bifenazate-diazene declined within a few days with less than 50% remaining after 14 days. Metabolite 4-hydroxybiphenyl was much more stable. In liver and kidney, bifenazate and bifenazate-diazene also declined rapidly with less than 50% of the spiked concentration remaining after 2 days. Metabolite 4-hydroxybiphenyl was reasonably stable in liver and kidney with 65–85% of the spiked concentration remaining after 9–11 weeks of storage.

The instability of bifenazate-diazene in bovine kidney matrix was noted during the analytical recovery testing. At time 0, the recoveries were treated in the same way as the storage test samples, i.e., 5–10 minutes were allowed after fortification for the fortification solvent to evaporate before the jars were capped and placed in the freezer. In day 2 and day 14 procedural recoveries, samples were extracted immediately after fortification.

Table 13. Freezer storage stability testing of fortified bifenazate, bifenazate-diazene and 4-hydroxybiphenyl in bovine tissues and milk (Jablonski, 1999, 7475). Samples were stored at freezer temperature -24 °C to -20 °C.

| Storage interval | Procedural recov % | Bifenazate + bifenazate-diazene, mg/kg | Procedural recov % | Bifenazate + bifenazate-diazene, mg/kg | Procedural recov % | 4-hydroxybiphenyl mg/kg |
|---|---|---|---|---|---|---|
| MILK, fortified at 0.20 mg/kg bifenazate | | | MILK, fortified at 0.20 mg/kg of bifenazate-diazene | | MILK, fortified at 0.20 mg/kg of 4-hydroxybiphenyl | |
| 0 days | 96% (n=6) | | 97% (n=6) | | 108% (n=6) | |
| 14 days | 100% | 0.188 | 97% | 0.194 | 104% | 0.201 |
| 42 days | 95% | 0.168 | 88% | 0.163 | 101% | 0.183 |
| 85 days | 86% | 0.160 | 85% | 0.157 | 102% | 0.183 |
| 202 days | 96% | 0.162 | 78% | 0.156 | 95% | 0.167 |
| MUSCLE, fortified at 0.20 mg/kg bifenazate | | | MUSCLE, fortified at 0.20 mg/kg of bifenazate-diazene | | MUSCLE, fortified at 0.20 mg/kg of 4-hydroxybiphenyl | |
| 0 days | 105% (n=6) | | 89% (n=6) | | 96% (n=6) | |
| 2 days | 83% | 0.100 | 83% | 0.12 | | |
| 14 days | 102% | 0.097 | 92% | 0.031 | 103% | 0.180 |
| 28 days | 98% | 0.042 | 90% | 0.000 | 93% | 0.155 |
| 86 days | | | | | 98% | 0.136 |
| | | 30% decline in 10 days | | 30% decline in 2 days | 30% decline in approx 100 days | |
| LIVER, fortified at 0.20 mg/kg bifenazate | | | LIVER, fortified at 0.20 mg/kg of bifenazate-diazene | | LIVER, fortified at 0.20 mg/kg of 4-hydroxybiphenyl | |
| 0 days | 98% (n=6) | | 82% (n=6) | | 95% (n=6) | |
| 2 days | 73% | 0.050 | 76% | 0.0146 | | |
| 14 days | 106% | 0.179 | 92% | 0.032 | 98% | 0.181 |
| 76 days | | | | | 98% | 0.175 |
| KIDNEY, fortified at 0.20 mg/kg bifenazate | | | KIDNEY, fortified at 0.20 mg/kg of bifenazate-diazene | | KIDNEY, fortified at 0.20 mg/kg of 4-hydroxybiphenyl | |
| 0 days | 91% (n=6) | | 57% (n=6) | | 91% (n=6) | |
| 2 days | 72% | 0.092 | 69% | 0.041 | | |
| 14 days | 95% | 0.121 | 79% | 0.003 | 99% | 0.155 |
| 63 days | | | | | 101% | 0.133 |

| Storage interval | Procedural recov % | Bifenazate + bifenazate-diazene, mg/kg | Procedural recov % | Bifenazate + bifenazate-diazene, mg/kg | Procedural recov % | 4-hydroxybiphenyl mg/kg |
|---|---|---|---|---|---|---|
| FAT, fortified at 0.20 mg/kg bifenazate | | | FAT, fortified at 0.20 mg/kg of bifenazate-diazene | | FAT, fortified at 0.20 mg/kg of 4-hydroxybiphenyl | |
| 0 days | 88% (n=6) | | 88% (n=6) | | 106% (n=6) | |
| 14 days | 75% | 0.123 | 75% | 0.148 | 103% | 0.168 |
| 36 days | 93% | 0.147 | 81% | 0.158 | 109% | 0.168 |
| 95 days | 104% | 0.141 | 82% | 0.156 | 107% | 0.151 |

Procedural recoveries are means of duplicate samples (except as stated otherwise) and concentrations in test samples are means of 4 replicates.

Labelled bifenazate was spiked into control samples of goat milk (0.04 and 0.25 mg/kg), muscle (0.25 mg/kg), fat (0.25 mg/kg), liver (2.0 mg/kg) and kidney (0.25 mg/kg) for freezer storage stability testing. Samples were analysed by HPLC-LSC methods, initially and after 9 months of storage below -10 °C (McClanahan and Bayus, 1999, 96-0064). Stability data are summarised in Table 14 and Table 15. Metabolites 4-methoxybiphenyl and bifenazate-diazene are recorded as a combined residue because of only partial separation on the HPLC systems used for analysis. In cases where some resolution was achieved, bifenazate-diazene was a substantial part of the combined residue and sometimes the major part of the residue.

Bifenazate was stable to the fortification and extraction procedure for milk, muscle, fat and liver (day 0 samples). For kidney, only 27% of the fortified bifenazate was recovered from the day 0 sample, with 40% and 5% appearing as 4-methoxybiphenyl + bifenazate-diazene and 4-hydroxybiphenyl respectively.

Bifenazate was stable in the fat samples for 9 months freezer storage. In milk after 9 months freezer storage, bifenazate had declined to 23% of its fortification level, with 60% appearing as 4-methoxybiphenyl + bifenazate-diazene. In muscle after 9 months freezer storage, bifenazate had declined to 8–10% of its fortification level, with 40–44% appearing as 4-methoxybiphenyl + bifenazate-diazene and 26–30% unextractable from the matrix.

In kidney after 9 months freezer storage, bifenazate had declined to 1.2% of its fortification level, with 46% appearing as 4-methoxybiphenyl + bifenazate-diazene and 25% unextractable from the matrix. In liver after 9 months freezer storage, bifenazate had declined to 9% of its fortification level, with 64% appearing as 4-methoxybiphenyl + bifenazate-diazene and 14% unextractable from the matrix.

Samples from the lactating goat metabolism study (McClanahan and Bayus, 1999, 96-0064) were analysed by HPLC before and after freezer storage of 15–187 days to test the stability of incurred residues (muscle 79 days; fat 15 days; liver 111 days; kidney 187 days; milk 81 days). The qualitative appearance of the initial and final chromatograms was reasonably similar for milk, fat, liver and kidney. Substantial changes were apparent for muscle.

Table 14. Freezer storage stability testing of fortified [$^{14}$C]bifenazate in tissues of lactating goats (McClanahan and Bayus, 1999, 96-0064). Samples were stored below -10 °C.

| | Concentration, mg/kg, expressed as parent | | | | | | | |
|---|---|---|---|---|---|---|---|---|
| | Loin muscle | | Leg muscle | | Omental fat | | Peri-renal fat | |
| Component | initial | 9 months | initial | 9 months | initial | 9 months | initial | 9 months |
| Total $^{14}$C residue (TRR) | 0.25 | 0.25 | 0.25 | 0.25 | 0.25 | 0.25 | 0.25 | 0.25 |
| Extracted residue | 0.25 | 0.17 | 0.24 | 0.17 | 0.25 | 0.24 | 0.24 | 0.24 |
| Unextractable | | 0.064 | | 0.074 | | 0.007 | | 0.006 |
| Bifenazate | 0.20 | 0.019 | 0.20 | 0.026 | 0.23 | 0.20 | 0.23 | 0.21 |
| 4-hydroxybiphenyl | nd | 0.009 | nd | 0.011 | nd | nd | nd | nd |
| 4-methoxybiphenyl + bifenazate-diazene | 0.037 | 0.10 | 0.032 | 0.11 | 0.012 | 0.028 | 0.009 | 0.027 |

nd: not detected. Detection limit approximately 0.0001 mg/kg.

Table 15. Freezer storage stability testing of fortified [$^{14}$C] bifenazate in tissues and milk of lactating goats (McClanahan and Bayus, 1999, 96-0064). Samples were stored below -10 °C.

| Component | Concentration, mg/kg, expressed as parent | | | | | |
|---|---|---|---|---|---|---|
| | Milk | | Liver | | Kidney | |
| | initial | 9 months | initial | 9 months | initial | 9 months |
| Total $^{14}$C residue (TRR) | 0.04 | 0.25 | 2.0 | 2.0 | 0.25 | 0.25 |
| Extracted residue | 0.042 | 0.23 | 1.98 | 1.65 | 0.25 | 0.17 |
| Unextractable | | 0.001 | | 0.28 | | 0.062 |
| Bifenazate | 0.034 | 0.057 | 1.7 | 0.18 | 0.067 | 0.003 |
| 4-hydroxybiphenyl | nd | 0.003 | nd | 0.045 | 0.013 | 0.010 |
| 4-methoxybiphenyl + bifenazate-diazene | 0.004 | 0.15 | 0.17 | 1.28 | 0.099 | 0.115 |

nd: not detected. Detection limit approximately 0.0001 mg/kg.

Bifenazate spiked into control samples of egg yolk, hen skin + fat and thigh muscle at approximately 0.10–0.13 mg/kg and liver at 0.60 mg/kg was tested for freezer storage stability. Samples were analysed by HPLC-LSC methods, initially and after 4.5–6 months of storage below -10 °C (McClanahan *et al.*, 1999, 96-0265). Stability data are summarised in Table 16. Bifenazate residues were stable in egg yolk and liver. In thigh muscle, 45% of the bifenazate disappeared, with 14% and 11% appearing as 4-hydroxybiphenyl and bifenazate-diazene respectively. In skin + fat, 97% of the bifenazate disappeared with 4%, 4% and 59% appearing as 4-hydroxybiphenyl, 4-methoxybiphenyl and bifenazate-diazene respectively.

Samples from the laying hen metabolism study (McClanahan *et al.*, 1999, 96-0265) were analysed by HPLC before and after freezer storage of 121–171 days to test the stability of incurred residues. The qualitative appearance of the initial and final chromatograms was reasonably similar for egg yolk, skin-with-fat and liver. Substantial changes were apparent for thigh muscle, but total residues in thigh muscle were very low (0.006 mg/kg).

Table 16. Freezer storage stability testing of fortified [$^{14}$C] bifenazate in tissues and eggs of laying hens (McClanahan *et al.*, 1999, 96-0265). Samples were stored below -10 °C.

| Component | Concentration, mg/kg, expressed as parent | | | | | | | |
|---|---|---|---|---|---|---|---|---|
| | Egg yolk | | Skin + fat | | Thigh muscle | | Liver | |
| | initial | 6 months storage | initial | 4.5 months storage | initial | 6 months storage | initial | 4.5 months storage |
| Total $^{14}$C residue | 0.13 | 0.13 | 0.11 | 0.10 | 0.13 | 0.13 | 0.61 | 0.61 |
| Extracted residue | 0.13 | 0.13 | 0.10 | 0.077 | 0.13 | 0.11 | 0.60 | 0.59 |
| Unextractable | | | nd | 0.013 | | 0.018 | | |
| Bifenazate | 0.12 | 0.12 | 0.095 | 0.003 | 0.12 | 0.067 | 0.57 | 0.54 |
| 4-hydroxy biphenyl | nd | nd | nd | 0.004 | nd | 0.017 | 0.005 | nd |
| 4-methoxy biphenyl | 0.002 | 0.001 | 0.001 | 0.004 | 0.002 | 0.002 | 0.021 | 0.009 |
| bifenazate-diazene | 0.008 | 0.006 | 0.004 | 0.056 | 0.006 | 0.013 | 0.012 | 0.009 |

nd: not detected. Detection limits for tissues and egg yolks were 0.005 and 0.003 mg/kg respectively.

## USE PATTERN

Bifenazate is a selective acaricide which controls the motile stage of mites either by direct contact or through contact with foliar residues. Bifenazate blocks or closes the gamma-aminobutyric acid

(GABA) activated chloride channels of susceptible pests resulting in over-excitation of the peripheral nervous system.

Bifenazate products are mixed with water and applied as foliar sprays or broadcast treatments using aerial or ground equipment equipped for conventional spraying on crops. Bifenazate is not systemic in action; therefore complete coverage of both upper and lower leaf surfaces is necessary for effective control. Bifenazate is effective for the control of a variety of mite species, especially spider mites, red mites, and grass mites. The Meeting received information on bifenazate registered uses in Australia, Japan, Netherlands and USA. In The Netherlands, bifenazate is registered for use only in floriculture and on nursery trees and perennials.

Table 17. Registered field uses of bifenazate in Australia, Japan and USA. Labels for the following uses were available to the Meeting.

| Crop | Country | Application | | | | | | | PHI days |
|------|---------|------|------|------------|-----------|----------------|--------|------|
| | | Form | Type | Rate kg ai/ha | Conc kg ai/hL | Min spray vol, L/ha | Max number | |
| Almonds | USA | 500 WP | foliar | 0.42-0.56 | | 470 | 1 | 7 |
| Apple | Australia | 480 SC | foliar | | 0.031 | 1000 | 1 | 7 |
| Apple | Japan | 200 SC | foliar | 0.27-1.4 | | | 1 | 7 |
| Apricot | Australia | 480 SC | foliar | | 0.031 | 1000 | 1 | 3 |
| Cherry | Japan | 200 SC | foliar | 0.27-1.4 | | | 1 | 14 |
| Cotton | USA | 480 SC | foliar, aerial application | 0.4-0.8 | | 47 | 1 | 60 |
| Cotton | USA | 480 SC | foliar, ground application | 0.4-0.8 | | 190 | 1 | 60 |
| Cucumber | Japan | 200 SC | foliar | 0.3-0.6 | | | 1 | 1 |
| Cucurbit vegetables[3] | USA | 500 WP | foliar | 0.42-0.56 | | 470 | 1 | 3 |
| Egg plant | Japan | 200 SC | foliar | 0.3-0.6 | | | 1 | 1 |
| Filbert | USA | 500 WP | foliar | 0.42-0.56 | | 470 | 1 | 14 |
| Fruiting vegetables[4] | USA | 500 WP | foliar | 0.42-0.56 | | 470 | 1 | 3 |
| Grapes | Japan | 200 SC | foliar | 0.27-1.4 | | | 1 | 21 |
| Grapes | USA | 500 WP | foliar | 0.42-0.56 | | 470 | 1 | 14 |
| Hops | USA | 500 WP | foliar | 0.42-0.84 | | 470 | 1 | 14 |
| Lime | Japan | 200 SC | foliar | 0.27-1.4 | | | 1 | 7 |
| Mandarin | Japan | 200 SC | foliar | 0.27-1.4 | | | 1 | 7 |
| Melon | Japan | 200 SC | foliar | 0.3-0.6 | | | 1 | 1 |
| Mint | USA | 500 WP | foliar | 0.42-0.84 | | 470 | 1 | 7 |
| Nectarine | Australia | 480 SC | foliar | | 0.031 | 1000 | 1 | 3 |
| Non-bearing crops | USA | 500 WP | foliar | 0.42-0.56 | | 470 | 1 | |
| Okra | USA | 500 WP | foliar | 0.42-0.56 | | 470 | 1 | 3 |
| Orange | Japan | 200 SC | foliar | 0.27-1.4 | | | 1 | 7 |
| Other tree nuts[5] | USA | 500 WP | foliar | 0.42-0.56 | | 470 | 1 | 7 |
| Peach | Australia | 480 SC | foliar | | 0.031 | 1000 | 1 | 3 |

[3] Cucurbit vegetables include cucumbers, muskmelon, pumpkin, squash and watermelon.

[4] Fruiting vegetables include eggplants, peppers (bell and non-bell) and tomatoes. Use only on tomatoes greater than 1 inch in diameter when mature.

[5] Other tree nuts include Beech nut, Brazil nut, Butternut, Cashew, Chestnut, Hickory nut, Macadamia nut.

| Crop | Country | Application | | | | | | | |
|------|---------|------|------|-------------|-------------|------------------|---------------|-----------|
| | | Form | Type | Rate kg ai/ha | Conc kg ai/hL | Min spray vol, L/ha | Max number | PHI days |
| Peach | Japan | 200 SC | foliar | 0.27-1.4 | | | 1 | 7 |
| Pear | Australia | 480 SC | foliar | | 0.031 | 1000 | 1 | 7 |
| Pear | Japan | 200 SC | foliar | 0.27-1.4 | | | 1 | 1 |
| Pecan | USA | 500 WP | foliar | 0.42-0.56 | | 470 | 1 | 14 |
| Pistachios | USA | 500 WP | foliar | 0.42-0.56 | | 470 | 1 | 14 |
| Plum | Australia | 480 SC | foliar | | 0.031 | 1000 | 1 | 3 |
| Plum | Japan | 200 SC | foliar | 0.27-1.4 | | | 1 | 2 |
| Pome fruit[6] | USA | 500 WP | foliar | 0.42-0.56 | | 470 | 1 | 7 |
| Stone fruit[7] | USA | 500 WP | foliar | 0.42-0.56 | | 470 | 1 | 3 |
| Strawberries | Japan | 200 SC | foliar | 0.27-1.4 | | | 1 | 1 |
| Strawberries | USA | 500 WP | foliar | 0.42-0.56 | | 940 | 2 | 1 |
| Tea | Japan | 200 SC | foliar | 0.4-0.8 | | | 1 | 14 |
| Tomato | Japan | 200 SC | foliar | 0.3-0.6 | | | 1 | 1 |
| Walnuts | USA | 500 WP | foliar | 0.42-0.56 | | 470 | 1 | 14 |
| Watermelon | Japan | 200 SC | foliar | 0.3-0.6 | | | 1 | 1 |

## RESIDUES RESULTING FROM SUPERVISED TRIALS

The Meeting received information on supervised field trials for bifenazate uses on the following crops.

| | | |
|---|---|---|
| Citrus fruits | citrus | Table 19 |
| Pome fruits | apples | Table 20 |
| | pears | Table 21 |
| Stone fruits | apricot | Table 22 |
| | peach | Table 23 |
| | plums | Table 24 |
| | cherries | Table 25 |
| Berry fruits | grapes | Table 26 |
| | strawberries | Table 27 |
| Tropical fruits | figs | Table 28 |
| Cucurbits | cantaloupe | Table 29 |
| | watermelon | Table 30 |
| | cucumber | Table 31 |
| | summer squash | Table 32 |
| Fruiting vegetables | tomatoes | Table 33 |
| | peppers | Table 34 |
| | egg plant | Table 35 |
| Tree nuts | almonds, pecans | Table 36 |
| Oil seeds | cotton seed | Table 37 |
| Herbs | mint | Table 38 |
| | hops | Table 39 |
| | tea | Table 40 |

---

[6] Pome fruit include apple, crabapple, pear, quince.

[7] Stone fruit include nectarines, peach, plums/prunes.

| Animal feeds | almond hulls, cotton gin trash | Table 41 |

Trials from Japan were available only in summary form.

Trials from USA and Canada were generally well documented with laboratory and field reports. Laboratory reports included method validation with procedural recoveries from spiking at residue levels similar to those occurring in samples from the supervised trials. Dates of analyses or duration of residue sample storage were also provided. Although trials included control plots, no control data are recorded in the tables except where residues in control samples exceeded the LOQ. Residue data are recorded unadjusted for recovery.

In most trials, duplicate field samples from an unreplicated plot were taken at each sampling time and were analysed separately. The mean of the two analytical results was taken as the best estimate of the residues in the plot and the means are recorded in the tables.

When residues were not detected they are shown as below the LOQ (e.g. < 0.01 mg/kg). Residues, application rates and spray concentrations have generally been rounded to two significant figures or, for residues near the LOQ, to one significant figure. Residue values from the trials conducted according to maximum GAP have been used for the estimation of maximum residue levels. Those results included in the evaluation are double underlined.

Conditions of the supervised residue trials were generally well reported in detailed field reports. Most trial designs used non-replicated plots. Most field reports provided data on the sprayers used, plot size, field sample size and sampling date.

Table 18. Summary of sprayers, plot size and field sample size in the US supervised trials.

| Crop | Country | Year | Sprayer | Plot size | Sample size |
|---|---|---|---|---|---|
| Almond | USA | 2001 | tractor-mounted airblast | 16-18 trees | 1 kg |
| Apple | USA | 1998 | tractor-mounted airblast | 180-560 m$^2$ | 24 fruit |
| Apricot | USA | 2002 | airblast sprayer | 16-20 trees | 2 kg |
| Cantaloupe | USA | 2000 | tractor-mounted boom, backpack | 37-74 m$^2$ | 2 kg |
| Cherries | USA | 2001 | airblast, handgun sprayer | 54-500 m$^2$ | |
| Cotton seed | USA | 1999, 2000 | tractor-mounted boom, CO$_2$ powered, backpack, ATV | 90-1200 m$^2$ | 0.5-15 kg |
| Cucumber | USA | 2000 | tractor-mounted boom, backpack boom | 30-60 m$^2$ | 12 fruits |
| Grapes | USA | 1998 | tractor-mounted airblast | 75-250 m$^2$ | 12 bunches |
| Hops | USA | 1999 | airblast | 140-260 m$^2$ | 4 kg green |
| Mint tops | USA | 2000, 2001 | backpack boom, ATV-mounted boom, tractor-mounted boom | 33-90 m$^2$ | |
| Peach | USA | 1998, 2002 | airblast sprayer | 12-17 trees | 24 fruits |
| Pears | USA | 1998 | airblast sprayer | 16-18 trees | 2 kg |
| Pecan | USA | 2001 | tractor-mounted airblast | 6 trees | 1 kg |
| Peppers | USA | 2000 | tractor-mounted boom, backpack boom | 14-60 m$^2$ | |
| Plums | USA | 1998 | tractor-mounted airblast | 15-16 trees | 24-50 fruits |
| Strawberries | USA | 1999 | backpack sprayers | 26-90 m$^2$ | 1 kg |
| Summer squash | USA | 2000 | tractor-mounted boom, backpack boom | 20-90 m$^2$ | |
| Tomato | USA | 2000, 2001 | tractor-mounted boom, backpack boom | 8-110 m$^2$ | 12-15 fruit |

Intervals of freezer storage between sampling and analysis were recorded for all trials and were compared with intervals in the freezer storage stability studies.

Table 19. Bifenazate residues in citrus fruit resulting from supervised trials in Japan.

| CITRUS FRUITS | Application | | | | | PHI | Commodity | Bifenazate + bifenazate-diazene | Ref |
|---|---|---|---|---|---|---|---|---|---|
| country, year (variety) | Form | kg ai/ha | kg ai/hL | water (L/ha) | no. | days | 1/ | as bifenazate, mg/kg | |
| MANDARIN | | | | | | | | | |
| Japan, 1997 (Nankan 20 gou) | 200 SC | 1.2 | 0.02 | 6000 | 4 | 7<br>14<br>30<br>45 | pu, pe<br>pu, pe<br>pu, pe<br>pu, pe | 0.02, 2.0<br>0.02, 0.90<br>0.02, 1.4<br>0.02, 1.2 | Report No 1 Tokushima |
| Japan, 1997 (Nitinan 1 gou) | 200 SC | 2.0 | 0.02 | 10000 | 4 | 7<br>14<br>30<br>45 | pu, pe<br>pu, pe<br>pu, pe<br>pu, pe | < 0.02, 3.9<br>0.02, 3.7<br>< 0.02, 3.1<br>< 0.02, 2.4 | Report No 2 Miyazaki |
| NATSUDAIDAI | | | | | | | | | |
| Japan, 1997 (Kawano-Natsudai) | 200 SC | 1.0 | 0.02 | 5000 | 1 | 7<br>14<br>30<br>45 | pu, pe, wf<br>pu, pe, wf<br>pu, pe, wf<br>pu, pe, wf | 0.03, 0.72, 0.24<br>0.02, 0.64, 0.20<br>0.02, 0.37, 0.14<br>0.03, 0.44, 0.14 | Report No 3, 4 Yamaguti |
| Japan, 1997 (Kawano-Natsudai) | 200 SC | 1.2 | 0.02 | 6000 | 1 | 7<br>14<br>30<br>45 | pu, pe, wf<br>pu, pe, wf<br>pu, pe, wf<br>pu, pe, wf | < 0.02, 0.61, 0.21<br>< 0.02, 0.38, 0.15<br>< 0.02, 0.30, 0.11<br>< 0.02, 0.07, 0.04 | Report No 3, 4 Ehime |
| LIME | | | | | | | | | |
| Japan, 1997 (Sudati lime) | 200 SC | 1.2 | 0.02 | 6000 | 4 | 7<br>14<br>30<br>45 | wf | 0.27<br>0.08<br>0.10<br>0.10 | Report No 5 Tokushima |
| Japan, 1997 (Kabosu lime) | 200 SC | 1.4 | 0.02 | 7000 | 4 | 7<br>14<br>21<br>28 | wf | 0.30<br>0.27<br>0.13<br>0.07 | Report No 5 Ooita |

1/ pu: pulp;   pe: peel;   wf: whole fruit.

Table 20. Bifenazate residues in apples resulting from supervised trials in USA and Japan.

| APPLE | Application | | | | | PHI | Commodity | Bifenazate + bifenazate-diazene | Ref |
|---|---|---|---|---|---|---|---|---|---|
| country, year (variety) | Form | kg ai/ha | kg ai/hL | water (L/ha) | no. | days | | as bifenazate, mg/kg 1/ 2/ | |
| USA (NY) 1998, Idared | 500 WP | 2.8 | | 470 | 1 | 7 | whole fruit | 1.3 | RGC-98107 GRL-11419 |
| USA (WA) 1998, Red Delicious | 500 WP | 2.8 | | 460 | 1 | 7 | whole fruit | 2.0 | DNJ-98107 GRL-11419 |
| USA (NY) 1998, Monroe | 500 WP | 0.56 | 0.12 | 470 | 1 | 7<br>14<br>21 | whole fruit | 0.058<br>0.014<br>0.014 | GRL-11346 RGC-98104 |
| USA (PA) 1998, Red Delicious | 500 WP | 0.56 | 0.12 | 480 | 1 | 7<br>14<br>21 | whole fruit | 0.58<br>0.36<br>0.084 | GRL-11346 RGC-98105 |
| USA (PA) 1998, Law Rome | 500 WP | 0.56 | 0.12 | 480 | 1 | 7<br>14<br>21 | whole fruit | 0.20<br>0.13<br>0.074 | GRL-11346 RGC-98112 |

| APPLE | Application | | | | | PHI | Commodity | Bifenazate + bifenazate-diazene | Ref |
|-------|------|------|------|------|------|------|------|------|------|
| country, year (variety) | Form | kg ai/ha | kg ai/hL | water (L/ha) | no. | days | | as bifenazate, mg/kg 1/ 2/ | |
| USA (GA) 1998, Golden Delicious | 500 WP | 0.55 | 0.12 | 470 | 1 | 7<br>14<br>21 | whole fruit | 0.16<br>0.052<br>0.086 | GRL-11346 RCP-98105 |
| USA (MI) 1998, Empire | 500 WP | 0.56 | 0.12 | 460 | 1 | 7<br>14<br>21 | whole fruit | 0.15    c 0.019<br>0.16<br>0.10 | GRL-11346 JGC-98097 |
| USA (MI) 1998, Red Max | 500 WP | 0.56 | 0.12 | 480 | 1 | 7<br>14<br>21 | whole fruit | 0.22<br>0.20<br>0.11 | GRL-11346 JGC-98098 |
| USA (CO) 1998, Golden Delicious | 500 WP | 0.56 | 0.12 | 460 | 1 | 7<br>14<br>21 | whole fruit | 0.23    c 0.068<br>0.20<br>0.016 | GRL-11346 SWF-98101 |
| USA (CA) 1998, Golden Delicious | 500 WP | 0.58 | 0.12 | 500 | 1 | 7<br>14<br>21 | whole fruit | 0.18<br>0.17<br>0.11 | GRL-11346 CLS-98105 |
| USA (WA) 1998, Red Delicious | 500 WP | 0.55 | 0.12 | 460 | 1 | 7<br>14<br>21 | whole fruit | 0.18<br>0.15<br>0.072 | GRL-11346 DNJ-98102 |
| USA (WA) 1998, Red Delicious | 500 WP | 0.56 | 0.12 | 470 | 1 | 7<br>14<br>21 | whole fruit | 0.37<br>0.15<br>0.17 | GRL-11346 DNJ-98103 |
| USA (OR) 1998, Jonagold | 500 WP | 0.54 | 0.12 | 470 | 1 | 7<br>14<br>21 | whole fruit | 0.17    c 0.01<br>0.13<br>0.078 | GRL-11346 DNJ-98104 |
| USA (OR) 1998, Gala | 500 WP | 0.55 | 0.12 | 470 | 1 | 7<br>14<br>21 | whole fruit | 0.049<br>0.017<br>0.024 | GRL-11346 DNJ-98120 |
| USA (NY) 1998, Empire | 500 WP | 0.56 | 0.12 | 470 | 1 | 3<br>7<br>14<br>20<br>30 | whole fruit | 0.11<br>0.19<br>0.13<br>0.13<br>0.15 | GRL-11346 RGC-98103 |
| USA (WA) 1998, Red Delicious | 500 WP | 0.56 | 0.12 | 470 | 1 | 3<br>7<br>14<br>21<br>28 | whole fruit | 0.48<br>0.38<br>0.36<br>0.25<br>0.22 | GRL-11346 DNJ-98105 |
| Japan, 1997 (Fuji) | 200 SC | 1.2 | 0.02 | 6000 | 4 | 7<br>14<br>21<br>28 | whole fruit | 0.28<br>0.43<br>0.11<br>0.13 | Report No 6 Akita |
| Japan, 1997 (Fuji) | 200 SC | 1.2 | 0.02 | 6000 | 4 | 7<br>14<br>21<br>30 | whole fruit | 0.62<br>0.32<br>0.13<br>0.09 | Report No 6 Nagano |
| Japan, 2003 (Tsugaru) | 200 SC | 1.2 | 0.02 | 6000 | 1 | 1<br>3<br>7 | whole fruit | 0.57<br>0.32<br>0.24 | Report No 24 Iwate |
| Japan, 2003 (Tsugaru) | 200 SC | 1.0 | 0.02 | 5000 | 1 | 1<br>3<br>7 | whole fruit | 0.82<br>0.39<br>0.26 | Report No 24 Fukushima |

1/ mean of duplicate field samples (USA trials)

2/ c: sample from control plot.

Table 21. Bifenazate residues in pears resulting from supervised trials in USA and Japan.

| PEAR | Application | | | | | PHI | Commodity | Bifenazate + bifenazate-diazene | Ref |
|---|---|---|---|---|---|---|---|---|---|
| country, year (variety) | Form | kg ai/ha | kg ai/hL | water (L/ha) | no. | days | | as bifenazate, mg/kg 1/ 2/ 3/ | |
| USA (NY) 1998, Bartlett | 500 WP | 0.55 | 0.12 | 470 | 1 | 7 14 21 | whole fruit | 0.10 0.036 0.025 | GRL-11418 RGC-98108 |
| USA (PA) 1998, Bartlett | 500 WP | 0.56 | 0.12 | 480 | 1 | 7 14 21 | whole fruit | 0.24 0.077 0.11 | GRL-11418 RGC-98113 |
| USA (CA) 1998, Bartlett | 500 WP | 0.55 | 0.13 | 440 | 1 | 7 14 21 | whole fruit | 0.14 0.034 0.025 | GRL-11418 CLS-98106 |
| USA (CA) 1998, Bartlett | 500 WP | 0.56 | 0.12 | 470 | 1 | 7 14 21 | whole fruit | 0.076 0.13 0.082 | GRL-11418 CLS-98107 |
| USA (WA) 1998, Bartlett | 500 WP | 0.54 | 0.12 | 450 | 1 | 7 14 21 | whole fruit | 0.16 0.12    c 0.014 0.12 | GRL-11418 DNJ-98108 |
| USA (WA) 1998, D'Anjou | 500 WP | 0.55 | 0.12 | 480 | 1 | 7 14 21 | whole fruit | 0.094 0.056    c 0.01 0.074 | GRL-11418 DNJ-98109 |
| USA (OR) 1998, Red Clapp | 500 WP | 0.55 | 0.12 | 460 | 1 | 7 14 21 | whole fruit | 0.097    c 0.01 0.095 0.043 | GRL-11418 DNJ-98110 |
| USA (WA) 1998, D'Anjou | 500 WP | 0.53 | 0.12 | 450 | 1 | 7 14 21 | whole fruit | 0.29 0.19 0.099 | GRL-11418 DNJ-98121 |
| Japan, 1998 (Housui) | 200 SC | 1.2 | 0.02 | 6000 | 4 | 7 14 21 28 | whole fruit | 0.45 0.36 0.11 0.12 | Report No. 7 Nagano |
| Japan, 1998 (Kousui) | 200 SC | 1.2 | 0.02 | 6000 | 4 | 7 14 21 28 | whole fruit | 0.44 0.31 0.09 0.06 | Report No. 7 Ooita |
| Japan, 2000 (Kousui) | 200 SC | 1.2 | 0.02 | 6000 | 1 | 1 3 7 | whole fruit | 0.42 0.26 0.32 | Report No. 14 Nagano |
| Japan, 2000 (Kousui) | 200 SC | 1.2 | 0.02 | 6000 | 1 | 1 3 7 | whole fruit | 0.82 0.90 0.57 | Report No. 14 Nagano |
| Japan, 2001 (Kousui) | 200 SC | 0.80 | 0.02 | 4000 | 1 | 1 3 7 | whole fruit | 0.54 0.34 0.28 | Report No. 19 Fukushima |
| Japan, 2001 (Housui) | 200 SC | 0.40 | 0.02 | 2000 | 1 | 1 3 7 | whole fruit | 0.32 0.26 0.18 | Report No. 19 Saitama |
| Japan, 2001 (Kousui) | 200 SC | 0.7 | 0.02 | 3500 | 1 | 1 3 7 | whole fruit | 0.56 0.50 0.15 | Report No. 19 Ishikawa |
| Japan, 2001 (Kousui) | 200 SC | 1.0 | 0.02 | 5000 | 1 | 1 3 7 | whole fruit | 0.10 0.24 0.11 | Report No. 19 Tokushima |

1/ mean of duplicate field samples (USA trials)

2/ c: sample from control plot.

3/ Pear samples from the 8 US trials recorded in this table spent 15-16 months in frozen storage between harvest and analysis.

Table 22. Bifenazate residues in apricots resulting from supervised trials in USA.

| APRICOT country, year (variety) | Application | | | | | PHI days | Commodity | Bifenazate + bifenazate-diazene as bifenazate, mg/kg 1/ | Ref |
|---|---|---|---|---|---|---|---|---|---|
| | Form | kg ai/ha | kg ai/hL | water (L/ha) | no. | | | | |
| USA (CA) 2002, Royal Blenheim | 500 WP | 0.56 | 0.12 | 460 | 1 | 1 3 7 14 21 | whole fruit | 0.49 0.36 <u>0.44</u> 0.349 0.21 | CLS-02-102 RP-02007 |
| USA (CA) 2002, Royal Rosa | 500 WP | 0.56 | 0.12 | 460 | 1 | 3 | whole fruit | <u>0.23</u> | CLS-02-103 RP-02007 |
| USA (CA) 2002, Royal Blenheim | 500 WP | 0.57 | 0.12 | 470 | 1 | 2 | whole fruit | <u>0.30</u> | CLS-02-104 RP-02007 |
| USA (WA) 2002, Rival | 500 WP | 0.57 | 0.12 | 480 | 1 | 3 | whole fruit | <u>0.73</u> | DJN-02-101 RP-02007 |
| USA (CA) 2002, Castlebright | 500 WP | 0.56 | 0.12 | 470 | 1 | 3 | whole fruit | <u>0.59</u> | CEJ-02-101 RP-02007 |

1/ mean of duplicate field samples

Table 23. Bifenazate residues in peaches resulting from supervised trials in USA and Japan.

| PEACH country, year (variety) | Application | | | | | PHI days | Commodity 2/ | Bifenazate + bifenazate-diazene as bifenazate, mg/kg 1/ | Ref |
|---|---|---|---|---|---|---|---|---|---|
| | Form | kg ai/ha | kg ai/hL | water (L/ha) | no. | | | | |
| USA (GA) 2002, Redskin | 500 WP | 0.56 | 0.12 | 470 | 1 | 3 | whole fruit | <u>0.44</u> | KHG-02-101 RP-02007 |
| USA (LA) 2002, Tex Royal | 500 WP | 0.56 | 0.12 | 480 | 1 | 3 | whole fruit | <u>0.45</u> | AWD-02-901 RP-02007 |
| USA (PA) 1998, Red Haven | 500 WP | 0.56 | 0.12 | 460 | 1 | 3 7 14 | whole fruit | <u>0.55</u> 0.33 0.19 | RGC-98-500 RP-02007 RP-98006 |
| USA (SC) 1998, Harvesters | 500 WP | 0.55 | 0.12 | 480 | 1 | 3 7 14 | whole fruit | <u>0.23</u> 0.14 0.12 | RCP-98-102 RP-02007 RP-98006 |
| USA (SC) 1998, Contender | 500 WP | 0.55 | 0.11 | 490 | 1 | 3 7 14 | whole fruit | <u>0.23</u> 0.18 0.14 | RCP-98-103 RP-02007 RP-98006 |
| USA (GA) 1998, Redskin | 500 WP | 0.56 | 0.12 | 470 | 1 | 3 7 14 | whole fruit | <u>0.17</u> 0.11 0.06 | RCP-98-104 RP-02007 RP-98006 |
| USA (MI) 1998, Red Haven | 500 WP | 0.55 | 0.12 | 450 | 1 | 3 7 14 | whole fruit | <u>0.22</u> 0.19 0.05 | JGC-98-099 RP-02007 RP-98006 |
| USA (TX) 1998, June Gold | 500 WP | 0.56 | 0.12 | 480 | 1 | 3 7 14 | whole fruit | <u>1.2</u> 1.0 0.73 | AWD-98-202 RP-02007 RP-98006 |

| PEACH country, year (variety) | Application Form | kg ai/ha | kg ai/hL | water (L/ha) | no. | PHI days | Commodity 2/ | Bifenazate + bifenazate-diazene as bifenazate, mg/kg 1/ | Ref |
|---|---|---|---|---|---|---|---|---|---|
| USA (CA) 1998, Carnival | 500 WP | 0.57 | 0.12 | 470 | 1 | 1<br>3<br>7<br>14<br>21 | whole fruit | 0.46<br>0.40<br>0.26<br>0.17<br>0.19 | CEJ-98-108<br>RP-02007<br>RP-98006 |
| USA (CA) 1998, O'Henry | 500 WP | 0.57 | 0.12 | 480 | 1 | 3<br>7<br>14 | whole fruit | 0.15<br>0.16<br>0.10 | CEJ-98-109<br>RP-02007<br>RP-98006 |
| USA (CA) 1998, Yodel | 500 WP | 0.57 | 0.13 | 440 | 1 | 3<br>7<br>14 | whole fruit | 0.13<br>0.10<br>0.03 | CLS-98-110<br>RP-02007<br>RP-98006 |
| USA (CA) 1998, Loadel | 500 WP | 0.56 | 0.12 | 470 | 1 | 4<br>7<br>14 | whole fruit | 0.26<br>0.12<br>0.09 | CLS-98-111<br>RP-02007<br>RP-98006 |
| Japan, 1998 (Akatsuki) | 200 SC | 0.80 | 0.02 | 4000 | 4 | 7<br>14<br>21<br>28 | pulp<br>pulp<br>pulp<br>pulp | 0.02<br>0.02<br>< 0.02<br>< 0.02 | Report No. 8<br>Fukushima |
| Japan, 1998 (Hakuhou) | 200 SC | 1.2 | 0.02 | 6000 | 4 | 7<br>14<br>21<br>28 | pulp<br>pulp<br>pulp<br>pulp | < 0.02<br>< 0.02<br>0.02<br>< 0.02 | Report No. 8<br>Tokushima |
| Japan, 2003 (Akatsuki) | 200 SC | 0.8 | 0.02 | 4000 | 1 | 1<br>3<br>7 | pu, pe<br>pu, pe<br>pu, pe | < 0.02, 8.8<br>< 0.02, 9.7<br>< 0.02, 3.4 | Report No. 25 & 26<br>Fukushima |
| Japan, 2003 (Akatsuki) | 200 SC | 1.4 | 0.02 | 7000 | 1 | 1<br>3<br>7 | pu, pe<br>pu, pe<br>pu, pe | < 0.02, 6.9<br>< 0.02, 6.0<br>< 0.02, 3.8 | Report No. 25 & 26<br>Nagano |

1/ mean of duplicate field samples for US trials.

2/ pu: pulp;　pe: peel.

Table 24. Bifenazate residues in plums resulting from supervised trials in USA and Japan.

| PLUMS country, year (variety) | Application Form | kg ai/ha | kg ai/hL | water (L/ha) | no. | PHI days | Commodity | Bifenazate + bifenazate-diazene as bifenazate, mg/kg 1/ | Ref |
|---|---|---|---|---|---|---|---|---|---|
| USA (CA) 2002, Fortune | 500 WP | 0.56 | 0.12 | 470 | 1 | 3 | whole fruit | 0.034 | CEJ-02-105<br>RP-02007 |
| USA (MI) 1998, Stanley | 500 WP | 0.56 | 0.12 | 460 | 1 | 3<br>7<br>14 | whole fruit | 0.13<br>0.08<br>0.05 | JGC-98-100<br>RP-02007<br>RP-98006 |
| USA (CA) 1998, Fortune | 500 WP | 0.57 | 0.12 | 480 | 1 | 4<br>7<br>14 | whole fruit | 0.01<br>< 0.01<br>< 0.01 | CEJ-98-110<br>RP-02007<br>RP-98006 |
| USA (CA) 1998, Simka | 500 WP | 0.56 | 0.12 | 470 | 1 | 3<br>7<br>14 | whole fruit | 0.04<br>0.02<br>0.01 | CEJ-98-111<br>RP-02007<br>RP-98006 |
| USA (CA) 1998, French prune | 500 WP | 0.56 | 0.12 | 470 | 1 | 1<br>3<br>7<br>14<br>21 | whole fruit | 0.07<br>0.06<br>0.04<br>0.04<br>0.02 | CLS-98-109<br>RP-02007<br>RP-98006 |

| PLUMS | Application | | | | | PHI | Commodity | Bifenazate + bifenazate-diazene | Ref |
|---|---|---|---|---|---|---|---|---|---|
| country, year (variety) | Form | kg ai/ha | kg ai/hL | water (L/ha) | no. | days | | as bifenazate, mg/kg 1/ | |
| USA (CA) 1998, Moyer | 500 WP | 0.56 | 0.13 | 450 | 1 | 3<br>7<br>14 | whole fruit | 0.04<br>0.04<br>0.02 | CLS-98-108<br>RP-02007<br>RP-98006 |
| USA (OR) 1998, Parsons | 500 WP | 0.56 | 0.12 | 470 | 1 | 1<br>3<br>7<br>14<br>21 | whole fruit | 0.04<br>0.03<br>0.02<br>0.02<br>0.01 | DNJ-98-112<br>RP-02007<br>RP-98006 |
| USA (WA) 1998, Friar | 500 WP | 0.57 | 0.12 | 470 | 1 | 3<br>7<br>14 | whole fruit | 0.04<br>0.03<br>0.01 | DNJ-98-111<br>RP-02007<br>RP-98006 |
| Japan, 2001 (Ooishiwase) | SC 200 | 0.80 | 0.02 | 4000 | 2 | 3<br>7<br>14 | whole fruit | 0.14<br>0.20<br>0.04 | Report No. 22 Fukushima |
| Japan, 2001 (Ooishiwase) | SC 200 | 1.0 | 0.02 | 5000 | 2 | 3<br>7<br>14 | whole fruit | 0.05<br>0.14<br>0.06 | Report No. 22 Nagano |

1/ mean of duplicate field samples for US trials.

Table 25.Bifenazate residues in cherries resulting from supervised trials in USA and Japan.

| CHERRIES | Application | | | | | PHI | Commodity | Bifenazate + bifenazate-diazene | Ref |
|---|---|---|---|---|---|---|---|---|---|
| country, year (variety) | Form | kg ai/ha | kg ai/hL | water (L/ha) | no. | days | | as bifenazate, mg/kg 1/ | |
| USA (CA) 2001, Kings | 500 WP | 0.57 | 0.03 | 1640 | 1 | 3 | pitted fruits | 0.29 | 07054.01-CA51 |
| USA (CA) 2001, Brooks | 500 WP | 0.56 | 0.06 | 940 | 1 | 4 | pitted fruits | 0.23 | 07054.01-CA93 |
| USA (CO) 2001, Montmorency | 500 WP | 0.55 | 0.04 | 1240 | 1 | 3 | pitted fruits | 1.6 | 07054.01-CO12 |
| USA (ID) 2001, Lambert | 500 WP | 0.56 | 0.06 | 950 | 1 | 3 | pitted fruits | 0.11 | 07054.01-ID09 |
| USA (ID) 2001, Montmorency | 500 WP | 0.56 | 0.06 | 930 | 1 | 3 | pitted fruits | 0.48 | 07054.01-ID10 |
| USA (MI) 2001, Cavalier | 500 WP | 0.57 | 0.10 | 570 | 1 | 3 | pitted fruits | 0.20 | 07054.01-MI20 |
| USA (MI) 2001, Emperor Francis | 500 WP | 0.58 | 0.10 | 580 | 1 | 3 | pitted fruits | 0.42 | 07054.01-MI21 |
| USA (MI) 2001, Montmorency | 500 WP | 0.56 | 0.06 | 940 | 1 | 3 | pitted fruits | 0.89 | 07054.01-MI22 |
| USA (MI) 2001, Montmorency | 500 WP | 0.55 | 0.06 | 930 | 1 | 3 | pitted fruits | 0.71 | 07054.01-MI23 |
| USA (MI) 2001, Montmorency | 500 WP | 0.57 | 0.06 | 950 | 1 | 3 | pitted fruits | 1.2 | 07054.01-MI24 |

| CHERRIES | Application | | | | | PHI | Commodity | Bifenazate + bifenazate-diazene | Ref |
|---|---|---|---|---|---|---|---|---|---|
| country, year (variety) | Form | kg ai/ha | kg ai/hL | water (L/ha) | no. | days | | as bifenazate, mg/kg 1/ | |
| USA (MI) 2001, Montmorency | 500 WP | 0.56 | 0.05 | 930 | 1 | 3 | pitted fruits | 0.81 | 07054.01-MI25 |
| USA (NJ) 2001, Montmorency | 500 WP | 0.57 | 0.06 | 1020 | 1 | 4 | pitted fruits | 0.18 | 07054.01-NJ17 |
| USA (OR) 2001, Bing | 500 WP | 0.57 | 0.05 | 1170 | 1 | 4 | pitted fruits | 0.27 | 07054.01-OR13 |
| USA (WA) 2001, Bing | 500 WP | 0.57 | 0.04 | 1420 | 1 | 3 | pitted fruits | 0.34 | 07054.01-WA22 |
| Japan, 1998 (Koukanishiki) | 200 SC | 1.2 | 0.02 | 6000 | 4 | 14 21 28 42 | whole fruits | 0.17 0.21 0.03 0.02 | Report No 9 Iwate |
| Japan, 1998 (Satonishiki) | 200 SC | 1.2 | 0.02 | 6000 | 4 | 14 21 28 42 | whole fruits | 0.45 0.30 0.06 0.20 | Report No 9 Fukushima |

1/ mean of duplicate field samples for US trials.

Table 26. Bifenazate residues in grapes resulting from supervised trials in USA and Japan.

| GRAPES | Application | | | | | PHI | Commodity | Bifenazate + bifenazate-diazene | Ref |
|---|---|---|---|---|---|---|---|---|---|
| country, year (variety) | Form | kg ai/ha | kg ai/hL | water (L/ha) | no. | days | | as bifenazate, mg/kg 1/ 2/ | |
| USA (NY) 1998, Seyval Blanc | 500 WP | 0.57 | 0.12 | 470 | 1 | 14 21 | whole fruit | 0.31 0.17 | 7545 RGC-98-110 |
| USA (NY) 1998, Concord | 500 WP | 0.56 | 0.12 | 470 | 1 | 14 21 | whole fruit | 0.11 0.08 | 7545 RGC-98-111 |
| USA (CA) 1998, Flame Seedless | 500 WP | 0.56 | 0.12 | 470 | 1 | 14 21 | whole fruit | 0.10 0.10 | 7545 CEJ-98-112 |
| USA (CA) 1998, Thomson seedless | 500 WP | 0.57 | 0.12 | 470 | 1 | 14 21 | whole fruit | 0.07 0.06 | 7545 CEJ-98-113 |
| USA (CA) 1998, Chenin Blanc | 500 WP | 0.58 | 0.13 | 450 | 1 | 14 21 | whole fruit | 0.05 0.02 | 7545 CEJ-98-114 |
| USA (CA) 1998, Carigane | 500 WP | 0.58 | 0.12 | 470 | 1 | 14 21 | whole fruit | 0.33 0.17 | 7545 CLS-98-112 |
| USA (CA) 1998, Palomino | 500 WP | 0.56 | 0.12 | 470 | 1 | 14 21 | whole fruit | 0.20 0.07 | 7545 CLS-98-113 |
| USA (CA) 1998, Sauvignon Blanc | 500 WP | 0.58 | 0.12 | 480 | 1 | 14 21 | whole fruit | 0.21 0.19 | 7545 CLS-98-114 |

| GRAPES country, year (variety) | Application | | | | | PHI days | Commodity | Bifenazate + bifenazate-diazene as bifenazate, mg/kg 1/ 2/ | Ref |
|---|---|---|---|---|---|---|---|---|---|
| | Form | kg ai/ha | kg ai/hL | water (L/ha) | no. | | | | |
| USA (CA) 1998, Cabernet Sauvignon | 500 WP | 0.62 | 0.12 | 500 | 1 | 14 21 | whole fruit | 0.17 0.14 | 7545 CLS-98-115 |
| USA (CA) 1998, Thomson seedless | 500 WP | 0.56 | 0.12 | 470 | 1 | 14 21 | whole fruit | 0.55 0.48 | 7545 CLS-98-116 |
| USA (WA) 1998, Riesling | 500 WP | 0.55 | 0.12 | 460 | 1 | 14 21 | whole fruit | 0.17 0.15    c 0.06 | 7545 DNJ-98-118 |
| USA (WA) 1998, White Riesling | 500 WP | 0.56 | 0.12 | 470 | 1 | 14 21 | whole fruit | 0.29 0.21 | 7545 DNJ-98-119 |
| USA (CA) 1998, Thomson Seedless | 500 WP | 2.8 | | 470 | 1 | 3 7 14 21 28 | whole fruit whole fruit whole fruit whole fruit whole fruit | 1.05 0.61 0.30 0.30 0.23 | RP-98007  CEJ-98-115 |
| USA (CA) 1998, Thomson Seedless | 500 WP | 2.9 | | 480 | 1 | 3 7 14 21 28 | whole fruit whole fruit whole fruit whole fruit whole fruit | 1.1 0.72 0.19 0.08 0.12 | RP-98007  CEJ-98-116 |
| Japan, 1997 (Delaware) indoor | SC 200 | 0.80 | 0.02 | 4000 | 3 | 7 14 21 30 45 | whole fruit | 1.7 1.0 0.65 1.0 1.2 | Report No. 17 Akita |
| Japan, 1997 (Kyohou) indoor | SC 200 | 0.80 | 0.02 | 4000 | 3 | 7 14 21 30 44 | whole fruit | 0.46 0.34 0.34 0.47 0.14 | Report No. 17 Saitama |
| Japan, 1999 (Kyohou) indoor | SC 200 | 0.80 | 0.02 | 4000 | 3 | 14 21 28 42 | whole fruit | 0.16 0.13 0.15 0.19 | Report No. 18 Nagano |
| Japan, 1999 (Delaware) indoor | SC 200 | 0.80 | 0.02 | 4000 | 3 | 14 21 28 42 | whole fruit | 2.2 1.1 0.88 0.68 | Report No. 18 Ishikawa |

1/ mean of duplicate field samples for US trials.

2/ c: sample from control plot.

Table 27. Bifenazate residues in strawberries resulting from supervised trials in USA and Japan.

| STRAWBERRY country, year (variety) | Application | | | | | PHI days | Commodity | Bifenazate + bifenazate-diazene as bifenazate, mg/kg 1/ 2/ | Ref |
|---|---|---|---|---|---|---|---|---|---|
| | Form | kg ai/ha | kg ai/hL | water (L/ha) | no. | | | | |
| USA (PA) 1999, Northeaster | 500 WP | 0.56 | 0.06 | 940 | 2 | 1 3 | whole fruit | 0.68 0.43 | GRL-11517 RGC-99001 |
| USA (GA) 1999, Chandler | 500 WP | 0.55 | 0.06 | 910 | 2 | 1 3 | whole fruit | 0.93 0.81 | GRL-11517 RCP-99100 |
| USA (FL) 1999, Camarosa | 500 WP | 0.61 | 0.04 | 1630 | 2 | 1 3 | whole fruit | 0.49 0.44 | GRL-11517 PAK-99003 |
| USA (IN) 1999, Tribute | 500 WP | 0.56 | 0.06 | 940 | 2 | 1 3 | whole fruit | 1.0 0.41 | GRL-11517 JGC-99001 |
| USA (CA) 1999, Selva | 500 WP | 0.54 | 0.06 | 900 | 2 | 1 3 | whole fruit | 0.63 0.62 | GRL-11517 CLS-99101 |
| USA (CA) 1999, PS-952 | 500 WP | 0.55 | 0.06 | 920 | 2 | 1 3 | whole fruit | 0.23 0.29 | GRL-11517 CLS-99102 |
| USA (CA) 1999, Camarosa | 500 WP | 0.56 | 0.06 | 930 | 2 | 1 3 | whole fruit | 0.44 3.2  3/ | GRL-11517 CEJ-99101 |
| USA (OR) 1999, Totem | 500 WP | 0.56 | 0.06 | 940 | 2 | 1 3 | whole fruit | 0.53 0.24 | GRL-11517 DNJ-99101 |
| Japan, 1998 (Nyohou) indoor | 200 SC | 0.40 | 0.02 | 2000 | 1 | 1 3 7 | whole fruit | 0.89 0.53 0.36 | Report No. 10 Gunma |
| Japan, 1998 (Nyohou) indoor | 200 SC | 0.50 | 0.02 | 2500 | 1 | 1 3 7 | whole fruit | 0.82 1.10 0.33 | Report No. 10 Mie |
| Japan, 2002 (Nyohou) indoor | 200 SC | 0.50 | 0.02 | 2500 | 2 | 1 3 7 | whole fruit | 2.0 1.1 0.67 | Report No. 23 Gifu |
| Japan, 2002 (Akihime) indoor | 200 SC | 0.50 | 0.02 | 2500 | 2 | 1 3 7 | whole fruit | 0.38 0.19 0.15 | Report No. 23 Mie |
| Japan, 2002 (Nyohou) indoor | 150 FT | 100 g/ 400 m3 ~ 1.1 kg/ha | | | 2 | 1 3 7 | whole fruit | 0.24 0.10 < 0.05 | Report No. 23 Gifu |
| Japan, 2002 (Akihime) indoor | 150 FT | 100 g/ 400 m3 ~ 1.1 kg/ha | | | 2 | 1 3 7 | whole fruit | 0.05 < 0.05 < 0.05 | Report No. 23 Mie |

1/ mean of duplicate field samples for US trials.

2/ c: sample from control plot.

3/ The two values at the 3-days PHI in report CEJ-99101 were 3.4 and 2.9 mg/kg. The authors of the report discounted the values as being due to analytical error, based on the reasons that the values were 7-8 times as high as in the other trials and also much higher than the 1-day sample from the same trial.

Table 28. Bifenazate residues in figs resulting from supervised trials in Japan.

| FIGS country, year (variety) | Application | | | | | PHI days | Commodity | Bifenazate + bifenazate-diazene as bifenazate, mg/kg | Ref |
|---|---|---|---|---|---|---|---|---|---|
| | Form | kg ai/ha | kg ai/hL | water (L/ha) | no. | | | | |
| Japan, 2003 (Masui-Dofin) | SC 200 | 0.60 | 0.02 | 3000 | 1 | 1 3 7 | whole fruit | 0.53 0.22 0.10 | Report No. 27 Aichi, Anjo |
| Japan, 2003 (Masui-Dofin) | SC 200 | 0.60 | 0.02 | 3000 | 1 | 1 3 7 | whole fruit | 0.55 0.30 0.14 | Report No. 27 Aichi, Nagakute |

Table 29. Bifenazate residues in cantaloupe resulting from supervised trials in Canada, USA and Japan.

| CANTALOUPE country, year (variety) | Application | | | | | PHI days | Commodity | Bifenazate + bifenazate-diazene as bifenazate, mg/kg 1/ 2/ | Ref |
|---|---|---|---|---|---|---|---|---|---|
| | Form | kg ai/ha | kg ai/hL | water (L/ha) | no. | | | | |
| USA (TX) 2000, Hy-Mark | 500 WP | 0.57 | 0.18 | 310 | 1 | 1 3 7 | fruit | 0.31 0.10 0.05 | 07510.00-TX27 |
| USA (GA) 2000, Vienna | 500 WP | 0.56 | 0.20 | 290 | 1 | 4 | fruit | 0.03 | 07510.00-GA15 |
| USA (TX) 2000, Primo | 500 WP | 0.55 | 0.17 | 320 | 1 | 2 | fruit | 0.16    c 0.01 | 07510.00-TX28 |
| USA (CA) 2000, Hymark | 500 WP | 0.54 | 0.18 | 300 | 1 | 3 | fruit | 0.04 | 07510.00-CA44 |
| USA (CA) 2000, Sol Real | 500 WP | 0.54 | 0.18 | 300 | 1 | 4 | fruit | 0.04 | 07510.00-CA45 |
| USA (WI) 2000, Super Star | 500 WP | 0.58 | 0.20 | 300 | 1 | 3 | fruit | 0.05 | 07510.00-WI04 |
| USA (CA) 2000, Ambrosia | 500 WP | 0.57 | 0.28 | 200 | 1 | 3 | fruit | 0.04 | 07510.00-CA43 |
| USA (NJ) 2000, Ambrosiaar | 500 WP | 0.57 | 0.14 | 400 | 1 | 3 | fruit | 0.08 | 07510.00-NJ35 |
| Japan, 1999 (Natsukei2gou) indoor | 200 SC | 0.40 | 0.02 | 2000 | 2 | 1 3 7 14 | pulp pulp pulp pulp | 0.04 < 0.02 < 0.02 < 0.02 | Report No. 16 Nagano |
| Japan, 1999 (Aruseinu) indoor | 200 SC | 0.40 | 0.02 | 2000 | 2 | 1 3 7 14 | pulp pulp pulp pulp | < 0.02 < 0.02 < 0.02 < 0.02 | Report No. 16 Ishikawa |

1/ mean of duplicate field samples for trials in USA and Canada.

2/ c: sample from control plot.

Table 30. Bifenazate residues in watermelon resulting from supervised trials in Japan.

| WATERMELON country, year (variety) | Application | | | | | PHI | Commodity | Bifenazate + bifenazate-diazene as bifenazate, mg/kg | Ref |
|---|---|---|---|---|---|---|---|---|---|
| | Form | kg ai/ha | kg ai/hL | water (L/ha) | no. | days | | | |
| Japan, 1998 (Benikodam) indoor | 200 SC | 0.40 | 0.02 | 2000 | 3 | 1<br>3<br>7<br>14<br>21 | pulp<br>pulp<br>pulp<br>pulp<br>pulp | 0.03<br>0.02<br>< 0.02<br>< 0.02<br>< 0.02 | Report No. 11 Ishikawa. |
| Japan, 1998 (Madabowl-2gou) indoor | 200 SC | 0.40 | 0.02 | 2000 | 3 | 1<br>3<br>7<br>14<br>21 | pulp<br>pulp<br>pulp<br>pulp<br>pulp | 0.02<br>0.02<br>< 0.02<br>< 0.02<br>< 0.02 | Report No. 11 Miyazaki.. |

Table 31. Bifenazate residues in cucumber resulting from supervised trials in Canada, USA and Japan.

| CUCUMBER country, year (variety) | Application | | | | | PHI | Commodity | Bifenazate + bifenazate-diazene as bifenazate, mg/kg 1/ | Ref |
|---|---|---|---|---|---|---|---|---|---|
| | Form | kg ai/ha | kg ai/hL | water (L/ha) | no. | days | | | |
| USA (MD) 2000, Regal (pickling cucumber) | 500 WP | 0.55 | 0.12 | 470 | 1 | 3 | fruit | < 0.01 | 07511.00-MD09 |
| USA (NJ) 2000, Dasher II | 500 WP | 0.56 | 0.14 | 400 | 1 | 3 | fruit | 0.22 | 07511.00-NJ20 |
| USA (FL) 2000, Sawan Slicer Cucumber | 500 WP | 0.57 | 0.14 | 400 | 1 | 3 | fruit | 0.08 | 07511.00-FL44 |
| USA (GA) 2000, Thunder, Hybrid Cucumber | 500 WP | 0.56 | 0.20 | 290 | 1 | 3 | fruit | < 0.01 | 07511.00-GA16 |
| USA (TX) 2000, Calypso pickling | 500 WP | 0.56 | 0.25 | 220 | 1 | 3 | fruit | 0.03 | 07511.00-TX29 |
| Canada (ON), 2000, Dasher II | 500 WP | 0.55 | 0.28 | 200 | 1 | 2 | fruit | 0.07 | 07511.00-ON04 |
| Canada (ON), 2000, Dasher II | 500 WP | 0.52 | 0.15 | 350 | 1 | 3 | fruit | 0.05 | 07511.00-ON03 |
| USA (CA) 2000, Thunder cucumber | 500 WP | 0.54 | 0.18 | 300 | 1 | 3 | fruit | 0.04 | 07511.00-CA73 |
| Japan, 2001 (Haruka) indoor | SC 200 | 0.50 | 0.02 | 2500 | 1 | 1<br>3<br>7 | fruit | 0.12<br>0.05<br>< 0.01 | Report No. 20 Nagano |

| CUCUMBER country, year (variety) | Application | | | | | PHI days | Commodity | Bifenazate + bifenazate-diazene as bifenazate, mg/kg 1/ | Ref |
|---|---|---|---|---|---|---|---|---|---|
| | Form | kg ai/ha | kg ai/hL | water (L/ha) | no. | | | | |
| Japan, 2001 (Suiseisessei-2gou) indoor | SC 200 | 0.61 | 0.02 | 3040 | 1 | 1 3 7 | fruit | 0.14 0.08 < 0.01 | Report No. 20 Miyazaki |

1/ mean of duplicate field samples for US trials.

## Table 32. Bifenazate residues in summer squash resulting from supervised trials in USA.

| SUMMER SQUASH country, year (variety) | Application | | | | | PHI days | Commodity | Bifenazate + bifenazate-diazene as bifenazate, mg/kg 1/ | Ref |
|---|---|---|---|---|---|---|---|---|---|
| | Form | kg ai/ha | kg ai/hL | water (L/ha) | no. | | | | |
| USA (NY) 2000, Yellow Crookneck | 500 WP | 0.54 | 0.13 | 410 | 1 | 4 | fruit | 0.04 | 07512.00-NY15 |
| USA (NJ) 2000, Early Yellow Straightneck | 500 WP | 0.56 | 0.14 | 390 | 1 | 3 | fruit | 0.34 | 07512.00-NJ21 |
| USA (FL) 2000, Gentry CS4 hybrid squash | 500 WP | 0.56 | 0.20 | 280 | 1 | 3 | fruit | 0.12 | 07512.00-FL45 |
| USA (MD) 2000, Puma | 500 WP | 0.55 | 0.12 | 475 | 1 | 3 | fruit | < 0.01 | 07512.00-MD02 |
| USA (TX) 2000, Senator | 500 WP | 0.57 | 0.25 | 224 | 1 | 3 | fruit | 0.06 | 07512.00-TX30 |
| USA (CA) 2000, Patty Groon Tint | 500 WP | 0.54 | 0.27 | 200 | 1 | 3 | fruit | 0.02 | 07512.00-CA46 |
| USA (IN) 2000, Aristocrat | 500 WP | 0.56 | 0.24 | 230 | 1 | 3 | fruit | 0.01 | 07512.00-IN03 |

1/ mean of duplicate field samples

## Table 33. Bifenazate residues in tomatoes resulting from supervised trials in Canada, USA and Japan.

| TOMATOES country, year (variety) | Application | | | | | PHI days | Commodity | Bifenazate + bifenazate-diazene as bifenazate, mg/kg 1/ | Ref |
|---|---|---|---|---|---|---|---|---|---|
| | Form | kg ai/ha | kg ai/hL | water (L/ha) | no. | | | | |
| USA (FL), 2000, FL47 | 500 WP | 0.56 | 0.12 | 470 | 1 | 3 | fruit | 0.19 | 07266.00-FL42 GRL-FR-11673 |
| USA (FL), 2000, Celebrity | 500 WP | 0.56 | 0.12 | 470 | 1 | 3 | fruit | 0.14 | 07266.00-FL43 GRL-FR-11673 |
| USA (GA), 2000, Mountain Springs | 500 WP | 0.57 | 0.20 | 290 | 1 | 1 2 7 14 | fruit | 0.27 0.11 0.03 0.02 | 07266.00-GA14 GRL-FR-11673 |

| TOMATOES country, year (variety) | Application | | | | | PHI days | Commodity | Bifenazate + bifenazate-diazene as bifenazate, mg/kg 1/ | Ref |
|---|---|---|---|---|---|---|---|---|---|
| | Form | kg ai/ha | kg ai/hL | water (L/ha) | no. | | | | |
| USA (OH), 2000, Hypeel 696 | 500 WP | 0.56 | 0.14 | 410 | 1 | 2 | fruit | 0.10 | 07266.00-OH12 GRL-FR-11673 |
| USA (CA), 2000, Rio Grande | 500 WP | 0.54 | 0.18 | 300 | 1 | 3 | fruit | 0.07 | 07266.00-CA39 GRL-FR-11673 |
| USA (CA), 2000, Yaqui | 500 WP | 0.54 | 0.18 | 300 | 1 | 3 | fruit | 0.03 | 07266.00-CA40 GRL-FR-11673 |
| USA (CA), 2000, Celebrity | 500 WP | 0.58 | 0.16 | 370 | 1 | 3 | fruit | 0.03 | 07266.00-CA159 GRL-FR-11673 |
| USA (CA), 2000, H9553 | 500 WP | 0.58 | 0.16 | 360 | 1 | 3 | fruit paste puree 2/ | 0.09 0.11 0.49 | 07266.00-CA41 GRL-FR-11673 |
| USA (CA), 2000, Shady Lady | 500 WP | 0.58 | 0.15 | 380 | 1 | 3 | fruit | 0.04 | 07266.00-CA42 GRL-FR-11673 |
| USA (CA), 2000, 3135 VF | 500 WP | 0.55 | 0.14 | 380 | 1 | 3 | fruit | 0.13 | 07266.00-CA107 GRL-FR-11673 |
| USA (NJ), 2000, FT 4010 | 500 WP | 0.57 | 0.14 | 400 | 1 | 3 | fruit | 0.04 | 07266.00-NJ19 GRL-FR-11673 |
| USA (TX), 2001, Better Boy greenhouse | SC | 0.56 | | 960 | 1 | 0 | fruit | 0.21 | 08035-01-TX-03 GRL 11805 |
| USA (NJ), 2001, Floralina greenhouse | SC | 1.1 | | 1140 | 1 | 0 0 | fruit washed fruit | 0.44 0.15 | 08035-01-NJ-04 GRL 11805 |
| USA (TN), 2001, Celebrity greenhouse | SC | 0.56 | | 1400 | 1 | 0 | fruit | 0.16 | 08035-01-TN-02 GRL 11805 |
| Canada (Ontario), 2000, 9478 | 500 WP | 0.52 | 0.15 | 350 | 1 | 3 | fruit | 0.29 | 07266.00-ON02 GRL-FR-11673 |
| Canada (Quebec), 2000, Aclaim | 500 WP | 0.20 | 0.025 | 800 | 1 | 3 | fruit | 0.02 | 07266.00-QC02 GRL-FR-11673 |
| Japan, 2001 (House-Momotarou) indoor | SC 200 | 0.50 | 0.02 | 2500 | 1 | 1 7 14 | fruit | 0.32 0.11 0.10 | Report No. 21 Ibaragi |
| Japan, 2001 (Rokusanmaru) indoor | SC 200 | 0.50 | 0.02 | 2500 | 1 | 1 7 14 | fruit | 0.11 0.07 0.03 | Report No. 21 Nagano |

1/ mean of duplicate field samples for US and Canadian trials.

2/ Processing factors: tomato paste 1.26 ; tomato puree 5.6.

Table 34. Bifenazate residues in bell peppers resulting from supervised trials in Canada and USA.

| PEPPERS country, year (variety) | Application | | | | | PHI days | Commodity | Bifenazate + bifenazate-diazene as bifenazate, mg/kg 1/ 2/ | Ref |
|---|---|---|---|---|---|---|---|---|---|
| | Form | kg ai/ha | kg ai/hL | water (L/ha) | no. | | | | |
| USA (FL), 2000, Camelot – bell pepper | 500 WP | 0.56 | 0.20 | 280 | 1 | 3 | fruit | <u>1.1</u> | 07552.00-FL41 GRL-FR-11660 |
| USA (GA), 2000, Ceystone – bell pepper | 500 WP | 0.56 | 0.12 | 480 | 1 | 2 | fruit | <u>0.52</u>    c 0.02 | 07552.00-GA13 GRL-FR-11660 |
| USA (OH), 2000, King Arthur – bell pepper | 500 WP | 0.57 | 0.14 | 420 | 1 | 2 | fruit | <u>0.15</u> | 07552.00-OH11 GRL-FR-11660 |
| USA (MD), 2000, Boynton Bell – bell pepper | 500 WP | 0.56 | 0.12 | 480 | 1 | 3 | fruit | <u>0.32</u> | 07552.00-MD13 GRL-FR-11660 |
| USA (CA), 2000, Ivan – bell pepper | 500 WP | 0.54 | 0.19 | 290 | 1 | 3 | fruit | <u>0.24</u> | 07552.00-CA37 GRL-FR-11660 |
| USA (CA), 2000, Jupiter – bell pepper | 500 WP | 0.55 | 0.18 | 310 | 1 | 3 | fruit | <u>0.15</u> | 07552.00-CA38 GRL-FR-11660 |
| Canada (Ontario), 2000, King Arthur – bell pepper | 500 WP | 0.61 | 0.20 | 310 | 1 | 3 | fruit | <u>0.23</u> | 07552.00-ON01 GRL-FR-11660 |
| USA (TX), 2000, Capistrano – bell pepper | 500 WP | 0.55 | 0.12 | 450 | 1 | 3 | fruit | <u>0.13</u> | 07552.00-TX32 GRL-FR-11660 |
| USA (TX), 2000, Tam Veracruz – Jalopeno non-bell pepper | 500 WP | 0.56 | 0.13 | 420 | 1 | 3 | fruit | <u>1.1</u>    c 0.08 | 07552.00-TX52 GRL-FR-11660 |
| USA (FL), 2000, Mesilla - non-bell pepper | 500 WP | 0.56 | 0.20 | 280 | 1 | 3 | fruit | <u>1.6</u>    c 0.01 | 07552.00-FL40 GRL-FR-11660 |
| USA (NM), 2000, Sandia - non-bell pepper | 500 WP | 0.54 | 0.17 | 320 | 1 | 3 | fruit | <u>0.54</u>    c 0.03 | 07552.00-NM16 GRL-FR-11660 |

1/ mean of duplicate field samples.

2/ c: sample from control plot.

Table 35. Bifenazate residues in eggplant resulting from supervised trials in Japan.

| EGGPLANT | Application | | | | | PHI | Commodity | Bifenazate + bifenazate-diazene as bifenazate, mg/kg | Ref |
|---|---|---|---|---|---|---|---|---|---|
| country, year (variety) | Form | kg ai/ha | kg ai/hL | water (L/ha) | no. | days | | | |
| Japan, 1997 (Kokuyou) indoor | 200 SC | 0.40 | 0.02 | 2000 | 1 | 1<br>3<br>7 | whole fruit | 0.41<br>0.33<br>0.09 | Report No. 15 Ibaragi |
| Japan, 2000 (Itifuji) indoor | 200 SC | 0.40 | 0.02 | 2000 | 1 | 1<br>3<br>7 | whole fruit | 0.35<br>0.16<br>0.02 | Report No. 15 Gunma |

Table 36. Bifenazate residues in tree nuts resulting from supervised trials in USA.

| TREE NUTS | Application | | | | | PHI | Commodity | Bifenazate + bifenazate-diazene as bifenazate, mg/kg 1/ 2/ | Ref |
|---|---|---|---|---|---|---|---|---|---|
| country, year (variety) | Form | kg ai/ha | kg ai/hL | water (L/ha) | no. | days | | | |
| ALMONDS | | | | | | | | | |
| USA (CA), 2001, NonPareil | 500 WP | 0.84 | 0.18 | 470 | 1 | 3<br>7 | kernel<br>kernel | 0.07<br>0.05 | RP-01002 CEJ-01-101 |
| USA (CA), 2001, NonPareil | 500 WP | 0.84 | 0.18 | 460 | 1 | 3<br>7 | kernel<br>kernel | 0.04<br>0.03 | RP-01002 CEJ-01-102 |
| USA (CA), 2001, Mission | 500 WP | 0.85 | 0.18 | 470 | 1 | 3<br>7 | kernel<br>kernel | 0.01<br>0.01 | RP-01002 CEJ-01-103 |
| USA (CA), 2001, Peerless | 500 WP | 0.85 | 0.18 | 470 | 1 | 3<br>7 | kernel<br>kernel | 0.04  3/<br>0.02  4// | RP-01002 CLS-01-101 |
| USA (CA), 2001, Butte | 500 WP | 0.84 | 0.17 | 490 | 1 | 3<br>7 | kernel<br>kernel | 0.11  c 0.025<br>0.10 | RP-01002 CLS-01-102 |
| PECANS | | | | | | | | | |
| USA (LA), 2001, Elliot, Summer and Kiowa | 500 WP | 0.85 | 0.18 | 480 | 1 | 14 | kernel | < 0.01 | RP-01001 AWD-01-901 |
| USA (OK), 2001, Natives | 500 WP | 0.84 | 0.17 | 490 | 1 | 15 | kernel | < 0.01 | RP-01001 AWD-01-902 |
| USA (NM), 2001, Western Schley | 500 WP | 0.89 | 0.18 | 500 | 1 | 14 | kernel | 0.014 | RP-01001 SWG-01-001 |
| USA (GA), 2001, Stewart | 500 WP | 0.85 | 0.19 | 460 | 1 | 14 | kernel | 0.013 | RP-01001 KHG-01-001 |
| USA (GA), 2001, Cape Fear | 500 WP | 0.84 | 0.18 | 470 | 1 | 14 | kernel | < 0.01 | RP-01001 KHG-01-002 |

1/ mean of 3 replicate field samples.

2/ c: sample from control plot.

3/ Almonds were dried for 9 days before hulling and shelling.

4/ Almonds were dried for 7 days before hulling and shelling.

Note: LOQ for almond hulls was 0.2 mg/kg.

Table 37. Bifenazate residues in cotton seed resulting from supervised trials on cotton in USA.

| COTTON SEED | Application | | | | | PHI | Commodity | Bifenazate + bifenazate-diazene as bifenazate, mg/kg 1/ 2/ | Ref |
|---|---|---|---|---|---|---|---|---|---|
| country, year (variety) | Form | kg ai/ha | kg ai/hL | water (L/ha) | no. | days | | | |
| USA (SC) 2000, DP 458 | 500 WP | 0.84 | 0.45 | 190 | 1 | 60 | cotton seed | < 0.01 | 012351 RCP-00-001 |
| USA (MI) 2000, DP 20B | 500 WP | 0.84 | 0.44 | 190 | 1 | 60 | cotton seed | 0.02 | 012351 RDH-00-101 |
| USA (TX) 2000, Stoneville 489 2BRB | 500 WP | 0.83 | 0.46 | 180 | 1 | 60 | cotton seed | < 0.01 | 012351 AWD-00-901 |
| USA (NM) 2000, Acala 1517-95 | 500 WP | 0.83 | 0.43 | 190 | 1 | 60 | cotton seed | 0.03 | 012351 SWF-00-204 |
| USA (NM) 2000, Acala 1517-95 | 500 WP | 0.84 | 0.44 | 190 | 1 | 60 | cotton seed | 0.06 | 012351 SWF-00-305 |
| USA (TX) 2000, Paymaster HS2326 | 500 WP | 0.84 | 0.46 | 180 | 1 | 65 | cotton seed | 0.04 | 012351 SWF-00-406 |
| USA (CA) 2000, Acala Maxxa | 500 WP | 0.83 | 0.45 | 190 | 1 | 60 | cotton seed | < 0.01 | 012351 CEJ-00-105 |
| USA (CA) 2000, Acala Riata RR | 500 WP | 0.86 | 0.45 | 190 | 1 | 61 | cotton seed | 0.03 | 012351 CEJ-00-106 |
| USA (SC) 1999, DP 458 | 500 WP | 0.85 | 0.45 | 190 | 1 | 61 | cotton seed | < 0.01 | RP-99008 RCP-99-101 |
| USA (MI) 1999, DPL 20B | 500 WP | 0.82 | 0.44 | 190 | 1 | 59 | cotton seed | < 0.01 | RP-99008 RDH-99-001 |
| USA (LA) 1999, ST474 | 500 WP | 0.84 | 0.44 | 190 | 1 | 60 | cotton seed | < 0.01 | RP-99008 AWD-99-903 |
| USA (LA) 1999, DPL 458 | 500 WP | 0.85 | 0.46 | 180 | 1 | 60 | cotton seed | < 0.01 | RP-99008 AWD-99-904 |
| USA (OK) 1999, PM 145 | 500 WP | 0.84 | 0.45 | 190 | 1 | 60 | cotton seed | < 0.01 | RP-99008 AWD-99-906 |
| USA (TX) 1999, PM 2200RR | 500 WP | 0.85 | 0.45 | 190 | 1 | 59 | cotton seed | < 0.01 | RP-99008 SWF-99-801 |
| USA (TX) 1999, Excess | 500 WP | 0.84 | 0.45 | 190 | 1 | 42 | cotton seed | 0.02 | RP-99008 SWF-99-901 |
| USA (TX) 1999, DP 2156 | 500 WP | 0.84 | 0.45 | 190 | 1 | 59 | cotton seed | 0.01 | RP-99008 SWF-99-10A |
| USA (NM) 1999, Acala 1517-95 | 500 WP | 0.84 | 0.44 | 190 | 1 | 60 | cotton seed | 0.06 | RP-99008 SWF-99-20A |

| COTTON SEED | Application | | | | | PHI | Commodity | Bifenazate + bifenazate-diazene as bifenazate, mg/kg 1/ 2/ | Ref |
|---|---|---|---|---|---|---|---|---|---|
| country, year (variety) | Form | kg ai/ha | kg ai/hL | water (L/ha) | no. | days | | | |
| USA (NM) 1999, Acala 1517-95 | 500 WP | 5.3 | | | 1 | 60 | cotton seed meal refined oil | 1.05 < 0.01 < 0.01 | RP-99008 SWF-99-20A |
| USA (CA) 1999, CB 232 | 500 WP | 0.82 | 0.45 | 180 | 1 | 61 | cotton seed | < 0.01 | RP-99008 CLS-99-109 |
| USA (CA) 1999, Acala Maxxa | 500 WP | 0.85 | 0.45 | 190 | 1 | 60 | cotton seed | 0.04 | RP-99008 CEJ-99-105 |
| USA (CA) 1999, GC 500 | 500 WP | 0.84 | 0.44 | 190 | 1 | 61 | cotton seed | 0.28 | RP-99008 CEJ-99-106 |
| USA (CA) 1999, GC 500 | 500 WP | 5.3 | | | 1 | 60 | cotton seed meal refined oil | 2.64 < 0.01 < 0.01 | RP-99008 CEJ-99-106 |

1/ mean of duplicate field samples.

2/ c: sample from control plot.

Table 38. Bifenazate residues in mint resulting from supervised trials in USA.

| MINT | Application | | | | | PHI | Commodity | Bifenazate + bifenazate-diazene as bifenazate, mg/kg 1/ 2/ | Ref |
|---|---|---|---|---|---|---|---|---|---|
| country, year (variety) | Form | kg ai/ha | kg ai/hL | water (L/ha) | no. | days | | | |
| USA (WA), 2000, Native spearmint | 500 WP | 0.82 | 0.17 | 480 | 1 | 8 | mint tops | 15.4 | 07386.00-WA49 GRL-FR-11712 |
| USA (WI), 2000, Scotch spearmint, H. Darling strain | 500 WP | 0.84 | 0.27 | 310 | 1 | 8 | mint tops | 12.9 | 07386.00-WI21 GRL-FR-11712 |
| USA (WI), 2000, Black Mitchem peppermint | 500 WP | 0.84 | 0.28 | 310 | 1 | 8 | mint tops | 18.1 | 07386.00-WI22 GRL-FR-11712 |
| USA (WA), 2000, Scotch spearmint | 500 WP | 0.83 | 0.19 | 440 | 1 | 7 | mint tops | 6.6 | 07386.00-WA20 GRL-FR-11712 |
| USA (WA), 2000, Native mint | 500 WP | 0.84 | 0.19 | 440 | 1 | 7 | mint tops | 6.4      c 0.01 | 07386.00-WA21 GRL-FR-11712 |
| USA (WA), 2001, Spearmint (Native) | 500 WP | 1.8 | | 290 | 1 | 7 7 | mint tops mint oil | 20 1.6 0.90 0.97 3/ | RP-01015 DNJ-01-101 |

1/ mean of duplicate field samples.

2/ c: sample from control plot.

3/ 3 processing runs.

Table 39. Bifenazate residues in hops resulting from supervised trials in USA.

| HOPS country, year (variety) | Application Form | kg ai/ha | kg ai/hL | water (L/ha) | no. | PHI days | Commodity 2/ | Bifenazate + bifenazate-diazene as bifenazate, mg/kg 1/ | Ref |
|---|---|---|---|---|---|---|---|---|---|
| USA (WA) 1999, Nugget | 500 WP | 0.85 | 0.18 | 470 | 1 | 13 | dried hops | 9.3 | RP-99006 DNJ-99-103 |
| USA (WA) 1999, Nugget | 500 WP | 0.85 | 0.18 | 470 | 1 | 14 | dried hops | 7.8 | RP-99006 DNJ-99-104 |
| USA (OR) 1999, Williamette | 500 WP | 0.85 | 0.18 | 470 | 1 | 14 | dried hops | 7.1 | RP-99006 DNJ-99-105 |

1/ mean of 3 replicate field samples.

2/ Samples of green hops from the field were taken to commercial hop dryers for drying at 52 °C and, one day later, they were removed as approximately 1 kg of dried hops.

Table 40. Bifenazate residues in tea resulting from supervised trials in Japan.

| TEA country, year (variety) | Application Form | kg ai/ha | kg ai/hL | water (L/ha) | no. | PHI days | Commodity 1/ | Bifenazate + bifenazate-diazene as bifenazate, mg/kg | Ref |
|---|---|---|---|---|---|---|---|---|---|
| Japan, 1998 (Ooiwase) | 200 SC | 0.80 | 0.02 | 4000 | 3 | 7 | leaves | 6.5 | Report No. 12 and 13 Shizuoka |
|  |  |  |  |  |  | 14 | leaves | 0.82 |  |
|  |  |  |  |  |  | 21 | leaves | < 0.1 |  |
|  |  |  |  |  |  | 7 | tea extract | 1.6 |  |
|  |  |  |  |  |  | 14 | tea extract | 0.22 |  |
|  |  |  |  |  |  | 21 | tea extract | < 0.1 |  |
| Japan, 1998 (Yabukita) | 200 SC | 0.80 | 0.02 | 4000 | 3 | 7 | leaves | 24 | Report No. 12 and 13 Fukuoka |
|  |  |  |  |  |  | 14 | leaves | 0.53 |  |
|  |  |  |  |  |  | 20 | leaves | 0.10 |  |
|  |  |  |  |  |  | 7 | tea extract | 5.4 |  |
|  |  |  |  |  |  | 14 | tea extract | 0.14 |  |
|  |  |  |  |  |  | 20 | tea extract | < 0.1 |  |

1/ Tea extract: Tea leaves (6 g) were mixed with 360 ml of distilled water at 100 °C for 5 minutes. After filtration, the tea extract was analysed.

Table 41. Bifenazate residues in feed commodities resulting from supervised trials in USA.

| FEED country, year (variety) | Application Form | kg ai/ha | kg ai/hL | water (L/ha) | no. | PHI days | Commodity | Bifenazate + bifenazate-diazene as bifenazate, mg/kg 1/ 2/ | Ref |
|---|---|---|---|---|---|---|---|---|---|
| ALMOND HULLS |  |  |  |  |  |  |  |  |  |
| USA (CA), 2001, NonPareil | 500 WP | 0.84 | 0.18 | 470 | 1 | 3 | hull | 4.4 | RP-01002 CEJ-01-101 |
|  |  |  |  |  |  | 7 | hull | 5.0 |  |
| USA (CA), 2001, NonPareil | 500 WP | 0.84 | 0.18 | 460 | 1 | 3 | hull | 7.3 | RP-01002 CEJ-01-102 |
|  |  |  |  |  |  | 7 | hull | 6.9 |  |

| FEED country, year (variety) | Application | | | | | PHI days | Commodity | Bifenazate + bifenazate-diazene as bifenazate, mg/kg 1/ 2/ | Ref |
|---|---|---|---|---|---|---|---|---|---|
| | Form | kg ai/ha | kg ai/hL | water (L/ha) | no. | | | | |
| USA (CA), 2001, Mission | 500 WP | 0.85 | 0.18 | 470 | 1 | 3<br>7 | hull<br>hull | 5.0<br>2.8 | RP-01002 CEJ-01-103 |
| USA (CA), 2001, Peerless | 500 WP | 0.85 | 0.18 | 470 | 1 | 3<br>7 | hull<br>hull | 3.9  3/<br>1.8  4/ | RP-01002 CLS-01-101 |
| USA (CA), 2001, Butte | 500 WP | 0.84 | 0.17 | 490 | 1 | 3<br>7 | hull<br>hull | 6.0<br>5.1 | RP-01002 CLS-01-102 |
| COTTON GIN TRASH | | | | | | | | | |
| USA (NM) 2000, Acala 1517-95 | 500 WP | 0.84 | 0.44 | 190 | 1 | 60 | gin trash | 3.8 | 012351   SWF-00-305 |
| USA (TX) 2000, Paymaster HS2326 | 500 WP | 0.84 | 0.46 | 180 | 1 | 65 | gin trash | 0.69 | 012351   SWF-00-406 |
| USA (CA) 2000, Acala Riata RR | 500 WP | 0.86 | 0.45 | 190 | 1 | 61 | gin trash | 2.5      c 0.02 | 012351   CEJ-00-106 |
| USA (SC) 1999, DP 458 | 500 WP | 0.85 | 0.45 | 190 | 1 | 61 | gin trash | 0.88 | RP-99008  RCP-99-101 |
| USA (LA) 1999, DPL 458 | 500 WP | 0.85 | 0.46 | 180 | 1 | 60 | gin trash | 1.3 | RP-99008  AWD-99-904 |
| USA (TX) 1999, PM 2200RR | 500 WP | 0.85 | 0.45 | 190 | 1 | 59 | gin trash | 0.07 | RP-99008  SWF-99-801 |
| USA (TX) 1999, Excess | 500 WP | 0.84 | 0.45 | 190 | 1 | 42 | gin trash | 0.46 | RP-99008  SWF-99-901 |
| USA (TX) 1999, DP 2156 | 500 WP | 0.84 | 0.45 | 190 | 1 | 59 | gin trash | 0.39 | RP-99008  SWF-99-10A |
| USA (NM) 1999, Acala 1517-95 | 500 WP | 0.84 | 0.44 | 190 | 1 | 60 | gin trash | 4.0 | RP-99008  SWF-99-20A |
| USA (NM) 1999, Acala 1517-95 | 500 WP | 5.3 | | | 1 | 60 | gin trash | 30 | RP-99008  SWF-99-20A |
| USA (CA) 1999, GC 500 | 500 WP | 0.84 | 0.44 | 190 | 1 | 61 | gin trash | 18 | RP-99008  CEJ-99-106 |
| USA (CA) 1999, GC 500 | 500 WP | 5.3 | | | 1 | 60 | gin trash | 110 | RP-99008  CEJ-99-106 |

1/ mean of duplicate samples

2/ c: sample from control plot.

3/ Almonds were dried for 9 days before hulling and shelling.

4/ Almonds were dried for 7 days before hulling and shelling.

Note: LOQ for almond hulls was 0.2 mg/kg.

## FATE OF RESIDUES IN STORAGE AND PROCESSING

### In processing

The Meeting received information on the fate of bifenazate residues during the juicing of apples, the drying of prunes, the production of grape juice and raisins, the production of tomato paste and puree, the production of cotton seed oil and the processing of mint tops.

Gaydosh (2000, GRL-11419) processed apples, from two bifenazate field trials with exaggerated (5×) application rates, into juice and wet pomace using a small-scale process. The process was suitable for 20–50 kg of apples. Apples were ground in a hammer-mill and the wet mash was collected into clothsacks on a hydraulic press. The sack was pressed at 2200–3000 psi for a minimum of 5 minutes and juice was collected. The wet pomace cake within the bags was sampled as wet pomace. Residues were measured on whole fruit, juice and wet pomace (Table 42).

Table 42. Bifenazate residues in apple juice and wet pomace from processing trials in USA (Gaydosh, 2000, GRL-11419).

| APPLE | Application | | | | | PHI | Commodity | Bifenazate + bifenazate-diazene as bifenazate, mg/kg 1/ 2/ | Ref |
|---|---|---|---|---|---|---|---|---|---|
| country, year (variety) | Form | kg ai/ha | kg ai/hL | water (L/ha) | no. | days | | | |
| USA (NY) 1998, Idared | 500 WP | 2.8 | | 470 | 1 | 7 7 | whole fruit whole fruit juice wet pomace | 1.3 3/ 0.89 4/ 0.20 c 0.14 1.6 | RGC-98107 GRL-11419 |
| USA (WA) 1998, Red Delicious | 500 WP | 2.8 | | 460 | 1 | 7 7 | whole fruit whole fruit juice wet pomace | 2.0 3/ 2.1 4/ 0.22 c 0.14 3.6 | DNJ-98107 GRL-11419 |

1/ mean of duplicate samples

2/ c control juice from untreated apples. On a second analysis of these juice samples, no residue was detected (< 0.005 mg/kg).

3/ field samples taken at the same time as the fruit for processing

4/ sampled at the processing laboratory

Korpalski and Puhl (2000, RP-98006) described the processing to dried prunes of plums treated in the field with bifenazate (Table 43). At each of two sites plums (approximately 22 kg) were harvested for drying at maturity 3 days after a bifenazate treatment. In the process, plums were washed, placed in mesh bags and dried in a drying tunnel for 18 to 27 hours at 71–88 °C simulating a commercial process. Approximately 7 and 9 kg of prunes were produced from 22 kg of fresh plums at the two sites.

Table 43. Bifenazate residues in prunes resulting from processing trials in USA.

| PLUMS | Application | | | | | PHI | Commodity | Bifenazate + bifenazate-diazene as bifenazate, mg/kg 1/ | Ref |
|---|---|---|---|---|---|---|---|---|---|
| country, year (variety) | Form | kg ai/ha | kg ai/hL | water (L/ha) | no. | days | | | |
| USA (CA) 1998, French prune | 500 WP | 0.56 | 0.12 | 470 | 1 | 3 3 | whole fruit whole fruit dried prune | 0.06 2/ 0.02 3/ 0.01 | CLS-98-109 RP-02007 RP-98006 |

| PLUMS | Application | | | | | PHI | Commodity | Bifenazate + bifenazate-diazene as bifenazate, mg/kg 1/ | Ref |
|---|---|---|---|---|---|---|---|---|---|
| country, year (variety) | Form | kg ai/ha | kg ai/hL | water (L/ha) | no. | days | | | |
| USA (OR) 1998, Parsons | 500 WP | 0.56 | 0.12 | 470 | 1 | 3 | whole fruit dried prune | 0.03  2/ < 0.01 | DNJ-98-112 RP-02007 RP-98006 |

1/ mean of duplicate samples.

2/ field samples taken at the same time as the fruit for processing..

3/ sampled at the processing laboratory.

Korpalski (1999, RP-98007) described the processing of field-treated grapes into juice and raisins.

Approximately 22 and 45 kg of grapes were available for processing in the two trials. Fresh grapes were removed from storage and fed into a crusher/stemmer. The grape pulp was collected and the stems were discarded. The collected grape pulp was heated in a steam kettle to 52–57°C for 10 minutes and then to 60–66°C for another 10 minutes. The heated grape pulp was pressed to separate the juice and wet pomace, which was discarded. The juice was filtered and placed in the freezer for analysis.

Grapes for raisins were placed on trays or paper to dry. After 7 days the raisin samples were turned. After 14 days the stems were removed and the raisin samples were placed in bags and sent to the laboratory.

Korpalski and Puhl (2002, RP-01015) described the processing of mint tops to mint oil. Approximately 11 kg of fresh mint tops were subject to steam distillation in a cooker to produce 33–43 mL of oil in the 4 process runs (1 control and 3 treated samples). The oil samples were filtered, refrigerated and sent to the laboratory.

Dorschner (2002, 07266) descibed the processing of tomatoes to produce paste and puree. Tomatoes were first cleaned and then soaked for 3 minutes in a dilute sodium hydroxide solution, then thoroughly rinsed. Tomatoes were then chopped and rapidly heated to about 80°C and skin and seeds were separated from juice. Juice was evaporated to produce a puree. Further concentration and addition of salt produced a paste that was heated to approximately 85°C and canned. Residue data are summarised in Table 45.

Belcher (2000, RP-99008) described the processing of harvested seed cotton to cotton seed and refined oil. The seed cotton was first dried and the burrs, sticks and other trash removed. Ginning then removed most of the lint and the ginned seed was further delinted to produce cotton seed containing approximately 3% lint. A mill cracked the seed and removed most of the hull material. Kernel material was then heated to 80–90°C for 15–30 minutes and then flaked and extruded with a flaking mill and steam treated. The material was then dried at 65–80°C ready for solvent extraction with hot hexane. Hexane was evaporated from the extracted material to produce meal. After hexane was removed from the oil by vacuum evaporation, the oil was refined. Residue data for the meal and oil are summarised in Table 45.

Table 44. Bifenazate residues in grapes and processed commodities resulting from supervised trials in USA.

| GRAPES | Application | | | | | PHI | Commodity | Bifenazate + bifenazate-diazene | Ref |
|---|---|---|---|---|---|---|---|---|---|
| country, year (variety) | Form | kg ai/ha | kg ai/hL | water (L/ha) | no. | days | | as bifenazate, mg/kg 1/ | |
| USA (CA) 1998, Thomson Seedless | 500 WP | 2.8 | | 470 | 1 | 14 14 14 | whole fruit juice raisins | 0.28 0.02  0.01 0.07  0.13 | RP-98007  CEJ-98-115 |
| USA (CA) 1998, Thomson Seedless | 500 WP | 2.9 | | 480 | 1 | 15 15 15 | whole fruit juice raisins | 0.115 0.02  0.02 0.40  0.34 | RP-98007  CEJ-98-116 |

1/ The residue on the whole fruit is the mean of duplicate field samples
The two residues for juice and raisins represent duplicate processing runs.

Table 45. Bifenazate residues in raw and processed commodities resulting from supervised trials on tomatoes, mint and cotton in USA.

| CROP | Application | | | | | PHI | Commodity | Bifenazate + bifenazate-diazene | Ref |
|---|---|---|---|---|---|---|---|---|---|
| country, year (variety) | Form | kg ai/ha | kg ai/hL | water (L/ha) | no. | days | | as bifenazate, mg/kg | |
| TOMATOES USA (CA), 2000, H9553 | 500 WP | 0.58 | 0.16 | 360 | 1 | 3 | fruit puree paste | 0.09 0.49 0.11 | 07266.00-CA41 GRL-FR-11673 |
| MINT USA (WA), 2001, Spearmint (Native) | 500 WP | 1.8 | | 290 | 1 | 7 7 | mint tops mint oil | 20 1.6  0.90  0.97 (3 processing runs) | RP-01015  DNJ-01-101 |
| COTTON USA (NM) 1999, Acala 1517-95 | 500 WP | 5.3 | | | 1 | 60 | cotton seed hulls meal refined oil | 1.05 0.11 < 0.01 < 0.01 | RP-99008  SWF-99-20A |
| COTTON USA (CA) 1999, GC 500 | 500 WP | 5.3 | | | 1 | 60 | cotton seed hulls meal refined oil | 2.64 0.92 < 0.01 < 0.01 | RP-99008  CEJ-99-106 |

Table 46. Summary of processing factors for bifenazate residues. The factors are calculated from the data recorded in tables in this section.

| Raw agricultural commodity (RAC) | Processed commodity | Calculated processing factors. | Mean or best estimate |
|---|---|---|---|
| Apple | wet pomace | 1.8, 1.7 | 1.8 |
| Apples | apple juice | 0.23, 0.10 | 0.17 |
| Cotton seed | cotton seed hulls | 0.105, 0.35 | 0.23 |
| Cotton seed | cotton seed meal | < 0.0095, < 0.0038 | < 0.0038 |
| Cotton seed | cotton seed refined oil | < 0.0095, < 0.0038 | < 0.0038 |
| Grapes | grape juice | 0.054, 0.17 | 0.11 |
| Grapes | raisins | 0.36, 3.2 | 3.2 |
| Mint tops | mint oil | 0.080, 0.045, 0.049 | 0.057 |

| Raw agricultural commodity (RAC) | Processed commodity | Calculated processing factors. | Mean or best estimate |
|---|---|---|---|
| Plums | dried prunes | 0.5, < 0.3 | 0.5 |
| Tomato | tomato paste | 1.26 | 1.3 |
| Tomato | tomato puree | 5.6 | 5.6 |

## RESIDUES IN ANIMAL COMMODITIES

### Farm animal feeding studies

The meeting received a lactating dairy cow feeding study, which provided information on likely residues resulting in animal tissues and milk from residues in the animal diet.

Groups of three lactating Holstein dairy cows (animals weighing 437–591 kg and 460–619 kg on days 1 and 28 respectively) were dosed once daily via gelatin capsule with bifenazate at 1 ppm (1×), 3 ppm (3×) and 10 ppm (10×) in the dry-weight diet, for 28 consecutive days (Wiedmann and Jablonski, 1999, 7474). Milk was collected twice daily for analysis and pooled from the morning and evening milkings. Butterfat and skim milk samples were taken on days 20 and 28. On day 29, within 24 hours of the final dose, the animals were slaughtered for tissue collection. Tissues collected for analysis were liver, kidney, perirenal fat, omental fat, round muscle and loin muscle. Animals consumed approximately 21–29 kg feed (approx 88% dry matter) each per day and produced approximately 16–24 kg milk per animal per day (means for each animal through the test period). Samples were analysed for bifenazate + bifenazate-diazene by HPLC-coulometer method 7473-98-0115CR-001 and for 4-hydroxybiphenyl and its sulphate conjugate by HPLC with fluorescence detection.

Residue samples were stored at freezer temperatures (below -10°C) awaiting extraction and analysis. Intervals of storage were: muscle, liver and kidney 1 day; fat 3–14 days; milk 5–112 days; butterfat 15–23 days and skim milk 95–103 days. Residues in muscle, liver and kidney decline very quickly, while the residues are much more stable in milk and fat matrices.

Residues of bifenazate + bifenazate-diazene did not exceed the LOQ (0.01 mg/kg) in loin muscle, round muscle, liver, milk or skim milk at the highest dosing level 10 ppm (Table 47). Residues were detected in the kidney of one animal at 0.01 mg/kg. Residues were present in omental and perirenal fat in the 3 ppm group (0.01–0.03 mg/kg) and the 10 ppm group (0.03–0.10 mg/kg), but not in the 1 ppm group. Residues were also present in butterfat from the 10 ppm group (0.01–0.03 mg/kg) but not from the 3 ppm group. Residues of 4-hydroxybiphenyl and its sulphate conjugate did not exceed the LOQ (0.01 mg/kg) in any sample of tissue, milk or butterfat.

Table 47. Residues in milk and tissues of lactating dairy cows (3 per group) dosed once daily via gelatin capsule with bifenazate at 1 ppm (1×), 3 ppm (3×) and 10 ppm (10×) in the dry-weight diet, for 28 consecutive days (Wiedmann and Jablonski, 1999, 7474).

| Substrate | Residues, mg/kg 1/ | | | | | |
|---|---|---|---|---|---|---|
| | Dosing, 1 ppm | | Dosing, 3 ppm | | Dosing, 10 ppm | |
| | bifenazate + bifenazate-diazene | 4-hydroxy biphenyl + sulphate conj | bifenazate + bifenazate-diazene | 4-hydroxy biphenyl + sulphate conj | bifenazate + bifenazate-diazene | 4-hydroxy biphenyl + sulphate conj |
| Loin muscle | | | | | < 0.01 (3) | < 0.01 (3) |
| Round muscle | | | | | < 0.01 (3) | < 0.01 (3) |
| Liver | | | | | < 0.01 (3) | < 0.01 (3) |
| Kidney | | | | | 0.01 < 0.01 (2) | < 0.01 (3) |
| Omental fat | < 0.01 (3) | < 0.01 (3) | 0.01 0.02 0.02 | < 0.01 (3) | 0.03 0.06 0.07 | < 0.01 (3) |

| Substrate | Residues, mg/kg 1/ | | | | | |
|---|---|---|---|---|---|---|
| | Dosing, 1 ppm | | Dosing, 3 ppm | | Dosing, 10 ppm | |
| | bifenazate + bifenazate-diazene | 4-hydroxy biphenyl + sulphate conj | bifenazate + bifenazate-diazene | 4-hydroxy biphenyl + sulphate conj | bifenazate + bifenazate-diazene | 4-hydroxy biphenyl + sulphate conj |
| Perirenal fat | < 0.01 (3) | < 0.01 (3) | 0.02 0.03 0.02 | < 0.01 (3) | 0.10 0.07 0.07 | < 0.01 (3) |
| Milk, days 1-28 | | | | | < 0.01 (84) | < 0.01 (84) |
| Butterfat day 20 | | | < 0.01 (3) | < 0.01 (3) | 0.01 (3) 2/ | < 0.01 (3) |
| Butterfat day 28 | | | < 0.01 (3) | < 0.01 (3) | 0.03 (2) 0.02 3/ | < 0.01 (3) |
| Skim milk day 20 | | | | | < 0.01 (3) | < 0.01 (3) |
| Skim milk day 28 | | | | | < 0.01 (3) | < 0.01 (3) |

1/ Concentration of bifenazate + bifenazate-diazene expressed as bifenazate. Concentration of 4-hydroxybiphenyl + sulphate conjugate expressed as 4-hydroxybiphenyl .

2/ Butterfat averaged 42.8% of milk sample due to separator failure. Therefore, residues are artificially low.

3/ Butterfat averaged 13% of milk sample.

## RESIDUES IN FOOD IN COMMERCE OR AT CONSUMPTION

No information was received on residues of bifenazate in food in commerce or at consumption.

## NATIONAL MAXIMUM RESIDUE LIMITS

Information was provided on national residue definitions for bifenazate.

*Australia (APVMA, 2006)*
Sum of bifenazate and bifenazate diazene (diazenecarboxylic acid, 2-(4-methoxy-[1,1'-biphenyl-3-yl] 1-methylethyl ester), expressed as bifenazate .

*Japan*
Commodities of plant origin and fat: Sum of bifenazate and isopropyl 2-(4-methoxybiphenyl-3-yl)diazenylformate expressed as bifenazate.

Other commodities: Sum of bifenazate, isopropyl 2-(4-methoxybiphenyl-3-yl)diazenylformate and 4-hydroxybiphenyl expressed as bifenazate.

*USA (USEPA, 2005)*
Combined residues of bifenazate and diazinecarboxylic acid, 2-(4-methoxy-[1,1'-biphenyl]-3-yl), 1-methylethyl ester (expressed as bifenazate) for food crop commodities and animal fats.

Combined residues of bifenazate; diazinecarboxylic acid, 2-(4-methoxy-[1,1'-biphenyl]-3-yl), 1-methylethyl ester (expressed as bifenazate); 1,1'-biphenyl, 4-ol; and 1,1'-biphenyl, 4-oxysulfonic acid (expressed as 1,1'-biphenyl, 4-ol) for meat and meat byptoducts.

## APPRAISAL

Bifenazate was considered for the first time by the present meeting. It is a selective acaricide which controls the motile stage of mites either by direct contact or through contact with foliar residues.

IUPAC: Isopropyl 2-(4-methoxybiphenyl-3-yl)hydrazinoformate

CAS: 1-methylethyl 2-(4-methoxy[1,1'-biphenyl]-3-yl)hydrazinecarboxylate

The Meeting received information on bifenazate metabolism and environmental fate, methods of residue analysis, freezer storage stability, national registered use patterns, supervised residue trials, fate of residues in processing and national MRLs. Australia and Japan submitted GAP information and labels to support MRLs for bifenazate.

## Animal metabolism

The Meeting received animal metabolism studies with bifenazate in rats, lactating goats and laying hens. Bifenazate $^{14}C$ labelled in the substituted phenyl ring was used in all of the metabolism studies.

Bifenazate is readily converted to bifenazate-diazene (isopropyl 2-(4-methoxybiphenyl-3-yl)diazenoformate) by mild oxidation. Primary metabolites are readily produced by removal of the side chain and by hydroxylation of the biphenyl rings. Glucuronide and sulphate conjugates are also produced.

When rats were orally dosed with labelled bifenazate it was readily absorbed followed by extensive metabolism and excretion. Parent bifenazate and the following metabolites were identified in excreta: bifenazate glucuronide, bifenazate-diazene, 4-hydroxy bifenazate, 4-hydroxy bifenazate-diazene, 4-hydroxybiphenyl and its sulphate conjugate, 4,4'-dihydroxybiphenyl and its glucuronate and sulphate conjugates, 4-methoxybiphenyl and 4-hydroxy-4'-methoxybiphenyl and its conjugates. (See the toxicology report for more details of laboratory animal metabolism)

When a lactating goat was orally dosed with labelled bifenazate for 4 consecutive days at 21 mg/animal/day, equivalent to 10 ppm in the feed, most of the administered $^{14}C$ was excreted in the faeces (47%) and urine (19.5%). $^{14}C$ recovery was borderline at 68%. Residues in milk and tissues plus blood accounted for 0.22% and 2.0% of the dose respectively.

Metabolite 4-hydroxybiphenyl sulphate was the major identified component of the residue in milk (41% of TRR), while bifenazate and bifenazate-diazene comprised 9%. In muscle, residue levels were low with 4-hydroxybiphenyl the major identified component. In the fat, bifenazate was the major component at 53–58% of TRR. Residue levels and patterns in omental and perirenal fats were quite similar.

In goat liver, only 10% of the TRR was extractable. Bifenazate + bifenazate-diazene and 4-hydroxybiphenyl glucuronide were the main identified components, each comprising about 1% of TRR. In goat kidney, 4-hydroxybiphenyl glucuronide and sulphate accounted for approximately 14% of TRR. Bifenazate + bifenazate-diazene comprised less than 2% of TRR. In both liver and kidney, some of the unextractable TRR was apparently bound to protein.

The concentration of parent compound + bifenazate-diazene was substantially higher in the fat than in the other tissues suggesting that bifenazate (+ bifenazate-diazene) is a fat-soluble compound. No information was available on the residue distribution into the fat of goat milk.

When laying hens were orally dosed with labelled bifenazate for 4 consecutive days at 1.3 mg/bird/day, equivalent to 10 ppm in the feed, most of the administered 14C was excreted in the faeces (84%). $^{14}$C recovery was approximately 85%. Residues in eggs and tissues accounted for 0.1% and 1.4% of the dose respectively. Residues were not detectable (< 0.005 mg/kg) in breast muscle and egg white.

The major identified residues in liver, skin + fat and egg yolk were 4-hydroxybiphenyl (0.013 mg/kg, 2% TRR), bifenazate-diazene (0.008 mg/kg, 17% TRR) and bifenazate (0.005 mg/kg, 20% TRR), respectively. The distribution of bifenazate and bifenazate-diazene in the tissues and egg yolk suggests fat solubility.

The metabolic pathways in goats and poultry were generally similar, but additional conjugates were identified in the goat.

### Plant metabolism

The Meeting received plant metabolism studies with bifenazate on oranges, apples, grapes, radish and cotton.

In plants, most of the resultant residue from the use of bifenazate was a surface residue. Parent bifenazate was the major component of the residue at shorter intervals and the major identified component at longer intervals after treatment. Bifenazate-diazene was usually also present, but at much lower levels than parent bifenazate. Very little of the residue was translocated to the roots from treated radish foliage.

When Valencia <u>orange</u> trees were treated with a single application of WP formulated [$^{14}$C]bifenazate (0.42 and 2.2 kg ai/ha), approximately 80% of the residue was on the fruit surface 43 days after treatment, with bifenazate + bifenazate-diazene constituting 83% of TRR in and on the fruit. Bifenazate-diazene oxide, 4-methoxybiphenyl and 3-hydroxy-4-methoxybiphenyl were identified as minor components. After the rinsed oranges were separated into peel and the peeled fruits were homogenized to pulp and juice, the TRR distribution was mostly into the peel (approx 19% of TRR in whole fruit) with 0.9% in the pulp and 1% in the juice. Bifenazate was the only identified component in the juice at 0.003 and 0.001 mg/kg from the 0.42 and 2.2 kg ai/ha treatments respectively.

The TRR in oranges declined substantially at longer harvest intervals of 184, 274 and 442 days. Bifenazate and bifenazate-diazene were identified as components of the residue even at the longer intervals.

When Granny Smith <u>apple</u> trees were treated with a single application of WP formulated [$^{14}$C]bifenazate (0.42 and 2.2 kg ai/ha), approximately 60% of the residue was on the fruit surface 101 days after treatment, with bifenazate + bifenazate-diazene constituting 38% of TRR in and on the fruit. Bifenazate-diazene oxide and 4-methoxybiphenyl were identified as minor components. After the rinsed apples were homogenized and centrifuged to produce pomace and juice, the TRR distribution was mostly into the pomace (approx 30% of TRR in whole fruit) with approx 10% into the juice. Parent bifenazate and identified metabolites were not detected (< 0.001 mg/kg) in the juice.

<u>Grape</u> vines (variety Thompson Seedless) were treated with a single foliar application of WP formulated [$^{14}$C]bifenazate at 0.56 and 1.1 kg ai/ha and grapes were harvested 30 days later at maturity. Approximately 97% of the residue was surface residue. Bifenazate + bifenazate-diazene accounted for 98% and 95% of the TRR for the 0.56 and 1.1 kg ai/ha treatments respectively.

<u>Radish</u> plants (variety French Breakfast) were sprayed with a single foliar application of WS formulated [$^{14}$C]bifenazate at 1.1 and 2.2 kg ai/ha and harvested 7 days later for analysis. Most of the $^{14}$C remained on the foliage (TRR 13 and 21 mg/kg) with little reaching the roots (0.0023 and 0.0043 mg/kg). The majority of the residue (60% and 80% TRR) remained on the surface. Bifenazate + bifenazate-diazene accounted for 57% and 71% of the TRR in and on the radish tops. A ring-hydroxylated bifenazate-diazene was identified as constituting approximately 1.3% of the TRR in radish tops.

After <u>cotton</u> plants (variety Maxxa), at late bloom to early boll set, were sprayed with a single foliar application of WP formulated [$^{14}$C]bifenazate (0.56 and 2.2 kg ai/ha), bifenazate and identifiable metabolites were present at very low levels (each < 0.001 mg/kg) in the cotton seed harvested 112 days after treatment. A high proportion (77–82%) of the cotton gin trash residue was extractable, with bifenazate approximately 50% of the extractable residue and bifenazate-diazene, bifenazate-diazene oxide and 4-methoxybiphenyl identified as minor residue components.

### Environmental fate in soil

The Meeting received information on crop rotational studies for bifenazate. Information on soil metabolism and field dissipation was not required because no bifenazate uses as seed treatments or on root crops were provided for evaluation.

In a confined rotational crop study in USA a loamy sand soil was treated directly with $^{14}$C labelled bifenazate at a rate equivalent to 0.56 kg ai/ha and allowed to age under greenhouse conditions prior to the sowing of the rotational crops. Crops of carrots, lettuce and wheat were sown into the treated soil in pots at intervals of 30, 125 and 360 days after treatment.

Immature lettuce was sampled at the 4-5 leaf stage. Immature carrot plants were sampled when carrots were approximately 6 mm in diameter. Wheat forage samples were taken approximately 5 weeks after sowing. The remainder of the crops were grown to maturity and subsequently harvested and analysed for $^{14}$C (TRR) content. Samples were extracted and, where extractable residues exceeded 0.01 mg/kg, they were analysed by HPLC. No parent compound or reference metabolite was observed (LOQ 0.01 mg/kg). The unextractable residual solids from the wheat straw and fractions from the wheat forage were subjected to acid, base and enzyme hydrolysis, but no parent bifenazate or recognizable metabolite was released.

### Methods of residue analysis

The Meeting received descriptions and validation data for analytical methods for residues of bifenazate in raw agricultural commodities, processed commodities, feed commodities, animal tissues, milk and eggs.

Because bifenazate and bifenazate-diazene are readily interconverted by mild oxidation and reduction conditions, the measured residue includes both compounds. The analytical methods use a mild reduction with ascorbic acid to convert the bifenazate-diazene residue to bifenazate before the measurement step. Residues are typically extracted with acetonitrile and water acidified with acetic acid. After a partition cleanup and reduction with ascorbic acid, the residue is analysed by HPLC with coulometric detection. The oxidative coulometric detection system is quite selective. Substituted hydrazines such as bifenazate are oxidised at 200 mV, but most sample matrix components are not.

LC-MS-MS has also been used in place of coulometric detection. The [M+H]$^+$ ion is used as the precursor ion for bifenazate. Transitions 301.1/198.1 (for quantification) and 301.1/170.1 are observed.

Numerous recovery data on a wide range of crop and animal commodity substrates and processed commodities were provided from validation testing of the methods, which showed that the methods were valid over the relevant concentration ranges. The validated LOQ was typically 0.01 mg/kg.

None of the tested multiresidue methods was suitable for the analysis of bifenazate and bifenazate-diazene.

Samples of apples and oranges from [$^{14}$C]bifenazate crop metabolism studies were extracted with acetonitrile + acetic acid and analysed by the HPLC-coulometer method and an HPLC-radiometric method. The HPLC-coulometer results were approximately 60% of those from the radiometric method.

Samples of fat and liver from a goat dosed orally for 4 consecutive days with [$^{14}$C]bifenazate at the equivalent of 20 ppm in the feed as in a goat metabolism study were analysed by the HPLC-coulometer method and by radiolabel measurement for bifenazate + bifenazate-diazene residues. Agreement was good for the fat tissue (0.043 and 0.045 mg/kg, radiolabel and enforcement respectively) while for the liver the level was too low for the enforcement method (0.0082 and < 0.01 mg/kg). Samples of milk and liver were hydrolysed with hydrochloric acid for 2 hours to convert the sulphate conjugate of 4-hydroxybiphenyl to the free metabolite for analysis. The analytical results for 4-hydroxybiphenyl, by a suggested enforcement method, were in good agreement with the radiolabel measurement for liver and in reasonable agreement for milk.

## *Stability of residues in stored analytical samples*

The Meeting received information on the freezer storage stability of residues of bifenazate and bifenazate-diazene in apples, apricots, cantaloupe, cherries, cotton seed, cotton seed hulls, cotton seed meal, cotton seed refined oil, egg yolk, fat, gin trash, grape juice, grapes, kidney, liver, milk, mint, muscle, oranges, peaches, peppers, plums, potatoes, poultry liver, poultry muscle, poultry skin + fat, prunes, tomato, tomato paste and tomato puree.

Bifenazate residues (measured as bifenazate + bifenazate-diazene) are not particularly stable in some substrates. Stability is improved where the commodity is stored unchopped and in processed commodities presumably where enzymes are denatured. Bifenazate residues are stable in fat and milk, but are particularly unstable in kidney. Bifenazate residues are unstable in potato tuber matrix to the extent that disappearance from spiked samples causes difficulty with analytical recovery testing.

In a number of substrates some losses appeared to occur at spiking or soon after, but these losses may not be relevant when assessing the stability of incurred residues.

Residues of bifenazate or bifenazate-diazene measured as the sum of bifenazate and bifenazate-diazene did not decline by more than 30% when spiked into the following substrates and stored in a freezer at temperatures below -18°C for the interval tested: homogenized tomato 6 months; homogenized peppers 6 months; homogenized mint tops 102 days; sliced plums 4 weeks; tomato paste 4 weeks; tomato puree 4 weeks; cottonseed refined oil 28 days; apples skin surface 224 days; grapes surface 224 days; peaches skin surface 223 days; homogenized oranges 186 days; grape juice 186 days; homogenized prunes 182 days; milk 202 days; fat 95 days.

Estimates were made of the time interval for a 30% decline of residues of bifenazate or bifenazate-diazene measured as the sum of bifenazate and bifenazate-diazene when spiked into the following substrates and stored in a freezer at temperatures below -18°C: homogenized cherries 2.6 months; homogenized cantaloupe 3.9 months; homogenized apples 106 days; homogenized grapes 22 days; homogenized peaches 92 days; muscle 10 days.

In some matrices, e.g. cotton seed, the stability data were variable and difficult to interpret precisely.

When bifenazate was spiked into control samples of egg yolk, hen skin + fat, thigh muscle and liver and stored for 6 months below -10 °C, residues were stable in egg yolk and liver. In thigh muscle, 45% of the bifenazate disappeared, with 14% and 11% appearing as 4-hydroxybiphenyl and bifenazate-diazene respectively. In skin + fat, 97% of the bifenazate disappeared with 4%, 4% and 59% appearing as 4-hydroxybiphenyl, 4-methoxybiphenyl and bifenazate-diazene respectively.

Samples from the laying hen metabolism study were analysed by HPLC before and after freezer storage of 121–171 days to test the stability of incurred residues. The qualitative appearance of the initial and final chromatograms were reasonably similar for egg yolk, skin-with-fat and liver. Substantial changes were apparent for thigh muscle, but total residues in thigh muscle were very low (0.006 mg/kg).

## Definition of the residue

The composition of the residue in the metabolism studies, the available residue data in the supervised trials, the toxicological significance of metabolites, the capabilities of enforcement analytical methods and the national residue definitions already operating all influence the decision on residue definition.

Parent compound and metabolite bifenazate-diazene are readily interconverted, so both should be included in the residue definition.

In crop residue situations, parent compound comprised a substantial part of the residue for commodities that were directly sprayed, so bifenazate and bifenazate-diazene should constitute the residue definition for crops.

In goat fat, poultry fats and egg yolks, the sum of bifenazate and bifenazate-diazene was the major identifiable residue.

In goat muscle, liver, kidney and milk, 4-hydroxybiphenyl and its conjugates constituted the main identifiable residue. However, 4-hydroxybiphenyl may arise from sources other than bifenazate uses. It is a mammalian[8] and fungal[9] metabolite of biphenyl, a post-harvest fungicide used on citrus. It is also an industrial chemical used in the rubber industry[10]. Origins of 4-hydroxybiphenyl other than bifenazate mean that it would not be useful as part of an enforcement residue definition.

In the animal metabolism studies, the concentration of bifenazate + bifenazate-diazene was higher in the fat than in other tissues. In the dairy cow feeding study, the residue of bifenazate + bifenazate-diazene partitioned into the butter fat at the highest dosing level. The octanol-water partition coefficient of bifenazate (log $P_{OW}$ = 3.5) also suggests that fat-solubility for the parent compound.

The Meeting recommended a residue definition for bifenazate for plants and animals.

Definition of the residue (for compliance with the MRL and for estimation of dietary intake): sum of bifenazate and bifenazate-diazene (diazenecarboxylic acid, 2-(4-methoxy-[1,1'-biphenyl-3-yl] 1-methylethyl ester), expressed as bifenazate. The residue is fat soluble.

## Results of supervised trials on crops

The Meeting received supervised trials data for bifenazate uses on citrus fruits (mandarin, natsudaidai, lime), pome fruits (apple, pear), stone fruits (apricot, peach, plum, cherry), berry fruits (grapes, strawberry), figs, cucurbit fruiting vegetables (cantaloupe, watermelon, cucumber, summer squash), fruiting vegetables (tomato, peppers, egg plant), tree nuts (almond, pecan), cotton and herbs (mint, hops, tea).

Trials from Japan were available only in summary form and could not be evaluated.

All other trials were from the USA. In most trials, duplicate field samples from an unreplicated plot were taken at each sampling time and were analysed separately. For the purposes of the evaluation, the mean of the two results was taken as the best estimate of the residue from the plot.

Labels (or translations of labels) were available from Australia, Japan describing the registered uses of bifenazate.

---

[8] Wiebkin P, Fry JR, Jones CA, Lowing RK and Bridges JW. 1978. Biphenyl metabolism in isolated rat hepatocytes: effect of induction and nature of the conjugates. *Biochemical Pharmacology* 27:1899-1907.

[9] Schwartz RD, Williams AL and Hutchinson. 1980. Microbial production 4,4'-dihydroxybiphenyl: biphenyl hydroxylation by fungi. *Appl. Environ. Microbiol.* 39:702-708.

[10] Merck Index. 1996. 12th Edition. 7459 *p*-phenylphenol.

*Pome fruits*

Bifenazate is registered in USA for use on pome fruit trees at 0.42–0.56 kg ai/ha with a PHI of 7 days.

In 14 US trials on apples in 1998 matching GAP, residues of bifenazate + bifenazate-diazene were: 0.049, 0.058, 0.16, 0.16, 0.17, 0.18, 0.18, 0.19, 0.20, 0.22, 0.23, 0.37, 0.38 and 0.58 mg/kg.

In eight trials on pears in USA in 1998 with conditions matching the registered use, residues of bifenazate + bifenazate-diazene were: 0.094, 0.097, 0.10, 0.13, 0.14, 0.16, 0.24 and 0.29 mg/kg. The Meeting noted that the pear samples had spent 15–16 months in frozen storage, which exceeded the proven frozen storage interval for apples (7–8 months) representing pome fruits. However, the residue levels appeared to be stable on the fruit surface for the interval tested and the residue trials were accepted as valid.

The Meeting decided to combine the apple and pear data to form a pome fruit crop group estimation (populations not significantly different – Mann-Whitney test). The combined pome fruit data (22 values), in rank order were: 0.049, 0.058, 0.094, 0.097, 0.10, 0.13, 0.14, 0.16, 0.16, 0.16, 0.17, 0.18, 0.18, 0.19, 0.2, 0.22, 0.23, 0.24, 0.29, 0.37, 0.38 and 0.58 mg/kg.

The Meeting estimated a maximum residue level and an STMR value for bifenazate in pome fruits, of 0.7 and 0.175 mg/kg respectively.

*Stone fruits*

Bifenazate is registered in USA for use on stone fruit trees at 0.42–0.56 kg ai/ha with a PHI of 3 days.

In five US trials on apricots in 2002 matching GAP, residues of bifenazate + bifenazate-diazene were: 0.23, 0.30, 0.44, 0.59 and 0.73 mg/kg.

In 12 US trials on peaches in 1998 and 2002 matching GAP, residues of bifenazate + bifenazate-diazene were: 0.13, 0.16, 0.17, 0.22, 0.23, 0.23, 0.26, 0.40, 0.44, 0.45, 0.55 and 1.2 mg/kg.

In eight US trials on plums in 1998 and 2002 matching GAP, residues of bifenazate + bifenazate-diazene were: 0.01, 0.03, 0.034, 0.04, 0.04, 0.04, 0.07 and 0.13 mg/kg.

In 14 US trials on cherries in 2001 matching GAP, residues of bifenazate + bifenazate-diazene were: 0.29, 0.23, 1.6, 0.11, 0.48, 0.20, 0.42, 0.89, 0.71, 1.2, 0.81, 0.18, 0.27 and 0.34 mg/kg. The data were on pitted cherries, but the Meeting accepted the data as valid for MRL setting.

The residue data from peaches, apricots and cherries appeared to be from similar populations and were combined for a stone fruits group MRL. Residues on plums appeared to be much lower (significantly different from peach and cherry residues – Mann-Whitney test) than on the other stone fruits and were not included in the data set for STMR estimation.

Residue data on stone fruits in rank order (median underlined) were: 0.11, 0.13, 0.16, 0.17, 0.18, 0.20, 0.22, 0.23, 0.23, 0.23, 0.23, 0.26, 0.27, 0.29, 0.30, 0.34, 0.40, 0.42, 0.44, 0.44, 0.45, 0.48, 0.55, 0.59, 0.71, 0.73, 0.81, 0.89, 1.2, 1.2 and 1.6 mg/kg.

The Meeting estimated a maximum residue level and an STMR value for bifenazate in stone fruits of 2 and 0.34 mg/kg respectively.

*Grapes*

Bifenazate is registered in USA for use on grape vines at 0.42–0.56 kg ai/ha with a PHI of 14 days.

In 12 US trials on grapes in 1998 matching GAP, residues of bifenazate + bifenazate-diazene were: 0.05, 0.07, 0.10, 0.11, 0.17, 0.17, 0.20, 0.21, 0.29, 0.31, 0.33 and 0.55 mg/kg.

The Meeting estimated a maximum residue level and an STMR value for bifenazate in grapes, of 0.7 and 0.185 mg/kg respectively.

*Strawberries*

Bifenazate is registered in USA for use on strawberries with two treatments at 0.42–0.56 kg ai/ha and a PHI of 1 day.

In seven US trials on strawberries in 1999 matching GAP, residues of bifenazate + bifenazate-diazene were: 0.29, 0.49, 0.53, 0.63, 0.68, 0.93 and 1.0 mg/kg.

One strawberry trial had produced values of 3.4 and 2.9 mg/kg for its 3 day sample and 0.44 as the mean of the 1 day samples. The authors of the report discounted the high values as being due to analytical error, but found no specific cause. The trial was not included in this appraisal because of doubts about its validity.

The Meeting estimated a maximum residue level and an STMR value for bifenazate in strawberries, of 2 and 0.63 mg/kg respectively.

*Fruiting vegetables, cucurbits*

Bifenazate is registered in USA for use on cucurbit vegetables at 0.42–0.56 kg ai/ha with a PHI of 3 days.

In eight US trials on cantaloupes in 2000 matching GAP, residues of bifenazate + bifenazate-diazene were: 0.03, 0.04, 0.04, 0.04, 0.05, 0.08, 0.10 and 0.16 mg/kg.

In eight US trials on cucumbers in 2000 matching GAP, residues of bifenazate + bifenazate-diazene were: < 0.01, < 0.01, 0.03, 0.04, 0.05, 0.07, 0.08 and 0.22 mg/kg.

In seven US trials on summer squash in 2000 matching GAP, residues of bifenazate + bifenazate-diazene were: < 0.01, 0.01, 0.02, 0.04, 0.06, 0.12 and 0.34 mg/kg.

The Meeting decided to combine the data from cantaloupes, cucumbers and summer squash to support a cucurbit fruiting vegetables group MRL (populations not significantly different – Mann-Whitney test).

Residue data from 23 trials on cucurbit fruiting vegetables in rank order (median underlined) were: < 0.01, < 0.01, < 0.01, 0.01, 0.02, 0.03, 0.03, 0.04, 0.04, 0.04, 0.04, <u>0.04</u>, 0.05, 0.05, 0.06, 0.07, 0.08, 0.08, 0.10, 0.12, 0.16, 0.22 and 0.34 mg/kg.

The Meeting estimated a maximum residue level and an STMR value for bifenazate in cucurbit fruiting vegetables of 0.5 and 0.04 mg/kg respectively.

*Fruiting vegetables, other than cucurbits*

Bifenazate is registered in USA for use on fruiting vegetables at 0.42–0.56 kg ai/ha with a PHI of 3 days.

In 12 US trials on tomatoes in 2000 matching GAP, residues of bifenazate + bifenazate-diazene were: 0.03, 0.03, 0.04, 0.04, 0.07, 0.09, 0.10, 0.11, 0.13, 0.14, 0.19 and 0.29 mg/kg.

The Meeting estimated a maximum residue level and an STMR value for bifenazate in tomatoes of 0.5 and 0.095 mg/kg respectively.

In eight US trials on bell peppers in 2000 matching GAP, residues of bifenazate + bifenazate-diazene were: 0.13, 0.15, 0.15, 0.23, 0.24, 0.32, 0.52 and 1.1 mg/kg.

The Meeting estimated a maximum residue level and an STMR value for bifenazate in sweet peppers of 2 and 0.235 mg/kg respectively.

In three US trials on non-bell peppers in 2000 matching GAP, residues of bifenazate + bifenazate-diazene were: 0.54, 1.1 and 1.6 mg/kg.

The Meeting noted that the residue data for non-bell peppers, a minor crop, were rather limited, but also noted that two values were equivalent to the high end of the bell pepper data with one slightly higher as expected.

The Meeting estimated a maximum residue level and an STMR value for bifenazate in chili peppers of 3 and 1.1 mg/kg respectively.

The Meeting noted that the registered use of bifenazate referred to the fruiting vegetables group but was unable to recommend a group MRL because the residue levels on the three crops were too different.

*Tree nuts*

Bifenazate is registered in the USA for use on almonds and other tree nuts (including beech nut, Brazil nut, butternut, cashew, chestnut, hickory nut and Macadamia nut) at 0.42–0.56 kg ai/ha with a PHI of 7 days and on filberts, pecans, pistachios and walnuts with a PHI of 14 days.

In five trials on almonds in USA in 2001, the application rate was 0.84 kg ai/ha, 50% higher than the GAP rate, but acceptable for trials on tree nuts. In five US trials on almonds in 2001 harvested 7 days after treatment, residues of bifenazate + bifenazate-diazene in almond kernels were: 0.01, 0.02, 0.03, 0.05 and 0.10 mg/kg.

The Meeting estimated a maximum residue level and an STMR value for bifenazate in almonds of 0.2 and 0.03 mg/kg respectively.

In five US trials on pecans in 2001 where the application rate (0.85 kg ai/ha) was 50% higher than the GAP rate and a PHI of 14 days, residues of bifenazate + bifenazate-diazene in pecan kernels were: < 0.01 (3), 0.013 and 0.014 mg/kg. The Meeting noted that the application rate was higher than GAP, but the residues were close to the LOQ and could be used for evaluation.

The Meeting agreed to extrapolate the almond data to the tree nuts group and recommended a maximum residue level and an STMR value for bifenazate in tree nuts of 0.2 and 0.03 mg/kg respectively.

*Cotton seed*

Bifenazate is registered in USA for use on cotton at 0.4–0.8 kg ai/ha with a PHI of 60 days.

In 19 US trials on cotton in 1999 and 2000 matching GAP, residues of bifenazate + bifenazate-diazene in cotton seed were: < 0.01 (10), 0.01, 0.02, 0.03, 0.03, 0.04, 0.04, 0.06, 0.06 and 0.28 mg/kg.

The Meeting estimated a maximum residue level and an STMR value for bifenazate in cotton seed of 0.3 and 0.01 mg/kg respectively.

*Mint*

Bifenazate is registered in USA for use on mint at 0.42–0.84 kg ai/ha with a PHI of 7 days.

In five US trials on mint in 2000 matching GAP, residues of bifenazate + bifenazate-diazene in mint tops were: 6.4, 6.6, 12.9, 15.4 and 18.1 mg/kg.

The Meeting estimated a maximum residue level and an STMR value for bifenazate in mint of 40 and 12.9 mg/kg respectively.

*Hops*

Bifenazate is registered in USA for use on hops at 0.42–0.84 kg ai/ha with a PHI of 14 days.

In three US trials on hops in 1999 matching GAP, residues of bifenazate + bifenazate-diazene in mint tops were: 7.1, 7.8 and 9.3 mg/kg.

The Meeting recognized that the database for hops was very limited. However, hops are a minor crop and the Meeting estimated a maximum residue level and an STMR value for bifenazate in hops of 20 and 7.8 mg/kg respectively.

*Cotton gin trash*

Bifenazate is registered in USA for use on cotton at 0.4–0.8 kg ai/ha with a PHI of 60 days.

Residues were measured on cotton gin trash in 9 of the previously mentioned cotton trials where the application rates and PHIs matched label rates. Residues of bifenazate + bifenazate-diazene in cotton gin trash were: 0.07, 0.39, 0.69, 0.88, 1.3, 2.5, 3.8, 4.0 and 18 mg/kg. No maximum residue level was estimated for dry cotton fodder (cotton gin trash) because it is not traded internationally.

*Almond hulls*

In five trials on almonds in USA in 2001, the application rate was 0.84 kg ai/ha, 50% higher than the GAP rate, but acceptable for trials on tree nuts. In five US trials on almonds in 2001, harvested 7 days after treatment, residues of bifenazate + bifenazate-diazene in almond hulls were: 1.8, 2.8, 5.0, 5.1 and 6.9 mg/kg.

The Meeting estimated a maximum residue level and an STMR value for bifenazate in almond hulls of 10 and 5.0 mg/kg respectively. The highest residue was 6.9 mg/kg.

**Fate of residues during processing**

The Meeting received information on the fate of bifenazate residues during the juicing of apples, the drying of prunes, the production of grape juice and raisins, the production of tomato paste and puree, the production of cotton seed oil and the processing of mint tops.

Apples from bifenazate field trials at exaggerated (5×) rates were ground in a hammer mill and the mash was collected in cloths sacks and pressed in a hydraulic press to produce the wet pomace and juice.

Plums were harvested 3 days after a bifenazate treatment, placed in mesh bags and dried in a drying tunnel for 18 to 27 hours at 71–88°C simulating a commercial process.

Fresh field-treated grapes were fed into a crusher/stemmer to produce a grape pulp that was heated in a steam kettle and then pressed to separate juice and wet pomace. Grapes for raisins were placed on trays or paper to dry with turning after 7 days. After 14 days the stems were removed to produce the raisins.

Approximately 11 kg of fresh mint tops were subject to steam distillation in a cooker to produce 33–43 mL of oil. The oil samples were then filtered and refrigerated.

Cotton seed was cracked and dried at 55–71°C to a kernel moisture level of 12%. After further heating, the kernel material was flaked, steam treated and extracted with hexane to produce meal and crude oil. Sodium hydroxide treatment of the crude oil produced the refined cotton seed oil.

Tomatoes were first cleaned and then soaked for 3 minutes in a dilute sodium hydroxide solution, then thoroughly rinsed. Tomatoes were then chopped and rapidly heated to about 80°C and skin and seeds were separated from juice. Juice was evaporated to produce a puree. Further concentration and addition of salt produced a paste that was heated to approximately 85°C and canned.

Calculated processing factors and the mean or best estimate are summarised in the following table.

| Raw agricultural commodity (RAC) | Processed commodity | Calculated processing factors (PF). | Median or best estimate PF |
|---|---|---|---|
| Apple | wet pomace | 1.8, 1.7 | 1.8 |
| Apples | apple juice | 0.23, 0.10 | 0.17 |
| Cotton seed | hulls | 0.105, 0.35 | 0.23 |
| Cotton seed | cotton seed meal | < 0.0095, < 0.0038 | < 0.0038 |
| Cotton seed | cotton seed refined oil | < 0.0095, < 0.0038 | < 0.0038 |
| Grapes | grape juice | 0.054, 0.17 | 0.11 |
| Grapes | raisins | 0.36, 3.2 | 3.2 |
| Plums | dried prunes | 0.5, < 0.3 | 0.5 |
| Tomato | tomato paste | 1.26 | 1.3 |
| Tomato | tomato puree | 5.6 | 5.6 |

The processing factors for wet apple pomace (1.8) and apple juice (0.17) were applied to the estimated STMR for pome fruits (0.175 mg/kg) to produce STMR-P values for wet apple pomace (0.32 mg/kg) and apple juice (0.030 mg/kg).

The processing factor for dried prunes (0.5) was applied to the median residue for plums (0.04 mg/kg) to produce an STMR-P value for dried prunes (0.02 mg/kg).

The processing factors for raisins (3.2) and grape juice (0.11) were applied to the estimated STMR for grapes (0.185 mg/kg) to produce STMR-P values for raisins (0.59 mg/kg) and grape juice (0.020 mg/kg). The processing factor for raisins (3.2) was applied to the grape residue data (highest value 0.55 mg/kg) to produce an estimated highest value for dried grapes (1.76 mg/kg).

The Meeting estimated a maximum residue level for bifenazate in dried grapes (= currants, raisins, sultanas) of 2 mg/kg.

The processing factors for tomato puree (5.6) and tomato paste (1.3) were applied to the estimated STMR for tomatoes (0.095 mg/kg) to produce STMR-P values for tomato puree (0.53 mg/kg) and tomato paste (0.13 mg/kg).

The processing factors for cotton seed hulls (0.23), cotton seed meal (< 0.0038) and cotton seed refined oil (< 0.0038) were applied to the estimated STMR for cotton seed (0.01 mg/kg) to produce STMR-P values for cotton seed hulls (0.0023 mg/kg), cotton seed meal (0.00004 mg/kg) and cotton seed refined oil (0.00004 mg/kg).

### Residues in animal commodities

*Farm animal feeding*

The meeting received a lactating dairy cow feeding study, which provided information on likely residues resulting in animal tissues and milk from residues in the animal diet.

Lactating Holstein cows were dosed with bifenazate at the equivalent of 1 (1×), 3 (3×) and 10 (10×) ppm in the dry-weight diet for 28 consecutive days. Milk was collected throughout and tissues were collected for residue analysis of bifenazate + bifenazate-diazene and metabolite 4-hydroxybiphenyl and its sulphate conjugate from animals slaughtered on day 29.

Residues of bifenazate + bifenazate-diazene did not exceed the LOQ (0.01 mg/kg) in loin muscle, round muscle, liver, milk or skim milk at the highest dosing level 10 ppm. Residues were detected in the kidney of one animal at 0.01 mg/kg.

Residues of 4-hydroxybiphenyl and its sulphate conjugate did not exceed the LOQ (0.01 mg/kg) in any sample of tissue, milk or butterfat.

Residues were present in omental and perirenal fat in the 3 ppm feeding group (0.01–0.03 mg/kg) and the 10 ppm feeding group (0.03–0.10 mg/kg), but not in the 1 ppm feeding group. Residues were also present in butterfat from the 10 ppm group (0.01–0.03 mg/kg) but not from the 3 ppm group.

The dairy cow feeding study confirms the fat-solubility of the residue, bifenazate + bifenazate-diazene and that fat is the target tissue.

### Farm animal dietary burden

The Meeting estimated the dietary burden of bifenazate in farm animals on the basis of the diets listed in Appendix IX of the FAO Manual. Calculation from highest residue and STMR-P values provides the levels in feed suitable for estimating MRLs, while calculation from STMR and STMR-P values for feed is suitable for estimating STMR values for animal commodities. The percentage dry matter is taken as 100% when the highest residue levels and STMRs are already expressed as dry weight.

*Estimated maximum dietary burden of farm animals*

| Commodity | CC | Residue mg/kg | Basis | DM % | Residue dw mg/kg | Diet content (%) | | | Residue contribution (mg/kg) | | |
|---|---|---|---|---|---|---|---|---|---|---|---|
| | | | | | | Beef cattle | Dairy cattle | Poultry | Beef cattle | Dairy cattle | Poultry |
| Almond hulls | AM | 6.9 | highest residue | 90 | 7.7 | | | | | | |
| Apple pomace, wet | AB | 0.32 | STMR-P | 40 | 0.800 | 40 | 20 | | 0.32 | 0.16 | |
| Cotton fodder, dry | AM | 18 | highest residue | 90 | 20.000 | 20 | 20 | | 4.00 | 4.00 | |
| Cotton seed | SO | 0.28 | highest residue | 88 | 0.318 | 25 | 25 | | 0.08 | 0.08 | |
| Cotton seed hulls | AM | 0.0023 | STMR-P | 90 | 0.003 | | | | | | |
| Cotton seed meal | | 0.00004 | STMR-P | 89 | 0.000 | | | 20 | | | 0.00 |
| Total | | | | | | *85* | *65* | *20* | *4.40* | *4.24* | *0.00* |

*Estimated mean dietary burden of farm animals*

| Commodity | CC | Residue mg/kg | Basis | DM % | Residue dw mg/kg | Diet content (%) | | | Residue contribution (mg/kg) | | |
|---|---|---|---|---|---|---|---|---|---|---|---|
| | | | | | | Beef cattle | Dairy cattle | Poultry | Beef cattle | Dairy cattle | Poultry |
| Almond hulls | AM | 5.0 | STMR | 90 | 5.6 | *10* | *10* | | 0.56 | 0.56 | |
| | | | STMR- | | 0.800 | *40* | *20* | | 0.32 | 0.16 | |
| Apple pomace, wet | AB | 0.32 | P | 40 | | | | | | | |
| Cotton fodder, dry | AM | 1.3 | median residue | 90 | 1.444 | *10* | *10* | | 0.14 | 0.14 | |
| Cotton seed | SO | 0.01 | STMR | 88 | 0.011 | *25* | *25* | | 0.00 | 0.00 | |
| | | | STMR- | | 0.003 | | | | | | |
| Cotton seed hulls | AM | 0.0023 | P | 90 | | | | | | | |
| | | | STMR- | | 0.000 | | | *20* | | | |
| Cotton seed meal | | 0.00004 | P | 89 | | | | | | | 0.00 |
| Total | | | | | | *85* | *65* | *20* | *1.02* | *0.86* | *0.00* |

### Animal commodities, MRL estimation

For MRL estimation, the high residues in the tissues were calculated by interpolating the maximum dietary burden between the relevant feeding levels from the dairy cow feeding study and using the highest tissue concentrations from individual animals within those feeding groups. The high residues for butterfat were calculated similarly except that the mean butterfat concentrations from the relevant groups were used instead of the highest individual values.

*Cattle*

The STMR values for the tissues, milk and butterfat were calculated by interpolating the STMR dietary burdens between the relevant feeding levels from the dairy cow feeding study and using the mean tissue and milk concentrations from those feeding groups.

In the table, dietary burdens are shown in round brackets (), feeding levels and residue concentrations from the feeding study are shown in square brackets [] and estimated concentrations related to the dietary burdens are shown without brackets.

| Dietary burden (ppm) Feeding level [ppm] | Milk | Butterfat | Muscle | Liver | Kidney | Fat |
|---|---|---|---|---|---|---|
| MRL | | | | | | |
| | mean | mean | highest | highest | highest | highest |
| MRL beef cattle (4.4) [3, 10] | | | < 0.01 [< 0.01, < 0.01] | < 0.01 [< 0.01, < 0.01] | < 0.01 [< 0.01, 0.01] | 0.044 [0.03, 0.10] |
| MRL dairy cattle (4.24) [3, 10] | < 0.01 [< 0.01, < 0.01] | 0.013 [< 0.01, 0.03] | | | | |
| STMR | | | | | | |
| | mean | mean | mean | mean | mean | mean |
| STMR beef cattle (1.02) [0, 1] | | | < 0.01 [0, < 0.01] | < 0.01 [0, < 0.01] | < 0.01 [0, < 0.01] | < 0.01 [0, < 0.01] |
| STMR dairy cattle (0.86) [0, 1] | < 0.01 [0, < 0.01] | < 0.01 [0, < 0.01] | | | | |

The Meeting estimated dietary burdens for bifenazate in dairy cows to be 4.24 and 0.86 ppm (maximum and mean) and for beef cattle to be 4.4 and 1.02 ppm (maximum and mean), which are all less than feeding levels that produced residues below LOQ (< 0.01 mg/kg) in the milk, muscle and liver.

The Meeting estimated a maximum residue level and an STMR value for bifenazate in milk of 0.01* and 0.01 mg/kg, respectively.

For kidney, there was one residue detection from three animals at the 10 ppm feeding level, so for a dietary burden of 4.4 ppm, the residue in kidney should not exceed 0.01 mg/kg. The kidney and liver residues were used to support an edible offal MRL recommendation.

The Meeting estimated a maximum residue level and an STMR value for bifenazate in mammalian edible offal of 0.01* and 0.01 mg/kg, respectively.

By interpolation, the highest residue in fat was estimated as 0.044 mg/kg, while the STMR value was below the LOQ (0.01 mg/kg).

The Meeting estimated a maximum residue level for bifenazate in mammalian meat of 0.05 (fat) mg/kg. The associated STMR values for muscle and fat were 0.01 and 0.01 mg/kg.

By interpolation, the highest residue in butterfat was estimated as 0.013 mg/kg, while the STMR value was below the LOQ (0.01 mg/kg). The Meeting noted that, in this experiment, the yield of butterfat averaged 13% of the milk sample, suggesting that this "butterfat" may have contained only about 33% lipid (if the milk contained 4% lipid). This would mean that the highest residue would be approximately 0.039 mg/kg, expressed on the lipid content.

The Meeting estimated a maximum residue level and an STMR value for bifenazate in milk fats of 0.05 and 0.01 mg/kg, respectively.

*Poultry*

The dietary burden for poultry, currently based only on cotton seed meal is very low and is essentially zero. According to the poultry metabolism study, residues in poultry tissues and eggs were very low even for a 10 ppm dietary burden. Bifenazate residues, with current uses, are therefore not anticipated to occur in poultry tissues and eggs.

The Meeting estimated maximum residue levels of 0.01* (fat), 0.01* and 0.01* for bifenazate in poultry meat, poultry offal and eggs, respectively. The Meeting also estimated STMR values of 0 mg/kg for bifenazate residues in poultry meat (muscle 0 mg/kg; fat 0 mg/kg), poultry edible offal and eggs.

## RECOMMENDATIONS

On the basis of the data from supervised trials, the Meeting concluded that the residue concentrations listed below are suitable for establishing MRLs and for assessing IEDIs.

Definition of the residue (for compliance with the MRL and for estimation of dietary intake): Sum of bifenazate and bifenazate-diazene (diazenecarboxylic acid, 2-(4-methoxy-[1,1'-biphenyl-3-yl] 1-methylethyl ester), expressed as bifenazate. The residue is fat soluble.

| CCN | Commodity | MRL, mg/kg | STMR or STMR-P, mg/kg |
|---|---|---|---|
| AM 0660 | Almond hulls | 10 | 5.0 |
| SO 0691 | Cotton seed | 0.3 | 0.01 |
| DF 0269 | Dried grapes (= Currants, Raisins, Sultanas) | 2 | 0.59 |
| MO 0105 | Edible offal (Mammalian) | 0.01* | 0.01 |
| PE 0112 | Eggs | 0.01* | 0 |
| VC 0045 | Fruiting vegetables, Cucurbits | 0.5 | 0.04 |
| FB 0269 | Grapes | 0.7 | 0.185 |
| DH 1100 | Hops, dry | 20 | 7.8 |
| MM 0095 | Meat (from mammals other than marine mammals) | 0.05 (fat) | 0.01 muscle 0.01 fat |
| FM 0813 | Milk fats | 0.05 | 0.01 |
| ML 0106 | Milks | 0.01* | 0.01 |
| HH 0738 | Mints | 40 | 12.9 |
| VO 0444 | Peppers, Chili | 3 | 1.1 |
| VO 0445 | Peppers, Sweet (including Pimento or pimiento) | 2 | 0.235 |
| FP 0009 | Pome fruits | 0.7 | 0.175 |
| PM 0110 | Poultry meat | 0.01* (fat) | 0 muscle 0 fat |
| PO 0111 | Poultry, Edible offal of | 0.01* | 0 |
| FS 0012 | Stone fruits | 2 | 0.34 |
| FB 0275 | Strawberry | 2 | 0.63 |
| VO 0448 | Tomato | 0.5 | 0.095 |
| TN 0085 | Tree nuts | 0.2 | 0.03 |
| JF 0226 | Apple juice | | 0.030 |
| | Apple pomace, wet | | 0.32 |
| | Cotton seed hulls | | 0.0023 |
| | Cotton seed meal | | 0.00004 |

| CCN | Commodity | MRL, mg/kg | STMR or STMR-P, mg/kg |
|---|---|---|---|
| OR 0691 | Cotton seed refined oil | | 0.00004 |
| DF 0014 | Plum, dried (prunes) | | 0.02 |
| JF 0269 | Grape juice | | 0.020 |
| | Tomato paste | | 0.13 |
| | Tomato puree | | 0.53 |

\* At or about the limit of quantification.

Note: Bifenazate is a fat-soluble compound. Previously, the milk MRL would have been marked with an F to indicate a procedure for calculating "MRLs" for processed dairy products. Currently, bifenazate MRLs for milk and milk fat are available to support "MRLs" for processed dairy products.

## DIETARY RISK ASSESSMENT

### Long-term intake

The evaluation of bifenazate resulted in recommendations for MRLs and STMR values for raw and processed commodities. Where data on consumption were available for the listed food commodities, dietary intakes were calculated for the 13 GEMS/Food Consumption Cluster Diets. The results are shown in Annex 3 of the 2006 JMPR Report.

The IEDIs in the thirteen Cluster Diets, based on estimated STMRs were 1–20% of the ADI (0-0.01 mg/kg bw). The Meeting concluded that the long-term intake of residues of bifenazate from uses that have been considered by the JMPR is unlikely to present a public health concern.

### Short-term intake

The Meeting decided that it was unnecessary to establish an ARfD. The Meeting concluded that the short-term intake of bifenazate residues is unlikely to present a public health concern.

## REFERENCES

**Author, Date, Title, Institute, Report Reference, Document No.**

APVMA, 2006, APVMA. 2006. Australian Pesticides and Veterinary Medicines Authority. Maximum Residue Limits: the MRL standard - maximum residue limits in food and animal feedstuff. Table 3. Residue definition. http://www.apvma.gov.au/residues/mrl03_August06.pdf

USEPA, 2005, USEPA. 2005. § 180.572 Bifenazate; tolerance for residues. *Federal Register*. 40 CFR Ch. I (7–1–05 Edition).

012351, Korpalski SJ, 2002, Korpalski SJ. 2002. UCC-D2341 4L-SC on cotton: Magnitude of the residue study. Ricerca, Inc., Lab Project ID 012351. Sponsor Study ID RP-00003. Crompton Co. Unpublished. [61]

07054, Corley J, 2003, Corley J. 2003. Bifenazate: magnitude of the residue on cherry, Volume 2 of 3, IR-4 Project, Center for Minor Crop pest Management, Technology Center of New Jersey, Rutgers, IR-4 PR No. 07054. Unpublished. [32]

07266, Dorschner KW, 2002, Dorschner KW. 2002. Bifenazate: Magnitude of the residue on tomato. IR-4 Project, Center for Minor Crop Pest Management, Technology Centre of New Jersey, Rutgers, Lab ID No. 07266.00-UCR02. Unpublished. [36]

07386.00-UCR06, Dorschner KW, 2002, Dorschner KW. 2002. Bifenazate: Magnitude of the residue on mint. Crompton, Co., Lab ID No. 07386.00-UCR06. IR-4 Project, Center for Minor Crop Pest Management, Technology Centre of New Jersey, Rutgers, IR-4 Study No. 07386. Unpublished. [38]

07510, Buckrell HM, 2000, Buckrell HM. 2000. Bifenazate: Magnitude of the residue on cantaloupe: Analytical Summary Report,. Uniroyal Chemical Co., Test Site project No. GRL-11658. IR-4 Headquarters, Technology Centre of New Jersey, Study No. Analytical Phase 07510.00-UCR03. Unpublished. [27]

07510, Dorschner KW, 2002, Dorschner KW. 2002. Bifenazate: Magnitude of the residue on cantaloupe. Volume 2 of 11, IR-4 Project, Center for Minor Crop Pest Management Technology Centre of New Jersey, Rutgers, IR-4 Study No. 07510. Unpublished. [33]

07511, Benstead JE, 2000, Benstead JE. 2000. Bifenazate: Magnitude of the residue on cucumber analytical summary report. Uniroyal Chemical Co., Test Site Project No., GRL-11661. IR-4 Headquarters, Technology Centre of New Jersey, Study No. Analytical Phase 07511.00-UCR04. Unpublished. [6]

07511, Dorschner KW, 2002, Dorschner KW. 2002. Bifenazate: Magnitude of the residue on cucumber. Crompton Co., Laboratory ID No. 07511.00-UCR04. IR-4 Project, Center for Minor Crop Pest Management, Technology Centre of New Jersey, Rutgers, IR-4 Study No. 07511. Unpublished. [34]

07512, Benstead JE, 2000, Benstead JE. 2000. Bifenazate: Magnitude of the residue on squash (summer) analytical summary report. Uniroyal Chemical Co., Test Site Project No. GRL-11659. IR-4 Headquarters, Technology Centre of New Jersey, Study No. Analytical Phase 07512.00-UCR05. Unpublished. [7]

07512, Dorschner KW, 2002, Dorschner KW. 2002. Bifenazate: Magnitude of the residue on squash (summer) Volume 4 of 11. Crompton Co., Laboratory ID No. 07512.00-UCR05. IR-4 Project, Center for Minor Crop Pest Management, Technology Centre of New Jersey, Rutgers, IR-4 Study No. 07512. Unpublished. [35]

07552.00-UCR01, Dorschner KW, 2002, Dorschner KW. 2002. Bifenazate: Magnitude of the residue on pepper (bell and non-bell. Crompton Co., Lab ID No. 07552.00-UCR01. IR-4 Project, Center for Minor Crop Pest Management, Technology Centre of New Jersey, Rutgers, IR-4 Study No. 07552. Unpublished. [37]

08035, Benstead JE, 2001, Benstead JE. 2001. Bifenazate: Magnitude of the residue on tomato (greenhouse) analytical summary report. Crompton Co., Test Site Project No. GRL-11805. IR-4 Headquarters, Technology Centre of New Jersey, Study No. Analytical Phase 08035.01-UCR01. Unpublished. [12]

08035, Samoil KS, 2002, Samoil KS. 2002. Bifenazate: Magnitude of the Residue on Tomato (Greenhouse. IR-4 Project, Center for Minor Crop Pest Management, Technology Centre of New Jersey, Rutgers, Laboratory ID No. 08035.01-UCR01. Unpublished. [80]

10495-1, Wiedmann JL, 2000, Wiedmann JL. 2000. UCC-DP2341 on cotton: magnitude of the residue and processing study. Document RP-99008, Ricerca 10495-1, USA. Unpublished. [4]

2000-097, McManus JP and DeMatteo V, 2001, McManus JP and DeMatteo V. 2001. Distribution and metabolism of $^{14}$C-bifenazate in grapes. Report No. 2000-097. Uniroyal Chemical Co., USA. Unpublished. [68]

2001-147, Charlton RB and Tecle B, 2002, Charlton RB and Tecle B. 2002. Distribution and metabolism of $^{14}$C-bifenazate in radishes, November 13, 2002. Study 2001-147 Crompton Corp. Unpublished. [31]

2003-016, Wood BJ, 2003, Wood BJ. 2003. Confirmatory method: analytical method validation for bifenazate and D3598 in bovine liver, kidney, milk and fat, and A1530 and A1530-Sulfate in bovine liver, kidney and milk. North Coast Laboratories, Ltd. Report No. 2003-016. Crompton Co. Unpublished. [103]

2005-013, Gupta K and Cassidy P, 2005, Gupta K and Cassidy P. 2005. Radiovalidation of bifenazate livestock analytical enforcement method. Ricerca, Inc., USA. Report No. 2005-013. Chemtura. Unpublished. [51]

217/27, Lewis CJ, 2001, Lewis CJ. 2001. ($^{14}$C)- Bifenazate: Photodegradation in sterile aqueous solution. Covance Laboratories, U.K., Report No. 217/27. Unpublished. [65]

45552, Hackert Anderson CR and Koch DA, 1999, Hackert Anderson CR and Koch DA. 1999. Multiresidue method testing for bifenazate and UCC-D3598. ABC Laboratories, Inc., Laboratory Study ID 45552. Uniroyal Chemical Company, Inc., Sponsor Study ID RP-99022. Unpublished. [1]

6337-95-0006-EF-001, Shah JF, 1997, Shah JF. 1997. A hydrolysis study of D2341. Ricerca, Inc., USA, Document No. 6337-95-0006-EF-001. Unpublished. [81]

6337-95-006-EF-001-001., Findak D, DiFrancesco D and Shah JF, 1999, Findak D, DiFrancesco D and Shah JF. 1999. A hydrolysis study of D2341. Ricerca, Inc., USA, Amendment Document No. 6337-95-006-EF-001-001. Unpublished. [42]

6381-95-0028-EF-001, Panthani AM and Hatzenbeler CJ, 1998, Panthani AM and Hatzenbeler CJ. 1998. Metabolism of [$^{14}$C]-D2341 in Citrus. Ricerca, Inc., USA, Report No. 6381-95-0028-EF-001. Unpublished. [70]

6394-95-0032-EF-001, Shah J, 1997, Shah J. 1997. Aqueous photolysis of [$^{14}$C] D2341 at pH 5. Ricerca, Inc., USA, Document No. 6394-95-0032-EF-001. Unpublished. [82]

6507-95-0124-EF-001, Findak DC, 2000, Findak DC. 2000. A confined rotational crop study with [$^{14}$C]D2341. Ricerca, Inc., USA. Report No. 6507-95-0124-EF-001. Crompton Co. Unpublished. [41]

666/079, Tremain S, 2003, Tremain S. 2003. Bifenazate pure: study report: Determination of vapour pressure. Safepharm Laboratories Limited, UK, Report No. 666/079. Unpublished.

6850-96-0101-EF-001, Panthani AM and Hatzenbeler CJ, 1998, Panthani AM and Hatzenbeler CJ. 1998. Metabolism of [$^{14}$C]D2341 in apples. Ricerca, Inc., USA, Report No. 6850-96-0101-EF-001. Unpublished. [71]

6998-97-0237-CR-001, Jablonski JE, 1998, Jablonski JE. 1998. Analytical method for the analysis of D2341 and D3598 in apples and citrus. Ricerca, Inc., USA. Report No. 6998-97-0237-CR-001. Crompton Co. Unpublished. [52]

6998-98-0051-CR-001, Jablonski JE, 1998, Jablonski JE. 1998. Validation of the crop residue method for D2341 and D3598 (combined method for apples and citrus, including radiovalidation. Ricerca, Inc., USA. Report No. 6998-98-0051-CR-001. Crompton Co. Unpublished. [53]

7137-97-0024-EF-001, Panthani AM and Hatzenbeler CJ, 2000, Panthani AM and Hatzenbeler CJ. 2000. Metabolism of [$^{14}$C]D2341 in cotton. Ricerca, Inc., USA, Report No. 7137-97-0024-EF-001. Unpublished. [72]

7473, Jablonski JE, 1999, Jablonski JE. 1999. Validation of the residue method for D2341, D3598 and A1530 in bovine tissues and milk.: Document 7473-98-0115-CR-001: 98221: Lab Project Number 7473. Ricerca, Inc. Crompton Co. Unpublished. [55]

7474, Wiedmann JL and Jablonski JE, 1999, Wiedmann JL and Jablonski JE. 1999. Meat and milk magnitude of the residue study in lactating dairy cows dosed with D2341 Technical. Ricerca, Inc. Lab Project Number: 7474-98-0186-CR-001: 7474: 109-032-10. Crompton Study Number 2002-002. Unpublished. [97]

7475, Jablonski JE, 1999, Jablonski JE. 1999. Stability of D2341 and metabolites D3598 and A1530 in bovine tissues and milk during freezer storage. Ricerca, Inc. Lab Project Number: 7475-98-0118-CR-001: 7475: 98222. Crompton Co. Unpublished. [56]

7545, Korpalski SJ, 1999, Korpalski SJ. 1999. UCC-D2341 50WP on grapes: Magnitude of the residue study. Ricerca, Inc., Lab Project ID 7545. Sponsor Study ID RP-98014. Uniroyal Chemical Company, Inc. Unpublished.

95236, McClanahan RH, 1998, McClanahan RH. 1998. Metabolism of [$^{14}$C]D2341 in rats. Ricerca, Inc., Document No. 6455-95-0089-EF-001. Uniroyal Project No. 95236. Uniroyal Chemical Company, Inc. Unpublished.

96-0064, McClanahan RH. and Bayus MA, 1999, McClanahan RH. and Bayus MA. 1999. Metabolism of [$^{14}$C]D2341 in lactating goats: Lab Project Number: 6802-96-0064-EF-001: 96-0064: 96047. Ricerca, Inc. Unpublished. [66]

96-0265, McClanahan RH, Shah JF and O'Meara HM, 1999, McClanahan RH, Shah JF and O'Meara HM. 1999. Metabolism of [$^{14}$C]D2341 in laying hens: Lab Project Number: 7034-96-0265-EF-001: 96-0265: 97005. Ricerca, Inc. Unpublished [67]

99214, Batorewicz W, 2000, Batorewicz W. 2000. Independent laboratory validation. Analytical method for the analysis of D2341 and D3598 in apples. Uniroyal Chemical Company, Inc., USA. Amended Report No. 99214. Unpublished. [2]

GRL FR-11291, Friedlander BT, 1998, Friedlander BT. 1998. The color of purified D2341 experimental miticide. Uniroyal Chemical Ltd., Canada, Report No. GRL FR-11291. Unpublished. [43]

GRL FR-11292-01, Friedlander BT, 1998, Friedlander BT. 1998. The physical state of purified D2341 experimental miticide. Uniroyal Chemical Ltd., Canada, Report No. GRL FR-11292-01. Unpublished. [45]

GRL FR-11293-01, Friedlander BT, 1998, Friedlander BT. 1998. The odor of purified D2341 experimental miticide. Uniroyal Chemical Ltd., Canada, Report No. GRL FR-11293-01. Unpublished. [44]

GRL-11346, Gaydosh KA, 2000, Gaydosh KA. 2000. UCC-D2341 50WP on apples: Magnitude of the residue and MOR decline study. Uniroyal Chemical Co., Lab Project ID GRL-11346. Sponsor Study ID RP-98004. Uniroyal Chemical Company, Inc. Unpublished. [46]

GRL-11418, Gaydosh KA, 2000, Gaydosh KA. 2000. Bifenazate 50WP on pears: Magnitude of the residue study, Uniroyal Chemical Co., Lab Project ID GRL-11418. Sponsor Study ID RP-98005. Uniroyal Chemical Company, Inc. Unpublished. [47]

GRL-11419, Gaydosh KA, 2000, Gaydosh KA. 2000. UCC-D2341 50WP on apples: processing study. Uniroyal Chemical Co., Lab Project ID GRL-11419. Sponsor Study ID RP-98013. Uniroyal Chemical Company, Inc. Unpublished. [49]

GRL-11517, Gaydosh KA, 2000, Gaydosh KA. 2000. UCC-D2341 50WP on strawberries: Magnitude of the residue study. Uniroyal Chemical Co., Lab Project ID GRL-11517. Sponsor Study ID RP-99007. Uniroyal Chemical Company, Inc. Unpublished.

GRL-11662, Buckrell HM, 2000, Buckrell HM. 2000. Validation of a working method for determination of combined D2341 and D3598 residues in cantaloupe. Project No. GRL-11662. Uniroyal Chemical Co. Unpublished. [23]

GRL-11664, Buckrell HM, 2000, Buckrell HM. 2000. Validation of a working method for determination of combined D2341 and D3598 residues in squash (summer. Project No. GRL-11664. Uniroyal Chemical Co. Unpublished. [24]

GRL-11666, Buckrell HM, 2000, Buckrell HM. 2000. Validation of a working method of combined D2341 and D3598 residues in tomatoes. Project No. GRL-11666. Uniroyal Chemical Co. Unpublished. [25]

GRL-11668, Benstead JE, 2000, Benstead JE. 2000. Validation of a working method for determination of combined D2341 and D3598 residues in peppers. Project No. GRL-11668. Uniroyal Chemical Co. Unpublished. [5]

GRL-11670, Buckrell HM, 2000, Buckrell HM. 2000. Validation of a working method for determination of combined D2341 and D3598 residues in cucumbers. Project No. GRL-11670. Uniroyal Chemical Co. Unpublished. [26]

GRL-11866, Benstead JE, 2001, Benstead JE. 2001. Validation of a working method for determination of combined bifenazate and UCC-D3598 in almond nutmeat and hulls, Laboratory Project No. GRL-11866, Crompton Co. Research Laboratories, Guelph, Ontario, Canada. Unpublished. [9]

GRL-11868, Buckrell HM, 2001, Buckrell HM. 2001. Validation of a working method for determination of combined bifenazate and UCC-D3598 residues in pecans, Laboratory Project No. GRL-11868, Crompton Co. Research Laboratories, Guelph, Ontario, Canada. Unpublished. [29]

GRL-11907, Riggs AS, 2004, Riggs AS. 2004. The solubility of bifenazate in water and pH 5 aqueous buffer. Crompton Co. Laboratories, Canada, Report No. GRL-11907. Unpublished.

GRL-11928, Wesley JE, 2002, Wesley JE. 2002. Validation of a working method for determination of combined D2341 and D3598 residues in plums. Study No. GRL-11928. Crompton Co. Unpublished. [92]

GRL-11929, Wesley JE, 2002, Wesley JE. 2002. Validation of a working method for determination of combined D2341 and D3598 residues in apricots. Study No. GRL-11929. Crompton Co. Unpublished. [90]

GRL-11930, Wesley JE, 2002, Wesley JE. 2002. Validation of a working method for determination of combined D2341 and D3598 residues in peaches. Study No. GRL-11930. Crompton, Co. Unpublished. [91]

GRL-11934, WesleyJE, 2002, WesleyJE. 2002. Two month freezer storage stability of bifenazate (D2341) residues in apricots. Study No. GRL-11934. Crompton Co. Unpublished. [93]

GRL-11936, Black HM, 2003, Black HM. 2003. Two month freezer storage stability of bifenazate (D2341) in potatoes. Report No: GRL-11936. Crompton Co./Cie. Guelph, Canada. Unpublished. [19]

GRL-11940, Black HM, 2002, Black HM. 2002. Validation of a working method for determination of combined D2341 and D3598 residues in strawberries. Study No. GRL-11940. Crompton Co. Unpublished. [15]

GRL-12057, Dunn NL, 2003, Dunn NL. 2003. The melting point of pure bifenazate. Crompton Co. Research Laboratories, Canada, Report No. GRL-12057. Unpublished. [39]

GRL-12061, Riggs AS, 2003, Riggs, AS. 2003. The partition coefficient (n-octanol/water) of bifenazate. Crompton Co. Laboratories, Canada, Report No. GRL-12061. Unpublished.

GRL-12062, Riggs AS, 2004, Riggs AS. 2004. The dissociation constant of purified bifenazate. Crompton Co. Research Laboratories, Canada, Report No. GRL-12062. Unpublished. [79]

GRL-12140, Black HM, 2004, Black HM. 2004. Four week freezer storage stability of bifenazate (D2341) in potatoes. Report No.GRL-12140. Crompton Co./Cie. Guelph, Canada. Unpublished. [20]

GRL-12171, Black HM, 2005, Black HM. 2005. Four week freezer storage stability of bifenazate (D2341) in plums. Report No. GRL-12171. Crompton Co./Cie. Guelph, Canada. Unpublished. [21]

GRL-12172, Black HM, 2005, Black HM. 2005. Four week freezer storage stability of bifenazate (D2341) in tomato paste and tomato puree. Report No. GRL-12172. Crompton Co./Cie. Guelph, Canada. Unpublished. [22]

GRL-FR-10806, Riggs AS, 1997, Riggs AS. 1997. The solubility of technical D2341 experimental miticide in organic solvents. Uniroyal Chemical Ltd., Canada, Report No. GRL-FR-10806. Unpublished. [74]

GRL-FR-11295, Riggs AS, 1998, Riggs, AS. 1998. Determination of the solubility of purified D2341 experimental miticide in various solvents. Uniroyal Chemical Co., Canada, Report No. GRL-FR-11295. Unpublished.

GRL-FR-11296, Stevenson WJ, 1998, Stevenson WJ. 1998. Determination of the density of purified D2341 experimental miticide. Uniroyal Chemical Ltd., Canada, Report No. GRL-FR-11296. Unpublished.

GRL-FR-11660, Benstead JE, 2000, Benstead JE. 2000. Bifenazate: Magnitude of the residue on pepper (bell and non-bell) analytical summary report. Uniroyal Chemical Co., Test Site Project No. GRL-FR-11660. IR-4 Headquarters, Technology Centre of New Jersey, Study No. Analytical Phase 07552.00-UCR01. Unpublished. [8]

GRL-FR-11663, Benstead JE, 2001, Benstead JE. 2001. Six month freezer storage stability of bifenazate (D2341) residues in canteloup: Final Report. Project Number: GRL-FR-11663. Crompton Co. Unpublished. [11]

GRL-FR-11667, Buckrell HM, 2001, Buckrell HM. 2001. Six month freezer storage stability of bifenazate (D2341) residues in tomato: Final Report. Project Number: GRL-FR-11667. Crompton Co. Unpublished. [30]

GRL-FR-11669, Benstead JE, 2001, Benstead JE. 2001. Six month freezer storage stability of bifenazate (D2341) residues in peppers. Project No. GRL-FR-11669. Uniroyal Chemical Co. Unpublished. [10]

GRL-FR-11673, Buckrell HM, 2000, Buckrell HM. 2000. Bifenazate: Magnitude of residue on tomato, analytical summary report. Uniroyal Chemical Co., Test Site Project No. GRL-FR-11673. IR-4 Headquarters, Technology Centre of New Jersey, Study No. Analytical Phase 07266.00-UCR02. Unpublished. [28]

GRL-FR-11712, Black HM, 2001, Black HM. 2001. Bifenazate: Magnitude of the residue on mint, analytical summary report. Crompton Co., Test Site Project No. GRL-FR-11712. IR-4 Headquarters, Technology Centre of New Jersey, Study No. Analytical Phase 07386.00-UCR06. Unpublished. [13]

GRL-FR-11853, Black HM, 2002, Black HM. 2002. Six month freezer storage stability of bifenazate (D2341) residues in cherries. Project GRL-FR-11853. Crompton document: Bifenazate - REGDOC 15296. Unpublished. [14]

GRL-FR-11911, Black HM, 2002, Black HM. 2002. Three month freezer storage stability of bifenazate (D2341) residues in mint. Study No. GRL-FR-11911. Crompton Co. Unpublished. [16]

RP-01001, Puhl J, 2002, Puhl J. 2002. UCC-D2341 50 WP on pecans: magnitude of the residue study. Crompton, Co. Lab Experiment No. GRL-11861. Uniroyal Chemical Company, Inc., Report No. RP-01001. Unpublished. [73]

RP-01002, Korpalski SJ, 2002, Korpalski SJ. 2002. UCC-D2341 50 WP on almonds: Magnitude of the residue study. Crompton Co., Lab Experiment No. GRL-11862. Sponsor Study ID RP-01002. Uniroyal Chemical Company, Inc. Unpublished. [60]

RP-01015, Korpalski SJ and Puhl RJ, 2002, Korpalski SJ. and Puhl RJ. 2002. UCC-D2341 50WP on mint: processing study: final report: Lab Project Number: RP-01015: 20.082: RP-01015-PWA01 Uniroyal Chemical Co., Inc., Ron Britt & Associates, and USDA/ARS-IAREC. Unpublished. [63]

RP-02007, Gaydosh KA, 2003, Gaydosh KA. 2003. Acramite®-50WS on peaches, plums, and apricots: Magnitude of the residue study. Crompton Co/Cie, Lab Experiment No. GRL-11951. Sponsor Study ID RP-02007. Crompton CO. Unpublished. [50]

RP-02009, Wood BJ, 2003, Wood BJ. 2003. Confirmatory method: residue analytical method validation for bifenazate and UCC-D3598 in peaches, raisins and almond. Study RP-02009. North Coast Laboratories, Ltd. Doc. No. 20.085 for Crompton Corp. Unpublished. [102]

RP-98006, Korpalski SJ and Puhl RJ, 2000, Korpalski SJ and Puhl RJ. 2000. UCC-D2341 50WP on stonefruit: Magnitude of the residue and processing study: Lab Project Number: RP-98006: RGC-98-500: RCP-98-102. Ricerca, Inc. Crompton Co. Unpublished. Note: Part 2 of 2 is in Box 5. [62]

RP-98007, Korpalski SJ, 1999, Korpalski SJ. 1999. UCC-D2341 50WP on grapes: processing and decline study. Ricerca, Inc., Lab Project ID 7573. Sponsor Study ID RP-98007. Uniroyal Chemical Company, Inc. Unpublished. [58]

RP-98018, Wiedmann JL, 1999, Wiedmann JL. 1999. Residue analytical method validation for bifenazate and metabolite in various fruit matrices, Department of Residue Analysis, Ricerca, Inc., Report No. 7543-98-0072-CR-002. Uniroyal Chemical Company, Inc., Study No. RP-98018. Unpublished. [96]

RP-98019, Wiedmann JL and Korpalski SJ, 1999, Wiedmann JL and Korpalski SJ. 1999. Stability of bifenazate and metabolite in fruit matrices during freezer storage, Department of Residue Analysis, Ricerca, Inc., Document No. 7546-98-0097-CR-001. Uniroyal Chemical Company, Inc., Study No. RP-98019. Unpublished. [98]

RP-99006, Korpalski SJ, 2000, Korpalski SJ. 2000. UCC-D2341 50WP on hops: Magnitude of the residue study. Uniroyal Chemical, Ltd., Lab Experiment No. GRL-11574. Sponsor Study ID RP-99006. Uniroyal Chemical company, Inc. Unpublished. [59]

RP-99008, Belcher TI, 2000, Belcher TI. 2000. UCC-D2341 50WP on cotton: Magnitude of the residue and processing study. Excel Research Services, Inc., Study No. ERS-99010. Uniroyal Study No. RP-99008. Uniroyal Chemical Company, Inc. Unpublished. [4]

## CROSS-REFERENCES

| Author | Document Code | Year |
|---|---|---|
| APVMA | | 2006 |
| Batorewicz W | 99214 | 2000 |
| Belcher TI | RP-99008 | 2000 |
| Benstead JE | 07511 | 2000 |
| Benstead JE | 07512 | 2000 |
| Benstead JE | 08035 | 2001 |
| Benstead JE | GRL-11668 | 2000 |
| Benstead JE | GRL-11866 | 2001 |
| Benstead JE | GRL-FR-11660 | 2000 |
| Benstead JE | GRL-FR-11663 | 2001 |
| Benstead JE | GRL-FR-11669 | 2001 |
| Black HM | GRL-11936 | 2003 |
| Black HM | GRL-11940 | 2002 |

| Author | Document Code | Year |
|--------|---------------|------|
| Black HM | GRL-12140 | 2004 |
| Black HM | GRL-12171 | 2005 |
| Black HM | GRL-12172 | 2005 |
| Black HM | GRL-FR-11712 | 2001 |
| Black HM | GRL-FR-11853 | 2002 |
| Black HM | GRL-FR-11911 | 2002 |
| Buckrell HM | 07510 | 2000 |
| Buckrell HM | GRL-11662 | 2000 |
| Buckrell HM | GRL-11664 | 2000 |
| Buckrell HM | GRL-11666 | 2000 |
| Buckrell HM | GRL-11670 | 2000 |
| Buckrell HM | GRL-11868 | 2001 |
| Buckrell HM | GRL-FR-11667 | 2001 |
| Buckrell HM | GRL-FR-11673 | 2000 |
| Charlton RB and Tecle B | 2001-147 | 2002 |
| Corley J | 07054 | 2003 |
| Dorschner KW | 07266 | 2002 |
| Dorschner KW | 07386.00-UCR06 | 2002 |
| Dorschner KW | 07510 | 2002 |
| Dorschner KW | 07511 | 2002 |
| Dorschner KW | 07512 | 2002 |
| Dorschner KW | 07552.00-UCR01 | 2002 |
| Dunn NL | GRL-12057 | 2003 |
| Findak D, DiFrancesco D and Shah JF | 6337-95-006-EF-001-001. | 1999 |
| Findak DC | 6507-95-0124-EF-001 | 2000 |
| Friedlander BT | GRL FR-11291 | 1998 |
| Friedlander BT | GRL FR-11292-01 | 1998 |
| Friedlander BT | GRL FR-11293-01 | 1998 |
| Gaydosh KA | GRL-11346 | 2000 |
| Gaydosh KA | GRL-11418 | 2000 |
| Gaydosh KA | GRL-11419 | 2000 |
| Gaydosh KA | GRL-11517 | 2000 |
| Gaydosh KA | RP-02007 | 2003 |
| Gupta K and Cassidy P | 2005-013 | 2005 |
| Hackert Anderson CR and Koch DA | 45552 | 1999 |
| Jablonski JE | 6998-97-0237-CR-001 | 1998 |
| Jablonski JE | 6998-98-0051-CR-001 | 1998 |
| Jablonski JE | 7473 | 1999 |
| Jablonski JE | 7475 | 1999 |
| Korpalski SJ | 012351 | 2002 |
| Korpalski SJ | 7545 | 1999 |
| Korpalski SJ | RP-01002 | 2002 |
| Korpalski SJ | RP-98007 | 1999 |
| Korpalski SJ | RP-99006 | 2000 |
| Korpalski SJ and Puhl RJ | RP-01015 | 2002 |
| Korpalski SJ and Puhl RJ | RP-98006 | 2000 |
| Lewis CJ | 217/27 | 2001 |
| McClanahan RH | 95236 | 1998 |
| McClanahan RH, Shah JF and O'Meara HM | 96-0265 | 1999 |
| McClanahan RH. and Bayus MA | 96-0064 | 1999 |
| McManus JP and DeMatteo V | 2000-097 | 2001 |
| Panthani AM and Hatzenbeler CJ | 6381-95-0028-EF-001 | 1998 |
| Panthani AM and Hatzenbeler CJ | 6850-96-0101-EF-001 | 1998 |
| Panthani AM and Hatzenbeler CJ | 7137-97-0024-EF-001 | 2000 |
| Puhl J | RP-01001 | 2002 |
| Riggs AS | GRL-11907 | 2004 |
| Riggs AS | GRL-12061 | 2003 |
| Riggs AS | GRL-12062 | 2004 |
| Riggs AS | GRL-FR-10806 | 1997 |
| Riggs AS | GRL-FR-11295 | 1998 |
| Samoil KS | 08035 | 2002 |
| Shah J | 6394-95-0032-EF-001 | 1997 |
| Shah JF | 6337-95-0006-EF-001 | 1997 |
| Stevenson WJ | GRL-FR-11296 | 1998 |
| Tremain S | 666/079 | 2003 |

| Author | Document Code | Year |
|---|---|---|
| USEPA | | 2005 |
| Wesley JE | GRL-11928 | 2002 |
| Wesley JE | GRL-11929 | 2002 |
| Wesley JE | GRL-11930 | 2002 |
| WesleyJE | GRL-11934 | 2002 |
| Wiedmann JL | 10495-1 | 2000 |
| Wiedmann JL | RP-98018 | 1999 |
| Wiedmann JL and Jablonski JE | 7474 | 1999 |
| Wiedmann JL and Korpalski SJ | RP-98019 | 1999 |
| Wood BJ | 2003-016 | 2003 |
| Wood BJ | RP-02009 | 2003 |

# BOSCALID (221)

*First draft prepared by Dr Yibing He, Institute for the Control of Agrochemicals, Beijing, China.*

## EXPLANATION

Boscalid is a new fungicide used to control a range of plant pathogens in broadacre and horticultural crops. It was advanced to the 2006 JMPR schedule of new compounds by the 38th session of the CCPR. The manufacturer has submitted information on physical and chemical properties, plant and animal metabolism, environmental fate, analytical mehods, storage stability, good agricultural practices, supervised field trials, processing and livestock feeding.

## IDENTITY

| | |
|---|---|
| ISO common name: | boscalid |
| Chemical name | |
| IUPAC name: | 2-Chloro-N-(4'-chlorobiphenyl-2-yl)nicotinamide |
| CAS: | 3-Pyridinecarboxamide, 2-chloro-N-(4'-chloro[1,1'-biphenyl]-2-yl)- |
| CAS Registry No: | 188425-85-6 |
| CIPAC No: | not assigned |
| Synonyms and trade names: | BAS 510 F, Nicobifen |
| Structural formula: | |

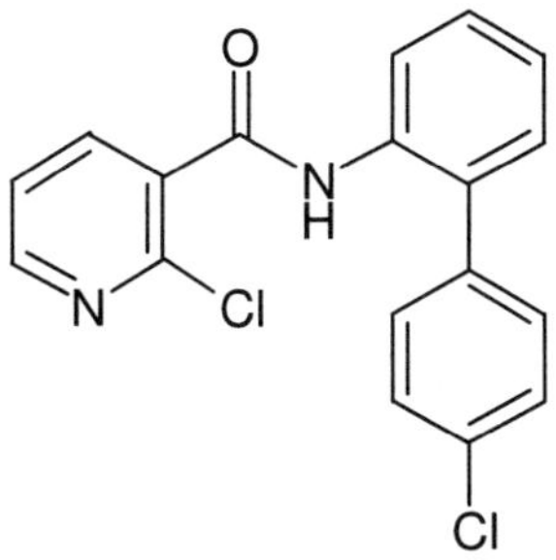

| | |
|---|---|
| Molecular formula: | $C_{18}H_{12}Cl_2N_2O$ |
| Molecular weight: | 343.21 |

## PHYSICAL AND CHEMICAL PROPERTIES

Pure active ingredient:

| Property | Result | Ref |
|---|---|---|
| Appearance (purity 99.7%) | White crystalline solid | Daum A., 1999, 1999/10991 |
| Odour (purity 99.7%) | Odourless | Daum A., 1999, 1999/10991 |
| Vapour pressure (purity 99.7%): | $7 \times 10^{-9}$ hPa at 20°C $2 \times 10^{-8}$ hPa at 25°C | Käestel R., 1999, 1999/10203 |
| Henry's law constant | $5.178 \times 10^{-8}$ kPa·m³/mol | Ohnsorge U., 2000, 2000/1001009 , |
| Boiling point (purity 99.7%) | Decomposition is observed at about 300 °C. | Daum A., 1999, 1999/10991 |

| Property | Result | Ref |
|---|---|---|
| Melting point (purity 99.7%) | 142.8 - 143.8°C | Daum A., 1999, 1999/ 10991 |
| Octanol-water partition coefficient: | 912 at a pH of 7.0 - 7.2 | Daum A., 1998, 1998/11082 |
| Solubility in water at 20°C: | 4.64 ± 0.06 mg/L (in deionized water at a pH of 6.0). There is no dissociation in water; therefore pH dependence on solubility is not applicable. | Daum A., 1998, 1998/10961 |
| Solubility in organic solvents at 20°C (purity of 99.4%): | Acetone: 160–200 g/L<br>Acetonitrile: 40–50 g/L<br>Dichloromethane: 200–250 g/L<br>N,N-Dimethylformamide: > 250 g/L<br>Ethyl acetate: 67–80 g/L<br>n-Heptane: < 10 g/L<br>Methanol: 40–50 g/L<br>1-Octanol: < 10 g/L<br>Olive oil: < 10 g/L<br>2-Propanol: < 10 g/L<br>Toluene: 20–25 g/L | Daum A., 1998, 1998/ 10953 |
| Relative density (purity 99.7%) | 1.381 g/cm$^3$ at room temperature | Käestel R., 1999, 1999/ 10203 |
| Dissociation constant in water | The test substance does not dissociate in deionized water (titration method, conc. of 20 mg/L [0.058 mmol/L], 20°C). Thus, no dissociation constant (pK$_a$) has been reported. | Daum A., 1998, 1998/10967 |
| Hydrolysis (sterile solution) | Boscalid is stable in aqueous solution in the dark at 50°C (pH 4, pH 7 and pH 9) for 5 days and at 25°C (pH 5, pH 7 and pH 9) for 30 days.<br><br>No DT$_{50}$ values were determined, neither for 50°C nor for 25°C, for they exceed twice the duration of the study. Therefore no DT$_{50}$ value for 20°C was calculated according to the Arrhenius equation. | Goetz von N., 1999, 1999/ 11285 |
| Photolysis in water | Boscalid is stable in direct aqueous photolysis for 15 days. In the dark control, likewise, no degradation was observed. No DT$_{50}$ value was determined for it exceeds twice the duration of the study. | Goetz N. von, 1999, 1999/11804 |

## FORMULATIONS

Boscalid is available in the following formulations:

| Formulation | Content of active ingredients | Trade names |
|---|---|---|
| WG | 500 g/kg boscalid | Cantus, Cantus WG, Filan, Pictor Pro |
| | 252 g/kg boscalid<br>128 g/kg pyraclostrobin | Bellis, Bellis 38 WG, Bellis WG,<br>Pristine, Cabrio C |
| | 136 g/kg boscalid<br>68 g/kg pyraclostrobin | Bellis Plus, Naria |
| WG | 267 g/kg boscalid | Signum, Signum 33 WG, Signum |

| Formulation | Content of active ingredients | Trade names |
|---|---|---|
| | 67 g/kg pyraclostrobin | WG |
| SE | 700 g/kg boscalid | Emerald, Endura, Lance WDG Fungicide, Cadence |
| SC | 200 g/L boscalid | Cantus Flüssig, Pictor |
| | 200 g/L dimoxystrobin | |
| | 200 g/L boscalid | Collis |
| | 100 g/L Kresoxim-methyl | |
| | 233 g/L boscalid | Tracker, Splice, Venture, Champion |
| | 67 g/L epoxyconazole | |

## METABOLISM AND ENVIRONMENTAL FATE

The Meeting received information on animal and plant metabolism and environmental fate studies which used boscalid labelled at the 3-pyridine carbon or uniformly labelled in the diphenyl moiety.

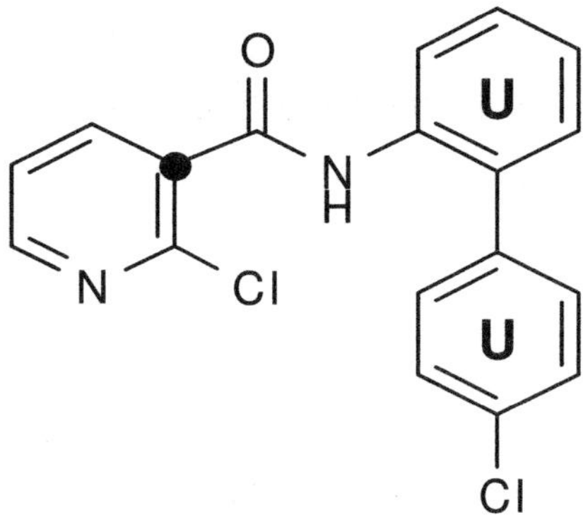

U = diphenyl label

● = 3-pyridine label

Structures, names and codes for boscalid and its metabolites in animal, plant and environmental fate studies are summarized below.

boscalid
2-chloro-N-(4'-chlorobiphenyl-2-yl)nicotinamide

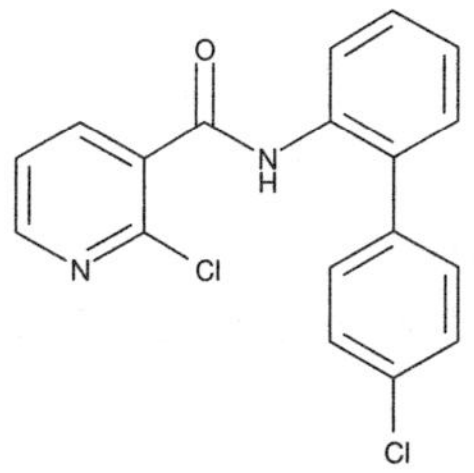

M510F01
2-chloro-N-(4'-chloro-5-hydroxybiphenyl-2-yl)nicotinamide

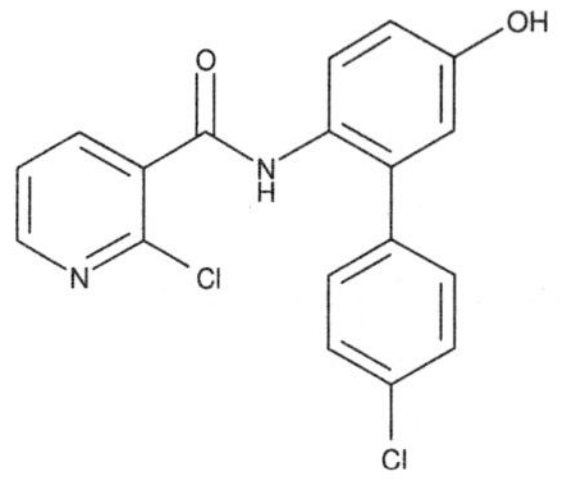

**M510F02**

4'-chloro-6-{[(2-chloro-3-pyridinyl)carbonyl]amino}biphenyl-3-yl glycopyranosiduronic acid

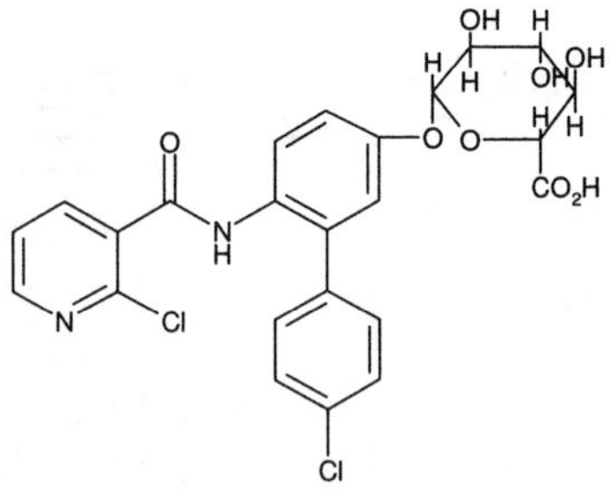

**M510F03**

4'-chloro-6-{[(2-chloro-3-pyridinyl)carbonyl]amino}biphenyl-3-yl hydrogen sulfate

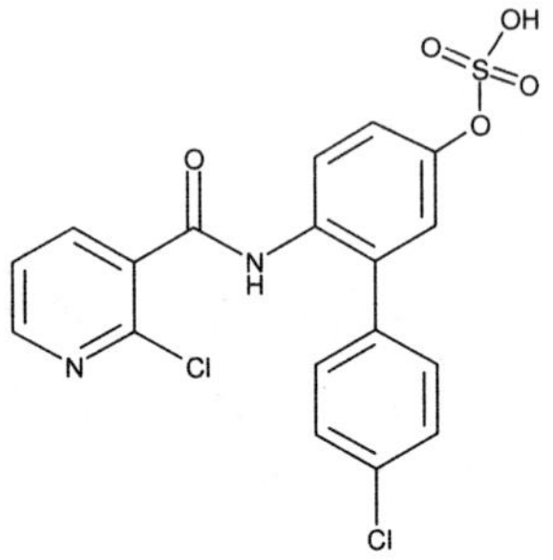

**M510F04**

*N*-acetyl(3-{[(4'-chlorobiphenyl-2-yl)amino]carbonyl}-2-pyridinyl)cysteine

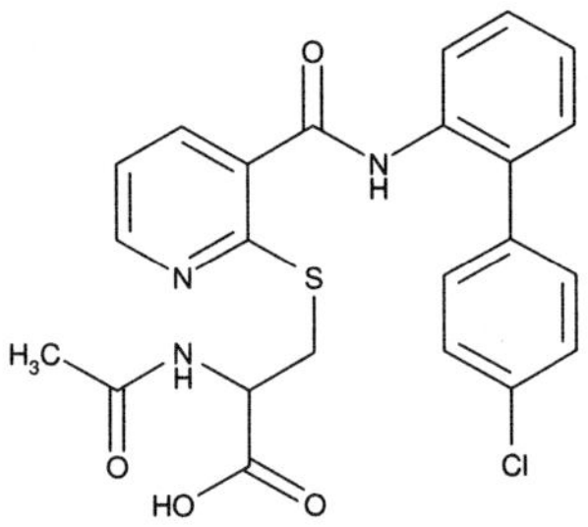

**M510F05**

(3-{[(4'-chlorobiphenyl-2-yl)amino]carbonyl}-2-pyridinyl)cysteine

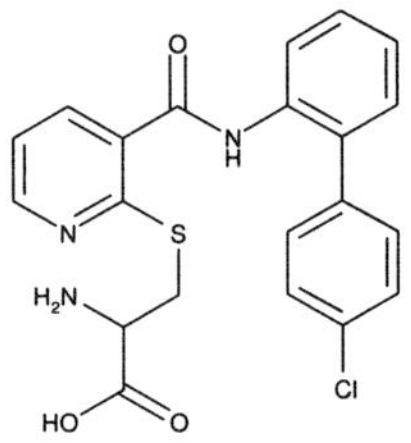

**M510F06**

*N*-(4'-chlorobiphenyl-2-yl)-2-sulfanylnicotinamide

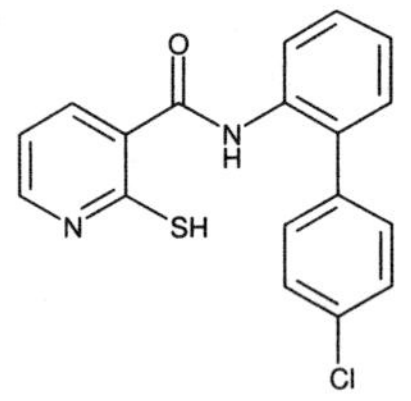

**M510F08**

*N*-(4'-chlorobiphenyl-2-yl)nicotinamide

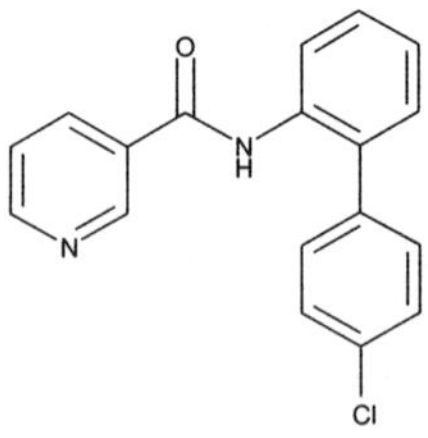

**M519F09**

*N*-biphenyl-2-yl-2-chloronicotinamide

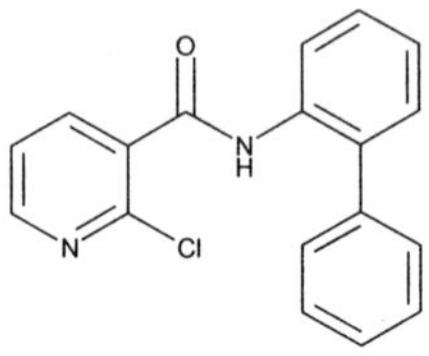

M510F10

2-chloro-N-(4'-chloro-?-hydroxybiphenyl-2-yl)nicotinamide

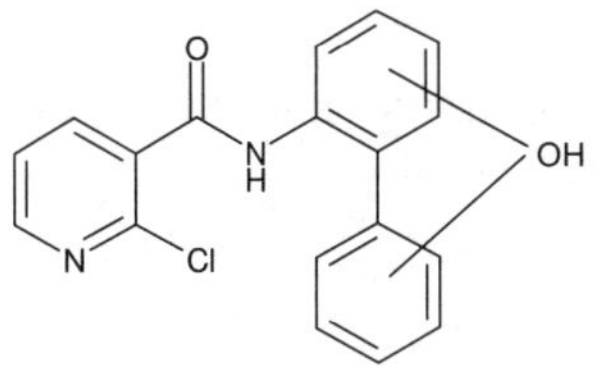

M510F11

N-(4'-chloro-?-hydroxybiphenyl-2-yl)-2-sulfanylnicotinamide

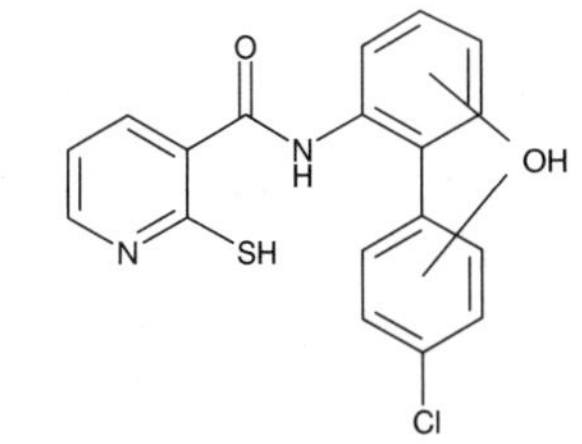

M510F12

N-(4'-chloro-?hydroxybiphenyl-2-yl)-2-methylsulfanylnicotinamide

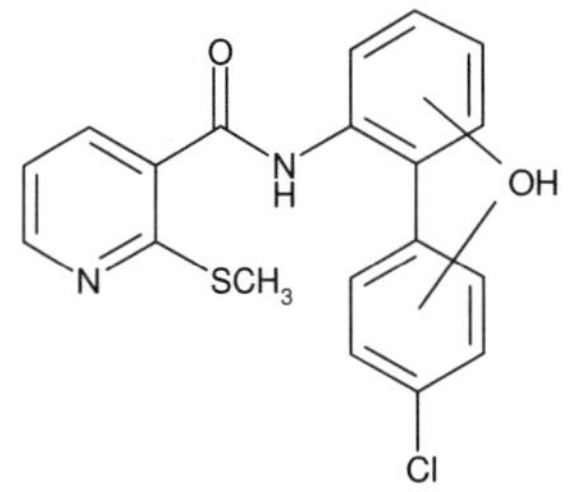

M510F13

3-{[(4'-chlorobiphenyl-2-yl)amino]carbonyl}-2-pyridinesulfinic acid

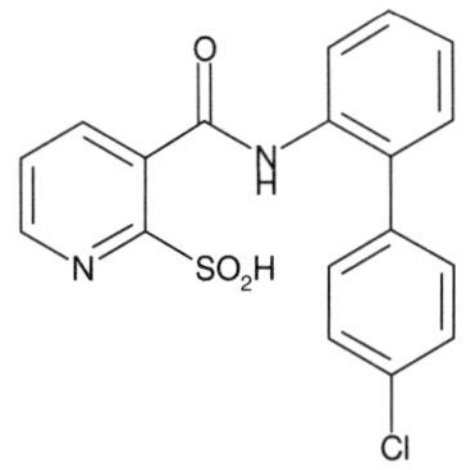

M510F14

2-chloro-N-(4'-chloro-?-dihydroxybiphenyl-2-yl)nicotinamide

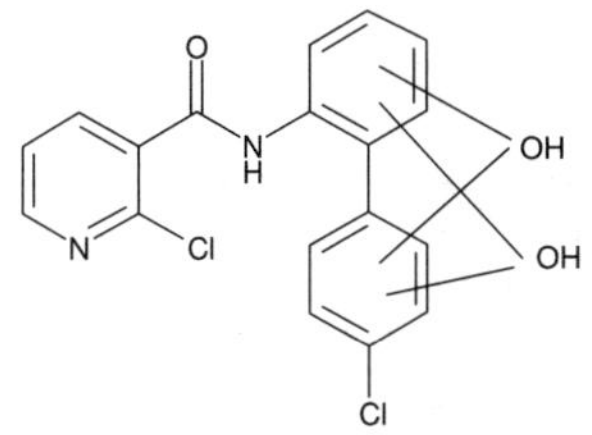

M510F15

2-chloro-N-(4'-chloro-?-sulfanylbiphenyl-2-yl)nicotinamide

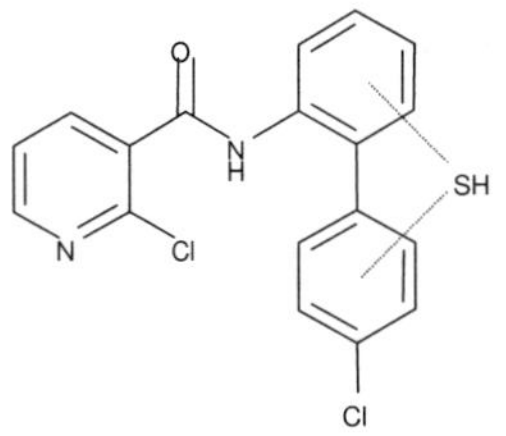

M510F16

2-chloro-N-(4'-chloro-?-hydroxy-?-methoxybiphenyl-2-yl)nicotinamide

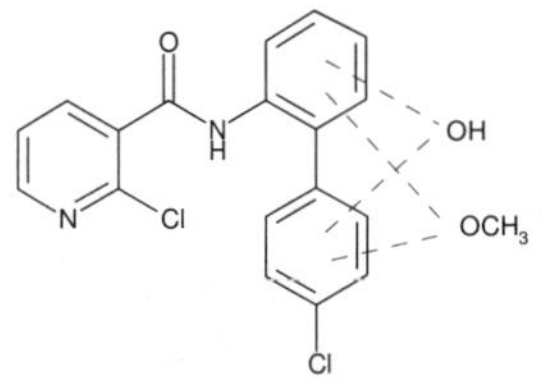

M510F18

[(3-{[(4'-chlorobiphenyl-2-yl)amino]carbonyl}-?-hydroxy-2-pyridinyl)o xy]acetic acid

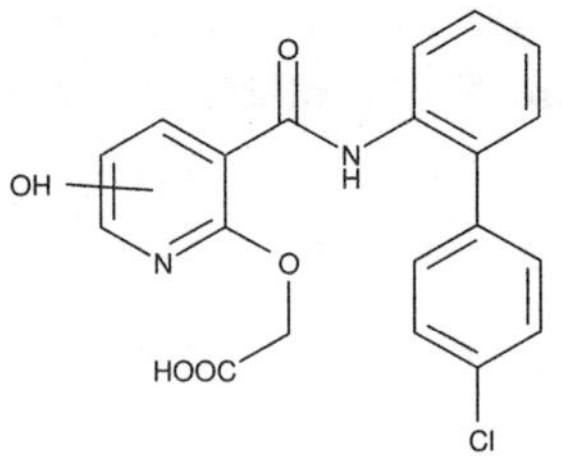

M510F19

[(3-{[(4'-chloro[1,1'-biphenyl]-2-yl)amino]carbonyl}-?-hydroxy-1-oxido -2-pyridinyl)oxy]acetic acid

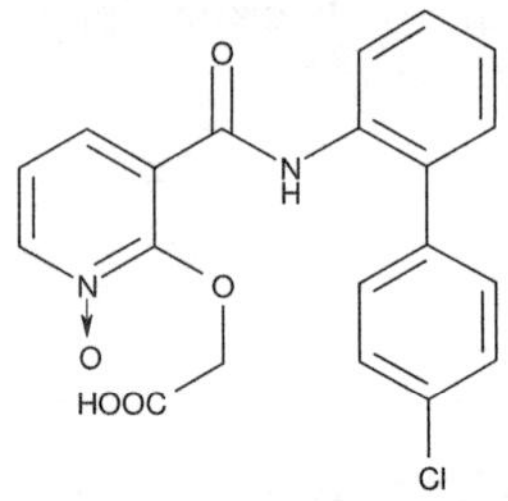

M510F20

2-chloro-N-(4'-chloro-?-hydroxy-?-methylsulfanylbiphenyl-2-yl)nicotin amide

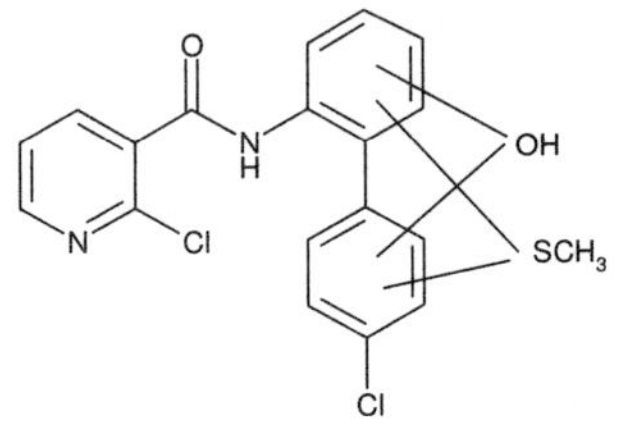

M510F22

(3-{[(4'-chloro-?-hydroxybiphenyl-2-yl)amino]carbonyl}-2-pyridinyl)cy steine

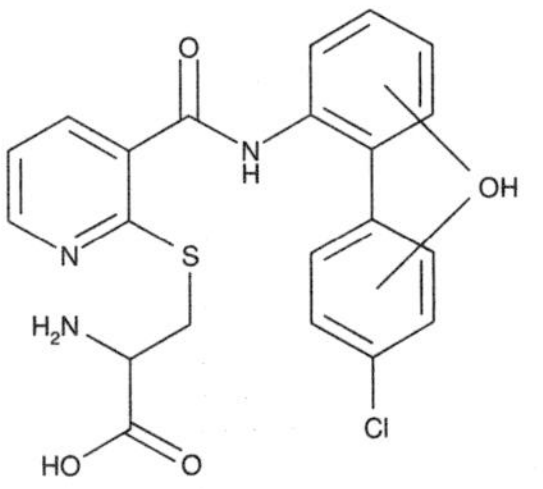

M510F23

(3-{[(4'-chlorobiphenyl-2-yl)amino]carbonyl}-?-hydroxy-2-pyridinyl)cy steine

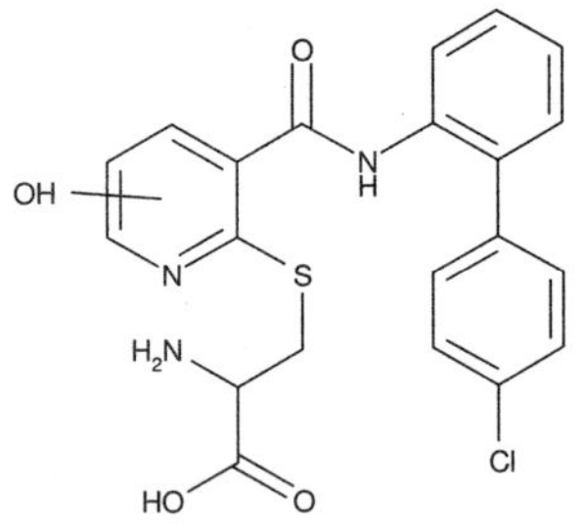

M510F28

N-acetyl(3-{[(4'-chloro-?-hydroxybiphenyl-2-yl)amino]carbonyl}-2-pyri dinyl)cysteine

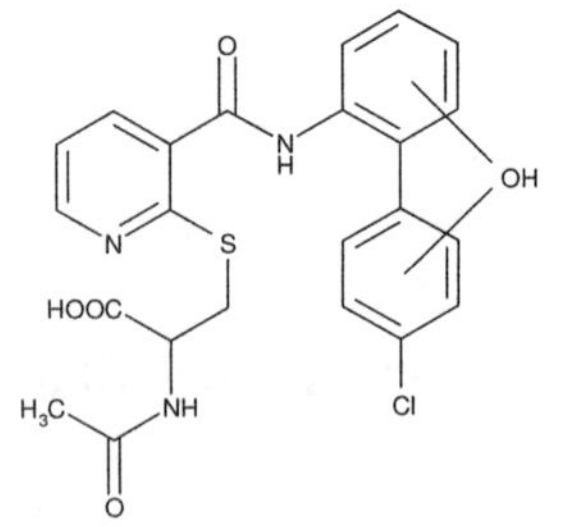

M510F29

2'-{[(2-chloro-3-pyridinyl)carbonyl]amino}biphenyl-4-yl glycopyranosiduronic acid

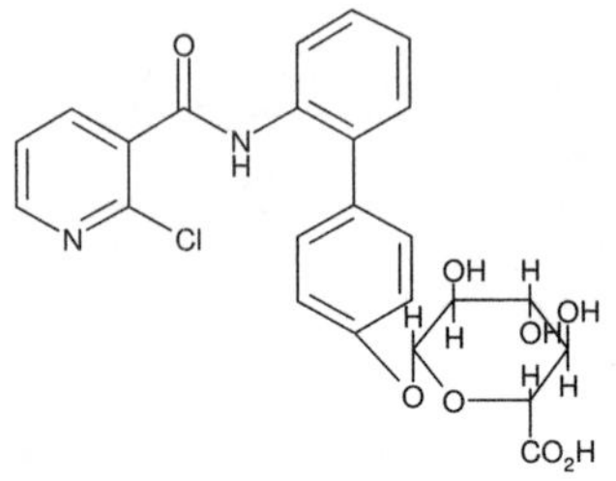

## M510F32

*N*-acetyl(4-chloro-2'-{[(2-chloro-3-pyridinyl)carbonyl]amino}biphenyl-?-yl)cysteine

## M510F33

*N*-acetyl(4-chloro-2'-{[(2-chloro-3-pyridinyl)carbonyl]amino}-?-hydroxybiphenyl-?-yl)cysteine

## M510F34

*N*-acetyl-3-[(4-chloro-2'-{[(2-chloro-3-pyridinyl)carbonyl]amino}biphenyl-?-yl)sulfinyl]alanine

## M510F39

4-chloro-2'-{[(2-methylsulfanyl-3-pyridinyl)carbonyl]amino}biphenyl-?-yl glycopyranosiduronic acid

## M510F40

2'-{[(2-chloro-3-pyridinyl)carbonyl]amino}-4-chloro-?-hydroxybiphenyl-?-yl glycopyranosiduronic acid

## M510F41

2'-{[(2-chloro-3-pyridinyl)carbonyl]amino}-4-chloro-?-methoxybiphenyl-?-yl glycopyranosiduronic acid

## M510F42

2'-{[(2-chloro-3-pyridinyl)carbonyl]amino}-4-chloro-?-methylsulfanylbiphenyl-?-yl glycopyranosiduronic acid

## M510F43
*N*-(4'-chlorobiphenyl-2-yl)-2-glutathionylnicotinamide

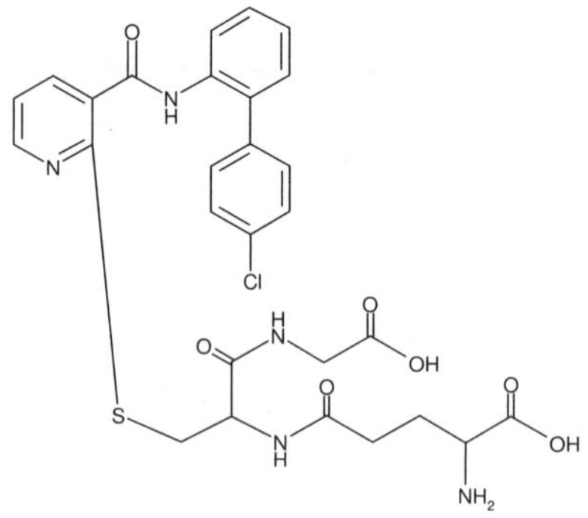

## M510F45
2-chloro-N-(4'-chloro-?-glutathionylbiphenyl-2-yl)nicotinamide

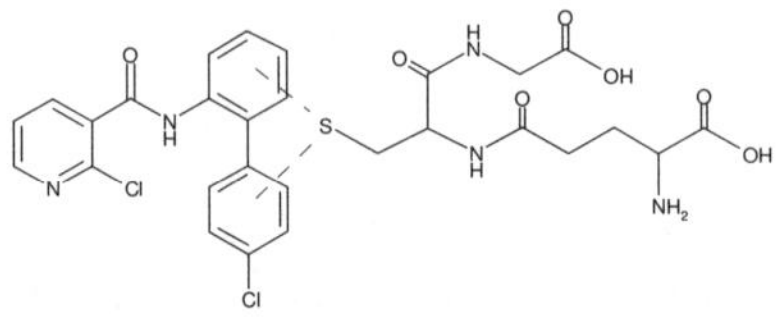

## M510F46
or isomer
$N^5$-(2-[(carboxymethyl)amino]-1-{[(5-(4-chlorophenyl)-4-{[(2-chloro-3-pyridinyl)carbonyl]amino}-6-hydroxy-2,4-cyclohexadien-1-yl)sulfanyl]methyl}-2-oxoethyl)glutamine

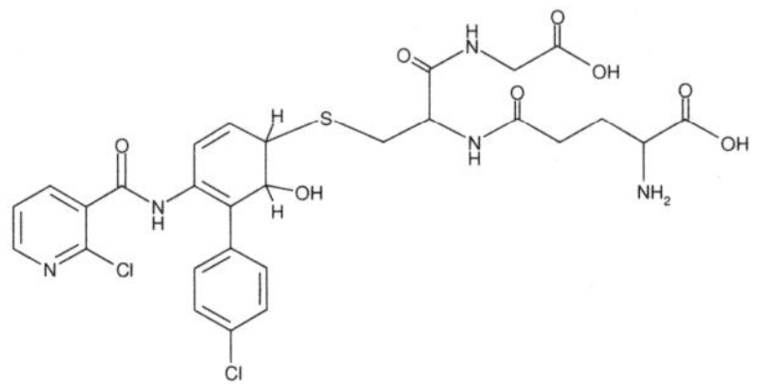

## M510F47
2-chloronicotinic acid

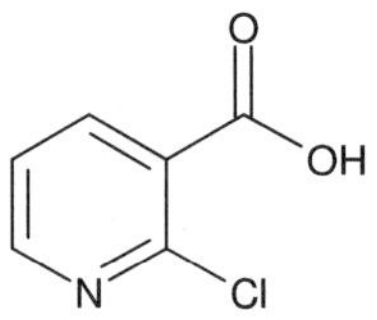

## M510F48
3-{[(4'-chloro-biphenyl-2-yl)amino]carbonyl}-2-pyridinyl
1-thiohexopyranosiduronic acid

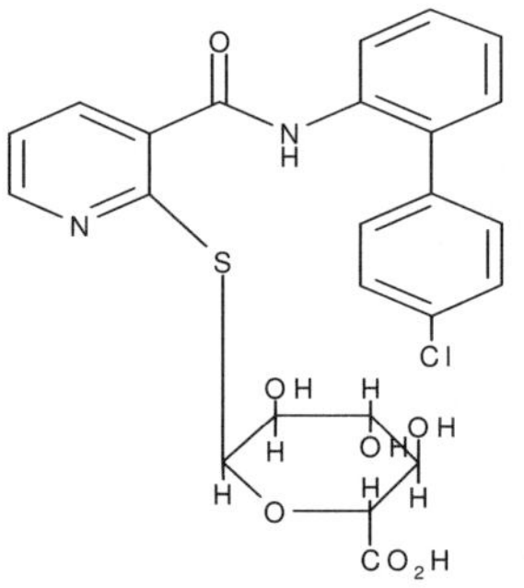

## M510F49
N-(4'-chlorobiphenyl-2-yl)-2-hydroxynicotinamide

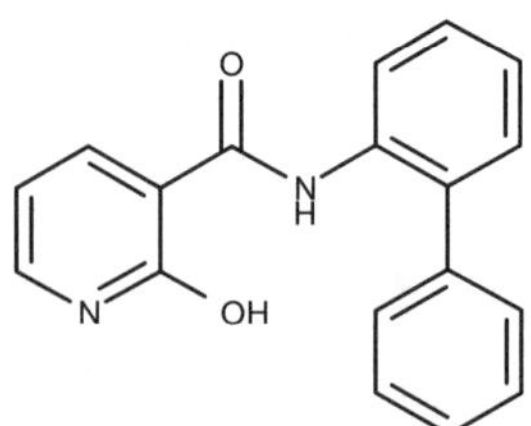

## M510F50
2-chloro-*N*-(4'-chlorobiphenyl-2-yl)-?-hydroxynicotinamide

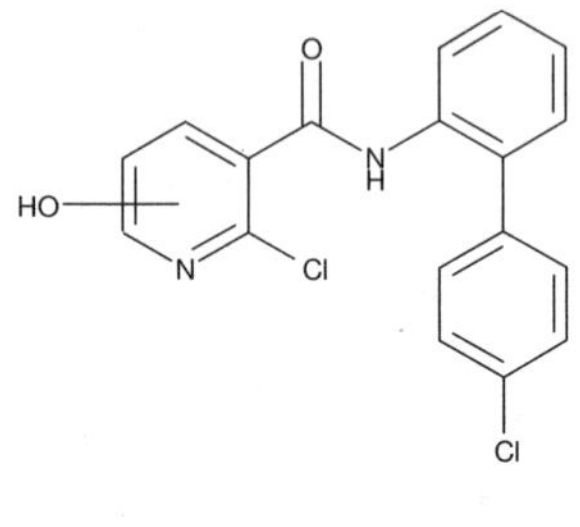

**M510F51**

*N*-(4'-chloro-5-hydroxybiphenyl-2-yl)-2-hydroxynicotinamide

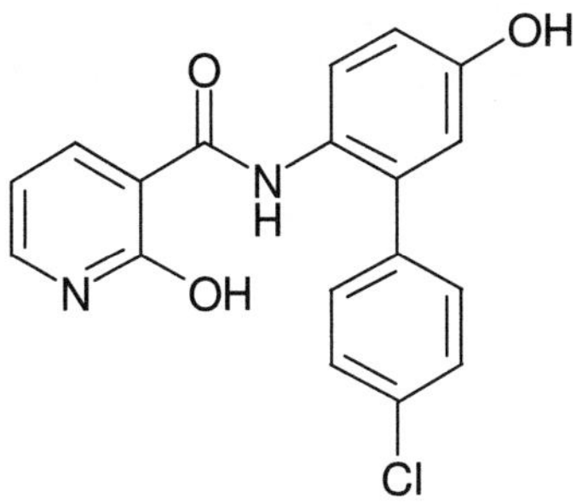

**M510F52**

4-chloro-2'-(formylamino)-biphenyl

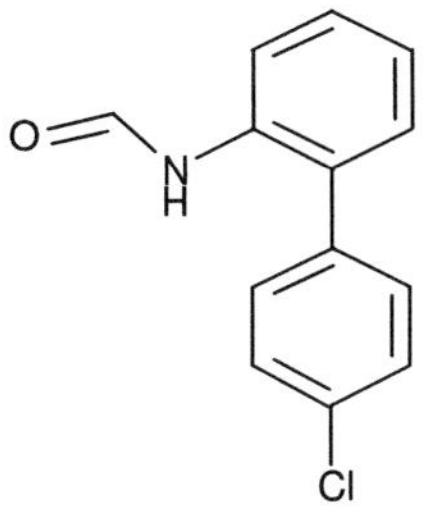

**M510F53**

4-chloro-2'-(acetylamino)-biphenyl

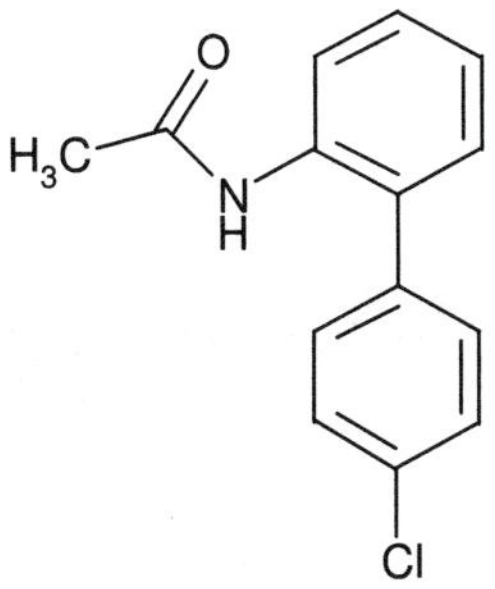

**M510F54**

2-chloro-N-(4'-chloro-?-sulfooxybiphenyl-2-yl)nicotinamide

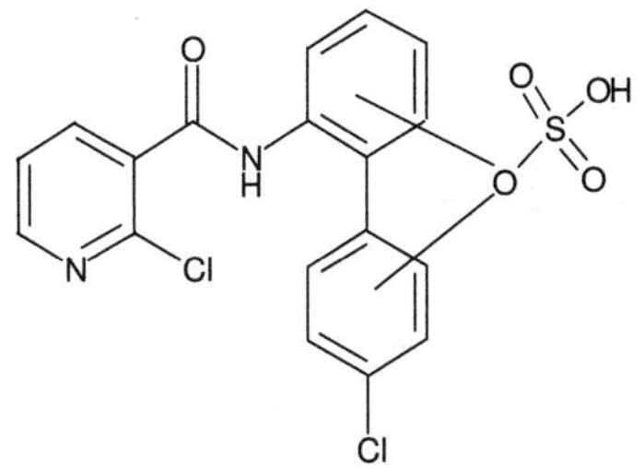

**M510F57**

or isomer

(5-(4-chlorophenyl)-4-{[(2-chloro-3-pyridinyl)carbonyl]amino}-6-hydroxy-2,4-cyclohexadien-1-yl)cysteine

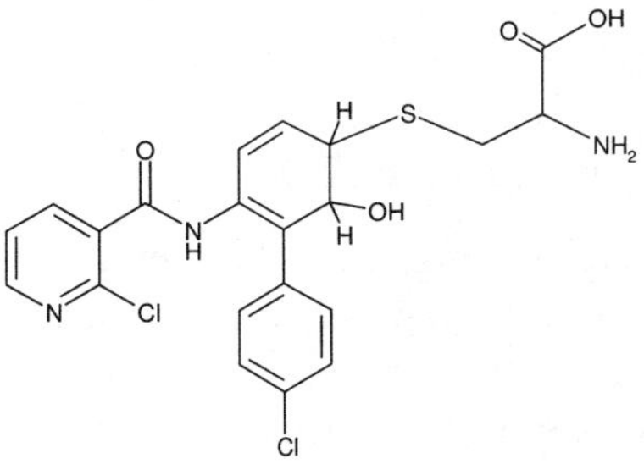

**M510F58**

(4-chloro-2'-{[(2-chloro-3-pyridinyl)carbonyl]amino}-?-hydroxybiphenyl-?-yl)cysteine

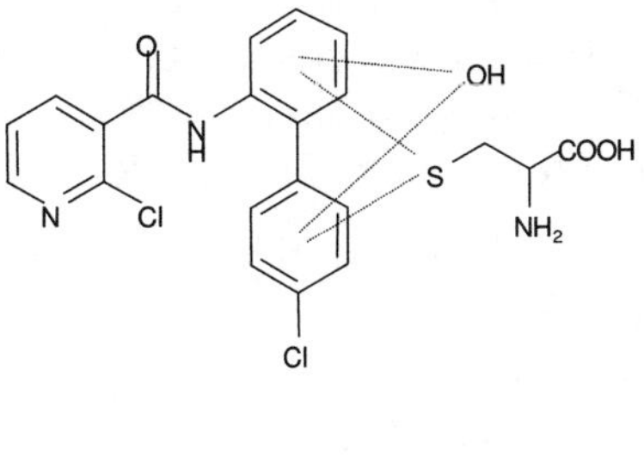

M510F59
2-chloro-N-(4'-chloro-?-hydroxybiphenyl-2-yl)nicotinamide

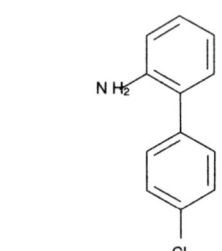

M510F61
4'-chloro-6-{[(2-chloro-3-pyridinyl)carbonyl]amino}biphenyl-?-yl
glycopyranosiduronic acid

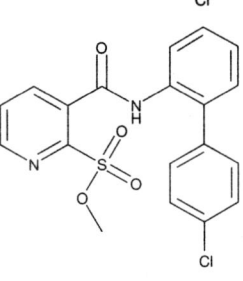

M510F62
4´-chlorobiphenyl-2-amine

M510F63
methyl 3-{[(4'-chloro[1,1'-biphenyl]-2-yl)amino]carbonyl}-2- pyridine
sulfonate

M510F64
4-chlorobenzoic acid

### Animal metabolism

The Meeting received animal metabolism studies with boscalid in lactating goats and laying hens. Two lactating goats (Bunte deutsche Edelziege) weighing 44.5 and 37.5 kg on the initial day were dosed orally once daily for 5 consecutive days, by gavage, with a syringe at about 65 mg/animal/day of [diphenyl-U-$^{14}$C] labelled boscalid, equivalent to 35 ppm in the feed (Leibold and Hoffmann, 2000, 2000/1012353; Fabian and Grosshans, 2000/1017221, 2001; based on a consumption of 1.9 kg/day feed). Milk was collected twice daily; a day's sample began in the afternoon after dosing and ended with the morning milking preceding the next dose. The animals were slaughtered 23 hours after the final dose for tissue collection. Recovery of administered $^{14}$C was 95% in Animal 1 and 98% in Animal 2.

The majority of the administered $^{14}$C was rapidly and almost completely excreted in the faeces (46% and 64%) and urine (24% and 44%). Radioactivity recovered from urine and faeces together with cage wash amounted to 94–95% of the total radioactivity recovered. Milk accounted for 0.06% and 0.15% of the administered $^{14}$C while tissues accounted for 0.46% and 0.66%. The distribution of the radiolabel and identified metabolites in milk and tissues are summarised in Table 1.

Parent compound and its hydroxylated metabolite M510F01, including the conjugate M510F02, were the main residues in milk and each of the tissues.

Table 1. Distribution of $^{14}C$ residue and identified metabolites in milk and tissues of lactating goats dosed orally for 5 days with 65 mg/animal/day of [diphenyl-U-$^{14}C$] labelled boscalid, equivalent to 35 ppm in the feed (Leibold and Hoffmann, 2000, 2000/1012353; Fabian and Grosshans, 2001, 2000/1017221).

| Metabolite code (Reg.-No. of reference substance) | Structure | Milk (pool) | Muscle | Fat | Kidney | Liver |
|---|---|---|---|---|---|---|
| | | mg/kg % TRR | mg/kg % TRR | mg/kg % TRR | mg/kg % TRR | mg/kg % TRR |
| Total $^{14}C$ residue | | 0.037 | 0.012 | 0.036 | 0.270 | 2.593 |
| Extracted residue | | 0.037 99.3 | 0.010 79.7 | 0.024 62.8 | 0.219 81.3 | 0.430 16.6 |
| Boscalid | 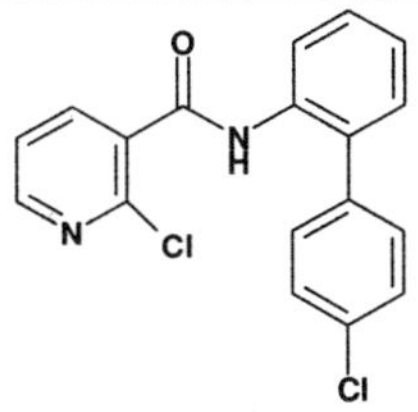 | 0.001 3.2 | 0.002 20.4 | 0.012 34.6 | 0.007 2.5 | 0.129 5.0 |
| M510F01 | 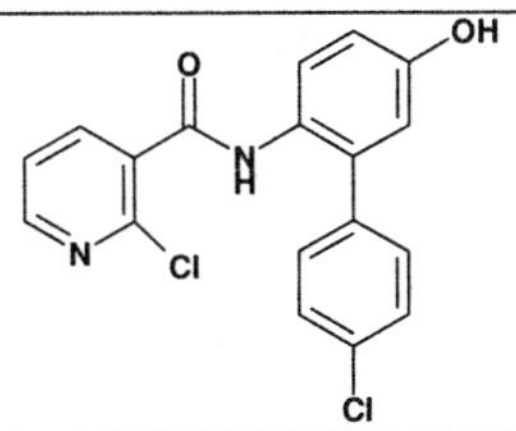 | 0.006 14.9 | 0.003 20.6 | 0.009 26.3 | 0.023 8.6 | 0.074 2.9 |
| M510F02 | 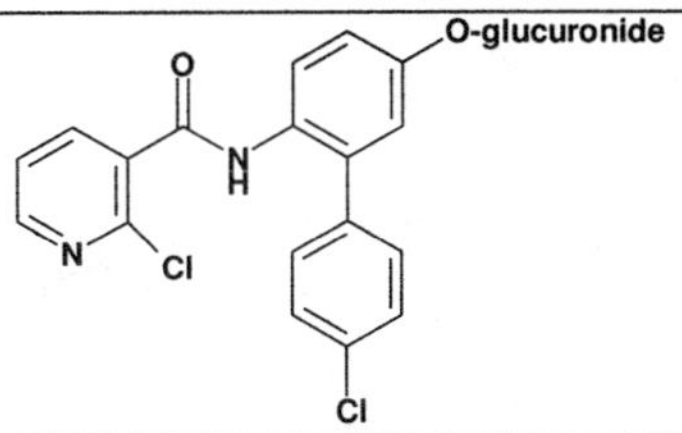 | 0.002 6.4 | 0.001 11.9 | n.d. | 0.136 50.3 | n.d. |

n.d.: not detected

Four analytes could be detected after microwave-treatment of liver samples using acetonitrile and acetic acid. M510F01, M510F49 and M510F51 originated from extractable residues. Bound residues of boscalid were cleaved under microwave treatment at the amide bond to form M510F53. The major residues in liver were bound residues (Table 2). M510F53 could also be detected as a minor residue in milk and is due to minor sulfur substituted metabolites.

Table 2. Characterisation of the residual radioactive in liver and minor metabolites in milk as M510F53 by microwave treatment with acetonitrile and acetic acid (Fabian and Grosshans, 2001, 2000/1017221).

| Sample | M510F53 [mg / kg] (% TRR) |
|---|---|
| Liver | 1.130 (43.6) |
| Milk | 0.004 (11.2) |

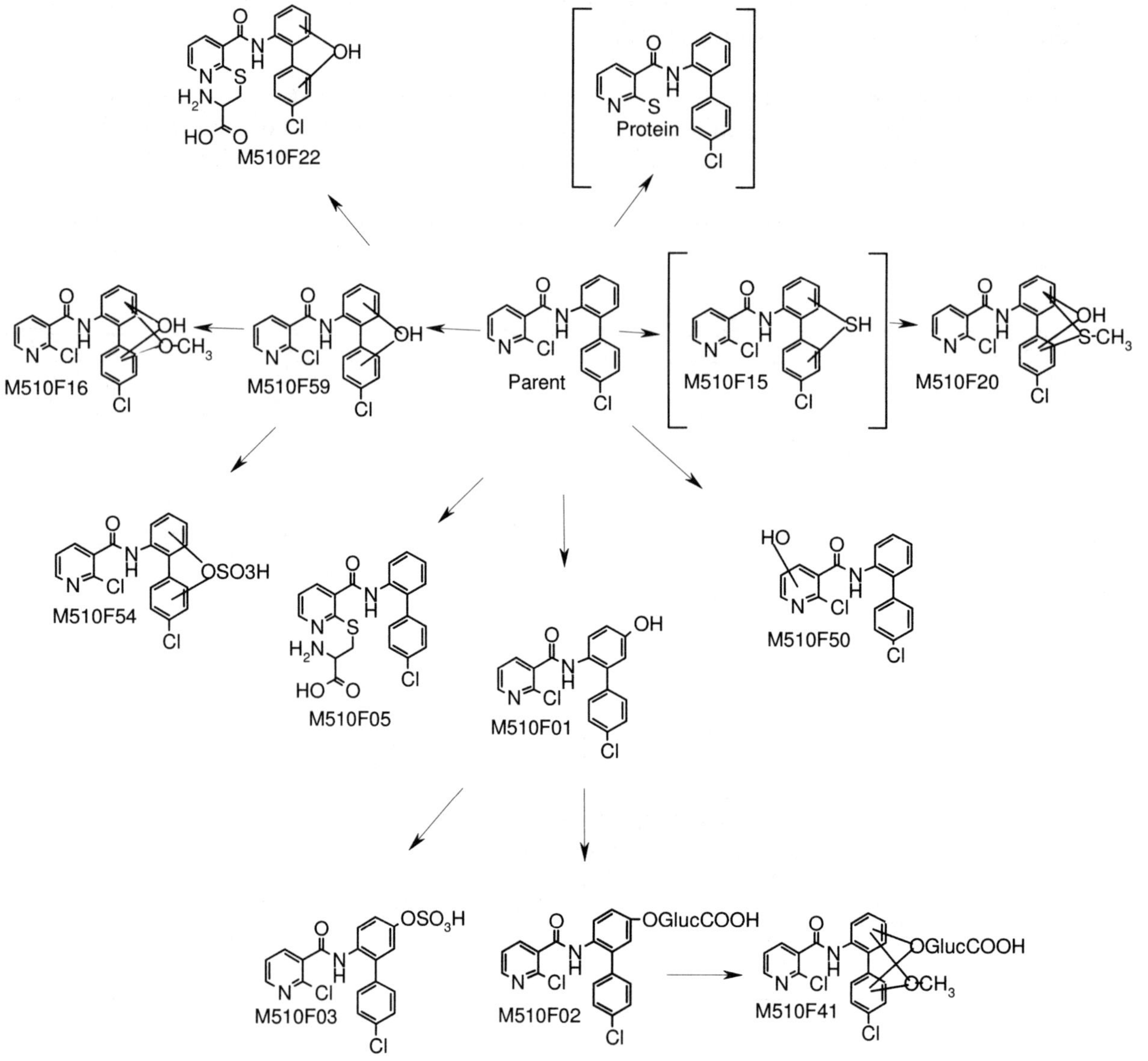

Figure 1. Proposed metabolic pathway for boscalid in lactating goats (Fabian and Grosshans, 2001, 2000/1017221).

Ten laying <u>hens</u> (body weight between 1.407 and 1.995 kg at study initiation) were administered encapsulated [diphenyl-U-$^{14}$C] labelled boscalid via a balling gun once daily for 10 consecutive days with 1.6 mg/bird/day, equivalent to 12.5 ppm in the feed (Nietschmann and Lam, 2000, 2000/5154; based on a consumption of 133 g/day feed). Individual eggs were collected twice daily and pooled by group starting on Study Day 0. The birds were slaughtered for tissue collection 21–23 hours after the final dose. Recovery of administered $^{14}$C was 98.3%.

The majority of the administered $^{14}$C was present in the excreta (97.7%). Eggs accounted for 0.115% of the administered $^{14}$C while tissues, liver, fat and muscle accounted for 0.046%. The distribution of the radiolabel and identified metabolites in eggs and tissues are summarised in Table 3.

The parent compound, boscalid and its hydroxylated metabolite M510F01, including the conjugates M510F02 and M510F54, were the main residues in eggs. Muscle had a very low residue level (0.0025 mg/kg) and therefore was not further investigated. The main residue in fat was identified as the parent compound.

Table 3. Distribution of $^{14}$C residue and identified metabolites in egg and tissues of laying hens administered encapsulated [diphenyl-U-$^{14}$C] labelled boscalid via a balling gun once daily for 10 consecutive days with 1.6 mg/bird/day, equivalent to 12.5 ppm in the feed (Nietschmann and Lam, 2000, 2000/5154).

| Metabolite code | Structure | Eggs (pool, day 2 - 10) | Fat | Liver[1] |
|---|---|---|---|---|
| | | mg/kg % TRR | mg/kg % TRR | mg/kg % TRR |
| Total $^{14}$C residue | | 0.058 | 0.025 | 0.169 |
| Extracted residue | | 0.053 92 | 0.023 93 | 0.020 12 |
| Boscalid | 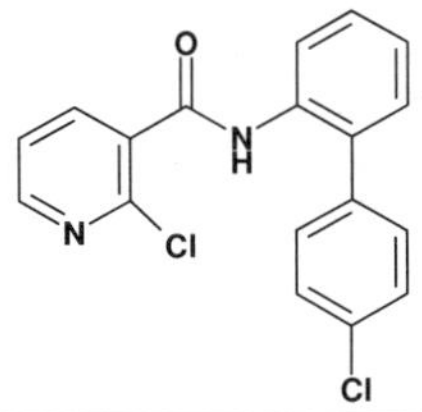 | 0.020 35 | 0.023 93 | n.d. |
| M510F01 | 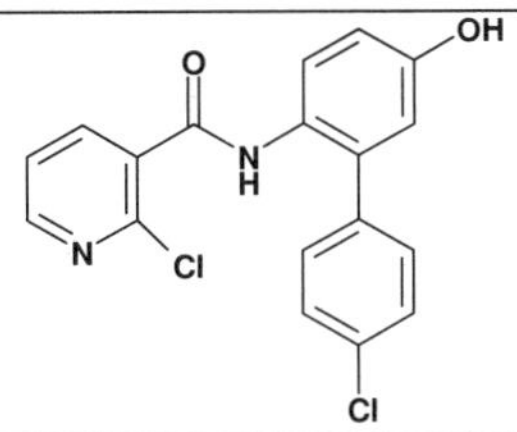 | 0.015 27 | n.d. | n.d. |
| M510F02 | 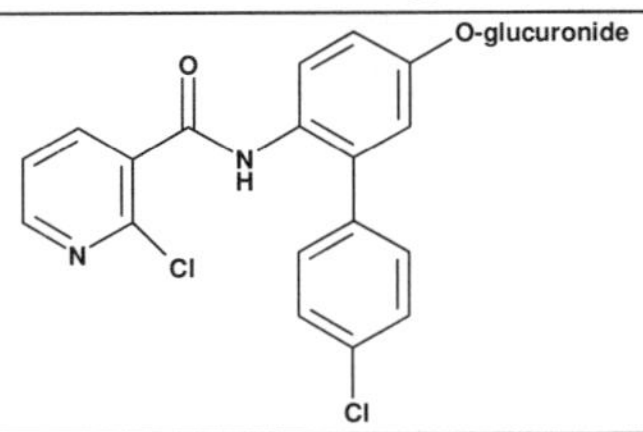 | 0.011 17 | n.d. | n.d. |
| M510F54 | 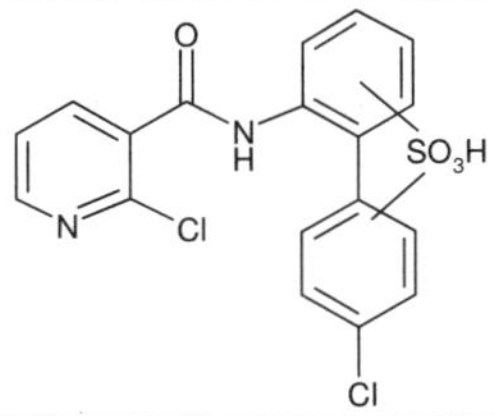 | 0.0013 1.9 | n.d. | n.d. |

n.d.: not detected

1) It was not possible to identify any metabolites in the liver acetonitrile extract.

For further characterisation of the residues of boscalid in liver, a specially developed microwave method was applied. Four metabolites were be detected following microwave treatment of liver samples using acetonitrile and formic acid. It was demonstrated, in detailed experiments, that the detected analytes M510F01, M510F49 and M510F51 originated from extractable residues. Bound residues of boscalid are cleaved under microwave treatment at the amide bond to form M510F52. The major residues in liver were bound residues that correspond to M510F52 (Table 4).

Table 4. Summary of metabolite identities and quantities in hen liver hydrolysate after microwave treatment with acetonitrile and formic acid (Nietschmann and Lam, 2000, 2000/5154).

| Metabolite code | Structure | Liver Hydrolysate mg/kg % TRR |
|---|---|---|
| M510F01 | 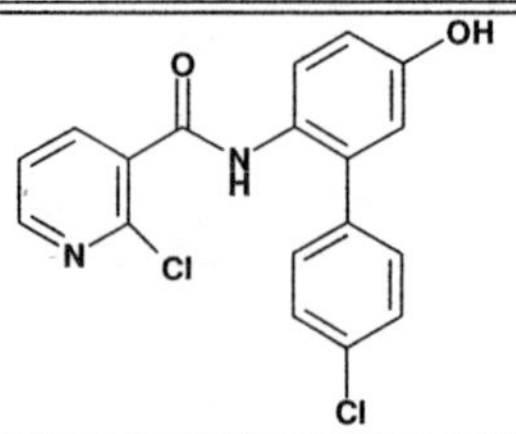 | 0.0094 5.55 |
| M510F49 | 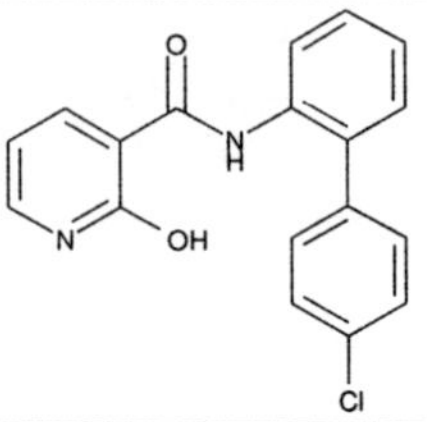 | 0.0214 12.71 |
| M510F51 | 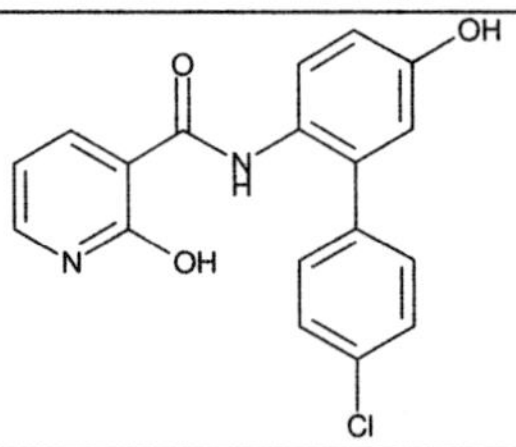 | 0.0366 21.69 |
| M510F52 | 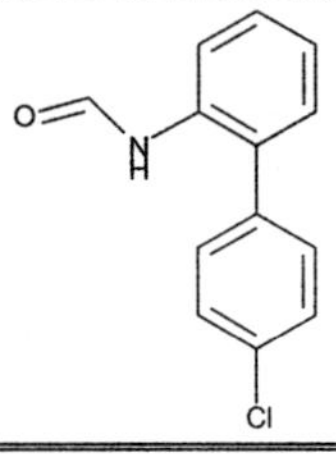 | 0.0710 42.09 |

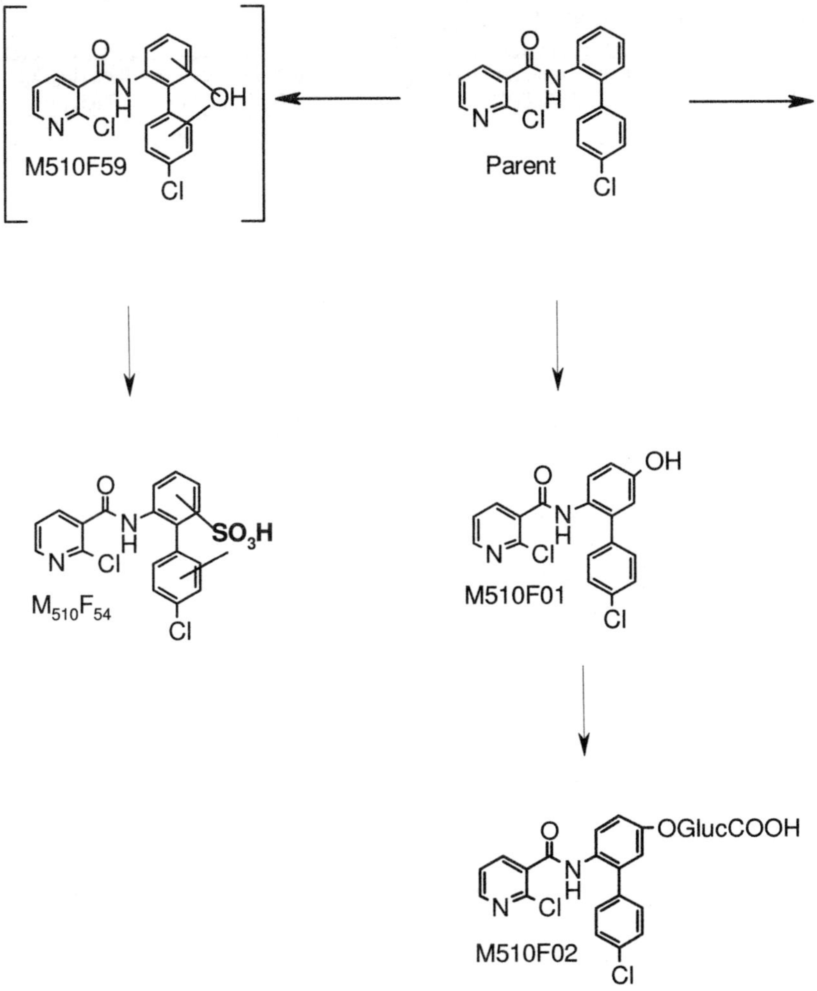

Figure 2. Proposed metabolic pathway for boscalid in laying hens (Nietschmann and Lam, 2000, 2000/5154).

No metabolism study was performed in pigs as the metabolic profile identified in rodents (rats) and ruminants (goats) did not differ significantly.

### Plant metabolism

The Meeting received plant metabolism studies for boscalid on grapes, lettuce and beans. In each crop tested, parent compound generally represented more than 90% of the total $^{14}$C residue and was essentially the only compound detected. The high extractability and the low non-extractable residues in various plant matrices demonstrate that boscalid showed almost no conversion to bound residues and therefore little or no presence of the corresponding carbohydrates, proteins or other natural products.

A grapevine (variety: Müller-Thurgau) metabolism study with $^{14}$C-boscalid (diphenyl- and pyridine-label) was performed with three spray applications, each with 800 g ai/ha, corresponding to the expected maximum recommended use rate (Rabe and Schlüter, 2001, 2000/1014860). The first application was carried out at the end of flowering, the second application 13 days later and the third 41 days after the second. Samples were taken at the maturity of grapes (45 days after the last treatment).

For all sample materials, the solvent extractability was very high. An HPLC analysis of the methanol and the water extracts of all matrices showed that most of the radioactive residues were unchanged parent (Table 5). Boscalid showed almost no metabolism to bound residues and thus no conversion to carbohydrates, proteins or other natural products.

Table 5. Summary of identified boscalid in grape samples after treatment with [14]C-boscalid (diphenyl- and pyridine- label) (Rabe and Schlüter, 2001, 2000/1014860).

| Matrices | Total radioactive residues [mg/kg] Diphenyl label (%TRR) | Total radioactive residues [mg/kg] Pyridine label (%TRR) |
|---|---|---|
| Grapes (at harvest) | 1.09 (92.7%) | 1.90 (92.2%) |
| Stalks (at harvest) | 11.9 (96.4%) | 19.1 (97.5%) |
| Leaves (at harvest) | 41.7 (95.6%) | 60.8 (96.1%) |

A lettuce (variety: Nadine) metabolism study with [14]C-boscalid (diphenyl- and pyridine-label) was performed with three spray applications, each with 700 g ai/ha, corresponding to the expected maximum recommended use rate in a greenhouse situation (Hamm, 1999, 1999/11240). The first application was performed 8 days after planting, the second and third applications 14 days later, respectively. The third treatment was done 18 days before harvest (PHI). Only one sampling was done 18 days after the third application.

The solvent extractability was high (about 99%) in all matrices. The two calculated TRRs of leaves treated with the two labels corresponded to 17.5 mg/kg (diphenyl label) and 17.6 mg/kg (pyridine label). The extractable radioactivity (ERR) was identified by HPLC and LC/MS/MS as unchanged parent only. Boscalid showed almost no conversion to bound residues such as carbohydrates, proteins or other natural products.

Table 6. Summary of identidied boscalid in lettuce samples after treatment with [14]C-boscalid (diphenyl- and pyridine- label) (Hamm, 1999, 1999/11240).

| Matrices | Total radioactive residues [mg/kg] Diphenyl label (%TRR) | Total radioactive residues [mg/kg] Pyridine label |
|---|---|---|
| Lettuce | 17.4 (99.3%) | 17.5 (99.3%) |

A bean metabolism study with [14]C-boscalid (diphenyl- and pyridine-label) was performed with three spray applications, each with 500g ai/ha, corresponding to the expected maximum recommended use rate in a glass house or in growth chambers (Veit, 2001, 2000/1014861). The first application was carried out at the beginning of flowering, the second application 8–10 days later and the third application 8–10 days after the second. The tests with the different labels were not done in the same time period. The pyridine label was applied at the end of February and the plants were grown in a growth chamber. The diphenyl label was applied in the beginning of March of the following year and the plants were grown in a glass house. Bean plant samples were collected directly after the last treatment. At 14/15 days (diphenyl/ pyridine label, respectively) after the last treatment, bean forage and green beans were harvested. The green beans were separated into pods and seeds and analysed individually to cover other bean varieties. At 53/51 days (diphenyl/pyridine label) after the last treatment, bean straw, dry pods and dry seeds were harvested.

The TRR values for the diphenyl label were, in all cases, higher than for the pyridine label. In all sample materials; the solvent extractability was generally > 80%. The fact that the two parts of the study were conducted a year apart may account for the differences in the TRR values. The pattern and the ratio of the TRR values of the two labels were similar for the different matrices. In plant samples, parent compound generally represented more than 90% of the total [14]C residue.

Table 7. Summary of identidied boscalid in beans samples after treatment with [14]C-boscalid (diphenyl- and pyridine- label) (Veit, 2001, 2000/1014861).

| Matrices (Days after last treatment DALT) | Total radioactive residues [mg/ kg] Diphenyl label (%TRR) | Total radioactive residues [mg/ kg] Pyridine label (%TRR) |
|---|---|---|
| Bean plant (0) | 48.7 (99.3) | 20.8 (98.1%) |
| Bean Forage (14/15) | 65.2 (98.6%) | 16.6 (98.4%) |

| Matrices (Days after last treatment DALT) | Total radioactive residues [mg/ kg] Diphenyl label (%TRR) | Total radioactive residues [mg/ kg] Pyridine label (%TRR) |
|---|---|---|
| Green Beans (14/15) | 0.999 (97.2%) | 0.071 (78.1) |
| Bean Pods (14/15) | 0.872 (96.7%) | 0.095 (87.0%) |
| Bean Seeds 14/15) | 0.173 (87.5%) | 0.043 (64.9%) |
| Bean Straw (53/51) | 120.9 (95.1%) | 87.7 (93.6%) |
| Bean Dry Pods (53/51) | 5.78 (94.5%) | 1.09 (79.7%) |
| Bean Dry Seeds (53/51) | 0.148 (72%) | 0.047 (36.9%) |

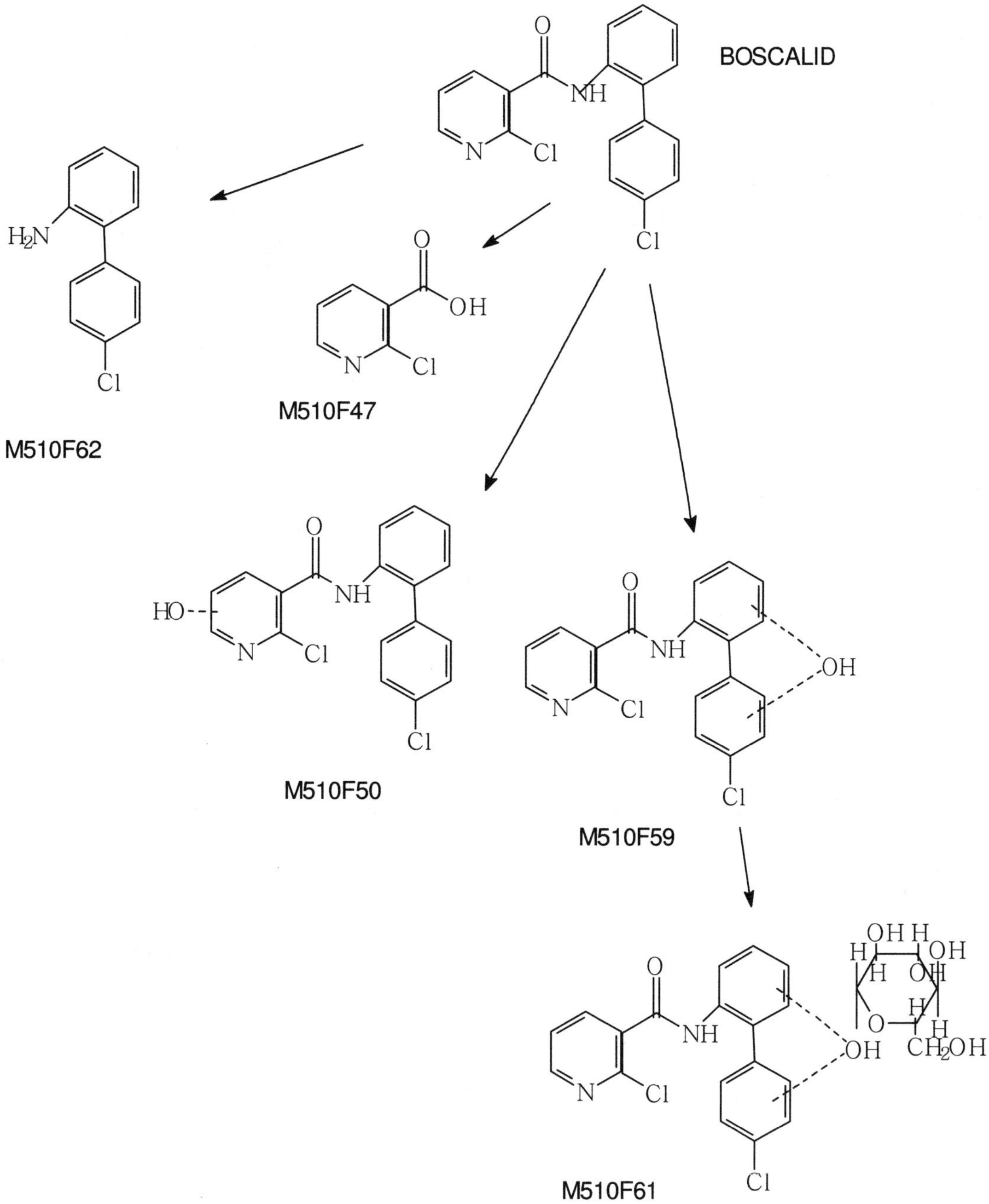

Figure 3. Proposed metabolic pathway for boscalid in beans.

### Environmental fate in soil

The Meeting received information on the environmental fate of boscalid in soil, including studies on aerobic and anaerobic soil metabolism, field dissipation and crop rotational studies. Because boscalid is used on peanuts and potatoes (the edible portion of which is in the soil), additional studies on aerobic soil metabolism and field dissipation are needed as well.

A study on the aerobic <u>soil metabolism</u> of [$^{14}$C] boscalid in a sandy loam soil showed that boscalid degraded slowly and identifiable metabolites were a minor part of the residue which in turn degraded almost at the same rate (Stephan, 1999, 1999/11807). The pyridine-labelled test compound was mineralized to $^{14}CO_2$ more quickly than the diphenyl-label test compound. No volatiles other than $^{14}CO_2$ were found. The non-extractable $^{14}$C in the soil treated with diphenyl labelled boscalid had begun to decline within 266 days and that in the soil treated with pyridine labelled boscalid increased continuously during the whole period of incubation.

*Aerobic soil metabolism*                                              Ref: Stephan, 1999, 1999/11807

Test material: [$^{14}$C]boscalid, diphenyl-labelled          Dose rate: 0.933 mg/kg dry soil
Sandy loam                              pH: 7.4               Organic carbon: 1.3%
Duration: 364 days                      Temp: 20°C           Moisture: 40% maximum water holding
                                                             capacity

Half-life of boscaild: 108 days
% boscalid remaining, day 364 = 16.7 %                       % mineralization, day 364 = 15.5 %

| Metabolites | Max (% of dose) | Day |
|---|---|---|
| M510F49 | 0.2 % | 57 |
| M510F50 | 0.1 % | 266 |
| Others | 1.0 % | 266 |
| $^{14}CO_2$ | 15.5% | 364 |
| Non-extractable $^{14}$C | 62.7 % | 266 |

Test material: [$^{14}$C]boscalid, pyridine-labelled          Dose rate: 1.022 mg/kg dry soil
Half-life of boscaild: 108 days
% boscalid remaining, day 364 = 17.3 %                       % mineralization, day 364 = 25.4 %

| Metabolites | Max (% of dose) | Day |
|---|---|---|
| M510F49 | 0.2 % | 93 |
| M510F50 | 0.1 % | 93 |
| Others | 1.0 % | 119 |
| $^{14}CO_2$ | 25.4% | 364 |
| Non-extractable $^{14}$C | 50.1% | 364 |

Another study on the aerobic soil metabolism of [diphenyl-U-$^{14}$C] boscalid in four different soils at different temperatures and soil moistures for 120 days showed again that boscalid degraded slowly and identifiable metabolites were a minor part of the residue which also mostly degraded relatively slowly (Ebert and Harder, 2000, 2000/1013279). The volatiles including $^{14}CO_2$ were not trapped in this study. The non-extractable $^{14}$C in the soil treated with diphenyl labelled boscalid had begun to decline within 266 days.

*Aerobic soil metabolism*                                    Ref: Ebert and Harder, 2000,
                                                                  2000/1013279

Test material: [$^{14}$C]boscalid, diphenyl-labelled          Dose rate: 1.0 mg/kg dry soil
Loamy sand                              pH: 5.6              Organic carbon: 2.5%
Duration: 120 days                      Temp: 20°C           Moisture: 40% maximum water holding
                                                             capacity

Half-life of boscalid: 384 days
% boscalid remaining, day 120 = 78.8%                        % mineralization, not measured

| Metabolites | Max (% of dose) | Day |
|---|---|---|
| Others | 0.7% | 60 |
| Non-extractable $^{14}$C | 16.8% | 120 |

| Duration: 120 days | pH: 5.6<br>Temp: 5°C | Organic carbon: 1.9%<br>Moisture: 40% maximum water holding capacity |
|---|---|---|

Half-life of boscalid: stable
% boscalid remaining, day 119 = 103.8%                                        % mineralization, not measured

| Metabolites | Max (% of dose) | Day |
|---|---|---|
| Others | 0.8% | 0 |
| Non-extractable $^{14}$C | 1.9% | 119 |

| Duration: 120 days | pH: 5.7<br>Temp: 30°C | Organic carbon: 2.18%<br>Moisture: 40% maximum water holding capacity |
|---|---|---|

Half-life of boscalid: 365 days
% boscalid remaining, day 120 = 84.4%                                         % mineralization, not measured

| Metabolites | Max (% of dose) | Day |
|---|---|---|
| Others | 1.6% | 120 |
| Non-extractable $^{14}$C | 13.0% | 120 |

| Duration: 120 days | pH: 5.9<br>Temp: 20°C | Organic carbon: 2.0%<br>Moisture: 20% maximum water holding capacity |
|---|---|---|

Half-life of boscalid: stable
% boscalid remaining, day 120 = 98.8%                                         % mineralization, not measured

| Metabolites | Max (% of dose) | Day |
|---|---|---|
| Others | 0.9% | 120 |
| Non-extractable $^{14}$C | 7.6% | 120 |

| Duration: 120 days, sterile | Temp: 20°C | Moisture: 40% maximum water holding capacity |
|---|---|---|

Half-life of boscalid: stable
% boscalid remaining, day 120 = 100.5%                                        % mineralization, day 120 = 0 %

| Metabolites | Max (% of dose) | Day |
|---|---|---|
| Others | 1.7% | 3 |
| Non-extractable $^{14}$C | 2.8% | 120 |

| Sandy loam<br>Duration: 120 days | pH: 7.0<br>Temp: 20°C | Organic carbon: 0.6%<br>Moisture: 40% maximum water holding capacity |
|---|---|---|

Half-life of boscalid: 376 days
% boscalid remaining, day 120 = 80.9 %                                        % mineralization, not measured

| Metabolites | Max (% of dose) | Day |
|---|---|---|
| Others | 0.5 % | 120 |
| Non-extractable $^{14}$C | 21.5 % | 120 |

| Loamy sand<br>Duration: 120 days | pH: 6.6<br>Temp: 20°C | Organic carbon: 1.0%<br>Moisture: 40% maximum water holding capacity |
|---|---|---|

Half-life of boscalid: 322 days
% boscalid remaining, day 120 = 77.7%                                         % mineralization, not measured

| Metabolites | Max (% of dose) | Day |
|---|---|---|
| Others | 0.3% | 91 |
| Non-extractable $^{14}$C | 16.6% | 120 |

Loam                              pH: 7.7                    Organic matter: 5.2%
Duration: 119 days                Temp: 20°C                 Moisture: 40% maximum water holding
                                                             capacity

Half-life of boscalid: 133 days
% boscalid remaining, day 119 = 53.6%                        % mineralization, not measured

| Metabolites | Max (% of dose) | Day |
|---|---|---|
| Others | 1.3% | 7 |
| Non-extractable $^{14}$C | 50.1% | 119 |

Two studies on the anaerobic soil metabolism of [$^{14}$C] boscalid in two soil types showed that boscalid degraded very slowly and identifiable metabolites were a minor part of the residue, except for M510F47 where a maximum of 6.7% formed in the study using pyridine-labelled boscalid (Staudenmaier and Schäfer, 2000, 2000/1014986; Staudenmaier, 2000, 2000/1014990). The non-extractable $^{14}$C in the soil treated with diphenyl labelled and pyridine labelled boscalid increased continuously during the whole period of the incubation.

*Anaerobic soil metabolism*                    Ref: Staudenmaier and Schäfer, 2000,
                                                    2000/1014986; Staudenmaier
                                                    2000/1014990

Test material: [$^{14}$C]boscalid, diphenyl-labelled          Dose rate: 1.0 mg/kg dry soil
Sandy loam                        pH: 7.2                    Organic carbon: 1.6%
Duration: 120 days                Temp: 20°C                 Moisture: 40% maximum water holding
                                                             capacity

Half-life of boscalid: 261 days
% boscalid remaining, day 120 = 73.6%                        % mineralization, day 120 = 0.1 %

| Metabolites | Max (% of dose) | Day |
|---|---|---|
| Others | 0.6% | 90 |
| Non-extractable $^{14}$C | 15.8% | 120 |

Test material: [$^{14}$C]boscalid, pyridine-labelled          Dose rate: 1.0 mg/kg dry soil
Sandy loam                        pH: 7.5                    Organic carbon: 1.7%
Duration: 120 days                Temp: 20°C                 Moisture: 40% maximum water holding
                                                             capacity

Half-life of boscalid: 345 days
% boscalid remaining, day 120 = 77.0%                        % mineralization, day 120 = 0.4 %

| Metabolites | Max (% of dose) | Day |
|---|---|---|
| M510F47 | 6.7% | 120 |
| Others | 0.8 % | 0 |
| $^{14}$CO$_2$ | 0.4% | 120 |
| Non-extractable $^{14}$C | 14.4% | 120 |

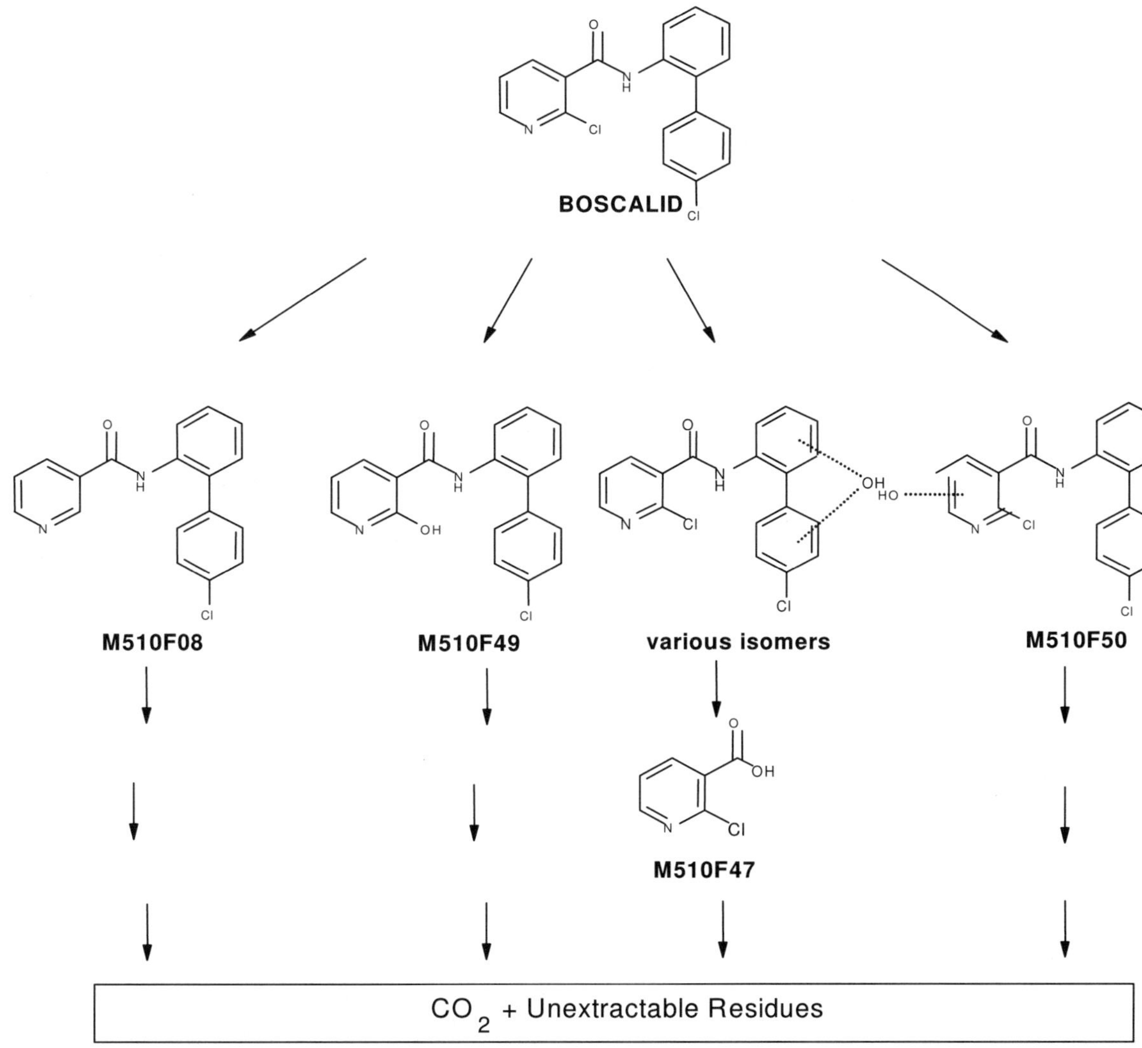

Figure 4. Proposed metabolic pathway for boscalid in soil.

The degradation behaviour (field dissipation) of boscalid in two different soils was investigated by Kellner and Keller (2000, 2000/1000123) under field conditions at two locations in Germany with three different application rates each on bare soil in 1997. Soil samples were taken at 9 sampling times up to 545 days and down to a maximum soil depth of 50 cm from the plots. The $DT_{50}$ values were shorter with higher application rates, and the $DT_{90}$ was not reached within one year, after application to bare soil. The highest amounts of boscalid were found in the top layer (0-10 cm) of soil (Table 8). Minor amounts were found in the 10 to 25 cm layer.

Table 8. Field dissipation of boscalid in 2 different soils in Germany in 1997 (Kellner and Keller, 2000, 2000/1000123).

| Trial | Applic rate, kg/ha | Initial conc, mg/kg | Boscalid as % of original concentration in 0-10 cm soil. | | | | | |
|---|---|---|---|---|---|---|---|---|
| | | | 30 days | 93 days | 176 days | 365 days | 449days | 544 days |
| Germany (Baden Württemberg) 1997 treat in April | Plot area 8.4 sq m. Silty loam: pH 7.5, 11.5% sand, 69.7% silt, 18.8% clay, 0.83% organic carbon. | | | | | | | |
| | 0.3 | 0.23 | 87% | 51% | 26% | 33% | 14% | 9.7% |
| | 0.6 | 0.63 | 63% | 21% | 15% | 21% | 12% | 8.9% |
| | 1.2 | 1.35 | 45% | 25% | 23% | 17% | 12% | 6.7% |

| Trial | Applic rate, kg/ha | Initial conc, mg/kg | Boscalid as % of original concentration in 0-10 cm soil. | | | | | |
|---|---|---|---|---|---|---|---|---|
| | | | 28 days | 97 days | 179 days | 367 days | 452 days | 545 days |
| Germany (Rheinland Pfalz) 1997 treat in April | Plot area 8.6 sq m. Silty sand: pH 5.4, 76.5% sand, 19.1% silt, 4.4% clay, 0.69% organic carbon. | | | | | | | |
| | 0.3 | 0.20 | 74% | 56% | 48% | 42% | 33% | 27% |
| | 0.6 | 0.41 | 78% | 62% | 43% | 53% | 35% | 26% |
| | 1.2 | 0.91 | 94% | 49% | 41% | 47% | 33% | 27% |

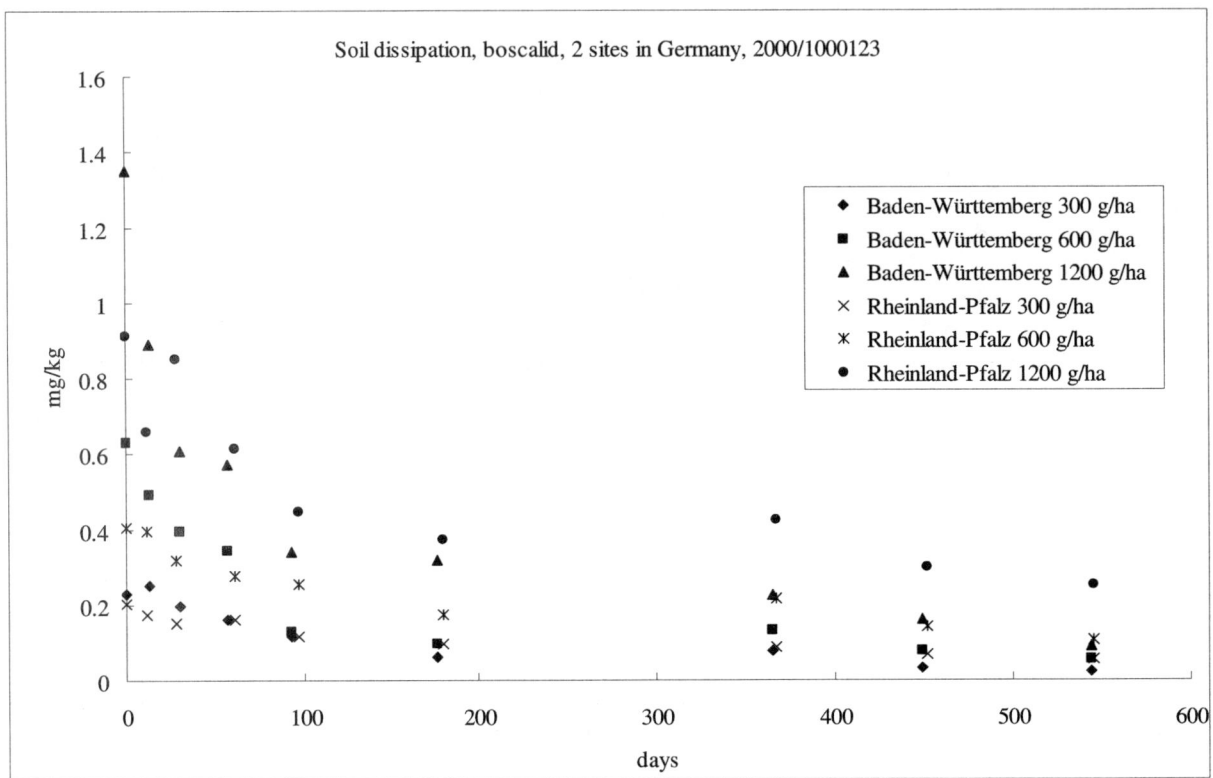

Figure 5. Disappearance of boscalid in soil at 2 sites in Germany at three different rates each on bare soil (Kellner and Keller, 2000, 2000/1000123).

The degradation behaviour of boscalid was also investigated, under field conditions on bare soil, at four locations in Europe in 1998 (Bayer and Grote 2001, 2000/1013295). Soil samples were taken on either 7 or 8 occasions, for up to 1 year and down to a maximum soil depth of 50 cm. The $DT_{50}$ values determined were shorter in Southern Europe than in Northern Europe, while the $DT_{90}$ was not reached one year after application to bare soil (see Table 9). The highest amounts of boscalid were found in the top layer (0-10cm) of the soils. Minor amounts were found in the 10 to 25 cm layer. No residues, above the limit of quantification, were found in the 25–50cm layer...

Table 9. Field dissipation of boscalid at four locations in Europe in 1998 (Bayer and Grote, 2001, 2000/1013295).

| Trial | Applic rate, kg/ha | Initial conc, mg/kg | Boscalid as % of original concentration in 0-10 cm soil. | | | | |
|---|---|---|---|---|---|---|---|
| | | | 30 days | 60 days | 98 days | 182 days | 349 days |
| Spain (Andalucid /Huelva) 1998 treat in May | Plot area 6.25 sq m. Sandy loam: pH 7.4, 58% sand, 23% silt, 19% clay, 0.6% organic carbon. | | | | | | |
| | 0.74 | 0.29 | 55% | 49% | 54% | 52% | 49% |

| Trial | Applic rate, kg/ha | Initial conc, mg/kg | Boscalid as % of original concentration in 0-10 cm soil. | | | | |
|---|---|---|---|---|---|---|---|
| | | | 30 days | 63 days | 99 days | 182 days | 356 days |
| Spain (Andalucid /Sevilla) 1998 | Plot area 6.25 sq m. Sandy loam: pH 7.7, 43% sand, 35% silt, 22% clay, 0.9% organic carbon. | | | | | | |
| treat in May | 0.76 | 0.30 | 97% | 102% | 56% | 39% | 55% |
| | | | | | | | |
| Sweden (Skane) | | | 31 days | 60 days | 101 days | 182 days | 352 days |
| 1998 | Plot area 9 sq m. Loamy sand: pH 5.9, 76% sand, 13% silt, 11% clay, 1.0% organic carbon. | | | | | | |
| treat in May | 0.80 | 0.32 | 107% | 138% | 178% | 114% | 115% |
| | | | | | | | |
| Germany | | | 30 days | 59 days | 97 days | 181 days | 357 days |
| (Schleswig-Holstein) 1998 | Plot area 18 sq m. Loamy sand: pH 8, 57% sand, 30% silt, 13% clay, 1.1% organic carbon. | | | | | | |
| treat in May | 0.78 | 0.52 | 71% | 70% | 59% | 31% | 42% |

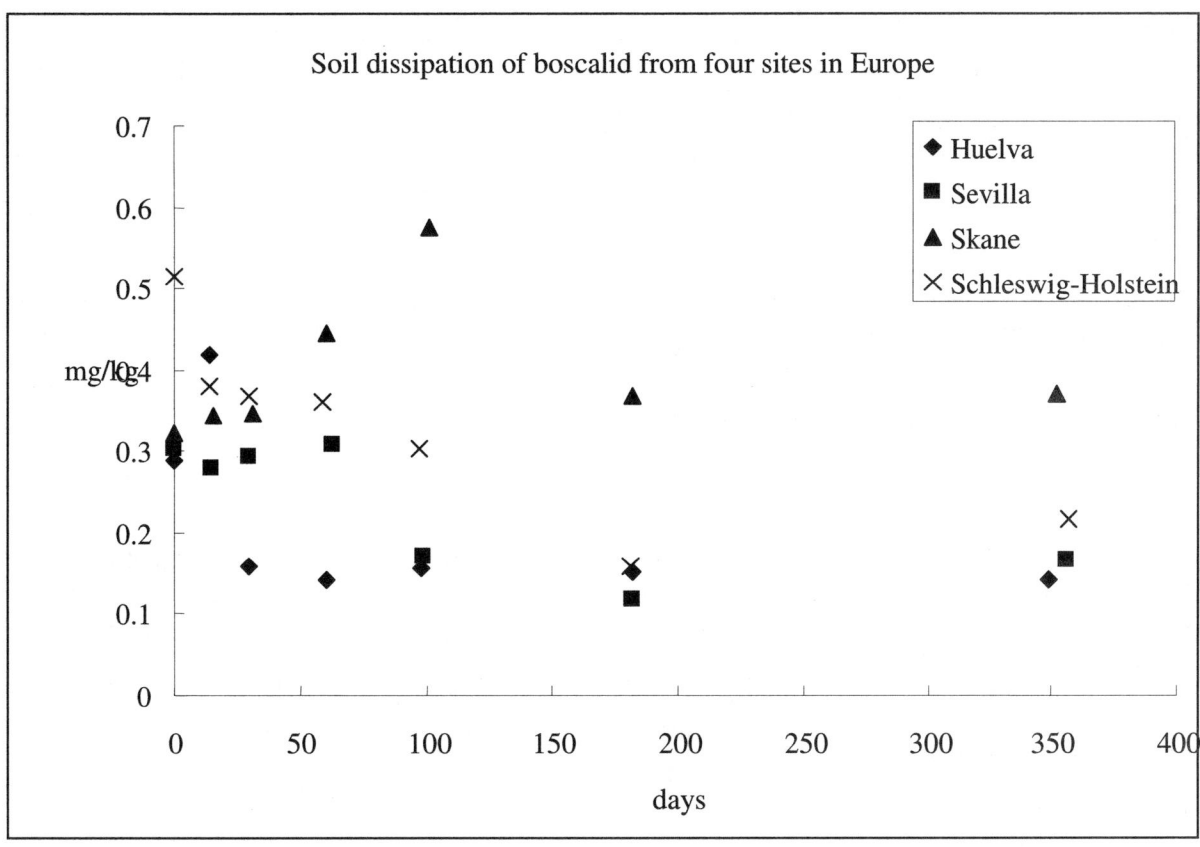

Figure 6. Disappearance of boscalid from soil at four sites in Europe after application to bare soil (Bayer and Grote, 2001, 2000/1013295).

### Rotational crop studies

Information on the fate of radiolabelled boscalid in a confined crop rotational study was made available to the Meeting.

A confined rotational crop study in Germany (Hamm and Veit, 2001, 2000/1014862) was conducted with [14]C-boscalid (diphenyl- and pyridine-label) with a single application of 2.1 kg ai/ha to bare soil (loamy sand, 0.9% organic matter, pH 6.7). Crops of lettuce, radish and wheat were sown into the treated soil at intervals of 30, 120, 270 and 365 days after the application. The crops were grown to maturity, harvested and analysed for [14]C content (Table 11). Samples were further examined by extraction and HPLC analysis. In all samples, unchanged $^{14}$C-boscalid was detected in various concentrations with the tendency for lower levels over the course of the study and with longer plant back intervals. One metabolite, M510F61, a glucoside of the hydroxylated parent compound, was found in various matrices, but not in lettuce, radish roots or wheat grains. The concentration was generally low, except in wheat straw.

Table 10. Total radioactive residues in soil samples after treatment with $^{14}$C-boscalid (pyridine and diphenyl label) (Hamm and Veit, 2001, 2000/1014862).

| Soil Samples | Boscalid Pyridine label TRR [mg/kg] | Boscalid Diphenyl label TRR [mg/kg] |
|---|---|---|
| After application | | |
| Plant back intervals (after soil aging and ploughing) | | |
| 30 DAT | 0.716 | 1.112 |
| 120 DAT | 0.648 | 0.813 |
| 270 DAT | 0.647 | n.d. |
| 365 DAT | 0.356 | 0.429 |
| After harvest of mature crops | | |
| Plant back interval: 30 DAT | | |
| Radish | n.d. | 0.731 |
| Lettuce | 0.545 | 0.747 |
| Wheat | 0.379 | 0.393 |
| Plant back interval: 120 DAT | | |
| Radish | 0.548 | 0.585 |
| Lettuce | 0.484 | 0.409 |
| Wheat | 0.386 | 0.506 |
| Plant back interval: 270 DAT | | |
| Radish | 0.377 | 0.521 |
| Lettuce | 0.321 | 0.436 |
| Wheat | 0.537 | 0.551 |
| Plant back interval: 365 DAT | | |
| Radish | n.d. | 0.460 |
| Lettuce | n.d. | 0.434 |
| Wheat | 0.125 | 0.343 |

n.d.: not determined.

Table 11. Investigation of the nature of the residues in rotational crops after treatment with $^{14}$C-boscalid (Hamm and Veit, 2001, 2000/1014862).

| Crop parts | TRR mg/kg | Extractable radioactive residue mg/kg (% TRR) | Unextractable radioactive residue mg/kg (% TRR) | Boscalid mg/kg (% TRR) | M510F61 mg/kg (% TRR) |
|---|---|---|---|---|---|
| Pyridine Label | | | | | |
| Plant back interval: 30 DAT | | | | | |
| lettuce leaf | 0.035 | 0.029 (83.1) | 0.007 (18.8) | 0.020 (58.5) | - |
| radish leaf | 0.343 | 0.317 (92.2) | 0.027 (7.8) | 0.301 (87.6) | 0.016 (4.6) |
| radish root | 0.048 | 0.040 (81.9) | 0.009 (19.3) | 0.030 (62.7) | - |
| wheat forage | 0.690 | 0.654 (94.7) | 0.047 (6.8) | 0.619 (89.8) | 0.024 (3.4) |
| wheat straw | 3.609 | 3.347 (92.8) | 0.351 (9.7) | 3.156 (87.5) | 0.102 (2.8) |

| Crop parts | TRR mg/kg | Extractable radioactive residue mg/kg (% TRR) | Unextractable radioactive residue mg/kg (% TRR) | Boscalid mg/kg (% TRR) | M510F61 mg/kg (% TRR) |
|---|---|---|---|---|---|
| wheat grain | 0.147 | 0.036 (24.7) | 0.130 (88.3) | 0.009 (6.1) | - |
| Plant back interval: 120 DAT | | | | | |
| lettuce leaf | 0.161 | 0.146 (90.8) | 0.015 (9.2) | 0.146 (90.8) | - |
| radish leaf | 0.211 | 0.187 (88.8) | 0.024 (11.2) | 0.172 (81.8) | 0.015 (7.0) |
| radish root | 0.038 | 0.031 (81.6) | 0.007 (18.4) | 0.023 (60.1) | 0.008 (21.5) |
| wheat forage | 0.433 | 0.384 (88.6) | 0.054 (12.5) | 0.379 (87.5) | - |
| wheat straw | 4.008 | 2.882 (71.9) | 1.293 (32.3) | 2.598 (64.8) | 0.117 (2.9) |
| wheat grain | 0.285 | 0.074 (26.2) | 0.260 (91.1) | 0.015 (5.3) | - |
| Plant back interval: 270 DAT | | | | | |
| lettuce leaf | 0.031 | 0.023 (74.5) | 0.008 (25.5) | 0.020 (65.1) | - |
| radish leaf | 0.125 | 0.108 (86.1) | 0.017 (13.9) | 0.104 (82.5) | 0.004 (3.6) |
| radish root | 0.017 | 0.013 (77.1) | 0.004 (22.9) | 0.009 (52.6) | - |
| wheat forage | 0.230 | 0.224 (97.3) | 0.006 (2.7) | 0.214 (92.8) | 0.005 (2.3) |
| wheat straw | 1.614 | 0.998 (61.8) | 0.703 (43.6) | 0.808 (50.0) | 0.071 (4.4) |
| wheat grain | 0.271 | 0.049 (17.9) | 0.260 (96.0) | 0.005 (1.9) | - |
| Plant back interval: 365 DAT | | | | | |
| lettuce leaf | 0.022 | 0.017 (76.1) | 0.005 (23.9) | 0.014 (61.6) | - |
| radish leaf | 0.113 | 0.103 (91.1) | 0.010 (8.9) | 0.088 (78.2) | 0.013 (11.2) |
| radish root | 0.066 | 0.060 (91.0) | 0.006 (9.0) | 0.060 (91.0) | - |
| wheat forage | 0.255 | 0.213 (83.5) | 0.042 (16.5) | 0.191 (74.7) | 0.008 (2.9) |
| wheat straw | 1.925 | 1.582 (82.1) | 0.437 (22.7) | 1.488 (77.3) | - |
| wheat grain | 0.148 | 0.029 (19.7) | 0.138 (93.2) | 0.006 (4.2) | - |
| Diphenyl Label | | | | | |
| Plant back interval: 30 DAT | | | | | |
| lettuce leaf | 0.050 | 0.047 (93.8) | 0.003 (6.2) | 0.047 (93.8) | - |
| radish leaf | 0.337 | 0.324 (96.1) | 0.013 (3.9) | 0.304 (90.2) | 0.020 (5.9) |
| radish root | 0.072 | 0.067 (93.1) | 0.005 (6.9) | 0.064 (89.6) | - |

| Crop parts | TRR mg/kg | Extractable radioactive residue mg/kg (% TRR) | Unextractable radioactive residue mg/kg (% TRR) | Boscalid mg/kg (% TRR) | M510F61 mg/kg (% TRR) |
|---|---|---|---|---|---|
| wheat forage | 1.575 | 1.531 (97.2) | 0.071 (4.5) | 1.472 (93.5) | 0.032 (2.0) |
| wheat straw | 9.826 | 9.214 (93.8) | 1.412 (14.4) | 7.991 (81.3) | 0.423 (4.3) |
| wheat grain | 0.166 | 0.059 (35.3) | 0.135 (81.6) | 0.028 (16.8) | - |
| Plant back interval: 120 DAT | | | | | |
| lettuce leaf | 0.084 | 0.075 (89.2) | 0.009 (10.8) | 0.072 (85.2) | - |
| radish leaf | 0.294 | 0.256 (87.2) | 0.046 (15.6) | 0.209 (71.2) | - |
| radish root | 0.052 | 0.043 (82.1) | 0.011 (21.3) | 0.035 (67.8) | 0.006 (10.9) |
| wheat forage | 0.980 | 0.909 (92.8) | 0.113 (11.5) | 0.846 (86.4) | 0.021 (2.1) |
| wheat straw | 3.912 | 3.704 (94.7) | 0.414 (10.6) | 3.311 (84.6) | 0.187 (4.8) |
| wheat grain | 0.243 | 0.062 (25.3) | 0.213 (87.8) | 0.023 (9.6) | - |
| Plant back interval: 270 DAT | | | | | |
| lettuce leaf | 0.067 | 0.063 (94.1) | 0.004 (5.9) | 0.063 (94.1) | - |
| radish leaf | 0.150 | 0.141 (94.3) | 0.009 (5.7) | 0.109 (73.1) | 0.032 (21.2) |
| radish root | 0.098 | 0.091 (92.8) | 0.007 (7.2) | 0.091 (92.8) | - |
| wheat forage | 0.562 | 0.512 (91.2) | 0.066 (11.7) | 0.352 (62.8) | 0.102 (18.1) |
| wheat straw | 3.226 | 2.865 (88.8) | 0.739 (22.9) | 2.283 (70.8) | 0.030 (0.9) |
| wheat grain | 0.023 | 0.013 (58.3) | 0.015 (64.6) | 0.008 (35.4) | - |
| Plant back interval: 365 DAT | | | | | |
| lettuce leaf | 0.028 | 0.022 (76.3) | 0.010 (37.2) | 0.016 (55.6) | - |
| radish leaf | 0.207 | 0.197 (95.2) | 0.018 (4.8) | 0.144 (69.4) | 0.032 (15.5) |
| radish root | 0.030 | 0.027 (89.9) | 0.003 (10.1) | 0.024 (78.4) | 0.001 (4.0) |
| wheat forage | 0.265 | 0.255 (96.1) | 0.018 (6.9) | 0.199 (75.0) | 0.026 (9.8) |
| wheat straw | 1.404 | 1.335 (95.1) | 0.151 (10.7) | 1.088 (77.6) | 0.025 (1.8) |
| wheat grain | 0.048 | 0.019 (40.3) | 0.036 (74.9) | 0.011 (23.6) | - |

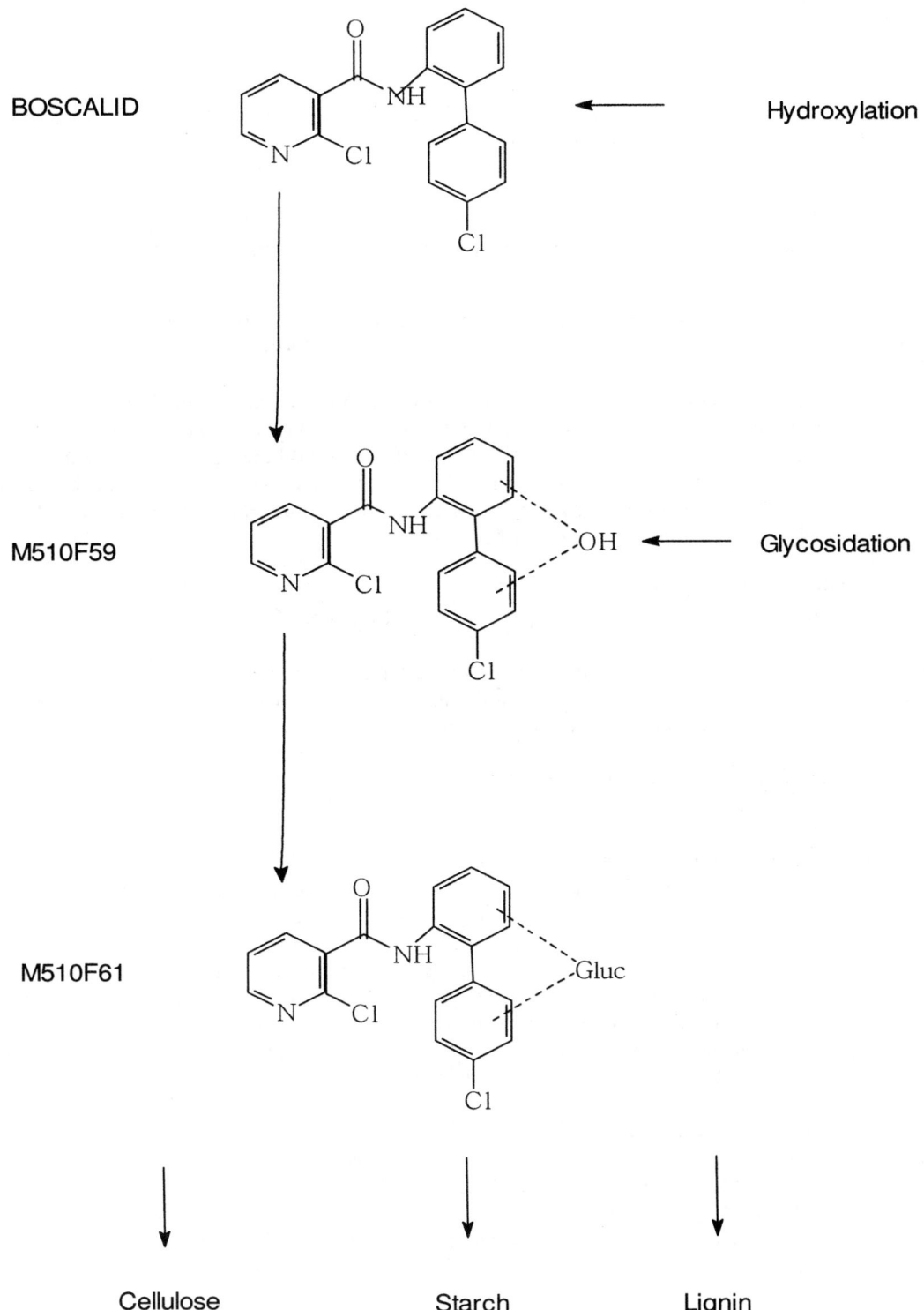

Figure 7. Proposed metabolic pathway for boscalid in succeeding crops (Hamm and Veit, 2001, 2000/1014862).

## RESIDUE ANALYSIS

### Analytical methods

The Meeting received descriptions and validation data for analytical methods for residues of boscalid in raw agricultural commodities, processed commodities, feed commodities, animal tissues, milk and eggs.

The methods rely on HPLC-UV, HPLC-MS/MS and GC-MSD for analysis of boscalid in the various matrices. Boscalid and its metabolites were determined and reported.

*Determination of boscalid in plant matrices* (Funk and Mackenroth, 2001, 2000/1012404)

| | | | |
|---|---|---|---|
| Analyte: | Boscalid | HPLC-MS/MS | Method 445/0 |
| LOQ: | 0.05 mg/kg in all matrices | | |
| Description | After extraction of the plant material with a methanol/water/hydrochloric acid mixture and subsequent centrifugation of an aliquot, transfer an aliquot of the supernatant into a culture tube containing water. For purification, perform liquid/liquid partitioning with cyclohexane. Evaporate the cyclohexane to dryness and dissolve the residue in methanol/water for HPLC-MS/MS quantification. For boscalid, the transition ions m/z = 343 -> 271 and m/z = 343 -> 307 can be used for quantification. | | |

*Cereal grains* (Funk, 2002, 2002/1004107)

| | | | |
|---|---|---|---|
| Analyte: | Boscalid | HPLC-MS/MS | Method 514/0 |
| LOQ: | 0.01 mg/kg | | |
| Description | Boscalid is extracted with a mixture of methanol, water and hydrochloric acid. An aliquot of the extract is centrifuged and partitioned against cyclohexane. The final determination of boscalid is performed by HPLC-MS/MS | | |

*Various plant materials* (Weeren and Pelz, 1999, 1999/11461)

| | | | |
|---|---|---|---|
| Analyte: | Boscalid | GC-MSD | DFG method S19 |
| LOQ: | Typically 0.01 mg/kg. Oilseed rape 0.02 mg/kg. | | |
| Description | The sample material is extracted with acetone/water 2+1 (v/v). In case of lemons, the pH value has to be adjusted to 7-8 by means of $NaHCO_3$. For liquid/liquid partition, ethyl acetate/cyclohexane (1+1) and sodium chloride is added and after repeated mixing, excess water is separated. The evaporated residue of an aliquot of the organic phase is cleaned up by gel permeation chromatography on Bio Beads S-X3 polystyrene gel using a mixture of ethyl acetate/cyclohexane (1+1) as eluant and an automated gel permeation chromatograph. The residue containing fraction is concentrated, followed by an additional clean-up by mini silica gel chromatography and analysed by gas chromatography using a fused silica capillary column (DB-5) and a mass selective detector (MSD). | | |

*White cabbage, rape (seed), hop and lettuce* (Reichert, 2001, 2000/1014886)

| | | | |
|---|---|---|---|
| Analyte: | Boscalid | GC-MSD | DFG method S19 |
| LOQ: | 0.01 mg/kg for white cabbage and lettuce, 0.02 mg/kg for rape seed, 0.05 mg/kg for hop. | | |
| Description | White cabbage, hop and lettuce are extracted with acetone, rape seed with acetone/acetonitrile. The extract is cleaned-up by means of gel permeation chromatography and additionally by means of a silica gel chromatography. The determination is performed by GC-MSD. | | |

*Plant matrices* (Abdel-Baky and Jones, 2001b, 2001/5001019)

| | | | |
|---|---|---|---|
| Analyte: | Boscalid | LC/MS/MS | Method 9908 |
| LOQ: | 0.05 mg/kg | | |
| Description | Boscalid are extracted with a 70:25:5 methanol:water:2N HCl mixture. An aliquot of the extract is removed and cleaned by either liquid/liquid partition or by Polar Plus C18 speedisk and/or silica speedisk column chromatography. If further cleaning is needed, the elute from the Polar Plus C18 speedisk column is then applied to a Silica gel speedisk column. The final chromatographic analysis of boscalid is determined by LC/MS/MS. The limit of quantitation of the method for boscalid is 0.05 mg/kg. | | |

*Plant and turf cloth matrices* (Abdel-Baky and Jones, 2001a, 2001/5000977)

| | | | |
|---|---|---|---|
| Analyte: | Boscalid | GC-MSD | Method D0008 |
| LOQ: | 0.05 mg/kg | | |
| Description | Boscalid was extracted from plant material (except oil) with a 70:25:5 methanol:water:HCl mixture. From oil, it was extracted by liquid/liquid partitioning with acetonitrile/hexane. An aliquot of the extract was removed and cleaned by liquid/liquid partitioning using iso-octane followed by further purification on a Silica gel speedisk column. The final chromatographic analysis of boscalid residues in plants was performed by gas chromatography with a mass selective detector (GC/MS). | | |

*Animal commodities* (Weeren and Pelz, 1999, 2000/1000221)

| | | | |
|---|---|---|---|
| Analyte: | Boscalid | GC-ECD | DFG method S19 |
| LOQ: | 0.01 mg/kg for milk, 0.02 mg/kg for meat, egg and fat. | | |
| Description | The sample material is extracted with acetone. Water is added beforehand in an amount that takes full account of the natural water content of the sample so that during extraction the acetone:water ratio remains constant at 2:1 (v:v). For liquid/liquid partition, ethyl acetate/cyclohexane (1+1) and sodium chloride is added and after repeated mixing, excess water is separated. The evaporated residue of an aliquot of the organic phase is cleaned up by gel permeation chromatography on Bio Beads S-X3 polystyrene gel using a mixture of ethyl acetate/cyclohexane (1+1) as eluant and an automated gel permeation chromatograph. The residue containing fraction is concentrated, additionally cleaned-up by mini silica gel chromatography and analysed by gas chromatography using a fused silica capillary column (DB-5) and an electron capture detector (ECD). | | |

*Animal commodities* (Grosshans, 2000a, 2000/1017223)

| | | | |
|---|---|---|---|
| Analyte: | Boscalid, M510F01 | HPLC-MS/MS | Method 471/0 |
| LOQ: | 0.01 mg/kg for milk, cream, egg, 0.0025 mg/kg for meat, fat, liver and kidney. | | |
| Description | A 25 g sample is extracted with methanol. An aliquot corresponding to a 5 g sample is taken for further work-up. The methanol extract is evaporated to dryness, redissolved in buffer solution and incubated with ß-glucuronidase / arylsulfatase to cleave the glucuronide M510F02 to M510F01. Then a liquid / liquid partition with ethyl acetate is carried out and the organic phase is purified on SPE C18 and if necessary on SPE silica gel columns. The final determination of the analytes boscalid and M510F01 is performed by HPLC/MS/MS. | | |

*Animal commodities* (Fabian,2000, 2000/1017224)

| | | | |
|---|---|---|---|
| Analyte: | Boscalid | GC-MSD | Method 476/0 |
| LOQ: | 0.01 mg/kg. | | |
| Description | The method 476 was developed to determine bound residues of boscalid in liver. It was also possible to apply the method on milk extracts to determine minor milk metabolites. Acetonitrile and concentrated acetic acid are added to the liver sample. In the case of milk, an acetonitrile extract is taken and concentrated acetic acid is added. The mixture is treated for 0.5 hours at 170°C in the microwave oven. The mixture is evaporated to a crude solution and a liquid / liquid partition against saturated sodium chloride solution is carried out. The pH is adjusted to about pH 12 with potassium hydroxide. A liquid / liquid partition with iso-octane is carried out. The organic phase is purified on SPE Silica (liver and milk) and C18 (liver), respectively. The final determination is performed by GC/MS. | | |

*Animal commodities* (Class, 2000, 2000/1017227)

| | | | |
|---|---|---|---|
| Analyte: | Boscalid, M510F01 | GC-ECD | DFG method S19 |
| LOQ: | 0.01 mg/kg for milk, 0.0025 mg/kg for meat, fat, liver kidney and egg. | | |
| Description | Residues of boscalid and its metabolite M510F01 (present also as a conjugate) are extracted from animal matrices by methanol. An aliquot of the filtered extract is concentrated and treated with enzymes (β-glucuronidase/arylsulfatase) for deconjugation of the metabolite. Water, acetone, sodium chloride and ethyl acetate/cyclohexane (1/1) are added to achieve a homogeneous partition of the analytes into the organic phase. An aliquot of the organic extract is cleaned-up by gel permeation chromatography followed by acetylation of the phenolic metabolite M510F01 with acetic anhydride and further fractionation on silica gel. The determination of boscalid and the acetylated metabolite M510F01 is achieved by gas chromatography with electron capture detection (GC/ECD). | | |

Recovery data of fortified samples in various matrices from the internal and independent laboratory validation testing are presented in Table 12.

Table 12. Validation data for analytical methods for the determination of boscalid residues in food (anonymous, 2006, 2006/1015800).

| Sample Matrix | Fortific. Level [mg/kg] | Average recovery [%] | RSD [%] | No. of analyses | Method | Ref |
|---|---|---|---|---|---|---|
| Apple, fruit | 0.05<br>0.5 | 94.9<br>88.1 | 1.8<br>13.3 | 5<br>5 | Method 445/0 | 2000/1012404 |
| Sour cherry, fruit | 0.05<br>0.5 | 91.3<br>86.1 | 4.9<br>3.5 | 5<br>5 | | |
| Grapes, fruit | 0.05<br>1.0 | 97.0<br>102.9 | 2.3<br>10.9 | 5<br>5 | | |
| Strawberry, fruit | 0.05<br>0.5 | 102.1<br>103.5 | 2.5<br>8.2 | 5<br>5 | | |
| Carrot, root | 0.05<br>0.5 | 87.3<br>85.3 | 2.9<br>4.6 | 5<br>5 | | |
| Onion, bulb | 0.05<br>0.5 | 101.0<br>98.4 | 5.7<br>5.9 | 5<br>5 | | |
| Tomato, fruit | 0.05<br>0.5 | 97.8<br>89.8 | 11.1<br>0.9 | 5<br>5 | | |
| Broccoli, plant w/o root | 0.05<br>0.5 | 108.8<br>94.3 | 18.5<br>2.2 | 5<br>5 | | |
| White cabbage, head | 0.05<br>0.5 | 89.8<br>92.8 | 9.3<br>3.8 | 5<br>5 | | |
| Leek, plant w/o root | 0.05<br>1.0 | 84.6<br>80.3 | 15.3<br>13.5 | 5<br>5 | | |
| Dwarf bean, pods w. seed | 0.05<br>0.5 | 95.9<br>92.7 | 6.5<br>3.8 | 5<br>5 | | |
| Oilseed rape, seed | 0.05<br>0.5 | 90.5<br>94.7 | 10.5<br>1.8 | 5<br>5 | | |
| Cereal, grain | 0.01<br>0.1<br>1.0 | 94.4<br>100.7<br>104.5 | 18.8<br>15.3<br>14.4 | 9<br>6<br>2 | Method 514/0 | 2002/1004107 |
| Tomato | 0.01<br>0.1 | 94<br>93 | 8.9<br>4.2 | 5<br>5 | DFG method S19 | 1999/1014886 |
| Lemon | 0.01<br>0.1 | 100 / 94 [1]<br>101 / 98 [1] | 8.9 / 8.2 [1]<br>6.2 / 1.4 [1] | 5 / 5 [1]<br>5 / 5 [1] | | |
| Wheat, grain | 0.01<br>0.1 | 93<br>83 | 5.8<br>4.5 | 5<br>5 | | |
| Oilseed rape, seed | 0.02<br>0.2 | 86<br>82 | 14<br>10 | 5<br>5 | | |
| White cabbage | 0.01<br>0.1 | 70<br>77 | 6<br>9 | 5<br>5 | DFG method S19 | 1999/1014886 |
| Oilseed rape, seed | 0.02<br>0.2 | 71<br>76 | 10<br>9 | 5<br>5 | | |
| Hop | 0.05<br>0.5 | 63<br>56 | 9<br>16 | 5<br>5 | | |
| Lettuce | 0.01<br>0.1 | 71<br>78 | 9<br>10 | 5<br>5 | | |
| Snap bean | 0.05<br>0.5 | 98<br>91 | 14<br>2 | 5<br>5 | Method 9908 | 2001/5001019 |
| Peanut, nutmeat | 0.05<br>0.5 | 85<br>92 | 8<br>6 | 5<br>5 | | |
| Canola, seed | 0.05<br>0.5 | 109<br>90 | 5<br>6 | 5<br>5 | | |

| Sample Matrix | Fortific. Level [mg/kg] | Average recovery [%] | RSD [%] | No. of analyses | Method | Ref |
|---|---|---|---|---|---|---|
| Canola, oil | 0.05<br>0.5 | 104<br>97 | 3<br>5 | 5<br>5 | | |
| Tomato | 0.05<br>0.5 | 113<br>90 | 3<br>3 | 5<br>5 | | |
| Lettuce | 0.05<br>0.5 | 112<br>89 | 8<br>5 | 5<br>5 | | |
| Canola, seed | 0.05<br>3.5 | 88<br>92 | 13<br>5 | 2<br>2 | Method D0008 | 2001/5000880 |
| Tomato, fruit | 0.05<br>1.0 | 94<br>83 | 4<br>6 | 2<br>2 | | |
| Milk | 0.01<br>0.1 | 78<br>91 | 8.6<br>3.3 | 5<br>5 | DFG method S19 | 2000/1000221 |
| Meat | 0.02<br>0.2 | 100<br>100 | 6.3<br>7.8 | 5<br>5 | | |
| Egg | 0.02<br>0.2 | 85<br>98 | 5.3<br>4.9 | 5<br>5 | | |
| Fat | 0.02<br>0.2 | 92<br>106 | 5.7<br>4.0 | 5<br>5 | | |
| Cow, milk | 0.01<br>0.1<br>0.01<br>0.1 | 86.0<br>88.7<br>88.4<br>84.9 | 3.6<br>7.7<br>5.8<br>8.6 | 5<br>5<br>5<br>5 | Method 471/0 | 2000/1017224 |
| Cow, cream | 0.01<br>0.1<br>0.01<br>0.1 | 72.2<br>89.9<br>89.5<br>94.2 | 1.5<br>4.7<br>1.7<br>2.3 | 5<br>5<br>5<br>5 | | |
| Cow, muscle | 0.025<br>0.25<br>0.025<br>0.25 | 86.4<br>94.5<br>89.3<br>86.3 | 4.0<br>1.5<br>2.1<br>1.4 | 5<br>5<br>5<br>5 | | |
| Cow, fat | 0.025<br>0.25<br>0.025<br>0.25 | 80.0<br>81.0<br>81.0<br>82.6 | 5.4<br>8.5<br>4.0<br>7.4 | 5<br>5<br>5<br>5 | | |
| Cow, kidney | 0.025<br>0.25<br>0.025<br>0.25 | 83.3<br>90.6<br>81.6<br>82.2 | 1.9<br>3.9<br>2.5<br>4.6 | 5<br>5<br>5<br>5 | | |
| Cow, liver | 0.025<br>0.25<br>0.025<br>0.25 | 86.7<br>96.0<br>90.9<br>91.5 | 6.3<br>8.7<br>10.3<br>6.2 | 5<br>5<br>5<br>5 | | |
| Hen, egg | 0.01<br>0.1<br>0.01<br>0.1 | 82.5<br>93.1<br>82.7<br>89.1 | 3.8<br>3.1<br>6.1<br>8.2 | 5<br>5<br>5<br>5 | | |
| Cow, milk | 0.01<br>0.1 | 95.0<br>100.0 | 4.5<br>4.2 | 5<br>5 | Method 476/0 | 2000/1017224 |
| Cow, liver | 0.05<br>0.5 | 91.0<br>97.9 | 2.7<br>3.3 | 5<br>5 | | |

| Sample Matrix | Fortific. Level [mg/kg] | Average recovery [%] | RSD [%] | No. of analyses | Method | Ref |
|---|---|---|---|---|---|---|
| Cow, milk | 0.01 | 82 | 12 | 5 | DFG method S19 | 2000/1017227 |
|  | 0.1 | 88 | 6 | 5 |  |  |
|  | 0.01 | 93 | 16 | 5 |  |  |
|  | 0.1 | 101 | 8 | 5 |  |  |
| Cow, muscle | 0.025 | 95 | 6 | 5 |  |  |
|  | 0.25 | 84 | 6 | 5 |  |  |
|  | 0.025 | 93 | 14 | 5 |  |  |
|  | 0.25 | 92 | 15 | 5 |  |  |
| Cow, fat | 0.025 | 105 | 15 | 5 |  |  |
|  | 0.25 | 91 | 10 | 5 |  |  |
|  | 0.025 | 85 | 9 | 5 |  |  |
|  | 0.25 | 86 | 17 | 5 |  |  |
| Cow, kidney | 0.025 | 93 | 13 | 5 |  |  |
|  | 0.25 | 89 | 9 | 5 |  |  |
|  | 0.025 | 87 | 9 | 5 |  |  |
|  | 0.25 | 99 | 4 | 5 |  |  |
| Cow, liver | 0.025 | 91 | 15 | 5 |  |  |
|  | 0.25 | 83 | 7 | 5 |  |  |
|  | 0.025 | 89 | 15 | 5 |  |  |
|  | 0.25 | 86 | 19 | 5 |  |  |
| Hen, egg | 0.025 | 97 | 17 | 5 |  |  |
|  | 0.25 | 89 | 4 | 5 |  |  |
|  | 0.025 | 80 | 14 | 5 |  |  |
|  | 0.25 | 78 | 12 | 5 |  |  |
| Cow, milk | 0.01 | 83.8 | 4 | 5 | DFG method S19 | 2000/1017226 |
|  | 0.1 | 95.4 | 4 | 5 |  |  |
|  | 0.01 | 97.3 | 11 | 5 |  |  |
|  | 0.1 | 107.3 | 5 | 5 |  |  |
| Cow, liver | 0.025 | 74.2 | 3 | 5 |  |  |
|  | 0.25 | 74.2 | 1 | 5 |  |  |
|  | 0.025 | 91.9 | 3 | 5 |  |  |
|  | 0.25 | 92.7 | 5 | 5 |  |  |
| Cow, milk | 0.01 | 79 | 1 | 2 | Method 471/0 | 2002/5002983 |
|  | 0.1 | 80 | 5 | 2 |  |  |
|  | 0.01 | 96 | 7 | 2 |  |  |
|  | 0.1 | 83 | 8 | 2 |  |  |
| Hen, egg | 0.01 | 91 | 7 | 2 |  |  |
|  | 0.1 | 89 | 3 | 2 |  |  |
|  | 0.01 | 93 | 4 | 2 |  |  |
|  | 0.1 | 89 | 1 | 2 |  |  |
| Cow, liver | 0.025 | 74 | 4 | 2 |  |  |
|  | 0.25 | 82 | 12 | 2 |  |  |
|  | 0.025 | 86 | 4 | 2 |  |  |
|  | 0.25 | 88 | 6 | 2 |  |  |

*Extraction efficiency of analytical methods*

Method No. 445/0 (Funk and Mackenroth, 2001, 2000/1012404) was used as the data generation method for fixing the maximum residue levels whereas the multi residue method 19 (Weeren and Pelz,

1999, 1999/11461) was proposed for monitoring purposes. As the extraction procedure used in method 445/0 and the multi residue method S19 slightly deviates from those used in the metabolism studies (Veit, 2001, 2000/1014861; Rabe and Schlüter, 2001, 2000/1014860; Hamm, 1999, 1999/11240; Hamm and Veit, 2001, 2000/1014862), [14]C-boscalid treated plant material was extracted according to these methods, the results of which are compared and summarized in Table 13 (Bross, 2001, 2001/1001739).

The results show that comparable or slightly higher amounts of radioactivity were released by extraction with methanol/water/hydrochloric acid (70/25/5) and acetone/water (70/30). The extractability with methanol/water/HCl ranged from 62.5% TRR (wheat straw) to 99.0% TRR (green bean). For acetone/water, the extractability ranged from 60.9% TRR (wheat grain) to 98.8% TRR (green bean). The HPLC metabolite profiles were comparable with the profiles obtained in the course of the metabolism studies.

Table 13. Comparison of extractability of boscalid obtained with different extraction solvents (Bross, 2001, 2001/1001739).

| Plant material | | Extraction results | | | | | |
|---|---|---|---|---|---|---|---|
| | | Metabolism study | | Method 445/0 | | Multi method S 19 | |
| | | mg/kg | % TRR | mg/kg | % TRR | mg/kg | % TRR |
| Green beans | TRR | 1.027 | 100 | 0.901 | 100 | 0.934 | 100 |
| | ERR | 1.010 | 98.3 | 0.892 | 99.0 | 0.923 | 98.8 |
| | RRR | 0.017 | 1.7 | 0.009 | 1.0 | 0.011 | 1.2 |
| Dry beans | TRR | 0.205 | 100 | 0.151 | 100 | 0.162 | 100 |
| | ERR | 0.165 | 80.5 | 0.129 | 85.4 | 0.141 | 87.0 |
| | RRR | 0.040 | 19.5 | 0.022 | 14.6 | 0.021 | 13.0 |
| Grapes | TRR | 1.181 | 100 | 1.291 | 100 | 1.185 | 100 |
| | ERR | 1.100 | 93.2 | 1.126 | 87.2 | 1.115 | 94.1 |
| | RRR | 0.081 | 6.8 | 0.165 | 12.8 | 0.070 | 5.9 |
| Lettuce | TRR | 0.067 | 100 | 0.069 | 100 | 0.071 | 100 |
| | ERR | 0.063 | 94.1 | 0.065 | 94.2 | 0.067 | 94.4 |
| | RRR | 0.004 | 5.9 | 0.004 | 5.8 | 0.004 | 5.6 |
| Wheat grain | TRR | 0.023 | 100 | 0.024 | 100 | 0.023 | 100 |
| | ERR | 0.008 | 35.4 | 0.015 | 62.5 | 0.014 | 60.9 |
| | RRR | 0.015 | 64.6 | 0.009 | 37.5 | 0.009 | 39.1 |
| Wheat straw | TRR | 3.226 | 100 | 3.177 | 100 | 3.176 | 100 |
| | ERR | 2.487 | 77.1 | 2.630 | 82.8 | 2.659 | 83.8 |
| | RRR | 0.739 | 22.9 | 0.546 | 17.2 | 0.516 | 16.2 |
| Radish roots | TRR | 0.098 | 100 | 0.095 | 100 | 0.101 | 100 |
| | ERR | 0.091 | 92.8 | 0.088 | 92.6 | 0.092 | 91.1 |
| | RRR | 0.007 | 7.2 | 0.007 | 7.4 | 0.008 | 7.9 |

### Stability of residues in stored analytical samples

The Meeting received information on the stability of residues of boscalid in wheat (green plant without roots, grain and straw), oil seed rape, sugar beet (roots), white cabbage (head), peach (fruit), peas, tomato paste, liver, milk and muscle.

The deep freeze stability of boscalid in different plant matrices such as wheat (green plant without roots, grain and straw), oil seed rape, sugar beet (roots), white cabbage (head), peach (fruit) and peas was investigated over a period of two years (Funk and Mackenroth, 2001, 2001/1015028). Untreated samples were fortified with 0.5 mg/kg boscalid. The samples were stored under the usual storage conditions for field samples (polyethylene bottle, -20°C). The samples were analysed with method No. 445/0 (Table 14).

Table 14. Storage stability of boscalid fortified at 0.5 mg/kg in various plant matrices (Funk and Mackenroth, 2001, 2001/1015028).

| Day | Boscalid found [1] [2] (mg/kg) | | | | | | | | | | | | | | | |
|-----|------|------|------|------|------|------|------|------|------|------|------|------|------|------|------|------|
| | Wheat plant | | Wheat grain | | Wheat straw | | Oil seed rape | | Sugar beet | | White cabbage | | Peach | | Pea | |
| 0 | 0.56 | 0.48 | 0.47 | 0.49 | 0.49 | 0.50 | 0.39 | 0.41 | 0.49 | 0.51 | 0.52 | 0.49 | 0.54 | 0.52 | 0.49 | 0.49 |
| 33 | 0.45 | 0.45 | 0.40 | 0.41 | 0.52 | 0.53 | 0.40 | 0.41 | 0.47 | 0.49 | 0.47 | 0.46 | 0.43 | 0.49 | 0.49 | 0.49 |
| 96 | 0.48 | 0.50 | 0.45 | - | 0.48 | 0.52 | 0.44 | 0.46 | 0.46 | 0.49 | 0.52 | 0.52 | 0.51 | 0.50 | 0.54 | 0.56 |
| 182 | 0.48 | 0.45 | 0.45 | 0.46 | 0.46 | 0.46 | 0.45 | 0.44 | 0.56 | 0.59 | 0.52 | 0.53 | 0.50 | - | 0.49 | 0.51 |
| 356 | 0.53 | 0.54 | 0.48 | 0.46 | 0.49 | 0.50 | 0.46 | 0.46 | 0.52 | 0.52 | 0.50 | 0.49 | 0.54 | 0.49 | 0.50 | 0.51 |
| 566 | 0.43 | 0.43 | 0.47 | 0.48 | 0.46 | 0.48 | 0.47 | 0.46 | 0.45 | 0.46 | 0.46 | 0.49 | 0.45 | 0.47 | 0.42 | 0.45 |
| 720 | 0.42 | 0.46 | 0.54 | 0.52 | 0.40 | 0.46 | 0.38 | 0.38 | 0.57 | 0.51 | 0.48 | 0.50 | 0.53 | 0.48 | 0.51 | 0.52 |
| Degradation after 24 months (%) | | | | | | | | | | | | | | | | |
| | 7.9 | | stable | | 10.9 | | 2.2 | | stable | | 2.7 | | 2.1 | | 3.7 | |

1) Corrected for individual procedural recovery.
2) Mean of two replicates.

A representative sample of tomato paste from a process fraction study was placed in a freezer (< -20°C) for 38.6 months. After storage the sample was analysed by method No. D9908. The average residue of boscalid found in the tomato paste sample from the analysis was 1.59 mg/kg. Re-analysis, after storage under deep freeze conditions, showed a mean residue of 1.30 mg/kg.

The freezer stability of boscalid and the metabolite M510F01 in animal matrices was also investigated over a period of 170 days (Grosshans, 2001, 2000/1017229). Untreated samples of muscle, liver and milk from cows were fortified with 0.5 mg/kg with boscalid and M510F01. These matrices were representative for the samples stored during the residue transfer study in cows. All samples were stored under the usual storage conditions for samples (polyethylene bottles, < -18°C). After 0 and approx. 60, 100 and 170 days, samples were analysed using method no. 471. Results are summarized in Table 15.

Table 15. Degradation of boscalid and M510F01 after 166/167 Days (Grosshans, 2001, 2000/1017229).

| Tissue | Boscalid [%] | M510F01 [%] |
|--------|--------------|-------------|
| Milk | 7.7 | -2.9 |
| Muscle | 1.8 | 8.5 |
| Liver | 5.2 | 4.2 |

## USE PATTERN

Boscalid is an anilide fungicide that inhibits mitochondrial respiration, thereby inhibiting spore germination, germ tube elongation, mycelial growth, and sporulation of pathogenic fungi on the leaf surface and is registered for use against a large number of fungi on a wide range of crops in many countries. Labels and English translations were available for all uses. Information on registered uses included in this monograph is generally limited to countries where supervised trials had been conducted, and is summarized in Table 16.

Table 16. Registered uses of boscalid on crops.

| Crop | Country | End-use product | F/G/P (a) | Method | No. per crop season min. max. | kg ai/hL (b) max. | Water L/ha per appl. Min. max. | kg ai/ha per applic. (b) min. max. | PHI [days] |
|---|---|---|---|---|---|---|---|---|---|
| Apple, pear | Belgium | 25.2% WG[1)] | F | Spraying | 3 | 0.045 | 300 | 0.134 | 7 |
| Apple, pear | United Kingdom | 25.2% WG | | Spraying | 4 | 0.013 – 0.067 | 300 – 1500 | 0.202 | 7 |
| Stone fruit | USA | 25.2% WG | F | Spray | 5 | NC | NS | 0.185 – 0.256 | 0 |
| Berry group | USA | 25.2% WG | | Spray | 4 | NC | NS | 0.326 – 0.406 | 0 |
| Grape (except Concord, Worden, Fredonia, Niagara and related varieties) | USA | 25.2% WG | | Spray | 6 | NC | NS | 0.14 – 0.185 | 14 |
| Grape (except Concord, Worden, Fredonia, Niagara and related varieties) | USA | 25.2% WG | | Spray | 5 | NC | NS | 0.14 – 0.22 | 14 |
| Grapes | USA | 70% WG | | Spray | 5 | NC | NS | 0.22 | 14 |
| Grapes | USA | 70% WG | | Spray | 3 | NC | NS | 0.392 | 14 |
| Grapes | Japan | WG 500g/kg | F | Spraying | 3 | 0.033-0.050 | 2000-7000 | 0.667-3.5 | 7 |
| Strawberry | Belgium | 26.7% WG[3)] | F/G | Spraying | 2 | 0.048 | 1000 | 0.481 | 3 |
| Strawberry (outdoor) | United Kingdom | 26.7% WG | F | Spraying | 2 | 0.024 – 0.048 | 1000 – 2000 | 0.481 | 3 |
| Strawberry (protected) | United Kingdom | 26.7% WG | G | Spraying | 2 | 0.024 – 0.107 | 450 – 2000 | 0.481 | 3 |
| Strawberry | Japan | WG 500g/kg | F/G | Spraying | 3 | 0.033-0.050 | 1000-3000 | 0.333-1.5 | 1 |
| Banana | USA | 50% WG | F | Spray | 4 | 0.500 – 0.833 (in the oil/water mixture) | 18 – 30 L oil/water mixture, where oil is added as a constant with always (5) 7 – 9 L | 0.150 | 0 |
| Bulb vegetables | USA | 25.2% WG | | Spray | 6 | NC | NS | 0.256 – 0.326 | 7 |
| Bulb vegetables | USA | 25.2% WG | | Spray | 6 | NC | NS | 0.326 | 7 |
| Onion | Japan | WG 500g/kg | F | Spraying | 3 | 0.033-0.050 | 1000-3000 | 0.333-1.5 | 1 |
| Leek | Belgium | 26.7% WG | F | Spraying | 2 | 0.100 | 400 | 0.400 | 14 |
| Leek | Netherlands | 26.7% WG | F | Spraying | 2 – 3 | 0.100 – 0.160 | 205 – 400 | 0.401 | 14 |
| Broccoli | USA | 70% WG | | Spray | 2 | NC | NS | 0.294 – 0.441 | 0 |
| Chinese broccoli (Gai lon) | USA | 70% WG | | Spray | 2 | NC | NS | 0.294 – 0.441 | 0 |
| Brussels sprouts | Belgium | 26.7% WG | | Spraying | 3 | 0.067 | 400 | 0.267 | 14 |
| Brussels sprouts | United Kingdom | 26.7% WG | F | Spraying | 3 | 0.027 – 0.133 | 200 – 1000 | 0.267 | 14 |
| Cauliflower | Belgium | 26.7% WG | | Spraying | 3 | 0.067 | 400 | 0.267 | 14 |

| Crop | Country | End-use product | F/G/P (a) | Application | | | | | | PHI [days] |
|------|---------|-----------------|-----------|-------------|--|--|--|--|--|------------|
| | | | | Method | No. per crop season min. max. | kg ai/hL (b) max. | Water L/ha per appl. Min. max. | kg ai/ha per applic. (b) min. max. | | |
| Cauliflower | United Kingdom | 26.7% WG | F | Spraying | 3 | 0.027 – 0.133 | 200 – 1000 | 0.267 | | 14 |
| Head cabbage | USA | 70% WG | | Spray | 2 | NC | NS | 0.294 – 0.441 | | 0 |
| Cucurbit vegetables | USA | 25.2% WG | | Spray | 4 | NC | NS | 0.22 – 0.326 | | 0 |
| Cucurbit vegetables | USA | 25.2% WG | | Spray | 4 | NC | NS | 0.326 | | 0 |
| Cucurbit vegetables | USA | 70% WG | | Spray | 4 | NC | NS | 0.319 | | 0 |
| Cucumber | Japan | WG 500g/kg | F/G | Spraying | 3 | 0.033-0.050 | 1000-3000 | 0.333-1.5 | | 1 |
| Melon | Germany | SC[4] 200g/L | F | Spraying | 1 – 3 | 0.008 – 0.017 | 600 – 1200 | 0.100 | | 3 |
| Melon | Japan | WG 500g/kg | F/G | Spraying | 3 | 0.033-0.050 | 1000-3000 | 0.333-1.5 | | 1 |
| Watermelon | Japan | WG 500g/kg | F/G | Spraying | 3 | 0.033-0.050 | 1000-3000 | 0.333-1.5 | | 1 |
| Pepper | USA | 70% WG | | Spray | 6 | NC | NS | 0.122 – 0.172 | | 0 |
| Eggplant | Japan | 500kg WG | F/G | Spraying | 3 | 0.033-0.050 | 1000-3000 | 0.333-1.5 | | 1 |
| Tomato | USA | 70% WG | | Spray | 6 | NC | NS | 0.122 – 0.172 | | 0 |
| Tomato | USA | 70% WG | | Spray | 2 | NC | NS | 0.44 – 0.613 | | 0 |
| Tomato | Japan | WG 500g/kg | F/G | Spraying | 3 | 0.033-0.050 | 1000-3000 | 0.333-1.5 | | 1 |
| Cabbage | Japan | 500g/kg WG | F | Spraying | 2 | 0.033 | 1000-3000 | 0.333-1.0 | | 7 |
| Brassica (outdoor crops of kale, collards (including spring greens), cabbage, leafy brassica crops grown for baby leaf production (i.e. crops harvested up to 8 true leaf stage), pak choi, choi sum and komatsuna | United Kingdom | 26.7% WG | F | Spraying | 3 | max. 0.133 | min. 200 | 0.267 | | 14 |
| Leaf herbs (outdoor and protected), leafy brassica crops (protected) grown for baby leaf production (i.e. Harvested up to 8 true leaf stage) | United Kingdom | 26.7% WG | F / G (protected) | Spraying | 2 | max. 0.133 | min. 200 | 0.401 | | 14 |
| Collards (kale; mizuna; mustard greens; mustard spinach; rape greens) | USA | 70% WG | | Spray | 2 | NC | NS | 0.294 – 0.441 | | 14 |

| Crop | Country | End-use product | F/G/P (a) | Application | | | | | PHI [days] |
|------|---------|-----------------|-----------|-------------|--|--|--|--|-----|
| | | | | Method | No. per crop season min. max. | kg ai/hL (b) max. | Water L/ha per appl. Min. max. | kg ai/ha per applic. (b) min. max. | |
| Chinese cabbage (bok choy) – leafy brassica greens | USA | 70% WG | | Spray | 2 | NC | NS | 0.294 – 0.441 | 14 |
| Lettuce (outdoor and protected) | Belgium | 26.7% WG | F / G | Spraying | 2 | 0.080 | 500 | 0.400 | 14 |
| Lettuce | Japan | WG 500g/kg | F/G | Spraying | 1 | 0.033-0.050 | 1000-3000 | 0.333-1.5 | 14 |
| Lettuce (outdoor and protected) | United Kingdom | 26.7% WG | F / G (protected) | Spraying | 2 | 0.045 – 0.200 | 200 – 900 | 0.400 | 14 |
| Lettuce | USA | 70% WG | F | Spray | 2 | NC | NS | 0.392 – 0.539 | 14 |
| Beans with pod | France | 50% WG | F | Spraying | 2 | 0.167 | 300 | 0.500 | 7 |
| Adzuki bean, dry | Japan | WG 500g/kg | F | Spraying | 3 | 0.033-0.050 | 1000-3000 | 0.333-1.5 | 7 |
| Beans with pod | Germany | 50% WG | F | Spraying | 2 | 0.167 | 300 – 600 | 0.500 | 7 |
| Beans with pod | Germany | 50% WG | G | Spraying | 2 | 0.167 | 300 – 600 | 0.500 | 7 |
| Common bean, dry | Japan | WG 500g/kg | F/G | Spraying | 2 | 0.033-0.050 | 1000-3000 | 0.333-1.5 | 21 |
| Succulent beans (Phaseolus spp., Vigna spp.) | USA | 70% WG | F | Spray | 2 | NC | NS | 0.392 – 0.539 | 7 |
| Succulent beans (Phaseolus spp., Vigna spp.) | USA | 70% WG | F | Spray | 2 | NC | NS | 0.294 | 7 |
| Succulent Peas (edible podded and succulent shelled) | USA | 70% WG | F | Spray | 2 | NC | NS | 0.392 – 0.539 | 7 |
| Soybean, immature seed (= edible podded) | USA | 70% WG | F | Spray | 2 | NC | NS | 0.392 – 0.539 | 7 |
| Beans, dried (Phaseolus spp., Vigna spp.) | USA | 70% WG | F | Spray | 2 | NC | NS | 0.392 – 0.539 | 21 |
| Beans, dried (Phaseolus spp., Vigna spp.) | USA | 70% WG | F | Spray | 2 | NC | NS | 0.294 | 21 |
| Dry beans (except soybeans) | USA | 25.2% WG | F | Spray | 2 | NC | NS | 0.176 – 0.264 | 21 |
| Dry beans (except soybeans) | USA | 25.2% WG | F | Spray | 2 | NC | NS | 0.264 – 0.441 | 21 |
| Peas – dry | USA | 70% WG | F | Spray | 2 | NC | NS | 0.392 – 0.539 | 21 |
| Peas – dry | USA | 70% WG | F | Spray | 2 | NC | NS | 0.294 | 21 |
| Soybean, dry | USA | 25.2% WG | | Spray | 2 | NC | NS | 0.141 – 0.282 | 21 |
| Soybean, dry | USA | 25.2% WG | | Spray | 2 | NC | NS | 0.221 – 0.282 | 21 |
| Soybean, dry | USA | 25.2% WG | | Spray | 2 | NC | NS | 0.282 | 21 |
| Soybean, dry | USA | 70% WG | | Spray | 2 | NC | NS | 0.172 – 0.270 | 21 |
| Soybean, dry | USA | 70% WG | | Spray | 2 | NC | NS | 0.270 | 21 |

| Crop | Country | End-use product | F/G/P (a) | Application | | | | | | PHI [days] |
|------|---------|-----------------|-----------|--------|------------------------------------|-------------------|------------------------------------|------------------------------------|---|
| | | | | Method | No. per crop season min. max. | kg ai/hL (b) max. | Water L/ha per appl. Min. max. | kg ai/ha per applic. (b) min. max. | |
| Soybean, dry | USA | 70% WG | | Spray | 2 | NC | NS | 0.270 – 0.540 | 21 |
| Carrot | USA | 25.2% WG | | Spray | 6 | NC | NS | 0.14 – 0.185 | 0 |
| Carrot | USA | 70% WG | | Spray | 5 | NC | NS | 0.22 | 0 |
| Potato | USA | 70% WG | | Spray | 4 | NC | NS | 0.122 – 0.22 | 30 |
| Potato | USA | 70% WG | | Spray | 2 | NC | NS | 0.27 – 0.49 | 30 |
| Potato | USA | 70% WG | | Spray | 4 | NC | NS | 0.122 – 0.22 | 30 |
| Potato | USA | 70% WG | | Spray | 2 | NC | NS | 0.27 – 0.49 | 30 |
| Cereals (wheat, barley, rye) | Germany | 233 g/L SC[2] | F | Spraying | 1 | 0.086 – 0.175 | 200 – 400 | 0.350 | Not given |
| Cereals (winter and spring wheat, winter and spring barley, oats) | United Kingdom | 233 g/L SC | F | Spraying | 2 | 0.175 | min. 200 L | 0.350 | Not given |
| Cereals (barley) | Germany | 233 g/L SC | F | Spraying | 2 | 0.086 – 0.175 | 200 – 400 | 0.350 | Not given |
| Cereals (rye) | Germany | 233 g/L SC | F | Spraying | 2 | 0.086 – 0.175 | 200 – 400 | 0.350 | Not given |
| Cereals (wheat) | Germany | 233 g/L SC | F | Spraying | 2 | 0.086 – 0.175 | 200 – 400 | 0.350 | Not given |
| Tree nuts (except almond, filbert, pecan) | USA | 25.2% WG | | Spray | 4 | NC | NS | 0.185 – 0.256 | 14 |
| Almond | USA | 25.2% WG | | Spray | 4 | NC | NS | 0.185 – 0.256 | 25 |
| Pecan | USA | 25.2% WG | | Spray | 4 | NC | NS | 0.185 – 0.256 | 14 |
| Pistachio | USA | 25.2% WG | | Spray | 4 | NC | NS | 0.185 – 0.256 | 14 |
| Filbert | USA | 25.2% WG | | Spray | 4 | NC | NS | 0.185 – 0.256 | 14 |
| Canola | USA | 70% WG | | Spray | 2 | NC | NS | 0.245 – 0.294 | 21 |
| Peanut | USA | 70% WG | | Spray | 3 | NC | NS | 0.319 – 0.49 | 14 |
| Peanut | USA | 70% WG | | Spray | 3 | NC | NS | 0.392 – 0.49 | 14 |
| Sunflower | USA | 70% WG | | Spray | 2 | NC | NS | 0.220 – 0.44 | 21 |
| Coffee | Brazil | 50% WG | | Spraying | 1 | 0.015 | 500 – 1000 | 0.075 | 45 |
| Hops (pending) | Germany | 25.2% WG | F | Spray | 3 | 0.019 | 600 – 2700 1.appl: 600 – 1200 2.appl: 1200 – 2300 3.appl: 2300 – 2700 | 0.114 – 0.504 | 21 |
| Hops (pending) | France | 25.2% WG | F | Spray | 3 | 0.019 | 600 – 2700 | 0.114 – 0.504 | 21 |

1)   contain boscalid (25.2%) and pyraclostrobin (12.8%) as coactive ingredients.
2)   contain boscalid (233 g/L) and epoxiconazole (67 g/L) as coactive ingredients.
3)   contain boscalid (26.7%) and pyraclostrobin (6.7%) as coactive ingredients.
4)   contain boscalid (200 g/L) and kresoxim-methyl (100 g/L) as coactive ingredients.
(a)   F = outdoor or field use, G = glasshouse, P = protected
(b)   Information given on active substance (as) refers to boscalid only
PHI = pre-harvest interval
NS = not specified in the label
NC = cannot be calculated as water volume is not specified on label

# RESIDUES RESULTING FROM SUPERVISED TRIALS

The Meeting received information on supervised field trials for boscalid uses on the following crops.

| | | |
|---|---|---|
| Apple: Belgium, Germany, France, Italy and the Netherlands | | Table 17 |
| Stone fruits: Canada, USA | | Table 18 |
| Raspberries, blueberries: USA | | Table 19 |
| Strawberry: Denmark, France, Germany, Great Britain, Greece, Italy, Japan, Spain, Sweden, and the Netherlands | | Table 20 |
| Grapes: USA, Japan | | Table 21 |
| Banana: Colombia, Costa Rica, Ecuador, Honduras, Martinique and Mexico | | Table 22 |
| Onion: USA, Japan | | Table 23 |
| Leek: Belgium, France, Germany, Great Britain and the Netherlands | | Table 24 |
| Broccoli: USA | | Table 25 |
| Cabbage: USA, Japan | | Table 26 |
| Cauliflower: Denmark, France, Germany, Great Britain and the Netherlands | | Table 27 |
| Brussels sprouts: Denmark, France, Germany, Great Britain, Sweden and the Netherlands | | Table 28 |
| Cucumbers: USA, Japan | | Table 29 |
| Cantaloupe: USA | | |
| Summer squash: USA | | |
| Melon: Italy, Japan, Spain | | |
| Watermelon: Japan | | |
| Tomatoes: USA, Japan | | Table 30 |
| Peppers: USA | | |
| Eggplant: Japan | | |
| Mustard greens: USA | | Table 31 |
| Lettuce: France, Germany, Japan, Spain, the Netherlands and USA | | Table 32 |
| Curly kale: Denmark, Great Britain, Sweden and the Netherlands | | Table 33 |
| Beans with pod: Denmark, France, Germany and Spain | | Table 34 |
| Peas, edible podded and succulent: USA | | Table 35 |
| Soybean, edible podded (immature): USA | | Table 36 |
| Beans without pod: USA | | Table 37 |
| Field beans: USA, Japan | | Table 38 |
| Peas dry: USA | | Table 39 |
| Soybean, dry (mature seed): USA | | Table 40 |
| Carrot: USA | | Table 41 |
| Radish: USA | | Table 42 |
| Potato: USA | | Table 43 |

Cereal grains: Belgium, Denmark, France, Germany, the Netherlands and UK          Table 44

Almond: USA                                                                       Table 45

Pecan nut: USA                                                                    Table 46

Pistachio: USA                                                                    Table 47

Canola: USA                                                                       Table 48

Sunflower: USA                                                                    Table 49

Peanuts: USA                                                                      Table 50

Coffee: Brazil                                                                    Table 51

Hops: Germany                                                                     Table 52

Where residues were not detected, they are reported as below the LOQ. Residue data, application rates and spray concentrations have generally been rounded to two significant figures or, for residues near the LOQ, to one significant figure. For trials, included control plots, no control data is given in the tables except where residues in control samples exceeded the LOQ. Residue data is recorded unadjusted for % recovery. Multiple results are recorded in the data tables where the trial design included replicate plots and where separate samples have been identified as being from these replicate plots. Results used to estimate STMRs are double underlined.

Trials were generally well documented with laboratory and field reports. Laboratory reports included method validation with procedural recoveries from spiking at residue levels similar to those occurring in samples from the supervised trials. Dates of analyses or duration of residue sample storage were also provided.

Conditions of the supervised residue trials were reported in detailed field reports. Most field reports provided data on the sprayers used, plot size, residue sample size and sampling date. In multiple applications, the application rate, spray concentration and water volume may not have been exactly the same for all applications; the recorded values in the supervised trials summary tables are for the final application.

Intervals of freezer storage between sampling and analysis were recorded for all trials and were covered by the conditions of the freezer storage stability studies.

Table 17. Results of residue trials with foliar treatment of boscalid conducted in apple (Raunft and Funk, 2001, 2001/1006135; Schulz, 2000, 2001/1000946; Raunft and Funk, 2001, 2001/1015029 Schulz, 2001, 2001/1015046 and Schulz, 2003, 2003/1001291).

| Country, Year Location (variety) | Application | | | | | PHI days | Commodity | Residues, mg/kg boscalid | Ref. Reg.DocID. (Trail No.) |
|---|---|---|---|---|---|---|---|---|---|
| | Formulation | kg ai/ha | kg ai/hL | Water L/ha | No. | | | | |
| Germany, 2000 Vehlefanz (Pinova) | SE | 0.20 | 0.020 | 1000 | 4 | 0 | fruit | 0.30 | 2001/1006135 (ACK/06/00) |
| | | | | | | 6 | | 0.15 | |
| | | | | | | 14 | | 0.14 | |
| | | | | | | 21 | | 0.14 | |
| | | | | | | 28 | | 0.11 | |
| Germany, 2000 Steiien a.H. (Jonagold) | SE | 0.20 | 0.020 | 1000 | 4 | 0 | fruit | 0.35 | 2001/1006135 (DU2/12/00) |
| | | | | | | 7 | | 0.36 | |
| | | | | | | 14 | | 0.27 | |
| | | | | | | 21 | | 0.24 | |
| | | | | | | 28 | | 0.16 | |

| Country, Year Location (variety) | Application | | | | | PHI days | Commodity | Residues, mg/kg boscalid | Ref. Reg.DocID. (Trail No.) |
|---|---|---|---|---|---|---|---|---|---|
| | Formulation | kg ai/ha | kg ai/hL | Water L/ha | No. | | | | |
| Germany, 2000 *Eschbach* (Braeburn) | SE | 0.20 | 0.020 | 1000 | 4 | 0 7 14 21 28 | fruit | 0.38 <u>0.32</u> 0.28 0.19 0.19 | 2001/1006135 (DU4/11/00) |
| France, 2000 *Cambrai* (Jonagold) | SE | 0.20 | 0.020 | 1000 | 4 | 0 6 13 21 28 | fruit | 0.55 <u>0.34</u> 0.31 0.17 0.15 | 2001/1000946 (X 00 62 03) |
| France, 2000 *St. Loup Terrier* (Jonagold) | SE | 0.20 | 0.020 | 1000 | 4 | 0 7 15 22 28 | fruit | 0.56 <u>1.24</u> 0.42 0.52 0.43 | 2001/1000946 (X 00 62 04) |
| France, 2000 *Buzet sur Baize* (Canada) | SE | 0.20 | 0.020 | 1000 | 4 | 0 7 14 21 28 | fruit | 0.38 <u>0.51</u> 0.39 0.28 0.22 | 2001/1000946 (X 00 62 05) |
| France, 2000 *Le Beugnon* (Golden) | SE | 0.20 | 0.020 | 1000 | 4 | 0 7 14 21 28 | fruit | 0.51 <u>0.42</u> 0.42 0.18 0.08 | 2001/1000946 (X 00 62 06) |
| Italy, 2000 *Ferrara* (Red Chief) | SE | 0.20 | 0.020 | 1000 | 4 | 0 7 13 20 27 | fruit | 0.36 <u>0.30</u> 0.19 0.20 0.22 | 2001/1000946 (0025R) |
| Italy, 2000 *Forli* (Royal Gala) | SE | 0.20 | 0.020 | 1000 | 4 | 0 8 14 22 28 | fruit | 0.36 <u>0.29</u> 0.24 0.14 0.12 | 2001/1000946 (0026R) |
| Belgium, 2001 *Kortenaken* (Jonagold) | SE | 0.20 | 0.020 | 1000 | 4 | 0 6 13 22 27 | fruit | 0.39 <u>0.37</u> 0.26 0.26 0.16 | 2001/1015029 (AGR/15/01) |
| Germany, 2001 *Stetten a. H.* (Golden Delicious) | SE | 0.20 | 0.020 | 1000 | 4 | 0 7 14 21 27 | fruit | 0.81 <u>0.55</u> 0.52 0.41 0.47 | 2001/1015029 (DU2/07/01) |
| France, 2001 *Chevire* (Golden Smoothy) | SE | 0.20 | 0.020 | 1000 | 4 | 0 8 14 20 28 | fruit | 0.42 0.38 0.35 <u>0.39</u> 0.20 | 2001/1015029 (FBM/02/01) |

| Country, Year Location (variety) | Application | | | | | PHI days | Commodity | Residues, mg/kg boscalid | Ref. Reg.DocID. (Trail No.) |
|---|---|---|---|---|---|---|---|---|---|
| | Formulation | kg ai/ha | kg ai/hL | Water L/ha | No. | | | | |
| Netherlands, 2001 *Groesbeek* (Elstar) | SE | 0.20 | 0.020 | 1000 | 4 | 0<br>8<br>13<br>21<br>29 | fruit | 0.24<br><u>0.42</u><br>0.25<br>0.26<br>0.15 | 2001/1015029 (AGR/16/01) |
| France, 2001 *Verquieres* (Ozar Gold) | SE | 0.20 | 0.020 | 1000 | 4 | 0<br>7<br>14<br>21<br>28 | fruit | 0.73<br><u>0.65</u><br>0.41<br>0.43<br>0.47 | 2001/1015046 (X 01 062 08) |
| France, 2001 *Verquieres* (Golden Delicious) | SE | 0.20 | 0.020 | 1000 | 4 | 0<br>7<br>14<br><br>21<br>28 | fruit | 0.60<br><u>0.53</u><br>0.51<br><br>0.35<br>0.35 | 2001/1015046 (X 01 062 09) |
| Italy, 2001 *Ferrara* (Red Chief) | SE | 0.20 | 0.020 | 1000 | 4 | 0<br>7<br>14<br>21<br>27 | fruit | 0.13<br><u>0.24</u><br>0.13<br>0.16<br>0.15 | 2001/1015046 (0148R) |
| Italy, 2001 *Ferrara* (Golden Delicious) | SE | 0.20 | 0.020 | 1000 | 4 | 0<br>7<br>14<br><br>21<br>28 | fruit | 0.27<br><u>0.20</u><br>0.18<br><br>0.20<br>0.18 | 2001/1015046 (0149R) |
| Italy, 2001 *Cesena* (Royal Gala) | SE | 0.20 | 0.020 | 1000 | 4 | 0<br>6<br>13<br>20<br>27 | fruit | 0.22<br><u>0.19</u><br>0.13<br>0.11<br>0.09 | 2001/1015046 (0150R) |
| Germany, 2003 *Vehlefanz* (Piros) | SE | 0.20 | 0.020 | 1000 | 4 | 0<br>8<br>15<br>21<br>28 | fruit | 0.37<br><u>0.29</u><br>0.16<br>0.16<br>0.17 | 2003/1001291 (ACK/11/03) Plot 2 |
| | WG | 0.20 | 0.020 | 1000 | 4 | 0<br>8<br>15<br>21<br>28 | fruit | 0.23<br><u>0.14</u><br>0.13<br>0.11<br>0.08 | Plot 3 |
| France, 2003 *Rottelsheim* (Golden) | SE | 0.20 | 0.020 | 1000 | 4 | 0<br>8<br>15<br>22<br>29 | fruit | 0.24<br><u>0.32</u><br>0.23<br>0.25<br>0.20 | 2003/1001291 (FAN/18/03) Plot 2 |
| | WG | 0.20 | 0.020 | 1000 | 4 | 0<br>8<br>15<br>22<br>29 | fruit | 0.42<br><u>0.24</u><br>0.19<br>0.20<br>0.15 | Plot 3 |

| Country, Year Location (variety) | Application | | | | | PHI days | Commodity | Residues, mg/kg boscalid | Ref. Reg.DocID. (Trail No.) |
|---|---|---|---|---|---|---|---|---|---|
| | Formulation | kg ai/ha | kg ai/hL | Water L/ha | No. | | | | |
| France, 2003 Bouloc (Star Kimson) | SE | 0.20 | 0.020 | 1000 | 4 | 0 | fruit | 0.92 | 2003/1001291 |
| | | | | | | 7 | | 0.86 | (FTL/15/03) |
| | | | | | | 14 | | 0.43 | Plot 2 |
| | | | | | | 21 | | 0.51 | |
| | | | | | | 28 | | 0.70 | |
| | WG | 0.20 | 0.020 | 1000 | 4 | 0 | fruit | 0.85 | Plot 3 |
| | | | | | | 7 | | 0.49 | |
| | | | | | | 14 | | 0.38 | |
| | | | | | | 21 | | 0.30 | |
| | | | | | | 28 | | 0.29 | |
| Italy, 2003 Piemonte (Golden Delicious) | SE | 0.20 | 0.020 | 1000 | 4 | 0 | fruit | 0.46 | 2003/1001291 |
| | | | | | | 7 | | 0.35 | (ITA/09/03) |
| | | | | | | 15 | | 0.39 | Plot 2 |
| | | | | | | 21 | | 0.32 | |
| | | | | | | 28 | | 0.16 | |
| | WG | 0.20 | 0.020 | 1000 | 4 | 0 | fruit | 0.55 | Plot 3 |
| | | | | | | 7 | | 0.43 | |
| | | | | | | 15 | | 0.17 | |
| | | | | | | 21 | | 0.20 | |
| | | | | | | 28 | | 0.17 | |

Table 18. Results of residue trials with foliar treatment of boscalid conducted in peach, plum and cherry (Wofford and Abdel-Baky, 2001, 2001/5000831; Leonard and Gooding, 2005, 2005/5000024).

| CROP Country, Year Location (variety) | Application | | | | | PHI days | Commodity | Residues, mg/kg boscalid | Ref. Reg.DocID. (Trial and treatment No.) |
|---|---|---|---|---|---|---|---|---|---|
| | Formulation | kg ai/ha | kg ai/hL[1] | Water L/ha | No. | | | | |
| PEACH | | | | | | | | | |
| USA, 1999 | WG | 0.26 | | | 5 | | fruit | | 2001/5000831 |
| PA, EPA Region 1 | | | 0.050 | 510-530 | | 0 | | 0.66 | (99107-02) |
| (Red Haven) | | | 0.013 | 2020-2050 | | 0 | | 0.75 | (99107-03) |
| SC, EPA Region 2 | | | 0.053 | 480-500 | | 0 | | 0.16 | (99108-02) |
| (Contender) | | | 0.013 | 1980-2050 | | 0 | | 0.19 | (99108-02) |
| GA, EPA Region 2 | | | 0.042 | 550-700 | | 0 | | 0.40 | (99109-02) |
| (Harmony) | | | 0.020 | 1120-1450 | | 0 | | 0.42 | (99109-03) |
| GA, EPA Region 2 | | | 0.051 | 490-520 | 5 | 0 | | 0.49 | (99110-02) |
| (June Gold) | | | | | | 7 | | 0.32 | |
| | | | | | | 14 | | 0.21 | |
| | | | | | | 21 | | 0.13 | |
| | | | | | | 28 | | 0.15 | |
| | | | 0.010 | 2500 | 5 | 0 | | 0.48 | (99110-03) |
| | | | | | | 7 | | 0.21 | |
| | | | | | | 14 | | 0.21 | |
| | | | | | | 21 | | 0.14 | |
| | | | | | | 28 | | 0.25 | |
| MI, EPA Region 5 | | | 0.041 | 620-650 | 5 | 0 | | 0.40 | (99111-02) |
| (Red Haven) | | | 0.014 | 1840-1900 | | 0 | | 0.33 | (99111-02) |
| TX, EPA Region 6 | | | 0.050 | 460-590 | | 0 | | 0.64 | (99112-02) |
| (Lauring) | | | 0.023 | 1010-1250 | | 0 | | 0.73 | (99112-03) |
| CA, EPA Region 10 | | | 0.037 | 670-730 | | 0 | | 0.52 | (99113-02) |
| (Red Sun) | | | 0.01 | 2500-2700 | | 0 | | 0.49 | (99113-03) |
| CA, EPA Region 10 | | | 0.028 | 930-950 | | 0 | | 0.48 | (99114-02) |
| (September Sun) | | | 0.014 | 1860-1910 | | 0 | | 0.19 | (99114-03) |
| CA, EPA Region 10 | | | 0.048 | 530-550 | | 0 | | 0.32 | (99115-02) |
| (Loadel) | | | 0.021 | 1250-1280 | | 0 | | 0.32 | (99115-03) |

| CROP Country, Year *Location* (variety) | Application | | | | | PHI days | Commodity | Residues, mg/kg boscalid | Ref. Reg.DocID. (Trial and treatment No.) |
|---|---|---|---|---|---|---|---|---|---|
| | Formulation | kg ai/ha | kg ai/hL[1] | Water L/ha | No. | | | | |
| USA, 2004 *WA, EPA region 11* (Snow king) | WG | 0.26 | 0.036 | 722-741 | 5 | 0 | fruit | 1.19 | 2005/5000024 (2004134) |
| *IL, EPA region 5* (Cresthaven) | | 0.26-0.27 | 0.033 | 772-827 | | | | 0.49 | (2004135) |
| *ON, EPA region 12* (Red Heaven) | | 0.25-0.26 | 0.030 | 844-868 | | | | 0.62 | (2004136) |
| *MN, EPA region 5* (Bailey Hardy) | | 0.25-0.26 | 0.027 | 920-952 | | | | 0.79 | (2004137) |
| PLUM | | | | | | | | | |
| USA, 2001 *MI, EPA Region 5* (Stanley) | WG | 0.26 | 0.037 | 690-730 | 5 | 0 7 14 21 28 | fruit | 0.57 0.55 0.40 0.29 0.23 | 2001/5000831 (99116-02) |
| | | | 0.012 | 2040-2160 | 5 | 0 7 14 21 28 | | 0.34 0.21 0.27 0.23 0.25 | (99116-03) |
| *CA, EPA Region 10* (July Rosu´s) | | | 0.050 0.012 | 510-530 2030-2160 | 5 | 0 0 | | 0.14 0.15 | (99117-02) (99117-03) |
| *CA, EPA Region 10* (Angelino) | | | 0.045 0.011 | 560-600 2260-2340 | | | | 0.17 0.32 | (99118-02) (99118-03) |
| *CA, EPA Region 10* (French Prune) | | | | 850-880 1600-1710 | | | | 0.09 0.10 | (99119-02) (99119-03) |
| *CA, EPA Region 10* (Howard Sun) | | | | 910-950 1890-1950 | | | | 0.24 0.25 | (99120-02) (99120-03) |
| *OR, EPA Region 12* (Parsons) | | | | 600-620 1790-1890 | | | | 0.08 0.11 | (99308-02) (99308-03) |
| USA, 2004 *ID, EPA region 11* (Empress) | WG | 0.26-0.27 | | 933-966 | 5 5 | 0 | fruit | 0.55 | 2005/5000024 (2004138) |
| *NS, EPA region 1[a]* (Blufre) | | 0.25-0.26 | | 625-645 | 5 | | | 0.70 | (2004139) |
| *ON, EPA region 12* (Yellow plum) | | 0.25-0.28 | | 838-921 | 5 | | | 0.46 | (2004140) |
| *MN, EPA region 5* (Alderman) | | 0.25-0.26 | | 920-952 | 5 | | | 0.17 | (2004141) |
| CHERRY (tart) | | | | | | | | | |
| USA, 2001 *NY, EPA Region 1* (Montmorency) | WG | 0.26 | 0.034 0.018 | 760 1420 | 5 | 0 | fruit | 1.64 1.42 | 2001/5000831 (99101-02) (99101-03) |
| *MI, EPA Region 5* (Montmorency) | | | 0.041 0.014 | 620-640 1820-1900 | | | | 1.31 1.51 | (99102-02) (99102-03) |
| *MI, EPA Region 5* (Montmorency) | | | 0.041 0.014 | 620-640 1820-1900 | | | | 1.09 1.21 | (99103-02) (99103-03) |
| CHERRY (sweet) | | | | | | | | | |
| USA, 2001 *MI, EPA Region 5* (Sommerset) | WG | 0.26 | 0.037 0.013 | 690-720 2010-2120 | 5 | 0 | fruit | 0.76 0.74 | 2001/5000831 (99104-02) (99104-03) |
| *CA, EPA Region 10* (Brooks) | | | 0.044 0.011 | 580-600 2290-2500 | | | | 0.64 1.00 | (99105-02) (99105-03) |
| *WA, EPA Region11* (Bing) | | | 0.055 0.014 | 470 1890-1920 | | | | 0.91 1.50 | (99106-02) (99106-03) |
| USA, 2004 *WA, EPA region 11* (Bing) | WG | 0.23-0.26 | 0.031 | 745-844 | 6 | 0 | fruit | 1.49 | 2005/5000024 (2004142) |

1) average value.

Table 19. Results of residue trials with foliar treatment of boscalid conducted in berries (Versoi and Abdel-Baky, 2000, 2000/5195; Leonard and Gooding, 2005, 2005/5000025).

| CROP Country, Year *Location* (variety) | Application | | | | | PHI days | Commodity | Residues, mg/kg boscalid | Ref. Reg.DocID. (Trial No.) |
|---|---|---|---|---|---|---|---|---|---|
| | Formulation | kg ai/ha | kg ai/hL[1] | Water L/ha | No. | | | | |
| **RED RASPBERRIES** | | | | | | | | | |
| USA, 1999 *NY, EPA Region 1* (Titau) | WG | 0.42 | 0.074 | 570 | 4 | 0 | fruit | <u>2.69</u> | 2000/5195 (99277) |
| | | | | | | 2 | | 2.32 | |
| | | | | | | 4 | | 1.93 | |
| | | | | | | 6 | | 1.58 | |
| | | | | | | 8 | | 1.25 | |
| *ON, EPA Region 12* (Meeker) | | 0.41 | 0.079 | 520 | 4 | 0 | | <u>1.49</u> | (99280) |
| *ON, EPA Region 12* (Tulamene) | | 0.41 | 0.079 | 520 | 4 | 0 | | <u>2.00</u> | (99281) |
| USA, 2004 *MN, EPA Region 5* (Nova) | WG | 0.42 | 0.074 | 570 | 4 | 0 | | <u>3.45</u> | 2005/5000025 (2004143) |
| *OR, EPA Region 12* (Caroline) | | 0.42 | 0.060 | 700 | 4 | 0 | | <u>2.44</u> | (2004144) |
| *QC,EPA Region 5B* (Kilarme) | | 0.40 | 0.057 | 600-800 | 4 | 0 | | <u>3.73</u> | (2004145) |
| **BLUEBERRIES** | | | | | | | | | |
| USA, 1999 *WI, EPA Region 5* (Blue Chop) | WG | 0.41 | 0.087 | 470 | | 0 | fruit | <u>1.16</u> | 2000/5195 (99278) |
| *OR, EPA Region 12* (Blue Crop) | | 0.41 | 0.043 | 950 | | 0 | | <u>0.84</u> | (99279) |
| *NY, EPA Region 1* (blue ray, blue crop) | | 0.41 | 0.071 | 580 | 4 | 0 | | <u>1.27</u> | (99328) |
| *WI, EPA Region 5* (Berkley) | | 0.43 | 0.090 | 480 | | 0 | | <u>1.26</u> | (99329) |
| *GA, EPA Region 2* (Tift blue) | | 0.41 | 0.079 | 520 | | 0 | | <u>1.46</u> | (99330) |
| *GA, EPA Region 2* (Climax) | | 0.41 | 0.079 | 520 | | 0 | | <u>2.34</u> | (99331) |
| USA, 2004 *PEI,EPA Region 1A* (wild blueberry) | WG | 0.41 | 0.15 | 270 | 4 | 0 | | <u>4.35</u> | 2005/5000025 (2004146) |
| *MI, EPA Region 5* (blue crop) | | 0.41 | 0.059 | 700 | | 0 | | <u>2.62</u> | (2004149) |
| *MI, EPA Region 5* (blue crop) | | 0.41 | 0.059 | 700 | | 0 | | <u>2.65</u> | (2004150) |
| *WI, EPA Region 5* (Elliot) | | 0.42 | 0.075 | 560 | | 0 | | <u>3.79</u> | (2004151) |
| *NS, EPA Region 1A* (not known) | | 0.41 | 0.059 | 700 | | 0 | | <u>6.78</u> | (2004198) |
| *NS, EPA Region 1A* (not known) | | 0.41 | 0.059 | 700 | | 0 | | <u>6.83</u> | (2004199) |

1) average value.

Table 20. Results of residue trials with foliar treatment of boscalid conducted in strawberries (Schulz, 2003, 2004/1015928; Johnston, 2004, 2004/7007479; Schulz, 2003, 2004/1015927; Reichert, 2004, 2005/1004969).

| CROP Country, Year Location (variety) | Application | | | | PHI days | Commodity | Residues, mg/kg boscalid | Ref. Reg.DocID. (Trial No.) |
|---|---|---|---|---|---|---|---|---|
| | Formulation | kg ai/ha | kg ai/hL | Water L/ha | No. | | | |
| **INDOOR** | | | | | | | | |
| Japan, 2002 Gifu (Meho) | WG | 1.25 | 0.06 | 2500 | 3 | 1 3 7 | fruit | 7.28 6.58 2.58 | na |
| Japan, 2002 Miyazaki (Toyonoka) | WG | 0.78 | 0.05 | 1565 | 3 | 1 3 7 | fruit | 2.04 0.89 0.82 | na |
| Spain, 2003 *Andalucía* (Camarosa) | WG | 0.48 | 0.048 | 1000 | 2 | 0 3 7 | fruit | 0.21 0.23 0.16 | 2004/1015928 (ALO/22/03) |
| Spain, 2003 *Andalucía* (Camarosa) | WG | 0.48 | 0.048 | 1000 | 2 | 0 3 7 | fruit | 0.36 0.27 0.12 | (ALO/23/03) |
| France, 2003 *Rhône-Alpes* (Dorselect) | WG | 0.48 | 0.048 | 1000 | 2 | 0 3 7 | fruit | 0.79 0.68 0.59 | (FBD/18/03) |
| Italy, 2003 *Piemonte* (Marmolada) | WG | 0.48 | 0.048 | 1000 | 2 | 0 3 7 | fruit | 1.17 0.46 0.34 | (ITA/18/03) |
| Spain, 2003 *Andalucía* (Camarosa) | WG | 0.48 | 0.048 | 1000 | 2 | 0 3 7 | fruit | 0.34 0.21 0.34 | 2004/7007479 (ALO/14/04) |
| Spain, 2003 *Andalucía* (Camarosa) | WG | 0.48 | 0.048 | 1000 | 2 | 0 3 7 | fruit | 0.80 0.49 0.25 | (ALO/15/04) |
| France, 2003 *Alsace* (Gariguette) | WG | 0.48 | 0.048 | 1000 | 2 | 0 3 7 | fruit | 0.64 0.57 0.30 | (FTL/10/04) |
| France, 2003 *Rhône-Alpes* (Marat des Bois) | WG | 0.48 | 0.048 | 1000 | 2 | 0 3 7 | fruit | 1.10 0.28 n.a. | (FBD/09/04) |
| Italy, 2003 *Piemonte* (Alba) | WG | 0.48 | 0.048 | 1000 | 2 | 0 3 7 | fruit | 0.39 0.31 0.24 | (ITA/07/04) |
| **OUTDOOR** | | | | | | | | |
| Germany, 2003 Nordrhein-Westf. (Elsanta) | WG | 0.48 | 0.048 | 1000 | 2 | 0 3 7 | fruit | 0.17 0.14 0.19 | 2004/1015927 (AGR/29/03) |
| Denmark, 2003 *Fuenen* (Elsanta) | WG | 0.48 | 0.048 | 1000 | 2 | 0 3 7 | fruit | 0.41 0.38 0.16 | (ALB/20/03) |
| France, 2003 *Alsace* (Dorselect) | WG | 0.48 | 0.048 | 1000 | 2 | 0 3 7 | fruit | 0.39 0.17 0.20 | (FAN/27/03) |
| France, 2003 *Midi-Pyrénées* (Gariguette) | WG | 0.48 | 0.048 | 1000 | 2 | 0 3 7 | fruit | 0.75 0.42 0.42 | (FTL/23/03) |
| France, 2003 *Rhône-Alpes* (Dorselect) | WG | 0.48 | 0.048 | 1000 | 2 | 0 3 7 | fruit | 2.03 1.74 0.71 | (FBD/17/03) |

| CROP Country, Year *Location* (variety) | Application | | | | | PHI days | Commodity | Residues, mg/kg boscalid | Ref. Reg.DocID. (Trial No.) |
|---|---|---|---|---|---|---|---|---|---|
| | Formulation | kg ai/ha | kg ai/hL | Water L/ha | No. | | | | |
| Great Britain, 2003 *Cambridgeshire* (Elsanta) | WG | 0.48 | 0.048 | 1000 | 2 | 0<br>3<br>7 | fruit | 0.47<br>0.27<br>0.24 | (OTA/23/03) |
| Italy, 2003 *Piemonte* (Marmolada) | WG | 0.48 | 0.048 | 1000 | 2 | 0<br>3<br>7 | fruit | 1.16<br>0.68<br>0.67 | (ITA/19/03) |
| Italy, 2003 *Piemonte* (Maya) | WG | 0.48 | | 1000 | 2 | 0<br>3<br>7 | fruit | 1.30<br>0.69<br>0.47 | (ITA/20/03) |
| Sweden, 2003 *Malmo* (Honeye) | WG | 0.48 | | 1000 | 2 | 0<br>3<br>7 | fruit | 0.23<br>0.15<br>0.15 | (HUS/13/03) |
| Germany, 2004 *Brandenburg* (Symphony) | WG | 0.48 | 0.048 | 1000 | 2 | 0<br>3<br>7 | fruit | 0.58<br>0.34<br>0.35 | 2005/1004969 (ACK/06/04) |
| France, 2004 *Rhône-Alpes* (Marat des Bois) | WG | 0.48 | 0.048 | 1000 | 2 | 0<br>3<br>7 | fruit | 0.53<br>0.89<br>0.49 | (FBD/08/04) |
| France, 2004 *Pays de la Loire* (Marat des Bois) | WG | 0.48 | 0.048 | 1000 | 2 | 0<br>3<br>7 | fruit | 0.39<br>0.29<br>0.41 | (FBM/04/04) |
| France, 2004 *Castelmaurou* (Gariguette) | WG | 0.48 | 0.048 | 1000 | 2 | 0<br>3<br>7 | fruit | 0.75<br>0.45<br>0.26 | (FTL/09/04) |
| Great Britain, 2004 *Seven Oaks, Kent* (Florence) | WG | 0.48 | 0.048 | 1000 | 2 | 0<br>3<br>7 | fruit | 0.69<br>0.55<br>0.48 | (OAT/06/04) |
| Greece, 2004 *Macedonia* (Siscape) | WG | 0.48 | 0.048 | 1000 | 2 | 0<br>3<br>7 | fruit | 1.44<br>1.87<br>0.77 | (GRE/10/04) |
| Italy, 2004 *Piemonte* (Maya) | WG | 0.48 | 0.048 | 1000 | 2 | 0<br>3<br>7 | fruit | 0.52<br>0.47<br>0.39 | (ITA/06/04) |
| Netherlands, 2004 *Limburg* (Elsanta) | WG | 0.48 | 0.048 | 1000 | 2 | 0<br>3<br>7 | fruit | 0.45<br>0.46<br>0.37 | (AGR/08/04) |

na: not available.

Table 21. Results of residue trials with foliar treatment of boscalid conducted in grapes (Haughey and Abdel-Baky, 2000, 2000/5228; Leonard and Gooding, 2005, 2005/5000023) and Japan.

| CROP Country, Year *Location* (variety) | Application | | | | | PHI days | Commodity | Residues, mg/kg boscalid | Ref. Reg.DocID. (Trial No.) |
|---|---|---|---|---|---|---|---|---|---|
| | Formulation | kg ai/ha | kg ai/hL | Water L/ha | No. | | | | |
| GRAPES | | | | | | | | | |
| Japan, 2002 Iwate (Beni-izu) | WG | 1.5 | 0.05 | 3000 | 3 | 7<br>14<br>21 | fruit | 4.30<br>3.95<br>3.92 | na |
| Japan, 2002 Nagano (Kyoho) | WG | 2.0 | 0.05 | 4000 | 3 | 7<br>14<br>21 | fruit | 5.20<br>4.16<br>3.84 | na |

| CROP Country, Year *Location* (variety) | Application | | | | PHI days | Commodity | Residues, mg/kg boscalid | Ref. Reg.DocID. (Trial No.) |
|---|---|---|---|---|---|---|---|---|
| | Formulation | kg ai/ha | kg ai/hL | Water L/ha | No. | | | |
| USA, 1999 *NY, EPA region 1* (Aurora) | WG | 0.41 | 0.087 | 470 | 3 | 14 | fruit | 2.97 | 2000/5228 (99122) |
| *CA, EPA region 10* (Emperor) | | 0.41 | 0.057 | 720 | 3 | 0 7 14 21 28 | fruit | 0.30 0.43 0.50 0.38 0.28 | (99124) |
| *CA, EPA region 10* (Flame Seedless) | | 0.41 | 0.055 | 740 | 3 | 14 | fruit | 0.34 | (99126) |
| *CA, EPA region 10* (Zinfandel) | | 0.41 | 0.084 | 490 | 3 | 14 | fruit | 1.50 | (99127) |
| *CA, EPA region 10* (Thompson Seedless) | | 0.41 | 0.059 | 700 | 3 | 14 | fruit | 0.65 | (99129) |
| *WA, EPA region 11* (White Riesling) | | 0.41 | 0.059 | 700 | 3 | 14 | fruit | 0.29 | (99132) |
| *NY, EPA region 1* (Seyval Blau) | | 0.41 | 0.043 | 950 | 3 | 14 | fruit | 2.08 | (99123) |
| *CA, EPA region 10* (Thompson Seedless) | | 0.41 | 0.028 | 1460 | 3 | 14 | fruit | 0.36 | (99125) |
| *CA, EPA region 10* (Zinfandel) | | 0.41 | 0.028 | 1460 | 3 | 14 | fruit | 1.38 | (99128) |
| *CA, EPA region 10* (Thompson Seedless) | | 0.41 | 0.029 | 1420 | 3 | 14 | fruit | 1.26 | (99130) |
| *CA, EPA region 10* (Thompson Seedless) | | 0.41 | 0.029 | 1420 | 3 | 14 | fruit | 0.65 | (99131) |
| *WA, EPA region 11* (White Riesling) | | 0.41 | 0.029 | 1420 | 3 | 14 | fruit | 0.34 | (99133) |
| USA, 2004 *ON, EPA region 5* (Concord) | WG | 0.41 | 0.041 | 980-1010 | 3 | 14 | fruit | 3.13 | 2005/5000023 (2004153) |
| *MN, EPA region 5* (Frontenac) | | 0.42 | 0.030 | 1420 | 3 | 14 | fruit | 0.92 | (2004154) |
| *MN, EPA region 5* (ES26-50) | | 0.41 | 0.044 | 940 | 3 | 14 | fruit | 1.83 | (2004155) |
| *IL, EPA region 5* (Chardonel) | | 0.42 | 0.052 | 780-830 | 3 | 13 14 | fruit | 2.28 | (2004156) |

na: not available.

Table 22. Results of residue trials with foliar treatment of boscalid conducted in banana (Wofford, 2004, 2004/5000480).

| CROP Country, Year *Location* (variety) | Application | | | | PHI days | Commodity | Residues, mg/kg boscalid | Ref. Reg.DocID. (Trial No.) |
|---|---|---|---|---|---|---|---|---|
| | Formulation | kg ai/ha | kg ai/hL | Water L/ha | No. | | | |
| USA, 2004 *Costa Rica* (Williams) | WG | 0.150 | 0.50 | 30 | 4 | 0 | wh. Fruit, bagged wh. Fruit unbag. pulp, bagged pulp, unbagged | ≤ 0.05 0.05 < 0.05 < 0.05 | 2004/5000480 (2003149) |

| CROP Country, Year *Location* (variety) | Application | | | | | PHI days | Commodity | Residues, mg/kg boscalid | Ref. Reg.DocID. (Trial No.) |
|---|---|---|---|---|---|---|---|---|---|
| | Formulation | kg ai/ha | kg ai/hL | Water L/ha | No. | | | | |
| *Costa Rica* (Williams) | | 0.150 | 0.50 | 30 | 4 | 0 | wh. Fruit, bagged<br>wh. Fruit unbag.<br>pulp, bagged<br>pulp, unbagged | ≤ 0.05<br>0.10<br>< 0.05<br>< 0.05 | (2003150) |
| *Ecuador* (Giant Cavendish) | | 0.150 | 0.60 | 25 | 4 | 0 | wh. Fruit, bagged<br>wh. Fruit unbag.<br>pulp, bagged<br>pulp, unbagged | ≤ 0.05<br>≤ 0.05<br>< 0.05<br>< 0.05 | (2003151) |
| *Ecuador* (Giant Cavendish) | | 0.150 | 0.60 | 25 | 4 | 0 | wh. Fruit, bagged<br>wh. Fruit unbag.<br>pulp, bagged<br>pulp, unbagged | ≤ 0.05<br>0.10<br>< 0.05<br>< 0.05 | (2003152) |
| *Colombia* (Gran Enano) | | 0.150 | 0.50 | 30 | 4 | 0 | wh. Fruit, bagged<br>wh. Fruit unbag.<br>pulp, bagged<br>pulp, unbagged | ≤ 0.05<br>≤ 0.05<br>< 0.05<br>< 0.05 | (2003153) |
| *Colombia* (Gran Enano) | | 0.148 | 0.50 | 30 | 4 | 0 | wh. Fruit, bagged<br>wh. Fruit unbag.<br>pulp, bagged<br>pulp, unbagged | ≤ 0.05<br>0.18<br>< 0.05<br>< 0.05 | (2003154) |
| *Honduras* (Ecuatoriano) | | 0.144 | 0.50 | 29 | 4 | 0 | wh. Fruit, bagged<br>wh. Fruit unbag.<br>pulp, bagged<br>pulp, unbagged | ≤ 0.05<br>≤ 0.05<br>< 0.05<br>< 0.05 | (2003155) |
| *Honduras* (Ecuatoriano) | | 0.144 | 0.50 | 29 | 4 | 0 | wh. Fruit, bagged<br>wh. Fruit unbag.<br>pulp, bagged<br>pulp, unbagged | ≤ 0.05<br>0.07<br>< 0.05<br>< 0.05 | (2003156) |
| *Guatemala* (Gran nane) | | 0.158 | 0.51 | 31 | 4 | 0 | wh. Fruit, bagged<br>wh. Fruit unbag.<br>pulp, bagged<br>pulp, unbagged | ≤ 0.05<br>0.09<br>< 0.05<br>< 0.05 | (2003157) |
| *Mexico* (Gran nane) | | 0.154 | 0.50 | 31 | 4 | 0 | wh. Fruit, bagged<br>wh. Fruit unbag.<br>pulp, bagged<br>pulp, unbagged | ≤ 0.05<br>0.11<br>< 0.05<br>< 0.05 | (2003158) |
| *Martinique* (Cavendish) | | 0.154 | 0.50 | 31 | 4 | 0 | wh. Fruit, bagged<br>wh. Fruit unbag.<br>pulp, bagged<br>pulp, unbagged | ≤ 0.05<br>0.07<br>< 0.05<br>< 0.05 | (2003159) |
| *Martinique* (Cavendish) | | 0.154 | 0.50 | 31 | 4 | 0 | wh. Fruit, bagged<br>wh. Fruit unbag.<br>pulp, bagged<br>pulp, unbagged | ≤ 0.05<br>≤ 0.05<br>< 0.05<br>< 0.05 | (2003160) |

Table 23. Results of residue trials with foliar treatment of boscalid conducted in onion (Versoi and Abdel-Baky, 2000, 2000/5207; Johnston and Jones, 2005, 2005/5000019).

| CROP Country, Year *Location* (variety) | Application | | | | | PHI days | Commodity | Residues, mg/kg boscalid[1] | Ref. Reg.DocID. (Trial No.) |
|---|---|---|---|---|---|---|---|---|---|
| | Formulation | kg ai/ha | kg ai/hL | Water L/ha | No. | | | | |
| **GREEN ONION** | | | | | | | | | |
| USA, 1999 *TX, EPA region 6* (Texas Early White) | WG | 0.34 | 0.16 | 210 | 6 | 7<br><br>7 | wh. Pl. w/o root<br><br>wh. Pl. w/o root | 2.73 | 2000/5207 (99177) |

| CROP Country, Year *Location* (variety) | Application | | | | PHI days | Commodity | Residues, mg/kg boscalid[1] | Ref. Reg.DocID. (Trial No.) |
|---|---|---|---|---|---|---|---|---|
| | Formulation | kg ai/ha | kg ai/hL | Water L/ha | No. | | | |
| CA, *EPA region 10* (Southport White 404) | | 0.34 | 0.12 | 280 | 6 | 7 | wh. Pl. w/o root | 2.39 | (99178) |
| | | | | | | 7 | wh. Pl. w/o root | | |
| CA, *EPA region 10* (K-99 Bunching) | | 0.34 | 0.12 | 280 | 6 | 7 | wh. Pl. w/o root | 1.13 | (99179) |
| | | | | | | 7 | wh. Pl. w/o root | | |
| USA, 2005 *ND; EPA region 5* (Lisbon) | WG | 0.33 | 0.18 | 180 | 6 | 7 | wh. Pl. w/o root | 2.01 | 2005/5000019 (R05104) |
| | | | | | | 7 | wh. Pl. w/o root | | |
| *QC; EPA region 5b* (Performer) | | 0.34 | 0.16 | 210 | 6 | 7 | wh. Pl. w/o root | 2.20 | (R05110) |
| | | | | | | 7 | wh. Pl. w/o root | | |
| ONION, bulb | | | | | | | | | |
| Japan, 2002 Hokkaido (Iomante) | WG | 0.75 | 0.050 | 1500 | 3 | 1 | bulb | 0.006 | na |
| | | | | | | 7 | | < 0.005 | |
| | | | | | | 14 | | < 0.005 | |
| Japan, 2002 Nagano (Kan 70) | WG | 0.75 | 0.050 | 1500 | 3 | 1 | bulb | 0.07 | na |
| | | | | | | 7 | | 0.03 | |
| | | | | | | 14 | | 0.006 | |
| USA, 1999 *PA, EPA region 1* (Stuttgarter) | WG | 0.34 | 0.16 | 210 | 6 | 7 | bulb | 0.11 | 2000/5207 (99171) |
| *TX, EPA region 6* (Cimarron) | | 0.33 | 0.17 | 200 | 6 | 7 | bulb | 0.10 | (99172) |
| *TX, EPA region 8* (Vega) | | 0.34 | 0.15 | 230 | 6 | 7 | bulb | 0.13 | (99173) |
| CA, *EPA region 10* (Red Cerole) | | 0.34 | 0.12 | 280 | 6 | 0 | bulb | 0.79 | (99174) |
| | | | | | | 0 | bulb | 0.70 | |
| | | | | | | 7 | bulb | 0.22 | |
| | | | | | | 14 | bulb | 0.05 | |
| | | | | | | 14 | bulb | 0.10 | |
| | | | | | | 21 | bulb | 0.08 | |
| | | | | | | 21 | bulb | 0.13 | |
| | | | | | | 28 | bulb | 0.17 | |
| | | | | | | 28 | bulb | 0.11 | |
| CA, *EPA region 10* (Blanco Duro) | | 0.34 | 0.12 | 280 | 6 | 7 | bulb | 0.93 | (99175) |
| OR, *EPA region 11* (Yellow Danver) | | 0.34 | 0.12 | 280 | 6 | 7 | bulb | 0.05 | (99176) |
| USA, 2005 *ND; EPA region 5* (Varsity) | WG | 0.33 | 0.18 | 180 | 6 | 7 | bulb | ≤ 0.05 | 2005/5000019 (R05105) |
| *ON; EPA region 5* (Hamlet) | | 0.34 | 0.13 | 260 | 6 | 7 | bulb | 0.78 | (R05106) |
| | | | | | | | bulb | | |
| *ON; EPA region 5* (Norstar) | | 0.34 | 0.13 | 260 | 6 | 7 | bulb | 0.92 | (R05107) |
| *QC; EPA region 5b* (Mountaineer) | | 0.34 | 0.13 | 220 | 6 | 7 | bulb | 2.61 | (R05109) |

na: not available.

1)   average values used for estimation of MRL.

Table 24. Results of residue trials with foliar treatment of boscalid conducted in leek (Raunft *et al.*, 2001, 2001/1006130; Raunft *et al.*, 2001, 2001/1006131; Schulz, 2004, 2004/1015937).

| CROP Country, Year *Location* (variety) | Application | | | | | PHI days | Commodity | Residues, mg/kg boscalid | Ref. Reg.DocID. |
|---|---|---|---|---|---|---|---|---|---|
| | Formulation | kg ai/ha | kg ai/hL | Water L/ha | No. | | | | |
| Belgium, 1999 *Scherpenheuvel* (Arbavas) | WG | 0.400 | 0.13 | 300 | 3 | 0 7 14 21 | pl. w/o root | 3.80 1.70 0.79 0.80 | 2001/1006130 (AGR/19/99) |
| Germany, 1999 *Wustrau* (Glorina) | WG | 0.400 | 0.13 | 300 | 3 | 0 7 14 20 | pl. w/o root | 2.05 1.30 0.90 0.42 | (ACK/09/99) |
| Germany, 1999 *Hochdorf* (Alaska) | WG | 0.400 | 0.13 | 300 | 3 | 0 7 14 21 | pl. w/o root | 4.74 2.80 1.90 1.19 | (DU2/14/99) |
| Netherland, 1999 *Ottersum* (Alaska) | WG | 0.400 | 0.13 | 300 | 3 | 0 7 14 21 | pl. w/o root | 4.16 2.52 0.93 0.78 | (AGR/18/99) |
| Belgium, 2000 *Halen* (Apollo) | WG | 0.400 | 0.13 | 300 | 3 | 0 7 14 20 | pl. w/o root | 5.32 1.56 1.09 1.31 | 2001/1006131 (AGR/08/00) |
| Netherland, 2000 *Siebengewald* (Porinto) | WG | 0.400 | 0.13 | 300 | 3 | 0 7 13 20 | pl. w/o root | 2.09 0.86 0.62 < 0.05 | (AGR/09/00) |
| Germany, 2000 *Horrenberg* (Verdea) | WG | 0.400 | 0.13 | 300 | 3 | 0 7 14 21 | pl. w/o root | 3.32 1.24 1.16 0.73 | (DU2/09/00) |
| Germany, 2000 *Lambsheim* (Duwina) | WG | 0.400 | 0.13 | 300 | 3 | 0 7 14 21 | pl. w/o root | 2.19 1.34 1.02 0.88 | (DU4/08/00) |
| UK, 2000 *Alcester* (Jolant) | WG | 0.400 | 0.13 | 300 | 3 | 0 7 14 21 | pl. w/o root | 3.07 1.06 0.58 0.46 | (OAT/10/00) |
| France (N), 2003 *Alsace* (Bleu de Solaise) | WG | 0.400 | 0.13 | 300 | 3 | 0 8 14 21 | pl. w/o root | 4.24 2.85 1.31 0.062 | 2004/1015937 (FAN1203) |
| France (N), 2003 *Pays de la Loire* (Sevilla) | WG | 0.400 | 0.13 | 300 | 3 | 0 7 14 20 | pl. w/o root | 4.98 4.07 2.30 1.60 | (FBM0603) |

Table 25. Results of residue trials with foliar treatment of boscalid conducted in broccoli (Wofford and Abdel-Baky, 2002, 2001/5002616).

| CROP Country, Year _Location_ (variety) | Application | | | | | PHI days | Commodity | Residues, mg/kg boscalid[1] | Ref. Reg.DocID. |
|---|---|---|---|---|---|---|---|---|---|
| | Formulation | kg ai/ha | kg ai/hL | Water L/ha | No. | | | | |
| USA, 2001 _TX, EPA region 6_ (not given) | WG | 0.448 | 0.24 | 190 | 2 | 0<br>3<br>7<br>10<br>14 | flower head + stem | <u>1.59</u><br>0.86<br>0.60<br>0.39<br>0.11 | 2001/5002616 (2001216) |
| _CA, EPA region 10_ (Marathon) | WG | 0.448 | 0.16 | 280 | 2 | 0<br>3<br>7<br>10<br>14 | flower head + stem | <u>0.98</u><br>0.28<br>0.29<br>0.20<br>0.19 | (2001217) |
| _CA, EPA region 10_ (Marathon) | WG | 0.448 | 0.16 | 280 | 2 | 0<br>3<br>7<br>10<br>14 | flower head + stem | 1.58<br><u>1.70</u><br>1.13<br>1.16<br>0.37 | (2001218) |
| _CA, EPA region 10_ (Marathon) | WG | 0.448 | 0.16 | 280 | 2 | 0<br>3<br>7<br>10<br>14 | flower head + stem | <u>2.70</u><br>1.72<br>1.26<br>0.88<br>0.81 | (2001219) |
| _CA, EPA region 10_ (greenbelt) | WG | 0.448 | 0.16 | 280 | 2 | 0<br>3<br>7<br>10<br>14 | flower head + stem | <u>0.81</u><br>0.68<br>0.49<br>0.36<br>0.20 | (2001220) |
| _OR, EPA region 10_ (Arcadia) | WG | 0.448 | 0.19 | 240 | 2 | 0<br>3<br>7<br>10<br>14 | flower head + stem | <u>1.45</u><br>0.76<br>0.22<br>0.25<br>0.09 | (2001221) |
| Germany, 2003 _Rhinweg 9_ (Marathon) | WG | 0.267 | 0.089 | 300 | 3 | 0<br>7<br>14<br>21 | Flower | 1.03<br>0.18<br><u>≤ 0.05</u><br>< 0.05 | 2004/1015910 (ACK/17/03) |
| France, 2003 _Rhone-Alpes_ (Milady) | WG | 0.267 | 0.089 | 300 | 3 | 0<br>7<br>14<br>20<br>28 | Flower | 2.99<br>0.54<br><u>0.20</u><br>0.06<br>< 0.05 | (FBD/15/03) |
| France, 2003 _Medi-Pyrenees_ (Coronado) | WG | 0.267 | 0.089 | 300 | 3 | 0<br>7<br>15<br>22 | Flower | 0.38<br>0.08<br><u>≤ 0.05</u><br>< 0.05 | (FTL/19/03) |

1) average values used for estimation of MRL.

Table 26. Results of residue trials with foliar treatment of boscalid conducted in cabbage (Wofford and Abdel-Baky, 2002, 2001/5002617).

| CROP Country, Year _Location_ (variety) | Application | | | | | PHI days | Commodity | Residues, mg/kg boscalid[1] | Ref. Reg.DocID. |
|---|---|---|---|---|---|---|---|---|---|
| | Formulation | kg ai/ha | kg ai/hL | Water L/ha | No. | | | | |
| Japan, 2005 Taitama (Ajio) | WG | 0.667 | 0.033 | 2000 | 2 | 1<br>7<br>14 | whole | 0.70<br>0.50<br>0.09 | na |

| CROP Country, Year *Location* (variety) | Application | | | | | PHI days | Commodity | Residues, mg/kg boscalid[1] | Ref. Reg.DocID. |
|---|---|---|---|---|---|---|---|---|---|
| | Formulation | kg ai/ha | kg ai/hL | Water L/ha | No. | | | | |
| Japan, 2005 Kochi (Kinkei 201 go) | WG | 0.667 | 0.033 | 2000 | 2 | 1 7 14 | whole | 2.16 0.80 0.19 | na |
| USA, 2001 PA, *EPA region 1* (Market Prize) | | 0.47 | 0.13 | 360 | 2 | 0 3 7 10 14 | fresh cabbage with wrapper leaves | 2.22 1.27 1.03 1.14 0.43 | 2001/5002617 (2001222) |
| NC, *EPA region 2* (Early Jersey Wakefield) | | 0.448 | 0.24 | 190 | 2 | 0 3 7 10 14 | fresh cabbage with wrapper leaves | 2.33 1.13 0.46 0.32 0.31 | (2001223) |
| FL, *EPA region 3* (Everlasting) | | 0.448 | 0.21 | 210 | 2 | 0 3 7 10 14 | fresh cabbage with wrapper leaves | 1.53 1.78 0.94 1.32 1.16 | (2001224) |
| MI, *EPA region 5* (Rinda) | | 0.448 | 0.19 | 240 | 2 | 0 3 7 10 14 | fresh cabbage with wrapper leaves | 0.73 0.29 0.35 0.22 0.16 | (2001225) |
| TX, *EPA region 6* (Pennant) | | 0.448 | 0.24 | 190 | 2 | 0 3 7 10 14 | fresh cabbage with wrapper leaves | 1.06 0.64 0.47 0.40 0.39 | (2001226) |
| CA, *EPA region 10* (Supreme Vantage) | | 0.437 | 0.13 | 330 | 2 | 0 3 3 7 7 10 10 14 14 | fresh cabbage with wrapper leaves | 0.64 0.67 0.81 0.31 0.34 0.48 0.39 0.57 0.70 | |

na: not available.

1) average values used for estimation of MRL.

Table 27. Results of residue trials with foliar treatment of boscalid conducted in cauliflower (Schulz, 2003, 2004/1015910; Johnston, 2005, 2004/7007476).

| CROP Country, Year *Location* (variety) | Application | | | | | PHI days | Commodity | Residues, mg/kg boscalid | Ref. Reg.DocID. (Trial No.) |
|---|---|---|---|---|---|---|---|---|---|
| | Formulation | kg ai/ha | kg ai/hL | Water L/ha | No. | | | | |
| Denmark, 2003 *Fuenen* (Aviso) | WG | 0.267 | 0.089 | 300 | 3 | 0 7 14 21 | cauliflower | 0.08 < 0.05 ≤ 0.05 < 0.05 | 2004/1015910 (ALB/16/03) |
| Netherlands, 2003 *Ottersum* (Aviron) | WG | 0.267 | 0.089 | 300 | 3 | 0 8 15 21 | cauliflower | < 0.05 < 0.05 ≤ 0.05 < 0.05 | (AGR/26/03) |

| CROP Country, Year Location (variety) | Application | | | | | PHI days | Commodity | Residues, mg/kg boscalid | Ref. Reg.DocID. (Trial No.) |
|---|---|---|---|---|---|---|---|---|---|
| | Formulation | kg ai/ha | kg ai/hL | Water L/ha | No. | | | | |
| France, 2003 Alsace (Aviso) | WG | 0.267 | 0.089 | 300 | 3 | 0 6 13 21 | cauliflower | < 0.05 < 0.05 ≤ 0.05 < 0.05 | (FBD/24/03) |
| UK, 2003 Bicester (Thallasa) | WG | 0.267 | 0.089 | 300 | 3 | 0 7 13 20 | cauliflower | 1.60 1.10 0.36 0.55 | (OAT/22/03) |
| Germany, 2004 Rheinland-Pfalz (Lecano) | WG | 0.267 | 0.089 | 300 | 3 | 0 7 15 21 | cauliflower | < 0.05 < 0.05 ≤ 0.05 < 0.05 | 2004/7007476 (DU4/04/04) |
| France, 2004 Alsace (Fremont) | WG | 0.267 | 0.089 | 300 | 3 | 0 7 14 21 | cauliflower | 0.55 0.17 < 0.05 0.06 | (FAN/09/04) |
| France, 2004 Midi-Pyrenees (Fremont) | WG | 0.267 | 0.089 | 300 | 3 | 0 6 13 21 | cauliflower | < 0.05 0.10 ≤ 0.05 < 0.05 | (FTL/11/04) |

Table 28. Results of residue trials with foliar treatment of boscalid conducted in Brussels sprouts (Klimmek and Schulz, 2004, 2004/1015912; Johnston, 2005, 2004/7007478).

| CROP Country, Year Location (variety) | Application | | | | | PHI days | Commodity | Residues, mg/kg boscalid | Ref. Reg.DocID. |
|---|---|---|---|---|---|---|---|---|---|
| | Formulation | kg ai/ha | kg ai/hL | Water L/ha | No. | | | | |
| Germany, 2003 Brandenburg (Ikarus) | WG | 0.267 | 0.089 | 300 | 3 | 0 8 13 20 | sprouts | 0.17 0.25 0.14 0.34 | 2004/1015912 |
| Denmark, 2003 Fuenen (Maximus) | WG | 0.267 | 0.089 | 300 | 3 | 0 7 14 21 | sprouts | 0.31 0.08 0.10 0.07 | |
| UK, 2003 Gloucestershire (Helemus) | WG | 0.267 | 0.089 | 300 | 3 | 0 7 14 21 | sprouts | < 0.05 < 0.05 ≤ 0.05 < 0.05 | |
| Netherlands, 2003 Limburg (Veloce) | WG | 0.267 | 0.089 | 300 | 3 | 0 6 13 20 | sprouts | 0.26 0.21 0.12 0.23 | |
| Sweden, 2003 Malmoe (Stallion) | WG | 0.267 | 0.089 | 300 | 3 | 0 8 15 22 | sprouts | 0.23 0.05 0.15 0.07 | |

| CROP Country, Year *Location* (variety) | Application | | | | PHI days | Commodity | Residues, mg/kg boscalid | Ref. Reg.DocID. |
|---|---|---|---|---|---|---|---|---|
| | Formulation | kg ai/ha | kg ai/hL | Water L/ha | No. | | | |
| Germany, 2004 *Baden-Württemberg* (Genius) | WG | 0.267 | 0.089 | 300 | 3 | 0<br>7<br>14<br>21 | sprouts | 0.50<br>0.41<br><u>0.40</u><br>0.27 | 2004/7007478 |
| France, 2004 *Alsace* (Cyrrus) | WG | 0.267 | 0.089 | 300 | 3 | 0<br>7<br>13<br>21 | sprouts | 0.18<br>0.16<br><u>0.16</u><br>0.14 | |
| UK, 2004 *Gloucestershire* (Genius) | WG | 0.267 | 0.089 | 300 | 3 | 0<br>7<br>13<br>21 | sprouts | 0.13<br>0.06<br><u>< 0.05</u><br>< 0.05 | |
| Sweden, 2004 *Malmoe* (Cyrrus) | WG | 0.267 | 0.089 | 300 | 3 | 0<br>7<br>15<br>21 | sprouts | 0.22<br>0.08<br>< 0.05<br><u>0.06</u> | |

Table 29. Results of residue trials with foliar treatment of boscalid conducted in cucurbits (Haughey and Abdel-Baky, 2001, 2001/5002593; Leonard, 2005, 2005/5000021; Heck *et al.*, 1999, 2000/1014874; Schroth, 2000, 2001/1009069).

| CROP Country, Year *Location* (variety) | Application | | | | PHI days | Commodity | Residues, mg/kg boscalid[1] | Ref. Reg.DocID. |
|---|---|---|---|---|---|---|---|---|
| | Formulation | kg ai/ha | kg ai/hL | Water L/ha | No. | | | |
| CUCUMBER | | | | | | | | |
| Japan, 2002 Gunma (Ona) | WG | 1.25 | 0.05 | 2500 | 3 | 1<br>3<br>7 | fruit | 1.00<br>0.56<br>0.26 | na |
| Japan, 2002 Chiba (Sharp 1) | WG | 1.00 | 0.05 | 2000 | 3 | 1<br>3<br>7 | fruit | 2.10<br>1.04<br>0.52 | na |
| USA, 2001 *NC, EPA region 2* (National Pickling) | WG | 0.35 | 0.12 | 290 | 4 | 0 | fruit | <u>0.14</u> | 2001/5002593 (2001262) |
| *GA, EPA region 2* (Long green imp) | | 0.35 | 0.13 | 275 | 4 | 0 | fruit | <u>0.14</u> | (2001263) |
| *FL, EPA region 3* (Poinsett 76) | | 0.35 | 0.12 | 290 | 4 | 0 | fruit | <u>0.13</u> | (2001264) |
| *MI, EPA region 5* (Marketmore) | | 0.35 | 0.18 | 200 | 4 | 0 | fruit | <u>0.07</u> | (2001265) |
| *WI, EPA region 5* (Eureka Hybrid) | | 0.35 | 0.18 | 200 | 4 | 0 | fruit | <u>0.05</u> | (2001266) |
| *TX, EPA region 6* (Straight 8) | | 0.35 | 0.09 | 400 | 4 | 0 | fruit | <u>0.07</u> | (2001267) |
| USA, 2004 *GA, EPA region 2* (Lightning) | WG | 0.34 | 0.13 | 260 | 4 | 0 | fruit | <u>0.12</u> | 2005/5000021 (2004123) |
| *CA, EPA region 10* (Poinsett 76 Armenian) | | 0.34 | 0.12 | 280 | 4 | 0 | fruit | <u>0.26</u> | (2004124) |
| *OR, EPA region 12* (Daytona) | | 0.34 | 0.12 | 280 | 4 | 0 | fruit | <u>0.07</u> | (2004125) |
| *QC, EPA region 5b* (Marketmore) | | 0.32-0.35 | 0.13 | 250-280 | 4 | 0 | fruit | <u>0.31</u> | (2004126) |

| CROP Country, Year *Location* (variety) | Application | | | | PHI days | Commodity | Residues, mg/kg boscalid[1] | Ref. Reg.DocID. |
|---|---|---|---|---|---|---|---|---|
| | Formulation | kg ai/ha | kg ai/hL | Water L/ha | No. | | | |
| **SUMMER SQUASH** | | | | | | | | |
| USA, 2001 *PA, EPA region 1* (Sunray) | WG | 0.35 | 0.18 | 200 | 4 | 0 | fruit | 0.12 | 2001/5002593 (2001274) |
| *NC, EPA region 2* (Straight neck early prolific) | | 0.34 | 0.12 | 280 | 4 | 0 | fruit | 0.14 | (2001275) |
| *FL, EPA region 3* (Yellow summer crookneck) | | 0.35 | 0.13 | 280 | 4 | 0 | fruit | 0.16 | (2001276) |
| *MI, EPA region 5* (Zucchini Elite) | | 0.34 | 0.15 | 220 | 4 | 0 | fruit | 0.31 | (2001277) |
| *CA, EPA region 10* (Sundance) | | 0.35 | 0.13 | 280 | 4 | 0 | fruit | 0.95 | (2001278) |
| USA, 2004 *MN, EPA region 5* (Monet) | WG | 0.34 | 0.14 | 235 | 4 | 0 | fruit | 0.11 | 2005/5000021 (2004129) |
| *QC, EPA region 5b* (Spineless Beauty) | | 0.34 | 0.12 | 280 | 4 | 0 | fruit | 0.27 | (2004130) |
| *OR, EPA region 12* (Golden Zucchini) | | 0.32-0.37 | 0.13 | 260-290 | 4 | 0 | fruit | 0.16 | (2004131) |
| *Prince Edw.Island EPA region 1A* (Rich Green) | | 0.34 | 0.12 | 280 | 4 | 0 | fruit | 0.19 | (2004132) |
| **MELON** | | | | | | | | |
| Japan, 2005 Shizuoka (Arus Miyabi) | WG | 3.0 | 0.050 | 6000 | 3 | 1 3 7 | fruit except peel | 0.03 0.02 0.02 | na |
| Japan, 2005 Ishikawa (Night Natsukei 2 go) | WG | 1.25 | 0.050 | 250 | 3 | 1 3 7 | fruit except peel | 0.006 0.006 < 0.005 | na |
| **(CANTALOUPE)** | | | | | | | | |
| USA, 2001 *GA, EPA region 2* (Edisto 47) | WG | 0.34 | 0.13 | 270 | 4 | 0 | fruit | 0.29 | 2001/5002593 (2001268) |
| *MI, EPA region 5* (Fire) | | 0.34 | 0.17 | 200 | 4 | 0 | fruit | 0.56 | (2001269) |
| *TX, EPA region 6* (Jumbo Hale´s best) | | 0.34 | 0.085 | 400 | 4 | 0 | fruit | 0.23 | (2001270) |
| *CA, EPA region 10* (Magnum PMR .45) | | 0.34 | 0.12 | 280 | 4 | 0 | fruit | 1.27 | (2001271) |
| *CA, EPA region 10* (Magnum .45) | | 0.34 | 0.12 | 280 | 4 | 0 | fruit | 0.39 | (2001272) |
| *CA, EPA region 10* (Mark) | | 0.34 | 0.12 | 280 | 4 | 0 | fruit | 0.71 | (2001273) |
| USA, 2004 *QC, EPA region 5b* (Athena) | WG | 0.32-0.34 | 0.13 | 240-260 | 4 | 0 | fruit | 0.57 | 2005/5000021 (2004127) |
| *MN, EPA region 5* (Primo) | | 0.32-0.34 | 0.13 | 240-260 | 4 | 0 | fruit | 0.14 | (2004128) |
| Spain, 1999 *Andalusia* (Cantaloupe) | SC | 0.100 | 0.010 | 1000 | 4 | 0 4 7 13 | fruit | < 0.05 ≤ 0.05 < 0.05 < 0.05 | 2000/1014874 (AC/21/99) |

| CROP<br>Country, Year<br>*Location*<br>(variety) | Application | | | | | PHI<br>days | Commodity | Residues,<br>mg/kg<br>boscalid[1] | Ref.<br>Reg.DocID. |
|---|---|---|---|---|---|---|---|---|---|
| | Formulation | kg ai/ha | kg ai/hL | Water<br>L/ha | No. | | | | |
| Spain, 1999<br><br>(Sirius) | SC | 0.100 | 0.010 | 1000 | 4 | 0<br>4<br>7<br>13 | fruit | < 0.05<br>≤ 0.05<br>< 0.05<br>< 0.05 | (AC/22/99) |
| Spain, 1999<br>*Andalusia*<br>(Makdimón F1) | SC | 0.100 | 0.010 | 1000 | 4 | 0<br>2<br>6<br>13 | fruit | < 0.05<br>≤ 0.05<br>< 0.05<br>< 0.05 | (ALO/13/99) |
| Spain, 1999<br>*Andalusia*<br>(Makdimón F1) | SC | 0.100 | 0.010 | 1000 | 4 | 0<br>3<br>7<br>14 | fruit | < 0.05<br>≤ 0.05<br>< 0.05<br>< 0.05 | (ALO/14/99) |
| Spain, 2000<br>*Andalucía*<br>(Sancho) | SC | 0.100 | 0.010 | 1000 | 4 | 0<br>3<br>3<br>3<br>7<br>14 | fruit<br>fruit<br>peel<br>flesh<br>fruit<br>fruit | 0.064<br>< 0.05<br>0.066<br>≤ 0.05<br>< 0.05<br>< 0.05 | 2001/1009069<br>(00S008R) |
| Spain, 2000<br>*Andalucía*<br>(Makdimon) | SC | 0.100 | 0.010 | 1000 | 4 | 0<br>3<br>3<br>3<br>7<br>14 | fruit<br>fruit<br>peel<br>flesh<br>fruit<br>fruit | 0.059<br>0.061<br>0.106<br>≤ 0.05<br>0.064<br>0.072 | (00S009R) |
| Italy, 2000<br>*Lombardia*<br>(Supermarket) | SC | 0.100 | 0.010 | 1000 | 4 | 0<br>3<br>3<br>3<br>7<br>14 | fruit<br>fruit<br>peel<br>flesh<br>fruit<br>fruit | < 0.05<br>< 0.05<br>< 0.05<br>≤ 0.05<br>< 0.05<br>< 0.05 | (00I010R) |
| Italy, 2000<br>*Piemonte*<br>(Supermarket) | SC | 0.100 | 0.010 | 1000 | 4 | 0<br>3<br>3<br>3<br>7<br>14 | fruit<br>fruit<br>peel<br>flesh<br>fruit<br>fruit | < 0.05<br>< 0.05<br>< 0.05<br>≤ 0.05<br>< 0.05<br>< 0.05 | (00I011R) |
| WATERMELON, INDOOR | | | | | | | | | |
| Japan, 2005<br>Ibaraki<br>(Benikodama) | WG | 1.5 | 0.050 | 3000 | 3 | 1<br>3<br>7 | fruit except peel | 0.02<br>0.04<br>0.04 | na |
| Japan, 2005<br>Isikawa<br>(Ajihimitsu) | WG | 1.0 | 0.050 | 2000 | 3 | 1<br>3<br>7 | fruit except peel | 0.04<br>0.02<br>0.02 | na |

na: not available

1) average values used for estimation of MRL.

Table 30. Results of residue trials with foliar treatment of boscalid conducted in fruiting vegetables (Haughey and Abdel-Baky, 2001, 2001/5000832; Johnston, 2005, 2005/5000016; Versoi and Abdel-Baky, 2000, 2000/5209).

| CROP<br>Country, Year<br>*Location*<br>(variety) | Application | | | | | PHI<br>days | Commodity | Residues,<br>mg/kg<br>boscalid[1] | Ref.<br>Reg.DocID. |
|---|---|---|---|---|---|---|---|---|---|
| | Formulation | kg ai/ha | kg ai/hL | Water<br>L/ha | No. | | | | |
| TOMATO | | | | | | | | | |
| Japan, 2002<br>Ibaraki<br>(House-Taro) | WG | 1.0 | 0.05 | 2000 | 3 | 1<br>3<br>7 | fruit | 0.85<br>0.51<br>0.65 | na |

| CROP Country, Year Location (variety) | Application | | | | | PHI days | Commodity | Residues, mg/kg boscalid[1] | Ref. Reg.DocID. |
|---|---|---|---|---|---|---|---|---|---|
| | Formulation | kg ai/ha | kg ai/hL | Water L/ha | No. | | | | |
| Japan, 2002 Kumamoto (House-Taro) | WG | 1.0 | 0.05 | 2000 | 3 | 1 3 7 | fruit | 1.09 0.56 0.34 | na |
| USA, 2000 PA, EPA region 1 (Mountain spring) | WG | 0.616 | 0.21 | 290 | 2 | 0 | fruit | 0.27 | 2001/5000832 (2000204) |
| NC, EPA region 2 (Homestead) | | 0.616 | 0.21 | 290 | 2 | 0 | fruit | 0.22 | (2000205) |
| FL, EPA region 3 (Sunny) | | 0.616 | 0.11 | 580 | 2 | 0 | fruit | 0.24 | (2000206) |
| FL, EPA region 3 (Asgrow FL-47) | | 0.616 | 0.11 | 580 | 2 | 0 | fruit | 0.21 | (2000207) |
| MI, EPA region 5 (Celebrity) | | 0.616 | 0.21 | 290 | 2 | 0 | fruit | 0.30 | (2000208) |
| CA, EPA region 10 (9557) | | 0.616 | 0.21 | 290 | 2 | 0 | fruit | 0.79 | (2000210) |
| CA, EPA region 10 (UF6203) | | 0.616 | 0.16 | 380 | 2 | 0 | fruit | 0.25 | (2000211) |
| CA, EPA region 10 (U 370) | | 0.616 | 0.13 | 470 | 2 | 0 | fruit | 0.59 | (2000212) |
| CA, EPA region 10 (La Roma Red) | | 0.616 | 0.13 | 470 | 2 | 0 | fruit | 0.17 | (2000213) |
| CA, EPA region 10 (3155) | | 0.616 | 0.27 | 230 | 2 | 0 | fruit | 0.92 | (2000214) |
| CA, EPA region 10 (Hypeel 108) | | 0.616 | 0.16 | 380 | 2 | 0 | fruit | 0.28 | (2000215) |
| CA, EPA region 10 (La Roma) | | 0.616 | 0.22 | 280 | 2 | 0 5 11 15 20 | fruit | 0.61 0.53 0.56 0.59 0.23 | (2000209) |
| USA, 2004 ND, EPA region 5 (Brandywine) | WG | 0.43 | 0.15 | 290 | 3 | 0 | fruit | 0.56 | 2005/5000016 (2004113) |
| ND, EPA region 5 (Early Girl Hybrid) | | 0.42 | 0.15 | 280 | 3 | 0 | fruit | 0.21 | (2004114) |
| MN, EPA region 5 (Sunbrite) | | 0.41 | 0.17 | 240 | 3 | 0 | fruit | 0.28 | (2004115) |
| MN, EPA region 5 (Sunbeam) | | 0.41 | 0.17 | 240 | 3 | 0 | fruit | 0.20 | (2004116) |
| ON, EPA region 5 (C337) | | 0.41 | 0.15 | 280 | 3 | 0 | fruit | 0.35 | (2004117) |
| ON, EPA region 5 (1069) | | 0.41 | 0.14 | 290 | 3 | 0 | fruit | 0.26 | (2004118) |
| QC, EPA region 5B (Pachero) | | 0.40 | 0.18 | 220 | 3 | 0 | fruit | 0.56 | (2004119) |
| EGGPANT | | | | | | | | | |
| Japan, 2002 Ibaraki (Senryo 2 go) | WG | 0.915 | 0.05 | 1830 | 3 | 1 3 7 | fruit | 0.61 0.45 0.11 | na |
| Japan, 2002 Nagano (Chikuyo) | WG | 1.0 | 0.05 | 2000 | 3 | 1 3 7 | fruit | 0.93 0.40 0.32 | na |
| PEPPER, non bell | | | | | | | | | |
| USA, 1999 OK, EPA region 8 (Big Chili) | WG | 0.168 | 0.086 | 130-260 | 6 | 0 | fruit | 0.83 | 2000/5209 (99349) |

| CROP Country, Year *Location* (variety) | Application | | | | | PHI days | Commodity | Residues, mg/kg boscalid[1] | Ref. Reg.DocID. |
|---|---|---|---|---|---|---|---|---|---|
| | Formulation | kg ai/ha | kg ai/hL | Water L/ha | No. | | | | |
| *TX, EPA region 8* (Anaheim TMR 23) | | 0.168 | 0.073 | 230 | 6 | 0 | fruit | 0.30 | (99350) |
| *NM, EPA region 9* (Sonora) | | 0.168 | 0.076 | 200-240 | 6 | 0 | fruit | 0.14 | (99351) |
| USA, 2004 *MN, EPA region 5* (Ancho Villa) | WG | 0.41 | 0.18 | 230 | 3 | 0 | fruit | 0.54 | 2005/5000016 (2004120) |
| PEPPER, bell | | | | | | | | | |
| USA, 1999 *GA, EPA region 2* (Camelot) | WG | 0.168 | 0.084 | 210-290 | 6 | 0 | fruit | 0.09 | 2000/5209 (99345) |
| *FL, EPA region 3* (California Wonder) | | 0.168 | 0.059 | 280-290 | 6 | 0 | fruit | 0.16 | (99346) |
| *IA, EPA region 5* (California Wonder) | | 0.168 | 0.084 | 160-240 | 6 | 0 | fruit | ≤ 0.05 | (99347) |
| *OK, EPA region 6* (California Wonder) | | 0.160 | 0.078 | 150-260 | 6 | 0 | fruit | 0.14 | (99348) |
| *CA, EPA region 10* (Torres) | | 0.168 | 0.044 | 380 | 6 | 0 | fruit | 0.30 | (99352) |
| *CA, EPA region 10* (Torres) | | 0.168 | 0.044 | 380 | 6 | 0 | fruit | 0.08 | (99353) |
| USA, 2004 *ON, EPA region 5* (Revolution) | WG | 0.41-0.51 | 0.15 | 290-340 | 3 | 0 | fruit | 0.18 | 2005/5000016 (2004121) |
| *QC, EPA region 5B* (Crusader) | | 0.38-0.42 | 0.19 | 220-245 | 3 | 0 | fruit | 1.35 | (2004122) |

na: not available.

1) average values used for estimation of MRL.

Table 31. Results of residue trials with foliar treatment of boscalid conducted in mustard greens (Wofford and Abdel-Baky, 2001, 2001/5003339; Leonard, 2005, 2005/7002859).

| CROP Country, Year *Location* (variety) | Application | | | | | PHI days | Commodity | Residues, mg/kg boscalid[1] | Ref. Reg.DocID. (Trial No.) |
|---|---|---|---|---|---|---|---|---|---|
| | Formulation | kg ai/ha | kg ai/hL | Water L/ha | No. | | | | |
| USA, 2001 *NC, EPA region 2* (Florida Broadleaf) | WG | 0.45 | 0.24 | 190 | 2 | 0<br>3<br>7<br>10<br>14 | green leaves | 25.9<br>17.3<br>17.9<br>13.1<br>12.9 | 2001/5003339 (2001307) |
| *MS, EPA region 4* (Florida Broadleaf) | WG | 0.45 | 0.38 | 120 | 2 | 0<br>3<br>7<br>10<br>14 | green leaves | 49.0<br>25.5<br>20.3<br>10.8<br>14.4 | (2001308) |
| *WI, EPA region 5* (Florida Broadleaf India Mustard) | WG | 0.45 | 0.24 | 190 | 2 | 0<br>3<br>7<br>10<br>14 | green leaves | 36.0<br>2.90<br>0.96<br>0.44<br>0.54 | (2001309) |

| CROP Country, Year *Location* (variety) | Application | | | | | PHI days | Commodity | Residues, mg/kg boscalid[1] | Ref. Reg.DocID. (Trial No.) |
|---|---|---|---|---|---|---|---|---|---|
| | Formulation | kg ai/ha | kg ai/hL | Water L/ha | No. | | | | |
| *TX, EPA region 6* (Southern Giant Curled Mustard) | WG | 0.45 | 0.16 | 290 | 2 | 0<br>3<br>7<br>10<br>14 | green leaves | 21.5<br>6.64<br>5.48<br>4.00<br>2.80 | (2001310) |
| *CA, EPA region 10* (Florida Broadleaf) | WG | 0.45 | 0.16 | 280 | 2 | 0<br>3<br>7<br>10<br>14 | green leaves | 41.1<br>24.9<br>12.6<br>19.5<br>6.04 | (2001311) |
| USA, 2004/05 *GA, EPA region 2* (Florida Broadleaf) | | 0.45 | 0.24 | 188 | 2 | 0 | green leaves | 15.6 | 2005/7002859 (2004110) |
| *FL, EPA region 3* (Southern Giant Curled Mustard) | | 0.45 | 0.28 | 158 | 2 | 0 | green leaves | 16.0 | (2004111) |
| *CA, EPA region 10* (Florida Broadleaf) | | 0.45 | 0.19 | 235 | 2 | 0 | green leaves | 28.1 | (2004112) |
| *GA, EPA region 2* (Southern Giant Curled Mustard) | | 0.45 | 0.35 | 130 | 2 | 0<br>3<br>7<br>14 | green leaves | 12.7<br>10.8<br>5.17<br>0.92 | (R05021) |
| *FL, EPA region 3* (Southern Giant Curled Mustard) | | 0.45 | 0.38 | 120 | 2 | 0<br>3<br>7<br>14 | green leaves | 12.7<br>1.63<br>1.31<br>0.45 | (R05022) |
| *CA, EPA region 10* (Florida Broadleaf) | | 0.45 | 0.15 | 300 | 2 | 3<br>7<br>15 | green leaves | 7.52<br>4.69<br>3.10 | (R05023) |

1) average values used for estimation of MRL.

Table 32. Results of residue trials with foliar treatment of boscalid conducted in lettuce (Haughey and Abdel-Baky, 2001, 2001/5000051; Beck *et al.*, 2001, 2001/1000998; Schulz, 2001, 2001/1000933; Beck *et al.*, 2001, 2001/1000999; Young and Atkinson, 2003, 2003/1001259).

| CROP Country, Year *Location* (variety) | Application | | | | | PHI days | Commodity | Residues, mg/kg boscalid[1] | Ref. Reg.DocID. |
|---|---|---|---|---|---|---|---|---|---|
| | Formulation | kg ai/ha | kg ai/hL | Water L/ha | No. | | | | |
| HEAD LETTUCE with wrapper leaves | | | | | | | | | |
| USA, 2000 *NC, EPA region 2* (Crisp head) | WG | 0.56 | 0.20 | 280 | 2 | 15 | head | 1.77 | 2001/5000051 (2000132) |
| *FL, EPA region 3* (Great Lakes) | WG | 0.56 | 0.20 | 280 | 2 | 13 | head | 2.73 | (2000133) |
| *CA, EPA region 10* (Empire) | WG | 0.56 | 0.15 | 380 | 2 | 14 | head | 0.98 | (2000135) |
| *CA, EPA region 10* (Salinas) | WG | 0.56 | 0.15 | 380 | 2 | 14 | head | 2.68 | (2000136) |
| *CA, EPA region 10* (Great Lakes) | WG | 0.56 | 0.12 | 470 | 2 | 14 | head | 5.42 | (2000137) |
| *AZ, EPA region 10* (Green Lightning) | WG | 0.56 | 0.20 | 280 | 2 | 14 | head | 3.18 | (2000138) |

| CROP Country, Year Location (variety) | Application | | | | | PHI days | Commodity | Residues, mg/kg boscalid[1] | Ref. Reg.DocID. |
|---|---|---|---|---|---|---|---|---|---|
| | Formulation | kg ai/ha | kg ai/hL | Water L/ha | No. | | | | |
| AZ, EPA region 10 (Diamond) | WG | 0.56 | 0.20 | 280 | 2 | 14 | head | <u>2.53</u> | (2000139) |
| CA, EPA region 10 (Bayview) | WG | 0.56 | 0.20 | 280 | 2 | 0 7 14 21 28 | head | 5.85 0.40 <u>0.11</u> 0.07 < 0.05 | (2000134) |
| HEAD LETTUCE without wrapper leaves | | | | | | | | | |
| USA, 2000 NC, EPA region 2 (Crisp head) | WG | 0.56 | 0.20 | 280 | 2 | 15 | head | 0.93 | 2001/5000051 (2000132) |
| FL, EPA region 3 (Great Lakes) | WG | 0.56 | 0.20 | 280 | 2 | 13 | head | 0.42 | (2000133) |
| CA, EPA region 10 (Empire) | WG | 0.56 | 0.15 | 380 | 2 | 14 | head | 0.17 | (2000135) |
| CA, EPA region 10 (Salinas) | WG | 0.56 | 0.15 | 380 | 2 | 14 | head | 0.41 | (2000136) |
| CA, EPA region 10 (Great Lakes) | WG | 0.56 | 0.12 | 470 | 2 | 14 | head | < 0.05 | (2000137) |
| AZ, EPA region 10 (Green Lightning) | WG | 0.56 | 0.20 | 280 | 2 | 14 | head | 0.72 | (2000138) |
| AZ, EPA region 10 (Diamond) | WG | 0.56 | 0.20 | 280 | 2 | 14 | head | 0.27 | (2000139) |
| AZ, EPA region 10 (Bayview) | WG | 0.56 | 0.20 | 280 | 2 | 0 7 14 21 28 | head | 0.32 0.09 < 0.05 < 0.05 < 0.05 | (2000134) |
| LEAF LETTUCE | | | | | | | | | |
| Japan, 2005 Hyogo (Smily) | WG | 1.0 | 0.050 | 2000 | 1 | 14 21 28 | leaves | 0.87 < 0.05 < 0.05 | na |
| Japan, 2005 Wakayama (Shizuka) | WG | 1.0 | 0.050 | 2000 | 1 | 14 21 28 | leaves | 0.89 2.29 0.16 | na |
| USA, 2000 NC, EPA region 2 (Salad Bowl) | WG | 0.56 | 0.20 | 280 | 2 | 14 | leaves | <u>9.55</u> | 2001/5000051 (2000140) |
| FL, EPA region 3 (Bibb) | WG | 0.56 | 0.20 | 280 | 2 | 13 | leaves | <u>0.74</u> | (2000141) |
| CA, EPA region 10 (Waldmans Green) | WG | 0.56 | 0.15 | 380 | 2 | 14 | leaves | <u>5.14</u> | (2000143) |
| CA, EPA region 10 (Red Salad Bowl) | WG | 0.56 | 0.15 | 380 | 2 | 14 | leaves | <u>1.60</u> | (2000144) |
| CA, EPA region 10 (Waldmans Green) | WG | 0.56 | 0.12 | 470 | 2 | 14 | leaves | <u>1.91</u> | (2000145) |
| AZ, EPA region 10 (Paris Island) | WG | 0.56 | 0.20 | 280 | 2 | 14 | leaves | <u>4.87</u> | (2000146) |
| AZ, EPA region 10 (Shining Star) | WG | 0.56 | 0.20 | 280 | 2 | 14 | leaves | <u>9.36</u> | (2000147) |
| CA, EPA region 10 (New Red Fire) | WG | 0.56 | 0.20 | 280 | 2 | 0 7 14 21 28 | leaves | 17.0 10.1 <u>1.63</u> 0.33 0.08 | (2000142) |

| CROP Country, Year *Location* (variety) | Application | | | | | PHI days | Commodity | Residues, mg/kg boscalid[1] | Ref. Reg.DocID. |
|---|---|---|---|---|---|---|---|---|---|
| | Formulation | kg ai/ha | kg ai/hL | Water L/ha | No. | | | | |
| LETTUCE, EUROPE OUTDOOR | | | | | | | | | |
| Germany, 1999 *Brandenburg* (Enrica) | WG | 0.400 | 0.080 | 500 | 2 | 0 7 14 21 | head | 10.3 1.09 0.09 < 0.05 | 2001/1000998 (ACK/06/99) |
| Germany, 1999 *Baden-Württemberg* (Einstein) | WG | 0.400 | 0.080 | 500 | 2 | 0 7 14 21 | head | 9.64 0.87 0.38 0.10 | (DU2/11/99) |
| Germany, 1999 *Rheinland-Pfalz* (Einstein) | WG | 0.400 | 0.080 | 500 | 2 | 0 5 13 20 | head | 20.5 5.54 0.64 0.30 | (DU4/08/99) |
| Spain, 1999 *Andalucía* (Yerga) | WG | 0.400 | 0.080 | 500 | 2 | 0 7 13 20 | head | 7.74 2.79 0.45 0.11 | (AC/14/99) |
| Spain, 1999 *Andalucía* (Yerga) | WG | 0.400 | 0.080 | 500 | 2 | 0 7 13 20 | head | 8.38 3.74 0.36 0.17 | (AC/15/99) |
| Spain, 1999 *Andalucía* (Little gem) | WG | 0.400 | 0.080 | 500 | 2 | 0 7 13 20 | head | 12.6 2.45 0.73 < 0.05 | (AC/16/99) |
| France, 1999 *Côte dÓr* (Titan) | WG | 0.400 | 0.080 | 500 | 2 | 0 7 14 21 | head | 13.8 1.27 0.65 0.13 | (FR4/01/99) |
| Netherlands, 1999 *Limburg* (Enrica) | WG | 0.400 | 0.080 | 500 | 2 | 0 6 14 22 | head | 6.67 1.24 0.33 0.30 | (AGR/13/99) |
| France, 1999 *Provence Alpes-Côte dÁzur* (Santa Fe) | WG | 0.400 | 0.13 | 300 | 2 | 0 7 14 21 | head | 12.4 1.56 0.21 < 0.05 | 2001/1000933 (X 99 62 11) |
| France, 1999 *Aquitaine* (Nadine) | WG | 0.400 | 0.13 | 300 | 2 | 0 7 14 21 | head | 14.3 2.30 0.50 0.06 | (X 99 62 12) |
| Germany, 2000 *Brandenburg* (Enrica) | WG | 0.400 | 0.080 | 500 | 2 | 0 6 14 22 | head | 6.17 0.45 1.58 < 0.05 | 2001/1000999 (ACK/03/00) |
| Germany, 2000 *Baden-Württemberg* (Einstein) | WG | 0.400 | 0.080 | 500 | 2 | 0 7 14 21 | head | 8.17 0.62 ≤ 0.05 < 0.05 | (DU2/05/00) |
| Spain, 2000 *Andalucía* (Yerga) | WG | 0.400 | 0.080 | 500 | 2 | 0 6 14 20 | head | 14.2 0.62 0.09 0.86 | (AC/15/00) |
| Spain, 2000 *Andalucía* (Yerga) | WG | 0.400 | 0.080 | 500 | 2 | 0 6 14 21 | head | 6.44 0.83 0.17 0.076 | (AC/16/00) |

| CROP Country, Year *Location* (variety) | Application | | | | | PHI days | Commodity | Residues, mg/kg boscalid[1] | Ref. Reg.DocID. |
|---|---|---|---|---|---|---|---|---|---|
| | Formulation | kg ai/ha | kg ai/hL | Water L/ha | No. | | | | |
| France, 2000 *Languedoc-Roussillon* (Rosny) | WG | 0.400 | 0.080 | 500 | 2 | 0<br>7<br>14<br>21 | head | 9.20<br>1.47<br>1.19<br>0.76 | (FR3/06/00) |
| France, 2000 *Bourgogne* (Titan) | WG | 0.400 | 0.080 | 500 | 2 | 0<br>7<br>13<br>20 | head | 6.38<br>0.28<br>0.15<br>< 0.05 | (FR4/06/00) |
| France, 2000 *Midi-Pyrénées* (Elselsa) | WG | 0.400 | 0.080 | 500 | 2 | 0<br>7<br>14<br>21 | head | 5.16<br>1.41<br>0.43<br>0.41 | (FR8/05/00) |
| Netherlands, 2000 *Limburg* (Flandria) | WG | 0.400 | 0.080 | 500 | 2 | 0<br>7<br>13<br>20 | head | 6.64<br>0.82<br>0.39<br>0.12 | (FR4/06/00) |
| LETTUCE, EUROPE INDOOR | | | | | | | | | |
| Germany, 2002 *Brandenburg* (Nadine) | WG | 0.400 | 0.080 | 500 | 2 | 0<br>7<br>13<br>22 | head | 29.3<br>2.51<br>0.72<br>< 0.05 | 2003/1001259 (ACK/03/02) |
| Netherlands, 2002 *Limburg* (Peter) | WG | 0.400 | 0.080 | 500 | 2 | 0<br>6<br>14<br>20 | head | 21.3<br>10.9<br>6.11<br>2.06 | (AGR/08/02) |
| Spain, 2002 *Andalucía* (Filipus) | WG | 0.400 | 0.080 | 500 | 2 | 0<br>7<br>14<br>21 | head | 14.9<br>4.18<br>5.96<br>3.21 | (ALO/04/02) |
| Spain, 2002 *Andalucía* (Candela) | WG | 0.400 | 0.080 | 500 | 2 | 0<br>7<br>14<br>21 | head | 29.9<br>3.46<br>0.37<br>0.05 | (AYE/03/02) |
| France, 2002 *Alsace* (Pantheon) | WG | 0.400 | 0.080 | 500 | 2 | 0<br>7<br>14<br>21 | head | 14.7<br>4.60<br>2.50<br>1.12 | (FAN/04/02) |
| France, 2002 *Rhone-Alps* (Grenobloise) | WG | 0.400 | 0.080 | 500 | 2 | 0<br>7<br>14<br>21 | head | 21.0<br>7.84<br>2.32<br>0.84 | (FBD/04/02) |
| France, 2002 *Pays del la Loire* (Angie) | WG | 0.400 | 0.080 | 500 | 2 | 0<br>7<br>14<br>21 | head | 9.92<br>9.73<br>5.63<br>4.81 | (FBM/02/02) |
| France, 2002 *Midi-Pyrénées* (Alexandria) | WG | 0.400 | 0.080 | 500 | 2 | 0<br>7<br>14<br>21 | head | 16.6<br>4.02<br>1.52<br>0.45 | (FTL/21/02) |

1) average values used for estimation of MRL.

Table 33. Results of residue trials with foliar treatment of boscalid conducted in curly kale (green cabbage) (Beck *et al.*, 2001, 2001/1001000; Beck and Benz, 2001, 2001/1001001).

| CROP Country, Year *Location* (variety) | Application | | | | | PHI days | Commodity | Residues, mg/kg boscalid | Ref. Reg.DocID. |
|---|---|---|---|---|---|---|---|---|---|
| | Formulation | kg ai/ha | kg ai/hL | Water L/ha | No. | | | | |
| Denmark, 2000 *Fuenen* (Winterbor) | WG | 0.400 | 0.13 | 300 | 4 | 0 | leaves | 7.04 | 2001/1001000 |
| | | | | | | 7 | | 7.77 | |
| | | | | | | 14 | | 0.67 | |
| | | | | | | 20 | | 1.09 | |
| UK, 2000 *Warwickshire* (Buffalo) | WG | 0.400 | 0.13 | 300 | 4 | 0 | leaves | 6.63 | |
| | | | | | | 8 | | 1.21 | |
| | | | | | | 14 | | 0.50 | |
| | | | | | | 21 | | 0.69 | |
| Netherlands, 1999 *Limburg* (Buffalo) | WG | 0.400 | 0.13 | 300 | 4 | 0 | leaves | 10.82 | 2001/1001001 |
| | | | | | | 7 | | 0.14 | |
| | | | | | | 15 | | 2.80 | |
| | | | | | | 22 | | 1.26 | |
| UK, 1999 *Lincolnshire* (Krypton) | WG | 0.400 | 0.13 | 300 | 4 | 0 | leaves | 0.34 | |
| | | | | | | 6 | | < 0.05 | |
| | | | | | | 13 | | 0.11 | |
| | | | | | | 21 | | 0.06 | |
| UK, 1999 *Worcestershire* (Winterbor) | WG | 0.400 | 0.13 | 300 | 4 | 0 | leaves | 3.33 | |
| | | | | | | 7 | | 0.91 | |
| | | | | | | 13 | | 0.55 | |
| | | | | | | 21 | | 0.43 | |
| Sweden, 1999 *Malmoe* (Arsis) | WG | 0.400 | 0.13 | 300 | 4 | 0 | leaves | 7.98 | |
| | | | | | | 7 | | 3.30 | |
| | | | | | | 14 | | 3.20 | |
| | | | | | | 21 | | 4.06 | |

Table 34. Results of residue trials with foliar treatment of boscalid conducted in bean (Treiber *et al.*, 2000, 2000/1014846; Treiber *et al.*, 2001, 2000/1014850; Perny, 2001, 2000/1014876; Treiber *et al.*, 2001, 2000/1014849).

| CROP Country, Year *Location* (variety) | Application | | | | | PHI days | Commodity | Residues, mg/kg boscalid | Ref. Reg.DocID. |
|---|---|---|---|---|---|---|---|---|---|
| | Formulation | kg ai/ha | kg ai/hL | Water L/ha | No. | | | | |
| BEAN (outdoor) | | | | | | | | | |
| D – Germany 1999 (Berggold) | WG | 0.500 | 0.17 | 300 | 3 | 0 | pl. w/o root | 20.2 | 2000/1014846 (ACK/03/99) |
| | | | | | | 3 | pods with seed | 0.71 | |
| | | | | | | 3 | shoots | 43.0 | |
| | | | | | | 7 | pods with seed | <u>0.47</u> | |
| | | | | | | 7 | shoots | 29.0 | |
| | | | | | | 15 | pods with seed | 0.47 | |
| | | | | | | 15 | shoots | 27.4 | |
| D – Germany 1999 (Primel) | WG | 0.500 | 0.17 | 300 | 3 | 0 | pl. w/o root | 14.0 | (DU2/07/99) |
| | | | | | | 3 | pods with seed | 0.63 | |
| | | | | | | 3 | shoots | 28.5 | |
| | | | | | | 7 | pods with seed | <u>0.26</u> | |
| | | | | | | 7 | shoots | 12.8 | |
| | | | | | | 14 | pods with seed | 0.18 | |
| | | | | | | 14 | shoots | 6.06 | |

| CROP Country, Year *Location* (variety) | Application | | | | | PHI days | Commodity | Residues, mg/kg boscalid | Ref. Reg.DocID. |
|---|---|---|---|---|---|---|---|---|---|
| | Formulation | kg ai/ha | kg ai/hL | Water L/ha | No. | | | | |
| D – Germany 1999 (Scuba) | WG | 0.500 | 0.17 | 300 | 3 | 0 | pl. w/o root | 14.5 | (D08/02/99) |
| | | | | | | 3 | pods with seed | 0.48 | |
| | | | | | | 3 | shoots | 20.4 | |
| | | | | | | 8 | pods with seed | 0.22 | |
| | | | | | | 8 | shoots | 22.2 | |
| | | | | | | 14 | pods with seed | 0.08 | |
| | | | | | | 14 | shoots | 6.50 | |
| DK – Denmark 1999 (Bonbon) | WG | 0.500 | 0.17 | 300 | 3 | 0 | pl. w/o root | 16.4 | (ALB/04/99) |
| | | | | | | 4 | pods with seed | 0.15 | |
| | | | | | | 4 | shoots | 10.5 | |
| | | | | | | 8 | pods with seed | 0.12 | |
| | | | | | | 8 | shoots | 11.4 | |
| | | | | | | 15 | pods with seed | 0.13 | |
| | | | | | | 15 | shoots | 6.75 | |
| D – Germany 2000 (Paulista) | WG | 0.500 | 0.17 | 300 | 3 | 0 | pl. w/o root | 30.1 | 2000/1014850 (DU2/10/00) |
| | | | | | | 3 | pods with seed | 0.89 | |
| | | | | | | 3 | shoots | 43.2 | |
| | | | | | | 8 | pods with seed | 0.67 | |
| | | | | | | 8 | shoots | 45.6 | |
| | | | | | | 14 | pods with seed | 0.53 | |
| | | | | | | 14 | shoots | 44.5 | |
| DK – Denmark 2000 (Bon-Bon) | WG | 0.500 | 0.17 | 300 | 3 | 0 | pl. w/o root | 22.9 | (ALB/02/00) |
| | | | | | | 3 | pods with seed | 0.84 | |
| | | | | | | 3 | shoots | 23.5 | |
| | | | | | | 7 | pods with seed | 0.50 | |
| | | | | | | 7 | shoots | 19.8 | |
| | | | | | | 14 | pods with seed | 0.28 | |
| | | | | | | 14 | shoots | 16.7 | |
| DK – Denmark 2000 (Bon-Bon) | WG | 0.500 | 0.17 | 300 | 3 | 0 | pl. w/o root | 24.7 | (ALB/03/00) |
| | | | | | | 3 | pods with seed | 0.81 | |
| | | | | | | 3 | shoots | 23.7 | |
| | | | | | | 7 | pods with seed | 0.83 | |
| | | | | | | 7 | shoots | 29.7 | |
| | | | | | | 14 | pods with seed | 0.43 | |
| | | | | | | 14 | shoots | 22.5 | |
| F – France 2000 (Flagrano) | WG | 0.500 | 0.17 | 300 | 3 | 0 | pl. w/o root | 18.1 | (FR2/02/00) |
| | | | | | | 4 | pods with seed | 0.53 | |
| | | | | | | 4 | shoots | 17.7 | |
| | | | | | | 7 | pods with seed | 0.26 | |
| | | | | | | 7 | shoots | 20.1 | |
| | | | | | | 14 | pods with seed | 0.29 | |
| | | | | | | 14 | shoots | 16.5 | |
| F – France 2000 (Masaï) | WG | 0.500 | 0.17 | 300 | 3 | 0 | pl. w/o root | 18.4 | 2000/1014876 (A0033 AN1) |
| | | | | | | 3 | pods with seed | 0.58 | |
| | | | | | | 3 | shoots | < 0.05 | |
| | | | | | | 7 | pods with seed | 0.53 | |
| | | | | | | 7 | shoots | 21.5 | |
| | | | | | | 14 | pods with seed | 0.31 | |
| | | | | | | 14 | shoots | 15.4 | |
| F – France 2000 (Booster) | WG | 0.500 | 0.17 | 300 | 3 | 0 | pl. w/o root | 13.9 | (A0033 TL2) |
| | | | | | | 3 | pods with seed | 0.71 | |
| | | | | | | 3 | shoots | 30.3 | |
| | | | | | | 7 | pods with seed | 0.95 | |
| | | | | | | 7 | shoots | 22.6 | |
| | | | | | | 15 | pods with seed | 0.65 | |
| | | | | | | 15 | shoots | 19.9 | |

| CROP Country, Year *Location* (variety) | Application | | | | | PHI days | Commodity | Residues, mg/kg boscalid | Ref. Reg.DocID. |
|---|---|---|---|---|---|---|---|---|---|
| | Formulation | kg ai/ha | kg ai/hL | Water L/ha | No. | | | | |
| F – France 2000 (Pissos) | WG | 0.500 | 0.17 | 300 | 3 | 0 | pl. w/o root | 10.6 | (A0033 SA1) |
| | | | | | | 3 | pods with seed | 0.74 | |
| | | | | | | 3 | shoots | 8.71 | |
| | | | | | | 6 | pods with seed | 0.62 | |
| | | | | | | 6 | shoots | 7.73 | |
| | | | | | | 13 | pods with seed | 0.29 | |
| | | | | | | 13 | shoots | 2.72 | |
| BEAN, (indoor) | | | | | | | | | |
| E – Spain 1999 (Primel) | WG | 0.500 | 0.17 | 300 | 3 | 0 | pl. w/o root | 78.3 | 2000/1014847 (AC/11/99) |
| | | | | | | 2 | pods with seed | 3.10 | |
| | | | | | | 2 | shoots | 51.8 | |
| | | | | | | 7 | pods with seed | 0.69 | |
| | | | | | | 7 | shoots | 30.3 | |
| | | | | | | 14 | pods with seed | 0.06 | |
| | | | | | | 14 | shoots | 18.5 | |
| E – Spain 1999 (Primel) | WG | 0.500 | 0.17 | 300 | 3 | 0 | pl. w/o root | 40.1 | (AC/12/99) |
| | | | | | | 2 | pods with seed | 2.63 | |
| | | | | | | 2 | shoots | 27.9 | |
| | | | | | | 8 | pods with seed | 0.29 | |
| | | | | | | 8 | shoots | 36.3 | |
| | | | | | | 15 | pods with seed | 0.06 | |
| | | | | | | 15 | shoots | 17.5 | |
| E – Spain 1999 (Festival RZ) | WG | 0.500 | 0.17 | 300 | 3 | 0 | pl. w/o root | 64.0 | (AC/13/99) |
| | | | | | | 3 | pods with seed | 0.49 | |
| | | | | | | 3 | shoots | 38.1 | |
| | | | | | | 7 | pods with seed | 0.28 | |
| | | | | | | 7 | shoots | 44.0 | |
| | | | | | | 14 | pods with seed | 0.14 | |
| | | | | | | 14 | shoots | 38.8 | |
| E – Spain 1999 (Emerite) | WG | 0.500 | 0.17 | 300 | 3 | 0 | pl. w/o root | 114.2 | (AC/23/99) |
| | | | | | | 3 | pods with seed | 0.56 | |
| | | | | | | 3 | shoots | 57.4 | |
| | | | | | | 7 | pods with seed | 0.28 | |
| | | | | | | 7 | shoots | 104.0 | |
| | | | | | | 14 | pods with seed | 0.10 | |
| | | | | | | 14 | shoots | 29.0 | |
| E – Spain 2000 (Primel) | WG | 0.500 | 0.17 | 300 | 3 | 0 | pl. w/o root | 34.6 | 2000/1014849 (AC/11/00) |
| | | | | | | 3 | pods with seed | 3.80 | |
| | | | | | | 3 | shoots | 83.3 | |
| | | | | | | 7 | pods with seed | 1.29 | |
| | | | | | | 7 | shoots | 71.8 | |
| | | | | | | 14 | pods with seed | 1.65 | |
| | | | | | | 14 | shoots | 121.3 | |
| E – Spain 2000 (Primel) | WG | 0.500 | 0.17 | 300 | 3 | 0 | pl. w/o root | 40.9 | (AC/12/00) |
| | | | | | | 3 | pods with seed | 1.46 | |
| | | | | | | 3 | shoots | 54.3 | |
| | | | | | | 7 | pods with seed | 1.67 | |
| | | | | | | 7 | shoots | 76.5 | |
| | | | | | | 14 | pods with seed | 1.13 | |
| | | | | | | 14 | shoots | 104.0 | |

| CROP Country, Year *Location* (variety) | Application | | | | | PHI days | Commodity | Residues, mg/kg boscalid | Ref. Reg.DocID. |
|---|---|---|---|---|---|---|---|---|---|
| | Formulation | kg ai/ha | kg ai/hL | Water L/ha | No. | | | | |
| E – Spain 2000 (Elda) | WG | 0.500 | 0.17 | 300 | 3 | 0 | pl. w/o root | 50.5 | (AC/13/00) |
| | | | | | | 3 | pods with seed | 0.95 | |
| | | | | | | 3 | shoots | 26.5 | |
| | | | | | | 8 | pods with seed | 0.06 | |
| | | | | | | 8 | shoots | 14.1 | |
| | | | | | | 14 | pods with seed | < 0.05 | |
| | | | | | | 14 | shoots | 15.1 | |
| E – Spain 2000 (Judia Dulce) | WG | 0.500 | 0.17 | 300 | 3 | 0 | pl. w/o root | 26.0 | (AC/14/00) |
| | | | | | | 3 | pods with seed | 0.96 | |
| | | | | | | 3 | shoots | 110.2 | |
| | | | | | | 7 | pods with seed | 0.61 | |
| | | | | | | 7 | shoots | 91.9 | |
| | | | | | | 14 | pods with seed | 0.46 | |
| | | | | | | 14 | shoots | 92.6 | |

Table 35. Results of residue trials with foliar treatment of boscalid conducted in peas (Haughey and Abdel-Baky, 2002, 2001/5003246).

| CROP Country, Year *Location* (variety) | Application | | | | | PHI days | Commodity | Residues, mg/kg boscalid[1] | Ref. Reg.DocID. |
|---|---|---|---|---|---|---|---|---|---|
| | Formulation | kg ai/ha | kg ai/hL | Water L/ha | No. | | | | |
| PEA, succulent, shelled | | | | | | | | | |
| USA, 2001 *PA, EPA region 1* (Wando) | WG | 0.57 | 0.17 | 340 | 2 | 7 | shelled pea | ≤ 0.05 | 2001/5003246 (2001252) |
| *NC, EPA region 2* (Green arrow) | | 0.56 | 0.20 | 280 | 2 | 7 | shelled pea | 0.06 | (2001253) |
| *GA, EPA region 2* (Progress No. 9) | | 0.57 | 0.57 | 100 | 2 | 7 | shelled pea | 0.24 | (2001254) |
| *MN, EPA region 5* (Knight) | | 0.56 | 0.33 | 170 | 2 | 7 | shelled pea | 0.15 | (2001256) |
| *MN, EPA region 5* (Top pod) | | 0.56 | 0.29 | 190 | 2 | 7 | shelled pea | 0.07 | (2001257) |
| *WI, EPA region 5* (Lazor) | | 0.57 | 0.30 | 190 | 2 | 8 | shelled pea | 0.37 | (2001258) |
| *WA, EPA region 11* (Case load) | | 0.56 | 0.29 | 190 | 2 | 7 | shelled pea | ≤ 0.05 | (2001260) |
| *QC, EPA region 5B* (XP 353) | | 0.57 | 0.24 | 240 | 2 | 7 | shelled pea | 0.19 | (2001338) |
| PEA, succulent, edible-podded | | | | | | | | | |
| USA, 2001 *MN, EPA region 5* (Casiadia) | WG | 0.57 | 0.36 | 160 | 2 | 6 | pods | 1.39 | 2001/5003246 (2001255) |
| *WA, EPA region 11* (Sugar snap) | | 0.56 | 0.29 | 190 | 2 | 7 | pods | 0.64 | (2001259) |
| *OR, EPA region 11* (SP704-3-8) | | 0.56 | 0.29 | 190 | 2 | 7 | pods | 0.97 | (2001261) |

1) average values used for estimation of MRL.

Table 36. Results of residue trials with foliar treatment of boscalid conducted in soybean (immature seed) (Leonard, 2003, 2002/5004272).

| CROP Country, Year Location (variety) | Application | | | | | PHI days | Commodity | Residues, mg/kg boscalid[1] | Ref. Reg.DocID. |
|---|---|---|---|---|---|---|---|---|---|
| | Formulation | kg ai/ha | kg ai/hL | Water L/ha | No. | | | | |
| USA, 2002 GA, EPA region 2 (NK RR S73-Z5) | WG | 0.56 | 0.18 | 270-370 | 2 | 5 14 | immature seed forage | ≤ 0.05 4.31 | 2002/5004272 (2002191) |
| VA, EPA region 2 (NK S53Q7 7B-1001) | WG | 0.58 | 0.19 | 310 | | 5 14 | immature seed forage | 0.06 4.48 | (2002192) |
| AR, EPA region 4 (AG4403) | WG | 0.56 | 0.27 | 210 | | 5 14 | immature seed forage | 1.18 16.0 | (2002193) |
| AR, EPA region 4 (AG5603) | WG | 0.56 | 0.20 | 280 | | 5 14 | immature seed forage | 0.05 2.98 | (2002194) |
| WI, EPA region 5 (BR2099RR) | WG | 0.56 | 0.23 | 240 | | 5 14 | immature seed forage | 0.08 1.62 | (2002195) |
| MN, EPA region 5 (BR2099RR) | WG | 0.56 | 0.23 | 240 | | 6 14 | immature seed forage | 0.20 6.59 | (2002196) |
| IA, EPA region 5 (SG 2531RR) | WG | 0.57 | 0.22 | 260 | | 5 14 | immature seed forage | ≤ 0.05 5.32 | (2002197) |
| IA, EPA region 5 (SG 2533RR) | WG | 0.57 | 0.22 | 260 | | 5 14 | immature seed forage | ≤ 0.05 1.18 | (2002198) |
| NE, EPA region 5 (Asgrow A2553) | WG | 0.56 | 0.29 | 190 | | 5 14 | immature seed forage | ≤ 0.05 6.67 | (2002199) |
| NE, EPA region 5 (Asgrow 2703) | WG | 0.56 | 0.29 | 190 | | 5 13 | immature seed forage | 0.09 7.94 | (2002200) |
| ND, EPA region 5 (Mycogen 5007) | WG | 0.56 | 0.20 | 280 | | 5 14 | immature seed forage | ≤ 0.05 5.43 | (2002201) |
| ND, EPA region 5 (Mycogen 5007) | WG | 0.56 | 0.20 | 280 | | 5 14 | immature seed forage | ≤ 0.05 8.70 | (2002202) |
| ND, EPA region 5 (Cropland RT0583) | WG | 0.56 | 0.23 | 240 | | 5 14 | immature seed forage | ≤ 0.05 4.73 | (2002203) |
| SD, EPA region 5 (Cropland RT0583) | WG | 0.56 | 0.23 | 240 | | 5 14 | immature seed forage | ≤ 0.05 8.25 | (2002204) |
| IL, EPA region 5 (B-T 441 CR) | WG | 0.56 | 0.24 | 210-260 | | 5 14 | immature seed forage | ≤ 0.05 3.05 | (2002205) |
| IL, EPA region 5 (Asgrow AG3302) | WG | 0.56 | 0.20 | 250-320 | | 5 14 | immature seed forage | ≤ 0.05 3.63 | (2002206) |
| QC, EPA region 5B (DKB07-51) | WG | 0.55 | 0.21 | 260 | | 5 14 | immature seed forage | ≤ 0.05 4.62 | (2002216) |

1) mean values used for estimation of MRL.

Table 37. Results of residue trials with foliar treatment of boscalid conducted in snap bean and lima bean (succulent shelled) (Haughey and Abdel-Baky, 2001, 2001/5000905).

| CROP Country, Year *Location* (variety) | Application | | | | PHI days | Commodity | Residues, mg/kg boscalid[1] | Ref. Reg.DocID. |
|---|---|---|---|---|---|---|---|---|
| | Formulation | kg ai/ha | kg ai/hL | Water L/ha | No. | | | |
| **BEAN, snap** | | | | | | | | |
| USA, 2000 *PA, EPA region 1* (Bush Bean Roma II) | WG | 0.57 | 0.20 | 280 | 2 | 7 | snap bean | 0.28 | 2001/5000905 (2000187) |
| *GA, EPA region 2* (Bush Bean Roma II) | | 0.56 | 0.19 | 260 | 2 | 7 | snap bean | 0.97 | (2000188) |
| *FL, EPA region 3* (Rhapsody) | | 0.57 | 0.22 | 260 | 2 | 7 | snap bean | 0.72 | (2000189) |
| *MN, EPA region 5* (Top Crop) | | 0.56 | 0.33 | 170 | 2 | 7 | snap bean | 0.54 | (2000190) |
| *WI, EPA region 5* (Bush Blue Lake 274 Green Bean) | | 0.56 | 0.29 | 190 | 2 | 0 3 7 10 14 | snap bean | 0.29 0.22 0.13 0.06 < 0.05 | (2000191) |
| *CA, EPA region 10* (Seville) | | 0.56 | 0.20 | 280 | 2 | 7 | snap bean | 0.42 | (2000192) |
| *ID, EPA region 11* (Tendergreen) | | 0.56 | 0.18 | 310 | 2 | 7 | snap bean | 0.36 | (2000193) |
| *NS, EPA region 1A* (Provider) | | 0.56 | 0.19 | 290 | 2 | 7 | snap bean | 0.41 | (2000194) |
| *QC, EPA region 5B* (Goldmine) | | 0.58 | 0.2 | 290 | 2 | 7 | snap bean | 0.52 | (2000195) |
| *QC, EPA region 5B* (Goldmine) | | 0.57 | 0.20 | 280 | 2 | 7 | snap bean | 0.46 | (2000196) |
| **LIMA BEAN** | | | | | | | | |
| USA, 2000 *NC, EPA region 2* (Early Thorogreen) | WG | 0.57 | 0.20 | 280 | 2 | 7 | lima bean | 0.47 | 2001/5000905 (2000197) |
| *GA, EPA region 2* (Henderson) | | 0.56 | 0.21 | 270 | 2 | 7 | lima bean | 0.07 | (2000198) |
| *GA, EPA region 2* (Nenagreen) | | 0.56 | 0.29 | 190 | 2 | 7 | lima bean | 0.07 | (2000199) |
| *WI, EPA region 5* (Henderson) | | | | | | 0 3 7 10 14 | | 0.10 0.08 0.07 0.08 0.07 | (2000200) |
| *CA, EPA region 10* (Henderson) | | 0.56 | 0.20 | 280 | 2 | 7 | lima bean | 0.08 | (2000201) |
| *CA, EPA region 10* (Fordhood 242) | | 0.57 | 0.20 | 280 | 2 | 7 | lima bean | ≤ 0.05 | (2000202) |
| *ID, EPA region 11* (Henderson) | | 0.58 | 0.20 | 290 | 2 | 7 | lima bean | ≤ 0.05 | (2000203) |

1) average values used for estimation of MRL.

Table 38. Results of residue trials with foliar treatment of boscalid conducted in bean (Haughey and Abdel-Baky, 2001, 2001/5000905).

| CROP Country, Year *Location* (variety) | Application | | | | | PHI days | Commodity | Residues, mg/kg boscalid[1] | Ref. Reg.DocID. |
|---|---|---|---|---|---|---|---|---|---|
| | Formulation | kg ai/ha | kg ai/hL | Water L/ha | No. | | | | |
| ADZUKI BEAN | | | | | | | | | |
| Japan, 2002 Hokkaido (Kitanootome) | WG | 0.75 | 0.050 | 1500 | 3 | 7 14 21 | dry bean | 0.13 0.07 0.06 | na |
| Japan, 2002 Yamagata (benidainagon) | WG | 0.75 | 0.050 | 1500 | 3 | 7 14 21 | dry bean | 0.14 0.08 0.06 | na |
| COMMON BEAN | | | | | | | | | |
| Japan, 2002 Hokkaido (Taishokintoki) | WG | 0.75 | 0.050 | 1500 | 3 | 7 14 21 | dry bean | 0.03 0.10 0.18 | na |
| Japan, 2002 Gifu (Nagauzura) | WG | 0.75 | 0.050 | 1500 | 3 | 7 14 21 | dry bean | 0.32 0.55 0.63 | na |
| Japan, 2002 Niigata (Kintoki) | WG | 0.75 | 0.050 | 1500 | 2 | 21 28 35 45 | dry bean | 0.34 0.30 0.17 0.06 | na |
| Japan, 2002 Gifu (Nagauzura) | WG | 0.75 | 0.050 | 1500 | 2 | 21 28 35 45 | dry bean | 0.44 0.45 0.29 0.10 | na |
| BEAN, dry | | | | | | | | | |
| USA, 2000 *ND, EPA region 5* (Pinto Topaz) | WG | 0.57 | 0.30 | 190 | 2 | 21 | dry bean | ≤ 0.05 | 2001/5000905 (2000177) |
| *ND, EPA region 5* (Maverick) | | 0.56 | 0.29 | 190 | 2 | 21 | dry bean | ≤ 0.05 | (2000178) |
| *MN, EPA region 5* (Pinto Topaz) | | 0.56 | 0.33 | 170 | 2 | 21 | dry bean | 0.06 | (2000179) |
| *WI, EPA region 5* (Dark red Kidney 126) | | 0.56 | 0.29 | 190 | 2 | 0 7 14 21 28 | dry bean | < 0.05 0.07 0.08 < 0.05 0.09 | (2000180) |
| *ND, EPA region 7* (Maverick) | | 0.56 | 0.40 | 140 | 2 | 21 | dry bean | ≤ 0.05 | (2000181) |
| *TX, EPA region 8* (Taylor Horticulture improved) | | 0.57 | 0.30 | 190 | 2 | 21 | dry bean | 0.12 | (2000182) |
| *CO, EPA region 9* (Pinto Bill Z) | | 0.56 | 0.56 | 100 | 2 | 21 | dry bean | 1.92 | (2000183) |
| *CA, EPA region 10* (Linden Light Red kidneys) | | 0.56 | 0.25 | 220 | 2 | 21 | dry bean | 0.37 | (2000184) |
| *ID, EPA region 11* (Pinto Apache) | | 0.56 | 0.20 | 280 | 2 | 21 | dry bean | ≤ 0.05 | (2000185) |
| *AB, EPA region 7[a]* (Pinto Othello) | | 0.55 | 0.28 | 200 | 2 | 21 | dry bean | 0.14 | (2000186) |

na: not available.

1) average values used for estimation of MRL.

Table 39. Results of residue trials with foliar treatment of boscalid conducted in peas (Haughey and Abdel-Baky, 2002, 2001/5003246).

| CROP Country, Year Location (variety) | Application | | | | | PHI days | Commodity | Residues, mg/kg boscalid[1] | Ref. Reg.DocID. |
|---|---|---|---|---|---|---|---|---|---|
| | Formulation | kg ai/ha | kg ai/hL | Water L/ha | No. | | | | |
| PEA, dry, shelled | | | | | | | | | |
| USA, 2001 ND, EPA region 5 (Athos) | WG | 0.58 | 0.61 | 95 | 2 | 20 | dry pea | 0.05 | 2001/5003246 (2001244) |
| AB, EPA region 14 (Admiral) | | 0.56 | 0.51 | 110 | 2 | 21 | dry pea | 0.31 | (2001248) |
| AB, EPA region 14 (Croma) | | 0.56 | 0.51 | 110 | 2 | 21 | dry pea | 0.17 | (2001249) |
| SK, EPA region 14 (Delta) | | 0.56 | 0.51 | 110 | 2 | 21 | dry pea | 0.39 | (2001250) |
| SK, EPA region 14 (Delta) | | 0.56 | 0.51 | 110 | 2 | 20 | dry pea | 0.23 | (2001251) |
| WA, EPA region 11 (Lazer) | | 0.56 | 0.29 | 190 | 2 | 21 | dry pea | 0.16 | (2001334) |
| WA, EPA region 11 (Estancia) | | 0.56 | 0.29 | 190 | 2 | 21 | dry pea | 0.12 | (2001335) |
| OR, EPA region 11 (Paso) | | 0.57 | 0.30 | 190 | 2 | 22 | dry pea | 0.11 | (2001336) |
| ID, EPA region 11 (09690470) | | 0.56 | 0.20 | 280 | 2 | 21 | dry pea | 0.09 | (2001337) |

1) average values used for estimation of MRL.

Table 40. Results of residue trials with foliar treatment of boscalid conducted in soybean (Lwonard, 2003, 2002/5004272).

| CROP Country, Year Location (variety) | Application | | | | | PHI days | Commodity | Residues, mg/kg boscalid[1] | Ref. Reg.DocID. |
|---|---|---|---|---|---|---|---|---|---|
| | Formulation | kg ai/ha | kg ai/hL | Water L/ha | No. | | | | |
| USA, 2002 GA, EPA region 2 (NK RR S73-Z5) | WG | 0.57 | 0.16 | 350 | 2 | 21 21 | mature seed hay | ≤ 0.05 1.29 | 2002/5004272 (2002191) |
| VA, EPA region 2 (NK S53Q7 7B-1001) | WG | 0.57 | 0.18 | 310 | 2 | 22 22 | mature seed hay | ≤ 0.05 2.28 | (2002192) |
| AR, EPA region 4 (AG4403) | WG | 0.56 | 0.27 | 210 | 2 | 20 20 | mature seed hay | ≤ 0.05 4.55 | (2002193) |
| AR, EPA region 4 (AG5603) | WG | 0.56 | 0.20 | 280 | 2 | 22 22 | mature seed hay | ≤ 0.05 2.12 | (2002194) |
| WI, EPA region 5 (BR2099RR) | WG | 0.56 | 0.23 | 240 | 2 | 21 21 | mature seed hay | ≤ 0.05 1.76 | (2002195) |
| MN, EPA region 5 (BR2099RR) | WG | 0.56 | 0.23 | 240 | 2 | 22 22 | mature seed hay | ≤ 0.05 2.78 | (2002196) |
| IA, EPA region 5 (SG 2531RR) | WG | 0.57 | 0.22 | 260 | 2 | 21 21 | mature seed hay | ≤ 0.05 4.79 | (2002197) |
| IA, EPA region 5 (SG 2533RR) | WG | 0.57 | 0.24 | 240 | 2 | 21 21 | mature seed hay | ≤ 0.05 1.99 | (2002198) |
| NE, EPA region 5 (Asgrow A2553) | WG | 0.56 | 0.29 | 190 | 2 | 21 21 | mature seed hay | ≤ 0.05 21.3 | (2002199) |
| NE, EPA region 5 (Asgrow 2703) | WG | 0.56 | 0.29 | 190 | 2 | 21 21 | mature seed hay | ≤ 0.05 1.38 | (2002200) |

| CROP Country, Year *Location* (variety) | Application | | | | | PHI days | Commodity | Residues, mg/kg boscalid[1] | Ref. Reg.DocID. |
|---|---|---|---|---|---|---|---|---|---|
| | Formulation | kg ai/ha | kg ai/hL | Water L/ha | No. | | | | |
| ND, EPA region 5 (Mycogen 5007) | WG | 0.56 | 0.19 | 290 | 2 | 21 21 | mature seed hay | ≤ 0.05 6.73 | (2002201) |
| ND, EPA region 5 (Mycogen 5007) | WG | 0.56 | 0.20 | 280 | 2 | 21 21 | mature seed hay | ≤ 0.05 7.08 | (2002202) |
| ND, EPA region 5 (Cropland RT0583) | WG | 0.56 | 0.23 | 240 | 2 | 21 21 | mature seed hay | ≤ 0.05 7.78 | (2002203) |
| SD, EPA region 5 (Cropland RT0583) | WG | 0.56 | 0.23 | 240 | 2 | 21 21 | mature seed hay | ≤ 0.05 11.3 | (2002204) |
| IL, EPA region 5 (B-T 441 CR) | WG | 0.56 | 0.20 | 260-300 | 2 | 20 20 | mature seed hay | ≤ 0.05 5.33 | (2002205) |
| IL, EPA region 5 (Asgrow AG3302) | WG | 0.56 | 0.22 | 200-310 | 2 | 21 21 | mature seed hay | ≤ 0.05 7.30 | (2002206) |
| QC, EPA region 5B (DKB07-51) | WG | 0.54 | 0.19 | 280 | 2 | 21 21 | mature seed hay | ≤ 0.05 3.57 | (2002216) |

1) average values used for estimation of MRL.

Table 41. Results of residue trials with foliar treatment of boscalid conducted in carrot (Versoi and Abdel-Baky, 2000, 2000/5208).

| CROP Country, Year *Location* (variety) | Application | | | | | PHI days | Commodity | Residues, mg/kg boscalid[1] | Ref. Reg.DocID. (Trial No.) |
|---|---|---|---|---|---|---|---|---|---|
| | Formulation | kg ai/ha | kg ai/hL | Water L/ha | No. | | | | |
| USA, 1999 FL, EPA region 3 (Choctaw) | WG | 0.19 | 0.069 | 240-310 | 6 | 0 | root w/o top | 0.19 | 2000/5208 (99180) |
| MN, EPA region 5 (Dundee) | WG | 0.38 | 0.22 | 170 | 3 | 0 | root w/o top | 0.18 | (99181) |
| TX, EPA region 6 (Mercury) | WG | 0.19 | 0.090 | 210 | 6 | 0 | root w/o top | 0.12 | (99182) |
| CA, EPA region 10 (Danvers Half Long) | WG | 0.38 | 0.14 | 280 | 3 | 0 | root w/o top | ≤ 0.05 | (99183) |
| CA, EPA region 10 (Danvers Half Long) | WG | 0.38 | 0.14 | 280 | 3 | 0 | root w/o top | 0.06 | (99184) |
| CA, EPA region 10 (Nance) | WG | 0.38 | 0.14 | 280 | 3 | 0 5 9 15 20 | root w/o top | 0.17 0.13 0.15 0.16 0.15 | (99185) |
| CA, EPA region 10 (Nantes) | WG | 0.38 | 0.14 | 380 | 3 | 0 | root w/o top | 0.34 | (99186) |
| ID, EPA region 11 (Danvers Half Long) | WG | 0.38 | 0.29 | 130 | 3 | 0 | root w/o top | 0.28 | (99187) |

1) average values used for estimation of MRL.

Table 42. Results of residue trials with foliar treatment of boscalid conducted in radish (Haughey and Abdel-Baky, 2001, 2001/50002572).

| CROP Country, Year Location (variety) | Application | | | | | PHI days | Commodity | Residues, mg/kg boscalid[1] | Ref. Reg.DocID. |
|---|---|---|---|---|---|---|---|---|---|
| | Formulation | kg ai/ha | kg ai/hL | Water L/ha | No. | | | | |
| USA, 1999 FL, EPA region 3 (Early Scarlet Globe) | WG | 0.38 | 0.14 | 280 | 3 | 0 | root | 0.40 0.38 | 2001/50002572 |
| | | | | | | 3 | | 0.34 0.25 | |
| | | | | | | 7 | | 0.14 0.18 | |
| | | | | | | 10 | | 0.12 0.12 | |
| | | | | | | 0 | top | 48.32 61.44 | |
| | | | | | | 3 | | 31.08 43.92 | |
| | | | | | | 7 | | 4.39 4.19 | |
| | | | | | | 10 | | 1.37 1.11 | |
| FL, EPA region 3 (Red Silk) | WG | 0.38 | 0.13 | 280-320 | 3 | 0 | root | 0.20 0.15 | |
| | | | | | | 3 | | 0.20 0.21 | |
| | | | | | | 7 | | 0.12 0.12 | |
| | | | | | | 10 | | 0.09 0.07 | |
| | | | | | | 0 | top | 33.04 26.2 | |
| | | | | | | 3 | | 6.46 5.79 | |
| | | | | | | 7 | | 4.08 4.87 | |
| | | | | | | 10 | | 2.33 2.99 | |
| MN, EPA region 5 (white Icicle) | WG | 0.38 | 0.22 | 170 | 3 | 0 | root | 0.06 0.08 | |
| | | | | | | 3 | | 0.15 0.12 | |
| | | | | | | 7 | | < 0.05 < 0.05 | |
| | | | | | | 10 | | < 0.05 < 0.05 | |
| | | | | | | 0 | top | 20.68 20.72 | |
| | | | | | | 3 | | 8.51 7.11 | |
| | | | | | | 7 | | 7.69 6.41 | |
| | | | | | | 10 | | 2.81 4.47 | |

| CROP Country, Year *Location* (variety) | Application | | | | | PHI days | Commodity | Residues, mg/kg boscalid[1] | Ref. Reg.DocID. |
|---|---|---|---|---|---|---|---|---|---|
| | Formulation | kg ai/ha | kg ai/hL | Water L/ha | No. | | | | |
| *CA, EPA region 10* (white Icicle) | WG | 0.38 | 0.12 | 280-330 | 3 | 0 | root | 0.10 0.12 | |
| | | | | | | 3 | | 0.11 0.06 | |
| | | | | | | 7 | | 0.08 0.09 | |
| | | | | | | 10 | | < 0.05 0.06 | |
| | | | | | | 0 | top | 25.76 24.44 | |
| | | | | | | 3 | | 13.84 11.18 | |
| | | | | | | 7 | | 10.46 10.54 | |
| | | | | | | 10 | | 7.78 6.56 | |
| *PA, EPA region 3* (Altaglobe) | WG | 0.38 | 0.11 | 350 | 3 | 0 | root | 0.61 0.60 | |
| | | | | | | 3 | | 0.43 0.46 | |
| | | | | | | 7 | | 0.20 0.23 | |
| | | | | | | 10 | | 0.15 0.16 | |
| | | | | | | 0 | top | 24.48 22.80 | |
| | | | | | | 3 | | 2.72 2.65 | |
| | | | | | | 7 | | 1.67 1.65 | |
| | | | | | | 10 | | 0.80 0.92 | |

1) average values used for estimation of MRL.

Table 43. Results of residue trials with foliar treatment of boscalid conducted in potato (Wofford and Abdel-Baky, 2001, 2001/5000879).

| CROP Country, Year *Location* (variety) | Application | | | | | PHI days | Commodity | Residues, mg/kg boscalid[1] | Ref. Reg.DocID. |
|---|---|---|---|---|---|---|---|---|---|
| | Formulation | kg ai/ha | kg ai/hL | Water L/ha | No. | | | | |
| POTATO | | | | | | | | | |
| USA, 2000 *PA, EPA region 1* (Andover) | WG | 0.50 | 0.15 | 340 | 2 | 30 | tuber | ≤ 0.05 | 2001/5000879 (2000161) |
| *NJ, EPA region 1* (Reba) | | 0.50 | 0.17 | 290 | 2 | 10 20 30 40 50 | tuber | < 0.05 < 0.05 ≤ 0.05 < 0.05 < 0.05 | (2000162) |
| *NC, EPA region 2* (Atlantic) | | 0.50 | 0.18 | 280 | 2 | 30 | tuber | ≤ 0.05 | (2000163) |
| *FL, EPA region 3* (Red Pontiac) | | 0.50 | 0.18 | 280 | 2 | 30 | tuber | ≤ 0.05 | (2000164) |
| *ND, EPA region 5* (Atlantic) | | 0.50 | 0.26 | 190 | 2 | 30 | tuber | ≤ 0.05 | (2000165) |

| CROP Country, Year *Location* (variety) | Application | | | | | PHI days | Commodity | Residues, mg/kg boscalid[1] | Ref. Reg.DocID. |
|---|---|---|---|---|---|---|---|---|---|
| | Formulation | kg ai/ha | kg ai/hL | Water L/ha | No. | | | | |
| *ND, EPA region 5* (Atlantic) | | 0.50 | 0.26 | 190 | 2 | 30 | tuber | ≤ 0.05 | (2000166) |
| *WI, EPA region 5* (Russet Burbank) | | 0.50 | 0.26 | 190 | 2 | 30 | tuber | ≤ 0.05 | (2000167) |
| *MN, EPA region 5* (Atlantic Newleaf) | | 0.50 | 0.26 | 190 | 2 | 30 | tuber | ≤ 0.05 | (2000168) |
| *CO, EPA region 9* (Norkotah) | | 0.50 | 0.21 | 240 | 2 | 30 | tuber | ≤ 0.05 | (2000169) |
| *CA, EPA region 10* (Russet) | | 0.50 | 0.18 | 280 | 2 | 10 20 30 40 50 | tuber | < 0.05 < 0.05 ≤ 0.05 < 0.05 < 0.05 | (2000170) |
| *OR, EPA region 11* (Russet Norkotah) | | 0.50 | 0.29 | 170 | 2 | 30 | tuber | ≤ 0.05 | (2000171) |
| *WA, EPA region 11* (Newleaf Plus) | | 0.50 | 0.26 | 190 | 2 | 30 | tuber | ≤ 0.05 | (2000309) |
| *ID, EPA region 11* (Russet Burbank) | | 0.50 | 0.18 | 280 | 2 | 30 | tuber | ≤ 0.05 | (2000173) |
| *ID, EPA region 11* (Russet Burbank) | | 0.50 | 0.18 | 280 | 2 | 30 | tuber | ≤ 0.05 | (2000174) |
| *WA, EPA region 11* (Newleaf Plus) | | 0.50 | 0.26 | 190 | 2 | 30 | tuber | ≤ 0.05 | (2000175) |
| *OR, EPA region 11* (Norchip) | | 0.50 | 0.25 | 200 | 2 | 30 | tuber | ≤ 0.05 | (2000176) |
| *WA, EPA region 11* (Newleaf Plus) | | 2.50 | 0.26 | 190 | 2 | 30 | tuber | < 0.05 | (2000175) |

1) average values used for estimation of MRL.

Table 44. Results of residue trials with foliar treatment of boscalid conducted in cereal grain (Raunft *et al.*, 2003, 2003/1009783; Johnston, 2005, 2005/5000151).

| CROP Country, Year *Location* (variety) | Application | | | | | PHI days | Commodity | Residues, mg/kg boscalid | Ref. Reg.DocID. |
|---|---|---|---|---|---|---|---|---|---|
| | Formulation | kg ai/ha | kg ai/hL | Water L/ha | No. | | | | |
| BARLEY | | | | | | | | | |
| UK; 2003 *Oxfordshire* (Cellar) | SC | 0.35 | 0.12 | 300 | 2 | 0 20 20 48 48 | plant w/o root ear culm grain straw | 12 1.1 5.3 0.03 13 | 2003/1009783 (OAT/15/03) |
| Netherlands, 2003 *Limburg* (Theresa) | SC | 0.35 | 0.12 | 300 | 2 | 0 7 7 61 61 | plant w/o root ear culm grain straw | 6.4 0.94 2.3 ≤ 0.01 0.51 | (AGR/13/03) |
| Germany; 2005 *Brandenburg* (Annabell) | SC | 0.35 | 0.12 | 300 | 2 | 0 19 19 54 54 | plant w/o root ear culm grain straw | 5.2 0.07 4.1 ≤ 0.01 2.5 | 2005/5000151 (ACK/07/05) |

| CROP Country, Year *Location* (variety) | Application | | | | PHI days | Commodity | Residues, mg/kg boscalid | Ref. Reg.DocID. |
|---|---|---|---|---|---|---|---|---|
| | Formulation | kg ai/ha | kg ai/hL | Water L/ha | No. | | | |
| Denmark; 2005 *Fuenen* (Chess) | SC | 0.35 | 0.12 | 300 | 2 | 0 | plant w/o root | 6.2 | (ALB/10/05) |
| | | | | | | 11 | ear | 3.3 | |
| | | | | | | 11 | culm | 4.2 | |
| | | | | | | 63 | grain | 0.19 | |
| | | | | | | 63 | straw | 5.8 | |
| France ; 2005 *Rhône-Alpes* (Orelie) | SC | 0.35 | 0.12 | 300 | 2 | 0 | plant w/o root | 18 | (FBD/28/05) |
| | | | | | | 13 | ear | 0.57 | |
| | | | | | | 19 | culm | 8.4 | |
| | | | | | | 61 | grain | 0.02 | |
| | | | | | | 61 | straw | 27 | |
| France ; 2005 *Midi-Pyrénées* (Prestige) | SC | 0.35 | 0.12 | 300 | 2 | 0 | plant w/o root | 7.7 | (FTL/17/05) |
| | | | | | | 7 | ear | 7.5 | |
| | | | | | | 7 | culm | 10 | |
| | | | | | | 44 | grain | 0.12 | |
| | | | | | | 44 | straw | 14 | |
| **WHEAT** | | | | | | | | | |
| Germany; 2003 *Brandenburg* (Colfiorito) | SC | 0.35 | 0.12 | 300 | 2 | 0 | plant w/o root | 12 | 2003/1009783 (ACK/08/03) |
| | | | | | | 13 | ear | 0.31 | |
| | | | | | | 13 | culm | 13 | |
| | | | | | | 52 | grain | 0.01 | |
| | | | | | | 52 | straw | 15 | |
| France ; 2003 *Pays de la Loire* (Isengrain) | SC | 0.35 | 0.12 | 300 | 2 | 0 | plant w/o root | 9.4 | (FBM/11/03) |
| | | | | | | 16 | ear | 0.51 | |
| | | | | | | 16 | culm | 3.9 | |
| | | | | | | 64 | grain | ≤ 0.01 | |
| | | | | | | 64 | straw | 3.1 | |
| France ; 2003 *Midi-Pyrénées* (Courtot) | SC | 0.35 | 0.12 | 300 | 2 | 0 | plant w/o root | 6.6 | (FTL/09/03) |
| | | | | | | 8 | ear | 6.8 | |
| | | | | | | 8 | culm | 7.8 | |
| | | | | | | 50 | grain | 0.06 | |
| | | | | | | 50 | straw | 7.9 | |
| Belgium; 2005 *Limburg* (Cadenza) | SC | 0.35 | 0.12 | 300 | 2 | 0 | plant w/o root | 7.2 | 2005/5000151 (AGR/18/05) |
| | | | | | | 22 | ear | < 0.05 | |
| | | | | | | 22 | culm | 4.7 | |
| | | | | | | 56 | grain | 0.01 | |
| | | | | | | 56 | straw | 3.0 | |
| France ; 2005 *Rhône-Alpes* (Karur) | SC | 0.35 | 0.12 | 300 | 2 | 0 | plant w/o root | 8.1 | (FBD/28/05) |
| | | | | | | 14 | ear | 0.19 | |
| | | | | | | 14 | culm | 3.7 | |
| | | | | | | 54 | grain | 0.03 | |
| | | | | | | 54 | straw | 5.3 | |
| France; 2005 *Pays de la Loire* (Royssac) | SC | 0.35 | 0.12 | 300 | 2 | 0 | plant w/o root | 6.3 | (FBM/10/05) |
| | | | | | | 23 | ear | 0.25 | |
| | | | | | | 23 | culm | 4.5 | |
| | | | | | | 64 | grain | 0.06 | |
| | | | | | | 64 | straw | 7.9 | |
| France ; 2005 *Midi-Pyrénées* (Estenia) | SC | 0.35 | 0.12 | 300 | 2 | 0 | plant w/o root | 6.1 | (FTL/18/05) |
| | | | | | | 9 | ear | 4.0 | |
| | | | | | | 9 | culm | 5.0 | |
| | | | | | | 50 | grain | 0.27 | |
| | | | | | | 50 | straw | 11 | |

| CROP Country, Year *Location* (variety) | Application | | | | | PHI days | Commodity | Residues, mg/kg boscalid | Ref. Reg.DocID. |
|---|---|---|---|---|---|---|---|---|---|
| | Formulation | kg ai/ha | kg ai/hL | Water L/ha | No. | | | | |
| UK ; 2005 *Oxfordshire* (Xi 19) | SC | 0.35 | 0.12 | 300 | 2 | 0 | plant w/o root | 4.1 | (OAT/15/05) |
| | | | | | | 17 | ear | < 0.05 | |
| | | | | | | 17 | culm | 2.7 | |
| | | | | | | 68 | grain | <u>0.01</u> | |
| | | | | | | 68 | straw | <u>5.8</u> | |

Table 45. Results of residue trials with foliar treatment of boscalid conducted in almond (Haughey and Abdel-Baky, 2000, 2000/5226; Johnston, 2004, 2004/5000223).

| CROP Country, Year *Location* (variety) | Application | | | | | PHI days | Commodity | Residues, mg/kg boscalid | Ref. Reg.DocID. (Trial No.) |
|---|---|---|---|---|---|---|---|---|---|
| | Formulation | kg ai/ha | kg ai/hL | Water L/ha | No. | | | | |
| USA, 1999 *CA, EPA region 10* (Monterey) | WG | 0.26 | 0.033 | 780-820 | 4 | 148 | | | 2000/5226 (99135) |
| | | | | | | | nutmeat | 0.03 | Treatment 2 |
| | | | | | | | hull | 1.36 | Treatment 3 |
| | | 0.26 | 0.013 | 1960-2150 | 4 | 148 | nutmeat | 0.02 | |
| | | | | | | | hull | 2.35 | |
| *CA, EPA region 10* (Non-pareil) | WG | 0.26 | 0.036 | 700-730 | 4 | 108 | | | (99136) |
| | | | | | | | nutmeat | 0.16 | Treatment 2 |
| | | | | | | | hull | 2.10 | Treatment 3 |
| | | 0.26 | 0.014 | 1830-1890 | 4 | 108 | nutmeat | 0.20 | |
| | | | | | | | hull | 1.62 | |
| *CA, EPA region 10* (Non-pareil) | WG | 0.26 | 0.028 | 910-930 | 4 | 116 | | | (99137) |
| | | | | | | | nutmeat | 0.13 | Treatment 2 |
| | | | | | | | hull | 1.10 | Treatment 3 |
| | | 0.26 | 0.011 | 2310-2360 | 4 | 116 | nutmeat | 0.16 | |
| | | | | | | | hull | 1.36 | |
| *CA, EPA region 10* (Non-pareil) | WG | 0.26 | 0.028 | 910-950 | 4 | 115 | | | (99138) |
| | | | | | | | nutmeat | 0.05 | Treatment 2 |
| | | | | | | | hull | 0.42 | Treatment 3 |
| | | 0.26 | 0.011 | 2320-2400 | 4 | 115 | nutmeat | 0.04 | |
| | | | | | | | hull | 0.82 | |
| *CA, EPA region 10* (Non-pareil) | WG | 0.26 | 0.040 | 640-660 | 4 | 120 | | | (99139) |
| | | | | | | | nutmeat | 0.04 | Treatment 2 |
| | | | | | | | hull | 1.65 | |
| | | | | | | 127 | nutmeat | 0.05 | |
| | | | | | | | hull | 1.72 | |
| | | | | | | 134 | nutmeat | 0.08 | |
| | | | | | | | hull | 1.86 | |
| | | | | | | 148 | nutmeat | 0.07 | |
| | | | | | | | hull | 2.63 | |
| | | | | | | 155 | nutmeat | 0.05 | |
| | | | | | | | hull | 1.43 | |
| | | 0.26 | 0.017 | 1480-1520 | 4 | 120 | nutmeat | 0.09 | Treatment 3 |
| | | | | | | | hull | 1.97 | |
| | | | | | | 127 | nutmeat | 0.08 | |
| | | | | | | | hull | 2.22 | |
| | | | | | | 134 | nutmeat | 0.11 | |
| | | | | | | | hull | 2.40 | |
| | | | | | | 148 | nutmeat | 0.11 | |
| | | | | | | | hull | 2.81 | |
| | | | | | | 155 | nutmeat | 0.08 | |
| | | | | | | | hull | 2.84 | |

| CROP Country, Year *Location* (variety) | Application | | | | | PHI days | Commodity | Residues, mg/kg boscalid | Ref. Reg.DocID. (Trial No.) |
|---|---|---|---|---|---|---|---|---|---|
| | Formulation | kg ai/ha | kg ai/hL | Water L/ha | No. | | | | |
| USA, 2003 | | | | | | | | | 2004/5000223 |
| *CA, EPA region 10* (Monterey) | WG | 0.26 | 0.045 | 580 | 4 | 25 | nutmeat hull | ≤ 0.05 <u>11.3</u> | (2003131) Treatment 2 |
| | WG | 0.26 | 0.020 | 1350 | 4 | 25 | nutmeat hull | ≤ 0.05 <u>11.9</u> | Treatment 3 |
| *CA, EPA region 10* (Non-pareil) | WG | 0.26 | 0.034 | 620-930 | 4 | 24 | nutmeat hull | ≤ 0.05 <u>3.45</u> | (2003132) Treatment 2 |
| | WG | 0.26 | 0.017 | 1380-1660 | 4 | 24 | nutmeat hull | ≤ 0.05 <u>6.78</u> | Treatment 3 |
| *CA, EPA region 10* (Butte) | WG | 0.26 | 0.037 | 700 | 4 | 26 | nutmeat hull | ≤ 0.05 <u>2.21</u> | (2002133) Treatment 2 |
| | WG | 0.26 | 0.014 | 1880 | 4 | 26 | nutmeat hull | ≤ 0.05 <u>5.41</u> | Treatment 3 |
| *CA, EPA region 10* (Non-pareil) | WG | 0.26 | 0.047 | 550 | 4 | 25 | nutmeat hull | ≤ 0.05 <u>2.64</u> | (2003134) Treatment 2 |
| | WG | 0.26 | 0.014 | 1850 | 4 | 25 | nutmeat hull | ≤ 0.05 <u>3.42</u> | Treatment 3 |
| *CA, EPA region 10* (Carmel) | WG | 0.26 | 0.035 | 750 | 4 | 25 | nutmeat hull | ≤ 0.05 <u>3.30</u> | (2003135) Treatment 2 |
| | WG | 0.26 | 0.013 | 1950 | 4 | 25 | nutmeat hull | ≤ 0.05 <u>3.91</u> | Treatment 3 |

Table 46. Results of residue trials with foliar treatment of boscalid conducted in pecan nut (Haughey and Abdel-Baky, 2001, 2000/5230).

| CROP Country, Year *Location* (variety) | Application | | | | | PHI days | Commodity | Residues, mg/kg boscalid | Ref. Reg.DocID. (Trial No.) |
|---|---|---|---|---|---|---|---|---|---|
| | Formulation | kg ai/ha | kg ai/hL | Water L/ha | No. | | | | |
| USA, 1999 | WG | | | | | | | | 2000/5230 |
| *GA, EPA region 2* (Stewart) | | 0.26 | 0.037 | 680-730 | 4 | 14 | nutmeat | ≤ 0.05 | (99332) Treatment 2 |
| | | 0.26 | 0.016 | 1650 | 4 | 14 | nutmeat | ≤ 0.05 | Treatment 3 |
| *GA, EPA region 2* (Sumner) | | 0.26 | 0.033 | 750-830 | 4 | 14 | nutmeat | ≤ 0.05 | (99333) Treatment 2 |
| | | 0.26 | 0.014 | 1800 | 4 | 14 | nutmeat | ≤ 0.05 | Treatment 3 |
| *MS, EPA region 4* (Kiowa) | | 0.26 | 0.036 | 620-830 | 4 | 14 | nutmeat | ≤ 0.05 | (99334) Treatment 2 |
| | | 0.26 | 0.019 | 1350 | 4 | 14 | nutmeat | ≤ 0.05 | Treatment 3 |
| *OK, EPA region 6* (Seedling) | | 0.26 | 0.042 | 620 | 4 | 14 | nutmeat | ≤ 0.05 | (99335) Treatment 2 |
| | | 0.26 | 0.019 | 1380 | 4 | 14 | nutmeat | ≤ 0.05 | Treatment 3 |
| *OK, EPA region 8* (Natives) | | 0.26 | 0.036 | 720 | 4 | 14 | nutmeat | ≤ 0.05 | (99336) Treatment 2 |
| | | 0.26 | 0.016 | 1650 | 4 | 14 | nutmeat | ≤ 0.05 | Treatment 3 |

Table 47. Results of residue trials with foliar treatment of boscalid conducted in pistachio (Haughey and Abdel-Baky, 2001, 2000/5229).

| CROP Country, Year Location (variety) | Application | | | | | PHI days | Commodity | Residues, mg/kg boscalid | Ref. Reg.DocID. |
|---|---|---|---|---|---|---|---|---|---|
| | Formulation | kg ai/ha | kg ai/hL | Water L/ha | No. | | | | |
| USA, 1999 | WG | | | | | | | | 2000/5229 |
| CA, EPA region 10 (Kerman) | | 0.26 | 0.042 | 620 | 4 | 14 | nutmeat | 0.19 | (99337) Treatment 2 |
| | | 0.26 | 0.017 | 1520 | 4 | 14 | nutmeat | 0.35 | Treatment 3 |
| CA, EPA region 10 (Calagucci) | | 0.26 | 0.044 | 500 | 4 | 14 | nutmeat | 0.45 | (99338) Treatment 2 |
| | | 0.26 | 0.019 | 1380 | 4 | 14 | nutmeat | 0.64 | Treatment 3 |
| CA, EPA region 10 (Kerman) | | 0.26 | 0.027 | 950 | 4 | 14 | nutmeat | $\leq 0.05$ | (99339) Treatment 2 |
| | | 0.26 | 0.014 | 1890 | 4 | 14 | nutmeat | $\leq 0.05$ | Treatment 3 |

Table 48. Results of residue trials with foliar treatment of boscalid conducted in canola (Versoi and Abdel-Baky, 2001, 2001/5000048).

| CROP Country, Year Location (variety) | Application | | | | | PHI days | Commodity | Residues, mg/kg boscalid | Ref. Reg.DocID. |
|---|---|---|---|---|---|---|---|---|---|
| | Formulation | kg ai/ha | kg ai/hL | Water L/ha | No. | | | | |
| USA, 2000 ND, EPA region 7 (Quantum) | WG | 0.43 | 0.24 | 180 | 2 | 0 | seed | 1.61 1.62 | 2001/5000048 |
| | | | | | | 10 | | 1.34 0.63 | |
| | | | | | | 21 | | 0.70 0.36 | |
| | | | | | | 30 | | 1.05 1.08 | |
| | | | | | | 40 | | 0.75 1.09 | |
| ND, EPA region 7 (Hyola 401) | WG | 0.45 | 0.24 | 190 | 2 | 21 | seed | 0.70 0.88 | |
| ND, EPA region 7 (Quantum) | WG | 1.35 | 0.64 | 190 | 2 | 21 | seed | 0.58 | |
| ID, EPA region 11 (Phoenix) | WG | 0.45 | 0.24 | 190 | 2 | 0 | seed | 1.23 1.34 | |
| | | | | | | 10 | | 0.43 0.45 | |
| | | | | | | 20 | | 0.45 0.55 | |
| | | | | | | 29 | | 0.46 0.45 | |
| | | | | | | 40 | | 0.22 0.17 | |
| AB, EPA region 14 (LG 3235) | WG | 0.45 | 0.45 | 100 | 2 | 22 | seed | 0.14 0.22 | |
| AB, EPA region 14 (LG 3235) | WG | 0.45 | 0.45 | 100 | 2 | 22 | seed | 0.20 0.19 | |
| AB, EPA region 14 (Agassiz) | WG | 0.45 | 0.39 | 115 | 2 | 22 | seed | 0.55 0.70 | |
| AB, EPA region 14 (Agassiz) | WG | 0.45 | 0.39 | 115 | 2 | 22 | seed | 3.42 2.99 | |
| AB, EPA region 14 (LG 3235) | WG | 0.45 | 0.23 | 200 | 2 | 21 | seed | 0.24 0.36 | |

| CROP Country, Year Location (variety) | Application | | | | | PHI days | Commodity | Residues, mg/kg boscalid | Ref. Reg.DocID. |
|---|---|---|---|---|---|---|---|---|---|
| | Formulation | kg ai/ha | kg ai/hL | Water L/ha | No. | | | | |
| AB, EPA region 14 (LG 3235) | WG | 0.45 | 0.23 | 200 | 2 | 20 | seed | 1.27 2.58 | |
| SK, EPA region 14 (45A71) | WG | 0.45 | 0.39 | 115 | 2 | 21 | seed | 0.78 1.23 | |
| SK, EPA region 14 (46A76) | WG | 0.45 | 0.39 | 115 | 2 | 20 | seed | 0.76 0.75 | |
| MB, EPA region 14 (Canterra 1867RR) | WG | 0.45 | 0.23 | 200 | 2 | 21 | seed | 1.74 0.85 | |
| MB, EPA region 14 (Canterra 1867RR) | WG | 1.35 | 0.68 | 200 | 2 | 21 | seed | 5.10 | |
| MB, EPA region 14 (LG3235) | WG | 0.45 | 0.23 | 200 | 2 | 21 | seed | 3.13 3.13 | |
| MB, EPA region 14 (Quest) | WG | 0.45 | 0.39 | 115 | 2 | 19 | seed | 0.19 0.44 | |
| MB, EPA region 14 (Quest) | WG | 1.36 | 1.18 | 115 | 2 | 20 | seed | 1.96 | |
| MB, EPA region 14 (45A51) | WG | 0.45 | 0.39 | 115 | 2 | 19 | seed | 0.30 0.34 | |
| MN, EPA region 5 (Golden Ready) | WG | 0.45 | 0.24 | 190 | 2 | 22 | seed | 0.91 0.92 | |
| MN, EPA region 5 (Golden Ready) | WG | 1.34 | 0.71 | 190 | 2 | 22 | seed | 2.15 | |

Table 49. Results of residue trials with foliar treatment of boscalid conducted in sunflower (Versoi and Abdel-Baky, 2002, 2001/5002552; Leonard, 2005, 2005/5000022).

| CROP Country, Year Location (variety) | Application | | | | | PHI days | Commodity | Residues, mg/kg boscalid[1] | Ref. Reg.DocID. |
|---|---|---|---|---|---|---|---|---|---|
| | Formulation | kg ai/ha | kg ai/hL | Water L/ha | No. | | | | |
| USA, 2001 ND, EPA region 5 (Cropland-CL-803) | WG | 0.45 | 0.19 | 240 | 2 | 21 | seed | 0.09 | 2001/50002552 (2001284) |
| ND, EPA region 5 (Interstate Seed) | WG | 0.45 | 0.24 | 190 | 2 | 21 | seed | 0.08 | (2001285) |
| ND, EPA region 7 (Interstate Seed) | WG | 0.45 | 0.24 | 190 | 2 | 21 | seed | 0.13 | (2001286) |
| SD, EPA region 7 (Mycogen 8377) | WG | 0.45 | 0.24 | 190 | 2 | 20 | seed | 0.23 | (2001287) |
| SD, EPA region 7 (Mycogen 8388) | WG | 0.45 | 0.24 | 190 | 2 | 20 | seed | 0.16 | (2001288) |
| TX, EPA region 8 (Triumph 567DW) | WG | 0.45 | 0.23 | 200 | 2 | 21 | seed | 0.16 | (2001289) |
| MB, EPA region 14 (Ag Canada 6111) | WG | 0.45 | 0.23 | 200 | 2 | 21 | seed | 0.45 | (2001290) |

| CROP Country, Year *Location* (variety) | Application | | | | | PHI days | Commodity | Residues, mg/kg boscalid[1] | Ref. Reg.DocID. |
|---|---|---|---|---|---|---|---|---|---|
| | Formulation | kg ai/ha | kg ai/hL | Water L/ha | No. | | | | |
| USA, 2004 *IL, EPA region 5* (Mycogen 8N429CL) | WG | 0.44 | 0.19 | 230 | 2 | 21 | seed | ≤ 0.05 | 2005/5000022 (2004152) |

\* Results of a confirmatory analysis on the same seed sample.

1) average values used for estimation of MRL.

Table 50. Results of residue trials with foliar treatment of boscalid conducted in peanut (Wofford and Abdel-Baky, 2001, 2001/5000870).

| CROP Country, Year *Location* (variety) | Application | | | | | PHI days | Commodity | Residues, mg/kg boscalid[1] | Ref. Reg.DocID. |
|---|---|---|---|---|---|---|---|---|---|
| | Formulation | kg ai/ha | kg ai/hL | Water L/ha | No. | | | | |
| PEANUT | | | | | | | | | |
| USA, 2000 *NC, EPA region 2* (NCV-11) | WG | 0.50 | 0.18 | 280 | 3 | 13 | nutmeat hay | 0.05 20.2 | 2001/5000870 (2000148) |
| *SC, EPA region 2* (Georgia Green) | WG | 0.50 | 0.28 | 180 | 3 | 14 | nutmeat hay | ≤ 0.05 24.2 | (2000149) |
| *SC, EPA region 2* (Georgia Green) | WG | 0.50 | 0.28 | 180 | 3 | 14 | nutmeat hay | ≤ 0.05 29.3 | (2000150) |
| *GA, EPA region 2* (Georgia Green) | WG | 0.50 | 0.18 | 280 | 3 | 7 14 21 28 35 7 14 21 28 35 | nutmeat hay | < 0.05 ≤ 0.05 < 0.05 < 0.05 < 0.05 5.78 7.79 3.87 5.38 3.90 | (2000151) |
| *FL, EPA region 3* (Georgia Green) | WG | 0.50 | 0.15 | 330 | 3 | 13 | nutmeat hay | ≤ 0.05 6.70 | (2000152) |
| *AL, EPA region 2* (Georgia Green) | WG | 0.50 | 0.42 | 120 | 3 | 15 | nutmeat hay | ≤ 0.05 5.84 | (2000153) |
| *GA, EPA region 2* (Agra Teck 201) | WG | 0.50 | 0.19 | 270 | 3 | 14 | nutmeat hay | ≤ 0.05 3.15 | (2000154) |
| *GA, EPA region 2* (Valencia) | WG | 0.50 | 0.19 | 270 | 3 | 14 | nutmeat hay | ≤ 0.05 28.4 | (2000155) |
| *GA, EPA region 2* (Georgia Green) | WG | 0.50 | 0.22 | 230 | 3 | 14 | nutmeat hay | ≤ 0.05 12.58 | (2000156) |
| *TX, EPA region 6* (Pronto) | WG | 0.50 | 0.26 | 190 | 3 | 14 | nutmeat hay | ≤ 0.05 - | (2000157) |
| *OK, EPA region 6* (Spanco) | WG | 0.50 | 0.36 | 140 | 3 | 13 | nutmeat hay | ≤ 0.05 6.65 | (2000158) |
| *OK, EPA region 8* | WG | 0.50 | 0.36 | 140 | 3 | 13 | nutmeat hay | ≤ 0.05 9.01 | (2000159) |

\* The confirmatory re-analyses of the same sample.

- Not available. Hay samples were not collected from this site in error.

1) average values used for estimation of MRL.

Table 51. Results of residue trials with foliar treatment of boscalid conducted in coffee (Dantas, 2001, 2001/1026859; Dantas, 2001, 2001/1026860; Dantas, 2001, 2001/1026861; Dantas, 2001, 2001/1026862).

| CROP Country, Year *Location* (variety) | Application | | | | | PHI days | Commodity | Residues, mg/kg boscalid | Ref. Reg.DocID. |
|---|---|---|---|---|---|---|---|---|---|
| | Formulation | kg ai/ha | kg ai/hL | Water L/ha | No. | | | | |
| Brazil, 2000 *Sao Paolo* (Catual vermelho e amarelo) | WG | 0.15 | 0.030 | 500 | 1 | 0 15 30 45 60 | coffee bean | < 0.05 < 0.05 < 0.05 ≤ 0.05 < 0.05 | 2001/1026859 |
| Brazil, 2000 *Sao Paolo* (Catual) | WG | 0.15 0.30 | 0.030 0.060 | 500 500 | 1 1 | 45 45 45 | coffee bean | ≤ 0.05 0.06 0.08 | 2001/1026860 |
| Brazil, 2000 *Romaria, MG* (Mundo Novo) | WG | 0.15 0.30 | 0.030 0.060 | 500 500 | 1 1 | 45 45 | coffee bean | ≤ 0.05 < 0.05 | 2001/1026861 |
| Brazil, 2000 *Araguari, MG* (Catual amarelo) | WG | 0.15 0.30 | 0.030 0.060 | 500 500 | 1 1 | 45 45 | coffee bean | ≤ 0.05 < 0.05 | 2001/1026862 |

Table 52. Results of residue trials with foliar treatment of boscalid conducted in hops (Schneider, 2002, 2001/1015050; Schneider, 2002, 2001/1015052; Schulz, 2004, 2003/1001292).

| CROP Country, Year *Location* (variety) | Application | | | | | PHI days | Commodity | Residues, mg/kg boscalid kg ai/ha | Ref. Reg.DocID. |
|---|---|---|---|---|---|---|---|---|---|
| | Formulation | kg ai/ha | kg ai/hL | Water L/ha | No. | | | | |
| Germany 2000 Niederlauterbach (Perle) | SE | 0.500 to 0.600 | 0.019 | 2700 to [a] 3200 | 3 | 0 13 20 26 20 26 | cone, green cone, green cone, green cone, green cone, dried cone, dried | 2.66 3.67 3.54 3.72 32.4 14.6 | 2001/1015050 |
| Germany 2000, Wolnzach (Hallertauer Magnum) | SE | 0.500 to 0.600 | 0.019 | 2700 to [a] 3200 | 3 | 0 13 20 26 20 26 | cone, green cone, green cone, green cone, green cone, dried cone, dried | 3.03 3.53 2.53 1.07 16.6 14.4 | |
| Germany 2000 Holzhof Gemeinde Au (Perle) | SE | 0.500 to 0.600 | 0.019 | 2700 to [a] 3200 | 3 | 0 13 20 26 20 26 | cone, green cone, green cone, green cone, green cone, dried cone, dried | 4.56 4.57 3.62 4.57 16.5 14.9 | |
| Germany 2000 Unterempfenbach (Perle) | SE | 0.500 to 0.600 | 0.019 | 2700 to [a] 3200 | 3 | 0 13 20 26 20 26 | cone, green cone, green cone, green cone, green cone, dried cone, dried | 12.2 5.89 5.46 1.77 20.4 33.6 | |

| CROP Country, Year *Location* (variety) | Application | | | | | PHI days | Commodity | Residues, mg/kg boscalid kg ai/ha | Ref. Reg.DocID. |
|---|---|---|---|---|---|---|---|---|---|
| | Formulation | kg ai/ha | kg ai/hL | Water L/ha | No. | | | | |
| Germany 2001, Hallertau (Perle) | SE | 0.420 to 0.500 | 0.018 | 2300 to [a)] 2700 | 3 | 0 | cone, green | 5.55 | 2001/1015052 |
| | | | | | | 14 | cone, green | 3.68 | |
| | | | | | | 21 | cone, green | 6.22 | |
| | | | | | | 28 | cone, green | 3.05 | |
| | | | | | | 21 | cone, dried | 10.4 | |
| | | | | | | 28 | cone, dried | 12.3 | |
| Germany 2001 Niederlauterbach (Perle) | SE | 0.420 to 0.500 | 0.018 | 2300 to [a)] 2700 | 3 | 0 | cone, green | 3.10 | |
| | | | | | | 14 | cone, green | 2.72 | |
| | | | | | | 21 | cone, green | 3.17 | |
| | | | | | | 28 | cone, green | 1.67 | |
| | | | | | | 21 | cone, dried | 5.69 | |
| | | | | | | 28 | cone, dried | 5.23 | |
| Germany 2001 Unterempfenbach (Perle) | SE | 0.420 to 0.500 | 0.018 | 2300 to [a)] 2700 | 3 | 0 | cone, green | 10.3 | |
| | | | | | | 14 | cone, green | 10.3 | |
| | | | | | | 21 | cone, green | 7.07 | |
| | | | | | | 28 | cone, green | 5.72 | |
| | | | | | | 21 | cone, dried | 36.6 | |
| | | | | | | 28 | cone, dried | 21.7 | |
| Germany 2001, Wolnzach (Hallertauer Magnum) | SE | 0.420 to 0.500 | 0.018 | 2300 to [a)] 2700 | 3 | 0 | cone, green | 4.51 | |
| | | | | | | 14 | cone, green | 2.30 | |
| | | | | | | 21 | cone, green | 2.89 | |
| | | | | | | 28 | cone, green | 1.97 | |
| | | | | | | 21 | cone, dried | 9.67 | |
| | | | | | | 28 | cone, dried | 6.46 | |
| Germany 2003, Bayern (Hallertauer Magnum) | SE | 1) 0.420 2+3) 0.500 | 0.018 0.019 | 2300 [b)] 2700 [c)] | 3 | 0 | cones | 13.0 | 2003/1001292 |
| | | | | | | 14 | cones | 3.91 | |
| | | | | | | 21 | cones | 5.83 | |
| | | | | | | 28 | cones | 8.04 | |
| | WG | 1) 0.428 2+3) 0.504 | 0.019 0.019 | 2300 [b)] 2700 [c)] | 3 | 0 | cones | 5.78 | |
| | | | | | | 14 | cones | 5.09 | |
| | | | | | | 21 | cones | 6.09 | |
| | | | | | | 28 | cones | 7.02 | |
| Germany 2003, Bayern (Spalter Select) | SE | 1) 0.420 2+3) 0.500 | 0.018 0.019 | 2300 [b)] 2700 [c)] | 3 | 0 | cones | 13.6 | 2003/1001292 |
| | | | | | | 14 | cones | 9.64 | |
| | | | | | | 21 | cones | 9.91 | |
| | | | | | | 28 | cones | 4.09 | |
| | WG | 1) 0.428 2+3) 0.504 | 0.019 0.019 | 2300 [b)] 2700 [c)] | 3 | 0 | cones | 22.7 | |
| | | | | | | 14 | cones | 18.4 | |
| | | | | | | 21 | cones | 7.55 | |
| | | | | | | 28 | cones | 10.5 | |
| Germany 2003, Bayern (Hall Tradition) | SE | 1) 0.420 2+3) 0.500 | 0.018 0.019 | 2300 [b)] 2700 [c)] | 3 | 0 | cones | 48.2 | |
| | | | | | | 14 | cones | 19.8 | |
| | | | | | | 21 | cones | 9.63 | |
| | | | | | | 28 | cones | 6.98 | |
| | WG | 1) 0.428 2+3) 0.504 | 0.019 0.019 | 2300 [b)] 2700 [c)] | 3 | 0 | cones | 15.3 | |
| | | | | | | 14 | cones | 12.4 | |
| | | | | | | 21 | cones | 21.1 | |
| | | | | | | 28 | cones | 6.48 | |

| CROP Country, Year *Location* (variety) | Application | | | | | PHI days | Commodity | Residues, mg/kg boscalid kg ai/ha | Ref. Reg.DocID. |
|---|---|---|---|---|---|---|---|---|---|
| | Formulation | kg ai/ha | kg ai/hL | Water L/ha | No. | | | | |
| Germany 2003, Bayern (Perle) | SE | 1) 0.420 2+3) 0.500 | 0.018 0.019 | 2300 [b] 2700 [c] | 3 | 0 14 21 28 | cones cones cones cones | 15.0 8.25 20.1 5.36 | |
| | WG | 1) 0.428 2+3) 0.504 | 0.019 0.019 | 2300 [b] 2700 [c] | 3 | 0 14 21 28 | cones cones cones cones | 15.5 6.27 16.1 12.8 | |

* average of multiple analysis

a) corresponding to 0.019 kg as/hL

b) corresponding to 0.0183 kg as/hL

[c] corresponding to 0.0185 kg as/hL

## FATE OF RESIDUES IN STORAGE AND PROCESSING

### *In processing*

The Meeting received information on the fate of boscalid residues during aqueous hydrolysis under conditions of pasteurisation, baking, brewing, boiling and sterilisation. Information was also provided on the fate of boscalid residues during the food processing of citrus, apples, plums, cherries, strawberries, grapes, white cabbage, gherkins, tomatoes, head lettuce, peas, soybeans, carrots, sugar beet, barley, winter wheat, corn, peanuts, sunflower, cotton seed, canola seed, mint and hops.

[$^{14}$C] boscalid, labelled in the diphenyl moiety, was dissolved in aqueous buffer solutions of different pHs (Scharf, 1998, 1998/10878). In order to simulate the process of pasteurisation, the test solutions were heated for 20 minutes at 90°C. For simulation of baking, brewing and boiling, the test substances were treated under reflux at 100°C for 60 minutes. For simulation of sterilisation, samples were treated at about 120°C in an autoclave for 20 minutes.

Boscalid was not degraded during the simulation of pasteurisation (pH 4, 90°C), baking, boiling, brewing (pH 5, 100°C) or during sterilisation (pH 6, 120°C) (Table 53).

Table 53. Recovery data after the simulation of processing (Scharf, 1998, 1998/10878).

| Process | Test conditions | Diphenyl-label, % TAR* after test |
|---|---|---|
| Pasteurisation | pH 4, 90 °C | 99.3 |
| Baking, brewing and boiling | pH 5, 100 °C | 100.2 |
| Sterilisation | pH 6, 120 °C | 99.1 |

* Total applied radioactivity

A study was performed testing boscalid in oranges, grapefruits and lemons in 2002 (Jordan, 2002, 2002/5002446). The trees were treated four times with boscalid at a rate of 0.34 kg ai/ha beginning 30 days before harvest with a spraying interval of 10 days. Citrus RAC samples were collected directly after the last application. They were separated into pulp and peel. All samples were analysed by method D9908 which led to an average recovery of 84 ± 9%. Residues found in orange fruit, separated into pulp and peel, showed significantly less residue in the pulp sample than in the whole fruit, thus confirming that the majority of the residue remains on the peel of the fruit (Table 54).

Table 54. Boscalid residues in citrus and processed fractions resulting from supervised trials in the USA (Jordan, 2002, 2002/5002446).

| CITRUS country, year, location (variety) | Form | Application kg ai/ha | kg ai/hL | water (L/ha) | no. | PHI days | Commodity | Residues, mg/kg boscalid | Ref |
|---|---|---|---|---|---|---|---|---|---|
| **Orange** | | | | | | | | | |
| USA, 2001, FL (Valencia) | WG | 1.34 | 0.095 | 1400-1410 | 4 | 0 | fruit pulp peel | 0.50, 0.68 0.10, 0.12 2.60, 3.97 | 2002/5002446 |
| USA, 2001, FL (Navel) | WG | 1.37 | 0.16 | 840-870 | 4 | 0 | fruit pulp peel | 0.18, 0.23 <0.05, <0.05 0.63, 0.69 | |
| USA, 2001, FL (Hamlin) | WG | 1.34 | 0.095 | 1400-1410 | 4 | 0 | fruit pulp peel | 0.54, 0.56 0.05, <0.05 2.57, 2.05 | |
| USA, 2001, FL (Hamlin) | WG | 1.36 | 0.16 | 810-840 | 4 | 0 | fruit pulp peel | 1.43, 1.36 0.20, 0.10 6.26, 4.55 | |
| USA, 2001, FL (Valencia, Swingle) | WG | 1.34 | 0.087 | 1520-1550 | 4 | 0 | fruit pulp peel | 0.47, 0.64 0.07, 0.09 2.33, 2.79 | |
| USA, 2001, FL (Pineapple) | WG | 1.33 | 0.16 | 800-820 | 4 | 0 | fruit pulp peel | 0.68, 1.19 0.08, 0.09 2.18, 2.57 | |
| USA, 2001, FL (Hamlin) | WG | 1.36 | 0.071 | 1860-1950 | 4 | 0 | fruit pulp peel | 0.33, 0.33 <0.05, <0.05 0.70, 0.76 | |
| USA, 2001, FL (Valencia) | WG | 1.34 | 0.19 | 690-700 | 4 | 0 | fruit pulp peel | 0.71, 0.27 <0.05, 0.06 1.72, 1.06 | |
| USA, 2001, TX (Everhard Navel) | WG | 1.36 | 0.058 | 2340-2380 | 4 | 0 | fruit pulp peel | 0.32, 0.24 <0.05, <0.05 0.68, 0.81 | |
| USA, 2001, CA (Navel) | WG | 1.34 | 0.040 | 3280-3390 | 4 | 0 | fruit pulp peel | 0.30, 0.24 <0.05, <0.05 0.98, 0.69 | |
| USA, 2001, CA (Navel) | WG | 1.34 | 0.18 | 720-730 | 4 | 0 | fruit pulp peel | 0.47, 0.31 0.05, 0.06 0.79, 0.85 | |
| USA, 2001, CA (Navel) | WG | 1.34 | 0.041 | 3240-3280 | 4 | 0 | fruit pulp peel | 0.35, 0.35 <0.05, <0.05 0.98, 1.16 | |
| USA, 2001, CA (Cutter) | WG | 1.33 | 0.18 | 720-730 | 4 | 0 | fruit pulp peel | 0.26, 0.24 <0.05, <0.05 0.42, 1.14 | |
| **Grapefruit** | | | | | | | | | |
| USA, 2001, FL (Flame) | WG | 1.34 | 0.080 | 1685 | 4 | 0 | fruit | 0.25, 0.27 | 2002/5002446 |
| USA, 2001, FL (White Marsh) | WG | 1.34 | 0.16 | 810-840 | 4 | 0 | fruit | 0.82, 0.85 | |
| USA, 2001, FL (Flame) | WG | 1.34 | 0.051 | 2540-2700 | 4 | 0 | fruit | 0.10, 0.06 | |
| USA, 2001, TX (Rio Red) | WG | 1.34 | 0.19 | 690 | 4 | 0 | fruit | 0.12, 0.12 | |
| USA, 2001, CA (Mello Gold) | WG | 1.34 | 0.16 | 830-850 | 4 | 0 | fruit | 0.15, 0.14 | |
| USA, 2001, CA (Oroblanco) | WG | 1.34 | 0.064 | 2080-2120 | 4 | 0 | fruit | 0.15, 0.10 | |

| CITRUS country, year, location (variety) | Application | | | | | PHI days | Commodity | Residues, mg/kg boscalid | Ref |
|---|---|---|---|---|---|---|---|---|---|
| | Form | kg ai/ha | kg ai/hL | water (L/ha) | no. | | | | |
| Lemon | | | | | | | | | |
| USA, 2001, FL (Bearss) | WG | 1.38 | 0.17 | 780-850 | 4 | 0 | fruit | 0.68, 0.66 | 2002/5002446 |
| USA, 2001, CA (Prior) | WG | 1.34 | 0.060 | 2200-2260 | 4 | 0 | fruit | 0.74, 0.60 | |
| USA, 2001, CA (Lisbon) | WG | 1.34 | 0.19 | 690-720 | 4 | 0 | fruit | 1.51, 0.97 | |
| USA, 2001, AZ (Lisbon) | WG | 1.32 | 0.18 | 720-730 | 4 | 0 | fruit | 0.52, 0.59 | |
| USA, 2001, AZ (Limonarie) | WG | 1.32 | 0.072 | 1830-1850 | 4 | 0 | fruit | 0.94, 0.81 | |

Four field trials were conducted in different representative apple growing areas in Germany to determine the residue level of boscalid in apples and processed fractions during the 2001 growing season (Schulz, 2001, 2001/1015047). Boscalid was applied four times with an application rate of 200 g ai/ha (38, 30, 22 and 14 days before the commercial harvest) resulting in a maximum seasonal target rate of 800 g ai/ha boscalid, in order to determine the magnitude of the residues of active ingredients in or on RACs. For analysis, apples were collected immediately after the last application and about 14 days later. The fruit was processed to the following products: wash water, washed apples, fresh pomace, dried pomace, thick juice, apple juice, the remainder of the straining process and apple sauce (Table 55).

Table 55. Boscalid residues in apples and processed fractions resulting from supervised trials in Germany (Schulz, 2001, 2001/1015047).

| APPLES country, year, location (variety) | Application | | | | | PHI days | Commodity | Residues, mg/kg boscalid | Ref |
|---|---|---|---|---|---|---|---|---|---|
| | Form | kg ai/ha | kg ai/hL | water (L/ha) | no. | | | | |
| Germany, 2001, Lower Saxony (Jona-gold) | SE | 0.6 | 0.060 | 1000 | 4 | 0 | fruit | 1.0 | 2001/1015047 |
| | | | | | | | wash water | 0.18 | |
| | | | | | | | fresh pomace | 3.9 | |
| | | | | | | | fresh juice | 0.08 | |
| | | | | | | 14 | fruit | 0.78 | |
| | | | | | | | wash water | 0.11 | |
| | | | | | | | washed apples | 0.33 | |
| | | | | | | | fresh pomace | 5.28 | |
| | | | | | | | dried pomace | 18.8 | |
| | | | | | | | thick juice | 0.23 | |
| | | | | | | | fresh juice | 0.05 | |
| | | | | | | | remainder of the straining process | 1.97 | |
| | | | | | | | apple sauce | 0.52 | |
| Germany, 2001, Thruingia (Remo) | SE | 0.6 | 0.060 | 1000 | 4 | 0 | fruit | 2.13 | |
| | | | | | | 14 | fruit | 2.27 | |
| | | | | | | | wash water | 0.18 | |
| | | | | | | | washed apples | 1.07 | |
| | | | | | | | fresh pomace | 14.48 | |
| | | | | | | | dried pomace | 31.01 | |
| | | | | | | | thick juice | 0.29 | |
| | | | | | | | fresh juice | 0.11 | |
| | | | | | | | remainder of the straining process | 5.32 | |
| | | | | | | | apple sauce | 1.89 | |

| APPLES country, year, location (variety) | Application | | | | | PHI days | Commodity | Residues, mg/kg boscalid | Ref |
|---|---|---|---|---|---|---|---|---|---|
| | Form | kg ai/ha | kg ai/hL | water (L/ha) | no. | | | | |
| Germany, 2001, | SE | 0.6 | 0.060 | 1000 | 4 | 0 | fruit | 1.24 | |
| | | | | | | | wash water | 0.12 | |
| | | | | | | | fresh pomace | 2.58 | |
| | | | | | | | fresh juice | 0.10 | |
| Schleswig-Holstein (Boskop) | | | | | | 14 | fruit | 0.58 | |
| | | | | | | | wash water | 0.07 | |
| | | | | | | | washed apples | 0.58 | |
| | | | | | | | fresh pomace | 4.79 | |
| | | | | | | | dried pomace | 11.62 | |
| | | | | | | | thick juice | 0.13 | |
| | | | | | | | fresh juice | < 0.05 | |
| | | | | | | | remainder of the straining process | 1.36 | |
| | | | | | | | apple sauce | 0.58 | |
| Germany, 2001, Hesse (Braeburn) | SE | 0.6 | 0.060 | 1000 | 4 | 0 | fruit | 0.54 | |
| | | | | | | 14 | fruit | 0.49 | |
| | | | | | | | wash water | 0.10 | |
| | | | | | | | washed apples | 0.41 | |
| | | | | | | | fresh pomace | 2.81 | |
| | | | | | | | dried pomace | 8.18 | |
| | | | | | | | thick juice | 0.13 | |
| | | | | | | | fresh juice | < 0.05 | |
| | | | | | | | remainder of the straining process | 1.07 | |
| | | | | | | | apple sauce | 0.56 | |

At one location in California in 1999 (Wofford and Abdel-Baky, 2001, 2000/5275), plum trees were treated according to different application schemes: on one plot, the GAP rate of 0.26 kg ai/ha was applied 5 times. An exaggerated treatment was done on a second plot with 5 applications at a rate of 1.3 kg ai/ha were used. Treatments started 28 days before expected harvest with a 7 day interval between the sprays. Plum fruit samples were collected directly after the last application. The 5× treated plums were processed to prunes according to typical commercial practices. All samples were analysed by method D9908 which led to an average recovery of 87 ± 4% for plums and 84% for prunes (Table 56).

Table 56. Boscalid residues in plums and processed fractions resulting from supervised trials in USA(Wofford and Abdel-Baky, 2001, 2000/5275).

| PLUMS country, year, location (variety) | Application | | | | | PHI days | Commodity | Residues, mg/kg boscalid | Ref |
|---|---|---|---|---|---|---|---|---|---|
| | Form | kg ai/ha | kg ai/hL | water (L/ha) | no. | | | | |
| USA, 1999, CA (French Plume) | WG | 0.77 | 0.034 | 2290 | 4 | 0 | unwashed plum | 0.68 | 2000/5275 |
| | | | | | | | | 1.09 | |
| | | | | | | | washed plums | 0.97 | |
| | | | | | | | | 0.43 | |
| | | | | | | | prunes | 0.51 | |
| | | | | | | | | 0.41 | |

Plum trees at four locations in typical plum growing areas in Germany were treated five times with boscalid at the double of the recommended rate, i.e., 0.401 kg ai/ha, in order to investigate the residues in plums and the processed fractions plum puree and prune production (Schulz and Scharm, 2000, 2001/1000936). The double rate was used to increase the probability of finding measurable residues in the processed fractions. Plums for residue analysis were sampled manually immediately after the last application once the spray had dried. Plums were also taken 7 days after the last application. These samples were used for residue analysis and for processing. The plums were processed to the following products: wash water, washed plums, the remainder of the straining process, condensed water, plum-puree, dipping water, and prunes (Table 57).

Table 57. Boscalid residues in plums and processed fractions resulting from supervised trials in Germany (Schulz and Scharm, 2000, 2001/1000936).

| PLUMS country, year, location (variety) | Application | | | | | PHI days | Commodity | Residues, mg/kg boscalid | Ref |
|---|---|---|---|---|---|---|---|---|---|
| | Form | kg ai/ha | kg ai/hL | water (L/ha) | no. | | | | |
| Germany, 2000 Rhineland-Palatine ("Elena" auf Fereley) | WG | 0.41 | 0.054 | 760 | 5 | 0 | fruit | 1.59 | 2001/1000936 |
| | | | | | | 7 | fruit | 1.55 | |
| | | | | | | | washed plums | 0.96 | |
| | | | | | | | wash water | 0.42 | |
| | | | | | | | remainder of straining process | 2.54 | |
| | | | | | | | condensed water | < 0.05 | |
| | | | | | | | puree | 2.97 | |
| | | | | | | | dipping water | 0.36 | |
| | | | | | | | prunes | 4.88 | |
| Germany, 2000 Hesse (Chrudimer) | WG | 0.41 | 0.040 | 1030 | | 0 | fruit | 0.52 | |
| | | | | | | 7 | fruit | 0.40 | |
| | | | | | | | washed plums | 0.32 | |
| | | | | | | | wash water | 0.07 | |
| | | | | | | | remainder of straining process | 1.85 | |
| | | | | | | | condensed water | < 0.05 | |
| | | | | | | | puree | 0.79 | |
| | | | | | | | dipping water | 0.05 | |
| | | | | | | | prunes | 1.12 | |
| Germany, 2000 Wiesbaden-Kloppenheim (Auerbacher) | WG | 0.41 | 0.040 | 1029 | | 0 | fruit | 0.46 | |
| | | | | | | 7 | fruit | 0.38 | |
| | | | | | | | washed plums | 0.22 | |
| | | | | | | | wash water | 0.15 | |
| | | | | | | | remainder of straining process | 1.43 | |
| | | | | | | | condensed water | < 0.05 | |
| | | | | | | | puree | 0.78 | |
| | | | | | | | dipping water | 0.09 | |
| | | | | | | | prunes | 1.39 | |
| Germany, 2000 Schleswig Holstein (Schraderhof) | WG | 0.41 | 0.040 | 1013 | | 0 | fruit | 0.33 | |
| | | | | | | 7 | fruit | 0.26 | |
| | | | | | | | washed plums | 0.37 | |
| | | | | | | | wash water | < 0.05 | |
| | | | | | | | remainder of straining process | 0.59 | |
| | | | | | | | condensed water | < 0.05 | |
| | | | | | | | puree | 0.40 | |
| | | | | | | | dipping water | < 0.05 | |
| | | | | | | | prunes | 0.63 | |

Cherry trees at four locations in typical cherry growing areas in Germany were treated five times at double of the recommended rate (0.401 kg ai/ha) with boscalid in order to investigate the residues in cherries and in the processed fractions from washed and canned cherry production (Schulz, 2000, 2001/1000938). Samples were taken twice. The first sampling was performed immediately after last application once the spray had dried. The second sampling was performed 7 days after last application. These samples were used for residue analysis and for processing. The cherries were processed to the following products: de-stemmed cherries (RAC), wash water, stones (not used for residue analyses), washed cherries (= de-stemmed, washed and de-stoned cherries), canned cherries, fruit syrup and cherry juice (Table 58).

Table 58. Boscalid residues in cherries and processed fractions resulting from supervised trials in Germany (Schulz, 2000, 2001/1000938).

| CHERRY country, year, location (variety) | Application | | | | | PHI days | Commodity | Residues, mg/kg boscalid | Ref |
|---|---|---|---|---|---|---|---|---|---|
| | Form | kg ai/ha | kg ai/hL | water (L/ha) | no. | | | | |
| Germany, 2000 Rhineland-Palatine (Schattenmorellen) | WG | 0.41 | 0.040 | 1032 | 5 | 0 | fruit | 1.56 | 2001/1000938 |
| | | | | | | 7 | fruit | 1.04 | |
| | | | | | | | washed cherries | 0.27 | |
| | | | | | | | wash water | 0.95 | |
| | | | | | | | canned cherries | 0.42 | |
| | | | | | | | fruit syrup | 0.10 | |
| | | | | | | | cherry juice | 0.29 | |
| Germany, 2000 Hesse (Haumüller) | WG | 0.42 | 0.031 | 986-1540 | 5 | 0 | fruit | 0.67 | |
| | | | | | | 7 | fruit | 0.29 | |
| | | | | | | | washed cherries | 0.35 | |
| | | | | | | | wash water | 0.16 | |
| | | | | | | | canned cherries | 0.25 | |
| | | | | | | | fruit syrup | 0.10 | |
| | | | | | | | cherry juice | 0.36 | |
| Germany, 2000 Schleswig-Holstein (Schattenmorellen) | WG | 0.40 | 0.040 | 1010 | 5 | 0 | fruit | 1.78 | |
| | | | | | | 7 | fruit | 1.03 | |
| | | | | | | | washed cherries | 0.60 | |
| | | | | | | | wash water | 0.74 | |
| | | | | | | | canned cherries | 0.56 | |
| | | | | | | | fruit syrup | 0.09 | |
| | | | | | | | cherry juice | 0.45 | |
| Germany, 2000 Schleswig-Holstein (Boseka) | WG | 0.40 | 0.040 | 996 | 5 | 0 | fruit | 2.17 | |
| | | | | | | 7 | fruit | 1.38 | |
| | | | | | | | washed cherries | 0.78 | |
| | | | | | | | wash water | 1.91 | |
| | | | | | | | canned cherries | 0.68 | |
| | | | | | | | fruit syrup | 0.18 | |
| | | | | | | | cherry juice | 0.43 | |

Strawberry plants at four locations, in typical strawberry growing areas in Germany, were treated four times with boscalid at double of the recommended rate (0.961 kg ai/ha) of boscalid in order to investigate the residues in strawberries and in the processed fractions canned strawberry and jam production (Scharm, 2000, 2001/1000937). Samples of strawberries for residue analysis were taken directly after the last application once the spray had dried. Strawberry samples were also collected 3 days after the last application. These samples were used for residue analysis and for processing. The strawberries were processed to the following products: washed strawberries, wash water, canned strawberries, fruit syrup, jam and distillate (Table 59).

Table 59. Boscalid residues in strawberries and processed fractions resulting from supervised trials in Germany (Scharm, 2000, 2001/1000937).

| STRAWBERRY country, year, location (variety) | Application | | | | | PHI days | Commodity | Residues, mg/kg boscalid | Ref |
|---|---|---|---|---|---|---|---|---|---|
| | Form | kg ai/ha | kg ai/hL | water (L/ha) | no. | | | | |
| Germany, 2000 Hesse (Symphony) | WG | 0.95 | 0.097 | 984 | 4 | 0 | fruit | 0.80 | 2001/1000937 |
| | | | | | | 3 | fruit | 0.79 | |
| | | | | | | | washed fruit | 0.20 | |
| | | | | | | | washing water | 0.23 | |
| | | | | | | | canned fruit | 0.27 | |
| | | | | | | | fruit syrup | 0.07 | |
| | | | | | | | jam | 0.11 | |
| | | | | | | | distillate | < 0.05 | |

| STRAWBERRY country, year, location (variety) | Application | | | | | PHI days | Commodity | Residues, mg/kg boscalid | Ref |
|---|---|---|---|---|---|---|---|---|---|
| | Form | kg ai/ha | kg ai/hL | water (L/ha) | no. | | | | |
| Germany, 2000 Thuringia (Elsanta) | WG | 1.01 | 0.096 | 1050 | 4 | 0 | fruit | 0.78 | |
| | | | | | | 3 | fruit | 1.05 | |
| | | | | | | | washed fruit | 0.41 | |
| | | | | | | | washing water | 0.33 | |
| | | | | | | | canned fruit | 0.41 | |
| | | | | | | | fruit syrup | 0.16 | |
| | | | | | | | jam | 0.25 | |
| | | | | | | | distillate | < 0.05 | |
| Germany, 2000 Schleswig-Holstein (Korona) | WG | 0.98 | 0.096 | 1019 | 4 | 0 | fruit | 1.17 | |
| | | | | | | 3 | fruit | 1.59 | |
| | | | | | | | washed fruit | 0.29 | |
| | | | | | | | washing water | 0.19 | |
| | | | | | | | canned fruit | 0.40 | |
| | | | | | | | fruit syrup | 0.11 | |
| | | | | | | | jam | 0.18 | |
| | | | | | | | distillate | < 0.05 | |
| Germany, 2000 Schleswig-Holstein (Elsanta) | WG | 0.98 | 0.096 | 1018 | 4 | 0 | fruit | 1.29 | |
| | | | | | | 3 | fruit | 1.22 | |
| | | | | | | | washed fruit | 0.34 | |
| | | | | | | | washing water | 0.22 | |
| | | | | | | | canned fruit | 0.39 | |
| | | | | | | | fruit syrup | 0.14 | |
| | | | | | | | jam | 0.29 | |
| | | | | | | | distillate | < 0.05 | |

A grape processing study was performed with boscalid in 2000 (Haughey and Abdel-Baky, 2001, 2001/5000065). At one location in California, grape vines were treated according to different application regimes: on one plot, the normal GAP was followed with 3 treatments at a rate of 0.41 kg ai/ha. An exaggerated treatment, 5× GAP rate, was done on a second plot with 3 applications at a rate of 2.08 kg ai/ha were used. Grape berry samples were collected at normal crop maturity, i.e., 14 days after the last application. The 5× treated grapes were processed to grape juice and raisins according to typical commercial practices (Table 60).

Table 60. Boscalid residues in grapes and processed juice and raisins resulting from supervised trials in USA (Haughey and Abdel-Baky, 2001, 2001/5000065).

| GRAPE country, year, location (variety) | Application | | | | | PHI days | Commodity | Residues, mg/kg boscalid | Ref |
|---|---|---|---|---|---|---|---|---|---|
| | Form | kg ai/ha | kg ai/hL | water (L/ha) | no. | | | | |
| USA, 2000, CA (Thompson Seedless) | WG | 2.08 | 0.18 | 1174 | 3 | 14 | fruit | 4.65 4.95 | 2001/5000065 |
| | | | | | | | raisins | 11.35 11.86 | |
| | | | | | | | juice | 1.84 2.15 | |

During the 1999 growing season four field trials, two each with red and white grape varieties, were conducted in different representative wine growing areas in Germany to determine the residue levels of boscalid in grapes and grape processing fractions (must, wine, pomace) (Meumann, 1999, 2000/1012412). Boscalid was applied 3 times 28 ± 2 days before normal harvest at application rates of about 0.64 kg ai/ha (average value). For the analysis grape samples were collected within 3 hours after the last application as well as 28 ± 2 days later. Fruit samples for processing of juice, wine and pomace were taken 28 ± 2 days after the last application (Table 61).

Table 61. Boscalid residues in grapes and processed fractions resulting from supervised trials in Germany (Meumann, 1999, 2000/1012412).

| GRAPE country, year, location (variety) | Application | | | | | PHI days | Commodity | Residues mg/kg boscalid | Ref |
|---|---|---|---|---|---|---|---|---|---|
| | Form | kg ai/ha | kg ai/hL | water (L/ha) | no. | | | | |
| Gremany, 1999 Baden-Wüttermberg (Spätburgunder) | WG | 0.63 | 0.090 | 612-803 | 3 | 0 | fruit | 1.13 | 2000/1012412 |
| | | | | | | 28 | fruit | 1.58 | |
| | | | | | | | wet pomace | 3.79 | |
| | | | | | | | must, cold | 0.51 | |
| | | | | | | | must, after mash heating | 0.14 | |
| | | | | | | | wine, from must, cold | 0.41 | |
| | | | | | | | wine from must, after mash heating | 0.13 | |
| Gremany, 1999, Baden-Wüttermberg (Mürttemberg) | WG | 0.64 | 0.090 | 610-810 | 3 | 0 | fruit | 1.49 | |
| | | | | | | 28 | fruit | 1.40 | |
| | | | | | | | wet pomace | 2.73 | |
| | | | | | | | must, cold | 0.68 | |
| | | | | | | | wine, from must, cold | 0.66 | |
| | | | | | | | must, after short time heating | 0.67 | |
| | | | | | | | wine from must, after short time heating | 0.64 | |
| Gremany, 1999, Rheinland Pfalz (Portugieser) | WG | 0.63 | 0.089 | 598-832 | 3 | 0 | fruit | 0.56 | |
| | | | | | | 28 | fruit | 0.50 | |
| | | | | | | | wet pomace | 1.30 | |
| | | | | | | | must, cold | 0.26 | |
| | | | | | | | must, after mash heating | 0.09 | |
| | | | | | | | wine, from must, cold | 0.18 | |
| | | | | | | | wine from must, after mash heating | 0.06 | |
| Gremany, 1999, Rheinland Pfalz (Riesling) | WG | 0.64 | 0.090 | 614-829 | 3 | 0 | fruit | 0.62 | |
| | | | | | | 28 | fruit | 0.58 | |
| | | | | | | | wet pomace | 1.98 | |
| | | | | | | | must, cold | 0.23 | |
| | | | | | | | wine, from must, cold | 0.20 | |
| | | | | | | | must, after short time heating | 0.26 | |
| | | | | | | | wine from must, after short time heating | 0.21 | |

White cabbage plants at four locations in typical cabbage growing areas in Germany were treated with boscalid to investigate the residues in the raw commodity and in processed fractions such as cooked white cabbage and sauerkraut (Schulz and Scharm, 2000, 2001/1000943). The cabbage plants was treated four times, at a rate of 1.2 kg ai/ha. The treatment rate was three times that of the recommended rate in order to increase the probability of finding residues. Samples of cabbage heads for residue analysis were taken after the last application once the spray had dried. Cabbage heads were also taken 14 to 15 days after the last application. These samples were used for residue analysis and for processing. The white cabbages were processed to the following products: Outer leaves, inner leaves, inner and outer stalks, cooked white cabbage head, boiled water, sauerkraut, juice of sauerkraut and pasteurised juice of sauerkraut (Table 62).

Table 62. Boscalid residues in white cabbages and processed fractions resulting from supervised trials in Germany (Schulz and Scharm, 2000, 2001/1000943).

| WHITE CABBAGE country, year, location (variety) | Application | | | | | PHI days | Commodity | Residues mg/kg boscalid | Ref |
|---|---|---|---|---|---|---|---|---|---|
| | Form | kg ai/ha | kg ai/hL | water (L/ha) | no. | | | | |
| Gremany, 2000 Thuringia (Megadon) | WG | 1.21 | 0.40 | 302 | 4 | 0 | head | 2.44 | 2001/1000943 |
| | | | | | | 15 | head | 0.12 | |
| | | | | | | | outer leaves | 1.87 | |
| | | | | | | | inner leaves | < 0.05 | |
| | | | | | | | inner/out stalks | 0.553 | |
| | | | | | | | cooked cabbage | < 0.05 | |
| | | | | | | | boiled water | < 0.05 | |
| | | | | | | | sauerkraut juice | < 0.05 | |
| | | | | | | | sauerkraut | < 0.05 | |
| | | | | | | | sauerkraut juice (pasteurized) | < 0.05 | |
| Gremany, 2000, Hesse (Transam) | WG | 1.20 | 0.40 | 299 | 4 | 0 | head | 2.80 | |
| | | | | | | 14 | head | 0.64 | |
| | | | | | | | outer leaves | 13.79 | |
| | | | | | | | inner leaves | < 0.05 | |
| | | | | | | | inner/out stalks | 1.14 | |
| | | | | | | | cooked cabbage | < 0.05 | |
| | | | | | | | boiled water | < 0.05 | |
| | | | | | | | sauerkraut juice | < 0.05 | |
| | | | | | | | sauerkraut | 0.14 | |
| | | | | | | | sauerkraut juice (pasteurized) | < 0.05 | |
| Gremany, 2000, Hesse (Transam) | WG | 1.22 | 0.40 | 304 | 4 | 0 | head | 6.81 | |
| | | | | | | 14 | head | 1.49 | |
| | | | | | | | outer leaves | 24.67 | |
| | | | | | | | inner leaves | < 0.05 | |
| | | | | | | | inner/out stalks | 0.05 | |
| | | | | | | | cooked cabbage | < 0.05 | |
| | | | | | | | boiled water | < 0.05 | |
| | | | | | | | sauerkraut juice | < 0.05 | |
| | | | | | | | sauerkraut | 0.13 | |
| | | | | | | | sauerkraut juice (pasteurized) | < 0.05 | |
| Gremany, 2000, Hesse (Transam) | WG | 1.24 | 0.40 | 311 | 4 | 0 | head | 5.90 | |
| | | | | | | 14 | head | 0.77 | |
| | | | | | | | outer leaves | 40.60 | |
| | | | | | | | inner leaves | < 0.05 | |
| | | | | | | | inner/out stalks | 0.09 | |
| | | | | | | | cooked cabbage | < 0.05 | |
| | | | | | | | boiled water | < 0.05 | |
| | | | | | | | sauerkraut juice | < 0.05 | |
| | | | | | | | sauerkraut | 0.08 | |
| | | | | | | | sauerkraut juice (pasteurized) | < 0.05 | |

During the 2000 growing season four field trials were conducted in different representative areas for cucumber (gherkins) cultivation in Germany, to determine the residue level of boscalid in cucumbers and resulting processed fractions (Scharm, 2000, 2001/1000942). The cucumber plants were treated four times, each time with about 0.30 kg ai/ha of boscalid, approximately double the GAP rate. The applications were performed at about 24, 17, 10 and 3 days before the expected harvest. For analysis, cucumbers were taken immediately after the last application and 3 days following. The cucumbers were processed to washed and canned cucumbers (Table 63).

Table 63. Boscalid residues in cucumbers and processed fractions resulting from supervised trials in Germany (Scharm, 2000, 2001/1000942).

| CUCUMBER country, year, location (variety) | Application | | | | | PHI days | Commodity | Residues mg/kg boscalid | Ref |
|---|---|---|---|---|---|---|---|---|---|
| | Form | kg ai/ha | kg ai/hL | water (L/ha) | no. | | | | |
| Gremany, 2000 Thuringia (Melody) | SC | 0.31 | 0.030 | 1041 | 4 | 0 | gherkins | 0.46 | 2001/1000942 |
| | | | | | | 3 | gherkins | 0.16 | |
| | | | | | | | gherkins[1] | 0.15 | |
| | | | | | | | washed gherkins | 0.11 | |
| | | | | | | | wash water | 0.09 | |
| | | | | | | | canned gherkins | 0.09 | |
| | | | | | | | brine | < 0.05 | |
| Gremany, 2000, Hesse (Pontomac) | SC | 0.31 | 0.030 | 1039 | 4 | 0 | gherkins | 0.14 | |
| | | | | | | 3 | gherkins | < 0.05 | |
| | | | | | | | gherkins[1] | 0.10 | |
| | | | | | | | washed gherkins | 0.05 | |
| | | | | | | | wash water | < 0.05 | |
| | | | | | | | canned gherkins | < 0.05 | |
| | | | | | | | brine | < 0.05 | |
| Gremany, 2000, Hesse (Musica) | SC | 0.31 | 0.030 | 1035 | 4 | 0 | gherkins | 0.23 | |
| | | | | | | 3 | gherkins | 0.11 | |
| | | | | | | | gherkins[1] | 0.13 | |
| | | | | | | | washed gherkins | 0.06 | |
| | | | | | | | wash water | < 0.05 | |
| | | | | | | | canned gherkins | 0.07 | |
| | | | | | | | brine | < 0.05 | |
| Gremany, 2000, Eich (Mauvin) | SC | 0.31 | 0.030 | 1018 | 4 | 0 | gherkins | 0.16 | |
| | | | | | | 3 | gherkins | 0.10 | |
| | | | | | | | gherkins[1] | 0.22 | |
| | | | | | | | washed gherkins | < 0.05 | |
| | | | | | | | wash water | < 0.05 | |
| | | | | | | | canned gherkins | 0.05 | |
| | | | | | | | brine | < 0.05 | |

1) for processing.

Tomato plants at four locations in California were treated with boscalid to determine the magnitude of the residues in processed tomato fractions. The tomato plants were treated twice, at a rate of 3 kg ai/ha, approximately five times the GAP rate (Haughey and Abdel-Baky, 2001, 2001/5000967). Whole tomato samples were collected 0 days after the last application. The tomato RAC samples from the control and 5× plots were harvested and processed according to simulated commercial practices into wet pomace, fresh canned juice, puree, paste, peeled tomatoes, canned tomatoes and peel (Table 64).

Table 64. Boscalid residues in tomatoes and processed fractions resulting from supervised trials in USA (Haughey and Abdel-Baky, 2001, 2001/5000967).

| TOMATO country, year, location (variety) | Application | | | | | PHI days | Commodity | Residues mg/kg boscalid | Ref |
|---|---|---|---|---|---|---|---|---|---|
| | Form | kg ai/ha | kg ai/hL | water (L/ha) | no. | | | | |
| USA, 2000, CA (UF6203) | WG | 3.09 | 1.09 | 283 | 2 | 0 | unwashed tomato RAC | 1.52 | 2001/5000967 |
| | | | | | | | wash water | 0.20 | |
| | | | | | | | washed tomatoes | 0.21 | |
| | | | | | | | wet pomace | 1.42 | |
| | | | | | | | canned juice | 0.19 | |
| | | | | | | | puree | 0.36 | |
| | | | | | | | tomato paste | 0.80 | |
| | | | | | | | peeled tomatoes | < 0.05 | |
| | | | | | | | canned tomatoes | < 0.05 | |
| | | | | | | | peel | 0.43 | |

| TOMATO country, year, location (variety) | Application | | | | | PHI days | Commodity | Residues mg/kg boscalid | Ref |
|---|---|---|---|---|---|---|---|---|---|
| | Form | kg ai/ha | kg ai/hL | water (L/ha) | no. | | | | |
| USA, 2000, CA (3155) | WG | 3.11 | 0.66 | 472 | 2 | 0 | unwashed tomato RAC | 0.71[1] | |
| | | | | | | | wash water | 0.14 | |
| | | | | | | | washed tomatoes | 0.67[1] | |
| | | | | | | | wet pomace | 1.54[1] | |
| | | | | | | | canned juice | 0.19 | |
| | | | | | | | puree | 0.52[1] | |
| | | | | | | | tomato paste | 1.59[1] | |
| | | | | | | | peeled tomatoes | - | |
| | | | | | | | canned tomatoes | - | |
| | | | | | | | peel | - | |
| USA, 2000, CA (Hypeel 108) | WG | 3.08 | 1.39 | 222 | 2 | 0 | unwashed tomato RAC | 0.96 | |
| | | | | | | | wash water | 0.34 | |
| | | | | | | | washed tomatoes | 0.16 | |
| | | | | | | | wet pomace | 0.82 | |
| | | | | | | | canned juice | 0.15 | |
| | | | | | | | puree | 0.23 | |
| | | | | | | | tomato paste | 0.79 | |
| | | | | | | | peeled tomatoes | < 0.05 | |
| | | | | | | | canned tomatoes | < 0.05 | |
| | | | | | | | peel | 0.34 | |
| USA, 2000, CA (La Roma) | WG | 3.08 | 0.82 | 374 | 2 | 0 | unwashed tomato RAC | 0.75 | |
| | | | | | | | wash water | 0.05 | |
| | | | | | | | washed tomatoes | 0.11 | |
| | | | | | | | wet pomace | 0.82 | |
| | | | | | | | canned juice | 0.07 | |
| | | | | | | | puree | 0.14 | |
| | | | | | | | tomato paste | 0.47 | |
| | | | | | | | peeled tomatoes | < 0.05 | |
| | | | | | | | canned tomatoes | 0.09 | |
| | | | | | | | peel | 0.44 | |

1) average of duplicate or triplicate analyses of a single sample.
- not collected.

Four field trials were conducted in representative head lettuce growing areas of Germany and Spain to determine the residue levels of boscalid (Raunft *et al.*, 2000, 2001/1006128). The lettuce plants were treated twice at a rate of 0.801 kg ai/ha, double GAP rate. For analysis, plant material was sampled within 3 hours after the second application and 14 days later. At the second sampling, additional lettuce was collected which was processed immediately in the field (Table 65).

Table 65. Boscalid residues in head lettuce and processed fractions resulting from supervised trials in Germany and Spain (Raunft *et al.*, 2000, 2001/1006128).

| HEAD LETTUCE country, year, location (variety) | Application | | | | | PHI days | Commodity | Residues mg/kg boscalid | Ref |
|---|---|---|---|---|---|---|---|---|---|
| | Form | kg ai/ha | kg ai/hL | water (L/ha) | no. | | | | |
| Spain, 2000 Andalucía (Yerga) | WG | 0.80 | 0.16 | 500 | 2 | 0 | head | 9.27 | 2001/1006128 |
| | | | | | | 14 | head | 1.05 | |
| | | | | | | | leaves (exterior) | 1.95 | |
| | | | | | | | leaves (exterior rinsed) | 1.32 | |
| | | | | | | | leaves (interior) | 0.06 | |
| | | | | | | | leaves (interior rinsed) | < 0.05 | |

| HEAD LETTUCE country, year, location (variety) | Application | | | | | PHI days | Commodity | Residues mg/kg boscalid | Ref |
|---|---|---|---|---|---|---|---|---|---|
| | Form | kg ai/ha | kg ai/hL | water (L/ha) | no. | | | | |
| Spain, 2000, Andalucía (Reina Verde) | WG | 0.82 | 0.16 | 511 | 2 | 0 | head | 9.81 | |
| | | | | | | 14 | head | 2.27 | |
| | | | | | | | leaves (exterior) | 2.52 | |
| | | | | | | | leaves (exterior rinsed) | 1.64 | |
| | | | | | | | leaves (interior) | < 0.05 | |
| | | | | | | | leaves (interior rinsed) | 0.07 | |
| Germany, 2000 Baden-Württemberg (Einstein) | WG | 0.79 | 0.16 | 492 | 2 | 0 | head | 21.01 | |
| | | | | | | 14 | head | 0.28 | |
| | | | | | | | leaves (exterior) | 0.35 | |
| | | | | | | | leaves (exterior rinsed) | 0.16 | |
| | | | | | | | leaves (interior) | 0.05 | |
| | | | | | | | leaves (interior rinsed) | < 0.05 | |
| Germany, 2000 Rheinland-Pfalz (Nadine) | WG | 0.79 | 0.16 | 492 | 2 | 0 | Head | 26.33 | |
| | | | | | | 14 | head | 0.07 | |
| | | | | | | | leaves (exterior) | 0.09 | |
| | | | | | | | leaves (exterior rinsed) | 0.06 | |
| | | | | | | | leaves (interior) | < 0.05 | |
| | | | | | | | leaves (interior rinsed) | < 0.05 | |

Four field trials were conducted in different pea cultivation areas in Germany to determine the residue levels of boscalid in peas and process fractions involved in canned pea production (Scharm, 2000, 2000/1014885). The peas were treated twice at a rate of 1.0 kg ai/, which is double the GAP rate. For analysis, shoots (whole plant without roots) were collected directly after the last application. In three trials, green seeds were sampled at 7–8 days after the last application and manually separated from the pods. In one trial, the seeds matured earlier than expected. These were harvested using a combine harvester at 7 days after the last application. All seed samples were used for residue analysis and processing. During the processing of peas, to produce canned peas, the following fractions were obtained: washed peas, wash water, cooked peas, boiled water, canned peas and vegetable stock (Table 66).

Table 66. Boscalid residues in peas and processed fractions resulting from supervised trials in Germany (Scharm, 2000, 2000/1014885).

| PEA country, year, location (variety) | Application | | | | | PHI days | Commodity | Residues mg/kg boscalid | Ref |
|---|---|---|---|---|---|---|---|---|---|
| | Form | kg ai/ha | kg ai/hL | water (L/ha) | no. | | | | |
| Germany, 2000 Hesse (Stok) | WG | 0.99 | 0.25 | 395 | 2 | 0 | shoots | 20.27 | 2000/1014885 |
| | | | | | | 7 | green seeds (RAC) | 0.14 | |
| | | | | | | | washed peas | 0.07 | |
| | | | | | | | wash water | 0.06 | |
| | | | | | | | cooked peas | < 0.05 | |
| | | | | | | | boiled water | < 0.05 | |
| | | | | | | | canned peas | < 0.05 | |
| | | | | | | | vegetable stock | < 0.05 | |
| Germany, 2000, Thuringia (Gloriosa) | WG | 0.88 | 0.25 | 353 | 2 | 0 | shoots | 5.09 | |
| | | | | | | 8 | green seeds (RAC) | < 0.05 | |
| | | | | | | | washed peas | < 0.05 | |
| | | | | | | | wash water | < 0.05 | |
| | | | | | | | cooked peas | < 0.05 | |
| | | | | | | | boiled water | < 0.05 | |
| | | | | | | | canned peas | < 0.05 | |
| | | | | | | | vegetable stock | < 0.05 | |

| PEA country, year, location (variety) | Application | | | | | PHI days | Commodity | Residues mg/kg boscalid | Ref |
|---|---|---|---|---|---|---|---|---|---|
| | Form | kg ai/ha | kg ai/hL | water (L/ha) | no. | | | | |
| Germany, 2000, Thuringia (Remus) | WG | 0.89 | 0.25 | 355 | 2 | 0 | shoots | 5.76 | |
| | | | | | | 8 | green seeds (RAC) | < 0.05 | |
| | | | | | | | washed peas | < 0.05 | |
| | | | | | | | wash water | < 0.05 | |
| | | | | | | | cooked peas | < 0.05 | |
| | | | | | | | boiled water | < 0.05 | |
| | | | | | | | canned peas | < 0.05 | |
| | | | | | | | vegetable stock | < 0.05 | |
| Germany, 2000 Hesse (Nitouch) | WG | 0.99 | 0.25 | 396 | 2 | 0 | shoots | 9.48 | |
| | | | | | | 7 | green seeds (RAC) | < 0.05 | |
| | | | | | | | washed peas | < 0.05 | |
| | | | | | | | wash water | < 0.05 | |
| | | | | | | | cooked peas | < 0.05 | |
| | | | | | | | boiled water | < 0.05 | |
| | | | | | | | canned peas | < 0.05 | |
| | | | | | | | vegetable stock | < 0.05 | |

A second processing study was conducted on peas in Germany and the Netherlands to determine the residue levels of boscalid in peas and process fractions involved in canned pea production (Schulz, 2003, 2004/1000750). The peas were treated twice, each time at a rate of 1.34 kg ai/ha, five times GAP rate. For analysis, whole plants, without roots, were collected directly after the last application. At 10 and 14 days after the last application respectively, seeds (RAC) were sampled. During the processing of peas to produce canned peas, the following fractions were obtained: washed peas, wash water, cooked peas, boiled water, canned peas and vegetable stock (Table 67).

Table 67. Boscalid residues in peas and processed fractions resulting from supervised trials in Germany and the Netherlands (Schulz, 2004, 2004/1000750).

| PEA country, year, location (variety) | Application | | | | | PHI days | Commodity | Residues mg/kg boscalid | Ref |
|---|---|---|---|---|---|---|---|---|---|
| | Form | kg ai/ha | kg ai/hL | water (L/ha) | no. | | | | |
| Germany, 2003 Brandenburg (Feltham) | WG | 1.34 | 0.34 | 400 | 2 | 0 | whole plant | 49.4 | 2004/1000750 |
| | | | | | | 10 | seeds | < 0.05 | |
| | | | | | | | washed peas | < 0.05 | |
| | | | | | | | wash water | < 0.05 | |
| | | | | | | | cooked peas | < 0.05 | |
| | | | | | | | boiled water | < 0.05 | |
| | | | | | | | canned peas | < 0.05 | |
| | | | | | | | vegetable stock | < 0.05 | |
| Germany, 2003, Brandenburg (Samish) | WG | 1.34 | 0.34 | 400 | 2 | 0 | whole plant | 70.5 | |
| | | | | | | 13 | seeds | < 0.05 | |
| | | | | | | | washed peas | < 0.05 | |
| | | | | | | | wash water | < 0.05 | |
| | | | | | | | cooked peas | < 0.05 | |
| | | | | | | | boiled water | < 0.05 | |
| | | | | | | | canned peas | < 0.05 | |
| | | | | | | | vegetable stock | < 0.05 | |
| Netherlands, 2003, Limburg (Remus) | WG | 1.34 | 0.34 | 400 | 2 | 0 | whole plant | 28.5 | |
| | | | | | | 13 | seeds | < 0.05 | |
| | | | | | | | washed peas | < 0.05 | |
| | | | | | | | wash water | < 0.05 | |
| | | | | | | | cooked peas | < 0.05 | |
| | | | | | | | boiled water | < 0.05 | |
| | | | | | | | canned peas | < 0.05 | |
| | | | | | | | vegetable stock | < 0.05 | |

| PEA country, year, location (variety) | Application | | | | | PHI days | Commodity | Residues mg/kg boscalid | Ref |
|---|---|---|---|---|---|---|---|---|---|
| | Form | kg ai/ha | kg ai/hL | water (L/ha) | no. | | | | |
| Germany, 2003 Nordehein-Westfalen (Misty) | WG | 1.34 | 0.34 | 400 | WG | 0 | whole plant | 22.1 | |
| | | | | | | 14 | seeds | < 0.05 | |
| | | | | | | | washed peas | < 0.05 | |
| | | | | | | | wash water | < 0.05 | |
| | | | | | | | cooked peas | < 0.05 | |
| | | | | | | | boiled water | < 0.05 | |
| | | | | | | | canned peas | < 0.05 | |
| | | | | | | | vegetable stock | < 0.05 | |

A processing study was performed testing boscalid in soybean in 2001 (Versoi and Malinsky, 2001, 2001/5002529). At one location in Nebraska, soybean plants were treated twice at a rate of 5.6 kg ai/ha of boscalid, 5 times the normal recommended rate. Treatments started about three weeks before expected harvest with a 7 day interval between sprays. Soybean seed samples were collected at normal crop maturity, which was about 14 days after the final application. The seed was processed to meal, hulls and refined oil simulating typical commercial practices (Table 68).

Table 68. Boscalid residues in soybean and processed fractions resulting from supervised trials in USA (Versoi and Malinsky, 2001, 2001/5002529).

| SOYBEAN country, year, location (variety) | Application | | | | | PHI days | Commodity | Residues mg/kg boscalid | Ref |
|---|---|---|---|---|---|---|---|---|---|
| | Form | kg ai/ha | kg ai/hL | water (L/ha) | no. | | | | |
| USA, 2001 Nebraska (?) | WG | 1.91 | | | 2 | 13 | seed RAC | 0.31 | 2001/5002529 |
| | | | | | | | hulls | 0.54 | |
| | | | | | | | meals | < 0.05 | |
| | | | | | | | refined oil | 0.13 | |

Carrot plants at four locations in typical growing areas in Germany were treated with boscalid to investigate the residues in raw carrots and in the processed fractions cooked carrot, canned carrot and carrot juice (Scharm, 2000, 2001/1000939). The carrot plants were treated three times, at a rate of 1.202 kg/ha of boscalid, which is three times the GAP rate. The applications were made at a 6 to 8 days interval, starting 42 days before the expected harvest. The last application was performed 28 days before expected harvest, i.e., at the GAP pre-harvest-interval. Samples were collected directly after the last application, once the spray had dried, and 28 days later. These samples were used for residue analysis and for processing. The carrots were processed into the following products: washed carrot, combined wash water, peeled carrot, carrot peel, cooked carrot, cooking liquid, carrot juice, pomace, canned carrot and vegetable stock (Table 69).

Table 69. Boscalid residues in carrots and processed fractions resulting from supervised trials in Germany (Scharm, 2001, 2001/1000939).

| CARROT country, year, location (variety) | Application | | | | | PHI days | Commodity | Residues mg/kg boscalid | Ref |
| | Form | kg ai/ha | kg ai/hL | water (L/ha) | no. | | | | |
|---|---|---|---|---|---|---|---|---|---|
| Germany, 2000 Schleswig-Holstein (Primecut 59) | WG | 1.21 | 0.30 | 402 | 3 | 0 | plant with roots | 19.95 | 2001/1000939 |
| | | | | | | 28 | plant with roots | 1.11 | |
| | | | | | | | washed carrot | 0.30 | |
| | | | | | | | wash water | 0.12 | |
| | | | | | | | topped/peeled | 0.07 | |
| | | | | | | | peel | 0.75 | |
| | | | | | | | cooked carrot | 0.05 | |
| | | | | | | | cooking liquid | < 0.05 | |
| | | | | | | | juice | < 0.05 | |
| | | | | | | | pomace | 0.07 | |
| | | | | | | | canned carrot | < 0.05 | |
| | | | | | | | vegetable stock | < 0.05 | |
| Germany, 2000, Rhineland-Palatinate (Majestro) | WG | 1.19 | 0.30 | 397 | 3 | 0 | plant with roots | 9.57 | |
| | | | | | | 28 | plant with roots | 0.35 | |
| | | | | | | | washed carrot | 0.15 | |
| | | | | | | | wash water | 0.07 | |
| | | | | | | | topped/peeled | < 0.05 | |
| | | | | | | | peel | 0.46 | |
| | | | | | | | cooked carrot | < 0.05 | |
| | | | | | | | cooking liquid | < 0.05 | |
| | | | | | | | juice | < 0.05 | |
| | | | | | | | pomace | 0.06 | |
| | | | | | | | canned carrot | < 0.05 | |
| | | | | | | | vegetable stock | < 0.05 | |
| Germany, 2000, Rhineland-Palatinate (Majestro) | WG | 1.22 | 0.30 | 405 | 3 | 0 | plant with roots | 7.90 | |
| | | | | | | 28 | plant with roots | 0.48 | |
| | | | | | | | washed carrot | 0.19 | |
| | | | | | | | wash water | 0.09 | |
| | | | | | | | topped/peeled | < 0.05 | |
| | | | | | | | peel | 0.72 | |
| | | | | | | | cooked carrot | < 0.05 | |
| | | | | | | | cooking liquid | < 0.05 | |
| | | | | | | | juice | < 0.05 | |
| | | | | | | | pomace | 0.08 | |
| | | | | | | | canned carrot | < 0.05 | |
| | | | | | | | vegetable stock | < 0.05 | |
| Germany, 2000, Rhineland-Palatinate (Majestro) | WG | 1.18 | 0.30 | 393 | 3 | 0 | plant with roots | 5.29 | |
| | | | | | | 28 | plant with roots | 0.39 | |
| | | | | | | | washed carrot | 0.14 | |
| | | | | | | | wash water | 0.15 | |
| | | | | | | | topped/peeled | < 0.05 | |
| | | | | | | | peel | 0.58 | |
| | | | | | | | cooked carrot | < 0.05 | |
| | | | | | | | cooking liquid | < 0.05 | |
| | | | | | | | juice | < 0.05 | |
| | | | | | | | pomace | 0.06 | |
| | | | | | | | canned carrot | < 0.05 | |
| | | | | | | | vegetable stock | < 0.05 | |

In 2001, a study was performed in the US and Canada testing boscalid on sugar beet (Haughey and Abdel-Baky, 2001, 2001/5003245). In addition to the residue trials (10 locations) which were done according to the recommended GAP (2 × 0.56 kg ai/ha of boscalid) one trial was performed in Texas using the 5× application rate (2 × 2.8 kg ai/ha) to increase the probability of finding residues. Applications were done 21 and 14 days before expected harvest. Sugar beet RAC samples (roots and tops) were collected 7 days after the last application. They were processed to dried pulp, molasses and refined sugar simulating typical commercial practices (Table 70).

Table 70. Boscalid residues in sugar beet and processed fractions resulting from supervised trials in USA and Canada (Haughey and Abdel-Baky, 2001, 2001/5003245).

| SUGAR BEET country, year, location (variety) | Application | | | | | PHI days | Commodity | Residues mg/kg boscalid | Ref |
|---|---|---|---|---|---|---|---|---|---|
| | Form | kg ai/ha | kg ai/hL | water (L/ha) | no. | | | | |
| USA, 2001, MI (ACH 1353) | WG | 0.56 | 0.28 | 201 | 2 | 7 | sugar beet, root | 0.07 0.09 | 2001/5003245 |
| | | | | | | | sugar beet tops | 4.28 4.19 | |
| | | | | | | 14 | sugar beet, root | 0.08 0.08 | |
| | | | | | | | sugar beet tops | 2.67 3.28 | |
| | | | | | | 21 | sugar beet, root | 0.08 0.07 | |
| | | | | | | | sugar beet tops | 2.11 1.64 | |
| USA, 2001, MN(Crystal 205) | WG | 0.57 | 0.34 | 168 | 2 | 7 | sugar beet, root | 0.19 0.23 | |
| | | | | | | | sugar beet tops | 3.31 3.49 | |
| | | | | | | 14 | sugar beet, root | 0.12 0.14 | |
| | | | | | | | sugar beet tops | 2.49 2.26 | |
| | | | | | | 21 | sugar beet, root | 0.09 0.10 | |
| | | | | | | | sugar beet tops | 2.18 2.12 | |
| | | | | | | 28 | sugar beet, root | 0.18 0.14 | |
| | | | | | | | sugar beet tops | 1.77 1.64 | |
| USA, 2001, MN (HM Resist RR) | WG | 0.56 | 0.30 | 187 | 2 | 7 | sugar beet, root | 0.08 < 0.05 | |
| | | | | | | | sugar beet tops | 3.15 3.08 | |
| | | | | | | 14 | sugar beet, root | < 0.05 0.05 | |
| | | | | | | | sugar beet tops | 2.64 2.68 | |
| | | | | | | 21 | sugar beet, root | < 0.05 < 0.05 | |
| | | | | | | | sugar beet tops | 2.69 2.64 | |
| USA, 2001, ND (Chrystal 222) | WG | 0.55 | 0.30 | 187 | 2 | 7 | sugar beet, root | < 0.05 0.08 | |
| | | | | | | | sugar beet tops | 3.67 4.73 | |
| | | | | | | 14 | sugar beet, root | < 0.05 0.06 | |
| | | | | | | | sugar beet tops | 4.03 2.92 | |
| | | | | | | 21 | sugar beet, root | < 0.05 < 0.05 | |
| | | | | | | | sugar beet tops | 2.24 3.27 | |

| SUGAR BEET country, year, location (variety) | Application | | | | | PHI days | Commodity | Residues mg/kg boscalid | Ref |
|---|---|---|---|---|---|---|---|---|---|
| | Form | kg ai/ha | kg ai/hL | water (L/ha) | no. | | | | |
| USA, 2001, WI (66283 Medium) | WG | 0.55 | 0.30 | 187 | 2 | 7 | sugar beet, root | 0.14 0.12 | |
| | | | | | | 14 | sugar beet, root | 0.09 0.14 | |
| | | | | | | | sugar beet tops | 3.60 4.29 | |
| | | | | | | 21 | sugar beet, root | 0.08 0.09 | |
| | | | | | | | sugar beet tops | 3.65 2.64 | |
| USA, 2001, ND (Chrystal 196) | WG | 0.55 | 0.23 | 243 | 2 | 7 | sugar beet, root | 0.07 < 0.05 | |
| | | | | | | | sugar beet tops | 15.15 21.30 11.40 19.60 | |
| | | | | | | 14 | sugar beet, root | < 0.05 < 0.05 | |
| | | | | | | | sugar beet tops | 14.61 21.50 13.86 19.80 | |
| | | | | | | 21 | sugar beet, root | < 0.05 < 0.05 | |
| | | | | | | | sugar beet tops | 15.51 15.90 19.83 26.90 | |
| USA, 2001, TX (Ranger) | WG | 0.55 | 0.30 | 187 | 2 | 7 | sugar beet, root | 0.06 < 0.05 | |
| | | | | | | | sugar beet tops | 8.00 7.24 | |
| | | | | | | 14 | sugar beet, root | < 0.05 0.05 | |
| | | | | | | | sugar beet tops | 3.68 6.24 | |
| | | | | | | 21 | sugar beet, root | < 0.05 < 0.05 | |
| | | | | | | | sugar beet tops | 6.69 4.70 | |
| | | 2.8 | 1.50 | 187 | 2 | 7 | sugar beet root RAC | 0.34 0.17 | |
| | | | | | | | dried pulp | 0.50 0.60 | |
| | | | | | | | molasses | 0.46 0.40 | |
| | | | | | | | refined sugar | 0.07 < 0.05 | |
| USA, 2001, CA (NB7R) | WG | 0.57 | 0.23 | 253 | 2 | 7 | sugar beet, root | 0.07 0.05 | |
| | | | | | | | sugar beet tops | 2.52 2.08 | |
| | | | | | | 14 | sugar beet, root | 0.07 0.10 | |
| | | | | | | | sugar beet tops | 1.62 1.04 | |
| | | | | | | 21 | sugar beet, root | 0.06 < 0.05 | |
| | | | | | | | sugar beet tops | 1.04 0.90 | |

| SUGAR BEET country, year, location (variety) | Application | | | | | PHI days | Commodity | Residues mg/kg boscalid | Ref |
|---|---|---|---|---|---|---|---|---|---|
| | Form | kg ai/ha | kg ai/hL | water (L/ha) | no. | | | | |
| USA, 2001, WA (Canyon) | WG | 0.56 | 0.30 | 187 | 2 | 7 | sugar beet, root<br><br>sugar beet tops | 0.08<br>0.08<br>6.24<br>7.72 | |
| | | | | | | 14 | sugar beet, root<br><br>sugar beet tops | 0.11<br>0.10<br>3.78<br>3.34 | |
| | | | | | | 21 | sugar beet, root<br><br>sugar beet tops | 0.12<br>0.08<br>3.98<br>2.82 | |
| Canada, 2001, MB (Bergen) | WG | 0.57 | 0.51 | 112 | 2 | 7 | sugar beet, root<br><br>sugar beet tops | < 0.05<br>< 0.05<br>22.41<br>17.51 | |
| | | | | | | 14 | sugar beet, root<br><br>sugar beet tops | 0.21<br>0.18<br>12.33<br>7.92 | |
| | | | | | | 21 | sugar beet, root<br><br>sugar beet tops | 0.17<br>0.14<br>8.85<br>7.44 | |

Four field trials were conducted in different representative cereal growing areas in Germany to determine the residue level of boscalid in barley and processed products during the 2002 growing season (Schulz, 2002, 2003/1000946). The barley was treated twice at a rate of 1.05 kg ai/ha, which is three times the GAP rate. The applications were made at 6 to 8 day intervals, starting at 42 days before expected harvest. Samples of barley grain were collected 40 to 59 days after the last application and used for the processing procedures. The material was subjected to two different processes simulating industrial practice: in a milling process, the consumer product pot barley and the waste fractions pearling dust and offal were obtained. In a malting and brewing process, the involved fractions malt germs, offal, steeping water, brewing malt, spent grain, condensed water, trub (flocs), beer yeast, and beer were investigated (Table 71).

Table 71. Boscalid residues in barley and processed fractions resulting from supervised trials in Germany (Schulz, 2003, 2003/1000946).

| BARLEY country, year, location (variety) | Application | | | | | PHI days | Commodity | Residues mg/kg boscalid | Ref |
|---|---|---|---|---|---|---|---|---|---|
| | Form | kg ai/ha | kg ai/hL | water (L/ha) | no. | | | | |
| Germany, 2002 Hesse (Scarlett) | SC | 1.06 | 0.35 | 303 | 2 | 59 | grain[1] | 1.98 | 2003/1000946 |
| | | | | | | | grain[2] | 1.93 | |
| | | | | | | | offal | 21.87 | |
| | | | | | | | pearling dust | 11.38 | |
| | | | | | | | pot barley | 0.71 | |

| BARLEY country, year, location (variety) | Application | | | | | PHI days | Commodity | Residues mg/kg boscalid | Ref |
|---|---|---|---|---|---|---|---|---|---|
| | Form | kg ai/ha | kg ai/hL | water (L/ha) | no. | | | | |
| | | | | | | | grain[3] | 1.87 | |
| | | | | | | | offal | 6.55 | |
| | | | | | | | brewing malt | 0.70 | |
| | | | | | | | malt germ | 0.99 | |
| | | | | | | | steeping water | 0.02 | |
| | | | | | | | brewing malt | 0.65 | |
| | | | | | | | spent grain | 0.54 | |
| | | | | | | | trub (flocs) | 0.89 | |
| | | | | | | | condensed water | < 0.01 | |
| | | | | | | | beer yeast | 0.29 | |
| | | | | | | | beer (cold) | 0.02 | |
| | | | | | | | beer (frozen) | 0.02 | |
| Germany, 2002 Hesse (Scarlett) | SC | 1.02 | 0.35 | 293 | 2 | 40 | grain[1] | 2.05 | |
| | | | | | | | grain[2] | 2.61 | |
| | | | | | | | offal | 34.14 | |
| | | | | | | | pearling dust | 12.61 | |
| | | | | | | | pot barley | 0.75 | |
| | | | | | | | grain[3] | 2.02 | |
| | | | | | | | offal | 4.34 | |
| | | | | | | | brewing malt | 0.90 | |
| | | | | | | | malt germ | 1.80 | |
| | | | | | | | steeping water | 0.02 | |
| | | | | | | | brewing malt | 0.88 | |
| | | | | | | | spent grain | 0.80 | |
| | | | | | | | trub (flocs) | 0.90 | |
| | | | | | | | condensed water | < 0.01 | |
| | | | | | | | beer yeast | 0.94 | |
| | | | | | | | beer (cold) | 0.03 | |
| | | | | | | | beer (frozen) | 0.03 | |
| Germany, 2002 Rhineland-Palatinate (Scarlett) | SC | 1.04 | 0.35 | 297 | 2 | 41 | grain[1] | 1.25 | |
| | | | | | | | grain[2] | 1.98 | |
| | | | | | | | offal | 26.69 | |
| | | | | | | | pearling dust | 12.55 | |
| | | | | | | | pot barley | 0.74 | |
| | | | | | | | grain[3] | 1.60 | |
| | | | | | | | offal | 5.00 | |
| | | | | | | | brewing malt | 0.93 | |
| | | | | | | | malt germ | 1.55 | |
| | | | | | | | steeping water | 0.02 | |
| | | | | | | | brewing malt | 0.79 | |
| | | | | | | | spent grain | 0.76 | |
| | | | | | | | trub (flocs) | 0.57 | |
| | | | | | | | condensed water | < 0.01 | |
| | | | | | | | beer yeast | 0.46 | |
| | | | | | | | beer (cold) | 0.03 | |
| | | | | | | | beer (frozen) | 0.03 | |
| Germany, 2002 Rhineland-Palatinate (Barke) | SC | 1.04 | 0.35 | 298 | 2 | 50 | grain[1] | 1.31 | |
| | | | | | | | grain[2] | 1.25 | |
| | | | | | | | offal | 16.00 | |
| | | | | | | | pearling dust | 9.16 | |
| | | | | | | | pot barley | 0.28 | |

| BARLEY country, year, location (variety) | Application | | | | | PHI days | Commodity | Residues mg/kg boscalid | Ref |
|---|---|---|---|---|---|---|---|---|---|
| | Form | kg ai/ha | kg ai/hL | water (L/ha) | no. | | | | |
| | | | | | | | grain[3] | 1.26 | |
| | | | | | | | offal | 5.47 | |
| | | | | | | | brewing malt | 0.65 | |
| | | | | | | | malt germ | 1.35 | |
| | | | | | | | steeping water | 0.01 | |
| | | | | | | | brewing malt | 0.51 | |
| | | | | | | | spent grain | 0.43 | |
| | | | | | | | trub (flocs) | 0.88 | |
| | | | | | | | condensed water | < 0.01 | |
| | | | | | | | beer yeast | 0.24 | |
| | | | | | | | beer (cold) | 0.02 | |
| | | | | | | | beer (frozen) | 0.02 | |

1) Transfer factor = residue in process fraction / residue in RAC

2) for residue analysis

3) starting material for pot barley production

4) starting material for malting/brewing process

Four field trials were conducted in different representative cereal growing areas in Germany to determine the residue level of boscalid in winter wheat and processed products (Renner, 2003, 2003/1000945). The wheat was treated twice at a rate of 1.05 kg/ha, which is three times the GAP rate. The applications were carried out at 42 days before expected harvest. Samples of wheat grain were collected 42 to 60 days after last application and used for the processing procedures. The material was subjected to processes simulating industrial practice producing wheat germ, coarse and total bran, flour (type 550) and wholemeal flour and bread (Table 72).

Table 72. Boscalid residues in winter wheat and processed fractions resulting from supervised trials in Germany (Renner, 2003, 2003/1000945).

| WINTER WHEAT country, year, location (variety) | Application | | | | | PHI days | Commodity | Residues mg/kg boscalid | Ref |
|---|---|---|---|---|---|---|---|---|---|
| | Form | kg ai/ha | kg ai/hL | water (L/ha) | no. | | | | |
| Germany, 2002 Gülzow-Wilheminenhof (Trifter) | SC | 1.06 | 0.35 | 306 | 2 | 53 | grain | 0.30 | 2003/1000945 |
| | | | | | | | middlings | 0.71 | |
| | | | | | | | coarse bran | 1.15 | |
| | | | | | | | total bran | 1.16 | |
| | | | | | | | toppings | 0.37 | |
| | | | | | | | flour type 550 | 0.07 | |
| | | | | | | | wholemeal flour | 0.33 | |
| | | | | | | | whole bread | 0.18 | |
| | | | | | | | wheat germs | 0.29 | |
| Germany, 2002 Nienburg (Cardos) | SC | 1.05 | 0.35 | 304 | 2 | 60 | grain | 0.36 | |
| | | | | | | | middlings | 1.32 | |
| | | | | | | | coarse bran | 1.85 | |
| | | | | | | | total bran | 1.96 | |
| | | | | | | | toppings | 0.81 | |
| | | | | | | | flour type 550 | 0.08 | |
| | | | | | | | wholemeal flour | 0.41 | |
| | | | | | | | whole bread | 0.27 | |
| | | | | | | | wheat germs | 0.57 | |
| Germany, 2002 Litzendorf (Ludwig) | SC | 1.05 | 0.34 | 305 | 2 | 42 | grain | 0.11 | |
| | | | | | | | middlings | 0.45 | |
| | | | | | | | coarse bran | 0.63 | |
| | | | | | | | total bran | 0.51 | |
| | | | | | | | toppings | 0.31 | |
| | | | | | | | flour type 550 | 0.05 | |
| | | | | | | | wholemeal flour | 0.20 | |
| | | | | | | | whole bread | 0.11 | |
| | | | | | | | wheat germs | 0.15 | |

| WINTER WHEAT country, year, location (variety) | Application | | | | | PHI days | Commodity | Residues mg/kg boscalid | Ref |
|---|---|---|---|---|---|---|---|---|---|
| | Form | kg ai/ha | kg ai/hL | water (L/ha) | no. | | | | |
| Germany, 2002 Motterwitz (Kanzler) | SC | 1.07 | 0.35 | 309 | 2 | 44 | grain | 0.17 | |
| | | | | | | | middlings | 0.41 | |
| | | | | | | | coarse bran | 0.57 | |
| | | | | | | | total bran | 0.56 | |
| | | | | | | | toppings | 0.24 | |
| | | | | | | | flour type 550 | 0.08 | |
| | | | | | | | wholemeal flour | 0.22 | |
| | | | | | | | whole bread | 0.15 | |
| | | | | | | | wheat germs | 0.22 | |

A study was performed in the US and Canada testing boscalid in sweet and field corn in 2001 (Versoi and Malinsky, 2001, 2001/5002624). In addition to the residue trials done according to the recommended GAP (2 × 0.56 kg ai/ha of boscalid at 30 locations) one trial was performed in Nebraska using a 5× application rate (2 × 2.8 kg ai/ha) to increase the probability of finding residues in the processed fractions. Applications were done 42 and 35 days before expected harvest. Corn samples were collected 7 and 14 days after the last application and processed into refined oil, starch, meal, grits and flour simulating typical commercial practices (see Table 73).

Table 73. Boscalid residues in corn and processed fractions resulting from supervised trials in USA and Canada (Versoi and Malinsky, 2001, 2001/5002624).

| CORN country, year, location (variety) | Application | | | | | PHI days | Commodity | Residues mg/kg boscalid | Ref |
|---|---|---|---|---|---|---|---|---|---|
| | Form | kg ai/ha | kg ai/hL | water (L/ha) | no. | | | | |
| USA, 2001, PA (Doebler's 642XP) | WG | 0.59 | 0.14 | 418 | 2 | 7 | fresh corn | < 0.05 < 0.05 | 2001/5002624 |
| | | | | | | 14 | | < 0.05 < 0.05 | |
| | | | | | | 21 | | < 0.05 < 0.05 | |
| | | | | | | 28 | | < 0.05 < 0.05 | |
| | | | | | | 35 | | < 0.05 < 0.05 | |
| | | 0.58 | 0.14 | 412 | 2 | 0 | forage | 4.86 4.31 | |
| | | | | | | 7 | | 4.65 4.14 | |
| | | | | | | 14 | | 3.07 3.36 | |
| | | | | | | 20 | | 4.14 4.10 | |
| | | | | | | 27 | | 3.30 3.64 | |

| CORN country, year, location (variety) | Application | | | | | PHI days | Commodity | Residues mg/kg boscalid | Ref |
|---|---|---|---|---|---|---|---|---|---|
| | Form | kg ai/ha | kg ai/hL | water (L/ha) | no. | | | | |
| | | 0.58 | 0.14 | 409 | 2 | 0 | grain | < 0.05 < 0.05 | |
| | | | | | | 7 | | < 0.05 < 0.05 | |
| | | | | | | 14 | | < 0.05 < 0.05 | |
| | | | | | | 20 | | < 0.05 < 0.05 | |
| | | | | | | 28 | | < 0.05 < 0.05 | |
| | | | | | | 0 | stover | 11.95 11.48 | |
| | | | | | | 7 | | 10.50 10.30 | |
| | | | | | | 14 | | 9.15 9.70 | |
| | | | | | | 20 | | 9.60 6.85 | |
| | | | | | | 28 | | 6.05 5.60 | |
| USA, 2001, PA (Argent) | WG | 0.57 | 0.14 | 407 | 2 | 7 | fresh corn | < 0.05 < 0.05 | |
| | | | | | | 7 | forage | 5.46 7.95 | |
| | | | | | | 14 | | 5.86 4.89 | |
| | | 0.57 | 0.14 | 406 | 1 | 7 | stover | 21.55 19.92 | |
| USA, 2001, NC (Pioneer) | WG | 0.50 | 0.18 | 280 | 2 | 6 | fresh corn | < 0.05 < 0.05 | |
| | | 0.56 | 0.20 | 283 | 2 | 7 | forage | 4.43 3.42 | |
| | | | | | | 14 | | 3.21 3.76 | |
| | | 0.56 | 0.20 | 281 | 2 | 7 | grain | < 0.05 < 0.05 | |
| | | | | | | 7 | stover | 18.30 19.45 | |
| USA, 2001, FL (Golden Queen) | WG | 0.57 | 0.20 | 286 | 2 | 6 | fresh corn | < 0.05 < 0.05 | |
| | | | | | | 6 | forage | 8.40 7.08 | |
| | | | | | | 14 | | 3.18 2.77 | |
| | | 0.56 | 0.20 | 281 | 2 | 7 | stover | 12.15 7.95 | |
| USA, 2001, IL (Burrus BX 789) | WG | 0.56 | 0.45 | 125 | 2 | 7 | fresh corn | < 0.05 < 0.05 | |
| | | 0.56 | 0.45 | 124 | 2 | 7 | forage | 11.00 14.69 | |
| | | | | | | 14 | | 7.27 9.65 | |
| | | 0.56 | 0.45 | 125 | 2 | 7 | grain | < 0.05 < 0.05 | |
| | | | | | | | stover | 11.25 10.40 | |

| CORN country, year, location (variety) | Application | | | | | PHI days | Commodity | Residues mg/kg boscalid | Ref |
|---|---|---|---|---|---|---|---|---|---|
| | Form | kg ai/ha | kg ai/hL | water (L/ha) | no. | | | | |
| USA, 2001, IL (Hamel H1166) | WG | 0.56 | 0.46 | 123 | 2 | 7 | fresh corn | < 0.05<br>< 0.05 | |
| | | 0.56 | 0.45 | 124 | 2 | 7<br><br>14 | forage | 5.09<br>4.01<br>3.91<br>3.91 | |
| | | 0.56 | 0.45 | 125 | 2 | 7 | grain<br><br>stover | < 0.05<br>< 0.05<br>14.15<br>12.75 | |
| USA, 2001, IL (Pioneer 33G26) | WG | 0.56 | 0.29 | 194 | 2 | 7 | fresh corn | < 0.05<br>< 0.05 | |
| | | 0.56 | 0.29 | 194 | 2 | 7<br><br>14 | forage | 7.59<br>7.81<br>1.80<br>3.00 | |
| | | 0.56 | 0.28 | 197 | 2 | 6 | grain<br><br>stover | < 0.05<br>< 0.05<br>8.75<br>10.60 | |
| USA, 2001, IL (Pioneer 33G26) | WG | 0.57 | 0.29 | 196 | 2 | 7 | fresh corn | < 0.05<br>< 0.05 | |
| | | 0.56 | 0.29 | 194 | 2 | 7<br><br>14 | forage | 7.16<br>7.43<br>5.11<br>5.95 | |
| | | 0.56 | 0.28 | 198 | 2 | 7 | grain<br><br>stover | < 0.05<br>< 0.05<br>10.35<br>9.25 | |
| USA, 2001, IA (Pioneer 34B23) | WG | 0.56 | 0.44 | 126 | 2 | 6 | fresh corn | < 0.05<br>< 0.05 | |
| | | 0.56 | 0.42 | 133 | 2 | 7<br><br>13 | forage | 5.63<br>3.59<br>2.32<br>2.08 | |
| | | 0.56 | 0.30 | 187 | 2 | 8 | grain<br><br>stover | < 0.05<br>< 0.05<br>3.29<br>3.18 | |
| USA, 2001, IA (Pioneer 33A14) | WG | 0.56 | 0.43 | 131 | 2 | 7<br><br>14 | forage | 3.13<br>5.10<br>4.26<br>4.15 | |
| | | 0.56 | 0.34 | 166 | 2 | 7 | grain<br><br>stover | < 0.05<br>< 0.05<br>7.65<br>6.70 | |
| USA, 2001, MO (Golden Harvest H-2552) | WG | 0.57 | 0.43 | 133 | 2 | 7<br><br>13 | forage | 6.06<br>6.44<br>3.99<br>7.41 | |
| | | 0.55 | 0.34 | 161 | 2 | 7 | grain<br><br>stover | < 0.05<br>< 0.05<br>10.25<br>6.95 | |

| CORN country, year, location (variety) | Application | | | | | PHI days | Commodity | Residues mg/kg boscalid | Ref |
|---|---|---|---|---|---|---|---|---|---|
| | Form | kg ai/ha | kg ai/hL | water (L/ha) | no. | | | | |
| USA, 2001, MN (DK C48-83) | WG | 0.56 | 0.36 | 154 | 2 | 7<br><br>14 | forage | 3.73<br>3.52<br>2.63<br>2.64 | |
| | | 0.56 | 0.32 | 175 | 2 | 7 | grain<br><br>stover | < 0.05<br>0.06<br>9.55<br>8.05 | |
| USA, 2001, MN (DK C48-83) | WG | 0.56 | 0.34 | 166 | 2 | 7<br><br>14 | forage | 1.90<br>2.68<br>1.99<br>1.90 | |
| | | 0.56 | 0.34 | 167 | 2 | 7 | grain<br><br>stover | < 0.05<br>< 0.05<br>8.20<br>13.15 | |
| USA, 2001, NE (Northrup King N58-D1) | WG | 0.56 | 0.30 | 187 | 2 | 7<br><br>13 | forage | 7.18<br>7.15<br>3.27<br>3.39 | |
| | | 0.56 | 0.30 | 187 | 2 | 7 | grain<br><br>stover | < 0.05<br>< 0.05<br>11.60<br>10.15 | |
| | | 2.81 | 1.50 | 187 | 2 | 7 | grain<br><br>refined oil, wet milled<br><br>starch<br><br>meal<br><br>refined oil, drt milled<br><br>grits<br><br>flour | < 0.05<br>0.05<br>0.27<br>0.33<br>< 0.05<br>< 0.05<br>< 0.05<br>< 0.05<br>< 0.05<br>< 0.05<br>< 0.05<br>< 0.05<br>< 0.05<br>< 0.05 | |
| USA, 2001, NE (Cargill M739) | WG | 0.56 | 0.30 | 188 | 2 | 7<br><br>13 | forage | 6.24<br>6.69<br>2.53<br>3.43 | |
| | | 0.56 | 0.30 | 189 | 2 | 7 | grain<br><br>stover | < 0.05<br>< 0.05<br>30.10<br>24.35 | |
| USA, 2001, WI (Pioneer 3753) | WG | 0.56 | 0.30 | 188 | 2 | 7<br><br>14 | forage | 5.58<br>6.05<br>2.01<br>1.87 | |
| | | 0.56 | 0.30 | 187 | 2 | 8 | grain<br><br>stover | < 0.05<br>< 0.05<br>19.00<br>13.25 | |
| USA, 2001, WI (Pioneer 3751) | WG | 0.57 | 0.30 | 190 | 2 | 7<br><br>14 | forage | 5.58<br>5.85<br>2.09<br>2.72 | |

| CORN country, year, location (variety) | Application | | | | | PHI days | Commodity | Residues mg/kg boscalid | Ref |
|---|---|---|---|---|---|---|---|---|---|
| | Form | kg ai/ha | kg ai/hL | water (L/ha) | no. | | | | |
| | | 0.57 | 0.30 | 189 | 2 | 8 | grain | 0.05 0.07 | |
| | | | | | | | stover | 17.75 24.55 | |
| USA, 2001, WI (DeKalb DKC42-22 YG) | WG | 0.57 | 0.30 | 192 | 2 | 7 | forage | 5.29 5.44 | |
| | | | | | | 14 | | 2.91 3.44 | |
| | | 0.56 | 0.30 | 188 | 2 | 8 | grain | 0.07 0.08 | |
| | | | | | | | stover | 26.65 22.35 | |
| USA, 2001, WI (Pioneer 38T27) | WG | 0.55 | 0.30 | 186 | 2 | 7 | forage | 5.40 5.31 | |
| | | | | | | 14 | | 2.79 2.64 | |
| | | 0.56 | 0.30 | 187 | 2 | 8 | grain | 0.07 0.06 | |
| | | | | | | | stover | 14.75 16.40 | |
| USA, 2001, MN (DeKalb DKC39-47) | WG | 0.56 | 0.30 | 188 | 2 | 0 | forage | 7.79 6.76 | |
| | | | | | | 7 | | 6.84 4.79 | |
| | | | | | | 14 | | 6.23 6.30 | |
| | | | | | | 21 | | 3.52 4.23 | |
| | | | | | | 28 | | 3.88 3.31 | |
| | | 0.56 | 0.30 | 188 | 2 | 0 | grain | < 0.05 < 0.05 | |
| | | | | | | 7 | | < 0.05 < 0.05 | |
| | | | | | | 14 | | < 0.05 < 0.05 | |
| | | | | | | 21 | | < 0.05 < 0.05 | |
| | | | | | | 28 | | < 0.05 < 0.05 | |
| | | | | | | 0 | stover | 7.93 8.38 | |
| | | | | | | 7 | | 5.63 4.51 | |
| | | | | | | 14 | | 3.81 4.06 | |
| | | | | | | 21 | | 2.60 2.60 | |
| | | | | | | 28 | | 2.57 2.03 | |
| USA, 2001, MN (DeKalb DKC39-47) | WG | 0.56 | 0.30 | 188 | 2 | 7 | forage | 6.08 4.50 | |
| | | | | | | 14 | | 3.80 3.55 | |
| | | 0.56 | 0.30 | 187 | 2 | 7 | grain | < 0.05 < 0.05 | |
| | | | | | | | stover | 8.70 10.10 | |

| CORN country, year, location (variety) | Application | | | | | PHI days | Commodity | Residues mg/kg boscalid | Ref |
|---|---|---|---|---|---|---|---|---|---|
| | Form | kg ai/ha | kg ai/hL | water (L/ha) | no. | | | | |
| Canada, 2001 Quebec (DK 44-22 bt) | WG | 0.54 | 0.24 | 228 | 2 | 7 | fresh corn | na < 0.05 | |
| | | 0.54 | 0.24 | 226 | 2 | 8 14 | forage | 4.61 5.32 1.62 3.29 | |
| | | 0.56 | 0.24 | 235 | 2 | 7 | grain stover | < 0.05 < 0.05 11.25 11.60 | |
| Canada, 2001 Quebec (DK 44-22 bt) | WG | 0.56 | 0.24 | 233 | 2 | 8 | fresh corn | < 0.05 < 0.05 | |
| | | 0.57 | 0.24 | 237 | 2 | 8 14 | forage | 3.34 2.87 2.76 1.89 | |
| | | 0.54 | 0.24 | 224 | 2 | 7 | grain stover | < 0.05 < 0.05 10.90 13.10 | |
| Canada, 2001 Quebec (DKC 44-41) | WG | 0.55 | 0.24 | 230 | 2 | 7 14 | forage | 5.44 3.70 3.51 5.55 | |
| | | 0.55 | 0.24 | 232 | 2 | 7 | grain stover | < 0.05 < 0.05 7.73 6.40 | |
| Canada, 2001 Quebec (DKC 359 RR) | WG | 0.56 | 0.24 | 233 | 2 | 7 14 | forage | 3.54 3.89 2.88 2.25 | |
| | | 0.58 | 0.24 | 244 | 2 | 6 | grain stover | < 0.05 < 0.05 15.45 15.80 | |
| USA, 2001, OK (NK 4242 BT) | WG | 0.56 | 0.46 | 123 | 2 | 6 | fresh corn | < 0.05 < 0.05 | |
| | | 0.52 | 0.42 | 124 | 2 | 6 13 | forage | 3.54 3.03 3.04 1.97 | |
| | | 0.55 | 0.43 | 128 | 2 | 7 | grain stover | < 0.05 < 0.05 6.20 5.25 | |
| Canada, 2001, Alberta (Sheeba) | WG | 0.55 | 0.28 | 197 | 2 | 7 7 14 | fresh corn forage | < 0.05 < 0.05 5.06 6.18 3.50 4.45 | |
| | | 0.57 | 0.28 | 202 | 2 | 7 | stover | 9.75 10.85 | |
| USA, 2001, CA (Silver Queen) | WG | 0.56 | 0.20 | 282 | 2 | 8 8 14 | fresh corn forage | < 0.05 < 0.05 8.06 9.24 3.38 9.45 | |

| CORN country, year, location (variety) | Application | | | | | PHI days | Commodity | Residues mg/kg boscalid | Ref |
|---|---|---|---|---|---|---|---|---|---|
| | Form | kg ai/ha | kg ai/hL | water (L/ha) | no. | | | | |
| | | 0.57 | 0.20 | 287 | 2 | 7 | stover | 37.10 21.45 | |
| USA, 2001, OR (Kandy Kris) | WG | 0.56 | 0.29 | 191 | 2 | 7 | fresh corn | < 0.05 < 0.05 | |
| | | | | | | 7 | forage | 8.04 4.78 | |
| | | | | | | 14 | | 5.36 2.72 | |
| | | 0.57 | 0.29 | 197 | 2 | 7 | stover | 8.30 9.95 | |
| USA, 2001, OR (Honey & Pearls Super Sweet) | WG | 0.57 | 0.20 | 284 | 2 | 7 | fresh corn | < 0.05 < 0.05 | |
| | | | | | | 7 | forage | 3.08 2.52 | |
| | | | | | | 14 | | 1.01 1.12 | |
| | | 0.57 | 0.20 | 287 | 2 | 7 | stover | 38.20 44.32 | |

A study was performed testing boscalid in peanuts in the USA (Wofford and Abdel-Baky, 2001, 2001/5000870). In addition to the residue trials done according to the recommended GAP (3 × 0.51 kg ai/ha of boscalid at 12 locations) one trial was performed in Georgia using a 3× application rate (3 × 1.5 kg ai/ha) to increase the probability of finding residues in the processed fractions. In this trial treatments began 42 days prior to harvest (digging) with retreatment intervals of 14 days and a PHI of 14 days. Peanuts were dug at normal crop maturity which was about 14 days after the last application. They were allowed to dry for 3–8 days and then separated in nutmeat and hay samples. Nutmeat was processed to meal, refined oil and bleached and treatment oil (Table 74).

Table 74. Boscalid residues in peanut and processed fractions resulting from supervised trials in USA (Wofford and Abdel-Baky, 2001, 2001/5000870).

| PEANUT country, year, location (variety) | Application | | | | | PHI days | Commodity | Residues mg/kg boscalid | Ref |
|---|---|---|---|---|---|---|---|---|---|
| | Form | kg ai/ha | kg ai/hL | water (L/ha) | no. | | | | |
| USA, 2000, NC (NCV-11) | WG | 0.51 | 0.18 | 282 | 3 | 13 | peanut nutmeat | < 0.05 < 0.05 | 2001/5000870 |
| | | | | | | | peanut hay | 22.17 18.22 | |
| USA, 2000, SC (Georgia Green) | WG | 0.51 | 0.29 | 177 | 3 | 14 | peanut nutmeat | < 0.05 < 0.05 | |
| | | | | | | | peanut hay | 25.61 22.84 | |
| USA, 2000, SC (Georgia Green) | WG | 0.52 | 0.29 | 178 | 3 | 14 | peanut nutmeat | < 0.05 < 0.05 | |
| | | | | | | | peanut hay | 27.77 30.76 | |

| PEANUT country, year, location (variety) | Application | | | | | PHI days | Commodity | Residues mg/kg boscalid | Ref |
|---|---|---|---|---|---|---|---|---|---|
| | Form | kg ai/ha | kg ai/hL | water (L/ha) | no. | | | | |
| USA, 2000, GA (Georgia Green) | WG | 0.51 | 0.19 | 274 | 3 | 7 | peanut nutmeat | < 0.05<br>< 0.05 | |
| | | | | | | 14 | peanut nutmeat | < 0.05<br>< 0.05 | |
| | | | | | | 21 | peanut nutmeat | < 0.05<br>< 0.05 | |
| | | | | | | 28 | peanut nutmeat | < 0.05<br>< 0.05 | |
| | | | | | | 35 | peanut nutmeat | < 0.05<br>< 0.05 | |
| | | | | | | 7 | peanut hay | 5.67<br>5.89 | |
| | | | | | | 14 | peanut hay | 8.78<br>6.79 | |
| | | | | | | 21 | peanut hay | 3.46<br>4.28 | |
| | | | | | | 28 | peanut hay | 3.91<br>6.84 | |
| | | | | | | 35 | peanut hay | 3.17<br>4.62 | |
| | | 1.53 | 0.56 | 275 | 3 | 14 | peanut nutmeat RAC | < 0.05<br>< 0.05 | |
| | | | | | | | meal | 0.30<br>0.47 | |
| | | | | | | | refined oil | 0.33<br>0.59 | |
| | | | | | | | bleached and deodorized oil | 0.24<br>0.45 | |
| USA, 2000, FL (Georgia Green) | WG | 0.50 | 0.15 | 331 | 3 | 13 | peanut nutmeat | < 0.05<br>< 0.05 | |
| | | | | | | | peanut hay | 6.23<br>7.17 | |
| USA, 2000, AL (Georgia Green) | WG | 0.51 | 0.41 | 125 | 3 | 15 | peanut nutmeat | < 0.05<br>< 0.05 | |
| | | | | | | | peanut hay | 5.43<br>6.25 | |
| USA, 2000, GA (Agra Teck 201) | WG | 0.50 | 0.41 | 122 | 3 | 14 | peanut nutmeat | < 0.05<br>< 0.05 | |
| | | | | | | | peanut hay | 3.68<br>2.61 | |
| USA, 2000, GA (Valencia) | WG | 0.51 | 0.19 | 273 | 3 | 14 | peanut nutmeat | < 0.05<br>< 0.05 | |
| | | | | | | | peanut hay | 28.08<br>28.79 | |
| USA, 2000, GA (Georgia Green) | WG | 0.51 | 0.23 | 225 | 3 | 14 | peanut nutmeat | < 0.05<br>< 0.05 | |
| | | | | | | | peanut hay | 10.11<br>15.04 | |
| USA, 2000, TX (Pronto) | WG | 0.47 | 0.27 | 172 | 3 | 14 | peanut nutmeat | < 0.05<br>< 0.05 | |
| | | | | | | | peanut hay | na<br>na | |
| USA, 2000, OK (Spanco) | WG | 0.50 | 0.38 | 132 | 3 | 13 | peanut nutmeat | < 0.05<br>< 0.05 | |
| | | | | | | | peanut hay | 7.21<br>6.08 | |
| USA, 2000, OK (Spanco) | WG | 0.50 | 0.38 | 133 | 3 | 13 | peanut nutmeat | < 0.05<br>< 0.05 | |
| | | | | | | | peanut hay | 6.68<br>11.33 | |

na: not available

A study was performed testing boscalid in sunflower in 2001 in USA and Canada (Versoi and Abdel-Baky, 2001, 2001/5002552). In addition to the trials conducted for residue analysis after use according to normal GAP, one trial was done to obtain processed products. Sunflowers were treated twice at a rate of 2.25 kg ai/ha of boscalid (5×GAP rate) beginning 28 days prior to harvest with a retreatment interval of 7 days and a PHI of 21 days. Sunflowers were harvested at normal crop maturity, about 21 days after the last application and were later processed according to simulated commercial procedures into meal and refined oil (Table 75).

Table 75. Boscalid residues in sunflower and processed fractions resulting from supervised trials in USA and Canada (Versoi and Abdel-Baky, 2001, 2001/5002552).

| SUNFLOWER country, year, location (variety) | Application | | | | | PHI days | Commodity | Residues mg/kg boscalid | Ref |
|---|---|---|---|---|---|---|---|---|---|
| | Form | kg ai/ha | kg ai/hL | water (L/ha) | no. | | | | |
| USA, 2001, ND (Cropland CL-803) | WG | 0.45 | 0.19 | 233 | 2 | 21 | sunflower seed RAC | < 0.05 0.13 | 2001/5002552 |
| USA, 2001, ND (Interstate Seed) | WG | 0.45 | 0.24 | 188 | 2 | 21 | sunflower seed RAC | < 0.05 0.11 | |
| USA, 2001, ND (Interstate Seed) | WG | 0.45 | 0.24 | 187 | 2 | 21 | sunflower seed RAC | 0.15 0.10 | |
| USA, 2001, SD (Mycogen 8377) | WG | 0.44 | 0.24 | 186 | 2 | 20 | sunflower seed RAC | 0.23 0.22 | |
| USA, 2001, SD (Mycogen 8388) | WG | 0.44 | 0.24 | 185 | 2 | 20 | sunflower seed RAC | 0.17 0.14 | |
| USA, 2001, TX (Triumph 567DW) | WG | 0.44 | 0.096 | 460 | 2 | 21 | sunflower seed RAC | 0.24 0.07 | |
| | | 2.19 | 0.48 | 456 | 2 | 21 | sunflower seed RAC sunflower, meal sunflower, refined oil | 4.95 2.40 0.09 0.08 0.08 0.06 | |
| Canada, 2001, Manitoba (Ag Canada 6111) | WG | 0.45 | 0.41 | 111 | 2 | 21 | sunflower seed RAC | 0.35 0.54 | |

In 2004, a processing study was performed testing boscalid in cotton in USA (Jordan and Jones, 2005, 2005/5000013). Boscalid was applied twice at a rate of 2.24 kg ai/ha of boscalid as an in-furrow application. Cotton seed samples were harvested at normal maturity which was 30 days after the last application. Cotton seed was processed to produce meal, hulls, crude and refined oil (Table 76).

Table 76. Boscalid residues in cotton seed and processed fractions resulting from supervised trials in USA (Jordan and Jones, 2005, 2005/5000013).

| COTTON country, year, location (variety) | Application | | | | | PHI days | Commodity | Residues mg/kg boscalid | Ref |
|---|---|---|---|---|---|---|---|---|---|
| | Form | kg ai/ha | kg ai/hL | water (L/ha) | no. | | | | |
| USA, 2004, AR (ST4793RR) | WG | 2.24 | 1.84 | 122 | 2 | 30 | undelinted cotton seed (RAC) meal hulls oil, crude oil, refined | 0.58 0.95 < 0.05 < 0.05 < 0.05 < 0.05 0.06 0.07 < 0.05 0.08 | 2001/5000013 |

A processing study was performed testing boscalid in canola in 2000 (Versoi and Abdel-Baky, 2001, 2001/5000049, 2001/5001064). At four locations in the US and Canada, canola was treated twice a rate of 1.34 kg ai/ha which is 3×GAP rate. Treatments were applied 26 and 21 days before expected harvest. Canola plant samples were collected at normal crop maturity which was about three weeks after the last application. The plants were allowed to dry for 0–7 days and subsequently processed to meal and refined oil, simulating typical commercial practices (Table 77).

Table 77. Boscalid residues in canola and processed fractions resulting from supervised trials in USA (Versoi and Abdel-Baky, 2001, 2001/5000049, 2001/5001064).

| CANOLA country, year, location (variety) | Application | | | | | PHI days | Commodity | Residues mg/kg boscalid | Ref |
|---|---|---|---|---|---|---|---|---|---|
| | Form | kg ai/ha | kg ai/hL | water (L/ha) | no. | | | | |
| USA, 2000, ND (Hyola 401) | WG | 1.35 | 0.72 | 187 | 2 | 21 | canola seed RAC | 0.72 | 2001/5000049 |
| | | | | | | | cleaned seed | 0.56 | 2001/5001064 |
| | | | | | | | expeller crude oil | 1.11 | |
| | | | | | | | solvent extracted crude oil | 1.05 | |
| | | | | | | | meal | 0.37 | |
| | | | | | | | refined oil | 1.39 | |
| | | | | | | | soapstock | 0.50 | |
| Canada, 2000, Manitoba (Canterra 1867RR) | WG | 1.33 | 0.67 | 198 | 2 | 21 | canola seed RAC | 1.87 | |
| | | | | | | | cleaned seed | 1.00 | |
| | | | | | | | expeller crude oil | 1.51 | |
| | | | | | | | solvent extracted crude oil | 1.26 | |
| | | | | | | | meal | 0.26 | |
| | | | | | | | refined oil | 1.35 | |
| | | | | | | | soapstock | 0.59 | |
| Canada, 2000, Manitoba (Canterra 1867RR) | WG | 1.36 | 1.21 | 112 | 2 | 20 | canola seed RAC | 2.28 | |
| | | | | | | | cleaned seed | 2.25 | |
| | | | | | | | expeller crude oil | 2.74 | |
| | | | | | | | solvent extracted crude oil | 2.49 | |
| | | | | | | | meal | 1.40 | |
| | | | | | | | refined oil | 2.61 | |
| | | | | | | | soapstock | 1.77 | |
| USA, 2000, MN (Golden Ready) | WG | 1.34 | 0.72 | 187 | 2 | 22 | canola seed RAC | 1.76 | |
| | | | | | | | cleaned seed | 1.57 | |
| | | | | | | | expeller crude oil | 2.00 | |
| | | | | | | | solvent extracted crude oil | 1.99 | |
| | | | | | | | meal | 1.11 | |
| | | | | | | | refined oil | 2.53 | |
| | | | | | | | soapstock | 1.13 | |

A study was performed testing boscalid in peppermint in 2001 (Versoi and Abdel-Baky, 2001, 2001/5002467). In addition to the trials conducted according to normal GAP, one trial was done to obtain processed products. Mint plants were treated four times at a rate of 2.25 kg ai/ha of boscalid (5×GAP rate) beginning 28 days prior to harvest with a retreatment interval of 7 days. Mint top samples were collected at 7 and about 14 days after the last application. After drying for two days, mint hay was processed, according to simulated commercial practices, into mint oil (Table 78).

Table 78. Boscalid residues in mint and processed fractions resulting from supervised trials in USA (Versoi and Abdel-Baky, 2001, 2001/5002467).

| MINT country, year, location (variety) | Application | | | | | PHI days | Commodity | Residues mg/kg boscalid | Ref |
|---|---|---|---|---|---|---|---|---|---|
| | Form | kg ai/ha | kg ai/hL | water (L/ha) | no. | | | | |
| USA, 2001, MI (Black Mitchum) | WG | 0.45 | 0.20 | 220 | 4 | 7 | mint tops | 29.25 | 2001/5002467 |
| | | | | | | | | 36.35 | |
| | | | | | | 14 | | 25.45 | |
| | | | | | | | | 25.25 | |
| USA, 2001, MI (Native) | WG | 0.45 | 0.21 | 218 | 4 | 7 | mint tops | 16.05 | |
| | | | | | | | | 15.40 | |
| | | | | | | 14 | | 12.30 | |
| | | | | | | | | 14.45 | |
| USA, 2001, OR (Native) | WG | 0.45 | 0.24 | 189 | 4 | 7 | mint tops | 7.00 | |
| | | | | | | | | 6.65 | |
| | | | | | | 14 | | 4.80 | |
| | | | | | | | | 5.20 | |
| | | 2.23 | 1.16 | 192 | 4 | 7 | mint hay | 81.00 | |
| | | | | | | | | 103.40 | |
| | | | | | | | mint oil | 0.08 | |
| | | | | | | | | 0.08 | |
| | | | | | | | | 0.19 | |
| | | | | | | | | 0.18 | |
| USA, 2001, ID (Native) | WG | 0.45 | 0.16 | 283 | 4 | 7 | mint tops | 31.90 | |
| | | | | | | | | 28.85 | |
| | | | | | | 14 | | 27.90 | |
| | | | | | | | | 28.65 | |
| USA, 2001, WA (Native) | WG | 0.45 | 0.24 | 186 | 4 | 7 | mint tops | 13.20 | |
| | | | | | | | | 13.35 | |
| | | | | | | 14 | | 13.55 | |
| | | | | | | | | 14.45 | |

Four field trials were conducted in different representative hops growing areas in Germany and the Netherlands to determine the residue level of boscalid in hops and processed fractions (Schulz, 2001, 2001/1015048; 2001/1015049). Boscalid was applied three times in total, once at 0.18 kg ai/ha and twice at 0.24 kg ai/ha (35, 28 and 22–21 days before the commercial harvest in three trials and 21, 18, and 15 days in one trial) resulting in a maximum seasonal target rate of 0.66 kg ai/ha, in order to determine the magnitude of the residues of active ingredients in or on RACs. For analysis, hops were taken immediately after the last application and 15-22 days following. The processing included the following fractions: drip dried cones, condensed water, trub (flocs), beer yeast, beer (cooled) and beer (frozen) (Table 79).

Table 79. Boscalid residues in hops and processed fractions resulting from supervised trials in Germany and the Netherlands (Schulz, 2001, 2001/1015048; 2001/1015049).

| HOPS country, year, location (variety) | Application | | | | | PHI days | Commodity | Residues mg/kg boscalid | Ref |
|---|---|---|---|---|---|---|---|---|---|
| | Form | kg ai/ha[1] | kg ai/hL[1] | water (L/ha)[1] | no. | | | | |
| Germany, 2001 Bavaria (Hersbrucker Spät) | WG | 0.57 | 0.019 | 3049 | 3 | 15 | cones green[2] | 5.42 | 2001/1015048 |
| | | | | | | | dried cones[3] | 20.79 | |
| | | | | | | | dried cones[4] | 18.12 | |
| | | | | | | | drip dried cones | 0.62 | |
| | | | | | | | condensed water | < 0.05 | |
| | | | | | | | trub | 0.43 | |
| | | | | | | | beer yeast | 0.24 | |
| | | | | | | | beer cooled | < 0.05 | |
| | | | | | | | beer frozen | < 0.05 | |

| HOPS country, year, location (variety) | Application | | | | PHI days | Commodity | Residues mg/kg boscalid | Ref |
|---|---|---|---|---|---|---|---|---|
| | Form | kg ai/ha[1] | kg ai/hL[1] | water (L/ha) [1] | no. | | | |
| Germany, 2001 Bavaria (Hersbrucker Spät) | WG | 0.57 | 0.019 | 3033 | 3 | 15 | cones green[2] dried cones[3] dried cones[4] drip dried cones condensed water trub beer yeast beer cooled beer frozen | 4.47 20.34 16.33 0.42 < 0.05 0.40 0.24 < 0.05 < 0.05 | |
| Netherlands, 2001 Reijmerstok (Tauer) | WG | 0.21 | 0.019 | 1115 | 3 | 0 21 | cones green[2] cones green[2] dried cones[3] dried cones[4] drip dried cones condensed water trub beer yeast beer cooled beer frozen | 3.99 2.09 5.89 4.24 0.24 < 0.05 0.12 < 0.05 < 0.05 < 0.05 | 2001/1015049 |
| Netherlands, 2001 Reijmerstok (Tauer) | WG | 0.21 | 0.019 | 1124 | 3 | | cones green[2] cones green[2] dried cones[3] dried cones[4] drip dried cones condensed water trub beer yeast beer cooled beer frozen | 4.18 1.62 6.77 3.86 0.25 < 0.05 0.12 0.06 < 0.05 < 0.05 | |

1) average values.

2) to be used for residue analysis.

3) to be used for processing to dried hops for residue analysis.

4) to be used for processing to dried hops for processing to beer.

Table 80. Summary of processing factors for boscalid residues. The factors are calculated from the data recorded in tables in this section.

| Raw agricultural commodity (RAC) | Processed commodity | Calculated processing factors. 1/ | Median |
|---|---|---|---|
| Orange | Pulp | < 0.09, 0.09, < 0.11, 0.11, < 0.14, 0.14(2), < 0.15, < 0.18, < 0.19, 0.19, < 0.20, < 0.24 | 0.14 |
| | Peel | 2.10, 2.21, 2.54, 2.66, 2.84, 3.06, 3.09, 3.12, 3.22, 3.87, 4.20, 4.61, 5.57 | 3.09 |
| Apples | Washed apples | 0.42, 0.47, 0.84, 1.00 | 0.66 |
| | Fresh pomace | 2.08, 3.90, 5.73, 6.38, 6.77, 8.26 | 6.06 |
| | Dried pomace | 13.66, 16.69, 20.03, 24.10 | 18.36 |
| | Thick juice | 0.13, 0.22, 0.27, 0.29 | 0.25 |
| | Fresh juice | 0.05, 0.06, 0.08(2), < 0.09, < 0.10 | 0.08 |
| | Apple Sauce | 0.67, 0.83, 1.00, 1.14 | 0.92 |
| Plums | Washed plum | 0.58, 0.62, 0.79, 0.80, 1.42 | 0.79 |
| | Puree | 1.54, 1.92, 1.98, 2.05 | 1.95 |
| | Prunes | 0.52, 2.42, 2.80, 3.15, 3.66 | 2.80 |
| Cherries | Washed cherries | 0.26, 0.57, 0.58, 1.21 | 0.58 |
| | Canned cherries | 0.40, 0.49, 0.54, 0.86 | 0.52 |
| | Fruit syrup | 0.10, 0.10, 0.13, 0.34 | 0.12 |
| | Cherry juice | 0.28, 0.31, 0.44, 1.24 | 0.38 |

| Raw agricultural commodity (RAC) | Processed commodity | Calculated processing factors. 1/ | Median |
|---|---|---|---|
| Strawberries | Washed strawberries | 0.54, 0.67, 0.69, 0.71 | 0.68 |
| | Canned strawberries | 0.62, 0.69, 0.90, 1.00 | 0.80 |
| | Fruit syrup | 0.22, 0.23, 0.27, 0.28 | 0.25 |
| | Jam | 0.37, 0.42, 0.45, 0.46 | 0.44 |
| | Distillate | < 0.08(2), < 0.13, < 0.17 | < 0.11 |
| Grapes | Raisins | 2.42 | 2.42 |
| | Wet pomace | 1.95, 2.40, 2.60, 3.41 | 2.50 |
| | Wine, from must, cold | 0.09, 0.34, 0.36, 0.47 | 0.35 |
| | Wine, from must after mash heating | 0.12, 0.26, 0.36, 0.46 | 0.31 |
| | Juice | 0.42 | 0.42 |
| White cabbage | Outer leaves | 15.58, 16.56, 21.55, 52.72 | 19.06 |
| | Inner leaves | < 0.03, < 0.06, < 0.08, < 0.42 | < 0.07 |
| | Inner and outer stalks | 0.03, 0.11, 1.78, 4.61 | 0.95 |
| | Cooked cabbage | < 0.03, < 0.06, < 0.08, < 0.42 | < 0.07 |
| | Sauerkraut | 0.09, 0.10, 0.22, < 0.42 | 0.16 |
| Gherkins | Washed gherkins | < 0.23, 0.46, 0.50, 0.73 | 0.48 |
| | Canned gherkins | 0.23, < 0.50, 0.54, 0.60 | 0.52 |
| Tomatoes | Washed tomatoes | 0.14, 0.15, 0.17, 0.94 | 0.16 |
| | Wet pomace | 0.85, 0.93, 1.09, 2.17 | 1.02 |
| | Canned juice | 0.09, 0.13, 0.16, 0.27 | 0.15 |
| | Puree | 0.19, 0.24(2), 0.73 | 0.24 |
| | Tomato paste | 0.53, 0.63, 0.82, 2.24 | 0.73 |
| | Peeled tomatoes | < 0.03, < 0.05, < 0.07 | < 0.05 |
| | Canned tomatoes | < 0.03, < 0.05, 0.12 | < 0.05 |
| | Peel | 0.28, 0.35, 0.59 | 0.35 |
| Head lettuce | Leaves (exterior) | 1.11, 1.25, 1.29, 1.86 | 1.27 |
| | Leaves (exterior rinsed) | 0.57, 0.72, 0.86, 1.26 | 0.79 |
| | Leaves (interior) | < 0.02, 0.06, 0.18, < 0.71 | 0.12 |
| | Leaves (interior rinsed) | 0.03, < 0.05, < 0.18, < 0.71 | < 0.12 |
| Green peas | Washed peas | 0.50, 1.0,1.0,1.0, 1.0, 1.0, 1.0, 1.0 | 1.0 |
| | Cooked peas | 1.0, 1.0, 1.0, 1.0, 1.0, 1.0, 1.0, 1.0 | 1.0 |
| | Canned peas | 1.0, 1.0, 1.0, 1.0, 1.0, 1.0, 1.0, 1.0 | 1.0 |
| | Vegetable stock | 1.0, 1.0, 1.0, 1.0, 1.0, 1.0, 1.0 | 1.0 |
| Soybean | Hulls | 1.74 | 1.74 |
| | meals | < 0.16 | < 0.16 |
| | Refined oil | 0.42 | 0.42 |
| Carrots | Washed carrots | 0.27, 0.36, 0.40, 0.43 | 0.38 |
| | Topped/peeled | 0.06, < 0.10, < 0.13, < 0.14 | < 0.12 |
| | Peel | 0.68, 1.31, 1.49, 1.50 | 1.40 |
| | Cooked carrot | 0.05, < 0.10, < 0.13, < 0.14 | < 0.12 |
| | Juice | < 0.05, < 0.10, < 0.13, < 0.14 | < 0.12 |
| | Pomace | 0.06, 0.15, 0.17(2) | 0.16 |
| | Canned carrot | < 0.05, < 0.10, < 0.13, < 0.14 | < 0.12 |
| | Vegetable stock | < 0.05, < 0.10, < 0.13, < 0.14 | < 0.12 |
| Sugar beets | Dried pulp | 2.16 | 2.16 |
| | Molasses | 1.88 | 1.88 |
| | Refined sugar | 0.24 | 0.24 |
| Barley | Offal | 2.2, 3.1, 3.5, 4.3, 11.3, 12.8, 13.1, 13.5 | 7.8 |
| | Pot barley | 0.22, 0.29, 0.37(2) | 0.33 |
| | Brewing malt for residue analysis | 0.37, 0.45, 0.52, 0.58 | 0.49 |
| | Malt germ | 0.53, 0.89, 0.97, 1.1 | 0.93 |
| | Spent grain | 0.29, 0.34, 0.40, 0.48 | 0.37 |
| | Trub (flocs) | 0.36, 0.45, 0.48, 0.70 | 0.47 |
| | Beer yeast | 0.16, 0.19, 0.29, 0.47 | 0.24 |
| | Beer (cold) | 0.01, 0.02, 0.02, 0.02 | 0.02 |
| | Beer (frozen) | 0.01, 0.02, 0.02, 0.02 | 0.02 |
| Wheat | Middlings | 2.37, 2.41, 3.67, 4.09 | 3.04 |
| | Coarse bran | 3.35, 3.83, 5.14, 5.73 | 4.49 |
| | Total bran | 3.29, 3.87, 4.64, 5.44 | 4.26 |
| | Toppings | 1.23, 1.41, 2.25, 2.82 | 1.83 |
| | Flour type 550 | 0.22, 0.23, 0.45, 0.47 | 0.34 |
| | Wholemeal flour | 1.10, 1.14, 1.29, 1.82 | 1.22 |

| Raw agricultural commodity (RAC) | Processed commodity | Calculated processing factors. 1/ | Median |
|---|---|---|---|
| | Wholemeal bread | 0.60, 0.75, 0.88, 1.00 | 0.82 |
| | Wheat germs | 0.97, 1.29, 1.36, 1.58 | 1.33 |
| Corn | Refined oil, wet milled | 6.0 | 6.0 |
| | Starch | 1.0 | 1.0 |
| | Meal | 1.0 | 1.0 |
| | Refined oil, dried milled | 1.0 | 1.0 |
| | Grits | 1.0 | 1.0 |
| | Flour | 1.0 | 1.0 |
| Peanut | Meal | >7.8 | -2/ |
| | Refined oil | >9.2 | - |
| | Bleached and deodorized oil | >6.9 | - |
| Sunflower | Meal | 0.02 | 0.02 |
| | Refined oil | 0.02 | 0.02 |
| Cotton | Meal | 0.07 | 0.07 |
| | Hulls | 0.07 | 0.07 |
| | Oil, crude | 0.08 | 0.08 |
| | Oil, refined | < 0.08(0.10) | < 0.08 |
| Canola | Cleaned seed | 0.53, 0.78, 0.89, 0.99 | 0.84 |
| | Expeller crude oil | 0.81, 1.14, 1.20, 1.54 | 1.17 |
| | Solvent extracted crude oil | 0.67, 1.09, 1.13, 1.46 | 1.11 |
| | Meal | 0.14, 0.51, 0.61, 0.63 | 0.56 |
| | Refined oil | 0.72, 1.14, 1.44, 1.93 | 1.29 |
| | Soapstock | 0.32, 0.64, 0.69, 0.78 | 0.67 |
| Mint | Mint oil | 0.002 | 0.002 |
| Hops | Drip dried cones | 0.02, 0.03, 0.04, 0.04 | 0.035 |
| | Trub (flocs) | 0.02, 0.02, 0.02, 0.02 | 0.02 |
| | Beer yeast | < 0.01, < 0.01, 0.01, 0.01 | 0.01 |
| | Beer (cooled) | < 0.002, < 0.002, < 0.01, < 0.01 | < 0.006 |
| | Beer (frozen) | < 0.002, < 0.002, < 0.01, < 0.01 | < 0.006 |

1/ 'Less-than' (<) values are derived from cases where residues were not detected in the processed commodity. The 'less-than' processing factor is then calculated from the LOQ of the analyte in the processed commodity and the residue in the RAC.

2/ Residues of boscalid were not detected in peanut, so a processing factor was not calculated and estimates in processed products were not made.

## RESIDUES IN ANIMAL COMMODITIES

### Direct animal treatments

Boscalid is not used for direct animal treatments.

### Farm animal feeding studies

#### Ruminants

The Meeting received a lactating dairy cow feeding study, which provided information on likely residues resulting in animal tissues and milk from residues in the animal diet.

Four Groups of lactating Holstein cows (animals weighing 510–735 kg initially) were dosed twice daily via transferring the contents of the appropriate vial onto the molasses sugar beet feed with boscalid at 1.5 (1×), 4.5 (3×) and 18 ppm (12×) in the dry-weight diet, for 28 consecutive days (Tilting, 2001, 2000/1017228) (Table 81). Milk was collected twice daily for analysis. Animals were sacrificed within 23 hours following the final dosing and tissue samples were taken, except for one cow of the 12× group which were sacrificed seven days after the final dose to determine residue levels post-dosing. Tissues collected for analysis were liver, kidney, fat and muscle. Animals consumed approximately 12.2–20.1 kg dry-weight feed each per day and produced at least 12.7 kg milk per animal per day (measured during dosing period). Samples were analysed by methods 471/0 and 476/0 (Tables 82, 83 and 84).

Table 81. Actual Dose Levels (Tilting, 2001, 2000/1017228).

| Cow (Number) | Nominal dose mg/kg feed | Actual dose: mg/animal/day | Actual concentration: mg/kg-bw | Actual concentration:** mg/kg-feed |
|---|---|---|---|---|
| 1068 (1) | 0 | 0 | 0 | 0 |
| 3191 (2) | 0 | 0 | 0 | 0 |
| 5337 (3) | 0 | 0 | 0 | 0 |
| Average group A | 0 | 0 | 0 | 0 |
| 996 (4) | 1.5 | 30 | 0.047 | 1.583 |
| 3541 (5) | 1.5 | 30 | 0.058 | 2.058 |
| 5180 (6) | 1.5 | 30 | 0.044 | 1.765 |
| Average group B | 1.5 | 30 | 0.050 | 1.802 |
| 3137 (7) | 4.5 | 90 | 0.161 | 5.56 |
| 3253 (8) | 4.5 | 90 | 0.130 | 7.15 |
| 3555 (9) | 4.5 | 90 | 0.176 | 5.02 |
| Average group C | 4.5 | 90 | 0.156 | 5.91 |
| 1226 (10) | 18 | 360 | 0.612 | 20.50 |
| 1938 (11) | 18 | 360 | 0.611 | 19.18 |
| 5070* (12) | 18 | 360 | - | - |
| 3048 (13) | 18 | 360 | 0.507 | 20.61 |
| 3083 (14) | 18 | 360 | 0.495 | 20.36 |
| Average group D | 18 | 360 | 0.556 | 20.16 |

\*      Animal withdrawn from study due to mastitis

\*\*    Based on average feed intake during dosing period

Table 82. Residues of boscalid and metabolite M510F01 in milk from lactating dairy cows dosed with boscalid at 1.5 (1×), 4.5 (3×) and 18 ppm (12×) in the dry-weight diet, for 28 consecutive days (Tilting, 2001, 2000/1017228). Reported values are means of 3 values for the 1.5 and 4.5 ppm groups and of 4 values in the 18 ppm group (Tilting, 2001, 2000/1017228).

| Day | 1.5 ppm group (1×) | | | 4.5 ppm group (3×) | | | 18 ppm group (12×) | | |
|---|---|---|---|---|---|---|---|---|---|
| | milk | skim milk | cream | milk | skim milk | cream | milk | skim milk | cream |
| | boscalid + M510F01, mg/kg | | | | | | | | |
| -3 | < 0.02 | | | < 0.02 | | | < 0.02 | | |
| | < 0.02 | | | < 0.02 | | | < 0.02 | | |
| | < 0.02 | | | < 0.02 | | | < 0.02 | | |
| | | | | | | | < 0.02 | | |
| | | | | | | | < 0.02 | | |
| 1 | n.a. | | | < 0.02 | | | 0.02 | | |
| | | | | < 0.02 | | | 0.02 | | |
| | | | | < 0.02 | | | < 0.02 | | |
| | | | | | | | 0.02 | | |
| | | | | | | | < 0.02 | | |
| 3 | n.a. | | | < 0.02 | | | 0.027 | | |
| | | | | < 0.02 | | | 0.041 | | |
| | | | | 0.02 | | | 0.039 | | |
| | | | | | | | 0.023 | | |
| | | | | | | | 0.047 | | |
| 6 | < 0.02 | | | < 0.02 | | | < 0.02 | | |
| | < 0.02 | | | < 0.02 | | | 0.025 | | |
| | < 0.02 | | | < 0.02 | | | 0.022 | | |
| | | | | | | | < 0.02 | | |
| | | | | | | | 0.031 | | |
| 9 | < 0.02 | | | < 0.02 | | | 0.030 | | |
| | < 0.02 | | | < 0.02 | | | 0.045 | | |
| | < 0.02 | | | < 0.02 | | | 0.036 | | |
| | | | | | | | < 0.02 | | |
| | | | | | | | 0.045 | | |

| Day | 1.5 ppm group (1×) | | | 4.5 ppm group (3×) | | | 18 ppm group (12×) | | |
|---|---|---|---|---|---|---|---|---|---|
| | milk | skim milk | cream | milk | skim milk | cream | milk | skim milk | cream |
| 12 | < 0.02 < 0.02 < 0.02 | | | < 0.02 < 0.02 < 0.02 | | | 0.025 0.032 0.032 0.023 0.032 | | |
| 15 | n.a. | | | < 0.02 0.021 < 0.02 | | | 0.034 0.042 0.051 0.026 0.042 | | |
| 18 | n.a. | | | < 0.02 < 0.02 0.023 | | | 0.096 0.055 0.021 0.036 0.055 | | |
| 21 | < 0.02 < 0.02 < 0.02 | < 0.02 < 0.02 < 0.02 | 0.033 0.055 0.035 | 0.02 < 0.02 < 0.02 | < 0.02 < 0.02 < 0.02 | 0.123 0.125 0.110 | 0.038 0.043 0.031 0.040 | 0.02 < 0.02 < 0.02 < 0.02 | 0.38 0.38 0.25 0.35 |
| 24 | n.a. | | | < 0.02 < 0.02 < 0.02 | | | 0.035 0.046 0.026 0.040 | | |
| 28 | < 0.02 < 0.02 < 0.02 | | | < 0.02 0.02 < 0.02 | | | 0.039 0.043 0.028 0.046 | | |
| 29 | n.a. | | | < 0.02 < 0.02 < 0.02 | | | n.a. | | |
| 32 | n.a. | | | n.a. | | | < 0.02 | | |
| 36 | n.a. | | | n.a. | | | < 0.02 | | |

n.a.: not analyzed.

Table 83. Residues of boscalid and metabolite M510F01 in tissues from lactating dairy cows dosed with boscalid at 1.5 (1×), 4.5 (3×) and 18 ppm (12×) in the dry-weight diet, for 28 consecutive days (Tilting, 2001, 2000/1017228). Reported values are means of 3 values for the 1.5 and 4.5 ppm groups and of 4 values in the 18 ppm group (Tilting, 2001, 2000/1017228).

| Tissue | 1.5 ppm group (1×) | | 4.5 ppm group (3×) | | 18 ppm group (12×) | |
|---|---|---|---|---|---|---|
| | individuals | mean | individuals | mean | individuals | mean |
| | boscalid + M510F01, mg/kg | | | | | |
| Muscle | < 0.05, < 0.05, < 0.05 | < 0.05 | < 0.05, < 0.05, < 0.05 | < 0.05 | < 0.05, < 0.05, 0.058 | 0.053 |
| Fat | 0.078, < 0.05, < 0.05 | 0.059 | 0.124, 0.109, 0.082 | 0.105 | 0.235, 0.292, 0.278 | 0.268 |
| Liver | < 0.05, < 0.05, < 0.05 | < 0.05 | 0.055, 0.51, 0.064 | 0.057 | 0.182, 0.170, 0.180 | 0.177 |
| Kidney | < 0.05, < 0.05, < 0.05 | < 0.05 | 0.071, 0.063, 0.088 | 0.074 | 0.318, 0.220, 0.169 | 0.236 |

Table 84. Residues of boscalid and metabolite M510F01 in milk and tissues from the depuration animal dosed with boscalid at 18 ppm (12×) in the dry-weight diet, for 28 consecutive days and then fed with control feed until slaughter on day 36 (Tilting, 2001, 2000/1017228).

| Matrix | Study day | boscalid, mg/kg | M510F01, mg/kg |
|---|---|---|---|
| Whole milk | 32 | < 0.01 | < 0.01 |
| Whole milk | 36 | < 0.01 | < 0.01 |
| Muscle | 36 | < 0.025 | < 0.025 |
| Fat | 36 | < 0.025 | < 0.025 |

| Matrix | Study day | boscalid, mg/kg | M510F01, mg/kg |
|--------|-----------|-----------------|----------------|
| Liver | 36 | < 0.025 | < 0.025 |
| Kidney | 36 | < 0.025 | < 0.025 |

## RESIDUES IN FOOD IN COMMERCE OR AT CONSUMPTION

No information was received on residues of boscalid in food in commerce or at consumption.

## NATIONAL MAXIMUM RESIDUE LIMITS

Information was provided on national residue definitions for boscalid.

Australia (APVMA, 2006)
Commodities of plant origin: boscalid.

Commodities of animal origin: Sum of boscalid, 2-chloro-N-(4'-chloro-5-hydroxybiphenyl-2-yl) nicotinamide including its conjugate, expressed as boscalid.

USA (USEPA, 2005)
Commodities of plant origin: boscalid.

Commodities of animal origin: Sum of boscalid, 2-chloro-N-(4'-chloro-5-hydroxybiphenyl-2-yl) nicotinamide including its conjugate, expressed as boscalid.

## APPRAISAL

Boscalid was considered for the first time by the present Meeting. It is an anilide fungicide that inhibits mitochondrial respiration, thereby inhibiting spore germination, germ tube elongation, mycelial growth, and sporulation of pathogenic fungi on the leaf surface, and is used against a broad spectrum of diseases in a wide range of crops.

2-Chloro-N-(4'-chlorobiphenyl-2-yl) nicotinamide

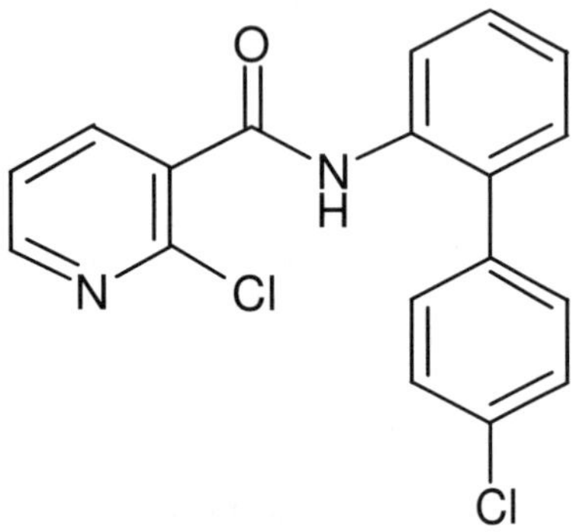

### Animal metabolism

The Meeting received animal metabolism studies for boscalid in lactating goats and laying hens.

When lactating goats were orally dosed with [diphenyl-U-$^{14}$C] labelled boscalid for 5 consecutive days at about 65 mg/animal/day, equivalent to 35 ppm in the feed, most of the administered $^{14}$C was excreted in the faeces (46% and 64%) and urine (23% and 44%). $^{14}$C recovery amounted to 94–95% of the total radioactivity. Milk and tissues accounted for 0.06%–0.15% and 0.46%–0.66% of the administered $^{14}$C respectively.

The major residues in muscle were boscalid, the hydroxylated compound M510F01 (2-chloro-N-(4'-chloro-5-hydroxybiphenyl- 2-yl)nicotinamide) and the glucuronic acid conjugate M510F02 (4'-chloro-6-{[(2-chloro-3-pyridinyl)carbonyl] amino}biphenyl-3- yl glycopyranosiduronic acid). These three metabolites were also detected in kidney. The main residues in fat were identified as boscalid and M510F01. The main residues in milk extracts were boscalid, M510F01, M510F02, and 4-chloro-2'-(acetylamino)-biphenyl (M510F53). Further characterisation of boscalid residues in liver demonstrated that the parent compound, M510F01, and M510F53 were the major residues. M510F53 originated from bound residues of parent compound during extraction with microwave treatment. Parent compound and its hydroxylated metabolite M510F01, including the conjugate M510F02 were the major residues in milk and each of the tissues.

When laying hens were orally dosed with [diphenyl-U-$^{14}$C] labelled boscalid for 10 consecutive days with 1.6 mg/bird/day, equivalent to 12.5 ppm in the feed, most of the administered $^{14}$C was excreted in the excreta (97.7%). Recovery of $^{14}$C amounted to 98.3% of the total radioactivity. Eggs and tissues accounted for 0.115% and 0.046% of the dose respectively.

The parent compound and its hydroxylated metabolite M510F01 were the main residues in eggs. Muscle had a very low residue levels and therefore was not further investigated. The main residue in fat was the parent compound.

Although there were similarities in the metabolic pathways in lactating goats and in poultry, there were also some differences, e.g., M510F53 in liver and milk was not identified in poultry tissues and eggs. The major residues in animal tissues are parent compound and M510F01 and M510F02. M510F54 was identified in chicken as a minor metabolite. M510F54 was found in liver and milk of lactating goats, but was not identified.

No metabolism study was performed in pigs since the metabolic patterns in rodents (rats) and ruminants (goats) did not differ significantly (See the toxicology report for more details of laboratory animal metabolism).

### Plant metabolism

The Meeting received plant metabolism studies with boscalid on grapes, lettuce and beans.

In each crop tested, parent compound generally represented more than 90% of the total $^{14}$C residue and showed almost no further metabolism to carbohydrates, proteins or other natural products.

When grapes plants were treated three times with $^{14}$C-boscalid (diphenyl- and pyridine-label), parent compound represented more than 90% of the total $^{14}$C residue in the samples taken at the maturity of grapes. An HPLC analysis of the methanol and the water extracts of all matrices showed that in grapes 92.7% of the TRR were represented by the unchanged parent compound for the diphenyl label and 92.2% for the pyridine label.

When lettuce plants were treated 3 times with $^{14}$C-boscalid (diphenyl- and pyridine-label), parent compound represented more than 90% of the total $^{14}$C residue in the samples taken 18 days after the last application.

When green beans were treated with 3 foliar applications, approximately 8–10 days apart, of $^{14}$C-boscalid (diphenyl- and pyridine-label), the majority of the residue associated with the fruit sampled 14/15 and 53/51 (diphenyl/pyridine label) days after the final application was mainly boscalid, accounting for about 90% of the residue.

### Environmental fate in soil

The Meeting received information on the environmental fate of boscalid in soil, including studies on aerobic soil metabolism, field dissipation and crop rotational studies.

When [$^{14}$C]-boscalid labelled in the diphenyl ring or the pyridine ring was incubated with four different soils under aerobic conditions in the dark at different temperatures and soil moistures, it degraded slowly and identifiable metabolites were a minor part of the residue and almost degraded at the same rate. The pyridine-label test compound was mineralized to $^{14}$CO$_2$ more quickly than the diphenyl-label test compound. No volatiles other than $^{14}CO_2$ were found. The soil metabolism and degradation studies described so far showed that boscalid is finally degraded in soil to $CO_2$ and bound residues.

Boscalid did not show a tendency to move into deeper layers of soil and was primarily detected in the top 10 cm soil layer during field dissipation trials (four different soils) of duration up to 12-18 months. Boscalid concentrations declined to half of their initial values in 28 days to 208 days. In all trials a $DT_{90}$ could not be reached within one year after application to bare soil.

In a confined rotational crop study in Germany, soil was treated directly with $[^{14}C]$-boscalid labelled in the diphenyl ring or the pyridine ring. Crops of lettuce, radish and wheat were sown into the treated soil at intervals of 30, 120, 270 and 365 days after treatment and were grown to maturity and harvested for analysis. The residues in the edible parts of succeeding crops destined for human consumption were low for lettuce and radish root, and slightly higher for wheat grain after all four plant back intervals. The major part of the residues was identified as parent. The concentration of boscalid in lettuce leaf ranged from 55.6–94.1% TRR, in radish leaf from 69.4–90.2% TRR, in radish root from 52.6–92.8% TRR and in wheat straw from 50.0–87.5% TRR. In wheat grain the concentration of parent was lower (1.9–35.4% TRR, $\leq 0.028$ mg/kg).

Field trials on rotational crop studies were not submitted prior to this Meeting and could not be used for evaluation.

## Methods of residue analysis

The Meeting received description and validation data for analytical methods for residues of boscalid in raw agricultural commodities, processed commodities, feed commodities as well as animal tissues, milk and eggs.

The methods rely on HPLC-UV, HPLC-MS/MS, GC-ECD and GC-MSD for analysis of boscalid in the various matrices. A multi-residue method with GC-ECD and GC-MSD suitable for enforcement for plant and animal commodities (LOQ values 0.0025–0.02 mg/kg) was adapted from an existing method (DFSG S19).

Numerous recovery data on a wide range of substrates were provided with validation testing of methods which showed that they were valid over the relevant concentration ranges.

## Stability of residues in stored analytical samples

The Meeting received information on the freezer storage stability of residues of boscalid in wheat (green plant without roots, grain and straw), oilseed rape, sugar beet (roots), white cabbage (head), peach (fruit), peas, tomato paste, liver, milk, muscle.

Residues were stable (less than 30% disappearance) in various plant matrices over a period of 2 years and longer. Storage data were available for some animal commodities for at least 5 months, and were suitable for showing stability of the residues in samples from these studies.

## Definition of the residue

The composition of the residue in the metabolism studies, the available residue data in the supervised trials, the toxicological significance of metabolites, the capabilities of enforcement analytical methods and the national residue definitions already operating all influence the decision for an appropriate residue definition.

The metabolism of boscalid was investigated in grapes, lettuce and beans. Unchanged parent compound formed the major part of the residue in these studies. The cleavage products M510F62 (chlorophenylaminobenzene) and M510F47 (chloronicotinic acid) and in addition hydroxy-parent and sugar conjugates were also identified in beans. However, all metabolites were of minor importance. Therefore only parent is included in the residue definition.

Metabolism studies performed on goats and hens show that residues in products of animal origin derive from the parent compound as well as from the hydroxylated metabolite M510F01 including its conjugates. M510F54 in chicken eggs was not found in lactating goats' tissues and milk, and M510F53 in liver and milk was not found in poultry tissues and eggs.

Ruminant feeding studies show that boscalid preferentially accumulates in cream as opposed to whole milk (concentration ratio is 9:1). The boscalid ratio between fat and muscle was about 5:1. In the metabolism studies of lactating goats, the ratio between fat and muscle was about 6:1. The log octanol-water partition coefficient was approximately 3 and also suggests that boscalid is likely to be a fat-soluble compound, although it does not accumulate in animal tissues, milk and eggs.

Based on the available comparative animal and plant metabolism studies, the Meeting recommended a residue definition for boscalid for plants and animals:

Definition of the residue (for compliance with the MRL for plant and animal commodities and for estimation of dietary intake for plant commodities): *boscalid.*

Definition of the residue (for estimation of dietary intake for animal commodities): sum of boscalid, 2-chloro-N-(4'-chloro-5-hydroxybiphenyl-2-yl)nicotinamide including its conjugate, expressed as boscalid.

The residue is fat soluble.

### Results of supervised trials on crops

The Meeting received supervised trial data for boscalid uses on apple, peach, plum, cherry, raspberries, blueberries, strawberry, grapes, banana, onion, leek, broccoli, cabbages, cauliflowers, Brussels sprouts, cucumbers, cantaloupe, summer squash, melons, tomatoes, peppers, mustard greens, lettuce, curly kale, beans, peas, soybeans, carrot, radish, potato, cereal grain (barley, wheat), almond, pecan nut, pistachio, canola, sunflower, peanuts and coffee.

Field studies on residues in follow-up and rotational crops were provided only at a late stage of the Meeting. In view of the wide range of crops in which boscalid residues may be present above the LOQ, the evaluation of residues deriving from direct application and through uptake from soil has to be assessed together, which was not possible due to late submission of the reports. Consequently, the Meeting could not make recommendations for residue levels in annual crops (onion, leek, broccoli, cabbages, cauliflower, Brussels sprouts, cucumbers, cantaloupe, summer squash, melons, strawberry, tomatoes, peppers, mustard greens, lettuce, curly kale, beans, peas, soybeans, carrot, radish, potato, canola, sunflower and peanuts). The supervised residue trials with direct application and field trials on rotational crops will be evaluated together at a future meeting when all results will become available.

Processing trials with boscalid were considered valid because the processing factors should not be influenced by higher residues than that achieved by GAP. It is common practice to apply a pesticide at an exaggerated rate in processing trials to achieve measurable levels in processed commodities.

Trials from Japan were available only in summary form and could not be evaluated.

### Apples

Boscalid has approval for use on <u>apple</u> in the UK for up to four applications at 0.202 kg ai/ha with a 7 day PHI. Supervised trials were conducted on apple trees conforming to UK GAP in Belgium (1), Germany (6), France (11), Italy (7), and the Netherlands (1) in 2000, 2001 and 2003. The residues in ranked order were: 0.14, 0.15, 0.19, 0.20, 0.24, 0.24, 0.29, 0.29, 0.30, 0.32, 0.32, 0.34, <u>0.36</u>, <u>0.37</u>, 0.39, 0.39, 0.42, 0.42, 0.43, 0.49, 0.51, 0.53, 0.55, 0.65, 0.86, 1.24 mg/kg.

The Meeting estimated a maximum residue level and an STMR value for boscalid in apples of 2 and 0.365 mg/kg respectively.

*Stone fruit*

Supervised trials were conducted on stone fruits in the USA (GAP: five applications at 0.256 kg ai/kg with a 0 day PHI) in 1999 and 2004. Twenty two trials were conducted on peach conforming to US GAP. The residues in ranked were: 0.16, 0.19, 0.19, 0.32, 0.32, 0.33, 0.40, 0.40, 0.42, 0.48, 0.48, 0.49, 0.49, 0.49, 0.52, 0.62, 0.64, 0.66, 0.73, 0.75, 0.79, 1.19 mg/kg.

Sixteen trials were conducted on plums conforming to US GAP in 2001 and 2004. The residues in ranked order, median underlined, were: 0.08, 0.09, 0.1, 0.11, 0.14, 0.15, 0.17, 0.17, 0.24, 0.25, 0.32, 0.34, 0.46, 0.55, 0.57, 0.70 mg/kg.

Thirteen trials were conducted on cherries conforming to US GAP in 2001 and 2004. The residues of boscalid in ranked order, median underlined, were: 0.64, 0.74, 0.76, 0.91, 1.0, 1.09, 1.21, 1.31, 1.42, 1.49, 1.5, 1.51, 1.64 mg/kg.

The data for peach was significantly different from those for plums. The data for peach and plums were also significantly different from those for cherry, and so could not be combined. The Meeting agreed to use the data from cherry and estimated a maximum residue level and an STMR value for boscalid in stone fruit of 3 and 1.21 mg/kg respectively.

*Berries and other small fruits*

Six supervised trials were conducted on raspberries in USA and Canada conforming to the GAP of the USA in 1999 and 2004 (GAP: four applications at 0.406 kg ai/kg with a 0 day PHI). The residues of boscalid in ranked order were: 1.49, 2.00, 2.44, 2.69, 3.45, 3.73 mg/kg.

Twelve supervised trials were conducted on blueberries in USA and Canada conforming to US GAP (0.406 kg ai/ha, four applications, 0 day PHI) in 1999 and 2004. The residues found in ranked order were: 0.84, 1.16, 1.26, 1.27, 1.46, 2.34, 2.62, 2.65, 3.79, 4.35, 6.78, 6.83 mg/kg.

The Meeting agreed that data populations for raspberries and blueberries could be combined, therefore, the residues of boscalid in ranked order, median underlined, were: 0.84, 1.16, 1.26, 1.27, 1.46, 1.49, 2.00, 2.34, 2.44, 2.62, 2.65, 2.69, 3.45, 3.73, 3.79, 4.35, 6.78, 6.83 mg/kg. The Meeting estimated a maximum residue level and STMR values for boscalid in berries and other small fruits except strawberry and grapes of 10 and 2.53 mg/kg respectively.

Boscalid is approved for use in strawberries in the UK at 0.481 kg ai/kg, a maximum of two applications with a 3 day PHI. Supervised trials conforming to UK GAP were conducted on strawberry in Demark, France, Germany, Greece, Italy, Spain, Sweden and the Netherlands in 2003 and 2004. At the GAP of the UK, residue values from outdoor trials in France were 0.20, 0.41, 0.42, 0.45, 0.89, 1.74 mg/kg; in Denmark 0.38 mg/kg; in Germany 0.19, 0.35 mg/kg; in Greece 1.87 mg/kg; in Italy 0.47, 0.68, 0.69 mg/kg; in Sweden 0.15 mg/kg; in the Netherlands 0.46 mg/kg; and those in the UK were 0.27, 0.55 mg/kg. In summary, residues of boscalid in strawberry from the 17 European trials in ranked order were: 0.15, 0.19, 0.20, 0.27, 0.35, 0.38, 0.41, 0.42, 0.45, 0.46, 0.47, 0.55, 0.68, 0.69, 0.89, 1.74, 1.87 mg/kg.

At UK GAP, the residue values from indoor trials in France were 0.28, 0.57, 0.68 mg/kg; in Italy were 0.31, 0.46 mg/kg; and in Spain were 0.23, 0.27, 0.34, 0.49 mg/kg. The ranked order of residue values were: 0.23, 0.27, 0.28, 0.31, 0.34, 0.46, 0.49, 0.57, 0.68 mg/kg.

The outdoor and indoor residue data appeared to be similar populations and were combined. The residues in ranked order were: 0.15, 0.19, 0.20, 0.23, 0.27 (2), 0.28, 0.31, 0.34, 0.35, 0.38, 0.41, 0.42, 0.45, 0.46 (2), 0.47, 0.49, 0.55, 0.57, 0.68 (2), 0.69, 0.89, 1.74 and 1.87 mg/kg.

No recommendation can be made for strawberries until the contribution of residues from direct application as well as uptake through the soil can be assessed.

Sixteen supervised trials were conducted on grapes in USA (GAP: 0.392 kg ai/ha, three applications with a 14 day PHI) in 1999. The residues in ranked order were: 0.33, 0.34 (2), 0.36, 0.50, 0.65 (2), 0.92, 1.26, 1.38, 1.50, 1.83, 2.08, 2.28, 2.97, 3.13 mg/kg.

The Meeting estimated a maximum residue level and an STMR value for boscalid in grapes of 5 and 1.09 mg/kg respectively.

## Banana

Twelve supervised trials conforming to the GAP of the USA (0.150 kg ai/ha, four applications with a 0 day PHI) were conducted on banana in Costa Rica, Panama, Guatemala, Honduras, Ecuador and Colombia in 2004. The residues in ranked order for bagged bananas were: < 0.05 (12) mg/kg. The residues in ranked order for unbagged bananas were: < 0.05 (5), 0.07, 0.07, 0.09, 0.10, 0.10, 0.11, 0.18 mg/kg. The ranked order of concentrations on banana pulp was: < 0.05 (12) mg/kg.

The unbagged and bagged residue data populations for whole fruit of banana were significantly different and could not be combined. The Meeting estimated a maximum residue level, based on unbagged bananas, and an STMR value, based on banana pulp, for boscalid in banana of 0.2 and 0.05 mg/kg respectively.

## Bulb vegetables

Data from five supervised trials on green onions was received from the USA (GAP: 0.326 kg ai/ha, six applications, 7 day PHI) in 1999. The residues in ranked order on green onions were: 1.13, 2.01, 2.20, 2.39, 2.73 mg/kg.

Ten supervised trials were conducted on bulb onions in the USA (maximum GAP: 0.326 kg ai/ha, six applications, 7 day PHI) in 1999. The residues in ranked order on bulb onions were: < 0.05, 0.05, 0.1, 0.11, 0.13, 0.22, 0.78, 0.92, 0.93, 2.61 mg/kg.

Eleven supervised trials were conducted on leek in Belgium in 1999 and 2000 (maximum GAP of two applications at 0.400 kg ai/ha with a 14 day PHI), France (no national GAP, that of the Netherlands used) in 2003, Germany in 1999 and 2000 (no national GAP, that of the Netherlands used), the Netherlands in 1999 and 2000 (three 3 applications at 0.410 kg ai/ha with a 14 day PHI), UK in 2000 (no national GAP, that of the Netherlands used). In summary, residues of boscalid in leek from the 11 European trials in ranked order were: 0.58, 0.62, 0.8, 0.9, 0.93, 1.02, 1.16, 1.31 (2), 1.90, 2.30 mg/kg.

No recommendation can be made for bulb vegetables until the contribution of residues from direct application as well as uptake through the soil can be assessed.

## Brassica

Six supervised trials were conducted on broccoli in the USA in 2001 (GAP: two applications at 0.441 kg ai/ha with a 0 day PHI). The residues on broccoli in ranked order were: 0.81, 0.98, 1.45, 1.59, 1.70 and 2.70 mg/kg.

Three supervised trials were conducted on broccoli one from Germany (no national GAP, the GAP for cauliflower from the UK used: three applications at 0.267 kg ai/ha with a 14 day PHI) and two trials from France (no national GAP, that of UK cauliflower used). The residues on broccoli in ranked order were: < 0.05, < 0.05 and 0.20 mg/kg.

The data populations for the USA and the EU were deemed significantly different and as such could not be combined.

Six supervised trials were conducted on cabbage in USA in 2001 (GAP: two applications at 0.441 kg ai/ha with a 0 day PHI). The residues on cabbage in ranked order were: 0.64, 0.73, 1.06, 1.78, 2.22, 2.33 mg/kg.

Boscalid is approved for use on cauliflower in the UK at 0.267 kg ai/ha, three applications with a 14 day PHI. Seven supervised trials were conducted on cauliflower in Denmark in 2003, France in 2003 and 2004, Germany in 2004, the Netherlands in 2003 and the UK in 2003conforming with UK GAP. The residues on cauliflower in ranked order were: < 0.05 (5), 0.06, 0.55 mg/kg.

Boscalid is approved for use on Brussels sprouts in the UK at 0.267 kg ai/ha, three applications with a 14 day PHI. Nine supervised trials were conducted on Brussels sprouts in Denmark in 2003, France in 2004, Germany in 2003 and 2004, the Netherlands in 2003, Sweden in 2003 and 2004 and the UK in 2003 and 2004 conforming to UK GAP. The residues on Brussels sprouts in ranked order were: < 0.05 (2), 0.06, 0.10, 0.15, 0.16, 0.23, 0.34 and 0.40 mg/kg.

The data for broccoli was significantly different from those for Brussels sprouts (Mann-Whitney test). The data for cauliflower was significantly different from those for Brussels sprouts. The data for cabbage was significantly different from those for Brussels sprouts. The data for cabbage were significantly different from those for cauliflower and so could not be combined. The data populations for broccoli against US GAP and cabbage were not significantly different. In summary, residues of boscalid in broccoli and cabbage from the 12 US trials in rank order were: 0.64, 0.73, 0.81, 0.98, 1.06, 1.45, 1.59, 1.70, 1.78, 2.22, 2.33 and 2.70 mg/kg.

No recommendation can be made for brassica vegetables until the contribution of residues from direct application as well as uptake through the soil can be assessed.

*Fruiting vegetables, cucurbits*

Ten supervised trials were conducted on cucumber in USA in 2001 and 2004 (GAP: four applications at 0.326 kg ai/ha with a 0 day PHI). The residues on cucumber in ranked order were: 0.05, 0.07 (3), 0.12, 0.13, 0.14 (2), 0.26 and 0.31 mg/kg.

Eight supervised trials were conducted on cantaloupe in USA in 2001 and 2004 (GAP: four applications at 0.326 kg ai/ha with a 0 day PHI). The residues on cantaloupe in ranked order were: 0.14, 0.23, 0.29, 0.39, 0.57, 0.56, 0.71 and 1.27 mg/kg.

Boscalid is approved for use in Germany on melons at 0.1 kg ai/ha, three applications with a 3 day PHI. Eight supervised trials, conforming to German GAP, were conducted on melons in Italy in 2000 and Spain in 1999 and 2000. The residues in ranked order were: < 0.05 (8) mg/kg.

The data populations for cantaloupe and melons were significantly different and could not be combined.

Nine supervised trials were conducted on Summer squash in USA in 2001 and 2004 (GAP: four applications at 0.326 kg ai/ha with a 0 day PHI). The residues in ranked order on summer squash were: 0.11, 0.12, 0.14, 0.16 (2), 0.19, 0.27, 0.31, 0.95 mg/kg.

The data for cucumber and summer squash appeared to be similar populations and could be combined. The data for cucumber was significantly different from those for cantaloupe. The data for cantaloupe was significantly different from those for summer squash.

No recommendation can be made for fruiting vegetables, cucurbits until the contribution of residues from direct application as well as uptake through the soil can be assessed.

*Fruiting vegetables other than cucurbits (except fungi, mushroom and sweet corn)*

Supervised trials were conducted on tomatoes in USA in 1999 and 2004 (GAP: two applications at 0.613 kg ai/ha with a 0 day PHI). Twelve trials were conducted at GAP. The residues in ranked order were: 0.17, 0.21, 0.22, 0.24, 0.25, 0.27, 0.28, 0.3, 0.59, 0.61, 0.79, 0.92 mg/kg.

Supervised trials were conducted on non-bell and bell peppers in USA in 1999 and 2004 (GAP: six applications at 0.172 kg ai/ha with a 0 day PHI). In six US trials on bell peppers in 1999 matching the maximum GAP, residues of boscalid were: < 0.05, 0.08, 0.09, 0.14, 0.16, 0.3 mg/kg. In three US trials on non-bell peppers in 1999 matching maximum GAP, residues of boscalid were: 0.14, 0.30 and 0.83 mg/kg.

The data populations for non-bell and bell peppers were not significantly different data and were combined. The residues on peppers in ranked order were: < 0.05, 0.08, 0.09, 0.14 (2), 0.16, 0.3 (2), 0.83 mg/kg.

The data populations for peppers and tomatoes were not significantly different and could be combined. The residues in ranked order were: < 0.05, 0.08, 0.09, 0.14 (2), 0.16, 0.17, 0.21, 0.22, 0.24, 0.25, 0.27, 0.28, 0.3 (3), 0.59, 0.61, 0.79, 0.83, 0.92 mg/kg.

No recommendation can be made for Fruiting vegetables other than cucurbits (except fungi, mushroom and sweet corn) until the contribution of residues from direct application as well as uptake through the soil can be assessed.

*Leafy vegetables*

Eleven supervised trials were conducted on Mustard greens in USA in 2001, 2004 and 2005 (GAP: two applications at 0.441 kg ai/ha with a 14 day PHI). Eight trials were conducted conforming to US GAP, with residues in ranked order of: of 0.45, 0.54, 0.92, 2.80, 3.1, 6.04, 12.9, 14.4 mg/kg.

Eight supervised trials were conducted on head lettuce and leafy lettuce in USA (GAP: two applications at 0.441 kg ai/ha with a 14 day PHI), respectively. The residues on head lettuce in ranked order were: 0.11, 0.98, 1.77, 2.53, 2.68, 2.73, 3.18, 6.15 mg/kg and on leafy lettuce were: 0.74, 1.60, 1.63, 1.91, 4.87, 5.14, 9.36 and 9.55 mg/kg.

The data populations for head lettuce and leaf lettuce were not significantly different and could be combined. The ranked order of concentrations was: 0.11, 0.74, 0.98, 1.6, 1.63, 1.77, 1.91, 2.53, 2.68, 2.73, 3.18, 4.87, 5.14, 5.42, 9.36 and 9.55 mg/kg.

Eighteen supervised trials were conducted on outdoor lettuce in France (6), Germany (5), the Netherlands (2), Spain (5) in 1999 and 2000 conforming to Belgian GAP of two applications at 0.4 kg ai/ha with a 14 day PHI. The residues on outdoor lettuce in ranked order were: < 0.05, 0.09, 0.15, 0.21, 0.33, 0.36, 0.38, 0.39, 0.43, 0.45, 0.50, 0.64, 0.65, 0.73, 0.76, 0.86, 1.19 and 1.58 mg/kg.

Eight supervised trials were conducted on indoor lettuce in France (4), Germany (1), the Netherlands (1) and Spain (2) in 2002 conforming to Belgian GAP. The residues on indoor lettuce in ranked order were: 0.37, 0.71, 1.52, 2.31, 2.50, 5.63, 5.96 and 6.11 mg/kg.

The outdoor and indoor residue data population for head lettuce in Europe were significantly different and could not be combined.

The US and European residue (indoor lettuce) data populations for head lettuce were not significantly different and could be combined. The residues on head lettuce in ranked order were: 0.11, 0.37, 0.72, 0.74, 0.98, 1.52, 1.6, 1.63, 1.77, 1.91, 2.32, 2.5, 2.53, 2.68, 2.73, 3.18, 4.87, 5.14, 5.42, 5.63, 5.96, 6.11, 9.36, 9.55 mg/kg.

Data was received from six supervised trials conducted on Curly kale in Denmark in 2000 (1), the Netherlands in 1999 (1), Sweden in 1999 (1) and UK in 1999 and 2000 (3) at 0.4 kg ai/ha (0.133 kg ai/hL). The trials did not conform to UK GAP of three applications at a rate of 0.267 kg ai/ha, i.e., treatments were made at a rate 50% higher than GAP with one extra application. As a result the data on Curly kale could not be evaluated (0.11, 0.50, 0.55, 0.67, 2.80, 3.20 mg/kg).

The combined data populations for lettuce were not significantly different from those for mustard greens and could be combined. The residues in ranked order were: 0.11, 0.37, 0.45, 0.54, 0.72, 0.74, 0.92, 0.98, 1.52, 1.60, 1.63, 1.77, 1.91, 2.32, 2.50, 2.53, 2.68, 2.73, 2.80, 3.10, 3.18, 4.87, 5.14, 5.42, 5.63, 5.96, 6.04, 6.11, 9.36, 9.55, 12.9, 14.4 mg/kg.

No recommendation can be made for leafy vegetables until the contribution of residues from direct application as well as uptake through the soil can be assessed.

*Legume vegetables*

Boscalid is approved for use in France with a maximum GAP of two applications at a rate of 0.5 kg ai/ha with a 7 day PHI. Eleven supervised trials were conducted on outdoor beans in Denmark 1999 and 2000 (3), France in 2000 (4), Germany in 1999 and 2000 (4) conforming to French GAP. The residues on field grown beans, with pod, were: 0.13, 0.22, 0.26, 0.29, 0.47, 0.50, 0.53, 0.62, 0.67, 0.83, 0.95 mg/kg.

Eight supervised trials were conducted on indoor beans in Spain in 1999 and 2000 conforming to French GAP. The residues on indoor beans, with pod, in ranked order were: 0.06, 0.28, 0.28, 0.29, 0.61, 0.69, 1.65, 1.67 mg/kg.

The outdoor and indoor residue data populations for beans, with pod, were not significantly different and could be combined. The residues on beans with pod in ranked order were: 0.06, 0.13, 0.22, 0.26, 0.28, 0.28, 0.29, 0.29, 0.47, 0.50, 0.53, 0.61, 0.62, 0.67, 0.69, 0.83, 0.95, 1.65, 1.67 mg/kg.

Eleven supervised trials were conducted on peas in USA in 2001 (maximum GAP: two applications at 0.539 kg ai/ha with a 7 day PHI) where peas were shelled in eight of the trials. The residues in ranked order on shelled peas (succulent seeds) were: < 0.05 (2), 0.06, 0.07, 0.15, 0.19, 0.24, 0.37 mg/kg. The residues on peas (pod and succulent seeds) in ranked order were: 0.64, 0.97, 1.39 mg/kg.

The data populations for shelled peas (succulent seeds) in the EU and peas (pods and succulent seed) were significantly different and could not be combined.

Seventeen supervised trials were conducted on soybean in USA in 2002 (maximum GAP: two applications at 0.539 kg ai/ha with a 7 day PHI). The residues on immature soybean in ranked order were: < 0.05 (11), 0.05, 0.06, 0.08, 0.09, 0.2, 1.18 mg/kg.

Ten supervised trials were conducted on snap beans in USA in 2000 (maximum GAP: two applications at 0.5 kg ai/ha with a 7 day PHI). The residues on snap bean (young pods) in ranked order were: 0.13, 0.28, 0.36, 0.41, 0.42, 0.46, 0.52, 0.54, 0.72, 0.97 mg/kg.

Seven supervised trials were conducted on Lima bean in USA in 2000 (maximum GAP: two applications at 0.5 kg ai/ha with a 7 day PHI). The residues on Lima bean (young pods and immature beans) in ranked order were: < 0.05 (2), 0.07 (2), 0.08 (2), 0.47 mg/kg.

The data for beans with pods were significantly different from those for succulent shelled peas. The data for beans with pods were significantly different from those for immature soybean. The data for snap beans were significantly different from those for lima beans and could not be combined. The data for snap beans were not significantly different from those for beans with pod. The data populations for peas with pod and beans with pod were not significantly different. In summary, residues of boscalid in beans with pods, peas with pods and snap beans from the 32 US trials in rank order were: 0.06, 0.13 (2), 0.22, 0.26, 0.28 (3), 0.29 (2), 0.36, 0.41, 0.42, 0.46, 0.47, 0.50, 0.52, 0.53, 0.54, 0.61, 0.62, 0.64, 0.67, 0.69, 0.72, 0.83, 0.95, 0.97 (2), 1.39, 1.65, 1.67 mg/kg.

No recommendation can be made for legume vegetables until the contribution of residues from direct application as well as uptake through the soil can be assessed.

*Pulses*

Ten supervised trials were conducted on beans in USA in 2000 (maximum GAP: two applications at 0.539 kg ai/ha with a 21 day PHI). The residues on dry beans in ranked order were: < 0.05 (4), 0.06, 0.09, 0.12, 0.14, 0.37, 1.92 mg/kg.

Nine supervised trials were conducted on peas in USA in 2000 (maximum GAP: two applications at 0.539 kg ai/ha with a 21 day PHI). The residues on dry peas in ranked order were: 0.05, 0.09, 0.11, 0.12, 0.16, 0.17, 0.23, 0.31, 0.46 mg/kg.

The data populations for dry beans and dry shelled peas were not significantly different and could be combined. In summary, residues of boscalid in dry beans and dry peas from the 19 US trials in rank order were: < 0.05 (4), 0.05, 0.06, 0.09, 0.09, 0.11, 0.12 (2), 0.14, 0.16, 0.17, 0.23, 0.31, 0.37, 0.39, 1.92 mg/kg.

Seventeen supervised trials were conducted on soybean in USA in 2002 (maximum GAP: two applications at 0.54 kg ai/ha with a 21 day PHI). The residues on dry soybean in ranked order were: < 0.05 (17) mg/kg.

No recommendation can be made for pulses until the contribution of residues from direct application as well as uptake through the soil can be assessed.

*Carrot*

Eight supervised trials were conducted on carrot in USA in 1999 (GAP: six applications at 0.185 kg ai/ha with a 0 day PHI). In two trials in 1999, matching GAP, residues of boscalid were: 0.19, 0.12 mg/kg. In six other trials only three applications were made but the total rate applied was equivalent and residue levels at PHI were similar. The Meeting agreed to combine the data at the same total application rate. The residues on carrot in ranked order were: < 0.05, 0.06, 0.12, 0.17, 0.18, 0.19, 0.28, 0.34 mg/kg.

No recommendation can be made for carrot until the contribution of residues from direct application as well as uptake through the soil can be assessed.

*Radish*

Five supervised trials were conducted on radish in USA in 1999 (no GAP). As no trials were conducted according to a GAP, the Meeting did not recommend a maximum residue level for radish.

*Potato*

Sixteen supervised trials were conducted on potato in USA in 2000 (GAP: two applications at 0.49 kg ai/ha with a 30 day PHI). The residues on potato in ranked order were: < 0.05 (16) mg/kg.

No recommendation can be made for potato until the contribution of residues from direct application as well as uptake through the soil can be assessed.

*Cereal grains*

Boscalid is approved for use on cereal grains in Germany (GAP: two applications at 0.35 kg ai/ha with no specified PHI). Data from six supervised trials were submitted on barley conforming to German GAP from Denmark from 2005 (1), France from 2005 (2), Germany from 2005 (1), the Netherlands from 2003 (1) and the UK from 2003 (1). The residues on barley in ranked order were: < 0.01 (2), 0.02, 0.03, 0.12, 0.19 mg/kg.

Eight supervised trials were submitted on wheat conforming to German GAP from Belgium from 2005 (1), France from 2003 and 2005 (5), Germany from 2003 (1), the UK from 2003 (1). The residues on wheat in ranked order were: < 0.01, 0.01 (3), 0.03, 0.06 (2), 0.27 mg/kg.

The data populations for barley and wheat were not significantly different and could be combined. In summary, residues of boscalid in barley and wheat from the 14 EU trials in ranked order were: < 0.01 (3), 0.01 (3), 0.02, 0.03 (2), 0.06 (2), 0.12, 0.19, 0.27 mg/kg.

*Tree nuts*

Boscalid is approved in the USA on tree nuts with four applications at 0.256 kg ai/ha with a 25 day PHI. Data was submitted from twenty supervised trials conducted on almond in USA between 1999 and 2003. In ten trials on almond from 2003 matching GAP, residues of boscalid were: < 0.05 (20) mg/kg.

Ten supervised trials were submitted on pecan nut from the USA from 1999 conforming to US GAP, i.e., four applications at 0.256 kg ai/ha with a 25 day PHI. The residues of boscalid in pecan nuts in ranked order were: < 0.05 (10) mg/kg.

Six supervised trials were conducted on pistachio nuts in USA in 1999 (maximum GAP: 0.256 kg ai/ha, four applications, 25 day PHI). In six US trials on pistachio in 1999 matching GAP, residues of boscalid, median residue underlined, were: < 0.05(2), <u>0.19</u>, <u>0.35</u>, 0.45, 0.64 mg/kg.

The data populations for almond and pistachio, and pecan nut and pistachio were significantly different and could not be combined.

The Meeting agreed to use the data from almond and pecan, and estimated a maximum residue level and an STMR value for boscalid in tree nuts except pistachio of 0.05 (*) and 0.05 mg/kg respectively.

The Meeting agreed to use the data from pistachio, and estimated a maximum residue level and an STMR value for boscalid in pistachio of 1 and 0.27 mg/kg respectively.

## Canola

Trials on canola were conducted in USA in 2000 (GAP: 0.294 kg ai/ha, two applications with a 21 day PHI). In the 16 US trials the application rate (0.45 kg ai/ha) was 50% higher than the GAP rate and at PHI of 20–22 days.

The Meeting noted that the application rate was 50% higher than GAP, and as such the residue values could not be used for evaluation.

## Sunflower

Eight supervised trials were conducted on sunflower in the USA according to GAP in 2001 (GAP: two applications at 0.44 kg ai/ha with a 21 day PHI). The residues in ranked order were: < 0.05, 0.08, 0.09, 0.13, 0.16, 0.16, 0.23, 0.45 mg/kg.

No recommendation can be made for sunflower until the contribution of residues from direct application as well as uptake through the soil can be assessed.

## Peanut

Twelve supervised trials were conducted on peanut in USA according to GAP in 2000 (GAP: three applications at 0.49 kg ai/ha with a 14 day PHI). The residues in peanut in ranked order were: < 0.05 (11), 0.05 mg/kg.

No recommendation can be made for peanut until the contribution of residues from direct application as well as uptake through the soil can be assessed.

## Coffee

Seven supervised trials were conducted on coffee in Brazil in 2000 (GAP: one application at 0.075 kg ai/ha with a 45 day PHI). In four of the trials the application rate of 0.15 kg ai/ha was two times that of the maximum GAP. The Meeting noted that while the application rate was higher than GAP, the residues were below the LOQ (0.05 mg/kg) and could be used for evaluation.

The Meeting estimated a maximum residue level and an STMR value for boscalid in coffee of 0.05 (*) and 0.05 mg/kg respectively.

## Animal feedstuffs

### Almond hull

See previous section on almond for GAP in USA. In ten supervised trials on almond at US GAP, residues of boscalid in almond hull in rank order, with median and highest residue values underlined, were: 2.21, 2.64, 3.30, 3.42, 3.45, 3.91, 5.41, 6.78, 11.3, 11.9 mg/kg (fresh weight).

Allowing for the standard 90% dry matter for almond hulls (*FAO Manual*, p. 147), the Meeting estimated a maximum residue level of 15 mg/kg and an STMR of 4.1 mg/kg for almond hulls (dry weight). A highest residue level of 13 mg/kg was estimated for calculating the dietary burden of farm animals.

### Straw and fodder (dry) cereal grains

In 10 trials on barley at German GAP, residues of boscalid in barley straw in rank order, median and highest residue underlined, were: 0.51, 2.5, 5.8, 13, 14, 27 mg/kg (fresh weight). See previous section on wheat for GAP in Europe. In eight trials on wheat at German GAP, residues of boscalid in wheat straw in rank order were: 3.0, 3.1, 5.3, 5.8, 7.9, 7.9, 11, 15 mg/kg (fresh weight).

*Peanut fodder*

See previous section on peanut for GAP in USA in 2000. Residues of boscalid in peanut hay in rank order were: 3.2, 5.8, 6.7, 6.7, 7.8, 9.0, 13, 20, 24, 28, 29 mg/kg.

No recommendation can be made for peanut fodder until the contribution of residues from direct application as well as uptake through the soil can be assessed.

*Soybean forage*

See previous section on soybean for GAP in USA in 2002. Seventeen supervised were conducted on soybean.

The Meeting noted that the PHIs were double that of the GAP, as such the residues could not be used for evaluation.

*Soybean fodder*

See previous section on soybean for GAP in USA in 2002. In 17 supervised trials on soybean at USA GAP, residues of boscalid in soybean hay in rank order were: 1.3, 1.4, 1.8, 2.0, 2.1, 2.3, 2.8, 3.6, 4.6, 4.8, 5.3, 6.7, 7.1, 7.3, 7.8, 11, 21 mg/kg.

## Fate of residues during processing

The Meeting received information on the fate of boscalid residues during aqueous hydrolysis under conditions of pasteurisation, baking, brewing and boiling and sterilisation. Information was also provided on the fate of boscalid residues during the food processing of citrus, apples, plum, cherries, strawberries, grapes, white cabbage, gherkins, tomatoes, head lettuce, peas, soybeans, carrots, sugar beet, barley, winter wheat, corn, peanuts, sunflower, cotton, canola seed, mint and hops.

Boscalid was not degraded during the simulation of pasteurisation (pH 4, 90°C) nor during simulated baking, boiling, brewing (pH 5, 100°C) or during sterilisation (pH 6, 120°C).

The processing factors for wet apple pomace (6.06) and apple juice (0.08) were applied to the estimated STMR for apple (0.365 mg/kg) to produce STMR-P values for wet apple pomace (2.2 mg/kg) and apple juice (0.03 mg/kg).

The processing factors for plum to dried prunes (2.80) and to puree (1.95) were applied to the estimated STMR for plums (0.205 mg/kg) to produce an STMR-P value for prunes (0.57 mg/kg) and puree (0.40 mg/kg).

The processing factors for raisins (2.42), wet pomace (2.50), wine (0.35) and juice (0.42) were applied to the estimated STMR for grapes (1.09 mg/kg) to produce STMR-P values for raisins (2.6 mg/kg), wet pomace (2.7 mg/kg), wine (0.38 mg/kg) and grape juice (0.46 mg/kg). The processing factor for raisins (2.4) was applied to the grape residue data (HR of 3.2 mg/kg) to produce an estimated highest value for dried grapes (7.8 mg/kg).

The Meeting estimated a maximum residue level for boscalid in dried grapes (= currants, raisins, sultanas) of 10 mg/kg.

## Residues in animal commodities

### Farm animal feeding

The Meeting received a lactating dairy cow feeding study which provided information on likely residues resulting in animal tissues and milk from residues in the animal diet.

Lactating Holstein cows were dosed with boscalid at the equivalent of 1.5 (1×), 4.5 (3×) and 18 (12×) ppm in the dry-weight diet for 28 consecutive days. Milk was collected twice daily for analysis. Animals were sacrificed within 23 hours after the final dosing, except for one cow of the 12× group which was sacrificed seven days after the final dose to determine residue levels post dosing.

No residues were detected in milk samples taken from the control and the 1× dose groups. In a few samples from the 3× dose group, residues just above the LOQ of 0.01 mg/kg for boscalid were detected, but no residues of M510F01 or M510F02 were observed. In the group average, residues were below the LOQ. In the 12× dose group, residues of boscalid occurred regularly from day one onward with residues reaching a plateau on day 14 with average residues between 0.04 mg/kg and 0.05 mg/kg. M510F53 was below LOQ (< 0.01 mg/kg) in milk from all three treatment groups.

In the tissues, the mean residues of boscalid at the 3 dosing levels were: muscle (< 0.05, < 0.05, < 0.05 mg/kg); fat (0.06, 0.11, 0.27 mg/kg); liver (< 0.05, 0.06, 0.18 mg/kg); kidney (< 0.05, 0.07, 0.24 mg/kg).

M510F53 was below LOQ (< 0.01 mg/kg) in liver from 1× and 3× dose groups, and up to 0.09 mg/kg from 12× dose group.

Residues depleted quickly from the milk of a high-dose animal after dosing was stopped, falling below LOQ (0.01 mg/kg) after 2 days. Residues fell to below the LOQ (< 0.05 mg/kg) in all tissues. It was shown by samples from the withdrawal animal that no residues in milk was observed two day after dosing had stopped and boscalid were rapidly excreted.

*Farm animal dietary burden*

The Meeting noted that field trials on rotational crops were provided at a late stage of the Meeting, and decided to estimate maximum residue levels and STMRs on annual crops that may lead to animal feeds at a future JMPR when all the data can be examined together. The Meeting was also informed that a new livestock feeding study is commencing in 2007. The Meeting decided to calculate the livestock dietary burden and estimate maximum residue levels and STMRs for animal commodities at a future JMPR meeting.

## RECOMMENDATIONS

On the basis of the data from supervised trials, the Meeting concluded that the residue concentrations listed below are suitable for establishing MRLs and for assessing IEDIs.

Definition of the residue (for compliance with the MRL for plant and animal commodities and for estimation of dietary intake for plant commodities): *boscalid.*

Definition of the residue (for estimation of dietary intake for animal commodities): *sum of boscalid, 2-chloro-N-(4'-chloro-5-hydroxybiphenyl-2-yl)nicotinamide including its conjugate, expressed as boscalid. The residue is fat soluble.*

| CCN | Commodity | MRL, mg/kg | STMR or STMR-P, mg/kg |
|-----|-----------|-----------|-----------|
| AM | Almond hulls | 15 | 4.1 |
| FP 0226 | Apple | 2 | 0.365 |
| | Apple pomace | | 2.2 |
| | Apple juice | | 0.03 |
| FI 0327 | Banana | 0.2 | 0.05 |
| FB 0018 | Berries and other small fruits [note 1] | 10 | 2.53 |
| SB 0716 | Coffee | 0.05* | 0.05 |
| FB 0269 | Grapes | 5 | 1.09 |
| DF 0269 | Dried grapes (= currants, raisins, sultanas) | 10 | 2.6 |
| | Wet pomace | | 2.7 |
| | Wine | | 0.38 |
| | Grape juice | | 0.46 |

| CCN | Commodity | MRL, mg/kg | STMR or STMR-P, |
|---|---|---|---|
| TN 0675 | Pistachio | 1 | 0.27 |
| FS 0012 | Stone fruit | 3 | 1.21 |
| DF 0014 | Prunes | | 0.57 |
| | Puree [note 2] | | 0.40 |
| TN 0085 | Tree nuts [note 3] | 0.05* | 0.05 |

\* At or about the limit of quantification.
Note 1 except strawberry and grapes.
Note 2 processed product of plum
Note 3 except pistachio.

## DIETARY RISK ASSESSMENT

### Long-term intake

The Meeting could not make any recommendation for residue levels in annual crops since field studies on residues in follow-up and rotational crops were provided only at a late stage of the Meeting. In view of the wide range of crops in which boscalid residues may be present above the LOQ, maximum residue levels could not be recommended for a large number of crops. The Meeting decided that the estimation of the long-term intake would not be realistic at this time. Consequently, the long-term intake will be estimated at a future meeting when the residues deriving from both direct application and those taken up from the soil in a rotational crop situation can be evaluated together.

### Short-term intake

The 2006 JMPR decided that an acute ARfD was unnecessary. The Meeting therefore concluded that the short-term intake of boscalid residues is unlikely to present a public health concern.

## REFERENCES

**Author, Date, Title, Institute, Report Reference, Document No.**

Abdel-Baky S and Jones J E. 2001. Boscalid: Method for Determination of BAS 500 F, BF 500-3 and BAS 510 F Residues in Plant Matrices using LC/MS/MS. BASF Corporation, Agricultural Products Center, NC 27709-3528, USA. BASF unpublished report method D9908, issued ?.02.2001. 2000/5001019

Abdel-Baky S and Jones J E. 2001. Boscalid: Validation of BASF Analytical Method D0008, Method for determining BAS 510 F in plant and turf cloth matrices using GC/MS and dislodgeable foliar samples using HPLC/UV. BASF Corporation, Agricultural Products Center, NC 27709-3528, USA. Unpublished report, issued ?.02.2001. 2001/5000977

Abdel-Baky S and Jones J E. 2001. Boscalid: Method for Determination of BAS 500F, BF 500-3 and BAS 510F Residues in Plant Matrices Using LC/MS/MS (Method No. D9908). BASF Corporation, Agricultural Products Center, NC 27709-3528, USA. BASF unpublished report 64276, issued 02.03.2001. 2001/5001019

Anonymous. 2006. Boscalid: JMPR Evaluation, Section 3 – Monograph, Part II. BASF the Chemical Company. BASF unpublished report, issued ?.04.2006. 2006/1015800

Bayer H and Grote C. 2001. Boscalid: Field soil dissipation of BAS 510 F (300 355) in formulation BAS 510 KA F (1998 – 1999). BASF AG, Agrarzentrum Limburgerhof; Limburgerhof; Germany Fed. Rep. BASF unpublished report EU/FA/051/98, issued 16.02.2001. 2000/1013295

Beck J and Benz A. 2001. Boscalid: Study on the residue behavior of BAS 500 F and BAS 510 F in brassicas after treatment with BAS 516 GA F under field conditions in Denmark, Germany, Great Britain, the Netherlands and Sweden, 1999. BASF AG, Agrarzentrum Limburgerhof; Limburgerhof; Germany Fed.Rep. BASF unpublished report NEU/FR/10/99, issued 22.05.2001. 2001/1001001

Beck J et al. 2001. Boscalid: Study on the residue behavior of BAS 500 F and BAS 510 F in lettuce after treatment with BAS 516 GA F under field conditions in Germany, France, the Netherlands and Spain, 1999. BASF AG, Agrarzentrum Limburgerhof; Limburgerhof; Germany Fed.Rep. BASF unpublished report EU/FR/04/99, issued 31.05.2001. 2001/1000998

Beck J et al. 2001. Boscalid: Study on the residue behavior of BAS 500 F and BAS 510 F in lettuce after treatment with BAS 516 GA F under field conditions in Germany, France, the Netherlands and Spain, 2000. BASF AG, Agrarzentrum

Limburgerhof; Limburgerhof; Germany Fed.Rep. BASF unpublished report EU/FR/04/00, issued 31.05.2001. 2001/1000999

Beck J *et al.* 2001. Boscalid: Study on the residue behavior of BAS 500 F and BAS 510 F in brassicas after treatment with BAS 516 GA F under field conditions in Denmark, Germany, Great Britain, the Netherlands and Sweden, 2000. BASF AG, Agrarzentrum Limburgerhof; Limburgerhof; Germany Fed.Rep. BASF unpublished report NEU/FR/10/00, issued 01.06.2001. 2001/1001000

Bross M. 2001. Boscalid: Investigations on the extractability of 14C-BAS 510 F residues from plant matrices. BASF AG, Agrarzentrum Limburgerhof; Limburgerhof; Germany Fed. Rep. BASF unpublished report 73479, issued 23.02.2001. 2000/101739.

Bross M. 2001. Boscalid: Investigation on the extractability of 14C-BAS 510F Residues from Plant Matrices. BASF AG, Agrarzentrum Limburgerhof; Limburgerhof; Germany Fed. Rep. BASF unpublished report 73497, issued 23.02.2001. 2001/1001739

Class T. 2001. Boscalid: Assessment and validation of the adapted multi-residue method DFG S19 for the determination of BAS 510 F and its metabolite M510F01 in animal matrices. PTRL Europe; Ulm; Germany Fed. Rep. BASF unpublished report P/B 453, issued 20.02.2001. 2000/1017227

Class T. 2004. Boscalid: Assessment and validation of the adapted multi-residue method DFG S19 for the determination of BAS 510 F and its metabolite M510F01 in animal matrices (Amendment). PTRL Europe; Ulm; Germany Fed. Rep. BASF unpublished report P/B 453, issued 03.02.2004. 2004/1005207

Dantas C. 2001. Boscalid: Estudo de residuo de BAS 510 F em cafe (grao) do Brasil. Laboratorio de Desenvolvimento de Produtos Agricolas da America Latina; Rio de Janeiro; Brazil. BASF unpublished report RFR-AR-464-01, issued 10.09.2001. 2001/1026859

Dantas C. 2001. Boscalid: Estudo de residuo de BAS 510 F em cafe (grao) do Brasil. Laboratorio de Desenvolvimento de Produtos Agricolas da America Latina; Rio de Janeiro; Brazil. BASF unpublished report RFR-AR-465-01, issued 10.09.2001. 2001/1026860

Dantas, C. 2001. Boscalid: Estudo de residuo de BAS 510 F em cafe (grao) do Brasil. Laboratorio de Desenvolvimento de Produtos Agricolas da America Latina; Rio de Janeiro; Brazil. BASF unpublished report RFR-AR-467-01, issued 10.09.2001. 2001/1026861

Dantas C. 2001. Boscalid: Estudo de residuo de BAS 510 F em cafe (grao) do Brasil. Laboratorio de Desenvolvimento de Produtos Agricolas da America Latina; Rio de Janeiro; Brazil. BASF unpublished report RFR-AR-466-01, issued 10.09.2001. 2001/1026862

Daum A. 1998. Boscalid: Determination of the solubility of BAS 510 F (Reg. No. 300355) pure active ingredient (PAI) in organic solvents at 20°C. BASF AG, Agrarzentrum Limburgerhof; Limburgerhof; Germany Fed. Rep. BASF unpublished report PCP04876, issued 07.09.1998. 1998/10953

Daum A. 1998. Boscalid: Determination of the pKa of Reg.No. 300 355 (BAS 510 F) in water at 20°C. BASF AG, Agrarzentrum Limburgerhof; Limburgerhof; Germany Fed. Rep. BASF unpublished report PCP04896, issued 10.09.1998. 1998/10967

Daum A. 1998. Boscalid: Determination of the octanol/water-partition coefficient of Reg.No. 300 355 (BAS 510 F) by HPLC. BASF AG, Agrarzentrum Limburgerhof; Limburgerhof; Germany Fed. Rep.BASF unpublished report PCP05017, issued 12.11.1998. 1998/11082

Daum A. 1998. Boscalid: Determination of the solubility of BAS 510 F (Reg.No. 300 355) in water at 20°C by column elution method and by HPLC. BASF AG, Agrarzentrum Limburgerhof; Limburgerhof; Germany Fed. Rep. BASF unpublished report PCP04888, issued 09.09.1998. 1998/10961

Daum A. 1999. Boscalid: Determination of the melting point and the appearance of Reg.N. 300355 (BAS 510F). BASF AG, Agrarzentrum Limburgerhof; Limburgerhof; Germany Fed. Rep. BASF unpublished report PCP05375, issued 16.08.1999. 1999/10991

Daussin S and Wofford J T. 2001. Boscalid: The magnitude of BAS 510 F residues in stone fruit. BASF Agro Research RTP; Research Triangle Park, NC 27709, USA. BASF unpublished report 63904, issued 23.01.2001. 2001/5000831

Ebert D and Harder U. 2000. Boscalid: The degradation behaviour of 14C-BAS 510 F in different soils (DT50/DT90). BASF AG, Agrarzentrum Limburgerhof; Limburgerhof; Germany Fed. Rep. BASF unpublished report 41860, issued 20.06.2000. 2000/1013279

Fabian E. 2001. Boscalid: The determination of BAS 510 F residues (as M510F53) in liver and milk by microwave treatment. BASF AG, Agrarzentrum Limburgerhof; Limburgerhof; Germany Fed. Rep. BASF unpublished report 96997, issued 09.03.2001. 2000/1017224

Fabian E. 2004. Boscalid: The validation of BASF method 476/0: The determination of BAS 510 F residues (as M510F53) in liver and milk by microwave (Amendment No.1). BASF AG, Agrarzentrum Limburgerhof; Limburgerhof; Germany Fed. Rep. BASF unpublished report 96997, issued 10.01.2004. 2003/1021923

Fabian E and Grosshans F. 2001. Boscalid: The metabolism of 14C – BAS 510 F in lactating goat. BASF AG, Agrarzentrum Limburgerhof; Limburgerhof; Germany Fed. Rep. BASF unpublished report 41837, issued 06.02.2001. 2000/1017221

Funk H. 2002. Boscalid: Parts of Technical Procedure: Method for the determination of BAS 510 F in wheat grain (BASF Method Number 514/0). BASF AG, Agrarzentrum Limburgerhof; Limburgerhof; Germany Fed. Rep. BASF unpublished report method 514/0, issued 24.04.2002. 2002/1004107

Funk H and Mackenroth C. 2001. Boscalid: Validation of BASF Method 445/0: Determination of BAS 510 in plant matrices. BASF AG, Agrarzentrum Limburgerhof; Limburgerhof; Germany Fed. Rep. BASF unpublished report 41840, issued 19.02.2001. 2000/1012404

Funk H and Mackenroth C. 2001. Boscalid: Investigation of the stability of residues of BAS 510 F in plant matrices under normal storage conditions. BASF AG, Agrarzentrum Limburgerhof; Limburgerhof; Germany Fed. Rep. BASF unpublished report 41851, issued 13.12.2001. 2001/1015028

Goetz N von. 1999. Boscalid: Hydrolysis of BAS 510 F. BASF AG, Agrarzentrum Limburgerhof; Limburgerhof; Germany Fed. Rep. BASF unpublished report 41874, issued 17.09.1999. 1999/11285

Goetz N von. 1999b. Boscalid: Aqueous photolysis of BAS 510 F. BASF AG, Agrarzentrum Limburgerhof; Limburgerhof; Germany Fed. Rep. BASF unpublished report 41875, issued 03.12.1999. 1999/11804

Grosshans F. 2001. Bascalid: The validation of BASF method 471/0: The determination of BAS 510 F and the metabolite M510F01 in animal matrices. BASF AG, Agrarzentrum Limburgerhof; Limburgerhof; Germany Fed. Rep. BASF unpublished report 42392, issued 20.02.2001. 2000/1017223

Grosshans F. 2001. Boscalid: Investigation of the stability of residues of BAS 510 F and M510F01 in sample materials of animal origin under usual storage conditions. BASF AG, Agrarzentrum Limburgerhof; Limburgerhof; Germany Fed. Rep. BASF unpublished report 42403, issued 20.02.2001. 2000/1017229

Grosshans F. 2004. Boscalid: The validation of BASF method 471/0: The determination of BAS 510 F and the metabolite M510F01 in animal matrices (Amendment No.1). BASF AG, Agrarzentrum Limburgerhof; Limburgerhof; Germany Fed. Rep. BASF unpublished report 42392, issued 10.01.2004. 2003/1021922

Grosshans F. 2004. Boscalid: The validation of BASF method 471/0: "The determination of BAS 510 F and the metabolite M510F01 in animal matrices. BASF AG, Agrarzentrum Limburgerhof; Limburgerhof; Germany Fed. Rep. BASF unpublished report 42392, issued 20.02.2001. 2004/5000681

Hamm R T. 1999. Boscalid: Metabolism of BAS 510 F in lettuce. BASF AG, Agrarzentrum Limburgerhof; Limburgerhof; Germany Fed. Rep. BASF unpublished report 44875, issued 21.09.1999. 1999/11240

Hamm R T and Veit P. 2001. Boscalid: Confined rotational crop study with 14C-BAS 510 F. BASF AG, Agrarzentrum Limburgerhof; Limburgerhof; Germany Fed. Rep. BASF unpublished report 42560, issued 27.02.2001. 2000/1014862

Haughey D W & Abdel-Baky S. 2000. Boscalid: The magnitude of BAS 510 F residue in almonds. BASF Agro Research RTP; Research Triangle Park, NC 27709; USA. BASF unpublished report 63902, issued 07.11.2000. 2000/5226

Haughey D W and Abdel-Baky S. 2000. Boscalid: The magnitude of BAS 510 F residues in grapes. BASF Agro Research RTP; Research Triangle Park, NC 27709; USA. BASF unpublished report 63892, issued 23.10.2000. 2000/5228

Haughey D W and Abdel-Baky S. 2000. Boscalid: The magnitude of BAS 510 F residues in pistachios. BASF Agro Research RTP; Research Triangle Park, NC 27709; USA. BASF unpublished report 63898, issued 05.01.2001. 2000/5229

Haughey D W and Abdel-Baky S. 2000. Boscalid: The magnitude of BAS 510 F residues in pecans. BASF Agro Research RTP; Research Triangle Park, NC 27709; USA. BASF unpublished report 63900, issued 05.01.2001. 2000/5230

Haughey D W and Abdel-Baky S. 2001. Boscalid: The magnitude of BAS 510 F residues in head and leaf lettuce. BASF Agro Research RTP; Research Triangle Park, NC 27709, USA. BASF unpublished report 64114, issued 22.02.2001. 2001/5000051

Haughey D W and Abdel-Baky S. 2001. Boscalid: The Magnitude of BAS 510 F Residues in Grape Processed Fractions. BASF Agro Research RTP; Research Triangle Park, NC 27709, USA. BASF unpublished report 46650, issued 22.02.2001. 2001/5000065

Haughey D W and Abdel-Baky S. 2001. Boscalid: The magnitude of BAS 510 F residues in tomatoes. BASF Agro Research RTP; Research Triangle Park, NC 27709, USA. BASF unpublished report 46786, issued 22.02.2001. 2001/5000832

Haughey D W and Abdel-Baky S. 2001. Boscalid: The magnitude of BAS 510 F residues in dry, snap, and lima beans. BASF Corporation, Agricultural Products Center, NC 27709-3528, USA. BASF unpublished report 46764, issued 09.03.2001. 2001/5000905

Haughey D W and Abdel-Baky S. 2001. Boscalid: The magnitude of BAS 510 F residues in tomato processed fractions. BASF Corporation, Agricultural Products Center, NC 27709-3528, USA. BASF unpublished report 46788, issued 16.03.2001. 2001/5000967

Haughey D W and Abdel-Baky S. 2001. Boscalid: The magnitude of BAS 510 F residues in radishes. BASF Corporation, Agricultural Products Center, NC 27709, USA. BASF unpublished report 66694, issued 20.12.2001. 2001/5002572

Haughey D W and Abdel-Baky S. 2001. Boscalid: The magnitude of BAS 510F residues in cucurbitsBASF Corporation Agro Research; Princeton NJ 08543-0400; USA. BASF unpublished report 66698, issued 20.12.2001. 2001/5002593

Haughey D W and Abdel-Baky S. 2002. Boscalid: The Magnitude of BAS 510 F Residues in Sugar Beet and Sugar Beet Processed Fractions. BASF Corporation, Agricultural Products Center, NC 27709, USA. BASF unpublished report 66706, issued 14.02.2002. 2001/5003245

Heck W et al. 2000. Boscalid: Study on the residue behavior of BAS 510 FD and Kresoxim-methyl in melons after treatment with BAS 517 00 F under greenhouse conditions in Spain, 1999. BASF AG, Agrarzentrum Limburgerhof; Limburgerhof; Germany Fed.Rep. BASF unpublished report ES/FR/04/99, issued 02.11.2000. 2000/1014874

Johnston R L. 2004. The magnitude of BAS 500 F and BAS 510 F residues in almonds with a 25 day PHI. BASF Agro Research RTP; Research Triangle Park, NC 27709; USA. BASF unpublished report 146467, issued 15.06.2004. 2004/5000223

Johnston R L. 2005. Study on the residue behaviour of BAS 510F and BAS 500 F in cauliflower and broccoli after application of BAS 516 00 F under field conditions in France (N & S), Germany, Sweden and Denmark, 2004. BASF Agro Research RTP; Research Triangle Park, NC 27709; USA. BASF unpublished report 168856, issued 17.02.2005. 2004/7007476

Johnston R L. 2005. Boscalid: Study on the residue behaviour of BAS 510 F and BAS 500 F in brussels sprouts after application of BAS 516 00 F under field conditions in France (N), England, Germany and Sweden, 2004. BASF Agro Research RTP; Research Triangle Park, NC 27709; USA. BASF unpublished report 168856, issued 17.02.2005. 2004/7007478

Johnston R L. 2005. Boscalid: Study on the residue behaviour of BAS 510 F and BAS 500 F in strawberries (glasshouse) after application of BAS 516 00 F under glasshouse conditions in France (S), Spain and Italy, 2004. BASF Agro Research RTP; Research Triangle Park, NC 27709, USA. BASF unpublished report 168898, issued 18.02.2005. 2004/7007479

Johnston R L. 2005. Boscalid: The magnitude of BAS 510F and BAS 500F residues in tomatoes and peppers following application of BAS 516 04F. BASF Agro Research RTP; Research Triangle Park, NC 27709; USA. BASF unpublished report 190540, issued 25.08.2005. 2005/5000016

Johnston R L and Jones J E. 2005. Boscalid: Magnitude of BAS 510 F and BAS 500 F residues in bulb vegetables after application of BAS 516 04F. BASF Agro Research RTP; Research Triangle Park, NC 27709; USA. BASF unpublished report 190537, issued 13.10.2005. 2005/5000019

Jordan J. 2001. Boscalid: Independent method validation of BASF analytical method D0008 entitled "Method for determining BAS 510 F in plant and turf cloth matrices using GC/MS and dislodgeable foliar samples using HPLC/UV". BASF Corporation, Agricultural Products Center, NC 27709-3528, USA. BASF unpublished report 64954, issued 22.03.2001. 2001/5000880

Jordan J. 2002. Boscalid: The Magnitude of BAS 500 02 F and BAS 510 02 F Residues in Citrus. BASF Corporation, Agricultural Products Center, NC 27709-3528, USA. BASF unpublished report 64978, issued 27.11.2002. 2002/5002446

Jordan J. 2003. Boscalid: BAS 510 Frozen Storage Stability in Treated Sample of Tomato Paste. BASF Corporation, Agricultural Products Center, NC 27709-3528, USA. BASF unpublished report 183754, issued 08.12.2003. 2003/5000538

Jordan J and Jones J E. 2005. Boscalid: he Magnitude of BAS 500 F and BAS 510 F Residues in Cotton Processed Fractions. BASF Corporation, Agricultural Products. BASF Agro Research RTP; Research Triangle Park, NC 27709; USA. BASF unpublished report 198163, issued 23.09.2005. 2005/5000013

Kampke-Thiel K. 2001. Boscalid: Independent laboratory validation of the adapted multi-residue method DFG S19 for the determination of BAS 510 F and its metabolite M510F01 in animal matrices (PTRL urope study No. P453G). BASF AG, Agrarzentrum Limburgerhof; Limburgerhof; Germany Fed. Rep. BASF unpublished report 42397, issued 09.03.2001. 2000/1017226

Kästel R. 1999. Boscalid: Physical properties of 300355 (PAI). BASF AG, Agrarzentrum Limburgerhof; Limburgerhof; Germany Fed. Rep. BASF unpublished report PCF 01962, issued 12.02.1999. 1999/10203

Kellner O and Keller W. 2000. Boscalid: Field soil dissipation of BAS 510 F (300 355) in formulation BAS 510 KB F (1997-1998). BASF AG, Agrarzentrum Limburgerhof; Limburgerhof; Germany Fed. Rep. BASF unpublished report DE/FA/047/97, issued 25.01.2000. 2000/1000123

Klimmek S and Schulz H. 2004. Boscalid: Study on the residue behaviour of BAS 510 F and BAS 500 F in Brussels sprouts after application of BAS 516 00 F under field conditions in the Netherlands, Germany, United Kingdom, Denmark and Sweden, 2003. Institut Fresenius, Chemische und Biologische Laboratorien GmbH; Taunusstein; Germany Fed. Rep. BASF unpublished report 168865, issued 18.06.2004. 2004/1015912

Leibold E and Hoffmann H D. 2000. Boscalid: 14C-BAS 510 F – Absorption, distribution and excretion after repeated oral administration in lactating goats. BASF AG; Ludwigshafen/Rhein; Germany Fed. Rep. BASF unpublished report 02B0426/976039, issued 24.05.2000. 2000/1012353

Leonard R C. 2003. Boscalid: Magnitude of BAS 500 F and BAS 510 F Residues in Soybean. BASF Agro Research RTP; Research Triangle Park, NC 27709; USA. BASF unpublished report 140578, issued 28.03.2003. 2002/5004272

Leonard R C. 2005. Boscalid: The magnitude of BAS 510F and BAS 500F residues in cucurbits. BASF Agro Research RTP; Research Triangle Park, NC 27709; USA. BASF unpublished report 190531, issued 22.03.2005. 2005/5000021

Leonard R C. 2005. Boscalid: The magnitude of BAS 510F and BAS 500F residues in sunflowers BASF Agro Research RTP; Research Triangle Park, NC 27709; USA. BASF unpublished report 190534, issued 21.03.2005. 2005/5000022

Leonard R C. 2005. Boscalid: The magnitude of BAS 510F and BAS 500F residues in stone fruit. BASF Agro Research RTP; Research Triangle Park, NC 27709, USA. BASF unpublished report 190555, issued 07.09.2005. 2005/5000024

Leonard R C. 2005. Boscalid: Study on the residue behavior of Boscalid and Epoxiconazole in cereals after treatment with BAS 549 00 F under field conditions in Denmark, Belgium, United Kingdom, Northern and Southern France and Germany, 2005. BASF Agro Research RTP; Research Triangle Park, NC 27709, USA. BASF unpublished report 164104, issued 14.09.2005. 2005/5000151

Leonard R C. 2005. Boscalid: Magnitude of BAS 510 F and BAS 500 F residues in mustard greens - 2004-2005 field study. BASF Agro Research RTP; Research Triangle Park, NC 27709, USA. BASF unpublished report 190528, issued 23.09.2005. 2005/7002859

Leonard R C and Gooding R F. 2005. Boscalid: The magnitude of BAS 510F and BAS 500F residues in grapes. BASF Agro Research RTP; Research Triangle Park, NC 27709, USA. BASF unpublished report 190543, issued 09.08.2005. 2005/5000023

Leonard R C and Gooding R F. 2005. Boscalid: The magnitude of BAS 510F and BAS 500F residues in berries. BASF Agro Research RTP; Research Triangle Park, NC 27709, USA. BASF unpublished report 190549, issued 07.22.2005. 2005/5000025

Leonard R C and Gooding R F. 2005. Boscalid: The magnitude of BAS 510 F and BAS 500 F residues in berries. BASF Agro Research RTP; Research Triangle Park, NC 27709, USA. BASF unpublished report 190549, issued 22.07.2005. 2005/5000144

Leonard R C and Gooding R F. 2005. Boscalid: The magnitude of BAS 510 F and BAS 500 F residues in berries (Amended final report). BASF Agro Research RTP; Research Triangle Park, NC 27709, USA. BASF unpublished report 190549, issued 25.08.2005. 2005/5000144

Mackenroth C. 2003. Boscalid: Summary of recoveries: Analytical method No. 514/0: Method for the determination of boscalid (BAS 510 F) in cereal grain. BASF AG, Agrarzentrum Limburgerhof; Limburgerhof; Germany Fed. Rep. BASF unpublished report method 514/0, issued ?.12.2003. 2003/1009787

Malinsky D S. 2002. Boscalid: Independent method validation of BASF analytical method 471/0 entitled: "Method for determination of BAS 510 F and the metabolite M510F01 in animal matrices. BASF Corporation, Agricultural Products Center, NC 27709-3528, USA. BASF unpublished report method 130730, issued 16.07.2002. 2002/5002983

Meumann H et al. 2000. Boscalid: Study on the residue behaviour of BAS 510 F in grape process fractions after treatment with BAS 510 01 F under field conditions in Germany, 1999. BASF AG, Agrarzentrum Limburgerhof; Limburgerhof; Germany Fed. Rep. BASF unpublished report DE/FV/04/99, issued 01.09.2000. 2000/1012412

Nietschmann D and Lam W W. 2000. Boscalid: Nature of residues of 14C-BAS 510 F in laying hens. BASF Corporation Agricultural Products Center; Research Triangle Park, NC 27709, USA. BASF unpublished report 98084, issued 29.06.2000. 2000/5154

Ohnsorge U. 2000. Boscalid: Henry's law constant for 300 355. BASF AG, Agrarzentrum Limburgerhof; Limburgerhof; Germany Fed. Rep. BASF AG, Agrarzentrum Limburgerhof; Limburgerhof; Germany Fed. Rep. BASF unpublished report APD/CP, issued 15.02.2000. 2000/1001009

Perny A. 2001. Boscalid: Residue study in beans following treatment with the preparation BAS 510 01 F under field conditions in France, 2000. ANADIAG Laboratoire; Haguenau ; France. BASF unpublished report R A0033, issued 06.02.2001. 2000/1014876

Rabe U and Schlüter H. 2001. Boscalid: Metabolism of 14C-BAS 510 F in grapevine. BASF AG, Agrarzentrum Limburgerhof; Limburgerhof; Germany Fed. Rep. BASF unpublished report 42389, issued 09.03.2001. 2000/1014860

Raunft E and Funk H. 2002. Boscalid : Study on the residue behavior of BAS 500 F and BAS 510 F in apples after application of BAS 516 01 F under field conditions in Belgium, Germany, France and the Netherlands, 2001. BASF AG, Agrarzentrum Limburgerhof; Limburgerhof; Germany Fed.Rep. BASF unpublished report 71133, issued 12.02.2001. 2001/1015029

Raunft E et al. 2003. Boscalid: Study on the residue behaviour of Boscalid (proposed) and Epoxiconazole in cereals after application of BAS 549 00 F under field conditions in France, Germany, the Netherlands and United Kingdom, 2003. BASF AG, Agrarzentrum Limburgerhof; Limburgerhof; Germany Fed.Rep. BASF unpublished report 164047, issued 17.12.2003. 2003/1009783

Raunft E, Benz A and Mackenroth C. 2001. Boscalid: Study on the residue behavior of BAS 500 F and BAS 510 F in leek after treatment with BAS 516 GA F under field conditions in Belgium, Germany and the Netherlands, 1999/2000. BASF AG, Agrarzentrum Limburgerhof; Limburgerhof; Germany Fed.Rep. BASF unpublished report NEU/FR/12/99, issued 23.05.2001. 2001/1006130

Raunft E, Funk H and Mackenroth C. 2001. Boscalid: Study on the residue behavior of BAS 500 F and BAS 510 F in head lettuce and its processing products after treatment with BAS 516 GA F under field conditions in Germany and Spain, 2000.

BASF AG, Agrarzentrum Limburgerhof; Limburgerhof; Germany Fed.Rep. BASF unpublished report NEU/FR/08/00, issued 08.06.2001. 2001/1006128

Raunft E, Funk H and Mackenroth C. 2001. Boscalid: Study on the residue behavior of BAS 500 F and BAS 510 F in leek after treatment with BAS 516 GA F under field conditions in Belgium, Germany, Great Britain and the Netherlands, 2000/2001. BASF AG, Agrarzentrum Limburgerhof; Limburgerhof; Germany Fed.Rep. BASF unpublished report NEU/FR/12/00, issued 31.05.2001. 2001/1006131

Raunft E, Funk H and Mackenroth C. 2001. Boscalid: Study on the residue behavior of BAS 500 F and BAS 510 F in apples after treatment with BAS 516 01 F under field conditions in Germany, 2000. BASF AG, Agrarzentrum Limburgerhof; Limburgerhof; Germany Fed.Rep. BASF unpublished report EU/FR/09/00, issued 28.02.2001. 2001/1006135

Reichert N. 2001. Boscalid: Independent laboratory validation of a method of analysis for the determination of BAS 510 F in white cabbage, rape (seed), hop and lettuce. Institut Fresenius, Chemische und Biologische Laboratorien GmbH; Taunusstein; Germany Fed. Rep. BASF unpublished report IF-100/35725-00, issued 15.02.2001. 2000/1014886

Reichert N. 2005. Boscalid: Study on the residue behaviour of BAS 500 F and BAS 510 F in strawberries after treatment with BAS 516 00 F under field conditions in Germany, England, Northern and Southern France, Greece, Netherlands and Italy, 2004. SGS Institut Fresenius GmbH; Taunusstein; Germany Fed. Rep. BASF unpublished report 168892, issued 15.02.2005. 2005/1004969

Renner G. 2003. Boscalid: Determination of the residues of BAS 510 F and Epoxiconazole in winter wheat processing products following double application of BAS 549 KA F in Germany. BioChem Agrar; Gerichshain; Germany Fed. Rep. BASF unpublished report 02 10 47 003, issued 26.06.2003. 2003/1000945

Scharf J. 1998. Boscalid: Hydrolysis of BAS 510 F at 90°C, 100°C, and 120°C. BASF AG, Agrarzentrum Limburgerhof; Limburgerhof; Germany Fed. Rep. BASF unpublished report 45505, issued 10.09.1998. 1998/10878

Scharm M. 2001. Boscalid: Determination of the residue of Reg.No. 300 355 in peas and processed products following reatment with BAS 510 01 F under field conditions in Germany 2000. Fresenius, Chem. Und biolog. Laboratorien; Taunusstein-Neuhof; Germany Fed. Rep. BASF unpublished report IF-100/13977-00, issued 08.01.2001. 2000/1014885

Scharm M. 2001. Boscalid: Determination of the residues of BA 510 F and BAS 500 F in plums and processed products following treatment with BAS 516 GA F under field conditions in Germany 2000. Institut Fresenius, Chemische und Biologische Laboratorien GmbH; Taunusstein; Germany Fed. Rep. BASF unpublished report IF-100/12669-00, issued 28.06.2001. 2001/1000937

Scharm M. 2001. Boscalid: Determination of the residues of BAS 510 F and BAS 500 F in carrots and processed products following treatment with BAS 516 GA F under field conditions in Germany 2000. Institut Fresenius, Chemische und Biologische Laboratorien GmbH; Taunusstein; Germany Fed. Rep. BASF unpublished report IF-100/13959-00, issued 31.05.2001. 2001/1000939

Scharm M. 2001. Boscalid: Determination of the residues of BAS 510 F and BAS 490 F in gherkins and processed products following treatment with BAS 517 00 F under field conditions in Germany 2000. Institut Fresenius, Chemische und Biologische Laboratorien GmbH; Taunusstein; Germany Fed. Rep. BASF unpublished report IF-100/13960-00, issued 08.06.2001. 2001/1000942

Schneider H. 2002. Boscalid: Determination of Residues of BAS 516 01 F in hops. Bayrische Landesanstalt für Bodenkultur und Pflanzenbau Freising; Germany Fed.Rep. BASF unpublished report Ho-BAS-2000, issued 14.02.2002. 2001/1015050

Schneider H. 2002. Boscalid: Determination of Residues of BAS 516 01 F in hops. Bayrische Landesanstalt für Bodenkultur und Pflanzenbau Freising; Germany Fed.Rep. BASF unpublished report Ho-BAS-2001, issued 14.02.2002. 2001/1015052

Schroth E. 2001. Boscalid: Determination of residues in/on melons (greenhouse) following multiple applications of BAS 517 00 F (Reg.No. 300 355, Reg.No. 242 009) under field conditions in Spain and Italy, 2000. Agrologia, SL; Palomares (Sevilla); Spain. BASF unpublished report 00/PF/003, issued 18.05.2001. 2001/1009069

Schulz H. 2001. Boscalid: Determination of the residues of BAS 510 F and BAS 500 F in lettuce following treatment with BAS 516 GA F under field conditions in southern France 1999. Institut Fresenius, Chemische und Biologische Laboratorien GmbH; Taunusstein; Germany Fed. Rep. BASF unpublished report IF-99/09279-00, issued 12.06.2001. 2001/1000933

Schulz H. 2001. Boscalid: Determination of residues of BAS 510 F and BAS 500 F in cherries and processed products following treatment with BAS 516 GA F under field conditions in Germany 2000. Institut Fresenius, Chemische und Biologische Laboratorien GmbH; Taunusstein; Germany Fed. Rep. BASF unpublished report IF-100/12501-00, issued 19.06.2001. 2001/1000938

Schulz H. 2001. Boscalid: Determination of the residues of BAS 500 F and BAS 510 F in apples following treatment with BAS 516 01 F under field conditions in Italy and France 2000. Institut Fresenius, Chemische und Biologische Laboratorien GmbH; Taunusstein; Germany Fed. Rep. BASF unpublished report IF-100/09103-00, issued 19.12.2001. 2001/1000946

Schulz H. 2002. Boscalid: Determination of the residues of BAS 500 F and BAS 510 F in apples following treatment with BAS 516 01 F under field conditions in Italy and Southern France 2001. Fresenius, Chem. Und biolog. Laboratorien; Taunusstein-Neuhof; Germany Fed.Rep. BASF unpublished report IF-101/10197-00, issued 19.02.2001. 2001/1015046

Schulz H. 2002. Boscalid: Determination of the residues of BAS 500 F and BAS 510 F in apples and processed products following treatment with BAS 516 01 F under field conditions in Germany 2001. Institut Fresenius, Chemische und Biologische Laboratorien GmbH; Taunusstein; Germany Fed. Rep. BASF unpublished report IF-101/14264-00, issued 06.03.2002. 2001/1015047

Schulz H. 2002. Boscalid: Determination of the residues of BAS 510 F and BAS 500 F in hops and processed products following treatment with BAS 516 01 F under field conditions in Germany 2001. Institut Fresenius, Chemische und Biologische Laboratorien GmbH; Taunusstein; Germany Fed. Rep. BASF unpublished report IF-101/23712-00, issued 05.03.2002. 2001/1015048

Schulz H. 2002. Boscalid: Determination of the residues of BAS 510 F and BAS 500 F in hops and processed products following treatment with BAS 516 01 F under field conditions in the Netherlands 2001. Institut Fresenius, Chemische und Biologische Laboratorien GmbH; Taunusstein; Germany Fed. Rep. BASF unpublished report IF-101/24112-00, issued 05.03.2002. 2001/1015049

Schulz H. 2003. Boscalid: Determination of the residues of Epoxiconazole and BAS 510 F in barley and processed products following treatment with BAS 549 KA F under field conditions in Germany 2002. Institut Fresenius, Chemische und Biologische Laboratorien GmbH; Taunusstein; Germany Fed. Rep. BASF unpublished report IF-02/00006864, issued 25.06.2003. 2003/1000946

Schulz H. 2004. Boscalid: Study on the residue behaviour of BAS 510 F and BAS 500 F in apples after application of either BAS 516 01 F or BAS 516 04 F under field conditions in France (North and South), Germany and Italy, 2003. Fresenius, Chem. Und biolog. Laboratorien; Taunusstein-Neuhof; Germany Fed.Rep. BASF unpublished report 165157, issued 06.02.2004. 2003/1001291

Schulz H. 2004. Boscalid: Study on the Residue Behaviour of Pyraclostrobin and Boscalid in Hops after treatment with BAS 516 01 F and BAS 516 04 F under field conditions in Germany, 2003. Insitut Fresenius Chemische Laboratorien AG; Taunusstein; Germany Fed.Rep. BASF unpublished report 165166, issued 04.02.2004. 2003/1001292

Schulz H. 2004. Boscalid: Study on the residue behaviour of Pyraclostrobin and Reg.No. 300 355 in peas and processed products after treatment with BAS 516 00 F under filed conditions in Germany and The Netherlands, 2003. Institut Fresenius, Chemische und Biologische Laboratorien GmbH; Taunusstein; Germany Fed. Rep. BASF unpublished report IF-03/00060684, issued 19.04.2004. 2004/1000750

Schulz H. 2004. Boscalid: Study on the residue behaviour of BAS 510 F and BAS 500 F in cauliflower and broccoli after application of BAS 516 00 F under field conditions in the Netherlands, Germany, United Kingdom, Denmark and France north and south, 2003. Institut Fresenius, Chemische und Biologische Laboratorien GmbH; Taunusstein; Germany Fed. Rep. BASF unpublished report 168853, issued 18.06.2004. 2004/1015910

Schulz H. 2004. Boscalid: Study on the residue behaviour of BAS 500 F and BAS 510 F in strawberries after application of BAS 516 00 F under field conditions in Germany, Italy, France (N & S), UK, Denmark and Sweden, 2003. Institut Fresenius, Chemische und Biologische Laboratorien GmbH; Taunusstein; Germany Fed. Rep. BASF unpublished report 168889, issued 24.09.2004. 2004/1015927

Schulz H. 2004. Boscalid: Study on the residue behaviour of BAS 500 F and BAS 510 F in strawberries after application of BAS 516 00 F under glasshouse conditions in Italy, France (S) and Spain, 2003. Institut Fresenius, Chemische und Biologische Laboratorien GmbH; Taunusstein; Germany Fed. Rep. BASF unpublished report 168895, issued 30.11.2004. 2004/1015928

Schulz H. 2004. Boscalid: tudy on the residue behaviour of BAS 510 F and BAS 500 F in leeks after application of BAS 516 00 F under field conditions in France North, 2003. Institut Fresenius, Chemische und Biologische Laboratorien GmbH; Taunusstein; Germany Fed. Rep. BASF unpublished report 157324, issued 28.09.2004. 2004/1015937

Schulz H and Scharm M. 2001. Boscalid: Determination of the residues of BAS 510 F and BAS 500 F in strawberries and processed products following treatment with BAS 516 GA F under field conditions in Germany 2000. Institut Fresenius, Chemische und Biologische Laboratorien GmbH; Taunusstein; Germany Fed. Rep. BASF unpublished report IF-100/14316-00, issued 13.06.2001. 2001/1000936

Schulz H and Scharm M. 2001. Boscalid: Determination of the residues of BAS 510 F and BAS 500 F in white cabbage and processed products following treatment with BAS 516 GA F under field conditions in Germany 2000. Institut Fresenius, Chemische und Biologische Laboratorien GmbH; Taunusstein; Germany Fed. Rep. BASF unpublished report IF-100/14317-00, issued 25.06.2001. 2001/1000943

Staudenmaier H. 2000. Boscalid: Anaerobic metabolism of BAS 510 F in soil (14C-pyridine-label). BASF AG, Agrarzentrum Limburgerhof; Limburgerhof; Germany Fed. Rep. BASF unpublished report 41858, issued 18.12.2000. 2000/1014990

Stephan A. 1999. Boscalid: Metabolism of BAS 510 F (14C-diphenyl and 14C-pyridin) in soil under aerobic conditions. BASF AG, Agrarzentrum Limburgerhof; Limburgerhof; Germany Fed. Rep. BASF unpublished report 42381, issued 21.12.1999. 1999/11807

Tilting N. 2001. Boscalid: Residues in milk and edible tissues following oral administration of BAS 510 F to lactating dairy cattle. BASF AG, Agrarzentrum Limburgerhof; Limburgerhof; Germany Fed. Rep. BASF unpublished report 42401, issued

09.03.2001. 2000/1017228

Treiber S et al. 2000. Boscalid: Study on the residue behavior of BAS 510 F in bush beans after treatment with BAS 510 01F under field conditions in Germany and Denmark, 1999. BASF AG, Agrarzentrum Limburgerhof; Limburgerhof; Germany Fed. Rep. BASF unpublished report EU/FA/03/99, issued 06.12.2000. 2000/1014846

Treiber S et al. 2001. Boscalid: Study on the residue behavior of BAS 510 F in bush- and climbing beans after treatment with BAS 510 01F under greenhouse conditions in Spain, 2000. BASF AG, Agrarzentrum Limburgerhof; Limburgerhof; Germany Fed. Rep. BASF unpublished report ES/FR/01/00, issued 12.05.2001. 2000/1014849

Treiber S et al. 2001. Boscalid: Study on the residue behavior of BAS 510 F in bush beans after treatment with BAS 510 01F under field conditions in Denmark, France and Germany, 2000. BASF AG, Agrarzentrum Limburgerhof; Limburgerhof; Germany Fed. Rep. BASF unpublished report EU/FA/03/00, issued 28.02.2001. 2000/1014850

Veit P. 2001. Boscalid: Metabolism of 14C-BAS 510 F in beans. BASF AG, Agrarzentrum Limburgerhof; Limburgerhof; Germany Fed. Rep. BASF unpublished report 42560, issued 09.02.2001. 2000/1014861
Versoi P and Abdel-Baky S. 2001. Boscalid: The Magnitude of BAS 510 F Residues in Canola. BASF Agro Research RTP; Research Triangle Park, NC 27709, USA. BASF unpublished report 58645, issued 05.02.2001. 2001/5000048

Versoi P and Abdel-Baky S. 2001. Boscalid: The Magnitude of BAS 510 F residues in canola seed processed fractions. BASF Agro Research RTP; Research Triangle Park, NC 27709, USA. BASF unpublished report 58646, issued 21.02.2001. 2001/5000049

Versoi P and Abdel-Baky S. 2001. Boscalid: The Magnitude of BAS 510 F Residues in Canola Seed Processed Fractions. BASF Corporation, Agricultural Products Center, NC 27709-3528, USA. BASF unpublished report 58646, issued 08.03.2001. 2001/5001064

Versoi P and Abdel-Baky S. 2001. Boscalid: The Magnitude of BAS 510 F and BAS 500 F Residues in Mint and Mint Processed Fractions. BASF Corporation, Agricultural Products Center, NC 27709, USA. BASF unpublished report 66700, issued 17.12.2001. 2001/5002467
Versoi P and Abdel-Baky S. 2002. Boscalid: The magnitude of BAS 510 F and BAS 500 F residues in sunflower and sunflower processed fractions. BASF Corporation, Agricultural Products Center, NC 27709, USA. BASF unpublished report 66710, issued 08.01.2002. 2001/5002552

Versoi P and Malinsky S. 2001. Boscalid: The Magnitude of BAS 500 F and BAS 510 F Residues in Soybean Processed Fractions. BASF Corporation, Agricultural Products Center, NC 27709-3528, USA. BASF unpublished report 66714, issued 12.12.2001. 2001/5002529

Versoi P and Malinsky S. 2002. Boscalid: The Magnitude of BAS 500 F and BAS 510 F Residues in Corn and Corn Grain Processed Fractions. BASF Corporation, Agricultural Products Center, NC 27709-3528, USA. BASF unpublished report 64554, issued 21.03.2002. 2001/5002624

Versoi P L and Abdel-baky S. 2000. Bascalid: Magnitude of BAS 510F residues in red raspberries and highbush blueberries. BASF Agro Research RTP; Research Triangle Park, NC 27709, USA. BASF unpublished report 63912, issued 04.08.2000. 2000/5195

Versoi P L and Abdel-Baky S. 2000. Boscalid: Magnitude of BAS 510F residues in dry bulb and green onions. BASF Agro Research RTP; Research Triangle Park, NC 27709; USA. BASF unpublished report 63894, issued 08.09.2000. 2000/5207

Versoi P L and Abdel-Baky S. 2000. Boscalid: The Magnitude of BAS 510F Residues in Carrots. BASF Agro Research RTP; Research Triangle Park, NC 27709; USA. BASF unpublished report 63896, issued 10.10.2000. 2000/5208

Versoi P L and Abdel-Baky S. 2000. Boscalid: The magnitude of BAS 510F residues in bell and chili peppers. BASF Agro Research RTP; Research Triangle Park, NC 27709; USA. BASF unpublished report 63918, issued 10.10.2000. 2000/5209

Weeren R D and Pelz S. 1999. Boscalid: Validation of DFG method S 19 for the determination of BAS 510 F in various plant materials. Dr. Specht & Partner, Chemische Laboratorien GmbH; Hamburg; Germany Fed. Rep. BASF unpublished report BAS09902V Az. M8020A/99, issued 19.10.1999. 1999/11461

Weeren R D and Pelz S. 1999. Boscalid: Validation of DFG method S 19 for the determination of BAS 505 F and BAS 510 F in animal matrices (milk, meat, egg, fat). Specht & Partner, Chemische Laboratorien GmbH, Hamburg, Germany Fed. Rep. BASF unpublished report BAS-9903V Az. T1170/99, issued 15.12.1999. 2000/1000221

Wofford J T. 2004. Boscalid: Magnitude of BAS 510 F residues in bananas for import tolerance. BASF Corporation Agro Research; Princeton NJ 08543-0400; USA. BASF unpublished report 179968, issued 21.07.2004. 2004/5000480

Wofford J T and Abdel-Baky S. 2001. Boscalid: The Magnitude of BAS 510 F Residues in Plum Processed Fractions. BASF Corporation, Agricultural Products Center, NC 27709-3528, USA. BASF unpublished report 63906, issued 09.01.2001. 2000/5275

Wofford J T and Abdel-Baky S. 2001. Boscalid: The magnitude of Bas 510 F residues in peanut raw agricultural commodities and peanut processed fractions. BASF Corporation, Agricultural Products Center, NC 27709; USA. BASF unpublished report 64106, issued 08.03.2001.2001/5000870

Wofford J T and Abdel-Baky S. 2001. Boscalid: The Magnitude of BAS 510 F Residues in Potato. BASF Corporation, Agricultural Products Center, NC 27709-3528, USA. BASF unpublished report 64126, issued 21.03.2001. 2001/5000879

Wofford J T and Abdel-Baky S. 2002. Boscalid: The magnitude of BAS 510 F residues in broccoli. BASF Corporation Agro Research; Princeton NJ 08543-0400; USA. BASF unpublished report 66702, issued 20.02.2002. 2001/5002616

Wofford J T and Abdel-Baky S. 2002. Boscalid: The magnitude of BAS 510 F residues in cabbage. BASF Corporation Agro Research; BASF Corporation Agro Research; Princeton NJ 08543-0400; USA. BASF unpublished report 66704, issued 20.02.2002. 2001/5002617

Wofford J T and Abdel-Baky S. 2002. Boscalid: Magnitude of BAS 510 F and BAS 500 F residues in dry and succulent peas. BASF Corporation, Agricultural Products Center, NC 27709, USA. BASF unpublished report 66696, issued 11.01.2002. 2001/5003246

Wofford J T and Abdel-Baky S. 2002. Boscalid: Magnitude of BAS 510 F and BAS 500 F residues in mustard greens – 2001 field study. BASF Corporation Agro Research; BASF Corporation Agro Research; Princeton NJ 08543-0400; USA. BASF unpublished report 67080, issued 26.02.2002. 2001/5003339

Young H and Atikinson S. 2003. Boscalid: Study on the residue behaviour of BA 510 F, BAS 500 F and BF 500-3 in lettuce (greenhouse) after application of BAS 516 00 F under field conditions in France North and South, Spain, the Netherlands and Germany, 2002. BASF plc, Agro Research; Gosport; United Kingdom BASF unpublished report 135163, issued 07.04.2003. 2003/1001259

# CHLORPYRIFOS (017)

*First draft was prepared by Arpad Ambrus, Hungary*

## EXPLANATION

Chlorpyrifos was last evaluated by the JMPR in 2004 when an ADI of 0-0.01 mg/kg bw/day and an ARfD of 0.1 mg/kg bw/day were established, and a number of maximum residue levels were estimated.

The cranberry industry performed a number of supervised trials within the Interregional Research Project No. 4 to provide data for the establishment of US tolerances for chlorpyrifos residues in cranberry. The relevant labels and reports of supervised trials were submitted for evaluation by the 2006 JMPR.

## RESIDUE ANALYSIS

### Analytical method

The harvested cranberry fruit samples were analyzed by a Dow Chemical method determining the total residues converted to 3,5,6-trichloro-2-pyridinol TCP) (McKellar)

Briefly, residues of chlorpyrifos in/on cranberry samples were extracted with methanol and 10% NaOH at 130°C in a pressure reaction bottle. The sample was cooled and residues were acidified with HCl, partitioned using diethyl ether, and eluted through an acidic alumina column using diethyl ether buffered to pH 6.5. Residues were then partitioned with 0.25 M sodium bicarbonate, acidified with HCl, partitioned into benzene, evaporated to dryness, and re-constituted in benzene prior to analysis by GC with a Ni 63 detector. Residues were detected and reported as TCP. Residues in chlorpyrifos equivalents were calculated by multiplying the TCP residue by 1.786. The method sensitivity (LOQ) was reported as 0.02 or 0.03 mg/kg TCP.

Concurrent recoveries of TCP in cranberries fortified at 0.20 or 0.25 mg/kg were 92–130% (110 ± 13.4, n=6). Control samples from each test site were analyzed. The apparent TCP residues were between < 0.03 and 0.06 mg/kg.

### Stability of pesticide residues in stored analytical samples

The maximum storage interval from harvest to extraction for analysis of the cranberry samples was 134 days. Storage conditions were specified as frozen only for a portion (approximately 4 months) of the total storage intervals used. Conditions prior to frozen storage were not reported. The 2004 JMPR reported that no loss of chlorpyrifos in frozen crop matrices was observed generally over one year. However about 20% of the residues decomposed in orange and orange juice samples within 170 days (FAO, 2005).

## USE PATTERN

In the USA, Lorsban 75WG, containing 75% active substance, can be applied to protect cranberry as shown in Table 1.

Table 1. Registered uses of Chlorpyrifos.

| Application | | | | | PHI |
|---|---|---|---|---|---|
| Method | No. | Interval, days | Water L/ha | Rate kg ai/ha | day |
| Broadcast foliar | Max 2 | Min. 10 | Min 140 | 2.24 | 60 |
| | | | | | |

## RESIDUES RESULTING FROM SUPERVISED TRIALS

A total of nine crop field trials on cranberries were conducted in 1981 in three geographical regions of the USA (Daussin 1982). At each test location, one untreated and at least two treated plots were established. Chlorpyrifos (EC 48%) was applied once at 1.68 or 3.36 kg ai/ha or twice at 0.84, 1.68, or 3.36 kg ai/ha for total seasonal rates of 1.68, 3.36, or 6.72 kg ai/ha. Retreatment intervals were 27–42 days. No adjuvants were used.

Cranberry fruit samples were harvested at 66–89 days following two applications of the test substance, or at 105 days following a single application of chlorpyrifos. Details regarding sampling methods were not reported. In each trial, samples submitted for analysis weighed 20 g only.

Cranberries were analyzed for combined residues of chlorpyrifos as TCP. The residues detected are summarized in Table 2.

Table 2. Residues in/on cranberry fruits derived from field trials with chlorpyrifos.

| Location (City, State) | Single Rate, kg ai/ha | Total Rate, (kg ai/ha) | DAT [a] | Residues mg/kg) TCP (Chlorpyrifos Equivalents) [b] | Average residues as chlorpyrifos equivalent [c] |
|---|---|---|---|---|---|
| East Wareham, MA (Plymouth County) | 1.68 | 1.68 | 105 | 0.02, 0.02, < 0.02 (0.04, 0.04, < 0.04) | 0.04 |
| | 3.36 | 3.36 | 105 | 0.02 (0.04) | 0.04 |
| | 1.68 | 3.36 | 66 | < 0.02, 0.11, 0.20, 0.39 (< 0.04, 0.20, 0.36, 0.70) | 0.42 |
| | 3.36 | 6.72 | 66 | 0.40, 0.50, 0.65, 0.80 (0.71, 0.89, 1.16, 1.43) | 1.05 |
| Long Beach, WA (Pacific County) | 0.84 | 1.68 | 89 | 0.14, 0.21, 0.24 (0.25, 0.38, 0.43) | 0.35 |
| | 1.68 | 3.36 | 89 | 0.19, 0.22, 0.42 (0.34, 0.39, 0.75) | 0.49 |
| | 3.36 | 6.72 | 89 | 0.54, 0.57, 0.60 (0.96, 1.02, 1.07) | 1.02 |
| Madison, WI (Dane County) | 1.68 | 3.36 | 77 | 0.31 (0.55) | 0.55 |
| | 3.36 | 6.72 | 77 | 0.25 (0.45) | 0.45 |

a   DAT = Days after treatment

b   TCP = 3,5,6-trichloro-2-pyridinol.  Chlorpyrifos equivalents = TCP value/0.56

c   As 20 g fruits were taken as one sample, the best estimate for the residue in composite samples is their average.

Residues in all control plots from the East Wareham, MA site were < 0.02 mg/kg. Apparent TCP residues were detected at 0.06, 0.03, and 0.03 mg/kg in the three controls from the Long Beach, WA site; and residues in the two controls from the Madison, WI site were at < 0.03 and 0.03 mg/kg.

## APPRAISAL

Chlorpyrifos was last evaluated by the JMPR in 2004 when an ADI of 0-0.01 mg/kg bw/day and an ARfD of 0.1 mg/kg bw/day were established, and a number of maximum residue levels were estimated. The 2004 JMPR defined the residue as chlorpyrifos for both compliance with MRLs and estimation of dietary intake.

Results of supervised trials, carried out on cranberry according the US registered uses were submitted for evaluation.

## *Results of supervised trials on crops*

Residues were detected and reported as 3,5,6-trichloro-2-pyridinol (TCP). Residues in chlorpyrifos equivalents were calculated by multiplying the TCP residue values by 1.786. The limit of quantification was reported as 0.02 or 0.03 mg/kg TCP.

Concurrent recoveries of TCP fortified in cranberries at 0.20 or 0.25 mg/kg spike levels were 92–130% (110 ± 13.4, n=6). Control samples from each test site were analyzed. The apparent TCP residues were between < 0.03 and 0.06 mg/kg.

The concurrent and previous stability studies reported by the 2004 JMPR suggest that the decrease of residues during storage was not significant.

Nine field trials on cranberries were carried out in three geographical region of USA. Chlorpyrifos (EC 48%) was applied once at 1.68 or 3.36 kg ai/ha or twice at 0.84, 1.68, or 3.36 kg ai/ha for total seasonal rates of 1.68, 3.36, or 6.72 kg ai/ha. None of the dosage rates corresponds to the maximum registered single dose of 2.24 kg ai/ha. Samples were collected at the registered PHI at one site. In other cases the PHI was much longer. Individual samples collected for analyses contained 20 g of fruit. In order to best represent the expected residues in composite samples, the averages of residues were calculated for each trial.

The residues expressed as chlorpyrifos derived from trials performed with ± 30% maximum rate are in rank order: 0.42, 0.49, and 0.55 mg/kg.

Taking into account the minimum data requirement (three trials) specified for commodities which are insignificant in trade and do not raise any intake concern (2004 JMPR Report, pp. 30-31), the Meeting estimated a maximum residue level of 1 mg/kg, HR of 0.55 mg/kg and an STMR of 0.49 mg/kg.

## RECOMMENDATION

On the basis of the data from supervised trials, the Meeting concluded that the residue levels listed below are suitable for establishing maximum residue limits and for dietary intake assessment.

Summary of recommendations for MRLs, STMRs and HRs for chlorpyrifos

| CCN | Commodity | MRL, mg/kg | | STMR or STMR-P, mg/kg | HR or HR/P mg/kg |
|-----|-----------|------------|----------|----------|----------|
| | | New | Previous | | |
| FB 0265 | Cranberry | 1 | | 0.49 | 0.55 |

## DIETARY RISK ASSESSMENT

### *Long-term intake*

The GEMS/Food Consumption Cluster Diets specifies the following long-term cranberry consumption (g/day/person) for various diets: A (0.1); D (0.3); F (0.6); M (2.5). The cranberry consumption in the other regions is nil.

The highest IEDI in the 13 GEMS/Food regional diets based on the estimated STMR was 0.2% of the maximum ADI (0.01 mg/kg bw).

The Meeting concluded that the long-term intake of residues of chlorpyrifos use on cranberry will not practically increase the intake of residues from other uses considered earlier by the JMPR.

*Short-term intake*

The GEMS/Food regional diet specifies the large portion sizes of cranberry of 3.53 g/kg bw for adults and 6.78 g/kg bw for children (both are from the USA).

The IESTIs of chlorpyrifos calculated on the basis of the large portion size and the estimated HR of 0.55 mg/kg are 1.9% and 3.7% of the ARfD for adults and children, respectively.

The Meeting concluded that the short-term intake of residues resulting from the use of chlorpyrifos on cranberry that have been considered by the JMPR is unlikely to present a public health concern.

## REFERENCES

**Author, Date, Title, Institute, Report Reference, Document No.**

Daussin S.1982. Petition Proposing a Tolerance for Chlorpyrifos for Use in Cranberries Production" Interregional Research Project No. 4, PR No. 0054Interregional Research Project No. 4, Rutgers University, The State University of New Jersey PR No. 0054

FAO.2005. Pesticide residues in food, Evaluations 2004, Plant Production and Protection Paper Plant Production and Protection Paper No. 182/1

McKellar R.L. Determination of Residues of 3,5,6-Trichloro-2-Pyridinol in Lima and Snapbean Forage and Beans by Gas Chromatography, Dow Chemical, ACR 71.19R.

# DIAZINON (022)

*First draft was prepared by Arpad Ambrus, Hungary*

## EXPLANATION

Diazinon was evaluated by the JMPR in 1993, 1999 and 2001 when an ADI of 0-0.002 mg/kg bw and ARfD of 0.03 mg/kg bw were established, and a number of maximum residue levels were estimated.

The cranberry industry performed a number of supervised trials within the Interregional Research Project No. 4 to provide data for the establishment of US tolerances for diazinon residues in cranberry. The relevant labels and reports of supervised trials were submitted for evaluation by the 2006 JMPR.

## RESIDUE ANALYSIS

### *Analytical methods*

The limit of quantification of the method (LOQ) was 0.01 mg/kg for each diazinon, diazoxon and 4-hydroxy-2-isopropyl-6-methylpyrimidine CGA-14128. No quantifiable residues were observed in the control samples. The recoveries obtained during the analysis of samples are summarised in Table 1.

Table 1. Summary of concurrent method recoveries of diazinon and its metabolites from cranberries

| Analyte | Spike level (mg/kg) | Sample size (n) | Type of Recovery | Recoveries (%)[a] | Mean ± std. dev. |
|---------|---------------------|-----------------|------------------|-------------------|------------------|
| Diazoxon | 0.016–0.40 | 7 | Concurrent | 70–00 | 82 ± 11 |
| Diazinon | 0.016–0.40 | 8 | Concurrent | 75–112 | 94 ± 14 |
| CGA-14128 | 0.016–0.40 | 7 | Concurrent | 106–125 | 114 ± 8 |

a Range of recoveries.

### *Stability of pesticide residues in stored analytical samples*

Samples were stored frozen from harvest to analysis for a maximum of 11 months. Freezer storage stability data included in the report indicated that residues of diazinon and CGA-14128 were stable for up to three months of frozen storage < -10 °C, and that residues of diazoxon were not (0% recovery at three months). The storage stability data previously submitted to JMPR (FAO 1994) indicated that diazinon and CGA-14128 residues were stable in/on frozen raw agricultural commodities for up to 26 months. Diazoxon was not stable (< 3 months).

## USE PATTERN

The diazinon is registered in various formulations in the USA: AG 500 (4 lb/gal EC, EPA Reg. No. 100-461), 50 W (50% WP, EPA Reg. 100-460), and 14 G (14% G, EPA Reg. No. 100-469) and diazinon 4E, containing 48% active substance (EPA Reg. No. 10163-100).

They can be applied to protect cranberry as shown in Table 2:

Table 2. Use pattern of diazinon on cranberry

| Application | | | | | | PHI |
|---|---|---|---|---|---|---|
| Method | No | Interval, days | Rate kg ai/hL | Water L/ha | Rate kg ai/ha | day |
| Ground | Max 6 | 14 | 1.6 | 140 | 2.24 | 7 |

| Application | | | | | | PHI |
|---|---|---|---|---|---|---|
| Method | No | Interval, days | Rate kg ai/hL | Water L/ha | Rate kg ai/ha | day |
| Chemigation | Max 6 | 14 | 1.6 | 140 | 2.24 | 7 |
|  | Max 3 | 14 | 0.06 | 1514 | 3.36 | 7 |

Do not apply more than 10.1 kg ai/ha per season.

## RESIDUES RESULTING FROM SUPERVISED TRIALS

During the 1988 and 1989 growing seasons, ten trials were conducted on cranberries in three regions of the USA (IR-4, 2006).

The cranberry crops (Early Black and Searles varieties) were grown and maintained according to typical agricultural practices for each geographical region.

At each test location, multiple broadcast foliar applications of diazinon were made using the 45% EC, 50% WP, and/or 14% G formulations at the rate of 3.36 kg ai/ha/application. Total seasonal application rates for all trials were 6.72, 13.45, and 20.17 kg ai/ha. Ground equipment was used at all trial locations except Carver, MA where applications were made aerially. When applied in water, spray volumes ranged from 46.8-2806 l/ha and no adjuvants were used. The granular formulation was applied dry followed by irrigation to wet the soil. Retreatment intervals were between 6 and 23 days.

Cranberry fruit samples were harvested at 7 or 31 days following the last application.

Four replicate samples were collected by hand from randomly chosen areas within the control and treated plots and placed in plastic bags.

The samples did not contain any detectable residues of diazoxon and CGA14128. The residue results from the supervised field trials are presented in Table 3.

Table 3. Residues of diazinon in/on cranberry fruits

| Location (City, State) | Trial Start Year | Variety | EP [b?] | Single Rate, kg ai/ha (in litre) | No. of appl. | DAT [a] | Residues (mg/kg) | |
|---|---|---|---|---|---|---|---|---|
|  |  |  |  |  |  |  | Diazinon | Average |
| East Wareham, MA (Plymouth County) | 1988 | Early Black | Diazinon (48% EC) | 3.36 (2806) | 4 | 7 | 0.17, 0.15, 0.07, 0.11 | 0.13 |
| Chatsworth, NJ (Burlington County) | 1988 | Early Black | Diazinon (14% G) | 3.36 | 2 | 7 | 0.04, 0.06, 0.02, 0.03 | 0.04 |
| Madison, WI (Dane County) | 1988 | Searles | Diazinon 48% EC | 3.36 (1871) | 4 | 7 | 0.08, 0.05, 0.07, 0.06 | 0.07 |
|  |  |  | Diazinon (14% G) | 3.36 (1871) | 2 | 31 | < 0.01, 0.02, < 0.01, < 0.01 | 0.01 |
|  |  |  | Diazinon (50% WP) | 3.36 (1871) | 4 | 7 | 0.06, 0.06, 0.04, 0.05 | 0.05 |
|  |  |  | Diazinon (50% WP + 14% G) | 3.36 or 6.72 (1871) | 4 | 7 | 0.15, 0.10, 0.17, 0.16 | 0.13 |
|  |  |  | Diazinon (48% EC + 14% G) | 3.36 or 6.72 (1871) | 4 | 7 | 0.05, 0.04, 0.07, 0.08 | 0.06 |
| Carver, MA (Plymouth County) | 1989 | Early Black | Diazinon [c] (48% EC) | 3.36 (46.8[c]) | 4 | 7 | < 0.01, 0.02, < 0.01, < 0.01 | 0.01 |
|  |  |  | Diazinon [c] (50% WP) | 3.36 (46.8[c]) | 4 | 7 | < 0.01, < 0.01, < 0.01, < 0.01 | < 0.01 |
| Bandon, Oregon (Coos County) | 1989 | McFarlin | Diazinon 14 G | 3.36 | 4 | 7 | 0.06, 0.10, 0.10, 0.06 | 0.08 |

a   DAT = Days after treatment.
b   Maximum combined residues reported in original study report.
c   These were aerial applications.

## APPRAISAL

Diazinon has been evaluated by the JMPR several times and a number of maximum residue levels were estimated. The residue was defined as diazinon for regulatory and dietary intake assessment purposes. The present Meeting established an ADI of 0-0.005 mg/kg bw but the ARfD of 0.03 mg/kg bw remained unchanged.

Results of supervised trials, carried out on cranberry according the US registered uses, were submitted for evaluation.

### Results of supervised residue trials

During the 1988 and 1989 growing seasons ten trials were conducted on cranberries in three regions of USA. Samples were collected at 7 day PHI following the total seasonal application rates of 6.72–13.45 (GAP or 1.3 × GAP), and 20.17 (double rate) kg ai/ha.

Samples were stored frozen from harvest to analysis for a maximum of 11 months. Storage stability data submitted to the 1993 JMPR indicated that diazinon and CGA-14128 (4-hydroxy-2-isopropyl-6-methylpyrimidine) residues were stable in/on frozen raw agricultural commodities (< -10° C) for up to 26 months. Diazoxon was not stable (< 3 months).

The harvested cranberry samples were analyzed by a method with an LOQ of 0.01 mg/kg for each residue component measured. The concurrent recoveries of diazoxon, diazinon and CGA-14128 from cranberries fortified at 0.016-0.40 mg/kg ranged between 70-100% (82 ± 11, n = 7), 75-112% (94 ± 14, n = 8), and 106-125% (114 ± 8, n = 7), respectively.

The average residues (mg/kg) from treatments carried out according to 1.5 times the individual treatment rate and at 1.3 times the total seasonal application rates specified in GAP were in rank order: < 0.01, 0.01, 0.04, 0.05, 0.07, 0.08 and 0.13mg/kg

The Meeting estimated a maximum residue level of 0.2 mg/kg, HR of 0.13 mg/kg and STMR of 0.05 mg/kg, respectively.

## RECOMMENDATION

On the basis of the data from supervised trials, the Meeting concluded that the residue levels listed below are suitable for establishing maximum residue limits and for dietary intake assessment.

Summary of recommendations for MRLs, STMRs and HRs for diazinon

| CCN | Commodity | MRL, mg/kg | | STMR or STMR-P, mg/kg | HR or HR/P mg/kg |
|-----|-----------|------------|-----------|--------|--------|
|     |           | New | Previous |        |        |
| FB 0265 | Cranberry | 0.2 |       | 0.05   | 0.13   |

## DIETARY RISK ASSESSMENT

### Long-term intake

As the ADI of the compound was revised, the long-term intake was recalculated using the current Codex MRLs or the estimated STMR values. The summaries of calculations are included in Annex 3 of the 2006 JMPR report. The intake of diazinon residues calculated based on the 13 regional diets ranged from 7 to 50% of the maximum ADI (0.005 mg/kg bw).

The Meeting concluded that the long-term intake of residues from the use of diazinon on the commodities considered by the CCPR or JMPR is unlikely to present a public health concern.

*Short-term intake*

Since the ARfD was not changed the IESTI was only calculated for cranberries:

The GEMS/Food regional diet specifies the large portion sizes of cranberry of 3.53 g/kg bw for adults and 6.78 g/kg bw for children (both are from the USA).

The IESTIs of diazinon calculated on the basis of the large portion size and the estimated HR of 0.13 mg/kg are 1.53% and 2.9% of the ARfD for adults and children, respectively.

The Meeting concluded that the short-term intake of residues resulting from the use of diazinon on cranberry that have been considered by the JMPR is unlikely to present a public health concern.

## REFERENCES

**Author, Date, Title, Institute, Report Reference, Document No.**

FAO. 1994. Plant production and Protection Paper No. 122 Pesticide Residues in Food - 1993 Evaluations, p. 39

FAO. 2000. Plant production and Protection Paper No. 153 Pesticide Residues in Food - 1999 Evaluations, p. 61

Interregional Research Project No. 4. 2006. Diazinon: Magnitude of Residue in or on Cranberry, The State University of New Jersey New Brunswick, New Jersey 08903 PR 3732

**DIMETHOATE (027)**

*First draft prepared by Dr Ursula Banasiak, Federal Institute for Risk Assessment, Berlin, Germany*

**EXPLANATION**

Dimethoate was evaluated by the JMPR several times from 1965 to 1994 and under the CCPR Periodic Review Programme in 1998. The compound was re-evaluated in 2003 for residues and toxicology. The 2003 Meeting recommended a number of MRLs and established an acute reference dose (ARfD) of 0.02 mg/kg bw.

The 38th session of the CCPR in 2006 decided to advance the MRL of dimethoate in barley of 2 mg/kg to step 8 (FAO/WHO, 2006), which was adapted by 29th Session of CAC as Codex Standard. Apart from that, the Committee was informed that new residue information for barley would be submitted to the JMPR (SCC 2006a and b).

**USE PATTERN**

Information on modified registered use for cereals was reported to the Meeting and is shown in Table 1.

Table 1. Registered uses for dimethoate.

| Crop G/F | Country | Form. ai g/L | Application | | | | | PHI days |
|---|---|---|---|---|---|---|---|---|
| | | | Method | Rate kg ai/ha | Spray conc. kg ai/hL | No. | | |
| Cereals F | Finland | EC, 400 | spraying, from leaf development (sprouting) to the end of tillering, before stem elongation (up to BBCH 30) | 0.16-0.32 | 0.08 – 0.16 | 1 | | application up to BBCH 30 |

F = outdoor or field

**RESIDUES RESULTING FROM SUPERVISED TRIALS ON CROPS**

The Meeting received new information on supervised field trials on barley. Residue data are reported for barley grains, straw and plant parts in Tables 2 - 4.

Where residues were not detected, data are recorded in the Tables as below the LOQ. Residue data, application rates and spray concentrations have generally been rounded to two significant figures or, for residues near the LOQ, to one significant figure. Although trials included control plots, no control data are recorded except where residues in control samples exceeded the LOQ. Residues are recorded unadjusted for procedural recoveries. Double-underlined residues are from treatments according to GAP.

Table 2. Dimethoate and omethoate residues in winter barley grains from supervised trials in Europe. Foliar spray treatment. Growth stage at last treatment BBCH 59 (Pollmann, 2006a and b).

| Location, year, variety, report No | Application | | | | | Commodity analysed | PHI (days) | Residues (mg/kg) | |
|---|---|---|---|---|---|---|---|---|---|
| | Form | kg ai/ha | Water L/ha | kg ai/hL | No. | | | Dimethoate | Omethoate |
| Bucknell, Oxfordshire, UK, 2005, Pearl, GB05W003R | EC | 0.35 | 307 | 0.11 | 1 | grain | 58 | <u>≤0.001</u> | <u>≤0.001</u> |
| Stratton Audley, Oxfordshire, UK, 2004, Regina, GB04W003R | EC | 0.32 | 283 | 0.11 | 1 | grain | 49 | <u>0.002</u> | <u>≤0.001</u> |

| Location, year, variety, report No | Application | | | | | Commodity analysed | PHI (days) | Residues (mg/kg) | |
|---|---|---|---|---|---|---|---|---|---|
| | Form | kg ai/ha | Water L/ha | kg ai/hL | No. | | | Dimethoate | Omethoate |
| Marest Damcourt, Picardie, France, 2005, Esterelle, F05W003R | EC | 0.34 | 296 | 0.11 | 1 | grain | 56 | <u><0.001</u> | <u><0.001</u> |
| Gertwiller, Alsace, France, 2004, Platine, F04W051R | EC | 0.38 | 218 | 0.17 | 1 | grain | 59 | <u><0.001</u> | <u><0.001</u> |
| Padborg, Sonderjylland Denmark, 2005, Vanessa, D05W003R | EC | 0.34 | 293 | 0.12 | 1 | grain | 53 | <u><0.001</u> | <u><0.001</u> |
| Padborg, Frosley, Bov, Denmark, 2004, Carola, D04W003R | EC | 0.35 | 305 | 0.11 | 1 | grain | 68 | <u><0.001</u> | <u><0.001</u> |
| Helmste, Niedersachsen, Germany, 2004, Passion, G04W067R | EC | 0.36 | 312 | 0.11 | 1 | grain | 63 | <u><0.001</u> | <u><0.001</u> |

Table 3. Dimethoate and omethoate residues in winter barley straw from supervised trials in Europe. Foliar spray treatment. Growth stage at last treatment BBCH 59 (Pollmann, 2006a and b).

| Location, year, variety, report No | Application | | | | | Commodity analysed | PHI (days) | Residues (mg/kg) | |
|---|---|---|---|---|---|---|---|---|---|
| | Form | kg ai/ha | Water L/ha | kg ai/hL | No. | | | Dimethoate | Omethoate |
| Bucknell, Oxfordshire, UK, 2005, Pearl, GB05W003R | EC | 0.35 | 307 | 0.11 | 1 | straw | 58 | <u><0.01</u> | <u><0.01</u> |
| Stratton Audley, Oxfordshire, UK, 2004, Regina, GB04W003R | EC | 0.32 | 283 | 0.11 | 1 | straw | 49 | <u>0.02</u> | <u><0.01</u> |
| Marest Damcourt, Picardie, France, 2005, Esterelle, F05W003R | EC | 0.34 | 296 | 0.11 | 1 | straw | 56 | <u><0.01</u> | <u><0.01</u> |
| Gertwiller, Alsace, France, 2004, Platine, F04W051R | EC | 0.38 | 218 | 0.17 | 1 | straw | 59 | <u><0.01</u> | <u><0.01</u> |
| Padborg, Sonderjylland Denmark, 2005, Vanessa, D05W003R | EC | 0.34 | 293 | 0.12 | 1 | straw | 53 | <u><0.01</u> | <u><0.01</u> |
| Padborg, Frosley, Bov, Denmark, 2004, Carola, D04W003R | EC | 0.35 | 305 | 0.11 | 1 | straw | 68 | <u><0.01</u> | <u><0.01</u> |
| Helmste, Niedersachsen, Germany, 2004, Passion, G04W067R | EC | 0.36 | 312 | 0.11 | 1 | straw | 63 | <u><0.01</u> | <u><0.01</u> |

Table 4. Dimethoate and omethoate residues in winter barley plants from supervised trials in Europe. Foliar spray treatment. Growth stage at last treatment BBCH 59 (Pollmann, 2006a and b).

| Location, year, variety, report No | Application | | | | | Commodity analysed | PHI (days) | Residues (mg/kg) | |
|---|---|---|---|---|---|---|---|---|---|
| | Form | kg ai/ha | Water l/ha | kg ai/hl | No. | | | Dimethoate | Omethoate |
| Stratton Audley, Oxfordshire, UK, 2004, Regina, GB04W003R | EC | 0.32 | 283 | 0.11 | 1 | plant | 0 | 7.0 | 0.02 |
| | | | | | | plant | 13 | 0.85 | 0.2 |
| | | | | | | plant | 21 | 0.45 | 0.13 |
| | | | | | | ears | 28 | <0.01 | <0.01 |
| | | | | | | rest of plant | 28 | 0.06 | 0.02 |
| | | | | | | ears | 35 | <0.01 | <0.01 |
| | | | | | | rest of plant | 35 | 0.03 | 0.01 |
| Gertwiller, Alsace, France, 2004, Platine, F04W051R | EC | 0.38 | 218 | 0.17 | 1 | plant | 0 | 7.2 | 0.02 |
| | | | | | | plant | 14 | 0.05 | 0.03 |
| | | | | | | plant | 20 | <0.01 | <0.01 |
| | | | | | | ears | 28 | <0.01 | <0.01 |
| | | | | | | rest of plant | 28 | <0.01 | <0.01 |
| | | | | | | ears | 35 | <0.01 | <0.01 |
| | | | | | | rest of plant | 35 | <0.01 | <0.01 |
| Padborg, Frosley, Bov, Denmark, 2004, Carola, D04W003R | EC | 0.35 | 305 | 0.11 | 1 | plant | 0 | 6.8 | 0.06 |
| | | | | | | plant | 13 | 0.49 | 0.25 |
| | | | | | | plant | 21 | 0.29 | 0.19 |
| | | | | | | ears | 27 | <0.01 | 0.02 |
| | | | | | | rest of plant | 27 | 0.31 | 0.16 |
| | | | | | | ears | 34 | <0.01 | <0.01 |
| | | | | | | rest of plant | 34 | 0.10 | 0.04 |
| Helmste, Niedersachsen, Germany, 2004, Passion, G04W067R | EC | 0.36 | 312 | 0.11 | 1 | plant | 0 | 7.3 | 0.02 |
| | | | | | | plant | 13 | 0.49 | 0.21 |
| | | | | | | plant | 20 | 0.26 | 0.12 |
| | | | | | | ears | 28 | <0.01 | <0.01 |
| | | | | | | rest of plant | 28 | 0.02 | 0.02 |
| | | | | | | ears | 34 | <0.01 | <0.01 |
| | | | | | | rest of plant | 34 | <0.01 | <0.01 |

## APPRAISAL

Dimethoate was evaluated by the JMPR on several occasions between 1965 and 1994 and under the CCPR Periodic Review Programme in 1998. The compound was re-evaluated in 2003 for residues and toxicology. The 2003 Meeting recommended a number of MRLs and established an acute reference dose (ARfD) of 0.02 mg/kg bw.

The 38th session of the CCPR in 2006 decided to advance the MRL of dimethoate in barley of 2 mg/kg to step 8, which was adopted by the 29th session CAC as a Codex standard. Apart from that, the Committee was informed that new residue information for barley would be submitted to the JMPR.

### Results of supervised residue trials

### Barley

Field trials conducted in northern Europe and involving one foliar application of dimethoate were made available to the Meeting.

GAP in Finland on cereals is for one application per season with a maximum rate of 0.32 kg ai/ha to be applied from leaf development to the end of tillering, before stem elongation. In seven trials on barley from northern Europe (two from the UK, two from France, two from Denmark and one from Germany) matching Finnish GAP residue levels of dimethoate were < 0.001 mg/kg in six trials and 0.002 mg/kg in one trial in grains. The omethoate residues in grains were < 0.001 (7) mg/kg.

The Meeting agreed that the MRL for barley recommended at the 1998 Meeting and advanced to step 8 by the 2006 CCPR was sufficient to accommodate the new residue data and the Finnish GAP, and that no further evaluation was required.

*Barley straw and fodder, dry*

Residues in barley straw and fodder, dry were evaluated by JMPR 1998. An MRL was not recommended but STMRs were estimated for dimethoate and omethoate separately. New field trials conducted in northern Europe and involving one foliar application of dimethoate were made available to the present Meeting.

GAP in Finland on cereals is for one application per season with a maximum rate of 0.32 kg ai/ha to be applied from leaf development to the end of tillering, before stem elongation. In 7 trials on barley from northern Europe (two from the UK, two from France, two from Denmark and one from Germany) matching Finish GAP, residue levels found were < 0.01 (6) and 0.02 mg/kg for dimethoate and < 0.01 (7) mg/kg for omethoate in straw.

The Meeting agreed that the STMRs of 0.495 mg/kg for dimethoate and 0.03 mg/kg for omethoate estimated by the 1998 JMPR for barley straw were sufficient to accommodate the new residue data and the Finnish GAP, and that no further evaluation was required.

## Further work

The Meeting noticed that the estimation of STMRs and HRs was not correct for some of the commodities calculated by the 2003 JMPR. The Meeting agreed to do a corrigendum based on the supervised residue trials data submitted to the 2003 JMPR. This should include consideration of new information on GAP and residue data.

## REFERENCES

**Author, Date, Title, Institute, Report Reference, Document No.**

FAO/WHO 2006. ALINORM 06/29/24. Joint FAO/WHO Food Standard Programme, Codex Alimentarius Commission, 29th Session, Geneva Switzerland, 3 – 7 July 2006. Report of the 38th Session of the Codex Committee on Pesticide Residues, Fortaleza, Brazil, 3 – 8 April 2006. para 57 and 58.

Pollmann, B. 2006 a. Residue Behaviour of Winter Barley after Application of Dimethoate 400 g/l EC – 4 Sites in Northern Europe 2004. Final Report amended on 15 Feb 2006, DTFDoc. No. 534-4305, 20044016/E1-FPWB. Trial codes of the testing facility: F04W051R, G04W067R, D04W003R, GB04W003R. Unpublished.

Pollmann, B. 2006 b. Residue Behaviour of Winter Barley after Application of Dimethoate 400 g/l EC – 4 Sites in Northern Europe 2005. Final Report amended on 14 March 2006, DTF Doc. No. 534-4306, 20054006/E1-FPWB. Trial codes of the testing facility: GB05W003R, F05W003R, D05W003R. Unpublished.

SCC 2006a. Submission of barley residue data – Dimethoate. SCC Project No.: 104 – 033. Submission by Dr. Monika Eder, SCC GmbH, Ring 1, D-55234 Wendelsheim, Germany, 13 April 2006. Unpublished.

SCC 2006b. Submission of Labels Dimethoate Finland and Ireland. SCC Project No.: 104 –003. Submission by Dr. Monika Hofer, SCC GmbH, Ring1, D-55234 Wendelsheim, Germany, 1 August 2006. Unpublished.

## CROSS-REFERENCES

| Author | Document Code | Year |
|---|---|---|
| Pollmann, B | D04W003R | 2006a. |
| Pollmann, B | D05W003R | 2006b. |
| Pollmann, B | F04W051R | 2006a. |
| Pollmann, B | F05W003R | 2006b. |
| Pollmann, B | G04W067R | 2006a. |
| Pollmann, B | GB04W003R | 2006a. |
| Pollmann, B | GB05W003R | 2006b. |

# DISULFOTON (74)

*First draft prepared by B.C. Ossendorp, Centre for Substances and*
*Integrated Risk Assessment, National Institute of Public Health and the Environment, The*
*Netherlands*

## EXPLANATION

Disulfoton, an insecticide/acaricide, was evaluated for residues within the periodic review programme by the 1991 JMPR. Additional residue information was evaluated by the 1994 and 1998 JMPR. An ADI of 0.0003 mg/kg bw was adopted in 1991, an acute reference dose (ARfD) of 0.003 mg/kg bw in 1996. Estimations of the short term intake (IESTI) in 2002 by WHO (CCPR 34, CX/PR 02/03) resulted in IESTIs exceeding the ARfD for broccoli, cabbage, cauliflower, lettuce, potato, Japanese radish and rice. In the meantime Codex MRLs for potato, Japanese radish and rice were withdrawn.

At CCPR 36, the Committee noted that the acute intake concerns had not been resolved even with the use of a probabilistic method (ALINORM 04/27/24, paragraph 106). The Committee returned the MRLs of broccoli, cabbages (head), cauliflower, lettuce head and leaf to Step 6 awaiting refinements in the acute dietary intake probabilistic methodology (ALINORM 04/27/24, paragraph 107).

At CCPR 37 the Committee decided to return all the MRLs currently at step 7 to step 6. Since this was the third time that the proposed MRLs were returned to Step 6 for intake concerns, the Committee also decided to request the JMPR to review GAPs that may result in lower MRL recommendations (ALINORM 05/28/24, paragraph 105).

The manufacturer submitted GAP data from Canada, Mexico, Japan and the USA on broccoli, cabbage, cauliflower, and lettuce.

Residue and GAP information was also submitted by Japan on Adzuki bean, burdock, cabbage, Chinese cabbage, cucumber, eggplant, common bean (pod), mitsuba, onion, Satsuma mandarin, pineapple, potato, Japanese radish, Chinese onion (rakkyo), soybean, Welsh onion, sugar cane, sweet pepper, tomato, upland wasabi and watermelon.

## Residue definition (from JMPR 1998)

Definition of the residue for compliance with MRLs and for the estimation of dietary intake: sum of disulfoton, demeton-S and their sulfoxides and sulfones expressed as disulfoton. Note that JMPR 1994 extensively discussed the inclusion of demeton-S.

## USE PATTERN

Disulfoton is a systemic pre-emergent and post-emergent insecticide used for the control of a variety of insect pests such as aphids, mites, leafhoppers, leafminers, nematodes, thrips and beetles. The active ingredient is formulated as granules, EC and SC formulations, and generally applied to the soil at planting as a soil injection, or at sowing in-furrow, as a side-dressing (i.e. at the side of the furrow), or as a broadcast spray. Foliar sprays may be applied after planting (pre-emergent) or at a post-emergence stage of growth. Information on registered use patterns was provided by the manufacturer and the Japanese government. The manufacturer provided the GAP data both in summarized form, as original labels, and in an English translation.(this could mean two things, depending on where the commas go – I've assumed 'both' means the summarised form and the original labels, and that the English translation is extra. The Japanese government provided only a summary table in English.

On the US labels, the relevant soil applications are expressed as ounce (oz) of product/1000 ft of row (36 inch row spacing for broccoli, cauliflower and cabbage, 20 inch row spacing for lettuce), with a maximum in pints per acre. This maximum is given in the GAP table.

Table 1. Registered uses of disulfoton supplied by the manufacturer.

| Crop | Country | Formulation (g ai/kg; g ai/L) | Application | | | | PHI, days |
|------|---------|------|------|------|------|------|------|
| | | | Method | Rate kg ai/ha | Spray conc, kg ai/hL | Number | |
| Broccoli | USA | EC 960 | Soil application: Transplanting beds: soil treatment before transplanting Field: soil injection/side dressing | 1.12 | | 1 | 14 |
| Broccoli | Canada | GR 15 | Soil application: band on each side of the seed furrow/side dressing | 1.12 | | 1 | 42 |
| Broccoli | Mexico | GR 10 | Soil application: Band to each side of the row, after transplanting small plants | 1 | | | |
| Cabbage | USA | EC 960 | Soil application: Transplanting beds: soil treatment before transplanting Field: soil injection/side dressing | 2.24 | | 1 | 42 |
| Cabbage | Canada | GR15 | Soil application: band on each side of the seed furrow/side dressing | 1.12 | | 1 | 42 |
| Cabbage | Mexico | GR10 | Soil application: Band to each side of the row, after transplanting small plants | 1 | | | |
| Cabbage | Japan | GR 5, GR 3 | Soil application: soil incorporation row/planting hole | 1.5-3 | | 1 | * |
| Cauliflower | USA | EC 960 | Soil application: Chemigation through low pressure irrigation systems Transplanting beds: soil treatment before transplanting Field: soil injection/side dressing | 1.12 | | 1 | 40 |
| Cauliflower | Canada | GR 15 | Soil application: Band/side dress | 1.12 | | 2 | 30 |
| Cauliflower | Mexico | GR 10 | Soil application: Band to each side of the row, after transplanting small plants | 1 | | | |
| Lettuce | USA | EC 960 | Soil application: Chemigation through low pressure irrigation systems Field application: soil | 1.12; 2.24 in heavy | | 1 | 60 |

| Crop | Country | Formulation (g ai/kg; g ai/L) | Application | | | | PHI, days |
|------|---------|------|------|------|------|------|------|
| | | | Method | Rate kg ai/ha | Spray conc, kg ai/hL | Number | |
| | | | injection/side dressing | organic soils | | | |
| Lettuce | Canada | GR 15 | Soil application: Band application at seeding | 1.12; 2.24 in heavy organic soils | | 1 | |
| Lettuce | Mexico | GR 10 | Soil application: Band or both sides of the sowing furrow | 1 | | | |

\* Application at transplanting , PHI ca. 60 days

Table 2. Registered uses of disulfoton supplied by the Japanese government.

| Crop | Formulation (g ai/kg; g ai/L) | Application | | | | PHI, days |
|------|------|------|------|------|------|------|
| | | Method | Rate kg ai/ha | Spray conc, kg ai/hL | Number | |
| Common bean | GR 5 | spreading, plant root, at development op fruit | 2.0 | | 1 | 60 |
| Broad beans(dry) | GR 5 | spreading, plant root, at development of fruit | 2.0 | | 1 | 60 |
| Burdock | GR 5 | spreading, row, at seeding | 2.0 | | 1 | n.a. |
| Cabbage | GR 3 | soil incorporation/ row, at transplanting | 1.8 | | 1 | n.a. |
| Cabbage | GR 5 | soil incorporation/ planting hole, at transplanting | 1.5-3.0 | | 1 | n.a. |
| Chinese cabbage | GR 3 | soil incorporation/ row, germination | 1.8 | | 1 | n.a. |
| Chinese cabbage | GR 5 | soil incorporation/ planting hole, at transplanting | 1.5-3.0 | | 1 | n.a. |
| Cucumber | GR 5 | spreading to planting hole, at transplanting | 1.25-2.5 | | 1 | n.a. |
| Cucumber | GR 2 | soil incorporation/ at transplanting | 1.2 | | 1 | n.a. |
| Eggplant | GR 5 | spreading, plant root, at transplanting | 1.0-2.0 | | 1 | n.a. |
| Eggplant | GR 2 | soil incorporation/ row, at leaf development | 1.2 | | 1 | n.a. |
| Common bean(pods and/or immature seeds) | GR 5 | spreading, planting hole, at seeding | 2.0 | | 1 | n.a. |
| Kidney bean | GR 5 | Spreading, at ripening of fruit and seed | 2.0 | | 1 | 60 |
| Mitsuba | | Spreading, at leaf development | 2.5 | | 1-3 | 90 |
| Onion | GR 5 | soil incorporation, at transplanting | 2.0 | | 1 | n.a. |
| Satsuma Mandarin (Unshu orange) | GR 5 | soil incorporation, at development of fruit | 7.5-10 | | 1-2 | 30 |
| Field pea(dry) | GR 5 | spreading, row, at leaf development | 1.5-2.5 | | 1 | 60 |

| Crop | Formulation (g ai/kg; g ai/L) | Application | | | | PHI, days |
|---|---|---|---|---|---|---|
| | | Method | Rate kg ai/ha | Spray conc, kg ai/hL | Number | |
| Pineapple | GR 5 | Spreading, at leaf development | 2.0-2.5 | | 1-3 | 120 |
| Potato | GR 5 | spreading, planting hole, at germination | 2.0 | | 1 | n.a. |
| Potato | GR 3 | soil incorporation, row, before germination (before transplanting) | 1.8-2.7 | | 1 | n.a. |
| Radish, Japanese | GR 3 | soil incorporation/ row, at seeding | 1.8 | | 1 | n.a. |
| Radish, Japanese | GR 5 | soil incorporation/ row, at transplanting | 1.5-2.0 | | 1 | n.a. |
| Onion , Chinese (Rakkyo) | GR 5 | soil incorporation/ row, at transplanting | 2.0 | | 1 | n.a. |
| Soybean | GR 5 | Spreading, at development of fruit | 1.5-3.0 | | 1 | 60 |
| Soybean | GR 2 | soil incorporation, at seeding | 1.2 | | 1 | n.a. |
| Adzuki bean (dry) | GR 5 | | 1.5-3.0 | | 1 | 60 |
| Onion , Welsh | GR 2 | soil incorporation, row, at transplanting | 1.2 | | 1 | n.a. |
| Onion , Welsh | GR 3 | soil incorporation , row, at transplanting | 0.9 | | 1 | n.a. |
| Onion , Welsh | GR 5 | soil incorporation, at transplanting | 1.5 | | 1 | n.a. |
| Sugar cane | GR 5 | soil incorporation, row, at germination | 3.0-4.5 | | 1 | n.a. |
| Sweet pepper | GR 3 | spreading, plant root, at leaf development | 1.8 | | 1 | 45 |
| Tomato | GR 5 | spreading, plant root, at transplanting | 1.5-3.0 | | 1 | n.a. |
| Tomato | GR 2 | spreading, plant root, at transplanting | 1.2 | | 1 | n.a. |
| Wasabi, upland | GR 3 | Spreading, at development of harvestable vegetative plant parts | 1.8 | | 1-2 | 60 |
| Watermelon | GR 5 | spreading, planting hole, at transplanting | 1.5-3.0 | | 1 | n.a. |

All refer to field and glasshouse use, except for burdock, pineapple, sugar cane, and wasabi which only have a field use. 'At transplanting' means at leaf development

## RESIDUE RESULTING FROM SUPERVISED TRIALS ON CROPS

Tables 3-24 show data from supervised trials provided by the Japanese government, on crops which have not previously been evaluated: Adzuki bean, burdock, cabbage, Chinese cabbage, cucumber, eggplant, common bean (pod), mitsuba, onion, Satsuma mandarin, pineapple, potato, Japanese radish, Chinese onion, soybean, Welsh onion, sugar cane, sweet pepper, tomato, wasabi and watermelon. Only summary data sheets were provided (ref Table XI.3), no reports on field and analytical parts of the trials were available. In all trials, it is stated that the residue definition is disulfoton-sulfon (P=S) + dimeton-thiosulfon (P=O). Residues are calculated as disulfoton. In accordance with Japanese legislation, all samples have been analyzed twice, once by an official institute and once by an in-house institute. In the tables, the mean of the duplicate analysis is given. The following summarizes information on residues resulting from supervised trials.

| Commodity | Table |
|---|---|
| Adzuki bean. | Table 3 |
| Burdock. | Table 4 |
| Cabbage | Table 5 |
| Chinese cabbage | Table 6 |
| Cucumber | Table 7 |
| Eggplant | Table 8 |
| Common bean (pod) | Table 9 |
| Mitsuba | Table 10 |
| Onion | Table 11 |
| Satsuma mandarin | Table 12 |
| Pineapple | Table 13 |
| Potato | Table 14 |
| Japanese radish, root | Table 15 |
| Japanese radish, leaves | Table 16 |
| Chinese onion | Table 17 |
| Soybean | Table 18 |
| Welsh onion | Table 19 |
| Sugar cane | Table 20 |
| Sweet pepper | Table 21 |
| Tomato | Table 22 |
| Wasabi | Table 23 |
| Watermelon | Table 24 |

Data on broccoli, cabbage, cauliflower and lettuce, which were reviewed in the 1991, 1994 and 1998 monographs, are interpreted in the light of current GAP in Tables 25-28.

| Broccoli | Table 25 |
|---|---|
| Cabbage | Table 26 |
| Cauliflower | Table 27 |
| Lettuce | Table 28 |

In conclusion, in table 29 a summary is given of broccoli, cabbage, cauliflower and lettuce data according to GAP, along with the corresponding HRs and %ARfD.

Table 3. Residues in Adzuki bean (whole commodity) after outdoor treatment in Japan.

| Location, year, (variety) | Form | No | kg ai/ha | method, timing | DAT | residues, mg/kg |
|---|---|---|---|---|---|---|
| Iwate, 1972, (Ohdate No. 2) | GR 5 | 1 | 1.5 | Germination (at seeding) | 128 | < 0.02 |
| idem | GR 5 | 1 | 3.0 | Germination (at seeding) | 128 | < 0.02 |
| Hokkaido, 1972, (Takaraazu) | GR 5 | 1 | 1.5 | Germination (at seeding) | 122 | < 0.02 |
| idem | GR 5 | 1 | 3.0 | Germination (at seeding) | 122 | < 0.02 |

Table 4. Residues in burdock (whole commodity) after outdoor treatment in Japan.

| Location, year, (variety) | Form | No | kg ai/ha | method, timing | DAT | residues, mg/kg |
|---|---|---|---|---|---|---|
| Hokkaido, 1989, (Yanagawariso) | GR 5 | 1 | 2.0 | Germination (at seeding) | 111 | < 0.02 |
| Saitama, 1989, (Isuke) | GR 5 | 1 | 2.0 | Germination (at seeding) | 140 | < 0.02 |

Table 5. Residues in cabbage (whole commodity after removal of obviously decomposed or withered leaves) after outdoor treatment in Japan.

| Location, year, (variety) | Form | No | kg ai/ha | method, timing | DAT | residues, mg/kg |
|---|---|---|---|---|---|---|
| Aomori, 1972, (Nagaokakohai-wase) | GR 5 | 1 | 2.0 | At leaf development | 57 | 0.074 |
| idem | GR 5 | 1 | 4.0 | At leaf development | 57 | 0.097 |
| Ymanashi, 1972, (Aozora) | GR 5 | 1 | 2.0 | At leaf development | 69 | 0.072 |
| idem | GR 5 | 1 | 4.0 | At leaf development | 69 | 0.063 |

Table 6. Residues in Chinese cabbage (whole commodity after removal of obviously decomposed or withered leaves) after outdoor treatment in Japan.

| Location, year, (variety) | Form | No | kg ai/ha | method, timing | DAT | residues, mg/kg |
|---|---|---|---|---|---|---|
| Ushiku, 1987, (Akogare) | GR 5 | 1 | 3.0 | At leaf development | 54 | < 0.02 |
| | | | | | 64 | < 0.02 |
| Nagano, 1987, (Taibyou 60) | GR 5 | 1 | 3.0 | At leaf development | 49 | 0.08 |
| | | | | | 59 | 0.04 |

Table 7. Residues in cucumber (whole commodity after removal of stems) after outdoor treatment in Japan.

| Location, year, (variety) | Form | No | kg ai/ha | method, timing | DAT | residues, mg/kg |
|---|---|---|---|---|---|---|
| Ushiku, 1987, (Tokiwahikari 3P) | GR 5 | 1 | 2.5 | At leaf development | 40 | < 0.02 |
| Nagano, 1987, (Nankyoku No. 1) | GR 5 | 1 | 2.5 | At leaf development | 26 | < 0.02 |

Table 8. Residues in eggplant (whole commodity after removal of stems and cap) after indoor treatment in Japan.

| Location, year, (variety) | Form | No | Interval (days) | g granule/plant | method, timing | DAT | residues, mg/kg |
|---|---|---|---|---|---|---|---|
| Nagasaki, 1971, (Naganasu) | GR 5 | 1 | - | 2 | At leaf development | 79 | < 0.02 |
| | | | | | | 116 | < 0.02 |
| | | | | | | 127 | < 0.02 |
| Idem | GR 5 | 2 | 96 | 2 | At leaf development | 20 | 0.018 |
| | | | | | | 31 | 0.024 |
| Ushiku, 1971, (Kurogishi) | GR 5 | 1 | - | 2 | At leaf development | 79 | < 0.02 |
| | | | | | | 111 | < 0.02 |
| | | | | | | 142 | < 0.02 |
| Idem | GR 5 | 2 | 50 | 2 | At leaf development | 29 | < 0.02 |
| | | | | | | 61 | < 0.02 |
| | | | | | | 92 | 0.024 |

Table 9. Residues in common bean (pod) (whole commodity) after indoor treatment in Japan

| Location, year, (variety) | Form | No | kg ai/ha | method, timing | DAT | residues P=S, mg/kg | residues P=O, mg/kg |
|---|---|---|---|---|---|---|---|
| Okinawa | GR 5 | 1 | 1.5 | At leaf development | 104 | 0.003 | < 0.008 |
| | | | | | 111 | 0.003 | < 0.008 |
| | GR 5 | 1 | 3.0 | At leaf development | 104 | 0.029 | < 0.008 |
| | | | | | 111 | 0.042 | < 0.008 |

Table 10. Residues in mitsuba (whole commodity) after indoor treatment in Japan.

| Location, year, (variety) | Form | No | kg ai/ha | method, timing | DAT | residues P=S, mg/kg | residues P=O, mg/kg |
|---|---|---|---|---|---|---|---|
| Ibaraki | GR 5 | 4 | 2.5 | At leaf development | 74 | 0.006 | 0.008 |
| | GR 5 | 9 | 2.5 | At leaf development | 100 | 0.013 | 0.060 |

Table 11. Residues in onion (whole commodity after removal of roots and adhering soil and whatever parchment skin is easily detached)) after indoor treatment in Japan.

| Location, year, (variety) | Form | No | kg ai/ha | method, timing | DAT | residues, mg/kg* |
|---|---|---|---|---|---|---|
| Ushiku, 1976, (Sensyuou) | GR 5 | 1 | 1 | At leaf development | 209 | < 0.02 |
| | GR 5 | 1 | 2 | At leaf development | 209 | < 0.02 |
| | | | | | 282 | < 0.02 |
| Hokkaido, 1976, (Sapporoou) | GR 5 | 1 | 1 | At leaf development | 125 | < 0.02 |
| | GR 5 | 1 | 2 | At leaf development | 125 | < 0.02 |

* at cross-check analysis in another laboratory: all residues < 0.01 mg/kg

Table 12. Residues in Satsuma mandarin (pulp) after indoor treatment in Japan.

| Location, year, (variety) | Form | No | kg ai/ha | method, timing | DAT | residues, mg/kg |
|---|---|---|---|---|---|---|
| Saga, 1971, (Wase) | GR 5 | 1 | 1 | At flowering | 155 | < 0.02 |
| | GR 5 | 2 | 1 | Last at development of fruit | 25 | < 0.02 |
| Ohsaka, 1971, (Common) | GR 5 | 1 | 1 | At flowering | 159 | < 0.02 |
| | GR 5 | 2 | 1 | Last at development of fruit | 47 | < 0.02 |

Table 13. Residues in pineapple (pulp) after outdoor treatment in Japan.

| Location, year, (variety) | Form | No | Interval (days) | kg ai/ha | method, timing | DAT | residues, mg/kg* |
|---|---|---|---|---|---|---|---|
| Okinawa, 1976, (Smooth kaien) | GR 5 | 3 | 32-33 | 2.5 | | 40 | 0.221 |
| | | | | | | 119 | 0.023 |
| | GR 5 | 4 | 32-33-26 | 2.5 | | 14 | 0.319 |
| | | | | | | 93 | 0.053 |

* Replicate samples and replicate analysis showing the mean of 2 replicate samples, each analyzed twice by two laboratories; a total of 8 numbers per mean.

Table 14. Residues in potato (whole commodity after removing tops, adhering soil removed) after outdoor treatment in Japan.

| Location, year, (variety) | Form | No | kg ai/ha | method, timing | DAT | residues, mg/kg |
|---|---|---|---|---|---|---|
| Nagasaki, 1971, (Tachibana) | GR 5 | 1 | 2.0 | Sprouting/germination | 85 | 0.042 |
| | | | | | 105 | 0.070 |
| | GR 5 | 1 | 4.0 | Sprouting/germination | 85 | 0.115 |
| | | | | | 105 | 0.098 |
| Ushiku, 1972, (Dansyaku) | GR 5 | 1 | 2.0 | Sprouting/germination | 102 | 0.091 |
| Idem | GR 5 | 1 | 4.0 | Sprouting/germination | 102 | 0.188 |

Table 15. Residues in Japanese radish (roots, whole commodity after removing tops, adhering soil removed) after outdoor treatment in Japan.

| Location, year, (variety) | Form | No | kg ai/ha | method, timing | DAT | residues* P=S, mg/kg | residues* P=O, mg/kg |
|---|---|---|---|---|---|---|---|
| Thiba, 1976, (Hayabutoriookura) | GR 5 | 1 | 1.0 | germination | 75 | 0.004 | < 0.004 |
| | | | | | 82 | 0.004 | < 0.004 |
| Idem | GR 5 | 1 | 1.5 | germination | 75 | 0.004 | < 0.004 |
| | | | | | 82 | 0.008 | 0.004 |
| Idem | GR 5 | 1 | 2.0 | germination | 75 | 0.004 | < 0.004 |
| | | | | | 82 | 0.005 | < 0.004 |
| Kanagawa, 1977, (Miuramiya) | GR 5 | 1 | 1.0 | germination | 102 | 0.014 | < 0.004 |
| | | | | | 112 | 0.009 | < 0.004 |
| Idem | GR 5 | 1 | 1.5 | germination | 102 | 0.013 | 0.007 |
| | | | | | 112 | 0.011 | 0.008 |
| Idem | GR 5 | 1 | 2.0 | germination | 102 | 0.019 | 0.008 |
| | | | | | 112 | 0.015 | 0.008 |

* replicate analysis showing the mean. - each sample was analyzed twice by two laboratories; a total of 4 numbers per mean. LOQ (lab 1) = 0.003 mg/kg for P=S and 0.005 or 0.01 mg/kg for P=O; LOQ (lab 2) = 0.004 mg/kg for both. The mean is calculated by assuming a residue at the LOQ when <LOQ; when all are <LOQ the lowest LOQ is stated.

Table 16. Residues in Japanese radish (leaves, whole commodity as usually marketed, after removal of obviously decomposed or withered leaves) after outdoor treatment in Japan.

| Location, year, (variety) | Form | No | kg ai/ha | method, timing | DAT | residues* P=S, mg/kg | residues* P=O, mg/kg |
|---|---|---|---|---|---|---|---|
| Thiba, 1976, (Hayabutoriookura) | GR 5 | 1 | 1.0 | germination | 75 | 0.005 | 0.032 |
| | | | | | 82 | 0.004 | 0.020 |
| Idem | GR 5 | 1 | 1.5 | germination | 75 | < 0.003 | 0.018 |
| | | | | | 82 | < 0.003 | 0.021 |
| Idem | GR 5 | 1 | 2.0 | germination | 75 | 0.005 | 0.030 |
| | | | | | 82 | 0.004 | 0.015 |
| Kanagawa, 1977, (Miuramiya) | GR 5 | 1 | 1.0 | germination | 102 | 0.009 | 0.046 |
| | | | | | 112 | 0.006 | 0.028 |
| Idem | GR 5 | 1 | 1.5 | germination | 102 | 0.014 | 0.054 |
| | | | | | 112 | 0.022 | 0.083 |
| Idem | GR 5 | 1 | 2.0 | germination | 102 | 0.026 | 0.114 |
| | | | | | 112 | 0.022 | 0.074 |

* replicate analysis showing the mean - each sample was analyzed twice by two laboratories; a total of 4 numbers per mean. LOQ (lab 1) = 0.003 mg/kg for P=S and 0.005 or 0.01 mg/kg for P=O; LOQ (lab 2) = 0.004 mg/kg for both. Mean is calculated by assuming a residue at the LOQ when <LOQ; when all are <LOQ the lowest LOQ is stated.

Table 17. Residues in Chinese onion (rakkyo) (whole commodity after removal of roots and adhering soil  and whatever parchment skin is easily detached) after outdoor treatment in Japan.

| Location, year, (variety) | Form | No | Interval (days) | kg ai/ha | method, timing | DAT | residues* P=S, mg/kg | residues* P=O, mg/kg |
|---|---|---|---|---|---|---|---|---|
| Fukui, 1985, (common) | GR 5 | 1 | - | 3.0 | germination | 282 | < 0.005 | 0.006 |
| | | | | | | 296 | < 0.005 | < 0.005 |
| Idem | GR 5 | 2 | 196 | 4.5 | Germination + at leaf development | 86 | < 0.005 | 0.014 |
| Idem | GR 5 | 1 | - | 4.5 | germination | 282 | < 0.005 | 0.007 |
| | | | | | | 296 | < 0.005 | 0.006 |

* replicate analysis showing the mean.

Table 18. Residues in soybean (dry) after outdoor treatment in Japan.

| Location, year, (variety) | Form | No | kg ai/ha | method, timing | DAT | residues, mg/kg |
|---|---|---|---|---|---|---|
| Fukushima, 1987, (Suzuyutaka) | GR 5 | 1 | 3.0 | flowering | 60 | < 0.02 |
| Nagano, 1987, (Hourei) | GR 5 | 1 | 3.0 | flowering | 60 | < 0.02 |

Table 19. Residues in Welsh onion (whole vegetable after removal of roots and adhering soil) after outdoor treatment in Japan.

| Location, year, (variety) | Form | No | kg ai/ha | method, timing | DAT | residues* P=S, mg/kg | residues* P=O, mg/kg |
|---|---|---|---|---|---|---|---|
| Nagano, 1976, (Shimonita) | GR 5 | 1 | 0.5 | germination | 102 | < 0.003 | 0.006 |
| Idem | GR 5 | 1 | 1.5 | germination | 102 | < 0.003 | < 0.005 |
| Idem | GR 5 | 1 | 0.5 | Leaf development | 122 | < 0.003 | 0.005 |
| Idem | GR 5 | 1 | 1.5 | Leaf development | 122 | < 0.003 | 0.006 |
| Gunma, 1976, (Ishikuranebukanegi) | GR 5 | 1 | 1.5 | Leaf development | 161 | < 0.003 | 0.016 |
| Idem | GR 5 | 1 | 3.0 | Leaf development | 161 | < 0.003 | 0.011 |
| Idem | GR 5 | 1 | 1.5 | germination | 247 | < 0.003 | < 0.005 |
| Idem | GR 5 | 1 | 3.0 | germination | 247 | < 0.003 | < 0.005 |

* replicate analysis showing the mean.

Table 20. Residues in sugar cane (stem after removal of skin) after outdoor treatment in Japan.

| Location, year, (variety) | Form | No | kg ai/ha | method, timing | DAT | residues* P=S, mg/kg | residues* P=O, mg/kg |
|---|---|---|---|---|---|---|---|
| Okinawa miyako, 1976, (NCO 310) | GR 5 | 1 | 3.0 | germination | 329 | < 0.002 | < 0.004 |
| | | | | | 519 | < 0.002 | < 0.004 |
| Idem | GR 5 | 1 | 4.5 | germination | 329 | < 0.002 | < 0.004 |
| | | | | | 519 | < 0.002 | < 0.004 |
| Okinawa yaeyama, 1976, (NCO 310) | GR 5 | 1 | 3.0 | germination | 200 | < 0.003 | 0.004 |
| | | | | | 297 | < 0.003 | < 0.004 |
| Idem | GR 5 | 1 | 4.5 | germination | 200 | 0.004 | 0.005 |
| | | | | | 297 | < 0.003 | < 0.004 |

* replicate analysis showing the mean - each sample was analyzed twice by two laboratories; a total of 4 numbers per mean. LOQ (lab 1) = 0.002 or 0.003 mg/kg for P=S and 0.004 or 0.02 mg/kg for P=O; LOQ (lab 2) = 0.004 mg/kg for both. The mean is calculated by assuming a residue at the LOQ when <LOQ; when all are <LOQ the lowest LOQ is stated.

Table 21. Residues in sweet pepper (whole commodity after removal of stems and cap) after outdoor treatment in Japan.

| Location, year, (variety) | Form | No | Interval (days) | kg ai/ha | method, timing | DAT | residues* P=S, mg/kg | residues* P=O, mg/kg |
|---|---|---|---|---|---|---|---|---|
| Ushiku, 1982, (Tosagreen B) | GR 5 | 1 | - | 2 | Maturity of fruit | 1 | 0.004 | < 0.008 |
| | | | | | | 7 | 0.006 | < 0.008 |
| | | | | | | 14 | < 0.004 | < 0.008 |
| | | | | | | 21 | < 0.004 | < 0.008 |
| Idem | GR 5 | 1 | - | 2 | Leaf development | 59 | 0.016 | 0.014 |

| Location, year, (variety) | Form | No | Interval (days) | kg ai/ha | method, timing | DAT | residues* P=S, mg/kg | residues* P=O, mg/kg |
|---|---|---|---|---|---|---|---|---|
| Idem | GR 5 | 2 | 59 | 2 | Maturity of fruit | 21 | 0.004 | < 0.008 |
| Kochi, 1982, (Original) | GR 5 | 1 | - | 2 | Maturity of fruit | 1 | < 0.004 | < 0.008 |
| | | | | | | 7 | 0.004 | < 0.008 |
| | | | | | | 14 | 0.009 | < 0.008 |
| | | | | | | 21 | 0.006 | < 0.008 |
| | GR 5 | 1 | - | 2 | Leaf development | 55 | 0.011 | 0.012 |
| | GR 5 | 2 | 25 | 2 | Maturity of fruit | 21 | 0.014 | 0.013 |

\* replicate analysis showing the mean.

Table 22. Residues in tomato (whole commodity after removal of stems and cap) after outdoor treatment in Japan.

| Location, year, (variety) | Form | No | Interval (days) | kg ai/ha | method, timing | DAT | residues* P=S, mg/kg | residues* P=O, mg/kg |
|---|---|---|---|---|---|---|---|---|
| Nagano, 1976, (Kyouryokubeiju) | GR 5 | 1 | - | 1 g GR/plant | Leaf development | 70 | < 0.003 | 0.005 |
| | | | | | | 100 | < 0.003 | < 0.005 |
| Idem | GR 5 | 1 | - | 2 g GR/plant | Leaf development | 70 | < 0.003 | 0.006 |
| | | | | | | 100 | < 0.003 | < 0.005 |
| Idem | GR 5 | 2 | 42 | 3 kg ai/ha, and 2 g GR/plant | Development of fruit (2x); flowering | 30 | 0.016 | 0.204 |
| | | | | | | 41 | 0.003 | 0.071 |
| | | | | | | 60 | < 0.003 | 0.008 |
| Koibuchi, 1976, (Hikari F1) | GR 5 | 1 | - | 1 g GR/plant | Leaf development | 38 | 0.003 | 0.076 |
| | | | | | | 64 | < 0.003 | 0.005 |
| | GR 5 | 1 | - | 2 g GR/plant | Leaf development | 38 | 0.008 | 0.140 |
| | | | | | | 64 | < 0.003 | 0.018 |
| | GR 5 | 2 | 42 | 3 kg ai/ha, and 2 g GR/plant | Leaf development | 33 | 0.006 | 0.062 |

\* replicate analysis showing the mean.

Table 23. Residues in upland wasabi (Shimane: top and root; Yamaguchi: top) after outdoor treatment in Japan.

| Location, year, (variety) | Form | No | kg ai/ha | DAT | residues* P=S, mg/kg | residues* P=O, mg/kg |
|---|---|---|---|---|---|---|
| Shimane | DP 3 | 1 | 1.8 | 14 | 0.72 | 0.20 |
| | | | | 30 | 0.14 | 0.05 |
| | | | | 60 | 0.02 | < 0.01 |
| | | | | 122 | < 0.01 | < 0.01 |
| Idem | DP 3 | 2 | 1.8 | 14 | 0.76 | 0.21 |
| | | | | 30 | 0.31 | 0.10 |
| | | | | 60 | 0.04 | 0.02 |
| | | | | 122 | < 0.01 | < 0.01 |
| Idem | DP 3 | 3 | 1.8 | 14 | 1.3 | 0.30 |
| | | | | 30 | 0.32 | 0.08 |
| | | | | 60 | 0.07 | 0.05 |
| | | | | 122 | 0.02 | < 0.01 |
| Yamaguchi, | GR 3 | 1 | 1.8 | 14 | 0.265 | 0.28 |
| | | | | 30 | 0.024 | 0.025 |
| | | | | 60 | < 0.004 | < 0.01 |
| | | | | 120 | < 0.004 | < 0.01 |
| | | | | 180 | < 0.004 | < 0.01 |

| Location, year, (variety) | Form | No | kg ai/ha | DAT | residues* P=S, mg/kg | residues* P=O, mg/kg |
|---|---|---|---|---|---|---|
| Idem | GR 3 | 2 | 1.8 | 14 | 0.382 | 0.54 |
| | | | | 30 | 0.070 | 0.095 |
| | | | | 60 | 0.006 | < 0.01 |
| | | | | 120 | < 0.004 | < 0.01 |
| | | | | 180 | < 0.004 | < 0.01 |
| Idem | GR 3 | 3 | 1.8 | 14 | 0.349 | 0.50 |
| | | | | 30 | 0.044 | 0.06 |
| | | | | 60 | 0.006 | < 0.01 |
| | | | | 120 | < 0.004 | < 0.01 |
| | | | | 180 | < 0.004 | < 0.01 |

* replicate analysis showing the mean.

Table 24. Residues in watermelon (whole commodity after removal of stems) after outdoor treatment in Japan.

| Location, year, (variety) | Form | No | kg ai/ha | method, timing | DAT | residues* P=S, mg/kg | residues* P=O, mg/kg |
|---|---|---|---|---|---|---|---|
| Kanagawa, 1976, (Shimao) | GR 5 | 1 | 1.43 | Leaf development | 79 | < 0.01 | 0.01 |
| Idem | GR 5 | 1 | 2.86 | Leaf development | 79 | < 0.01 | 0.04 |
| Ushiku, 1976, (Kodama) | GR 5 | 1 | 1.5 | Leaf development | 78 | < 0.01 | < 0.01 |
| | GR 5 | 1 | 3.0 | Leaf development | 78 | < 0.01 | < 0.01 |

* replicate analysis showing the mean.

Table 25. Residue interpretation table for disulfoton residues in broccoli. GAP and trial conditions are compared for treatments considered valid for MRL and STMR estimation.

| Crop | Country | Use pattern Form. | kg ai/ha | No of appl Note | PHI days | Trial | disulfoton mg/kg |
|---|---|---|---|---|---|---|---|
| Broccoli | Canadian GAP | GR | 1.12 | 1 | 42 | | |
| No relevant trials | | | | | | | |
| Broccoli | US GAP | GR/EC | 1.12 | 1 | 14 | | |
| Broccoli | US 1987 | EC | 1.12 + 1.68 | 1 + 1 | 14 | M 91147[a] | < 0.02 |
| Broccoli | US 1987 | EC | 1.12 + 1.68 | 1 + 1 | 14 | M 91147[a] | 0.05 |
| Broccoli | US 1987 | EC | 1.12 + 2.04 | 1 + 1 | 15 29 | M 91147[a] | 0.03 0.11 |
| Broccoli | US 1987 | GR+EC | 1.12 + 2.04 | 1 + 1 | 14 | M 91147[a] | 0.03 |
| Broccoli | US 1987 | GR+EC | 1.12 + 2.04 | 1 + 1 | 15 29 | M 91147[a] | 0.06 0.09 |
| Broccoli | US 1987 | GR | 1.12 + 2.04 | 1 + 1 | 14 | M 91147[a] | ≤ 0.02 |
| Broccoli | Mexican GAP | GR | 1.0 | 1 | ns | Apply after transplant of small plants | |
| No relevant trials | | | | | | | |

a One soil spray followed by one side-dress spray.

Table 26. Residue interpretation table for disulfoton residues in cabbage. GAP and trial conditions are compared for treatments considered valid for MRL and STMR estimation.

| Crop | Country | Application Form. | kg ai/ha | No of appl | PHI days | Trial | disulfoton mg/kg |
|---|---|---|---|---|---|---|---|
| Cabbage, head | Canadian GAP | GR | 1.12 | 1 | 42 | | |
| Cabbage, head | US 1987 | EC | 1.12 | 2 | 42 | M 91152[a] | ≤ 0.02 |
| Cabbage, head | US 1987 | GR+EC | 1.12 + 1.30 | 1 + 1 | 39 | M 91152[a] | ≤ 0.02 |
| Cabbage, head | US 1987 | GR | 1.12 + 1.48 | 1 + 1 | 42 | M 91152[a] | ≤ 0.02 |

| Crop | Country | Form. | Application | | | | Trial | disulfoton |
| | | | kg ai/ha | No of appl | PHI days | | mg/kg |
|---|---|---|---|---|---|---|---|
| Cabbage, head | US GAP | EC | 1.12 – 2.24 | 1 | 42 | | |
| Cabbage, head | US 1987 | EC | 1.12 + 1.72 | 1 + 1 | 42 | M 91152 [a] | 0.17 |
| Cabbage, head | US 1987 | EC | 1.12 + 1.72 | 1 + 1 | 41 | M 91152 [a] | 0.02 |
| Cabbage, head | US 1987 | GR | 1.12 + 1.72 | 1 + 1 | 42 | M 91152 [a] | 0.17 |
| Cabbage, head | US 1987 | GR | 1.12 + 1.72 | 1 + 1 | 41 | M 91152 [a] | 0.03 |
| Cabbage, head | US 1987 | EC | 1.12 + 1.72 | 1 + 1 | 42 | M 91152 [a] | 0.06 |
| Cabbage, head | US 1987 | EC | 1.12 + 1.72 | 1 + 1 | 42 | M 91152 [a] | 0.08 |
| Cabbage | Mexican GAP | GR | 1.0 | 1 | ns | Apply after transplant of small plants | |
| No relevant GAP | | | | | | | |
| Cabbage, head | Japan GAP | GR | 1.5-3.0 | 1 | ns | Apply at seeding or transplanting | |
| Cabbage, head | Japan 1972 | GR | 2 | 1 | 57 | N 27/72 (1991) | 0.073 |
| Cabbage, head | Japan 1972 | GR | 4 | 1 | 57 | N 27/72 (1991) | 0.096 |
| Cabbage, head | Japan 1972 | GR | 2 | 1 | 69 | N 28/72 (1991) | 0.072 |
| Cabbage, head | Japan 1972 | GR | 4 | 1 | 69 | N 28/72 (1991) | 0.063 |

[a] One soil broadcast followed by one side-dress spray.

Table 27. Residue interpretation table for disulfoton residues in cauliflower. GAP and trial conditions are compared for treatments considered valid for MRL and STMR estimation.

| Crop | Country | Form. | Use pattern | | | | Trial | disulfoton |
| | | | kg ai/ha | No of appl | PHI days | | mg/kg |
|---|---|---|---|---|---|---|---|
| Cauliflower | US GAP | EC/GR | 1.12 | 1 | 40 | | |
| Cauliflower | US 1987 | EC | 1.12 | 3 | 40 | M 91154 [a] | 0.31 |
| Cauliflower | US 1987 | EC | 1.12 + 1.68 | 1 + 2 | 40 | M 91154 | < 0.01 |
| Cauliflower | US 1987 | EC | 1.12 + 1.01 | 1 + 2 | 38 | M 91154 | < 0.01 |
| Cauliflower | US 1987 | EC | 1.12 | 3 | 38-43 | M 91154 | < 0.01 |
| Cauliflower | US 1987 | GR | 1.12 | 3 | 40 | M 91154 | 0.04 |
| Cauliflower | US 1987 | GR | 1.12 + 1.68 | 1 + 2 | 40 | M 91154 | < 0.01 |
| Cauliflower | US 1987 | GR | 1.12 | 3 | 40 | M 91154 | 0.01 |
| Cauliflower | US 1987 | GR | 1.12 + 1.01 | 1 + 2 | 38 | M 91154 | 0.01 |
| Cauliflower | Canadian GAP | EC/GR | 1.12 | 2 (i=21) | 30 | | |
| Cauliflower | US 1987 | EC | 1.12 | 3 | 30 40 | M 91154 | 0.01 0.31 |
| Cauliflower | US 1987 | EC | 1.1 + 1.7 | 1 + 2 | 30 | M 91154 | < 0.01 |
| Cauliflower | US 1987 | EC | 1.1 + 1.0 | 1 + 2 | 28-30 | M 91154 | < 0.01 |
| Cauliflower | US 1987 | EC | 1.12 | 3 | 28-30 | M 91154 | < 0.01 |
| Cauliflower | US 1987 | GR | 1.12 | 3 | 30 40 | M 91154 | 0.02 0.04 |
| Cauliflower | US 1987 | GR | 1.12 | 3 | 30 40 | M 91154 | < 0.01 0.01 |
| Cauliflower | US 1987 | GR | 1.12 + 1.01 | 1 + 2 | 28 | M 91154 | 0.05 |
| Cauliflower | Mexican GAP | GR | 1.0 | 1 | ns | Apply after transplant of small plants | |
| No relevant trials | | | | | | | |

[a] One broadcast application followed by two site-dress sprays.

Table 28. Disulfoton residues in lettuce (leaf and head), resulting from soil application in supervised trials.

| Crop | Country | Form. | Use pattern | | | | Trial | disulfoton |
| | | | kg ai/ha | kg ai/hL | No of appl. | PHI days | | mg/kg |
|---|---|---|---|---|---|---|---|---|
| Lettuce | Canadian GAP | GR | 1.12-2.25[a] | | 1 | ns | Apply at seeding time only | |
| Lettuce, leaf | US 1985 | EC | 2.23 | | 1 | 59 | M 91473 (1991) | < 0.03 |

| Crop | Country | Form. | Use pattern | | | | Trial | disulfoton mg/kg |
| | | | kg ai/ha | kg ai/hL | No of appl. | PHI days | | |
|---|---|---|---|---|---|---|---|---|
| Lettuce, leaf | US 1985 | EC | 2.44 | | 1 | 50 | M 91473 (1991) | 0.56 |
| Lettuce, leaf | US 1985 | EC | 2.9 | | 1 | 59 | M 91473 (1991) | 0.10 |
| Lettuce | Mexican GAP | GR | 1.0 | | 1 | ns | Apply at seeding time only | |
| Lettuce, leaf | US 1985 | EC | 1.1 | | 1 | 61 | M 91473 (1991) | 0.64 |
| Lettuce, leaf | US 1985 | EC | 1.1 | | 1 | 61 | M 91473 (1991) | ≤ 0.03 |
| Lettuce, leaf | US 1995 | EC | 1.1 | 1.8 | 1 | 60 | 107520 (1998) FCA-DI004-95H | 1.1 |
| Lettuce, leaf | US 1995 | EC | 1.1 | 0.29 | 1 | 90 | 107520 (1998) 458-DI005 -95H | ≤ 0.05 |
| Lettuce, leaf | US 1996 | EC | 1.14 | 2.3 | 1 | 73 | 107520 (1998) VBL-DO006-95H | 0.58 |
| Lettuce, head | US 1995 | EC | 1.1 | 1.8 | 1 | 97 | 107520 (1998) FCA-DI001-95H | 0.22 |
| Lettuce, head | US 1995 | EC | 1.2 | 0.25 | 1 | 116 | 107520 (1998) 458-DI002 -95H | ≤ 0.05 |
| Lettuce, head | US 1995 | EC | 1.21 | 2.0 | 1 | 62 | 107520 (1998) VBL-DI003-95H | ≤ 0.05 |
| Lettuce | US GAP | EC | 1.12-2.24 | | 1 | 60 | | |
| Same as Canada GAP | | | | | | | | |

ns = not specified

[a] Use the higher rate on heavy organic soils

Table 29. Summary of disulfoton residue data according to GAP, corresponding HRs and %ARfD.

| Crop | GAP | form. | kg ai/ha | No of appl | PHI days | Residue data | n | HR | % ARfD gen pop | % ARfD children |
|---|---|---|---|---|---|---|---|---|---|---|
| Broccoli | US | GR/EC | 1.12 | 1 | 14 | < 0.02, < 0.02, 0.03, 0.05, 0.09, 0.11 | 6 | 0.11 | 60% | 120% |
| Cabbage | Canada | GR | 1.12 | 1 | 42 | < 0.02, < 0.02, < 0.02 | 3 | 0.02 | 10% | 30% |
| Cabbage | US | EC | 1.12-2.24 | 1 | 42 | 0.02, 0.03, 0.06, 0.08, 0.17, 0.17 | 6 | 0.17 | 110% | 260% |
| Cabbage | Japan | GR | 1.5-3.0 | 1 | | 0.063, 0.072, 0.073, 0.096 | 4 | 0.096 | 60% | 150% |
| Cauliflower | US | EC/GR | 1.12 | 1 | 40 | < 0.01, < 0.01, < 0.01, < 0.01, 0.01, 0.01, 0.04, 0.31 | 8 | 0.31 | 150% | 380% |
| Cauliflower | Canada | EC/GR | 1.12 | 2 | 30 | < 0.01, < 0.01, < 0.01, 0.01, 0.04, 0.05, 0.31 | 7 | 0.31 | 150% | 380% |
| Lettuce | Mexico | GR | 1.0 | 1 | Apply at seeding | < 0.03, < 0.05, < 0.05, < 0.05, 0.22, 0.58, 0.64, 1.1 | 8 | 1.1 | 380% | 570% |
| Lettuce | Canada | GR | 1.12-2.25 | 1 | Apply at seeding | < 0.03, 0.10, 0.56 | 3 | 0.56 | 180% | 280% |

## RESIDUES IN FOOD IN COMMERCE OR AT CONSUMPTION

The results from different monitoring programs in Europe and the US are summarised in Table 30. The results show that in lettuce no residues were found in the EU. In the USDA Pesticide Data Program only low residues were detected in very few samples of broccoli or lettuce during the last eleven years. Each year from 1996–2003, many samples imported to the US or grown in the US were analysed by the FDA for residues of disulfoton. Only in three samples given in Table 30 were residues detected (Andersch, 2006).

Table 30. Monitoring data for disulfoton.

| Monitoring program | Origin | Year | Crop | Total samples analysed | Samples with residues | Residues (mg/kg) |
|---|---|---|---|---|---|---|
| EU Monitoring program | Austria | 2001 | Lettuce | 12 | 0 | |
| | Belgium | | | 27 | 0 | |
| | Finland | | | 42 | 0 | |
| | France | | | 343 | 0 | |
| | Greece | | | 16 | 0 | |
| | Italy | | | 18 | 0 | |
| | Norway | | | 88 | 0 | |
| | Spain | | | 45 | 0 | |
| USDA's Pesticide Data Program | USA | 1994 | Broccoli | 673 | 0 | |
| | | 1994 | lettuce | 691 | 0 | |
| | | 1999 | Lettuce | 185 | 0 | |
| | | 2000 | Lettuce | 740 | 1 | 0.015 |
| | | 2001 | Broccoli | 720 | 2 | 0.008, 0.026 |
| | | 2001 | Lettuce | 554 | 3 | 0.007-0.015 |
| | | 2002 | Broccoli | 737 | 0 | |
| FDA's Enforcement Monitoring Program | Mexico | 1997 | Broccoli | na | 1 | 0.2 mg/kg disulfoton sulfone, 0.1 mg/kg demeton-S sulfone |
| | USA | 1999 | Cabbage | na | 1 | 0.1 mg/kg demeton-S sulfone |
| | USA | 2001 | Lettuce | na | 1 | 0.1 mg/kg demeton-S sulfone |

na – Exact number not available

## APPRAISAL

Disulfoton, an insecticide acaricide, was evaluated for residues within the periodic review programme by the 1991 JMPR. Additional residue information was evaluated by the JMPR in 1994 and 1998. An ADI of 0.0003 mg/kg bw and an acute RfD of 0.003 mg/kg bw were established in 1991 and in 1996, respectively. Estimations of the short-term intake (IESTI) in 2002 by WHO (CCPR 34, CX/PR 02/03) resulted in IESTIs exceeding the acute reference dose (ARfD) of 0.003 mg/kg bw for broccoli, cabbage, cauliflower, lettuce, potato, Japanese radish and rice. In the interim Codex MRLs for potato, Japanese radish and rice were withdrawn.

At the 36th Session of CCPR, the Committee noted that the acute intake concerns had not been resolved even with the use of a probabilistic method (ALINORM 04/27/24, paragraph 106). The Committee returned the MRLs of broccoli; cabbages, head; cauliflower; lettuce head and leaf to Step 6 awaiting refinements in the acute dietary intake probabilistic methodology (ALINORM 04/27/24, paragraph 107).

At the 37th Session of CCPR the Committee decided to return all the MRLs currently at step 7 to step 6. Since this was the third time that the proposed MRLs were returned to Step 6 for intake concerns, the Committee also decided to request the JMPR to review alternate GAPs that may result in lower MRL recommendations (ALINORM 05/28/24, paragraph 105).

The manufacturer submitted GAP information from Canada, Mexico, Japan and the USA on broccoli, cabbage, cauliflower, and lettuce.

Residue and GAP information were also submitted by Japan on Adzuki bean, burdock, cabbage, Chinese cabbage, cucumber, eggplant, common bean (pod), mitsuba, onion, Satsuma mandarin, pineapple, potato, Japanese radish, Chinese onion (rakkyo), soybean, Welsh onion, sugar cane, sweet pepper, tomato, upland wasabi and water melon.

***Results of supervised residue trials on crops***

Data from supervised trials which have not previously been evaluated on Azuki bean, burdock, cabbage, Chinese cabbage, cucumber, egg plant, common bean (pod), mitsuba, onion, Satuma mandarin, pineapple, potato, Japanese radish, Chinese onion, soybean, Welsh onion, sugar cane, sweet pepper, tomato, wasabi and water melon were provided by the Japanese government.

Trials from Japan were available only in summary form and could not be evaluated.

Data on broccoli, cabbage, cauliflower and lettuce which were reviewed in the 1991, 1994 and 1998 monographs were interpreted in the light of current GAP. The Meeting noted that current GAP was the same as that recorded in 1998.

*Broccoli*

In USA, disulfoton may be used on broccoli in the field as a soil injection or side dressing at 1.1 kg ai/ha with a PHI of 14 days.

In Canada, disulfoton may be used on broccoli in the field as a soil-application band or side dressing at 1.1 kg ai/ha with a PHI of 42 days.

In Mexico, disulfoton may be used on broccoli in the field as a soil-application band at 1 kg ai/ha after transplanting small plants.

In six US field trials where disulfoton was used as a soil spray (pre-plant) at 1.1 kg ai/ha and a side-dressing at 1.7–2.0 kg ai/ha, with PHIs of 14–29 days, residues in rank order were < 0.02, < 0.02, 0.03, 0.05, 0.09 and 0.11 mg/kg. The Meeting noted that the field trial application rates considerably exceeded the GAP rates, so the trial data could not be used.

The Meeting noted that a highest residue of 0.11 mg/kg for broccoli would be associated with IESTI values of 60% and 120% of the current ARfD (0.003 mg/kg).

None of the residue data relating to available alternative GAP suggests a lower maximum residue level to replace the current proposal of 0.1 mg/kg disulfoton on broccoli.

*Cabbage*

In USA, disulfoton may be used on cabbage in the field as a soil injection or side dressing at 2.2 kg ai/ha with a PHI of 42 days. In six US field trials where disulfoton was used as a soil spray (pre-plant) at 1.1 kg ai/ha and a side-dressing at 1.7 kg ai/ha, with PHIs of 41–42 days, residues in rank order were 0.02, 0.03, 0.06, 0.08, 0.17 and 0.17 mg/kg.

Six trials were insufficient for estimation of an HR. The Meeting noted that a highest residue of 0.17 mg/kg for head cabbage would be associated with IESTI values of 110% and 260% of the current ARfD (0.003 mg/kg).

In Canada, disulfoton may be used on cabbage in the field as a soil-application band or side dressing at 1.1 kg ai/ha with a PHI of 42 days. In one US field trial where disulfoton was used twice as a band at 1.1 kg ai/ha and a PHI of 42 days, the residue was < 0.02 mg/kg. In two US field trials where disulfoton was used as a soil spray (pre-plant) at 1.1 kg ai/ha and a side-dressing at 1.3–1.5 kg ai/ha, with PHIs of 39 and 42 days, residues were < 0.02 and < 0.02 mg/kg.

Three trials were insufficient for estimation of an HR.

In Japan, disulfoton may be used on cabbage in the field by soil incorporation in the row or in the planting furrow at 1.5–3 kg ai/ha. In four Japanese field trials where disulfoton was used twice at 2 or 4 kg ai/ha and with PHIs of 57–69 days, the residues in rank order were 0.063, 0.072, 0.073 and 0.096 mg/kg.

Four trials were insufficient for estimation of an HR.

In Mexico, disulfoton may be used on cabbage in the field as a soil-application band at 1 kg ai/ha after transplanting small plants. No residue data were available at this GAP.

None of the residue data relating to available alternative GAP suggests a lower maximum residue level to replace the current proposal of 0.2 mg/kg disulfoton on head cabbage.

*Cauliflower*

In Canada, disulfoton may be used twice on cauliflower in the field as a soil injection or side dressing at 1.1 kg ai/ha with a PHI of 30 days.

In four US field trials on cabbage where disulfoton was used three times as a side-dressing at 1.1 kg ai/ha, with PHIs of 28-40 days, residues in rank order were < 0.01, 0.01, 0.04 and 0.31 mg/kg.

In three US field trials on cabbage where disulfoton was used as a soil spray (pre-plant) at 1.1 kg ai/ha and with 2 side-dressings at 1.0 or 1.7 kg ai/ha, with PHIs of 28–30 days, residues in rank order were < 0.01, < 0.01, and 0.05 mg/kg. Residues in rank order for the seven trials were: < 0.01, < 0.01, < 0.01, 0.01, 0.04, 0.05 and 0.31 mg/kg.

The 1991 JMPR assumed that the value of 0.31 mg/kg 40 days after the last application was an outlier (JMPR Residue Evaluations, 1991, page 325). On re-examination of the study report (Di-Syston - Magnitude of residue on cauliflower, Report M 91154, 1987), the Meeting noted that there was no documented reason for classifying the residue value of 0.31 mg/kg as invalid. The Meeting therefore included the value in the evaluation.

From this use, the Meeting estimated an HR of 0.31 mg/kg for disulfoton on cauliflower. The Meeting noted that an HR of 0.31 mg/kg for cauliflower would be associated with IESTI values of 150% and 380% of the current ARfD (0.003 mg/kg).

In Mexico, disulfoton may be used on cauliflower in the field as a soil-application band at 1 kg ai/ha after transplanting small plants. No residue data were available at this GAP.

In USA, disulfoton may be used on cauliflower in the field as a soil injection or side dressing at 1.1 kg ai/ha with a PHI of 40 days.

In four US field trials where disulfoton was used three times as a side-dressing at 1.1 kg ai/ha, with PHIs of 38–43 days, residues in rank order were < 0.01, 0.01, 0.04 and 0.31 mg/kg. In four US field trials where disulfoton was used as a soil spray (pre-plant) at 1.1 kg ai/ha and two side-dressings at 1.0 or 1.7 kg ai/ha, with PHIs of 38 and 40 days, residues in rank order were < 0.01, < 0.01, < 0.01 and 0.01 mg/kg. Residues in rank order for the eight trials were: < 0.01, < 0.01, < 0.01, < 0.01, 0.01, 0.01, 0.04 and 0.31 mg/kg.

None of the residue data relating to available alternative GAP suggests a lower maximum residue level to replace the current proposal of 0.05 mg/kg disulfoton on cauliflower. On the contrary, the inclusion of 0.31 mg/kg as a valid residue for disulfoton use on cauliflower suggests that the recommended maximum residue level on cauliflower should be adjusted to a higher level.

From this use, the Meeting estimated HR and STMR values of 0.31 and 0.01 mg/kg respectively for disulfoton on cauliflower.

The Meeting estimated a maximum residue level of 0.5 mg/kg for disulfoton on cauliflower to replace the current recommendation of 0.05 mg/kg.

*Lettuce*

In Canada, disulfoton may be used on lettuce as a band application at seeding at 1.1–2.2 kg ai/ha. In three US leaf lettuce trials where disulfoton was used at 2.2–2.9 kg ai/ha and the lettuce were harvested 50–59 days later, disulfoton residues were < 0.03, 0.10 and 0.56 mg/kg.

Three trials were insufficient for estimation of an HR.

In USA, disulfoton may be used on lettuce in the field as a soil injection or side dressing at 2.2 kg ai/ha with a PHI of 60 days. The US and Canadian GAPs were accepted as essentially equivalent.

In Mexico, disulfoton may be used on lettuce as a band or furrow application at seeding at 1 kg ai/ha. In eight US lettuce trials where disulfoton was used at 1.1–1.2 kg ai/ha and the lettuce were harvested 60–116 days later, disulfoton residues were < 0.03, < 0.05, < 0.05, < 0.05, 0.22, 0.58, 0.64 and 1.1 mg/kg.

From this use, the Meeting estimated an HR of 1.1 mg/kg for disulfoton on lettuce. The Meeting noted that an HR of 1.1 mg/kg for lettuce would be associated with IESTI values of 180% and 280% of the current ARfD (0.003 mg/kg).

None of the residue data relating to available alternative GAP suggests a lower maximum residue level to replace the current proposal of 1 mg/kg disulfoton on head and leaf lettuce.

## RECOMMENDATIONS

On the basis of the data from supervised trials, the Meeting concluded that the residue concentrations listed below are suitable for establishing MRLs and for assessing intakes.

Residue definition: *Sum of disulfoton, demeton-S and their sulphoxides and sulphones, expressed as disulfoton.*

| CCN | Commodity | MRL, mg/kg | Previous MRL mg/kg | STMR mg/kg | HR mg/kg |
|-----|-----------|-----------|--------------------|-----------|----------|
| VB 0404 | Cauliflower | 0.5 | 0.05 | 0.01 | 0.31 |

## DIETARY RISK ASSESSMENT

### Long-term intake

The estimates of long-term intake for disulfoton (ADI 0-0.0003 mg/kg bw) in 2002 for the five regional diets were 10–120% of the ADI[1]. The STMR of 0.01 mg/kg for cauliflower is unchanged, as a result the estimates of long-term intake are unchanged.

### Short-term intake

The International Estimated Short-Term Intakes (IESTI) was calculated for cauliflowers. An ARfD of 0.003 mg/kg bw has been established by the JMPR. The IESTI represented 180% and 280% of the ARfD for the general population and children respectively. The information provided to the JMPR precludes an estimate that the dietary intakes calculated for cauliflowers would be below the acute reference dose.

## REFERENCES

**Author, Date, Title, Institute, Report Reference, Document No.**

Andersch, I, 2006, Disulfoton (74). Registered Uses in Broccoli, Cabbage, Cauliflower and Lettuce, Residues from Supervised Trials, IESTI Calculations and Monitoring results. Bayer Crop Science AG, Monheim, Germany, Report No. M-2666618-01-1,

---

[1] WHO. 2002. Dietary exposure in relation to MRL setting. Codex Committee on Pesticide Residues. CCPR, 34th Session. Document CX/PR 02/03.

Anonymous, 2005, Summary of Good Agricultural Practices for pesticide uses. Disulfoton. Japan. Bayer Crop Science K.K. 1-6-5, Marunouchi, Chiyodaku, Tokyo, Table XI. 2.,

Anonymous, 2005, Residues data summary from supervised trials. Disulfoton. Japan. Bayer CropScience K.K. 1-6-5, Marunouchi, Chiyodaku, Tokyo, Table XI.3.,

Andersch, I, 2006, Disulfoton (74). Registered Uses in Broccoli, Cabbage, Cauliflower and Lettuce, Residues from Supervised Trials, IESTI Calculations and Monitoring results. Bayer Crop Science AG, Monheim, Germany, Report No. M-2666618-01-1,

# ENDOSULFAN (032)

*First draft prepared by Mr Bernard Declerq, Epinay sur Orge, France*

## EXPLANATION

Endosulfan is a synthetic cyclodiene non-systemic insecticide and acaricide with both contact and stomach activity. It is widely used in agriculture to control a range of insects and mites on a broad spectrum of crops. It has been evaluated several times by the JMPR, the initial evaluation for residues being in 1967 and the latest in 1993. The 1998 JMPR established an ADI and an acute reference dose for endosulfan of 0-0.006 mg/kg bw and 0.02 mg/kg bw respectively. It was listed under the periodic re-evaluation programme in the $36^{th}$ session of the CCPR for residue review by the 2006 JMPR.

The current Meeting received information from the manufacturer on physical and chemical properties, metabolism studies on plants and animals, environmental fate in soil, crop rotation, analytical methods, supervised trial data, processing studies, feeding studies, residues in food in commerce or at consumption and national maximum residues limits, residues in animal commodities as well as use patterns for Australia and The Netherlands.

## IDENTITY

| | |
|---|---|
| Common Name: | Endosulfan (BSI, ANSI and ISO approved) |
| Chemical Name: | |
| IUPAC: | 6,7,8,9,10,10-hexachloro-1,5,5ª,6,9,9ª-hexahydro-6,9-methano-2,4,3-benzo-dioxathiepin-3-oxide |
| CAS: | 6,9-methano-2,4,3-benzodioxathiepin,6,7,8,9,10,10-hexachloro-1,5,5ª,6,9,9ª-hexahydro-3-oxide |
| FAO Specification (including year of publication): | CP/228 |
| CAS registry number: | 115-29-7 |
| CIPAC number: | 89 |
| Synonyms and trade names: | See Names and other metabolites |
| Molecular formula: | $C_9H_6Cl_6O_3S$ |
| Molecular mass: | 406.96 g/mol |
| Structural Formula: | |

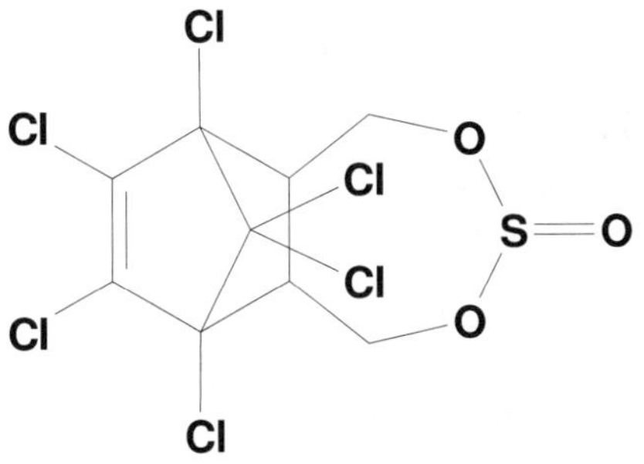

Endosulfan consists of 2 isomers that differ in the configuration of the 7-membered dioxothiepin-oxide ring. These isomers are known as alpha endosulfan and beta endosulfan. The ratio of alpha endosulfan and beta endosulfan is approximately 2:1.

## PHYSICAL AND CHEMICAL PROPERTIES

Pure active ingredient:

| Property | Result | Reference |
|---|---|---|
| Purity | > 99% endosulfan ratio 2.1 α endosulfan/β endosulfan | |
| Melting point | α endosulfan: 109.2°C, β endosulfan: 213.3 °C Mean 83.3 °C | Albrecht & Kappes 1974b, Smeykal H 2001a |
| Boiling point | 290 to 350°C. | Smeykal H 2001b, Röchling, Rexer, K and Maier 1990 |
| Relative density | 1.745 g / cm$^3$ at 20 °C | Albrecht, Kappes and Maier1974 |
| Vapour pressure | Mixture alpha, beta - endosulfan: 1.7 x 10$^{-3}$ Pa at 25 °C | OECD 104 |
| | α - endosulfan: 1.9 x 10$^{-3}$ Pa, β- endosulfan: 9.2 x 10$^{-5}$ Pa at 25 °C | Sarafin, R 1987 |
| Henry's law constant | α- endosulfan: 1.48 Pa x m$^3$ x mol$^{-1}$, β endosulfan: 0.07 Pa x m$^3$ x mol$^{-1}$ at 24 °C | Weller O, 1990b |
| Colour | Colourless. | Kappes A, 1974a |
| Physical state | Solid. | |
| Odour | Odourless | Kappes A, 1974b |
| UV/VIS IR, NMR, MS spectra | α - endosulfan | |
| | UV molecular extinction ε =7.00 x 10$^3$ (mol$^{-1}$ x cm$^{-1}$) at 212 nm | Wink O, 1985a |
| | IR wave number cm$^{-1}$: 3440 (OH, water), 2936 (C-H), 1605 (C=C), 1192 (S=O), 793 (C-Cl), 754 (C-C), 702 (S-O) | Sarafin R 1985b |
| | NMR : Multiplet 3.44 ppm, doublet-doublet 3.94 ppm and 4.77 ppm | Sarafin R 1985a |
| | β - endosulfan | |
| | UV molecular extinction ε =7.06 x 10$^3$ (mol$^{-1}$ x cm$^{-1}$) at 212 nm | Wink O, 1985b |
| | IR wave number cm-1: 3440 (OH, water), 2953 (C-H), 1607 C=C), 1194 (S=O),779(C-Cl), 745 (C-C), 691 (S-O) | Sarafin R 1985c |
| | NMR : Multiplet 3.15 ppm, doublet-multiplet 4.12 ppm, doublet 5.08 ppm | Sarafin 1985d |
| | endosulfan | |
| | MS: m/z 407, 339, 323, 307, 295, 277, 207, 195,159 | Sarafin R, & Winterscheidt, G 1985a, b |
| Solubility in water | α - endosulfan: 0.33 mg/L, β - endosulfan: 0.32 mg/L (pH 5, 22°C) | Sarafin, R. and Aßhauer, J.1987b, Görlitz, G.1990 |
| Solubility in organic solvents | n-hexane                           2.40 g/100 mL | OECD 116, Rexer K,1981, Huth G, 1996, Görlitz G, and Eyrich U.1986 |
| | ethanol                            6.50 g/100 mL | |
| | dichloromethane            > 20 g/100 mL | |
| | ethyl acetate, toluene     > 20 g/100 mL | |
| n-octanol/water coefficient | α - endosulfan 55500; log Pow= 4.74, | Sarafin R, Aßhauer J. 1987a; internal method. |
| | β - endosulfan 61400; log Pow=4.79 (pH 5.1, 22 °C). | |
| Dissociation in water | No indication of dissociation of endosulfan | Weller, O.1990a |
| Hydrolysis stability (25°C) | pH 5 > 200 days for α - endosulfan and β - endosulfan | Göerlitz,G and Rutz,U 1989 |
| | pH 7= 19 days for α - endosulfan and = 10.7 days for β - endosulfan. | |
| | pH 9= 0.26 and 0.17 days for α- endosulfan and β- endosulfan respectively | |
| | | |
| Estimated photochemical oxidative degradation | Reacts through photolysis (main product is endosulfan sulfate) | Paarlar, H. 1988 |

Technical material:

| Property | Result | Reference |
|---|---|---|
| Purity | 96.5%. | |
| Relative density | 1.745 g/cm$^3$ CIPAC 1, 1970 | Albrecht and Kappes K 1974 |
| Colour | Cream to tan mainly beige | Rexer K and Maier1990/43649 |
| Physical state | Flakes with tendency to agglomerate | Albrecht and Rexer K 1982/24344 |
| Odour | Odour like sulphur dioxide | Rexer K and Albrecht 1982/24333 |
| Flammability | | |
| Autoflammability | No spontaneous ignition up to 400°C. Internal method. | Rexer, K. and Albrech 1982c; Huth G 1996 |
| Flash point | Above 40°C | Rexer, K and Albrecht 1982b |
| Explosive properties | No potential for explosivity | Rexer, K and Albrecht 1982c |
| Oxidising properties | Not considered an oxidising agent. | Klais,O,Rexer,K,1995 |

## FORMULATIONS

The main formulations were EC at the concentration between 30 and 50%.(w/w);

WP formulations were between 30 and 50%.(w/w);

SC, CS, GR and powder formulations were also present in the market.

Endosulfan in EC and ULV formulations could be mixed with deltamethrin or dimethoate.

*Names and codes of metabolites*

| Common Name | Structural Formula | Company Code | Chemical Name |
|---|---|---|---|
| endosulfan (parent) | isomeric mixture of alpha- and beta-endosulfan in the ratio of approx. alpha 2-to beta −1 | AE F002671 (former code: Hoe 002671) | 6,7,8,9,10,10-hexachloro-1,5,5a,6,9,9a-hexahydro-6,9-methano-2,4,3-benzo-dioxathiepine-3-oxide (IUPAC) |
| alpha-endosulfan (asymmetric twist forms that only can be distinguished in the crystalline phase by X-ray spectrometry) (parent) | first twist form 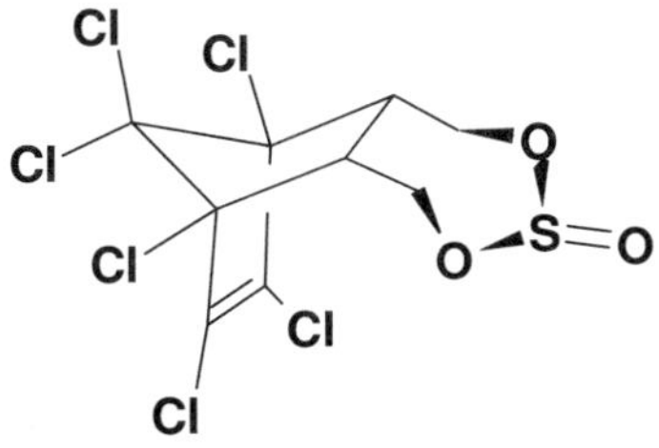<br><br>second twist form 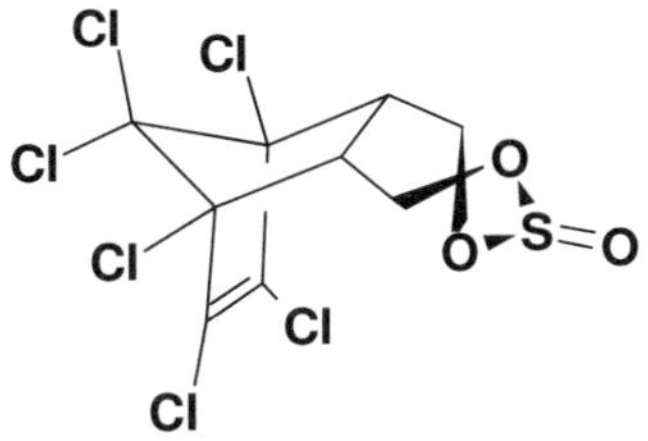<br> | AE F052618 (former code: Hoe 052618) | 6,7,8,9,10,10-hexachloro-1,5,5a,6,9,9a-hexahydro-6,9-methano-2,4,3-benzo-dioxathiepine-3-oxide-alpha-isomer (IUPAC) |
| beta-endosulfan (symmetric) (parent) | 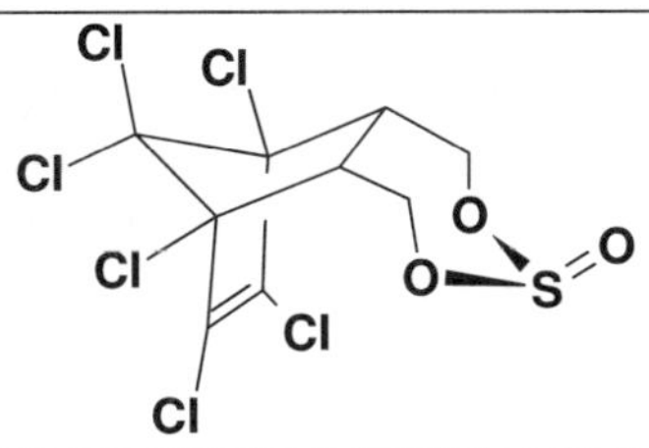 | AE F052619 (former code: Hoe 052619) | 6,7,8,9,10,10-hexachloro-1,5,5a,6,9,9a-hexahydro-6,9-methano-2,4,3-benzo-dioxathiepine-3-oxide-beta-isomer (IUPAC) |

| Common Name | Structural Formula | Company Code | Chemical Name |
|---|---|---|---|
| endosulfan sulfate<br><br>(plant, rat, cow, soil, natural water) | Cl Cl Cl Cl Cl Cl O O—S=O O | AE F051827<br>(former code: Hoe 051327) | 6,7,8,9,10,10-hexachloro-1,5,5a,6,9,9a-hexahydro-6,9-methano-2,4,3-benzo-dioxathiepine-3,3-dioxide (IUPAC) |
| endosulfan diol<br><br>(plant, rat, cow, surface water) | Cl Cl Cl Cl Cl Cl OH OH | AE F051329<br>(former code: Hoe 051329) | 1,4,5,6,7,7-hexachloro-bicyclo[2.2.1]hept-5-ene-2,3-dimethanol (IUPAC) |
| endosulfan ether<br><br>(rat, cow) | Cl Cl Cl Cl Cl Cl O | AE F051330<br>(former code: Hoe 051330) | 4,5,6,7,8,8-hexachloro-1,3,3a,4,7,7a-hexahydro-4,7-methano-isobenzofuran (IUPAC) |
| endosulfan hydroxy ether<br><br>(rat, cow) | Cl Cl Cl Cl Cl Cl O HO | AE F051326<br>(former code: Hoe 051326) | 4,5,6,7,8,8-hexachloro-1,3,3a,4,7,7a-hexahydro-4,7-methano-isobenzofuran-1-ol (IUPAC) |
| endosulfan lactone<br><br>(rat, cow) | Cl Cl Cl Cl Cl Cl O O | AE F051328<br>(former code: Hoe 051328) | 4,5,6,7,8,8-hexachloro-1,3,3a,4,7,7a-hexahydro-4,7-methano-isobenzofuran-1-one (IUPAC) |
| endosulfan hydroxyl carboxylic acid<br><br>(plant, surface water) | Cl Cl Cl Cl Cl Cl OH COOH | AE 0365278<br><br>(instable, therefore stored as Na salt, AE F114151) | 1,4,5,6,7,7-hexachloro-3-hydroxymethyl-bicyclo[2.2.1]hept-5-ene-2-carboxylic acid<br>and sodium 1,4,5,6,7,7-hexachloro-3-hydroxymethyl-bicyclo[2.2.1]hept-5-ene-2-carboxylate) (IUPAC) |

| Common Name | Structural Formula | Company Code | Chemical Name |
|---|---|---|---|
| endosulfan dihydroxy ether (rat) | 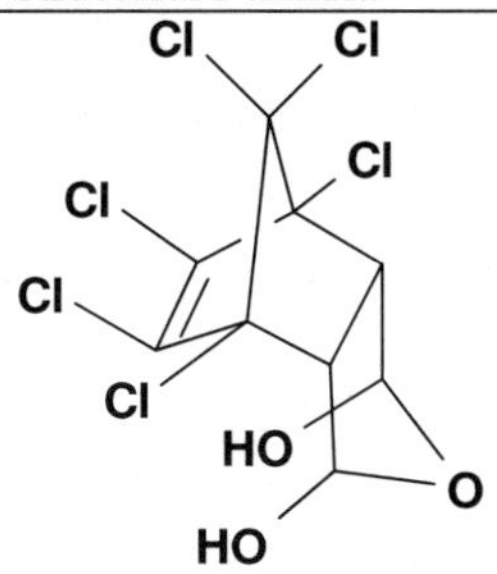 | AE 0035655 (intermediate, detected by MS of polar residues in the rat) | 4,5,6,7,8,8-hexachloro-1,3,3a,4,7,7a-hexahydro-4,7-methano-isobenzofuran-1,3-diol |
| endosulfan dihydroxy ether disulfate (rat) | 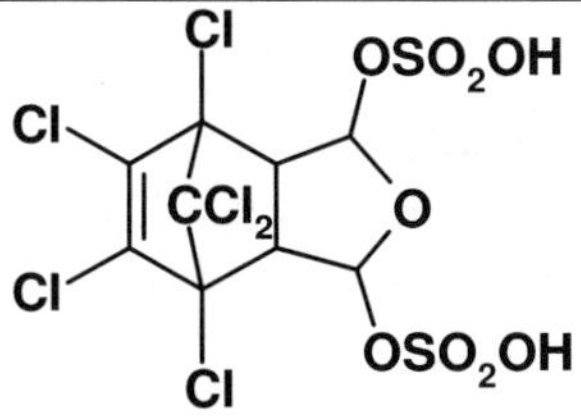 | - | 4,5,6,7,8,8-hexachloro-1,3,3a,4,7,7a-hexahydro-4,7-methano-isobenzofuran-1,3-diyl-bis(hydrogen sulfate) |
| endosulfan diol monosulfate (rat) | 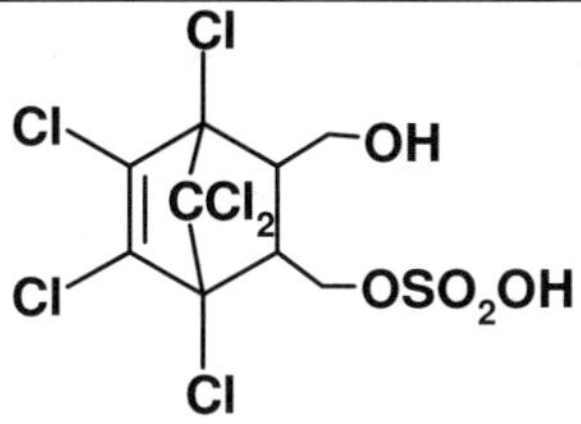 | - | [1,4,5,6,7,7-hexachloro-3-(hydroxymethyl)bicyclo[2.2.1]-hept-5-en-2-methyl]-hydrogen sulfate |
| endosulfan hydroxy ether sulfate (rat) | 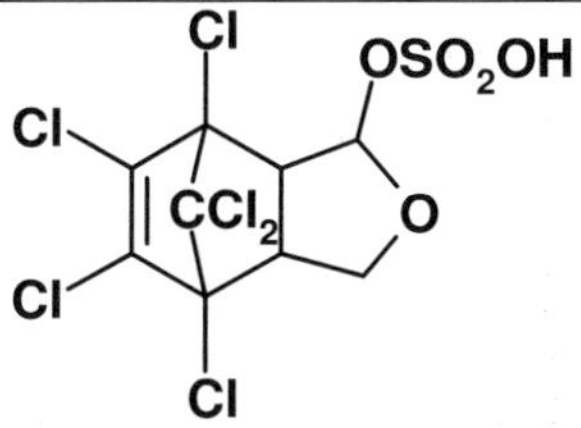 | - | 4,5,6,7,8,8-hexachloro-1,3,3a,4,7,7a-hexahydro-4,7-methano-isobenzofuran-1-yl-hydrogen sulfate |

## METABOLISM AND ENVIRONMENTAL FATE

The metabolism and distribution in livestock and plants of endosulfan was investigated using endosulfan labelled with [14]C as shown below or alternatively with the [14]C at the remaining carbon atoms of the 6-membered ring only.

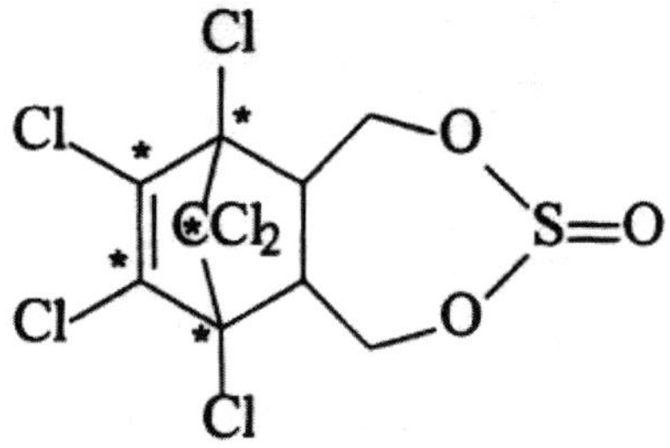

*position of radiolabel

### Animal metabolism

The Meeting received animal metabolism studies for endosulfan on rats, lactating cows, lactating sheep, and laying hens.

*Rats*

Studies on laboratory animal metabolism (rats) were evaluated by the WHO panel of the 1998 JMPR. Some studies not previously submitted are summarised below

Needham and Gutierrez-Giulianotti (1997) studied the metabolism of endosulfan by administration of a single oral dose of nominally either 1 or 6 mg $^{14}$C endosulfan/kg body weight (bw). The majority of the oral dose was excreted in the faeces (70–90%) and urine (9–20%) as polar metabolites. The highest tissue residues were found in the kidneys (1.3 and 1.4 mg/kg equivalent) and liver (0.2 and 0.3 mg/kg-equivalent) for male and female rats. The residues in fat were generally lower (0.02 and 0.1–0.2 mg/kg equivalent in males and females).

Later Needman (2001) investigated the nature of metabolites in faeces and urine. The sulfate conjugate of endosulfan dihydroxyether, endosulfan diol sulfate conjugate, endosulfan dihydroxyether, two isomers of endosulfan dihydroxyether sulfate conjugates and endosulfan dihydroxyether disulfate conjugate were identified in urine. Faeces contained endosulfan diol sulfate conjugate, hydroxyl endosulfan ether sulfate conjugate, two isomers of endosulfan dihydroxyether sulfate conjugates, and endosulfan dihydroxyether disulfate conjugate. These metabolites accounted for approximately 2.1–8.6% of the dose in the urine and a further 5.5–8.6% of the dose in the acetonitrile extract of the 0-24 hour faeces sample.

In a toxicokinetics study (Needham *et al.*, 1998) in the rat following repeated daily oral administration of 1 mg/kg/bw for 28 days, endosulfan sulfate was the major component in the fat. Metabolism of endosulfan involved either sulfoxidation to endosulfan-sulfate, a fat-soluble metabolite, followed by desulfatation to the diol, or direct hydrolysis to the diol followed by oxidation to the ether, the hydroxy-ether, the dihydroxy ether, and to the main metabolite in urine and faeces, the lactone.

The metabolites identified in this study were: alpha and beta endosulfan, endosulfan sulfate, endosulfan diol, endosulfan ether, endosulfan diol sulfate, endosulfan dihydroxy ether sulfate, endosulfan hydroxy ether, endosulfan lactone, endosulfan dihydroxy ether disulfate and endosulfan dihydroxy ether.

*Lactating cows*

The metabolism and distribution of $^{14}$C endosulfan (98% purity; ratio alpha:beta = 2) was investigated by Leah and Reynolds (1996) in a lactating cow (450 kg) following repeated oral administration of $^{14}$C-endosulfan at a mean daily dose of 288 mg/kg, equivalent to 0.64 mg/kg/bw/day for five consecutive days.

The dose was equivalent to 22 ppm in the diet. Urine and faeces were collected daily; milk was collected twice daily. At slaughter, liver, kidney, renal fat, subcutaneous fat and muscle (psoas and hindquarter) were sampled and the residues identified and quantified.

In milk, the TRR was detectable after 6 hours, reaching a maximum of 0.17 mg/kg eq 102 hours after the first dose.

TRR at 120 hours after initial dose and approximately 22 hours after final dose were generally low with the exception of the liver where the TRR were found to be 3.57 mg/kg eq tissue. TRR in fat were as follows: omental fat (1.28 mg/kg eq), renal fat (0.84 mg/kg eq) and subcutaneous fat (0.305 mg/kg eq). Residue levels in kidney (0.785 mg/kg eq) were comparable with levels in fat; residue levels in muscle (0.052 mg/kg eq) were lower.

For kidney 3.5% of TRR was extracted. A further 52% of the total $^{14}$C residue (unextracted) was released by successive enzymatic incubations and mild acid/ base hydrolysis. Analysis was by TLC and HPLC. The final unextracted radioactivity accounted for 1.3%.

A similar fractionation for liver indicated that only 30% of TRR was associated with lipids and that the majority of the residue was associated with more polar fractions such as proteins (24%) and the perchloric acid fraction (32%). This suggests that a large proportion of the residue is hydrophilic in nature.

Overall 38% of TRR in liver was extractable. A further 60% of the total $^{14}$C residue (unextracted) was released by successive enzymatic incubations and mild acid/base hydrolysis. The final unextracted radioactivity was approximately 2.3%.

For fat (omental, renal and subcutaneous) and milk TRR was extracted by hexane. No further extraction was performed.

For muscle 69% of TRR was extracted with hexane. The unextracted radioactivity was released by enzymatic incubation and showed very low level of radioactivity which was not identified.

Table 1. Distribution and characterisation of $^{14}$C in tissues and milk of cow.

| | Tissue | | | | | | |
|---|---|---|---|---|---|---|---|
| | Liver | Kidney | Muscle (psoas) | Omental fat | Renal fat | Subcut. fat | Milk (54 hrs) |
| TRR (mg/kg eq) | 3.572 | 0.785 | 0.052 | 1.278 | 0.840 | 0.305 | 0.147 |
| % Extracted | 37.98 | 43.49 | 69.24 | 95.70 | 96.46 | 81.34 | 95.32 |
| % Released[1] | 53.87 | 52.24 | 13.86 | - | - | - | |
| % Identified | 87.80 | 86.03 | 65.87 | 83.64 | 83.83 | 67.79 | 88.57 |
| % α-Endosulfan | 2.72 | - | 15.14 | - | - | - | - |
| % β-Endosulfan | - | 3.08 | - | - | - | - | - |
| % Endosulfan sulfate ( mg/kg) | 27.16 (0.97) | 12.58 (0.10) | 50.73 (0.026) | 82.08 (1.05) | 83.83 (0.70) | 67.79 (0.21) | 88.57 (0.13) |
| % Endosulfan lactone | 9.16 | 6.86 | - | - | - | - | |
| % Endosulfan diol | 6.48 | 4.96 | - | - | - | - | - |
| % Hydroxy Endosulfan ether | - | 4.77 | - | - | - | - | |
| % Endosulfan ether | 6.76 | 2.13 | - | - | - | - | - |
| % Polar | 35.49 | 51.65 | - | 1.56 | - | - | - |
| % BLA | 0 | 0 | 15.65 | 0.32 | 1.86 | 0 | 3.22 |
| unextracted (final) | 2.23 | 1.27 | 15.8 | 4.30 | 3.50 | 19.75 | 4.68 |
| loss | 9.97 | 10.49 | 2.68 | 7.5 | 10.77 | 3.75 | 3.50 |

1) = By enzyme catalysed or enzyme catalysed + acid/base hydrolysis

Following administration of [$^{14}$C]-endosulfan at a dose rate equivalent to 22 ppm in the diet for five consecutive days, equivalent to 0.64 mg/kg body weight/day, radioactivity was detected in all edible tissues at between 0.052 and 3.57 mg/kg eq. The major metabolite identified in all tissues was endosulfan sulfate at levels ranging from 68 to 84% in fat and up to 88% in milk. Endosulfan lactone was found in kidney and liver tissue, indicating that the endosulfan is readily cleaved following administration to a dairy cow.

*Lactating sheep*

The metabolism and distribution of endosulfan was investigated in studies by Gorbach, *et al.* in 1968 using $^{14}$C endosulfan (98% purity alpha/ beta=2 labelled in methylene group) and by Gorbach in 1965 in sheep following repeated oral administration of cold-endosulfan.

In the first study, a single dose of 0.3 mg/kg of $^{14}$C endosulfan was administered to two East-Friesian milk sheep each weighing approximately 50 kg. Milk samples were drawn twice daily every morning and afternoon (for up to 22 days); urine and faeces samples were collected once daily.

Approximately 90% of the administered $^{14}$C material was excreted in the urine and faeces. Endosulfan-diol and endosulfan-hydroxy-ether, but not parent, were found in urine while endosulfan was the majority of the residue in faeces.

In milk 1% of the TRR was found, with the entire radioactivity appearing to be endosulfan sulfate. The highest concentration of radioactivity found in the milk was 0.15 mg/kg eq at 24 hours after administration. Radioanalysis showed that 88% of the $^{14}$C in milk was present in cream.

In the second study (cold study) three Merino sheep, each weighing approximately 50 kg, were given a gelatine capsule containing 15 mg cold endosulfan daily for 26 days (0.3 mg/bw/day). Endosulfan sulfate was the primary residue found in milk (0.02–0.1 mg/kg).

*Laying hens*

Reynolds (1996) investigated the metabolism and distribution of $^{14}$C-endosulfan in laying hens. Six laying hens were orally dosed with [$^{14}$C]-endosulfan (98% purity; ratio alpha/beta=2) at 1.36 mg/animal/day (11 ppm in diet) for twelve consecutive days. At slaughter, liver, abdominal and subcutaneous fat, skin, skeletal muscle and undeveloped eggs were removed for determination of the distribution and magnitude of [$^{14}$C]-endosulfan related residues.

In egg yolks and whites TRR was detectable within 48 hours of the administration of the initial dose, with radioactivity in egg yolks continuing to rise until plateau at day 10 (0.85 mg/kg eq). The residues in egg whites were an order of magnitude lower with a maximum concentration of 0.013 mg/kg eq seen on day 6 of dosing. In undeveloped eggs the mean concentration of endosulfan-derived residues was 0.77 mg/kg eq.

TRR in tissues were generally low with the highest residues seen in the subcutaneous and abdominal fat at 0.88 and 0.97 mg/kg eq, respectively. The residues in skin and liver were slightly lower at 0.69 and 0.47 mg /kg eq, respectively. The value for muscle was much lower at 0.028 mg/kg eq.

Following the first dose of [$^{14}$C]-endosulfan, elimination was fairly rapid with 51% of the administered dose recovered within the first twenty-four hours of dosing. The overall mean daily recovery was 87% over the twelve-day study period indicating that the majority of the dose was excreted.

In muscle 64% of the TRR was extractable by hexane following partition with acetonitrile and was identified and quantified. Some 36% of the total [$^{14}$C] residue in muscle was unextractable and no further investigation was conducted.

The level of endosulfan and its metabolites in egg whites was low reaching a plateau of 0.013 mg/kg eq by day 6 of dosing. About 65% of the TRR (0.008 mg/kg eq) was organoextracted (acetonitrile). No further analysis was performed.

In abdominal and subcutaneous fat the TRR after extraction was quantified (97% and 98%, respectively) and identified. The loss and unextracted radioactivity were low (approximately 1–2%).

In the case of skin, the TRR after extraction was 94%, of which 90% was identified. The loss and unextracted radioactivity were approximately 4–6%.

Over 90% of the total [$^{14}$C] residue in liver was extractable. Subsequent enzymatic digestion permitted further quantitation of the nature of the residues.

Almost 60% of the radioactivity in liver was identified as endosulfan and its metabolites, of which the major metabolite was identified as endosulfan sulfate (46%). Approximately 23% of the total 14C residue was polar in nature. Approximately 9% was accounted for in procedural losses due partly to the particulate nature of some of the samples and another 9% of the radioactivity remained unextractable. For egg yolks 92% of TRR was extractable. Procedural losses and unextractable radioactivity was approximately 8% for each.

Isolation and identification of the residues in the tissues and excreta is summarised below. The results were the mean value of six birds.

Table 2. Distribution of $^{14}C$ endosulfan in tissues; eggs and excreta of hens.

| % Identified / characterised | | | | | | | | | |
|---|---|---|---|---|---|---|---|---|---|
| Tissue | TRR (mg/kg eq) | % Extracted | % α endosulfan | % β endosulfan | % endosulfan sulfate | % endo-sulfan lactone | % endo-sulfan diol | % polar | % BLA* |
| Egg yolks | 0.853 | 92.17[a] | 4.74 | 1.32 | 46.38 | 1.82 | - | 20.18[b] | 0.89 |
| Egg whites | 0.013 | 64.91 | - | - | - | - | - | - | 64.91 |
| Skin | 0.651 | 94.43 | 11.66 | 4.79 | 51.32 | 4.50 | - | 17.55 | 0.52 |
| Subcut. fat | 0.875 | 98.21 | 16.15 | 8.90 | 61.10 | 5.03 | - | 4.71 | 0.51 |
| Abdom. fat | 0.974 | 97.04 | 16.77 | 7.80 | 65.45 | 4.99 | - | - | 0.56 |
| Liver | 0.466 | 91.41 | 0.99 | 1.60 | 45.62 | 6.27 | 4.25 | 23.27[c] | - |
| Muscle | 0.028 | 64.40 | 6.52 | 4.40 | 35.83 | 3.51 | - | 4.71 | 8-15 |
| Excreta | NA | 60.62 | 4.90 | 6.25 | 0.79 | - | 1.43 | 46.61 | - |

a = Each polar component was > 0.03 ppm

b = Hydrolysed to identified metabolites

c = Unknown metabolite less polar than parent compound 6.03% total

NA = Not applicable

*BLA = Below level of analysis

Following dosing of [$^{14}C$]-endosulfan at a dose rate equivalent to 11 ppm in the diet for 12 consecutive days, endosulfan-derived radioactivity was detectable in all edible tissues at between 0.013 and 0.97 mg/kg eq. The major metabolite identified in tissues (excluding egg white) was endosulfan sulfate, along with a small percentage of unchanged α - and β- endosulfan as well as some products of hydrolysis and oxidation of endosulfan, namely endosulfan diol (in liver) and endosulfan lactone.

The primary residues found in animal tissues were the parent, endosulfan, both alpha and beta isomers, and to a larger extent, endosulfan sulfate. The metabolism studies are consistent with the view that the parent is converted to the sulfate *in situ*, the latter of which is more likely to accumulate in tissues than the parent molecule. While liver seems to be the primary target organ for metabolism, the above residue components are clearly present in significant amounts in fat. It is likely that the high presence of these metabolites in fat is consistent with endosulfan being a fat-soluble pesticide.

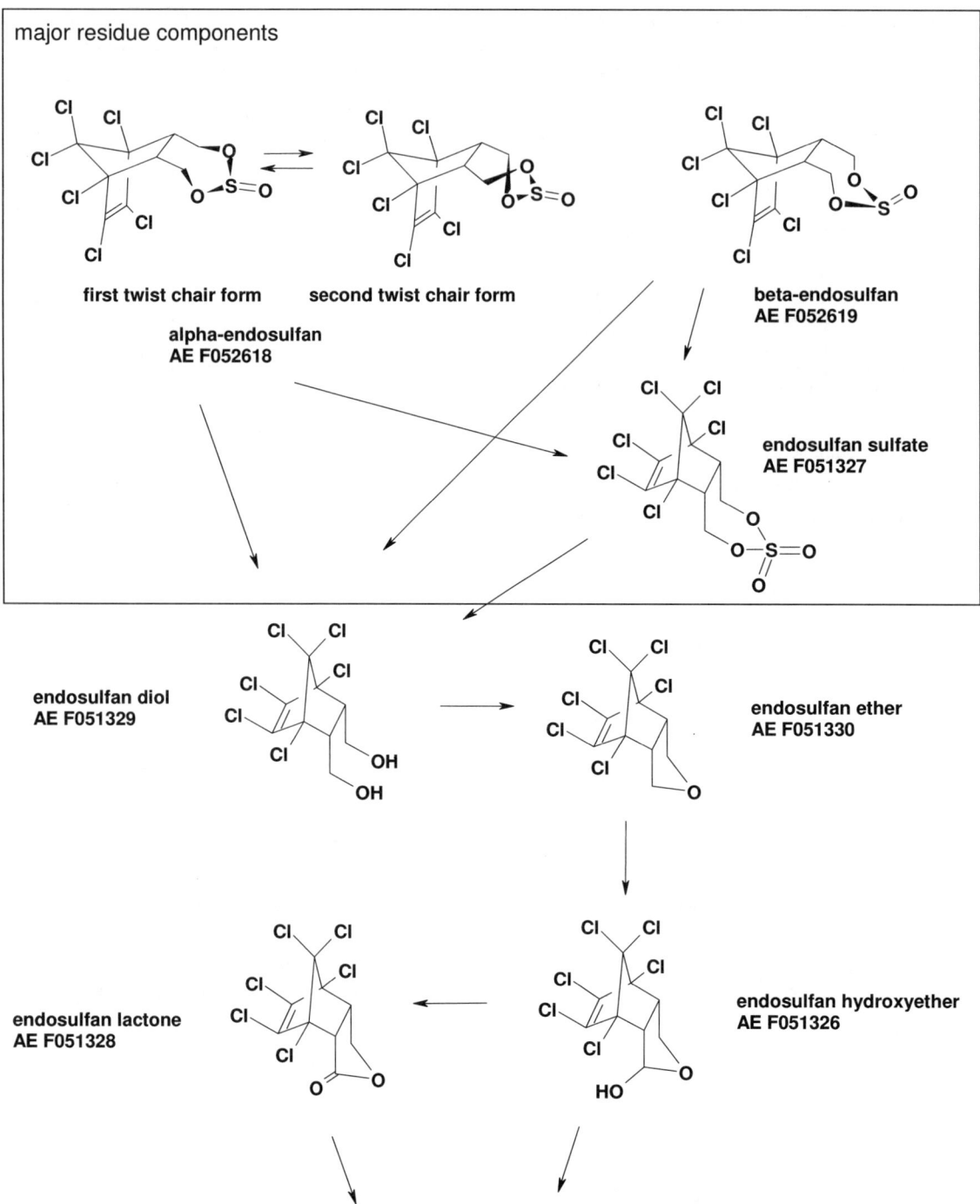

Figure 1. Metabolic pathway of endosulfan in the lactating cow.

## Plant metabolism

The Meeting received information on the fate of endosulfan after application to tomato, cucumber, apple, sugar beet and soybean.

*Tomato*

The study was performed by Buerkle *et al.* ( 1990). Young tomato plants *(Lycopersicon lycopersicum,* variety: *Fruchta)* with green fruit were treated three times with 6,7,8,9,10-U-$^{14}$C-labelled endosulfan (98% purity; alpha:beta = 2 ) at intervals of 7 days, each time at an application rate equivalent to 635 g ai/ha. The active substance was formulated as an emulsifiable concentrate (35EC) with an actual concentration of 32.9% (w/v).

After treatment, the plants were grown outdoors, but protected from rain by a glass roof. Leaf and fruit were sampled 8, 13, 21, 27, 42, and 48 days after the last treatment. In addition, leaf samples were taken one or two days after each treatment. Samples were extracted and quantified by liquid scintillation counting (LSC). Identification was achieved through the use of high performance liquid chromatography (HPLC) using co-eluted reference standards and HPLC with mass spectrometric detection (HPLC/MS).

The total radioactive residues decreased from 0.35 to 0.03 mg/kg eq in the fruit (day 8 to 48 after the last treatment) and 27 to 13 mg/kg eq in the leaf (day 2 to 48). The isomeric ratio ($\alpha/\beta$-endosulfan) in the leaf rinsate decreased from the original alpha 2/ beta 1 to alpha 0.54/ beta 1, two days after the third application.

Table 3. Characterisation of 6, 7, 8, 9, 10-u-$^{14}$C-endosulfan in% of TRR after treatment at 0.635 kg ai/ha on tomato and identification of metabolites.

| Plant part | Days after the 3rd treatment | Alpha-endosulfan | Beta-endosulfan | endosulfan sulfate | endosulfan diol | Polar metabolite fraction | Non-extracted |
|---|---|---|---|---|---|---|---|
| Leaves | 2 | 14.4 | 26.6 | 15.05 | 0.3 | 23.5 | 8.8 |
| | 8 | 12.2 | 18.2 | 21.7 | - | 26.8 | 8.5 |
| | 13 | 0.9 | 1.8 | 10.8 | - | 46.4 | 7.2 |
| | 21 | 2.3 | 6.3 | 16.7 | 0.6 | 46.1 | 15.0 |
| | 42 | 1.0 | 4.3 | 12.8 | 0.3 | 46.0 | 16.3 |
| | 48 | 1.5 | 5.2 | 11.2 | - | 46.5 | 26.4 |
| Fruit | 27* | 75 (a + b) isomer | | 15 | | | 6.8 |

* Fruit analysis could not be conducted with samples at later dates due to a significant decrease of the total residues at later sampling intervals.

The polar metabolite fraction extracted from leaves contained predominantly endosulfan diol as the aglycone, which was conjugated as the sulfate (releasable with arylsulfatase).

Following repeated application of endosulfan, approximately 90% of the total radioactive residues could be extracted from tomato fruit and consisted of the parent isomers, $\alpha$- and $\beta$-endosulfan and the metabolite endosulfan sulfate. These components therefore have to be considered as the relevant endosulfan residues in tomatoes. The leaf material contained trace amounts of free and considerable amounts of conjugated endosulfan diol.

*Apple*

A study on apple trees *(Malus sylvestris var. domestica,* variety: *Elstar)* was performed by Schwab (1995). A young apple tree was treated with 5a, 9a - $^{14}$C-labelled formulated endosulfan (98% purity; alpha:beta =2 ) at a rate which corresponded to 1.5 kg ai/ha. The active substance was formulated as a 35EC. Thirteen (13) near-mature apples were present on the tree. The tree was protected from rain by a glass roof. Two to five apples and several leaves were sampled at 0, 7, 14, and 21 days after treatment. The apples were first rinsed with acetone and then extracted. Leaves of the last sampling date were extracted as for the apples, but without prior rinsing. Quantitation of radioactivity was achieved by LSC and identification conducted by HPLC and TLC using co-chromatographed reference standards and gas chromatography with mass spectrometric detection (GC/MS).

The total radioactive residues decreased from 81 to 25 mg/kg eq in the leaves from day 0 to day 21 and varied in the range 0.44–1.37 mg/kg eq in the apples but remained stable. The ratio of apple extract/rinse increased from about 1:1 (day 0) to 3:1 (day 27). The composition of residues in leaves and fruits are shown in Table 4.

Table 4. Fate of characterisation of 5a, 9a, [14]C-endosulfan in% of TRR after treatment at 1.5 kg ai/ha on apple and identification of metabolites.

| Plant part | Days after treatment | Alpha-endosulfan | Beta-endosulfan | endosulfan sulfate | endosulfan diol | Non extracted |
|---|---|---|---|---|---|---|
| Leaves | 21 | 7.6 | 28.3 | 49.6 | 0.9 | 9.9 |
| Apples* | 0 | 54.3 | 43.1 | - | - | 3.8 |
|  | 7 | 49.7 | 44.0 | 0.9 | - | 4.3 |
|  | 14 | 47.9 | 43.4 | 1.5 | - | 2.9 |
|  | 21 | 50.7 | 43.1 | 1.5 | - | 2.0 |

* The corresponding components in rinse and extract are given after summation.

Following application of endosulfan approximately 90% of the TRR could be extracted from the apples. These residues consisted almost exclusively of the parent isomers, α and β-endosulfan and to a very low extent the metabolite endosulfan sulfate. These findings suggest only the parent endosulfan being the relevant residue in the edible portion. In the leaves, endosulfan sulfate occurred as the major metabolite accounting for about 50% of the TRR. Only a trace of endosulfan diol was detected.

*Cucumber*

A study on cucumbers *(Cucumis sativus,* variety: *Melani* F,) was performed by Buerkle in 1995. Small fruits were treated three times with 5a,9a-[14]C -labelled endosulfan (98% purity; alpha:beta = 2 ) at intervals of 7 days, each time at an application rate equivalent to 530 g ai/ha (470–537 g ai/ha; of a 350EC). After treatment, the plants were grown outdoors and protected from rain by a glass roof. Leaves and fruit were sampled 0, 3, 7, and 14 days after the last treatment. The sample analysis began no later than 3 weeks after sampling. Both the leaf and fruit samples were macerated and extracted without prior rinsing. The resulting organic extract was analysed for non-polar residue components. The aqueous phase was analysed for conjugated metabolites. Quantitation of radioactivity was achieved by LSC and identification conducted by HPLC and TLC using co-chromatographed reference standards and gas chromatography with mass spectrometric detection (GC/MS).

The total radioactive residues in the leaves decreased from 185 mg /kg eq to 52 mg/kg from 0 to 14 days after the last treatment. The corresponding levels in the fruit decreased from 0.23 to 0.18 mg/kg eq. The composition of residues was determined at the last sampling time, since the metabolism was extensive at this date. The residues in cucumbers following 14 days after the third treatment with endosulfan are presented in Table 5.

Table 5. Identification of metabolites resulting from 5a, 9a-[14]C-endosulfan in% of TRR and mg/kg after treatment at 530g ai/ha on cucumber and a PHI of 14 days.

| Components | mg/kg | % of TRR |
|---|---|---|
| α -endosulfan | 0.026 | 14.5 |
| β -endosulfan | 0.026 | 14.6 |
| endosulfan sulfate | 0.038 | 21.4 |
| Sum of 3 non-polar fractions | 0.017 | 9.5 |
| Non-polar fraction after hydrolysis* | 0.005 | 2.7 |
| Sum of 2 polar fractions after hydrolysis | 0.029 | 15.9 |
| Non-extracted | 0.020 | 11.4 |
| Loss during work-up | 0.018 | 10.2 |
| Sum | 0.180 | 100 |

* This fraction is assumed to contain the aglycones after cleavage of the conjugates

Following three treatments with endosulfan about 75% of the total radioactive residues could be extracted from the cucumbers. The major components α- and β endosulfan and endosulfan sulfate contributed approximately 50% of the total radioactive residues. The portion of conjugated metabolites was negligible, since hydrolysis of the polar fraction did not release a significant portion of apolar compounds. Several small fractions with different polarity were detected, each of them amounting to < 0.05 mg /kg eq. In contrast, the leaves contained 9.5% endosulfan diol and 24% hydroxy endosulfan carboxylic acid, mostly conjugated as glycosides, besides the previously mentioned components in cucumbers. The proportion of non-extracted TRR was low (about 10% in cucumber and 17% in the leaves).

*Sugar beet*

A study in sugar beet was performed by Selzer in 2001. The nature of residues, following treatment with (6,7,8,9,10 $^{14}$C)endosulfan (98% purity; alpha:beta =2), was investigated.

The test item was applied as an emulsifiable concentrate, containing 33% ai (w/w). Plants were treated twice at a 21 day interval by spraying at a rate of 630 g ai/ha at each application. At both applications the plants were in the growth stage 39 (BBCH code). Plant samples were taken shortly after the 1$^{st}$ and 2$^{nd}$ applications (14/21 days after 1st application) and at maturity (4 weeks after the 2$^{nd}$ application).

Plant material at maturity was divided into leaves without leaf base ("leaves"), leaf base, and tap root with swollen stem base without leaf base ("roots") prior to analysis. The residues in leaves and roots were investigated separately. Leaves were rinsed with acetonitrile/water. Aliquots of the rinsed leaves and root samples were immediately analysed for TRR and for the composition of the residues by extraction followed by radio-HPLC. Representative samples were investigated with TLC as a second independent method.

Analysis of mature plants revealed that the concentration of radioactivity was highest in leaves (5.7 mg/kg eq), whereas in roots only very low residues were found (0.09 mg/kg eq), indicating only slight translocation from leaves to roots.

In leaves, more than 93% of TRR were extractable. In total, 52% of the TRR was identified in the leaves. A further 33% of the TRR was characterised as polar radioactivity, of which approximately 8% of TRR proved to be conjugates of endosulfan diol. By a stepwise treatment of the non-extractable residues (6.7% of TRR) with enzymes and acid/alkali, most of this residue (4% of TRR) could be released.

In roots, 93% of TRR (0.081 mg/kg eq) was extractable leaving 6.6% of TRR (0.006 mg/kg eq) as non-extractable. Re-extraction of the extract with dichloromethane resulted in the partition between the aqueous phase which was not further analysed due to the low amount of radioactivity (9.2% of TRR, 0.008 mg/kg eq) and the organic phase (79.2% of TRR, 0.069 mg/kg eq).

The organo-soluble residues in roots were almost completely identified (75% of TRR) with the exception of a minor metabolite (3.9% of TRR, 0.003 mg /kg eq) which was suggested to be de-chlorinated endosulfan sulfate by HPLC-MS.

Alpha endosulfan, beta endosulfan and endosulfan sulfate were the major radioactivity components in all plant parts. Endosulfan sulfate was the major metabolite of the extractable radioactivity in all cases.

The following summary table gives an overview on the total radioactive residues in leaves 21 days after the 1st treatment (representative for single application) and also of the TRR in leaves and roots at the final harvest 28 days after the 2nd application.

Table 6. Identification of metabolites resulting from [6,7,8,9,10-$^{14}$C]-endosulfan treatment at 630g ai/ha on sugar beet.

| | % of TRR | Concentration [mg/kg eq] |
|---|---|---|
| Leaves (without leaf base), 21 days after 1st application, intermediate sample | | |
| | 100 | 2.02 |
| Rinsing | 16.29 | 0.33 |
| HPLC Analysis | | |

| | | |
|---|---|---|
| Endosulfan diol | 0.18 | < 0.01 (0.004) |
| β-endosulfan | 8.03 | 0.16 |
| α-endosulfan | 4.54 | 0.09 |
| Extractable | 67.46 | 1.36 |
| (aqueous phase) Endosulfan diol | 8.23 | 0.17 |
| Endosulfan sulfate | 21.72 | 0.44 |
| β-endosulfan | 3.96 | e0.08 |
| α-endosulfan | 3.58 | 0.07 |
| Total sum identified | 50.24 | 1.02 |
| Non-extractable residues | 16.25 | 0.33 |
| Releasable by enzyme/ NaOH treatment | 5.64 | 0.11 |
| Leaves (without leaf base), 28 days after 2nd application, final harvest | | |
| | 100 | 5.73 |
| Rinsing | 8.72 | 0.50 |
| Endosulfan diol | 0.20 | 0.01 |
| Endosulfan sulfate | 0.25 | 0.01 |
| β-endosulfan | 4.70 | 0.27 |
| α-endosulfan | 1.95 | 0.11 |
| Extractable | 84.60 | 4.85 |
| Aqueous phase | 32.73 | 1.87 |
| HPLC Analysis after hydrolysis of the aqueous phase | | |
| Endosulfan diol AEF051329* | 8.11 | 0.46 |
| Organic phase | 46.68 | 2.67 |
| Endosulfan diol | 0.91 | 0.05 |
| Endosulfan sulfate | 33.09 | 1.89 |
| β-endosulfan | 6.90 | 0.39 |
| α-endosulfan | 3.77 | 0.21 |
| Total sum identified | 51.77 | 2.96 |
| Non-extractable residues | 6.69 | 0.38 |
| Releasable by enzyme/ NaOH treatment | 3.97 | 0.23 |
| * released by acid hydrolysis indicating acid labile conjugates of Endosulfan diol in the aqueous phase; this value is not included in the total sum identified in the table | | |
| Roots (beets), 28 days after 2nd application, final harvest | | |
| | % of TRR | Concentration [mg/kg eq] |
| | 100 | 0.087 |
| Extractable | 93.38 | 0.081 |
| Aqueous phase | 9.19 | 0.006 |
| Organic phase | 79.15 | 0.069 |
| HPLC Analysis | | |
| Endosulfan sulfate | 59.65 | 0.052 |
| β-endosulfan | 4.06 | 0.004 |
| α-endosulfan | 11.60 | 0.01 |
| Total sum identified | 75.32 | 0.066 |
| Non-extractable residues | 6.62 | 0.006 |

## Soybean

A study in soybeans was performed by Mislankar and Tull in 2003. The soybeans were treated with two applications of 530 g ai/ha, equivalent to the annual maximum of 1060 g ai/ha (6,7,8,9,10 [14]C) labelled as previously described (alpha:beta =2). Applications were made at forage stage 61 days before harvest and at hay stage 38 days before harvest.

Soybean (seed) samples were analysed in duplicate at maturity to determine the metabolic profile of the endosulfan. Soybeans (forage and hay) were also analysed at intermediate time points to establish the trend in metabolism. These time points included Day 0 (immediately post-treatment at R5, forage stage) and day 23 (prior to the second application at R6.5, hay stage).

The residue in the day zero soybean forage was recovered by acetonitrile surface wash and acetonitrile extraction. Residue remaining in the extracted fibre was measured by combustion. Residues in hay and beans were extracted; the remaining fibre was combusted to determine non-extractable radioactive residue. Metabolites in the extracted component were identified by retention time comparison with authentic standards. The identities were confirmed by MS.

The total residue in forage samples at day 0 was 21.6 mg/kg eq, but after 23 days (prior to second application) this had declined to 0.54 mg/kg eq. The residue at final harvest in beans (Day 61) was 0.47 mg/kg eq.

The majority of residue at all time points was solvent extractable. In forage at day 0, the residue remained principally on the leaf surface and 16 mg/kg eq was recovered in the wash. The rest (4.95 mg/kg eq) was extractable with acetonitrile, for a total of 21.2 mg/kg eq.

In hay samples 87.0% of the TRR was extractable with acetonitrile (acetonitrile:water = 80:20). Similarly, in beans, 95.0% of the TRR was extractable with acetonitrile.

Table 7. Distribution of $^{14}$C-endosulfan as% of TRR and mg/kg eq in the different parts of the plant.

| Harvest time | Total residue (mg/kg eq) | Surface Wash Residue | | Extractable Residue | | Non extractable residue | |
|---|---|---|---|---|---|---|---|
| | | % TRR | mg/kg eq | % TRR | mg/kg eq | % TRR | mg/kg eq |
| Day 0 Forage | 21.5 | 75.4 | 16.3 | 23.0 | 4.9 | 1.6 | 0.34 |
| Day 23 Hay | 0.54 | NA | NA | 87.0 | 0.47 | 13.1 | 0.068 |
| Day 61 Beans | 0.47 | NA | NA | 94.5 | 0.45 | 5.6 | 0.026 |

Table 8. Characterisation of 6, 7, 8, 9, 10-u-$^{14}$C-endosulfan in% of the TRR after treatment at 0.53 kg ai/ha on soybean.

| Plant part | Days after the last treatment | extractable | Alpha-endosulfan | Beta-endosulfan | endosulfan sulfate | total identified | Non-extracted |
|---|---|---|---|---|---|---|---|
| forage | 0 | 98.4 | 61.2 | 36.8 | 0.3 | 98.3 | 1.6 |
| hay | 23 | 87 | 1.3 | 4 | 51.2 | 56.5 * | 13.1 |
| beans | 61 | 94.5 | 1.5 | 5.3 | 78.4 | 85.2 | 5.6 |

*) Remainder of extracted material consisted of multiple water soluble components, none of which accounted for more than 0.05 ppm or 10% of TRR.

Non-extracted radioactivity was very low in forage and beans and it accounted for a maximum of 2% and 6.0%, respectively. In one hay sample non-extracted radioactivity accounted for 13.1% of the TRR. This was further subjected to hydrolysis with strong acid and base. The non-extracted radioactivity in hay accounted for 2% of the TRR.

The predominant residues found in plant material examined in these metabolism studies include parent endosulfan, both alpha and beta isomers, as well as endosulfan sulfate.

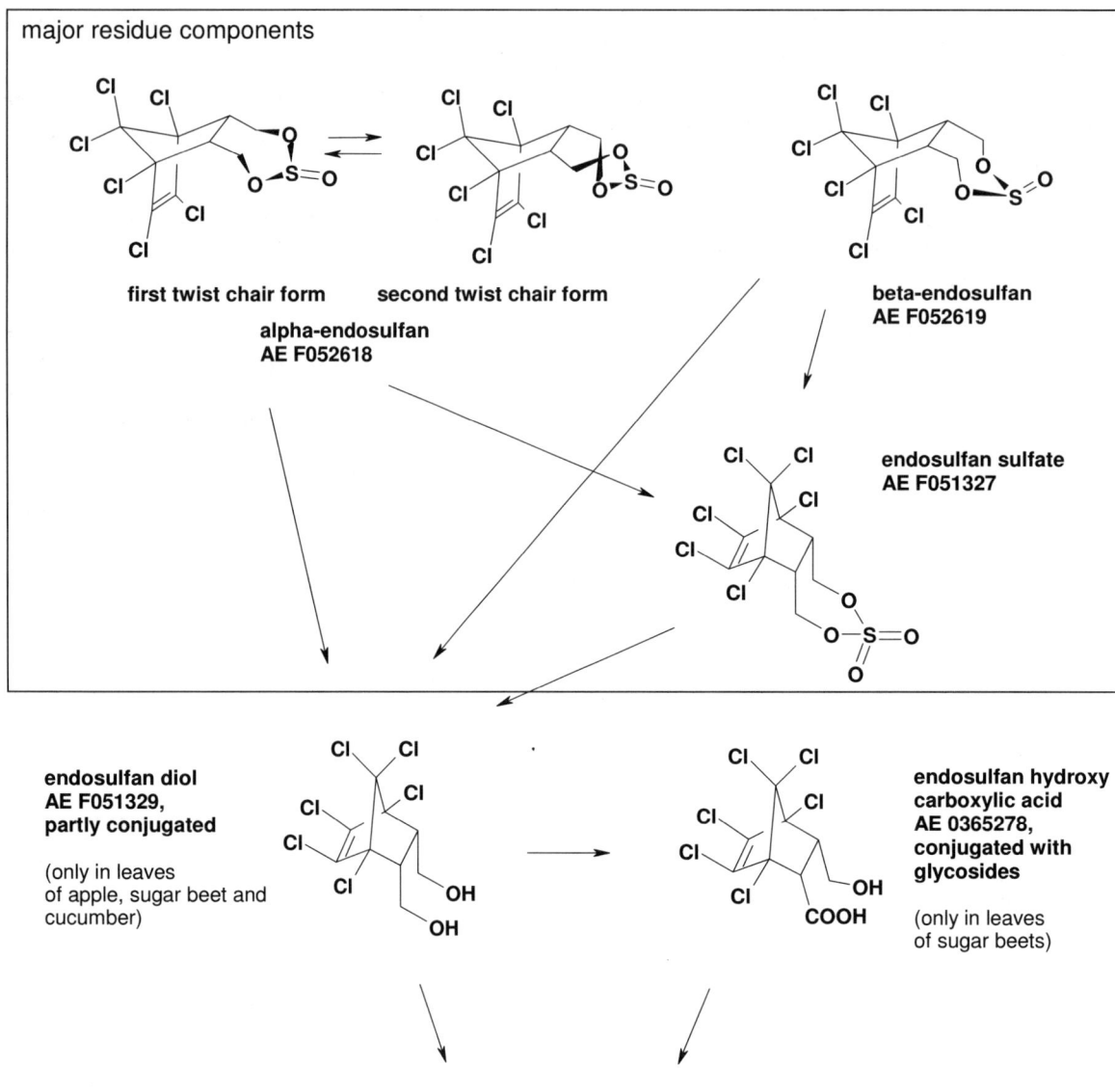

Figure 2. Metabolic pathway of endosulfan in plants (tomato, apple, sugar beet, cucumber and soybean).

## Environmental fate in soil

The Meeting received information on the aerobic degradation, soil accumulation, hydrolysis, degradation in water and sediment systems and rotational crops.

### Aerobic degradation (fate)

The aerobic degradation of endosulfan in soil was investigated in three laboratory studies with the [14]C-labelled parent isomers in the original ratio alpha-to-beta of approx. 2 to 1. An extra study was conducted with the major soil metabolite endosulfan sulfate.

In the first study, the isomeric endosulfan mixture (5a,9a-[14]C, dioxothiepin-labelled) was applied to a silt loam and a loamy sand soil at a dose corresponding to a field rate of 2.63 kg ai/ha (Gildemeister and Jordan, 1984). After 60 days of incubation at $22 \pm 2°C$ in the dark the composition of radioactive residues was similar in both soils consisted of a low contribution of the parent isomers (more beta than alpha isomer) while endosulfan sulfate was the main metabolite. Approximately 25–50% of applied radioactivity was not extractable.

A second study was conducted with 6,7,8,9,10-[14]C labelled (bicyclic ring labelled) endosulfan (98% purity; alpha:beta = 2) in a sandy loam soil at $28 \pm 2°C$ with an application rate corresponding to 2.6 kg ai/ha (Stumpf et al., 1988). Under comparable conditions as described above, the residues were also similar with a significant decrease of the parent, while endosulfan sulfate was the major soil metabolite found.

In a third soil metabolism study, 6,7,8,9,10-[14]C labelled (bicyclic ring labelled) endosulfan was applied to five different soils (sandy loam SLV, loamy sand LS 2.2, silt loam SL2, sandy loam F821, sandy loam SLG; soil types according to USDA) and incubated for a maximum of 365 days at $21 \pm 2°C$ in the darkness (Stumpf et al., 1995). In all soils the major metabolite was endosulfan sulfate which was subsequently degraded to the minor metabolites endosulfan diol, endosulfan lactone and other polar degradates, all of which amounting to less than 10% of applied radioactivity. The non-extractable residues amounted to 10–34% of applied radioactivity at the end of the study. The mineralization was observed to be very low in this study.

While the microbial activity significantly decreased after 3–4 months of incubation, the study does not represent field conditions.

Subsequently a special soil metabolism study was conducted with endosulfan sulphate (6,7,8,9,10 -[14]C-labelled, bicyclic ring labelled) to four different soils (sandy loam LS 2.2, silty clay loam HE, loam SP, silt loam SLS) for a total incubation period of up to 365 days at $20 \pm 2°C$ (Schnoeder, 2002 a,b,). The application rate of endosulfan sulfate corresponded to 0.84 kg ai/ha.

The mineralization rate at the end of the study, 365 days after application, was significant: 35% of applied radioactivity in LS 2.2, 16.7% in HE, 5% in SP and 23.4% in SLS indicating the complete degradation of the chlorinated ring structure. The non-extractable residues amounted to 25 – 33% of the applied radioactivity after 365 days of incubation. The expected soil metabolites endosulfan diol, endosulfan lactone and endosulfan hydroxy carboxylic acid were not detected at any time. However, a new polar metabolite was detected at 2–16% of the applied endosulfan sulfate in the four soils. This metabolite could not be identified but could be characterised as being more polar than the identified metabolites and containing at least one carboxylic group.

Table 9. Identification of metabolites resulting from [6,7,8,9,10-U-[14]C]-endosulfan treatment in soil.

| Name of metabolite | Max. concentration (% of AR) | Report |
|---|---|---|
| Endosulfan sulfate | 45.1% at day 30<br>34.7% at day 30<br>51-68% at days 59-120<br>in 4 soils and 77% at day 365 in 1 soil*) | Gildemeister, Jordan (1984) A29680<br>Stumpf (1988) A39429<br>Stumpf et al. (1995) A53618 |
| Endosulfan diol | Sporadic (0.3%) | Stumpf (1988) A39429 |
| Endosulfan diol | 0.2-8.5% at days 120 to 365*) | Stumpf et al. (1995) A53618 |
| Endosulfan lactone | 1.0% at day 30 | Gildemeister, Jordan (1984) A29680 |
| Endosulfan lactone | 0.3-2.5% at days 59 to 240 | Stumpf et al. (1995) A53618 |
| Endosulfan ether | 0.8% at days 8 and 30 | Gildemeister, Jordan (1984) A29680 |
| Endosulfan ether | Sporadic (0.6%) | Stumpf (1988) A39429 |

* Artificially increased portion due to drastic decrease of the biomass after approx. 4 months.

The degradation of endosulfan in soil is initiated through oxidation resulting in formation of the main metabolite, endosulfan sulphate.

Subsequent microbially induced hydrolysis of the sulfite diester (of endosulfan) and sulfate diester (of endosulfan sulfate) leads to ring opening of the 7-membered ring and formation of endosulfan diol. The endosulfan diol can be condensed to endosulfan ether (minor pathway) or oxidised to endosulfan hydroxy carboxylic acid and its condensation product endosulfan lactone. The chlorinated bicyclic carbon skeleton was shown to be completely degraded by considerable formation of labelled carbon dioxide in the soil metabolism study with ring labelled endosulfan sulfate.

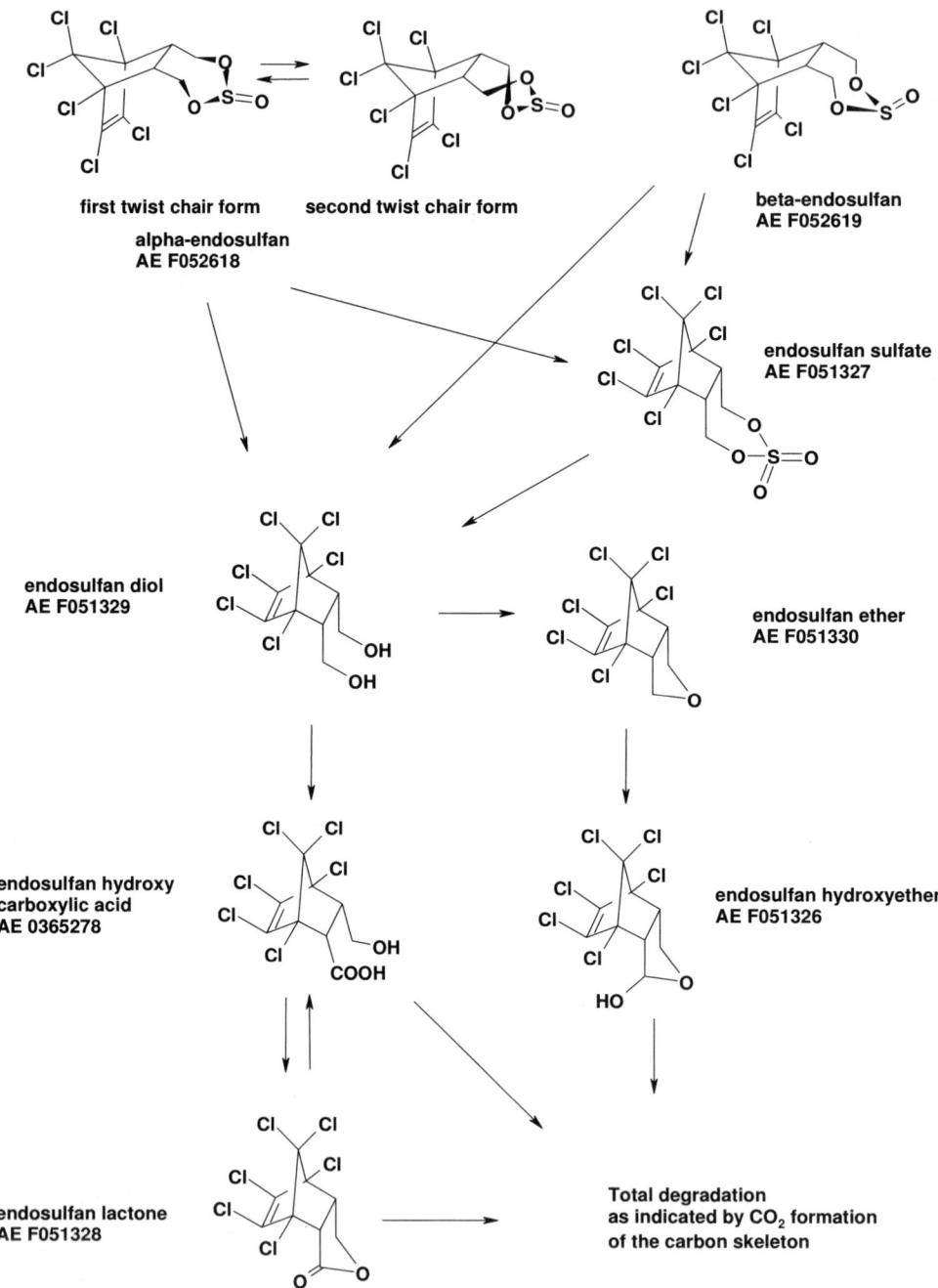

Figure 3. Proposed metabolic pathways for the degradation of endosulfan in soil and surface water.

*Aerobic degradation (rate)*

*Laboratory conditions*

Soil samples were taken from fields (Stumpf *et al.*, 1995). Sieved (< 2mm) 50 or 100 g samples (dry weight) were adjusted to the intended moisture content (usually 40% of the maximum water holding capacity). The radioactive test substance was dissolved and aliquoted into the samples at a rate representing the maximum field application rate. The treated soil samples were stored in the dark at the intended temperature. During storage, radiolabelled $CO_2$ and other volatiles were trapped. At selected time periods the complete soil samples (usually two replicates per sampling time) were extracted with a mixture of acetonitrile and water. The extracted radioactivity was measured by LSC, with identification proceeding with liquid chromatography (TLC, reversed phase HPLC) through co-chromatography with authentic reference substances and by MS.

Laboratory studies were provided which investigated the metabolic pathway of endosulfan degradation. The half lives were calculated for the individual isomers and the sum of the endosulfan isomers assuming single first order kinetics. In addition, a half life was also given for the sum of parent endosulfan and the main soil metabolite endosulfan sulfate, as this metabolite is generally included in the residue definition.(Table 10).

The soil metabolism studies in the laboratory primarily served to disclose the metabolites of endosulfan in soil rather than to derive a realistic figure of the degradation half lives. This was due to the partial decrease of the microbial biomass and degradation capacity during incubation periods longer than 3 to 4 months. As a consequence, the degradation half lives determined in field dissipation studies are more realistic and reliable (see Table 11).

Table 10. Degradation half lives of endosulfan under laboratory conditions (days).

| Soil | Incub. Temp. (°C) | DT50 α-endosulfan | DT50 β-endosulfan | DT50 α +β endosulfan - | DT50 endosulfan sulfate | Report |
|---|---|---|---|---|---|---|
| silt loam | 22 ± 2 | | | 25.6*) | | Gildemeister, Jordan (1984) A29680 |
| loamy sand | | | | 37.6*) | | |
| sandy loam | 28 ± 2 | 23 | 58 | 37 | 100-150 | Stumpf (1988) A39429 |
| sandy loam | 21 ± 2 | 12 | 158**) | 98 | | Stumpf et al. (1995) A53618 |
| loamy sand | | 39 | 264**) | 128**) | | |
| silt loam | | 19 | 132**) | 90 | | |
| sandy loam | | <10 | 108**) | 92 | | |
| sandy loam | | 14 | 115**) | 80 | | |
| sandy loam***) | 20 ± 2 | | | | 117 | Schnoeder (2002) C019647 and C020629 (fortified with field fresh soil every 3 months) |
| silty clay loam | | | | | 138 | |
| loam | | | | | 412****) | |
| silt loam | | | | | 134 | |

*) Recalculated from the decreasing residue levels given in the report

**) The DT50 values for beta and partly for parent sum of Endosulfan are artificially increased, as the microbial biomass and consequently the degradation capacity of the soil samples significantly decreased after 3-4 months of incubation. Therefore, DT50 values from field dissipation studies are more reliable.

***) Soil types according to the USDA

****) Outlier, because the DT50 of Endosulfan sulfate was determined as 75.2 days in a field trial, which was conducted at that field, where this soil was sampled for the lab study.

*Field conditions*

In all field trials, the predominant portion of endosulfan and endosulfan sulfate residues were detected in the upper 5 cm soil layer indicating very low mobility in soil. Relevant trials are summarised in Table 11.

Table 11. Rate of degradation of α +β endosulfan and endosulfan sulfate in soil under field conditions.

| Appl. rate [kg ai/ha] | Time schedule | Soil type* | | DT50 (days) | DT90 (days) | Report |
|---|---|---|---|---|---|---|
| α +β endosulfan | | | | | | |
| 0.84 | 7/ 2000 - 4/ 2001 | loam pH 5.8 | cotton (application before emergence) | 7.4**) | 24.6**) | Hardy (2001) C015651: Spain |
| 0.84 | 6/ 2000 – 6/ 2001 | silt soil pH 7.7 | cotton (application before emergence) | 21**) | 70**) | Balluff (2001) C018180 Greece |
| 5x0.56 with 7-day intervals | 8/ 1987 – 3/ 1989 | sandy soil pH 4.1-6.4 | tomato bare ground | 75.9***) 89.6***) | 252***) 298***) | Hacker (1989) A42193Georgia, USA |
| 2x1.68 with a 29-day interval | June 1987 – Dec. 1988 | loam – clay loam pH 6.7-6-9 | cotton bare ground | 92.9***) 89.5***) | 309***) 207.5***) | Mester (1990) A42997 California, USA |
| Appl. rate [kg ai/ha] | Time schedule | Soil type* | | DT50 (days) | DT90 (days) | Report |
| 2x1.68 with a 29-day interval | July 1990 – Jan. 1992 | loamy sand pH 6.8-7.8 | cotton bare ground | alpha: 6-7 beta: 19-63 alpha: 6-11 beta: 23-36 | | Czarnecki & Mayasich (1992) A51819 California, USA |
| endosulfan sulfate | | | | | | |
| 0.84 | July 2000 - April 2001 | loam pH 5.8 | cotton (application before emergence) | 75.2 | 249.7 | Hardy (2001) C015651 Spain |
| 0.84 | June 2000 – June 2001 | silt soil pH 7.7 | cotton (application before emergence) | 161 or 46.8****) | 536 or 156****) | Balluff (2001) C018180 Greece |

*)  Soil type according to USDA

**) Re-calculated assuming simple first order kinetics evaluation without the first day after application to exclude loss by volatilisation

***) Unexpected long DT values because the latest application was made late in season when cool fall and winter temperatures dominate (explanation by the author of report).

****) Alternative DT values depending on the evaluation: Inclusion of the slightly increased sulfate level at day 277 results in DT50/90 = 161/536 days, exclusion of the day 277 results in DT50/90 = 46.8/156 days.

The parent endosulfan (alpha and beta isomers in the ratio of approx. 2 to 1) degraded in soil with a half life in the range of 25–40 days at a temperature of 22°C according to the lab study of Gildemeister and Jordan (1984). In the field, however, the degradation half life was shortened to 7–21 days under southern European summer conditions (Hardy, 2001 and Balluff, 2001). However, at colder autumn and winter temperatures, the half life was increased to 75–93 days (Hacker, 1989 and Mester, 1990).

It appears that the alpha isomer degrades slightly faster (with a half life of 6–11 days) than the beta isomer (with a half life of 19–36 days) in the field (Czarnecki and Mayasich, 1992).

The main metabolite endosulfan sulfate degraded in soil with a half life of 117–138 days at 20°C in the laboratory (Schnoeder, 2002). In the field, a degradation half life was in the range of 75 to 161 days under southern European summer conditions (Hardy, 2001 and Balluff, 2001).

*Soil accumulation*

Endosulfan was applied to an apple orchard in the Netherlands three times per year with 14–35 day intervals over four subsequent years at a rate of 0.71 kg ai/ha (Tiirmaa *et al.*, 1993). The soil of the orchard, a loamy clay with a pH of 6.6–6.8, was repeatedly sampled up to 1 year after the last application within and between the tree rows.

In spite of showing a great variation of residues in the individual soil samples, mean and plateau levels could be derived. Between the tree rows, these levels amounted to approx. 0.1 mg total residues (alpha + beta endosulfan + endosulfan sulfate + endosulfan diol) per kg soil during the first year and approx. 0.2 mg/kg total residues between the second and the fourth year of application in the upper 10 cm soil layer followed by a decrease after termination of the applications. In the tree rows, the mean residue levels amounted to 0.15 mg total residues/kg during the first year and to approximately 0.3 mg total residues/kg between the second and the fourth year of application in the upper 10 cm soil layer followed by a decrease after termination.

The residues consisted predominantly of endosulfan sulfate, a smaller proportion of beta endosulfan and very low levels of alpha endosulfan. Endosulfan diol only occasionally appeared at the level of the LOQ (0.01 mg/kg). Most of the residues were detected in the upper 5 cm layer. Residues below 10 cm were negligible.

In summary, this multi-year study showed only a slight increase, from the residue level of the first year, to form a relatively constant plateau level in the subsequent years, even in northern Europe with cold to moderate temperatures. This level again decreased following termination of the application.

In biologically active soil, endosulfan was degraded via oxidation to the main soil metabolite endosulfan sulfate, followed by hydrolysis to the minor metabolite endosulfan diol which was subsequently further oxidised. The chlorinated bicyclic carbon skeleton has been shown to be completely degraded by stoichiometric production of carbon dioxide formed from the ring carbon atoms.

Under warm to moderate conditions in the field (southern Europe), endosulfan degraded moderately. A detailed review indicates that the degradation half life of alpha endosulfan is shorter than that of the beta isomer. The main soil metabolite endosulfan sulfate is more persistent than the isomers of the parent, and degrades with a half life of approximately 75–161 days, depending on the evaluation conditions. Other metabolites only appear at low levels in soil and are deemed to be not relevant.

*Hydrolysis*

Two studies (Gorlitz, 1982 and 1989) were submitted on the abiotic hydrolysis of alpha and beta endosulfan. The parent isomers were incubated in sterile buffer solution at constant temperatures. The results are given in Table 12.

Soil hydrolysis would only play a role under very alkaline and moist conditions where the half life of hydrolysis is shorter than the half life of microbial degradation in soil. In the summer, the upper soil layer can dry out and hydrolysis (and microbial degradation) is reduced. Irrigation and precipitation can accelerate both degradation processes, but in soil, microbial degradation is more prominent.

Table 12. Half lives of sterile hydrolysis of alpha and beta endosulfan at different pH values.

| Isomer | T in °C | pH 5 | pH 7 | pH 9 | Report |
|---|---|---|---|---|---|
| alpha<br>beta | 22 | > 1 year<br>> 1 year | 22 days<br>17 days | 7.0 hours<br>5.1 hours | Goerlitz & Kloeckner, 1982, A31069 |
| alpha | 25 | > 200 days | 19 days | 6.2 hours | Goerlitz & Rutz, 1988, |

| Isomer | T in °C | pH 5 | pH 7 | pH 9 | Report |
|--------|---------|------|------|------|--------|
| beta   |         | > 200 days | 10.7 days | 4.1 hours | A40003 |

Endosulfan-diol was identified as the only hydrolysis product.

### Degradation in water/sediment systems

Two studies on the behaviour of radiolabelled endosulfan in water-sediment systems were submitted. In the first study, endosulfan was incubated in two natural water-sediment systems at a pH of 7.3 and 7.8 at $22 \pm 2°C$ for 51 days (Gildemeister, 1983 and addendum Stumpf, 1990). Endosulfan rapidly disappeared from the water body by adsorption to the sediment and oxidation to endosulfan sulfate and hydrolysis to endosulfan diol. Finally endosulfan hydroxy carboxylic acid was formed.

In the second study, radiolabelled endosulfan was incubated at a pH range of 7.2–8.2 also in two natural water-sediment systems up to 120 days at $20 \pm 2°C$ (Jonas, 2002). This study yielded similar results. A rapid disappearance from the water body was observed with the concomitant formation of the above mentioned metabolites. This was followed by formation of a moderate portion of residues which were non-extractable from the sediment (8.7–19.0% of the applied radioactivity (AR) after 120 days of incubation) and partial mineralization (1.5% of AR).

The maximum portion of the metabolites and the day of reaching the maximum are shown in Table 13.

Table 13. Identification of metabolites of endosulfan in water and sediment. The incubation day of maximum appearance is added in brackets.

| Name of metabolite | Max. concentration (% of AR/day of max.) | Report |
|--------------------|------------------------------------------|--------|
| Endosulfan sulfate | water body:24.1 (8) and 57.6 (0)<br>sediment:16.2 (8) and 12.2 (16) | Gildemeister, (1983) A31182 and Stumpf, (1990), A44231 |
|                    | water body:8 (3) and 5.3 (58)<br>sediment:0 (120) and 22.3 (120) | Jonas, (2002) C022921, |
| Endosulfan diol    | sporadically<br>water body:2.4 (51) and 2.3 (32)<br>sediment:not detectable | Gildemeister, (1983) A31182 and Stumpf, (1990), A44231 |
|                    | water body:35.0 (2) and 23.6 (3)<br>sediment:41.5 (10) and 12.4 (10) | Jonas, (2002) C022921 |
| Endosulfan lactone | water body: 1.2 (4 and 8) and 1.6 (2)<br>sediment:not detectable | Gildemeister, (1983) A31182 and Stumpf, (1990), A44231 |
|                    | water body:0.1 (93) and 0.3 (120)<br>sediment:3.1 (58) and 0.7 (120) | Jonas, (2002) C022921 |
| Endosulfan hydroxy carboxylic acid | water body:24.7 (16) and 28.4 (16)<br>sediment:4.0 (32) and 3.1 (51) | Gildemeister, (1983) A31182 and Stumpf, (1990), A44231 |
|                    | water body:32.9 (93) and 44.3 (93)<br>sediment:not detectable | Jonas, (2002) C022921 |

The major metabolites of endosulfan in surface water were: endosulfan sulfate, endosulfan diol and endosulfan hydroxy carboxylic acid. In the sediment, the major metabolites of endosulfan were endosulfan sulfate and endosulfan diol.

### Rate of degradation in water and sediment

Different half lives could be derived for the disappearance of endosulfan and its metabolites from the water body and the total water-sediment system. The dissipation rates of the parent isomers can be directly derived from the basic water sediment reports (Gildemeister, 1983 and addendum Stumpf, 1990; Jonas, 2002). For degradation rates of the metabolites, a kinetic modelling of the residue levels in water and sediment of the study of Jonas was conducted (Hammel, 2004).

The resulting dissipation (transfer to another compartment, i.e., adsorption of dissolved residues by the sediment) and degradation (disappearance by chemical/microbial transformation) rates are given in the table below.

Table 14. Dissipation and degradation half lives of endosulfan and its metabolites in water and sediment.

| System | Temp (°C) | pH | Dissipation: DT50 (α +β endosulfan) from the water body | Degradation:DT50 (α +β endosulfan) in the total system | Report |
|---|---|---|---|---|---|
| α +β endosulfan | | | | | |
| River Main<br>Gravel pit | 22 ± 2<br>22 ± 2 | 7.3<br>7.8 | < 1 day<br>< 1day | 12 days<br>10 days | Gildemeister, (1983) A31182 and Stumpf, (1990), A44231 |
| Krempe (stream in a marsh area) Ohlau (stream surrounded by pastures) | 20 ± 2 | 7.5 – 8.2<br><br>7.2 – 8.2 | 0.7 days<br><br>1.6 days | 3.3 days<br><br>15.8 days | Jonas (2002), C022921 |
| Kinetic modelling of numerically combined data of Krempe and Ohlau | | | | | |
| Endosulfan | 20 ± 2 | 7.2 – 8.2 | | 38.0 days | Hammel (2004), C042131 |
| Endosulfan sulfate | 20 ± 2 | 7.2 – 8.2 | | 53.2 days | Hammel (2004), C042131 |
| Endosulfan diol | 20 ± 2 | 7.2 – 8.2 | | 29.5 days | Hammel (2004), C042131 |
| Endosulfan hydroxy carboxylic acid | 20 ± 2 | 7.2 – 8.2 | | 106.4 days | Hammel (2004), C042131 |

Endosulfan was stable under abiotic hydrolysis in acid conditions. However, the half life through hydrolysis decreased significantly as the pH-value increased.

In a surface-water system, endosulfan disappeared very rapidly from the water body by partition to the sediment with a dissipation half life of < 1 to 1.6 days. The microbial degradation in the total water-sediment system occurred at a moderate rate, the half life dependant on the evaluation method used. Assuming single first order degradation kinetics, the degradation half life was in the range 3.3–15.8 days. Using a more complex partition and degradation model (ACSL) the degradation half life determined was 38 days for the combined parent isomers.

In summary endosulfan was degraded in natural surface water and sediment via the major metabolites endosulfan sulfate, endosulfan diol and endosulfan hydroxy carboxylic acid. These metabolites were degraded with half lives between 29.5 and 106.4 days, with the formation of a small amount of non-extractable residues and carbon dioxide.

## Residues in succeeding crops

The Meeting received endosulfan residues studies in succeeding crops. The study by Krebs *et al.*, (1986) investigated three crops spinach, radish and carrots which were sown into a field plot directly after application of endosulfan.

The formulation was sprayed on three field plots (4 x 25 m²) in 1985 at an application rate equivalent to 18.8 L of formulation/ha. The study was conducted at an exaggerated rate of 6.6 kg ai/ha which is between 2 to 6× that of the current maximum application rate. The pesticide was incorporated in the plough layer (0 - 25 cm) by grubbing, ploughing and agitation with a Rotavapor. Immediately after this tillage, spinach *(Spinacia oleracea,* variety *Matador),* carrot *(Daucus carota ssp. sativa,* variety *Duwicker)* and little radish *(Raphanus sativus, var. sativa,* variety *Sora)* were sown with a hand seed drill and grown till harvest. Plant samples and soil from the corresponding plot (0 - 20 cm) were taken at the following intervals: spinach and little radish: 0, 28, and 42 days after treatment and sowing; carrot: 0, 106, and 133 days after soil treatment and sowing. The samples were analysed for α- and β-endosulfan and endosulfan sulfate using the residue method (AL 9/83 of Hoechst AG).

Table 15. Total endosulfan residues in mg/kg on crops rotated immediately after soil treatment at a rate of 6.6 kg ai/ha.

| Crop | Days after soil treatment and sowing | Residues in soil [mg/kg] | Residues in Leaf [mg/kg] | Residues in Root /tuber[mg/kg] |
|---|---|---|---|---|
| Spinach | 0 | 0.24 | - | - |
| | 28 | 0.13 | 0.19 | - |
| | 42 | 0.44 | 0.16 | - |
| Carrot | 0 | 0.58 | - | - |
| | 106 | 0.62 | 0.2 | 0.3 |
| | 133 | 0.96 | 0.14 | 0.18 |
| Little radish | 0 | 0.96 | - | - |
| | 28 | 0.54 | 0.06 | 0.052 |
| | 42 | 0.66 | 0.025 | 0.05 |

The total endosulfan residues in soil varied considerably even at day 0 probably due to incomplete incorporation in the soil. Nevertheless, a general finding could be observed. The residues in the plants were always lower than the corresponding soil residues (with one exception: spinach on day 28, probably due to an outlier of the corresponding soil residue).

With the data given above, uptake factors were estimated, i.e., the ratio of residues between soil and plant. The results of this calculation are given in Table 16.

Table 16. Uptake factors for Endosulfan (ratio of the residues in crop/soil).

| Crop | Days after Application | Leaf | Root/Tuber |
|---|---|---|---|
| Spinach | 42 | 0.36 | - |
| Carrot | 133 | 0.14 | 0.19 |
| Little radish | 42 | 0.04 | 0.08 |

In summary, the endosulfan residues taken up by root and leafy crops, which were sown immediately after soil treatment at a 6× application rate, were generally lower than the corresponding residues in soil.

## RESIDUE ANALYSIS

### *Analytical methods*

The Meeting received descriptions and validation data of analytical methods for endosulfan (α and β) and the metabolite endosulfan sulfate residues in crops, animal commodities, soil and water. Special methods were made available that could determine further minor metabolites such as endosulfan diol, endosulfan lactone or endosulfan hydroxyether.

The principle of most methods involves a solvent extraction step such as acetone, acetonitrile followed by different matrix dependant clean up steps such as GPC, Florisil or silica gel column chromatography. The final determination is carried out by GC mostly with ECD detection.

The Dutch multiresidue method MRM-1 (anonymous) has a reported limit of quantitation (LOQ) of 0.02 mg/kg for each component. This method was also adopted as AL 60/86 (Werner *et al.*, 1987) for the analysis of soil and body fluids. For soil a lower limit of the practical working range of 0.01 mg/kg was validated (Seefeld, 1990).

As a confirmatory method, GC-MS detection can be used as described in Multi-residue method 1 in Part 1 of "Analytical methods for Pesticide Residue in Foodstuffs (Working Group for the Development and Improvement of Residue-analytical Methods, 1996).

The method described in report 95-0061 (Huff and Winkler, 1997) is recommended for the analysis of animal tissues.

Table 17. Summary of analytical method.

| Sample material | | Method of analysis | LOQ | Reference |
| Sample type | Matrix | | | Author, Year |
|---|---|---|---|---|
| Plant | | | | |
| Potato | tuber | extraction acetone or acetone+dichloromethane+petroleum ether detection GC ECD | 0.02 mg/kg α-endosulfan, β-endosulfan and endosulfan sulfate | Martens, R. 1998a Analytical method and validation for the determination of residues of endosulfan and deltamethrin by GC. Date: 8/24/98. Hoechst Schering AgrEvo GmbH, Frankfurt am Main. Agredoc No.: C000413. Report No: DGMF01/97-0. Unpublished. Amendment: Martens, R. 1998b. Deltamethrin, endosulfan, AE F032640, AE F002671.Date: 11/30/98.Agredoc No: C001652. Amendment to Report No: DGMF01/97-0. R. Unpublished. |
| Peach | fruit |
| Onions | vegetable |
| Rape seed | oilseed |
| Cucumber | fruiting vegetable | extraction and detection as above | 0.02 mg/kg α-endosulfan, β-endosulfan endosulfan sulfate | Martens, R. 1998c. Validation of analytical method DGM F01/97-0 for residues of endosulfan and deltamethrin in cucumber, orange, melon and tomato. Date: 11/18/1998. Hoechst Schering AgrEvo GmbH, Frankfurt am Main. Agredoc No.: C001152. Report No.: CR97/027. |
| Orange | fruit |
| Tomato | fruiting vegetable |
| Melon | fruiting vegetable |
| Wheat | cereals ( grain) | extraction and detection as above | 0.02 mg/kg α-endosulfan, β-endosulfan endosulfan sulfate | Martens, R. 2000b. Validation of analytical method DGM F01/97-0 for dry crops (grain). Date: 4/3/2000. Hoechst Schering AgrEvo GmbH, Frankfurt am Main. Agredoc No.: C006935. Report No.: CR99/025. Unpublished |
| Citrus | fruit | extraction and detection as above | 0.02 mg/kg α-endosulfan, β-endosulfan endosulfan sulfate | Martens, R. 2000a. Data generation and enforcement method for residues on plant material by GC. Date: 3/28/00. Aventis CropScience. Agredoc No.: C007949. Report No.: DGM F01/97-1. Unpublished. |
| Peach | fruit |
| Tomato | fruiting vegetable |
| Sugar beet | Root, tuber |
| Oranges | fruit | extraction and detection as above | 0.02 mg/kg α-endosulfan, β-endosulfan endosulfan sulfate | Haines, B.K., 2001. Independent Laboratory Validation for the Determination of Residues of deltamethrin in Lettuce, Oranges, Milk and Fat, and endosulfan in Lettuce and Oranges Using Method DGM F01/97-1. Date: 3.29.01. Xenos Laboratories, Inc., Ottawa, Ontario, Canada. Agredoc No.: B003259 |
| Lettuce | leafy vegetable |
| Cotton | oilseed | extraction  cyclohexane+ethyl acetate GC-ECD GC-NPD | 0.02 mg/kg α-endosulfan, β-endosulfan endosulfan sulfate | Wrede, A. 2002. Analytical method and validation of endosulfan in cotton by GC-MSD Code: AE F002671. Date: 5/21/2002. Aventis CropScience GmbH, Frankfurt am Main. Germany. Agredoc No.: C022533. Report No.: AM 02/03 |
| Grape Tomato potato | processed product | extraction acetone/wate65/35 partition dichloromethane petroleum ether clean up florisil GC/ECD | 0.25 mg/kg α-endosulfan, β-endosulfan endosulfan sulfate | Winkler, D.Λ. 1997. Freezer Storage Stability of Endosulfan (α ,β and Sulfate) On Crop Raw Agricultural Commodities and Processed Commodities. Date: 9/29/97. Agredoc No.: A57831. Report No.: EN-CAS # 95-0072. Unpublished |

| Sample material | | Method of analysis | LOQ | Reference Author, Year |
|---|---|---|---|---|
| Sample type | Matrix | | | |
| Animal matrices | | | | |
| Milk | | extraction acetone+dichloromethane+petroleum ether detection GC ECD | 0.02 mg/kg α-endosulfan, β-endosulfan endosulfan sulfate | Haines, B.K., 2001. see above |
| Milk Egg Fat, Liver Muscle | | extraction petroleum ether+ acetone 75/25 detection GC ECD Confirmation GC/MSD | 0.025 mg/kg α-endosulfan, β-endosulfan endosulfan sulfate | Huff, D.K. and Winkler, D.A. 1997. Validation of the analytical method in animal tissues, egg (white and yolk) and dairy matrices based upon FDA pesticide analytical manual, volume I multi-residue methodology for the determination of endosulfan (alpha, beta and sulfate). Date: 11/13/97. EN-CAS Analytical Laboratories, Winston-Salem, NC. Agredoc No.: A57847. Report No.: 95-0061. Reg. No.: 44427601 |
| Milk, Egg Liver and Muscle | | GC/ECD | 0.25 mg/kg α-endosulfan, β-endosulfan endosulfan sulfate | Winkler, D.A. 1998a. Freezer Storage Stability of Endosulfan (α ,β and Sulfate) On Animal Tissue and Dairy Matrices. Date: 6/22/98. Agredoc No.: A67512. Report No.: EN-CAS # 96-0046. Unpublished |
| Soil | | extraction with acetone partition with dichloromethane GC/ECD | 0.01 mg/kg α-endosulfan, β-endosulfan endosulfan sulfate | Seefeld, F. 1990. Validation report. Analysis of endosulfan residues in soil. Date: 12/10/90. Biologische Zentralanstalt Berlin, Kleinmachnow. Agredoc No.: C008891. Report No.: Oec11/90 |

## Plant matrices

The method of analysis (Martens, 1998a, b) is based on the Dutch multiresidue method MRM-1, and describes the quantification of residues of endosulfan. The relevant residue of endosulfan consists of α-endosulfan, β-endosulfan and endosulfan sulfate. Basic validation data for potatoes, peaches, onions and rape seed were reported as well as recovery data from supervised residue trials where a modified version of this method was applied.

In the basic validation, all compounds were extracted from matrices with acetone followed by dichloromethane/petroleum ether (1/1 v/v). Later, for the analysis of samples from field trials, this extraction was simplified by taking a mixture of acetone/dichloromethane/petroleum ether (1:1:1 v/v/v). The extract was centrifuged and cleaned-up via gel permeation chromatography (GPC) and mini silica gel column. Analytes were determined by GC with ECD detection.

Table 18. Basic validation, recoveries in potato (tuber), peach, onions and rape seed.

| Analyte | Fortification Level (mg/kg) | Mean Recovery (%) | RSD (%) | N | Overall mean recovery, RSD |
|---|---|---|---|---|---|
| Potatoes tuber | | | | | |
| α-endosulfan | 0.02 | 84 | 4.3 | 5 | 84% |
| | 0.2 | 85 | 5.2 | 5 | RSD 4.5 |
| β-endosulfan | 0.02 | 89 | 4.4 | 5 | 89% |
| | 0.2 | 90 | 8.2 | 5 | RSD 6.3 |
| Endosulfan sulfate | 0.02 | 78 | 2.9 | 5 | 86% |
| | 0.2 | 94 | 5.3 | 5 | RSD 11.1 |
| Peaches | | | | | |
| α-endosulfan | 0.02 | 91 | 14.9 | 5 | 91% |
| | 0.2 | 92 | 4.3 | 5 | RSD 10.3 |
| β-endosulfan | 0.02 | 85 | 8.2 | 5 | 90% |
| | 0.2 | 96 | 3.4 | 5 | RSD 8.8 |
| Endosulfan sulfate | 0.02 | 86 | 7.8 | 5 | 92% |
| | 0.2 | 98 | 2.6 | 5 | RSD 10 |

| Analyte | Fortification Level (mg/kg) | Mean Recovery (%) | RSD (%) | N | Overall mean recovery, RSD |
|---|---|---|---|---|---|
| Onions | | | | | |
| α-endosulfan | 0.02 | 80 | 7.0 | 5 | 82% |
| | 0.2 | 83 | 3.4 | 5 | RSD 5.5 |
| β-endosulfan | 0.02 | 80 | 3.8 | 5 | 83% |
| | 0.2 | 86 | 2.8 | 5 | RSD5 |
| Endosulfan sulfate | 0.02 | 83 | 7.5 | 5 | 87% |
| | 0.2 | 90 | 4.2 | 5 | RSD 7 |
| Rape seed | | | | | |
| α-endosulfan | 0.02 | 78 | 6.4 | 5 | 86% |
| | 0.2 | 94 | 1.7 | 5 | RSD 10.5 |
| β-endosulfan | 0.02 | 87 | 2.9 | 5 | 89% |
| | 0.2 | 91 | 3.1 | 5 | RSD3.8 |
| Endosulfan sulfate | 0.02 | 81 | 12.2 | 5 | 88% |
| | 0.2 | 95 | 3.0 | 5 | RSD11.4 |

The report provides further validation of the method DGM F01/97-0 (Martens) for cucumber, orange, melon and tomato. All compounds were extracted from matrices with acetone followed by dichloromethane/petroleum ether (1:1 v/v). After centrifugation and cleanup via GPC and mini silica-gel column, the analytes were determined by GC/ECD.

Table 19. Basic validation, recoveries in cucumber, orange, melon and tomato.

| Analyte | Fortification Level (mg/kg) | Mean Recovery(%) | RSD (%) | N | Overall mean recovery, RSD |
|---|---|---|---|---|---|
| Cucumber | | | | | |
| α-endosulfan | 0.02 | 106 | 4.3 | 5 | 101% |
| | 0.2 | 95 | 2.0 | 5 | RSD 6.5 |
| β-endosulfan | 0.02 | 109 | 4.9 | 5 | 103% |
| | 0.2 | 97 | 2.8 | 5 | RSD 7.5 |
| Endosulfan sulfate | 0.02 | 107 | 4.2 | 5 | 101% |
| | 0.2 | 96 | 2.3 | 5 | RSD 6.7 |
| Orange peel | | | | | |
| α-endosulfan | 0.02 | 72 | 3.2 | 5 | 76% |
| | 0.2 | 80 | 3.0 | 5 | RSD 6.4 |
| β-endosulfan | 0.02 | 84 | 5.1 | 5 | 83% |
| | 0.2 | 83 | 3.6 | 5 | RSD 4.1 |
| Endosulfan sulfate | 0.02 | 104 | 13.1 | 5 | 96% |
| | 0.2 | 88 | 5.9 | 5 | RSD 13.6 |
| Orange pulp | | | | | |
| α-endosulfan | 0.02 | 67 | 2.7 | 5 | 73% |
| | 0.2 | 79 | 4.1 | 5 | RSD9.3 |
| β-endosulfan | 0.02 | 75 | 5.1 | 5 | 78% |
| | 0.2 | 82 | 7.9 | 5 | RSD 7.7 |
| Endosulfan sulfate | 0.02 | 88 | 3.8 | 5 | 90% |
| | 0.2 | 92 | 6.7 | 5 | RSD 5.7 |
| Melon peel | | | | | |
| α-endosulfan | 0.02 | 93 | 6.7 | 5 | 95% |
| | 0.2 | 96 | 3.7 | 5 | RSD 5.2 |
| β-endosulfan | 0.02 | 93 | 3.3 | 5 | 95% |
| | 0.2 | 96 | 4.8 | 5 | RSD 4.5 |
| Endosulfan sulfate | 0.02 | 100 | 6.3 | 5 | 103% |
| | 0.2 | 105 | 2.1 | 5 | RSD 5.2 |
| Melon fruit | | | | | |
| α-endosulfan | 0.02 | 90 | 5.5 | 4 | 95% |
| | 0.2 | 102 | 4.5 | 3 | RSD 8.2 |
| β-endosulfan | 0.02 | 91 | 3.5 | 5 | 96% |
| | 0.2 | 102 | 7.1 | 4 | RSD 7.5 |
| Endosulfan sulfate | 0.02 | 91 | 1.9 | 5 | 97% |
| | 0.2 | 104 | 5.9 | 4 | RSD 8 |

| Analyte | Fortification Level (mg/kg) | Mean Recovery(%) | RSD (%) | N | Overall mean recovery, RSD |
|---|---|---|---|---|---|
| Tomato | | | | | |
| α-endosulfan | 0.02 | 85 | 0.7 | 3 | 88% |
| | 0.2 | 91 | 2.3 | 5 | RSD 4.5 |
| β-endosulfan | 0.02 | 88 | 2.6 | 5 | 91% |
| | 0.2 | 94 | 3.8 | 5 | RSD 4.9 |
| Endosulfan sulfate | 0.02 | 79 | 5.3 | 5 | 89% |
| | 0.2 | 98 | 2.2 | 5 | RSD11.7 |

This report (Martens, 2000) contains validation data of the method DGM F01/97-0 for dry crops (grain). All compounds were extracted from matrices with acetone/dichloromethane/petroleum ether (1:1:1 v/v). After centrifugation and cleanup via GPC and mini silica-gel column, the analytes were determined by GC/ECD. The LOQs for α-endosulfan, β-endosulfan and endosulfan sulfate were established at 0.02 mg/kg. There were no interferences in control samples.

The specificity of the method was demonstrated by a confirmatory technique using a GC column with a different stationary phase a medium-polarity (50% phenyl-50% methyl polysiloxane column) as opposed to the original method and validations which used a DB-1 or the equivalent EC-1, a non-polar 100% dimethyl polysiloxane column. Recoveries using the DB-17 column were in the range of 68–107% and control samples showed no apparent residues.

Table 20. Recoveries and RSD at 0.02 and 1 mg/kg for endosulfan α, β and sulfate in grain.

| Analyte | Fortification Level (mg/kg) | Mean Recovery(%) | RSD(%) | N | Overall mean recovery, RSD |
|---|---|---|---|---|---|
| α-endosulfan | 0.02 | 88 | 3 | 7 | 88% |
| | 1 | 89 | 11 | 7 | RSD 4.5 |
| β-endosulfan | 0.02 | 89 | 5 | 7 | 91% |
| | 1 | 96 | 13 | 7 | RSD 4.9 |
| Endosulfan sulfate | 0.02 | 88 | 7 | 7 | 89% |
| | 1 | 98 | 10 | 7 | RSD11.7 |

The report (Martens, 2000) describes a new version of analytical method DGM F01/97-0 where some modifications with regard to optimisation were incorporated. Endosulfan residues are extracted from the matrices with a mixture of acetone/dichloromethane/petroleum ether (1/1/1). After centrifugation and clean-up via GPC (gel permeation chromatography) and mini silica-gel column, the analytes were determined by GC/ECD.

Table 21. Recoveries and RSD for endosulfan α, β and sulfate in citrus, tomato, peach and sugarbeet.

| Analyte | Fortification Level (mg/kg) | N | Overall mean recovery, RSD | RSD (%) | STUDY |
|---|---|---|---|---|---|
| Mandarin peel | | | | | |
| α-endosulfan | 0.02/1 | 2+2 | 93 | 8 | ER 98 ECS 741 |
| β-endosulfan | 0.02/1 | 2+2 | 109 | 13 | ER 98 ECS 741 |
| Endosulfan sulfate | 0.02/1 | 2+2 | 113 | 6 | ER 98 ECS 741 |
| Mandarin pulp | | | | | |
| α-endosulfan | 0.02/1 | 2+2 | 101 | 2 | ER 98 ECS 741 |
| β-endosulfan | 0.02/1 | 2+2 | 90 | 17 | ER 98 ECS 741 |
| Endosulfan sulfate | 0.02/1 | 2+2 | 96 | 5 | ER 98 ECS 741 |
| Orange peel | | | | | |
| α-endosulfan | 0.02/1 | 1+2 | 98 | 7 | ER 98 ECS 740 |
| β-endosulfan | 0.02/1 | 1+2 | 108 | 13 | ER 98 ECS 740 |
| Endosulfan sulfate | 0.02/1 | 1+2 | 113 | 5 | ER 98 ECS 740 |

| Analyte | Fortification Level (mg/kg) | N | Overall mean recovery, RSD | RSD (%) | STUDY |
|---|---|---|---|---|---|
| Orange pulp | | | | | |
| α-endosulfan | 0.02 | 2 | 91 | 12 | ER 98 ECS 740 |
| β-endosulfan | 0.02 | 2 | 93 | 0 | ER 98 ECS 740 |
| Endosulfan sulfate | 0.02 | 2 | 98 | 1 | ER 98 ECS 740 |
| Tomato | | | | | |
| α-endosulfan | 0.02/0.5 | 2+1 | 90 | 17 | ER 98 ECS 752 |
| β-endosulfan | 0.02/0.5 | 2+1 | 103 | 21 | ER 98 ECS 752 |
| Endosulfan sulfate | 0.02/0.5 | 2+1 | 97 | 8 | ER 98 ECS 752 |
| Peach | | | | | |
| α-endosulfan | 0.02/1/2 | 3+1+2 | 88 | 15 | ER 98 ECS 754 |
| β-endosulfan | 0.02/1/2 | 3+1+2 | 91 | 19 | ER 98 ECS 754 |
| Endosulfan sulfate | 0.02/1/2 | 3+1+2 | 89 | 21 | ER 98 ECS 754 |
| Sugar beet leaves with head | | | | | |
| α-endosulfan | 0.02/0.5/5 | 1+1+1 | 82 | 6 | ER 98 ECS 746 |
| β-endosulfan | 0.02/0.5/5 | 1+1+1 | 82 | 6 | ER 98 ECS 746 |
| Endosulfan sulfate | 0.02/0.5/5 | 1+1+1 | 92 | 8 | ER 98 ECS 746 |
| Sugar beet roots | | | | | |
| α-endosulfan | 0.02 | 2 | 114 | 5 | ER 98 ECS 746 |
| β-endosulfan | 0.02 | 2 | 107 | 11 | ER 98 ECS 746 |
| Endosulfan sulfate | 0.02 | 2 | 105 | 15 | ER 98 ECS 746 |

Method DGM F01/97-1 (Haines B.K 2001) is a modification of method DGM F01/97-0 for which validation data was provided in the previously summarised reports. Both are derived from the multiresidue method MRM-1. The main difference with method DGM F01/97-0 is that a single extraction step is performed employing a mixture of acetone/dichloromethane/petroleum ether (1/1/1). Some validation data for various matrices were reported in the previously summarised reports in which these modifications were already introduced. Quantitation was performed by GCD/ECD. For each matrix type, five replicates fortified at the LOQ and five replicates fortified at 10 × LOQ were analysed. Two blank samples were also analysed for each matrix. Endosulfan was successfully validated in lettuce and oranges.

Table 22. Recoveries and RSD at 0.02 and 1 mg/kg for endosulfan α, β and sulfate in lettuce, orange and milk.

| Analyte | Fortification Level (mg/kg) | Mean Recovery (%) | RSD(%) | Mean Recovery (%) | RSD(%) | N |
|---|---|---|---|---|---|---|
| Lettuce | | | | | | |
| | | Set #1 | | Set # 2 | | |
| α-endosulfan | 0.02 | 122 | 7.47 | 101 | 4.47 | 5 |
| | 0.2 | 81.2 | 7.17 | 98.1 | 10.9 | 5 |
| β-endosulfan | 0.02 | 102 | 6.15 | 108 | 4.62 | 5 |
| | 0.2 | 75.2 | 7.97 | 108 | 11.8 | 5 |
| Endosulfan sulfate | 0.02 | 99.1 | 4.22 | 99.6 | 6.88 | 5 |
| | 0.2 | 80.7 | 8.99 | 110 | 14.8 | 5 |
| Orange | | | | | | |
| | | Set #1 | | Set # 2 | | |
| α-endosulfan | 0.02 | 105 | 9.97 | NA | NA | 5 |
| | 0.2 | 87.4 | 6.58 | NA | NA | 5 |
| β-endosulfan | 0.02 | 116 | 6.95 | NA | NA | 5 |
| | 0.2 | 94.8 | 7.39 | NA | NA | 5 |
| Endosulfan sulfate | 0.02 | 92.2 | 4.99 | NA | NA | 5 |
| | 0.2 | 90.7 | 7.68 | NA | NA | 5 |

| Analyte | Fortification Level (mg/kg) | Mean Recovery (%) | RSD(%) | Mean Recovery (%) | RSD(%) | N |
|---|---|---|---|---|---|---|
| Milk | | | | | | |
| | | Set #1 | | Set # 2 | | |
| α-endosulfan | 0.02 | 101 | 4.28 | NA | NA | 5 |
| | 0.2 | 96.2 | 2.94 | NA | NA | 5 |
| β-endosulfan | 0.02 | 112 | 4.77 | NA | NA | 5 |
| | 0.2 | 103 | 3.44 | NA | NA | 5 |
| Endosulfan sulfate | 0.02 | 102 | 5.13 | NA | NA | 5 |
| | 0.2 | 101 | 3.53 | NA | NA | 5 |

The method was developed by Wrede (2002). Residues of α-endosulfan, β-endosulfan and endosulfan sulfate were extracted with cyclohexane/ethyl acetate (1:1, v/v) from the different matrices. After different clean-up steps, e.g. GPC and silica gel column, analytes were determined by GC-MSD. The LOQ was 0.02 mg/kg in cotton bolls, lint and seed for each analyte. Apparent levels in controls were < 30% of the LOQ. The specificity of the method was demonstrated by a confirmatory technique using GC-MS.

Table 23. Recoveries in cotton matrices.

| Analyte | Matrix | Fortification Level (mg/kg) | Mean Recovery% | SD (a)% | RSD (a)% | n | Overall Mean recovery, SD/RSD |
|---|---|---|---|---|---|---|---|
| α- endosulfan (AE F052618) | Bolls | 0.02 | 69 | 15 | 22 | 9 | 68 |
| | | 0.20 | 60 | 6 | 9 | 5 | 13/19 |
| | | 1.0 | 81 | 3 | 3 | 2 | |
| | Lint | 0.02 | 81 | 9 | 11 | 7 | 80 |
| | | 0.20 | 79 | 5 | 6 | 5 | 7/9 |
| | Seed | 0.02 | 81 | 19 | 23 | 9 | 75 |
| | | 0.20 | 65 | 3 | 5 | 6 | 17/22 |
| β-endosulfan (AE F052619) | Bolls | 0.02 | 80 | 12 | 15 | 9 | 79 |
| | | 0.20 | 72 | 9 | 13 | 5 | 12/15 |
| | | 1.0 | 93 | 1 | 2 | 2 | |
| | Lint | 0.02 | 90 | 8 | 9 | 7 | 90 |
| | | 0.20 | 89 | 7 | 8 | 5 | 7/8 |
| | Seed | 0.02 | 81 | 10 | 13 | 9 | 83 |
| | | 0.20 | 86 | 13 | 15 | 6 | 11/14 |
| Endosulfan sulfate (AE F051327) | Bolls | 0.02 | 93 | 11 | 11 | 9 | 89 |
| | | 0.20 | 82 | 10 | 13 | 5 | 11/12 |
| | | 1.0 | 87 | 8 | 10 | 2 | |
| | Lint | 0.02 | 99 | 17 | 17 | 7 | 98 |
| | | 0.20 | 96 | 9 | 10 | 5 | 14/14 |
| | Seed | 0.02 | 97 | 14 | 15 | 9 | 95 |
| | | 0.20 | 93 | 12 | 13 | 6 | 13/14 |

a) RSD = SD/Mean Recovery x 100%

*Animal matrices*

The report by Huff and Winkler (1997) describes the validation of an optimised analytical method for the determination of endosulfan (alpha, beta and sulfate) residues in animal tissues, eggs (whites and yolks) and dairy matrices based on methodologies described in the FDA Pesticide Analytical Manual. The report validates the method by determining endosulfan (alpha, beta and sulfate) in animal tissues, eggs (whites and yolks) and dairy matrices treated with [14]C-endosulfan, and additionally establishes a GC/MSD confirmatory procedure for determining endosulfan (alpha, beta and sulfate) residues in animal tissues, egg (whites and yolks) and dairy matrices.

Samples of animal tissues (except liver) were blended with 75:25 petroleum ether/acetone, sodium sulfate, and Celite and then filtered. The evaporated sample was then dissolved in petroleum ether and partitioned with acetonitrile (ACN). The ACN was combined with water and saturated NaCl and partitioned against petroleum ether. The sample was evaporated to dryness, then dissolved in hexane and loaded onto a Florisil SPE column. The eluate was analysed by GC/ECD.

Samples of liver were treated the same way except that the aqueous ACN/brine was partitioned with petroleum ether.

Milk and egg white samples were treated with ethanol and potassium oxalate and extracted with diethyl ether and petroleum ether. The top layer was removed, washed and then passed over sodium sulphate prior to evaporation to dryness and further purified with a Florisil SPE column.

Egg yolk samples were blended with 75:25 petroleum ether/acetone, sodium sulfate and Celite, then filtered. The filtrate was evaporated, dissolved in petroleum ether and passed over alumina-N. The alumina-N eluate was partitioned with can and then petroleum ether. The sample was evaporated, dissolved in hexane and loaded onto a Florisil SPE column.

The mean and standard deviations for method validation recovery results are shown in the tables. The method is reliable and reproducible for the determination of endosulfan residues in beef muscle, liver, heart, kidney, fat and poultry liver, muscle and fat, milk egg whites and egg yolks.

Table 24. Mean recoveries and standard deviation in animal matrices.

| Matrix | Mean Recovery and Standard Deviation[1] | | |
| | Alpha endosulfan | Beta endosulfan | Endosulfan sulfate |
|---|---|---|---|
| Method Validation Results | | | |
| Beef Muscle | 83% ± 3.0% (n=15) | 82% ± 3.2% (n=15) | 87% ± 3.4% (n=15) |
| Beef Liver | 87% ± 7.6% (n=15) | 85% ± 7.5% (n=15) | 88% ± 5.5% (n=15) |
| Beef Fat | 84% ± 7.2% (n=15) | 87% ± 4.5% (n=15) | 94% ± 3.2% (n=15) |
| Milk | 80% ± 10% (n=15) | 86% ± 11% (n=15) | 85% ± 11% (n=15) |
| Egg Whites | 78% ± 4.9% (n=15) | 86% ± 6.9% (n=15) | 85% ± 8.9% (n=15) |
| Egg Yolks | 80% ± 5.7% (n=15) | 81% ± 5.8% (n=15) | 86% ± 6.4% (n=15) |
| [14]C Method Validation Results | | | |
| Beef Muscle | 77% ± 6.5% (n=6) | 82% ± 6.2% (n=6) | 81% ± 5.9% (n=6) |
| Beef Liver | 76% ± 6.0% (n=6) | 83% ± 2.0% (n=6) | 89% ± 4.3% (n=6) |
| Beef Heart | 80% ± 7.8% (n=6) | 86% ± 2.6% (n=6) | 81% ± 3.7% (n=6) |
| Beef Kidney | 86% ± 7.2% (n=6) | 86% ± 8.4% (n=6) | 87% ± 6.4% (n=6) |
| Renal fat | 63% ± 23% (n=6) | 94% ± 7.2% (n=6) | 89% ± 9.8% (n=6) |
| Omental Fat | 72% ± 8.2% (n=6) | 75% ± 4.5% (n=6) | 73% ± 5.3% (n=6) |
| Poultry Liver | 98% ± 18% (n=6) | 96% ± 14% (n=6) | 100% ± 11% (n=6) |
| Poultry Muscle | 94% ± 4.3% (n=6) | 91% ± 1.9% (n=6) | 90% ± 2.9% (n=6) |
| Poultry Fat | 52% ± 26% (n=6) | 66% ± 19% (n=6) | 80% ± 8.8% (n=6) |
| Milk | 77% ± 1.6% (n=6) | 81% ± 3.1% (n=6) | 84% ± 5.4% (n=6) |
| Egg Whites | 88% ± 7.9% (n=6) | 88% ± 3.1% (n=6) | 88% ± 3.0% (n=6) |
| Egg Yolks | 76% ± 3.3% (n=6) | 76% ± 3.7% (n=6) | 76% ± 3.6% (n=6) |

1: For Method Validation there were 5 fortifications at 0.025 ppm, 5 at 0.25 ppm and 5 at 0.70 ppm; for [14]C Method Validation there were 3 fortifications at 0.025 ppm and 3 fortifications at 0.70 ppm, except for egg yolk, where there were 2 fortifications at 0.025 ppm and 3 fortifications at 0.70 ppm.

The method completely accounted for all incurred endosulfan derived residues in animal tissues, eggs (whites and yolks) and dairy matrices with the minor exception of poultry liver.

Table 25. Recoveries of endosulfan sulfate in animal matrices and alpha and beta endosulfan in poultry fat and egg yolks.

| Matrix | EN-CAS ID# | ppm Expected | | | % Recovery of Expected [14]C Residue | | |
| | | Alpha | Beta | Sulfate | Alpha | Beta | Sulfate |
|---|---|---|---|---|---|---|---|
| Beef Muscle | EN8169A | < 0.025 | < 0.025 | 0.026 | -- | -- | 112 |
| | EN8169B | < 0.025 | < 0.025 | 0.026 | -- | -- | 85 |
| Beef Liver | EN8170A | < 0.025 | < 0.025 | 0.0969 | -- | -- | 76 |
| | EN8170B | < 0.025 | < 0.025 | 0.969 | -- | -- | 79 |
| Beef Heart | EN8172A | < 0.025 | < 0.025 | < 0.025 | -- | -- | 174[a] |
| | EN8172B | < 0.025 | < 0.025 | < 0.025 | -- | -- | 191[a] |
| Beef Kidney | EN8173A | < 0.025 | < 0.025 | 0.065 | -- | -- | 122 |
| | EN8173B | < 0.025 | < 0.025 | 0.065 | -- | -- | 98 |
| Renal Fat | EN8174A | < 0.025 | < 0.025 | 0.704 | -- | -- | 138 |
| | EN8174B | < 0.025 | < 0.025 | 0.704 | -- | -- | 138 |

| Matrix | EN-CAS ID# | ppm Expected | | | % Recovery of Expected [14]C Residue | | |
|--------|-----------|--------|------|---------|--------|------|---------|
| | | Alpha | Beta | Sulfate | Alpha | Beta | Sulfate |
| Omental Fat | EN8175A | < 0.025 | < 0.025 | 1.049 | -- | -- | 119 |
| | EN8175B | < 0.025 | < 0.025 | 1.049 | -- | -- | 129 |
| Poultry Liver | EO6586A | < 0.025 | < 0.025 | 0.213 | -- | -- | 41 |
| | EO6586B | < 0.025 | < 0.025 | 0.213 | -- | -- | 32 |
| Poultry Muscle | EN8179A | < 0.025 | < 0.025 | < 0.025 | -- | -- | -- |
| | EN8179B | < 0.025 | < 0.025 | < 0.025 | -- | -- | -- |
| Poultry Fat | EN8180A | 0.137 | 0.076 | 0.637 | 105 | 93 | 98 |
| | EN8180B | 0.137 | 0.076 | 0.637 | 114 | 95 | 102 |
| Milk | EO5090A | < 0.025 | < 0.025 | 0.130 | -- | -- | 151 |
| | EO5090B | < 0.025 | < 0.025 | 0.130 | -- | -- | 80 |
| Egg Whites | EN8176A | < 0.025 | < 0.025 | < 0.025 | -- | -- | -- |
| | EN8176B | < 0.025 | < 0.025 | < 0.025 | -- | -- | -- |
| Egg Yolks | EN8177A | 0.040 | < 0.025 | 0.396 | 98 | | 116 |
| | EN8177B | 0.040 | < 0.025 | 0.396 | 108 | | 79 |

a. Percent recovery elevated due to samples being quantitated below screening level.

## Soil

The endosulfan and the metabolite endosulfan-sulfate were extracted from the soil with acetone. After dilution with brine and clean-up by liquid-liquid partition with dichloromethane and a silica gel column, determination was carried out by GC/ECD.

Table 26. Recovery values determined for residues of Endosulfan in soil.

| Analyte | Fortification Level (mg/kg) | Recoveries (mg/kg) | Recoveries (%) | RSD[1] (%) | LOD (mg/kg) | LOQ (mg/kg) |
|---------|------|------|------|------|------|------|
| Soil DEU89 I 71721 | | | | | | |
| AE F052618 (alpha endosulfan) | UTC | 0.0001 | | 201 | | |
| | 0.01 | 0.0091 | 90.3 | 16.7 | 0.004 | 0.01 |
| | 1.0 | 0.895 | 89.5 | 15.9 | | |
| AE F052619 (beta endosulfan) | UTC | 0.0006 | | 183 | | |
| | 0.01 | 0.0099 | 93.4 | 19.7 | 0.006 | 0.01 |
| | 1.0 | 0.870 | 87.0 | 15.0 | | |
| AE F051327 (Endosulfan sulfate) | UTC | 0.0001 | | 332 | | |
| | 0.01 | 0.010 | 99.8 | 19.1 | 0.005 | 0.01 |
| | 1.0 | 0.857 | 85.7 | 17.0 | | |
| Soil DEU89 I 71741 | | | | | | |
| AE F052618 (alpha endosulfan) | UTC | 0.0004 | | 142 | | |
| | 0.01 | 0.080 | 75.8 | 12.3 | 0.004 | 0.01 |
| | 1.0 | 0.784 | 78.4 | 14.1 | | |
| AE F052619 (beta endosulfan) | UTC | 0.0005 | | 156 | | |
| | 0.01 | 0.0083 | 78.6 | 11.9 | 0.004 | 0.01 |
| | 1.0 | 0.777 | 77.7 | 20.1 | | |
| AE F051327 (Endosulfan sulfate) | UTC | 0.0003 | | 195 | | |
| | 0.01 | 0.0085 | 81.7 | 10.7 | 0.004 | 0.01 |
| | 1.0 | 0.757 | 75.6 | 21.1 | | |

## Specific methods

Hong Li (1999) used the method XAM-53 described in the report for storage stability for endosulfan on sugar beet leaves for the determination of endosulfan alpha, beta, sulfate, lactone and diol residues in wheat grain, forage and straw, and sugar beet roots and tops. Residues were extracted from crops partitioned with hexane and the organic phase dried with sodium sulfate and concentrated. The sample is cleaned up using a silica gel SPE conditioned with hexane. Alpha and beta endosulfan, endosulfan sulfate and lactone were analysed by GC/ECD. The diol was derivatised with N-methyl-N-trimethylsilyl-trifluoroacetamide and analysed using GC/ECD. The LOQ was 0.05 mg/kg. The method validation showed recoveries in sugarbeet leaves of 79, 78 and 79% for endosulfan lactone spiked at 0.05 mg/kg, and of 77, 81 and 74% for endosulfan diol spiked at 0.05 mg/kg.

Gardner and Snowdon (1995) used extraction by acetone followed by partitioning with dichloromethane. After purification on silica solid phase the determination was by gas chromatography with electron capture detection. The recovery on melon (peel and pulp) and grapes was above 90% for alpha, beta and sulfate endosulfan.

Idstein *et al.* (1995) used extraction by acetone/water following by partitioning with dichloromethane. Final quantification was done by GC/ECD. The LOQ was 0.01 mg/kg. The recoveries obtained with alpha, beta endosulfan and endosulfan sulfate on potatoes were 74, 81 and 84% respectively at the 0.01 mg/kg fortification level.

Werner *et al.* (1987) investigated the analysis of endosulfan-diol and endosulfan-lactone in soil, water and urine and endosulfan and endosulfan-sulfate in soil, water, urine and plant material. Samples were extracted, cleaned up and then analysed for α and ß endosulfan and endosulfan sulfate with GC/ECD. In the case of endosulfan-diol the cleaned final extract was derivatised (silylated) before being quantified. This method was validated in Seefeld (1990.

## Stability of pesticide residues in stored analytical samples

The deep freeze stability (Winkler, 1997, 1998b) of endosulfan (alpha, beta and the sulfate) on raw agricultural commodities (RACs) and processed commodities (PCs) were investigated when stored frozen for 18 months. Control samples were fortified at 0.25 mg/kg with endosulfan (alpha, beta and the sulfate) and stored at approximately < -10°C. Unfortified control samples were stored frozen under the same conditions and one unfortified control and two freshly fortified controls were analysed concurrently with stored fortification samples at each analysis interval to determine procedural recovery. The method in FDA Pesticide Analytical Manual (PAM) (Volume 1, Sections 302, 303 and 304, 1994 Edition) was used for the analyses.

Endosulfan was stable for 18 months in RAC matrices (grape, potato, tomato, cantaloupe and lettuce) and PC matrices (grape juice, potato flakes, potato flakes, potato wet peel, tomato paste and tomato puree), and for at least 12 months in grape raisin. The recovery ranges for the stored fortifications are shown in the following tables for the average fresh fortification recovery.

Table 27. Summary of endosulfan stored fortification results in crop RAC and in crop PC.

| Storage Months | Matrix | Alpha | Beta | Sulfate | Matrix | Alpha | Beta | Sulfate |
|---|---|---|---|---|---|---|---|---|
| 3 | Grape | 89, 97 | 94, 102 | 91, 97 | Grape | 105, 113 | 111, 118 | 110, 118 |
| 6 | | 87, 91 | 88, 91 | 89, 94 | Raisin | 100, 113 | 107, 120 | 107, 119 |
| 9 | | 89, 90 | 92, 91 | 92, 92 | | 98, 100 | 99, 94 | 108, 98 |
| 12 | | 92, 88 | 75, 70 | 94, 87 | | 92, 99 | 68, 68 | 94, 93 |
| 18 | | 93, 91 | 100, 93 | 102, 94 | | NA | NA | NA |
| 3 | Potato | 91, 90 | 98, 102 | 89, 92 | Potato | 78, 81 | 84, 88 | 87, 91 |
| 6 | | 96, 89 | 97, 91 | 99, 91 | Flakes | 83, 86 | 93, 98 | 93, 103 |
| 9 | | 77, ,84 | 81, 88 | 80, 85 | | 82, 86 | 60, 63 | 92, 94 |
| 12 | | 94, 93 | 96, 94 | 95, 93 | | 84, 87 | 91, 93 | 101, 102 |
| 18 | | 54, 57 | 59, 61 | 62, 63 | | 68, 69 | 75, 74 | 80, 80 |
| 3 | Tomato | 90, 95 | 93, 97 | 89, 92 | .Tomato | 94, 91 | 100, 99 | 92, 93 |
| 6 | | 90, 102 | 91, 102 | 92, 104 | Paste | 88, 90 | 94, 96 | 97, 98 |
| 9 | | 93, 93 | 95, 95 | 96, 95 | | 85, 86 | 90, 93 | 90, 92 |
| 12 | | 100, 98 | 103, 101 | 107, 103 | | 77, 76 | 64, 58 | 82, 77 |
| 18 | | 79, 88 | 81, 91 | 80, 95 | | 95, 102 | 97, 106 | 99, 108 |
| 3 | Cantaloupe | 87, 87 | 91, 92 | 88, 87 | Potato | 99, 92 | 96, 88 | 98, 85 |
| 6 | | 91, 94 | 92, 91 | 92, 92 | Wet Peel | 90, 93 | 87, 89 | 91, 96 |
| 9 | | 91, 98 | 91, 85 | 93, 86 | | 72, 67 | 76, 81 | 75, 79 |
| 12 | | 94, 98 | 95, 100 | 95, 99 | | 92, 92 | 71, 70 | 90, 89 |
| 18 | | 81, 102 | 81, 103 | 78, 98 | | 97, 112 | 97,117 | 109, 92 |

| Storage Months | Matrix | Alpha | Beta | Sulfate | Matrix | Alpha | Beta | Sulfate |
|---|---|---|---|---|---|---|---|---|
| 3 | Lettuce | 93, 94 | 98, 97 | 94, 92 | Tomato | 100, 98 | 100, 95 | 100, 96 |
| 6 | | 89, 90 | 88, 87 | 92, 91 | Puree | | | |
| 9 | | 80, 81 | 83, 82 | 86, 83 | | 82, 87 | 95, 103 | 93, 98 |
| 12 | | 87, 87 | 90, 90 | 93, 94 | | 89, 91 | 65, 67 | 85, 86 |
| 18 | | 86, 104 | 86, 109 | 84, 112 | | 81, 105 | 85, 113 | 81, 114 |
| 3 | Grape Juice | 97, 95 | 100, 101 | 104, 102 | | | | |
| 6 | | 98, 81 | 103, 102 | 107, 105 | | | | |
| 9 | | 90, 98 | 76, 78 | 102, 103 | | | | |
| 12 | | 92, 89 | 92, 98 | 96, 99 | | | | |
| 18 | | 79, 88 | 81, 91 | 80, 95 | | | | |

Control samples were fortified at 0.50 mg/kg with endosulfan lactone or endosulfan diol, (Diot and Kieken, 2004) and stored at about -18°C. Unfortified control samples were stored frozen under the same conditions and one unfortified control and two freshly fortified controls were analysed concurrently with stored fortification samples at each analysis interval to determine procedural recovery. The method of analysis used was XAM-53; the LOQ for both endosulfan lactone and endosulfan diol is 0.05 mg/kg.

Apparent residues of endosulfan lactone and endosulfan diol in stored control sugar beet leaves samples were below 10% of the spiking level in spiked samples. Stored samples spiked with endosulfan lactone were stable for 18 months, but then showed a decrease of 33% over the final 6 months of the 2 year storage period. Endosulfan diol was stable for 24 months in sugar beet leaves.

Table 28. Endosulfan stored fortification results in sugar beet leaves.

| Storage Interval (Days) | Mean Recovery,% | Mean Concurrent Recovery,% | Mean Recovery,% | Mean Concurrent Recovery,% |
|---|---|---|---|---|
| | endosulfan lactone | | endosulfan diol | |
| 0 | 74 | 87 | 80 | 81 |
| 91 | 68 | 77 | 80 | 80 |
| 174 | 74 | 86 | 110 | 111 |
| 287 | 80 | 103 | 94 | 96 |
| 365 | 62 | 87 | 90 | 95 |
| 553 | 62 | 92 | 94 | 94 |
| 749 | 48 | 80 | 64 | 76 |

Samples of beetroot, lemons and leafy lettuce were analysed after six months of storage (Bodnaruk 2001). The analysis was conducted by extraction with acetone following by partition with dichloromethane; no further clean up was conducted.

The results are including in the following table.

Table 29. Endosulfan stability during 6 months of storage at −20°C.

| Storage Months | Matrix | Alpha mg/kg | Beta mg/kg | Sulfate mg/kg | total mg/kg | % degradation |
|---|---|---|---|---|---|---|
| 0 | Leafy lettuce | 1.60 | 1.50 | 0.31 | 3.40 | |
| 6 | | 1.50 | 1.20 | 0.28 | 3.00 | -13 |
| 0 | | 0.34 | 0.39 | 0.33 | 1.00 | |
| 6 | | 0.42 | 0.39 | 0.36 | 1.20 | +20 |
| 0 | | 0.15 | 0.14 | 0.19 | 0.48 | |
| 6 | | 0.14 | 0.12 | 0.19 | 0.45 | -6 |
| 0 | Lemons | 0.10 | 0.14 | 0.012 | 0.25 | |
| 6 | | 0.11 | 0.15 | 0.01 | 0.27 | +8 |
| 0 | | 0.049 | 0.10 | 0.008 | 0.16 | |
| 6 | | 0.059 | 0.13 | 0.006 | 0.20 | +24 |
| 0 | Lemons | 0.052 | 0.12 | 0.007 | 0.17 | |
| 6 | | 0.052 | 0.12 | 0.007 | 0.18 | +6 |

| Storage Months | Matrix | Alpha mg/kg | Beta mg/kg | Sulfate mg/kg | total mg/kg | % degradation |
|---|---|---|---|---|---|---|
| 0 | Beetroot | 0.18 | 0.11 | 0.10 | 0.39 | |
| 6 | | 0.53 | 0.30 | 0.15 | 0.98 | +125 |
| 0 | | 0.10 | 0.11 | 0.11 | 0.32 | |
| 6 | | 0.084 | 0.082 | 0.026 | 0.25 | -21 |
| 0 | | 0.062 | 0.063 | 0.075 | 0.20 | |
| 6 | | 0.051 | 0.062 | 0.085 | 0.20 | 0 |

### *Stability of pesticide residues in stored analytical samples (animal tissue)*

Control samples were fortified at 0.25 mg/kg with endosulfan (alpha, beta and sulfate) and stored at < 10°C (Winkler, 1998a). Unfortified control samples were stored frozen under the same conditions and one unfortified control and two freshly fortified controls were analysed concurrently with stored fortification samples at each analysis interval to determine procedural recovery. Recovery results were corrected for the average recovery of the corresponding fresh fortification samples. The analytical method used was derived from PAM 1, Sections 303/304 (see A57847).

The analysis results indicated that endosulfan was stable for 12 months in animal tissues (beef muscle and liver), egg (whites and yolks) and milk. The overall fresh procedural recoveries for all matrices ranged from 63% to 104% for endosulfan (alpha, beta and sulfate), with the exception of 4 recoveries, ranging from 52% to 59%, shown as outliers. The recovery ranges for the stored fortifications are shown in the following tables for the average fresh fortification recovery.

Table 30. Endosulfan stored fortification results in animal tissues.

| Matrix | Storage Interval (Months) | % Recovery Range for Stored Fortifications (Uncorrected) | | |
|---|---|---|---|---|
| | | Alpha | Beta | Sulfate |
| Beef Muscle | 3 | 86, 82 | 88, 82 | 89, 82 |
| | 6 | 87, 82 | 86, 79 | 90, 82 |
| | 9 | 77, 76 | 81, 81 | 84, 84 |
| | 12 | 82, 95 | 87, 101 | 86, 101 |
| Beef Liver | 3 | 76, 75 | 75, 74 | 75, 73 |
| | 6 | 70, 69 | 71, 73 | 69, 72 |
| | 9 | 59, 57 | 75, 74 | 83, 82 |
| | 12 | 82, 81 | 94, 90 | 118, 111 |
| Egg Whites | 3 | 76, 79 | 78, 81 | 83, 84 |
| | 6 | 66, 66 | 71, 70 | 84, 81 |
| | 9 | 67, 68 | 74, 74 | 80, 82 |
| | 12 | 61, 70 | 62, 70 | 78, 83 |
| Egg Yolks | 3 | 82, 74 | 84, 77 | 64, 74 |
| | 6 | 78, 78 | 83, 84 | 83, 84 |
| | 9 | 72, 83 | 73, 83 | 67, 75 |
| | 12 | 72, 63 | 64, 50 | 76, 58 |
| Milk | 3 | 83, 73 | 83, 81 | 80, 74 |
| | 6 | 86, 83 | 87, 86 | 86, 85 |
| | 9 | 75, 74 | 77, 76 | 76, 75 |
| | 12 | 98, 104 | 103, 114 | 104, 115 |

## USE PATTERN

Information on registered uses made available to the meeting is shown in Table 31.

Table 31. Registered uses of endosulfan on crops.

| Crop | Country | Formulation | | Application | | | | PHI |
|------|---------|------|------|--------|------|-------|------|-----|
| | | Type | Conc | Method | Rate g/hL | Water L/ha | Rate kg ai/ha | day |
| Apple | Australia | EC | 350 g/L | Spray | 66.5 | | NS | 28 |
| Apple | Canada | WP | 50% | | | | 0.65 | 15 |
| Apple | Canada, BC | WP | 50% | | | 4333- 4500 | 0.75-1.12 | 15 |
| Apple | Canada, E. | WP | 50% | | | 4500 to NS | 2.25-3.38 | 15 |
| Apple | Central America | EC | 350 g/L | | | | 0.7 | 21 |
| Apple | Chile | EC | 320 g/L | | 3.2 to 9.6 | 1500- 2000 | | 25 |
| Apple | Chile | EC | 320 g/L | | 12..8-16 | 2000- 2500 | | 25 |
| Apple | Chile | EC | 320 g/L | | 19 to 25 | 2000- 2500 | | 25 |
| Apple | Chile | WP | 48.25% | | 75-96 | | | NS |
| Apple | China | EC | 350 g/L | | | 1500- 2000 | | NS |
| Apple | Japan | WP | 48% | Spray | 0.03% | NS | | 30 |
| Apple | Japan | EC | 30% | Spray | 0.03-0.04% | NS | | 30 |
| Apple | Japan | EC | 30% | Dip | 0.006% | NS | | 120 |
| Apple | Namibia | SC | 475 g/L | Spray | | 300 to 3500 | 1.18-1.66 | 14 |
| Apple | S. Arabia | EC | 350 g/L | Spray | 35-52.5 | NS | | 28 |
| Apple | S. Africa | SC | 475 g/L | Spray | | 300 to 3500 | 1.18-1.66 | 14 |
| Apple | USA, exc CA | EC | 360 g/L | Spray | | 4675 to NS | 0.55- 2.80 | 21 |
| Apple | USA, exc CA | WSB | 50% | Spray | | 4675 (air 187) | 2.80 | 21 |
| Apple | USA-CA | EC WSB | 360 g/L 50% | Spray | | 4675 (air 187) | 0.55- 2.80 | 30 |
| Apple | Zimbabwe | SC | 47.5% | Spray | 23-368 | NS | | 14 |
| Apple, Custard | Australia | EC | 350 g/L | Spray | 52.5-70 | | NS | 7 |
| Avocados | Australia | EC | 350 g/L | Spray | | 52.5-70 | 0.735 | 14 |
| Bananas | Australia | EC | 350 g/L | Spray | | 52.5 | | 14 |
| Bean | Angola | SC | 475 g/L | Spray | | NS | 0.36-0.72 | NS |
| Bean, common, adzuki, faba, mung. | Australia | EC | 350 g/L | Spray | | 50 to NS | 0.17-0.35 | NR |
| Bean | Canada | EC | 400 g/L | Spray | | NS | 0.6-1.0 | 2 |
| Beans | Central America | EC | 350 g/L | Spray | | NS | 0.5-0.7 | 4 |
| Bean | Chile | EC | 320 g/L | Spray | | 400 to NS | 0.24-0.4 | 14 |
| Bean, broad | Chile | WP | 48.2% | Spray | | NS | 0.48-0.72 | 7 |
| Bean, French | Chile | WP | 48.2% | Spray | | NS | 0.48-0.72 | 7 |
| Bean | Japan | EC | 30% | Spray | 0.1-0.25% | NS | | 14 |
| Bean | Myanmar | EC | 350 g/L | Spray | | 233-420 | 0.44- 0.67 | 14 |
| Bean, incl kidney | Namibia | SC | 475 g/L | Spray | | 30 (air) to NS | 0.75 to 1.5 | 2 |
| Bean, kidney | Peru | EC | 355 g/L | Spray | 88.5 | NS | | 21 |
| Beans, incl kidney beans | South Africa | SC | 475 g/L | Spray | | 30 to NS | 0.35-0.72 | 2 |
| Beans, | South Africa | SC | 475 g/L | Spray | | 30 to NS | 0.35-0.72 | NS |
| Bean | USA, exc CA | EC | 360 g/L | Spray | | 93(air 185) | 0.56-1.12 | 3 |
| Bean | USA, exc CA | WSB | 50% | Spray | | 93.5 (air 9.31) | 1.12- 2.24 | 3 |
| Bean | USA-CA | EC | 360 g/L | Spray | | 93.5 (air 9.31) | 0.56-1.12 | 3 |
| Bean | USA-CA | WSB | 50% | Spray | | 93.5 (air 9.31) | 1.12-2.24 | 3 |
| Bean | Zimbabwe | EC | 35% | Spray | | 200 | 0.49 | 2 |
| Bean | Zimbabwe | SC | 47.5% | Spray | | 1000 | 0.57 | 2 |
| beetroot | Australia | EC | 350 g/L | Spray | 66.5 | | 0.735 | 14 |
| Brassicas | N. Zealand | EC | 350 g/L | Spray | | NS | 0.70 | 14 |
| Broccoli | Australia | EC | 350 g/L | Spray | 66.5 | | 0.73 | 7 |
| Broccoli | Canada | EC | 400 g/L | Spray | | NS | 0.6-0.8 | 7 |
| Broccoli | Canada | WP | 50% | Spray | | 1000 to NS | 0.5-0.88 | 7 |

| Crop | Country | Formulation | | Application | | | | PHI |
|------|---------|------|------|--------|-----------|-----------|-----------|-----|
| | | Type | Conc | Method | Rate g/hL | Water L/ha | Rate kg ai/ha | day |
| Broccoli | Central America | EC | 350 g/L | Spray | | NS | 0.52-0.7 | 7 |
| Broccoli | USA, exc CA | EC | 360 g/L | Spray | | 93 (air 186) | 0.83-1.12 | 7 |
| Broccoli | USA, exc CA | WSB | 50% | Spray | | 93 (air 9.3) | 0.84-1.1 | 7 |
| Broccoli | USA-CA | EC | 360 g/L | Spray | | 93 (air 9.3) | 1.33 qt/A | 7 |
| Broccoli | USA-CA | WSB | 50% | Spray | | 93 (air 9.3) | 0.84-1.12 | 7 |
| Br. Sprts | Canada | EC | 400 g/L | Spray | | NS | 0.52-0.7 | 7 |
| Br. Sprts | Central America | EC | 350 g/L | Spray | | NS | 0.52-0.7 | 7 |
| Br. Sprts | Namibia | SC | 475 g/L | Spray | | 30 to NS | 0.47 | 7 |
| Br. Sprts | South Africa | SC | 475 g/L | Spray | | 30 to NS | 0.47 | 7 |
| Br. Sprts | USA, exc CA | EC | 360 g/L | Spray | | 93 (air 186) | 0.72- 1.12 | 14 |
| Br. Sprts | USA, exc CA and USA-CA | WSB | 50% | Spray | | 93 (air 9.3) | 1.65-2.2 | 14 |
| Br. Sprts | USA-CA | EC | 360 g/L | Spray | | 93(air 9.3) to NS | 1.12 | 14 |
| Cabbage | Australia | EC | 350 g/L | Spray | 66.5 | | 0.735 | 7 |
| Cabbage | Canada | EC | 400 g/L | Spray | | NS | 0.6-0.8 | 7 |
| Cabbage | Canada | WP | 50% | Spray | | 1000 to NS | 0.5-0.88 | 7 |
| Cabbage | Central America | EC | 350 g/L | Spray | | NS | 0.52-0.7 | 7 |
| Cabbage | Chile | WP | 48.2% | Spray | | NS | 0.5-0.72 | NS |
| Cabbage | Japan | EC | 30% | Spray | 0.03-0.06% | NS | | 7 |
| Cabbage, Chinese | Japan | EC | 30% | Spray | 0.03-0.06% | NS | | 30 |
| Cabbage | Japan | EC | 35% | Spray | 0.03-0.07% | NS | | 7 |
| Cabbage | N.Zealand | EC | 350 g/L | Spray | | NS | 0.42-0.7 | 14 |
| Cabbage | Turkey | EC | 360 g/L | Spray | | NS | 0.72 | 14 |
| Cabbage | USA, exc CA | EC WSB | 360 g/L 50% | Spray Spray | | 93 (air 9.3) | 0.82-1.12 | 7 |
| Cabbage | USA-CA | EC WSB | 360 g/L 50% | Spray Spray | | 93 (air 9.3) | 0.82-1.12 | 7 |
| Cauliflower | Australia | EC | 350 g/L | Spray | 66.5 | | 0.735 | 7 |
| Cauliflower | Canada | EC | 400 g/L | Spray | | NS to NS | 0.6-0.8 | 7 |
| Cauliflower | Canada | WP | 50% | Spray | | 1000 to NS | 0.5-0.87 | 7 |
| Cauliflower | Central America | EC | 350 g/L | Spray | | NS | 0.42-0.7 | 7 |
| Cauliflower | Japan | EC | 35% | Spray | 0.03-0.07% | NS | | 7 |
| Cauliflower | N Zealand | EC | 350 g/L | Spray | | NS | 0.42-0.7 | 14 |
| Cauliflower | USA, exc CA | EC | 360 | Spray | | 93 (air 186) to NS | 0.83-1.120 | 14 |
| Cauliflower | USA, exc CA | WSB | 50% | Spray | | 93 (air 9.3) | 0.82-1.12 | 14 |
| Cauliflower | USA-CA | EC WSB | 360 50% | Spray Spray | | 93 (air 9.3) | 1.33 qt/A 0.82-1.12 | 14 |
| Carrots | Australia | EC | 350 g/L | Spray | 66.5 | | | 14 |
| Cashew | Australia | EC | 350 g/L | Spray | 70 | | | 14 |
| Celery | Australia | EC | 350 g/L | Spray | 66.5 | | | 7 |
| Celery | Canada | EC | 400 g/L | Spray | | NS | 0.8 | 14 |
| Celery | Canada | WP | 50% | Spray | | NS | 0.875 | 14 |
| Celery | Central America* | EC | 350 g/L | Spray | | NS | 0.52-0.70 | 4 |
| Celery | USA, exc CA | EC WSB | 360 g/L 50% | Spray Spray | | 93 (air 187) to NS | 0.56-1.12 | 4 |
| Celery | USA, exc CA | EC WSB | 360 g/L 50% | Spray | | 93 (air 187) to NS | 0.56 | 7 |
| Celery | USA-CA | EC WSB | 360 g/L 50% | Spray | | 93 (air 9.3) | 0.56-1.12 | 4 |
| Celery | USA-CA | EC WSB | 360 g/L 50% | Spray | | 93 (air 9.3) | 0.66 qt/A 0.56 | 7 |
| Cereals | Australia | EC | 350 g/L | Spray | | | 0.17-035 | NR |
| Chayote | Australia | EC | 350 g/L | | 66.5-70 | | 0.735 | 3 |
| Cherry | Canada, E/W | WP | 50% | Spray | | 4333- 4500 | 1.62-2.25 | 15 |
| Cherry | Chile | WP | 48.2% | Spray | 70-96 | NS | | NS |
| Cherry | Japan | EC | 35% | Spray | 0.04-0.07% | NS | 0.43-0.7 | GS |

| Crop | Country | Formulation | | Application | | | | PHI |
|------|---------|------|------|--------|------|------|------|-----|
| | | Type | Conc | Method | Rate g/hL | Water L/ha | Rate kg ai/ha | day |
| Cherry | Namibia | SC | 475 g/L | Spray | | 300 to 3500 | 1.18-1.66 | 14 |
| Cherry | S. Africa | SC | 475 g/L | Spray | | 300 to 3500 | 1.18-1.66 | 14 |
| Cherry | USA, exc CA | EC | 360 g/L | Spray | | 2500-3115 | 2.24-2.76 | 21 |
| Cherry | USA, exc CA | EC | 360 g/L | Spray | 120 | NS | | NS |
| Cherry | USA, exc CA | WSB | 50% | Spray | | NS | 2.24-2.75 | 21 |
| Cherry | USA, exc CA | WSB | 50% | Spray | 32 | NS | | 21 |
| Cherry | USA-CA | EC | 360 | Spray | | 3812 (air 187) | 2.24-2.76 | 21 |
| Cherry | USA-CA | WSB | 50% | Spray | | 3812 -4676 (air 187) | 2.24-2.75 | 21 |
| Cherry | USA, exc CA | WSB | 50% | dip | 600 | NS | | GS |
| Cherry | USA-CA | EC | 360 | dip | 600 | NS | | GS |
| Cherry | USA-CA | WSB | 50% | dip | 600 | NS | | GS |
| Citrus | Angola | SC | 475 g/L | Spray | 47-107 | NS | | NS |
| Citrus | Australia | EC | 350 g/L | Spray | 3.5-10.5 | NS | | 3 |
| Citrus ( thrips) | Central America | EC | 350 g/L | Spray | 35-70 | NS | | NS |
| Citrus | Central America | EC | 350 g/L | Spray | 35-52 | NS | | NS |
| Citrus | Chile | WP | 48.2% | Spray | 72-96.5 | NS | | NS |
| Citrus | Morocco | EC | 330 g/L | Spray | 57 | NS | | 30 |
| Citrus | Morocco | CS | 330 g/L | Spray | 57 | NS | | NS |
| Citrus | Mozambique | SC | 47.5% | Spray | 1-107 | NS | | NS |
| Citrus | S. Arabia | EC | 350 g/L | Spray | 35-52 | NS | | NS |
| Citrus | S. Africa | SC | 475 g/L | Spray | 47-107 | NS | | NS |
| Citrus, no bearing | USA, exc CA USA-CA | EC | 360 | Spray | | 4676 to NS (air 9.31) | 2.76 | >12 mo |
| Citrus, no bearing | USA, exc CA USA-CA | WP WSB | 50% 50% | Spray Spray | | NS 4676 | 2.5 1.26 | >12 mo >12 mo |
| Cocoa | Brazil | EC | 350 g/L | Spray | | 400 to 600 | 0.35-0.52 | 30 |
| Cocoa | Cameroon | EC | 350 g/L | Spray | | NS | 0.26 | 28 |
| Cocoa | Ivory Coast | EC/CS | 280 g/L | Spray | | 40 to NS | 0.21/0.24 | NS |
| Cocoa | Ivory Coast | EC | 500 g/L | Spray | | 40 to NS | 0.25 | 4 |
| Cocoa | Malaysia | EC | 33% | Spray | | NS | 0.15 | 28 |
| Cocoa | Nigeria | EC | 280 g/L | Spray | | NS | 0.21 | NS |
| Coffee | Brazil | EC | 350 g/L | Spray | | 100 to 250 | 0.52-0.70 | 70 |
| Coffee | Cameroon | EC | 350 g/L | Spray | | NS | 0.52 | 28 |
| Coffee | Central America | EC | 350 g/L | Spray | | 200 to 600 | 0.52-0.70 | 21 |
| Coffee | Cuba | EC | 350 g/L | soil | | NS | 0.52-.61 | 30 |
| Coffee | Cuba | EC | 482g/kg | Spray | | NS | 0.48-0.72 | NS |
| Coffee | Ecuador | EC | 350 g/L | Spray | | NS | 0.63-0.70 | 14 |
| Coffee | Namibia | SC | 475 g/L | Spray | 47.5 | 30 to NS | | 14 |
| Coffee | Peru | EC | 355 g/L | Spray | 105 | NS | | 21 |
| Coffee | S. Africa | SC | 475 g/L | Spray | 47.5 | 30 to NS | | 14 |
| Coffee | Sudan | EC | 500 g/L | Spray | NS | NS | | NS |
| Coffee | Thailand | EC | 350 g/L | Spray | 70- 87.5 | NS | | 7 to 14 |
| Coffee | Zimbabwe | EC/ /SC | 35,35 47,5% | Spray | 245 | NS | | 21 |
| Cotton | Angola | SC | 475 g/L | Spray | | NS | 0.24-0.83 | NS |
| Cotton | Australia | EC | 350 g/L | Spray | | NS | 0.73 | 56 |
| Cotton | Benin | EC | 330 g/L | Spray | | 10 | 0.66 | NS |
| Cotton | Brazil | EC | 350 g/L | Spray | | 100 to 250 | 0.35-0.875 | 30 |
| Cotton | Burkina | EC | 350 g/L | Spray | | 10 | 0.35 | NS |
| Cotton | Central America | EC | 350 g/L | Spray | | NS | 0.52-0.70 | 0 |
| Cotton | China | EC | 350 g/L | Spray | | NS | 0.052-0.084 | NS |
| Cotton | Greece, Cyprus | WP | 470 g/kg | Spray | 71-94 | NS | | 60 |
| Cotton | Ecuador | CE | 350 g/L | Spray | | NS | 0.35-0.875 | 14 |
| Cotton | Ethiopia | ULV | 250 g/L | Spray | | NS | 0.625-0.75 | 7 to 14 |
| Cotton | Ethiopia | EC | 350 g/L | Spray | | 20 to 300 | 0.70-0.78 | 7 to 14 |
| Cotton | India | NS | 29.7% | Spray | | NS | | NS |
| Cotton | Iran | EC | 350 g/L | Spray | | NS | 1.05 | 15 |

| Crop | Country | Formulation | | Application | | | | PHI |
|------|---------|------|------|--------|-----------|-----------|-----------|-----|
| | | Type | Conc | Method | Rate g/hL | Water L/ha | Rate kg ai/ha | day |
| Cotton | Ivory Coast | CS | 330 g/L | Spray | | NS | 0.66 | NS |
| Cotton | Madagascar | EC | 352 g/L | Spray | | NS | 0.85 | 21 |
| Cotton | Mali | EC | 500 g/L | Spray | | 10 | 0.50 | NS |
| Cotton | Morocco | EC/CS | 330 g/L | Spray | | NS | 0.66 | 30 |
| Cotton | Mozambique | SC | 47.5% | soil | | NS | 0.04-0.71 | NS |
| Cotton | Myanmar | EC | 350 g/L | Spray | | 233—420 | 0.52-0.7 | 14 |
| Cotton | Namibia | SC | 475 g/L | Spray | | 280 to NS | 0.24-0.83 | 35 |
| Cotton | Namibia | SC | 475 g/L | air | | 280 to NS | 0.47-0.71 | 35 |
| Cotton | Pakistan | EC | 320 g/L | Spray | | NS | 0.4 | 3 |
| Cotton | Pakistan | EC | 352 g/L | Spray | | NS | 0.7-1 | NS |
| Cotton | Peru | EC | 355 g/L | Spray | 88 | NS | | 21 |
| Cotton | South Africa | EC | 352 g/L | Spray | | NS | 0.85 | 21 |
| Cotton | South Africa | SC | 475 g/L | Spray | | 30 to NS | 0.24-0.83 | 35 |
| Cotton | South Africa | SC | 475 g/L | Air | | 30 to NS | 0.47-0.71 | 35 |
| Cotton | Southern EU | SC | 330 g/L | Spray | 0.1% | 700-1000 | | 21 |
| Cotton | Spain | EC | 350 g/L | Spray | 0.15 -0.3% | NS | | 21 |
| Cotton | Spain | EC | 350 g/L | Spray | 52-70 | NS | | 21 |
| Cotton | Sudan | EC | 500 g/L | Spray | | NS | 0.90 | 28 |
| Cotton | Sudan | ULV | 500 g/L | Spray | | NS | 0.84 | NS |
| Cotton | Sudan | EC | 50% | G, A | | NS | 0.90 | NS |
| Cotton | Thailand | EC | 350 g/L | Spray | | NS | 0.21 | 7 to 14 |
| Cotton | Togo | CS | 330 g/L | Spray | | 10 to NS | 0.66 | NS |
| Cotton | Turkey | EC | 360 g/L | Spray | | NS | 0.72 | 14 |
| Cotton | Turkey | WP | 32.9% | Seed | 0.93/100 kg | NS | | NA |
| Cotton | USA, exc CA | EC | 360 g/L | Spray | | 93 (air 187) | 0.41-1.12 | NS |
| Cotton | USA, exc CA | WSB | 50% | Spray | | 93 (air 9.3) | 0.56-1.68 | NS |
| Cotton | USA-CA | EC | 360 g/L | Spray | | 93 (air 9.3) | 0.41-1.12 | NS |
| Cotton | Venezuela | CE | 350 g/L | Spray | | NS | 0.35-0.70 | 14 |
| Cotton | Zimbabwe | EC | 35% | Spray | 333 | 50 to 150 | | NS |
| Cotton | Zimbabwe | EC | 35% | Spray | 150 | 25 to 100 | | NS |
| Cotton | Zimbabwe | EC | 35% | ULV | 100-162 | 1.7 to 5 | | NS/0 |
| Cotton | Zimbabwe | EC EC | 35% 35% | Spray, Air | 5100 | 5 to 15 | | NS/0 |
| Cotton | Zimbabwe | EC | 35% | Spray | 499 | 25 to 100 | | 0 |
| Cotton | Zimbabwe | WP WP | 50% 50% | Spray spray | 330 500 | 50 to 150 35 to 100 | | NS |
| Cotton | Zimbabwe | WP | 50% | Soil | | 20to 200 | 2 | NS |
| Cotton | Zimbabwe | SC | 47.5% | Spray | 500 | 50 to 150 | | NS |
| Cotton | Zimbabwe | SC | 47.5% | Spray | 475 | 35 to 100 | | NS |
| Cotton | Zimbabwe | SC | 47.5% | Spray, air | 125.5 | 5 to 15 | | NS |
| Crucifers | Namibia | SC | 475 g/L | Spray | 95 | 30 to NS | | 7 |
| Crucifers | S. Africa | SC | 475 g/L | Spray | 95 | 30 to NS | | 7 |
| Crucifers | Zimbabwe | SC | 47.5% | Spray | | 1000- 2000 | 0.47-0.95 | 7 |
| Cucumber | Australia | EC | 350 g/L | Spray | 66.5-70 | | 0.735 | 3 |
| Cucumber | Canada | EC | 400 g/L | Spray | 50-60 | NS | | 2 |
| Cucumber | Canada | WP | 50% | Spray | | | 0.50-0.75 | 2 |
| Cucumber | Canada | EC | 400 g/L | Spray | | NS | 0.60 | 2 |
| Cucumber | Canada | WP | 50% | Spray | | NS | 0.50-0.55 | 2 |
| Cucumber | Central America | EC | 350 g/L | Spray | | NS | 0.52-0.70 | 0 |
| Cucumber | Chile | WP | 48.2% | Spray | | NS | 0.48-0.73 | NS |
| Cucumber | Japan | EM | 30% | Spray | 0.03-0.06% | NS | | 1 |
| Cucumber | USA, exc CA | EC | 360 g/L | Spray | | 93 (air 187) | 0.55- 1.12 | 2 |
| Cucumber | USA, exc CA JSA -CA | WSB | 50% | Spray | | 93 (air 9.3) | 0.56-1.12 | 2 |
| Cucumber | USA-CA | EC | 360 g/L | Spray | | NS | 0.55 to 112 | 2 |
| Cucurbits | Australia | EC | 350 g/L | Spray | 66.5-70 | | 0.735 | 3 |
| Cucurbits | Namibia | SC | 475 g/L | Spray | 47.5 | 30 to NS | | 1 |
| Cucurbits | S. Africa | SC | 475 g/L | Spray | 47.5 | 30 to NS | | 1 |
| Cucurbits | Zimbabwe | SC | 47.5% | Spray | 47.5 | NS | | 0 |

| Crop | Country | Formulation | | Application | | | | PHI |
|------|---------|------|------|------|------|------|------|------|
| | | Type | Conc | Method | Rate g/hL | Water L/ha | Rate kg ai/ha | day |
| Gooseberries | Australia | EC | 350 g/L | | 66.5 | | 0.735 | 7 |
| Eggplant | Australia | EC | 350 g/L | | 66.5 | | 0.735 | 7 |
| Grape | Canada | WP | 50% | Spray | | 3000 to NS | 1.50 | 30 |
| Grape | Central America | EC | 350 g/L | Spray | | NS | 0.52-0.70 | 7 |
| Grape | Chile | WP | 48.25% | Spray | 72-96.5 | NS | | NS |
| Grape | Croatia | EC | 350 g/L | Spray | 0.15-0.2% | NS | | NS |
| Grape | Japan | EC | 30% | spray | 0.3-0.6% | NS | | Before germin'n |
| Grape | Namibia | SC | 475 g/L | Spray | | 500 to 1500 | 0.24-0.72 | 14 to 42 |
| Grape | Namibia | SC | 475 g/L | Spray | | 500 to 1500 | 0.30-0.89 | 14 to 42 |
| Grape | S. Africa | SC | 475 g/L | Spray | | 500 to 1500 | 0.30-0.89 | 14-42 |
| Grape | Turkey | EC | 360 g/L | Spray | 54 | NS | | 14 |
| Grape | Turkey | WP | 32.9% | Soil | | NS | 0.75 | 15 |
| Grape | Turkey | EC | 360 g/L | Spray | 54 | NS | | 14wp |
| Grape | USA, exc CA | EC | 360 g/L | Spray | | 2100-2800 | 1.25 -2.24 | 7 |
| Grape | USA, exc CA | WSB | 50% | Spray | | NS | 1.12-1..68 | 7 |
| Grape | USA-CA | EC | 360 g/L | Spray | | NS | 1.25-2.24 | 7 |
| Grape | USA-CA | WSB | 50% | Spray | | 1870-2800 (air 187) | 1.12-1.68 | 7 |
| Grapefruit | Australia | EC | 350 g/L | spray | 3.5-10.5 | | | 3 |
| Guavas | Australia | EC | 350 g/L | spray | 52.5-70 | | | 7 |
| Hazelnut | Poland | EC | 350 g/L | Spray | | 2000- 2500 | 0.875 | NS |
| Hazelnut | Spain | EC | 350 g/L | Spray | 0.30% | NS | | 30 |
| Hazelnut | Spain | EC | 350 g/L | Spray | 70 | NS | | 30 |
| Hazelnut | Turkey | EC | 360 g/L | Spray | 54 | NS | | 14 |
| Hazelnut | Turkey | WP | 32.9% | Spray | | NS | 3 | 15 |
| Kiwi | Australia | EC | 350 g/L | Spray | 52.5-70 | | | 14 |
| Lemons | Australia | EC | 350 g/L | Spray | 3.5-10.5 | | | 3 |
| Longans | Australia | EC | 350 g/L | Spray | 70 | | | 7 |
| Loquats | Australia | EC | 350 g/L | Spray | 70 | | | 28 |
| Linseed | Australia | EC | 350 g/L | Spray | | | 0.175-0.35 | NR |
| Lupins | Australia | EC | 350 g/L | Spray | | | 0.175-0.35 | NR |
| Lychees | Australia | EC | 350 g/L | Spray | 52.5-70 | | | 7 |
| Macadamia | Australia | EC | 350 g/L | Spray | 52.5-70 | | 0.525 | 2 |
| Mandarins | Australia | EC | 350 g/L | Spray | 3.5-10.5 | | | 3 |
| Mangoes | Australia | EC | 350 g/L | Spray | 52-70 | | | 7 |
| Marrow | Australia | EC | 350 g/L | Spray | 66.5-70 | | 0.735 | 3 |
| Melon | Australia | EC | 350 g/L | Spray | 66.5-70 | | 0.735 | 3 |
| Melon | Canada | EC | 400 g/L | Spray | | NS | 0.60 | 2 |
| Melon | Canada | WP | 50% | Spray | | NS | 0.50-0.55 | 2 |
| Melon | Central America | EC | 350 g/L | Spray | | NS | 0.52-0.70 | 0 |
| Melon | Chile | EC | 320 g/L | Spray | | 400 to NS | 0.24-0.40 | 7 |
| Melon | Chile | WP | 48.25% | Spray | | NS | 0.48-0.73 | NS |
| Melon | Japan | GR | 3.3% | Spray | | NS | 1.5 | GS |
| Melon | Turkey | EC | 360 g/L | Spray | | | 0.36-1.08 | 14 |
| Melon | USA, exc CA | EC | 360 g/L | Spray | | 93 (air 187) | 0.563-1.12 | 2 |
| Melon | USA, exc CA USA-CA | WSB | 50% | Spray | | 93(air 9.3) | 0.56-1.12 | 2 |
| Melon | USA-CA | EC | 360 g/L | Spray | | NS | 0.56-1.12 | 2 |
| Melon | Venezuela | CE | 350 g/L | Spray | | NS | 0.52 | 7 |
| Okra | Australia | EC | 350 g/L | Spray | 66.5 | | 0.735 | 7 |
| Orange | Australia | EC | 350 g/L | Spray | 3.5-10.5 | | 0.35-0.735 | 3 |
| Papaya | Australia | EC | 350 g/L | Spray | 52.5 | | 0.175-0.735 | 7 |
| Passion fruit | Australia | EC | 350 g/L | Spray | 52.5-70 | | | 14 |
| Peach | Canada | EC | 400 g/L | Spray | 70 | NS | | 15 |
| Peach | Canada | EC | 400 g/L | Dip tree | 600 | NS | | NA |
| Peach | Canada, E/W | WP | 50% | Spray | 75 | | | 15 |
| Peach | Canada, E/W | WP | 50% | Spray | 37-50 | | 1.6-2.25 | 15 |
| Peach | Central America | EC | 350 g/L | Spray | | NS | 0.52-0.70 | 21 |

| Crop | Country | Formulation | | Application | | | | PHI |
|------|---------|------|------|--------|--------------|-------------|----------------|-----|
|      |         | Type | Conc | Method | Rate g/hL | Water L/ha | Rate kg ai/ha | day |
| Peach | Chile | EC | 320 g/L | Spray | 8-10 | 1500- 2000 | | 25 |
| Peach | Chile | EC | 320 g/L | Spray | 13-16 | 2000- 2500 | | 25 |
| Peach | Chile | EC | 320 g/L | Spray | 20-25 | 2000- 2500 | | 25 |
| Peach | Chile | WP | 48.25% | Spray | 72-96.5 | NS | | NS |
| Peach | Japan | EC | 30% | Trunk | 2% | NS | | Post hvst |
| Peach | Namibia | SC | 475 g/L | Spray | 2.5 to 3.5 | 300 to 3500 | 1.19-1.66 | 14 |
| Peach | S. Africa | SC | 475 g/L | Spray | 2.5 to 3.5 | 300 to 3500 | 1.19-1..66 | 14 |
| Peach | USA, exc CA | EC | 360 | Spray | | 2500- 3130 | 2.24-2.76 | 21 |
| Peach | USA, exc CA | EC | 360 | Spray | | 4675 | to 2.76 | 30 |
| Peach | USA, exc CA USA-CA | EC | 360 | dip | 600 | NS | | NA GS |
| Peach | USA, exc CA | WSB | 50% | Spray | | NS | 2.24-2.50 | 21 |
| Peach | USA, exc CA | WSB | 50% | Spray | | NS | 2.24-2.50 | 30 |
| Peach | USA, exc CA | WSB | 50% | dip | | NS | 2.24-2.50 | GS |
| Peach | USA-CA | EC | 360 g/L | Spray | | 93, (air 187) | to 2.76 | 30 |
| Peach | USA-CA | WSB | 50% | Spray | | 3740-4675 (air 187) | 2.24-2.50 | 30 |
| Peach | USA-CA | WSB | 50% | dip | 1.21 | NS | | GS |
| Peach | Zimbabwe | EC | 35% | Spray | 35 | NS | | 10 |
| Pear | Australia | EC | 350 | Spray | 52.5-66.5 | | 0.735 | 28 |
| Pear | Canada, BC | WP | 50% | Spray | | 4333- 4500 | 1.62-2.25 | 15 |
| Pear | Canada, E. | WP | 50% | Spray | | 4500 to NS | 2.25-3.37 | 15 |
| Pear | Central America | EC | 350 g/L | Spray | | NS | 0.52-0.70 | 21 |
| Pear | Chile | EC | 320 g/L | Spray | 8-9.6 | 1500- 2000 | | 25 |
| Pear | Chile | EC | 320 g/L | Spray | 12.8-16 | 2000- 2500 | | 25 |
| Pear | Chile | EC | 320 g/L | Spray | 19- 25 | 2000-2500 | | 25 |
| Pear | Chile | WP | 48.25% | Spray | 72.2-96.5 | NS | | NS |
| Pear | Cyprus | EC | 350 g/L | Spray | 52-122 | NS | | 21 |
| Pear | Greece | WP | 470 g/kg | Spray | 70-94 | NS | | 30 |
| Pear, Asian | Japan | WP | 48% | Spray | 0.032 | NS | | 30 |
| Pear, Asian | Japan | EC | 30% | Spray | 0.03-0.06% | NS | | 30 |
| Pear | Namibia | SC | 475 g/L | Spray | | 300 to 3500 | 1.18-1.66 | 14 |
| Pear | S. Arabia | EC | 350 g/L | Spray | 35-52 | NS | | 28?? |
| Pear | S. Africa | SC | 475 g/L | Spray | | 300 to 3500 | 1.18-1.66 | 14 |
| Pear | USA, exc CA | EC | 360 g/L | Spray | | 1870-4675 (air 187) | 2.24-2.76 | 7 |
| Pear | USA, exc CA | EC | 360 g/L | Spray | 30-60 | NS | | NS |
| Pear | USA, exc CA | WSB | 50% | Spray | | 2800(air 93) | 2.24-2.73 | 7 |
| Pear | USA, exc CA | EC | 360 g/L | soil | 180-270//hl | 1870-3740 | | Prior to bloom |
| Pear | USA, exc CA | WSB | 50% | soil | 0.06% | 1870-3740 | | GS |
| Pear | USA, exc CA | WSB | 50% | Spray | 0.03-0.06% | NS | | GS |
| Pear | USA-CA | EC | 360 g/L | Spray | | 4675 | 2.24-2.76 | 7 |
| Pear | USA-CA | EC WSB | 360 g/L 50% | Soil | 600 0.06% | 1870-3740 1870-4675 | | prior to bloom |
| Pear | USA-CA | EC | 360 g/L | Spray | 0.02-0.05% | NS | | GS |
| Pear | USA-CA | WSB | 50% | Spray | | 3700-4675 air 93 | 2.24-2.76 | 7 |
| Pear | USA-CA | WSB | 50% | Spray | 0.03-0.06% | NS | | GS |
| Peas field, cow, chick, pigeon | Australia | EC | 350 g/L | Spray | | | 0.175-0.35 | NR |
| Pecan | Australia | EC | 350 g/L | Spray | 52.5 | | 0.98 | 14 |
| Pepper sweet | Australia | EC | 350 g/L | Spray | 66.5 | | 0.735 | 3 |
| Peppers G | Canada | EC | 400 g/L | Spray | 500-600 | NS | | 2 |
| Peppers | Canada | EC | 400 g/L | Spray | | NS | 0.6-1.13 | 2 |
| Peppers | Canada | WP | 50% | Spray | | NS | 0.5-1.12 | 2 |
| Peppers | Cyprus, Greece | WP | 470 g/kg | Spray | 70-94 | NS | | GS |
| Peppers | Cyprus | EC | 350 g/L | Spray | 52-122 | NS | | |
| Peppers | Greece | EC | 350 g/L | Spray | 52-122 | NS | | 1 |

| Crop | Country | Formulation | | Application | | | | PHI |
|------|---------|-------------|------|--------|------------|-----------|------------|-----|
| | | Type | Conc | Method | Rate g/hL | Water L/ha | Rate kg ai/ha | day |
| Peppers | USA, exc CA | EC | 360 g/L | Spray | | 93 (air 187) | 0.56-1.12 | 4 |
| Peppers | USA, exc CA | EC | 360 g/L | Spray | | As above | 0.56 | 1 |
| Peppers | USA, exc CA | WSB | 50% | Spray | | 93 (air 9.3) | 0.56 | 1 |
| Peppers | USA, exc CA | WSB | 50% | Spray | | As above | 0.56-1.12 | 4 |
| Peppers | USA-CA | EC | 360 g/L | Spray | | 93 (air 9.3) | 0.56 | 1 |
| Peppers | USA-CA | EC | 360 g/L | Spray | | As above | 0.56 to 1.12 | 4 |
| Peppers | USA-CA | WSB | 50% | Spray | | As above | 0.56 | 1 |
| Peppers | USA-CA | WSB | 50% | Spray | | As above | 0.56-1.12 | 4 |
| Persimmons | Australia | EC | 350 g/L | Spray | 52.5-70 | | 0.735 | 7 |
| Pineapple | Central America | ECF | 350 g/L | Spray | | 200 to 600 | 1.4-2.1 | 7 |
| Pineapple | Namibia | SC | 475 g/L | Spray | | 30 to NS | 1.187 | 5 mo |
| Pineapple | S. Africa | SC | 475 g/L | Spray | | 30 to NS | 1.187 | 5 mo |
| Pineapple | USA-CA | EC | 360 g/L | Spray | | 93 (air 187) | 1.66-2.24 | 7 |
| Pistachio | Australia | EC | 350 g/L | Spray | 70 | NS | 0.735 | 14 |
| Pomegranate | Australia | EC | 350 g/L | Spray | 70 | NS | 0.735 | 14 |
| Pome fruit | Australia | EC | 350 g/L | Spray | 52.5-66.5 | NS | | 28 |
| Potato | Australia | EC | 350 g/L | Spray | | NS | 0.735 | 14 |
| Potato | Austria | WP | 32.9% | Spray | | NS | 0.19-0.26 | 35 |
| Potato | Canada | EC | 400 g/L | Spray | | NS | 0.6-0.8 | 0 |
| Potato | Canada | WP | 50% | Spray | | NS | 0.5-0.75 | 0 |
| Potato | Central America | EC | 350 g/L | Spray | | NS | 0.52-0.70 | 0 |
| Potato | Chile | EC | 320 g/L | Spray | | 400 to NS | 0.24-0.40 | 7 |
| Potato | Chile | WP | 48.25% | Spray | | NS | 0.48-0.72 | 7 |
| Potato | Iran | EC | 350 g/L | Spray | | NS | 0.35-0.70 | 15 |
| Potato | Japan | EMU | 30% | Spray | 0.03-0.04% | NS | | 7 |
| Potato | New Zealand | EC | 350 g/L | Spray | | NS | 0.70 | 0 |
| Potato | Peru | EC | 355g/L | Spray | 88 | NS | | 21 |
| Potato | Turkey | EC | 360 g/L | Spray | 108 | NS | | 14 |
| Potato | USA | EC | 360 g/L | Spray | | 93 (air 187) | 0.56 to 1.12 | 1 |
| Potato | USA, exc CA | EC | 360 g/L | Chemigation | | NS | | 1 |
| Potato | USA, | WSB | 50% | Spray | | 93 (air 9.3) | 0.56-1.12 | 1 |
| Potato | Zimbabwe | EC | 35% | Spray | | 225 to NS | 0.70 | 14 |
| Pumpkins | Australia | EC | 350 g/L | Spray | 66.5-70 | | 0.735 | 3 |
| Rape seed | Australia | EC | 350 g/L | | | | 0.175-0.35 | NR |
| Rambutans | Australia | EC | 350 g/L | Spray | 70 | | | 7 |
| Safflower | Australia | EC | 350 g/L | Spray | | | 0.175-0.35 | NR |
| Sapodillas | Australia | EC | 350 g/L | Spray | 70 | | | 14 |
| Soybean | Australia | EC | 350 g/L | Spray | | 50 to NS | 0.175-0.35 | NR |
| Soybean | Central America | EC | 350 g/L | | | NS | 0.52-0.70 | 0 |
| Soybean | Chile | WP | 48.25% | | | NS | 0.48-0.73 | NS |
| Soybean | Iran | EC | 350 g/L | | | NS | 1.05 | 15 |
| Soybean | Zimbabwe | EC | 35% | | | 200 to NS | 0.10-0.70 | 21 |
| Soybean | Zimbabwe | MO | 35% | | | 200 | 0.10-0.35 | 21 |
| Soybean | Brazil | EC | 350 g/L | | | 100 to 250 | 0.17-0.52 | 30 |
| Squash | Australia | EC | 350 g/L | Spray | 66.5-70 | | 0.735 | 3 |
| Squash | Canada | EC | 400 g/L | Spray | | NS | 0.60 | 2 |
| Squash | Canada | WP | 50% | Spray | | NS | 0.5-0.55 | 2 |
| Squash, summer | USA, exc CA | EC | 360 g/L | Spray | | 93 (air 187) | 0.56 to 1.12 | 2 |
| Squash, summer | USA, exc CA | WSB | 50% | Spray | | 93 (air 9.3) | 0.56-1.12 | 2 |
| Squash, summer | USA-CA | EC | 360 g/L | Spray | | NS | 0.56-1.12 | 2 |
| Squash, summer | USA-CA | WSB | 50% | Spray | | 93 (air 9.3) | 0.56-1.12 | 2 |
| Sunflower | Australia | EC | 350 g/L | Spray | | | 0.175-0.35 | NR |
| Sugarbeet | Canada | EC | 400 g/L | Spray | | NS | 0.8-1.10 | 45 |
| Sugarbeet | Chile | WP | 48.25% | Spray | | NS | 0.48-0.72 | NS |
| Sugarbeet | Japan | EC | 30% | Spray | 0.03-0.07% | NS | | 30 |
| Sw. potato | Australia | EC | 350 g/L | Spray | | NS | 0.735 | 14 |
| Sw. potato | Japan | EC | 30% | Spray | 0.06-0.07% | NS | | 7 |
| Sw. potato | Japan | GR | 3,3% | Spray | | NS | 1.32-1.9 | 7 |
| Sw. potato | Japan | DP | 5% | Spray | | NS | 0.15-0.20 | 7 |

| Crop | Country | Formulation | | Application | | | | PHI |
|------|---------|------|------|------|------|------|------|------|
| | | Type | Conc | Method | Rate g/hL | Water L/ha | Rate kg ai/ha | day |
| Sw. potato | USA, exc CA | EC | 360 g/L | Spray | | 93 (air 187) | 0.56 -1.12 | 1 |
| Sw. potato | USA, exc CA | EC | 360 g/L | Soil | | 93 (air 187) | 1.12-2.24 | NS |
| Sw. potato | USA, exc CA | WSB | 50% | Spray | | 93 (air 9.3) | 0.563-1.12 | 1 |
| Sw. potato | USA-CA | EC | 360 g/L | Spray | | 93 (air 9) to NS | 0.56 to NS | 1 |
| Sw. potato | USA-CA | WSB | 50% | Spray | | 93 (air 9.3) | 0.56 | 1 |
| Tamarillo | Australia | EC | 350 g/L | Spray | 70 | | | 7 |
| Taro | Australia | EC | 350 g/L | Spray | 70 | | | 14 |
| Tea | China | EC | 350 g/L | Spray | | NS | 0.06-0.08 | NS |
| Tea | Japan | EC | 30% | Spray | 0.04-0.06% | NS | | GS |
| Tea | Malaysia | EC | 33% | Spray | | NS | 0.59 | 30 |
| Tomato | Angola | EC | 475g/L | Spray | | 500-1500 | 0.23-0.70 | NS |
| Tomato | Australia | EC | 350 g/L | Spray | 66.5 | NS | 0.735 | 3 |
| Tomato | Canada | G | 400 g/L | Spray | 50-60 | NS | | 2 |
| Tomato | Canada | EC | 400 g/L | Spray | | NS | 0.60-1.10 | 2 |
| Tomato | Canada | WP | 50% | Spray | | NS | 0.50-1.12 | 2 |
| Tomato G | Canada | G | 50% | Spray | | NS | 0.50-0.75 | 2 |
| Tomato | Central America | EC | 350 g/L | Spray | | NS | 0.52-0.70 | 1 |
| Tomato | Chile | EC | 320 g/L | Spray | | 400 to NS | 0.24-0.40 | 5 |
| Tomato | Chile | WP | 48.25% | Spray | | NS | 0.48-0.72 | NS |
| Tomato | Cyprus | WP | 470 g/kg | Spray | 70-94 | NS | | 4 |
| Tomato | Cyprus | EC | 350 g/L | Spray | 52-122 | NS | | 1 |
| Tomato | Ecuador | CE | 350 g/L | Spray | | NS | 0.35-0.52 | 14 |
| Tomato | Greece | EC | 350 g/L | Spray | 52-122 | NS | | 1 |
| Tomato | Greece | WP | 470 g/kg | Spray | 70-94 | NS | | 4 |
| Tomato | Morocco | CS | 330 g/L | Spray | 57 | NS | | NS |
| Tomato | Japan | EC | 30% | Spray | 0.03-0.06% | NS | | 14 |
| Tomato | Japan | EC | 35% | Spray | 0.035% | NS | | 14 |
| Tomato | Mozambique | SC | 47.5% | Spray | | 62.5 -375 | 0.47-0.72 | NS |
| Tomato | Namibia | SC | 475 g/L | Spray | 47.5 | 30 to NS | | 1 |
| Tomato | N. Zealand | EC | 350 g/L | Spray | | NS | 0.42-0.70 | 2 |
| Tomato | South Africa | SC | 475 g/L | Spray | | 62 to 1500 | 0.24-0.71 | 1 |
| Tomato | Southern EU | SC | 330 g/L | Spray | | NS | < 0.53 | 3 |
| Tomato | Spain | EC | 350 g/L | Spray | 0.15-0.30% | NS | | 3 |
| Tomato | Spain | EC | 350 g/L | Spray | 52-122 | NS | | 3 |
| Tomato F/G | USA, exc CA | EC | 360 g/L | Spray | | 93 (air 187) | 0.56 to 1.12 | 2 |
| Tomato F/G | USA, exc CA | EC | 360 g/L | Spray | | As above | | 2 |
| Tomato F/G | USA, exc CA | WP | 50% | Spray | | 93-187 (air 9) | 0.56-1.12 | 2 |
| Tomato F/G | USA, exc CA | WP | 50% | Spray | 58 | As above | | 2 |
| Tomato F/G | USA-CA | EC | 360 g/L | Spray | | 935-1870 | 0.56 to 1.12 | 2 |
| Tomato F/G | USA-CA | EC | 360 g/L | Spray | 2.3 qt /378l | As above | | 2 |
| Tomato F/G | USA-CA | WP | 50% | Spray | | 93 (air 9.3) | 0.56-1.12 | 2 |
| Tomato F/G | USA-CA | WP | 50% | Spray | 58 | As above | | 2 |
| Tomato | Venezuela | EC | 350 g/L | Spray | | NS | 0.35-0.70 | 7 |
| Tomato | Venezuela | SC | 350 g/L | Spray | | NS | 0.35 | NS |
| Tomato | Zimbabwe | EC/ | 35 -30% | Spray | 35-67 | NS | | 7 |
| Vegetables | Algeria | EC | 350 g/L | Spray | 52 | NS | | 15 |
| Vegetables | Cameroon | EC | 350 g/L | Spray | | NS | 0.26- | 28 |
| Vegetables | Ecuador | EC | 350 g/L | Spray | | NS | 0.35-0.52 | 14 |
| Vegetables | Ethiopia | ULV | 250 g/L | Spray | | NS | 0.61-0.75 | 7 -14 |
| Vegetables | Ethiopia | EC | 350 g/L | Spray | | 20 to 300 | 0.70 | 7 -14 |
| Vegetables | Morocco | EC | 330 g/L | Spray | 57 | NS | | 30 |
| Vegetables | Sudan | EC | 500 g/L | Spray | NS | NS | | NS |
| Vegetables | Turkey | EC | 360 g/L | Spray | | NS | 0.54 | 14 |
| Vegetables | Turkey | WP | 32.9% | Seed | 5 g//kg seed | NS | | NA |
| Vegetables | Turkey | WP | 32.9% | Soil app | | NS | 5 | NS |
| Vegetables | Turkey | WP | 32.9% | Bait | | NS | 26 | NS |
| Vegetables | Turkey | WP | 32.9% | Spray | | NS | 0.50-0.66 | 15 |
| Vegetables | Turkey | EC | 360 g/L | Bait | 350ml/500g | NS | | 3wp |

| Crop | Country | Formulation | | Application | | | | PHI |
|------|---------|------|------|--------|-------------|--------------|-----------------|-----|
| | | Type | Conc | Method | Rate g/hL | Water L/ha | Rate kg ai/ha | day |
| Vegetables | Turkey | EC | 360 g/L | Spray | | NS | 0.35-1.08 | 7wp |
| Vegetables | Venezuela | CE | 350 g/L | Spray | | NS | 0.52-0.87 | 21 |
| Watermelon | Chile | EC | 320 g/L | Spray | | 400 to NS | 0.24-0.40 | 7 |
| Watermelon | Cyprus, Greece | WP | 470 g/kg | Spray | 70-94 | NS | | 7 |
| Watermelon | Greece | EC | 350 g/L | Spray | 52-122 | NS | | 1 |
| Watermelon | Japan | EC | 30% | Spray | 0.03-0.06% | NS | | 14 |
| Watermelon | Japan | GR | 3.3% | Spray | | NS | 1-2 | GS |

## RESIDUES RESULTING FROM SUPERVISED TRIALS

The Meeting received information on supervised trials for the following commodities:

| Commodity | Table |
|-----------|-------|
| Lemon | 33 |
| Mandarin | 34 |
| Orange | 35 |
| Apple | 36 |
| Pear | 37 |
| Cherry | 38 |
| Apricot | 39 |
| Nectarine | 40 |
| Peach | 41 |
| Grapes | 42 |
| Avocado | 43 |
| Custard apple | 44 |
| Litchi | 45 |
| Mango | 46 |
| Papaya | 47 |
| Persimmon | 48 |
| Pineapple | 49 |
| Cabbage, head | 50 |
| Cabbage, Savoy | 51 |
| Broccoli | 52 |
| Brussels sprouts | 53 |
| Cauliflower | 54 |
| Cucumber | 55 |
| Melons (Australia) | 56 |
| Melons (Europe, USA) | 57 |
| Squash, summer | 58 |
| Zucchini | 59 |
| Peppers, sweet (Spain, USA) | 60 |
| Peppers, sweet (Australia) | 61 |
| Tomato (field) | 62 |
| Tomato (indoor) | 63 |
| Eggplant | 64 |
| Sweet corn | 65 |
| Beans | 66 |
| Peas | 67 |
| Soybean | 68 |
| Beetroot | 69 |
| Carrot | 70 |

| Commodity | Table |
|---|---|
| Potato | 71 |
| Sweet potato | 72 |
| Sugar beet | 73 |
| Celery | 74 |
| Rhubarb | 75 |
| Hazelnut | 76 |
| Macadamia | 77 |
| Cotton | 78 |
| Cocoa | 79 |
| Coffee | 80 |
| Tea | 81 |
| Sugar beet leaves and head | 82 |
| Forage and vines beans | 83 |
| Pea hay | 84 |
| Cocoa shell | 85 |
| Cotton lint | 86 |

Trials were well documented. The residues were expressed for the total compound endosulfan and in the majority of cases for alpha, beta and endosulfan sulfate. Laboratory reports included method validation and recoveries with spiking at residue levels. Dates of analysis or duration of analysis were reported.

In cases where the result was reported as ND, the LOQ of 0.02 mg/kg was usually applicable. For the total in the case of the LOQ of each metabolite, the residue was reported according to the FAO guideline as indicated in the Table 32 below.

Table 32. Reported residue where one or more of the analytes was present at the relevant LOQ.

| alpha endosulfan | beta endosulfan | endosulfan sulfate | residue total |
|---|---|---|---|
| < 0.0X | < 0.0X | < 0.0X | < 0.0X |
| < 0.0X | 0.10 | 0.05 | 0.15 |
| < 0.0X | < 0.0Y | 0.03 | 0.03 |
| < 0.0X | < 0.0Y | < 0.0Z | < maximum but < 0.05 |
|  |  |  |  |
| alpha endosulfan | beta endosulfan | endosulfan sulfate | residue total |
| < 0.005 | < 0.005 | < 0.005 | < 0.005 |
| < 0.01 | < 0.01 | < 0.01 | < 0.01 |
| < 0.02 | < 0.02 | < 0.02 | < 0.02 |
| < 0.05 | < 0.05 | < 0.05 | < 0.05 |

In the case where the LOQs were different, the higher level was taken into account with a limit of 0.05 mg/kg

No conversion of the content of endosulfan sulfate to endosulfan was introduced due to the fact that the molecular weights are almost identical.

Indices were "c" for calculated, "d" for determined and "co" for control trial for the raw material when the trials were conducted for processed studies.

Table 33. Endosulfan residues in lemons resulting from supervised trials in Australia.

| LEMONS | Application | | | | PHI | Sample | Residues of alpha, beta | Total |
| Country, Year of trial *Variety* Report | Form. | Method | g ai./hL | L/ha | No. | days | analysed | endosulfan, endosulfan sulfate, mg/kg | Residues, mg/kg |
|---|---|---|---|---|---|---|---|---|---|
| Australia 2000 Beverford *lisbon* N° 1/10/543 | EC | spray | 10.5 | 2000 | 4 | 0 1 3 7 | fruit fruit fruit fruit | 0.10, 0.14, 0.012 0.049, 0.10, 0.008 0.052, 0.12, 0.007 0.038,0.14, 0.012 | 0.25 0.16 0.17 <u>0.19</u> |
| Australia 2000 Beverford *lisbon* N° 1/10/543 | EC | spray | 21 | 2000 | 4 | 0 3 0 | fruit fruit fruit | 0.36, 0.35, 0.013 0.21, 0.47, 0.024 0.007,< 0.005,< 0.005 | 0.72 0.70 0.007co |
| Australia 2000 Koah *lisbon* N° 1/10/543 | EC | spray | 10.5 | 1428 | 4 | 0 1 3 7 | fruit fruit fruit | 0.064, 0.069, 0.021 0.014, 0.021, 0.013 0.006, 0.014,0.013 < 0.005, 0.008, 0.013 | 0.15 0.05 <u>0.03</u> 0.02 |
| Australia 2000 Koah *lisbon* N° 1/10/543 | EC | spray | 21. | 1428 | 4 | 0 +3 +3 0co | fruit peel flesh fruit | 0.068, 0.072, 0.021 0.063, 0.20, 0.10 < 0.005,< 0.005, < 0.005 0.019, 0.012,0.005 | 0.16 .0.36 < 0.005 0.04 |
| Australia 2000 Montacute *lisbon* N° 1/10/543 | EC | spray | 10.5 | 2000 | 4 | 0 1 3 7 | fruit fruit fruit fruit | 0.082, 0.095., 0.016 0.058, 0.099, 0.021 0.045, 0.090, 0.029 0.024, 0.074,0.033 | 0.19 0.18 <u>0.16</u> 0.13 |
| Australia 2000 Montacute *lisbon* N° 1/10/543 | EC | spray | 21 | 2000 | 4 | 0 3 | fruit fruit | 0.18,0.23, 0.028 0.088, 0.22, 0.034 | 0.44 0.34 |

co control trial

Table 34. Endosulfan residues in mandarins resulting from supervised trials in Spain, Italy, Greece, and Australia.

| MANDARINS | Application | | | | PHI | Sample | Residues of alpha, beta | Total |
| Country, Year of trial *Variety* Report | Form. | Method | g ai/hL | L/ha | No. | days | analysed | endosulfan, endosulfan sulfate, mg/kg | Residues, mg/kg |
|---|---|---|---|---|---|---|---|---|---|
| Spain1999 *Clemenules* C016672 *ER 99 ECS 751* | CS* | Spray Interval 12d BBCH 79 | 1.05 | 3000 (0.035% ai) | 2 | 0 22 | Fruit Fruit | 0.46, 0.25, 0.042 0.089, 0.063, 0.048 | 0.75 0.20 |
| Spain1999 *Clemenules* C016672 *ER 99 ECS 751* | CS* | Spray Interval 15d BBCH 79 | 1.05 | 3000 (0.035% ai) | 2 | 0 21 | Fruit Fruit | 0.48, 0.29, 0.047 0.20, 0.15, 0.11 | 0.82 0.46 |
| Spain1999 *Clemenules* C016672 *ER 99 ECS 751* | CS* | Spray Interval 14 d BBCH 79 | 1.05 | 3000 (0.035% ai) | 2 | 0 21 | Fruit Fruit | 0.32, 0.21, 0.066 0.088, 0.079, 0.087 | 0.60 0.25 |

| MANDARINS | Application | | | | | PHI days | Sample analysed | Residues of alpha, beta endosulfan, endosulfan sulfate, mg/kg | Total Residues, mg/kg |
|---|---|---|---|---|---|---|---|---|---|
| Country, Year of trial _Variety_ Report | Form. | Method | g ai/hL | L/ha | No. | | | | |
| Italy1999 _Oroval_ C016672 _ER 99 ECS 751_ | CS* | Spray Interval 13 d BBCH 79 | 1.05 | 3000 (0.035% ai) | 2 | 0 | Fruit | 0.52. 0.32, 0.046 | 0.89 |
| | | | | | | 21 | Fruit | 0.24, 0.16, 0.044 | 0.44 |
| Italy1999 _Comune_ C016672 _ER 99 ECS 751_ | CS* | Spray Interval15 d BBCH 79, 81 | 1.05 | 3000 (0.035% ai) | 2 | 0 | Fruit | 0.62, 0.41, 0.13 | 1.16 |
| | | | | | | 20 | Fruit | 0.29, 0.26, 0.17 | 0.72 |
| Spain1998 _Clemenules_ C005296 _ER 98 ECS 751_ | CS* | Spray Interval 14 d BBCH 75, 81 | 1.05 | 3000 (0.035% ai) | 2 | 0 | Peel | 2.4, 1.4, 0.11 | 3.91 |
| | | | | | | 7 | Peel | 2.3, 1.4, 0.21 | 3.91 |
| | | | | | | 14 | Peel | 1.6, 0.92, 0.21 | 2.73 |
| | | | | | | 20 | Peel | 0.79, 0.54, 0.24 | 1.57 |
| | | | | | | 29 | Peel | 0.65, 0.43, 0.27 | 1.30 |
| | | | | | | 0 | Pulp | < 0.02, ND, ND | < 0.02 |
| | | | | | | 7 | Pulp | < 0.02, < 0.02, ND | < 0.02 |
| | | | | | | 14 | Pulp | < 0.02, ND, ND | < 0.02 |
| | | | | | | 20 | Pulp | ND, ND, ND | < 0.02 |
| | | | | | | 29 | Pulp | < 0.02, ND, ND | < 0.02 |
| | | | | | | 0 | Fruit[c] | 0.56, 0.33, 0.03 | 0.92 |
| | | | | | | 7 | Fruit[c] | 0.58, 0.36, 0.06 | 1.0 |
| | | | | | | 14 | Fruit[c] | 0.41, 0.24, 0.06 | 0.71 |
| | | | | | | 20 | Fruit[c] | 0.2, 0.14, 0.06 | 0.40 |
| | | | | | | 29 | Fruit[c] | 0.17, 0.12,0.08 | 0.37 |
| Spain1998 _Ortanique_ C005296 _ER 98 ECS 751_ | CS* | Spray Interval 14 d BBCH 75, 79 | 1.05 | 3000 (0.035% ai) | 2 | 0 | Peel | 1.8, 0.97, 0.15 | 2.92 |
| | | | | | | 7 | Peel | 2.1, 1.3, 0.22 | 3.62 |
| | | | | | | 14 | Peel | 0.95, 0.68, 0.26 | 1.89 |
| | | | | | | 20 | Peel | 0.80, 0.51, 0.22 | 1.53 |
| | | | | | | 29 | Peel | 0.59, 0.46, 0.27 | 1.32 |
| | | | | | | 0 | Pulp | < 0.02, ND, ND | < 0.02 |
| | | | | | | 7 | Pulp | < 0.02, ND, ND | < 0.02 |
| | | | | | | 14 | Pulp | ND, ND, ND | < 0.02 |
| | | | | | | 20 | Pulp | ND, ND, ND | < 0.02 |
| | | | | | | 29 | Pulp | ND, ND, ND | < 0.02 |
| | | | | | | 0 | Fruit[c] | 0.42, 0.23, 0.04 | 0.69 |
| | | | | | | 7 | Fruit[c] | 0.47, 0.29, 0.06 | 0.82 |
| | | | | | | 14 | Fruit[c] | 0.22, 0.16, 0.06 | 0.44 |
| | | | | | | 20 | Fruit[c] | 0.18, 0.12, 0.05 | 0.35 |
| | | | | | | 29 | Fruit[c] | 0.13,0.11, 0.06 | 0.30 |
| Spain1998 _Clemenules_ C005296 _ER 98 ECS 751_ | CS* | Spray Interval 14 d BBCH 75, 79 | 1.05 | 3000 (0.035% ai) | 2 | 0 | Peel | 1.8,1.1, 0.13 | 3.03 |
| | | | | | | 7 | Peel | 1.5,0.86, 0.19 | 2.55 |
| | | | | | | 14 | Peel | 0.83, 0.50, 0.19 | 1.52 |
| | | | | | | 22 | Peel | 0.65, 0.44, 0.26 | 1.35 |
| | | | | | | 29 | Peel | 0.33, 0.24, 0.20 | 0.77 |
| | | | | | | 0 | Pulp | < 0.02, ND, ND | < 0.02 |
| | | | | | | 7 | Pulp | ND, ND, ND | < 0.02 |
| | | | | | | 14 | Pulp | < 0.02, ND, ND | < 0.02 |
| | | | | | | 22 | Pulp | ND, ND, ND | < 0.02 |
| | | | | | | 29 | Pulp | < 0.02, ND, ND | < 0.02 |
| | | | | | | 0 | Fruit[c] | 0.47, 0.29, 0.04 | 0.80 |
| | | | | | | 7 | Fruit[c] | 0.41, 0.24, 0.06 | 0.71 |
| | | | | | | 14 | Fruit[c] | 0.22, 0.13, 0.06 | 0.41 |
| | | | | | | 22 | Fruit[c] | 0.17, 0.12, 0.07 | 0.36 |
| | | | | | | 29 | Fruit[c] | 0.09, 0.07, 0.06 | 0.22 |

| MANDARINS Country, Year of trial *Variety* Report | Form. | Method | g ai/hL | L/ha | No. | PHI days | Sample analysed | Residues of alpha, beta endosulfan, endosulfan sulfate, mg/kg | Total Residues, mg/kg |
|---|---|---|---|---|---|---|---|---|---|
| Italy1998 *Clementino Comun* C005296 *ER 98 ECS 751* | CS* | Spray Interval 14 d BBCH 81, 81 | 1.05 | 3000 (0.035% ai) | 2 | 0 | Peel | 2.6, 1.4, 0.33 | 4.33 |
| | | | | | | 7 | Peel | 1.6, 0.88, 0.25 | 2.73 |
| | | | | | | 14 | Peel | 1.6, 0.91, 0.30 | 2.81 |
| | | | | | | 21 | Peel | 0.86, 0.55, 0.24 | 1.65 |
| | | | | | | 28 | Peel | 1.1, 0.69, 0.25 | 2.04 |
| | | | | | | 7 | Pulp | 0.03, 0.02, ND | 0.05 |
| | | | | | | 14 | Pulp | 0.04, 0.03, ND | 0.07 |
| | | | | | | 21 | Pulp | 0.05, 0.03, ND | 0.08 |
| | | | | | | 28 | Pulp | 0.06, 0.04, ND | 0.10 |
| | | | | | | 0 | Fruit[c] | 0.74, 0.4, 0.10 | 1.24 |
| | | | | | | 7 | Fruit[c] | 0.51, 0.28, 0.08 | 0.87 |
| | | | | | | 14 | Fruit[c] | 0.58, 0.34, 0.11 | 1.03 |
| | | | | | | 21 | Fruit[c] | 0.26, 0.16, 0.07 | 0.49 |
| | | | | | | 28 | Fruit[c] | 0.34, 0.22, 0.08 | 0.64 |
| Italy1998 *Oroval* C005296 *ER 98 ECS 751* | CS* | Spray Interval 14 d BBCH 79, 81 | 1.05 | 3000 (0.035% ai) | 2 | 0 | Peel | 2.6, 1.5, 0.11 | 4.21 |
| | | | | | | 7 | Peel | 2.1, 1.2, 0.21 | 3.51 |
| | | | | | | 14 | Peel | 1.2, 0.75, 0.28 | 2.23 |
| | | | | | | 21 | Peel | 0.43, 0.33, 0.25 | 1.01 |
| | | | | | | 28 | Peel | 0.15, 0.12, 0.13 | 0.40 |
| | | | | | | 0 | Pulp | 0.03, < 0.02, ND | 0.03 |
| | | | | | | 7 | Pulp | 0.03, < 0.02, ND | 0.03 |
| | | | | | | 14 | Pulp | 0.05, 0.02, ND | 0.07 |
| | | | | | | 21 | Pulp | 0.02, < 0.02, ND | 0.02 |
| | | | | | | 28 | Pulp | < 0.02, < 0.02, ND | < 0.02 |
| | | | | | | 0 | Fruit[c] | 0.72, 0.41, 0.04 | 1.17 |
| | | | | | | 7 | Fruit[c] | 0.68, 0.38, 0.07 | 1.13 |
| | | | | | | 14 | Fruit[c] | 0.41, 0.25, 0.1 | 0.76 |
| | | | | | | 21 | Fruit[c] | 0.12, 0.09, 0.07 | 0.28 |
| | | | | | | 28 | Fruit[c] | 0.04, 0.04, 0.04 | 0.12 |
| Spain1998 *Clemenules* C003107 *ER 98 ECS 741* | 33EC | Spray Interval 14 d BBCH 75, 81 | 1.05 | 3000 (0.035% ai) | 2 | 0 | Peel | 1.4, 1.4, 0.48 | 3.28 |
| | | | | | | 0 | Pulp | < 0.02, < 0.02, ND | < 0.02 |
| | | | | | | 20 | Peel | 0.06, 0.21, 0.52 | 0.79 |
| | | | | | | 20 | Pulp | ND, ND, ND | < 0.02 |
| | | | | | | 0 | Fruit[c] | 0.35, 0.35, 0.12 | 0.82 |
| | | | | | | 20 | Fruit[c] | 0.02, 0.06, 0.14 | 0.22 |
| Spain1998 *Ortanique* C003107 *ER 98 ECS 741* | 33EC | Spray Interval 14 days BBCH 75, 79 | 1.05 | 3000 (0.035% ai) | 2 | 0 | Peel | 1.10, 0.96, 0.24 | 2.30 |
| | | | | | | 0 | Pulp | < 0.02, ND, ND | < 0.02 |
| | | | | | | 20 | Peel | 0.04, 0.15, 0.36 | 0.55 |
| | | | | | | 20 | Pulp | ND, ND, ND | < 0.02 |
| | | | | | | 0 | Fruit[c] | 0.28, 0.24, 0.07 | 0.59 |
| | | | | | | 20 | Fruit[c] | 0.02, 0.04, 0.10 | 0.16 |
| Italy1998 *Clementino comune* C003107 *ER 98 ECS 741* | 33EC | Spray Interval 14 d BBCH 81, 81 | 1.05 | 3000 (0.035% ai) | 2 | 0 | Peel | 1.10, 1.20, 0.49 | 2.79 |
| | | | | | | 0 | Pulp | < 0.02, < 0.02, ND | < 0.02 |
| | | | | | | 21 | Peel | 0.17, 0.34, 0.58 | 1.09 |
| | | | | | | 21 | Pulp | < 0.02, < 0.02, ND | < 0.02 |
| | | | | | | 0 | Fruit[c] | 0.27, 0.29, 0.12 | 0.68 |
| | | | | | | 21 | Fruit[c] | 0.05, 0.09, 0.14 | 0.28 |
| Italy1998 *Oroval* C003107 *ER 98 ECS 741* | 33EC | Spray Interval 14 d BBCH 79, 81 | 1.05 | 3000 (0.035% ai) | 2 | 0 | Peel | 1.10, 1.10, 0.42 | 2.62 |
| | | | | | | 0 | Pulp | 0.03, 0.03, ND | 0.06 |
| | | | | | | 21 | Peel | 0.13, 0.29, 0.58 | 1.00 |
| | | | | | | 21 | Pulp | < 0.02, < 0.02, ND | < 0.02 |
| | | | | | | 0 | Fruit[c] | 0.38, 0.38, 0.15 | 0.91 |
| | | | | | | 21 | Fruit[c] | 0.05, 0.10, 0.19 | 0.34 |

| MANDARINS | Application | | | | | PHI | Sample | Residues of alpha, beta | Total |
|---|---|---|---|---|---|---|---|---|---|
| Country, Year of trial *Variety* Report | Form. | Method | g ai/hL | L/ha | No. | days | analysed | endosulfan, endosulfan sulfate, mg/kg | Residues, mg/kg |
| Spain1997 *Clemenvilla* C001465 *ER 97 ECS741* | 35EC | Spray Interval 14 d BBCH 77, 78 | 1.05 | 3000 (0.035% ai) | 2 | 0 | Peel | 1.10, 0.89, 0.69 | 2.68 |
| | | | | | | 0 | Pulp | ND, < 0.02, < 0.02 | < 0.02 |
| | | | | | | 7 | Peel | 0.28, 0.47, 1.10 | 1.85 |
| | | | | | | 7/14 | Pulp | ND, ND, < 0.02 | < 0.02 |
| | | | | | | 14 | Peel | 0.17, 0.31, 0.94 | 1.42 |
| | | | | | | 21 | Peel | 0.12, 0.19, 1.20 | 1.51 |
| | | | | | | 21 | Pulp | ND, ND, < 0.02 | < 0.02 |
| | | | | | | 28 | Peel | 0.04, 0.05, 0.34 | 0.48 |
| | | | | | | 28 | Pulp | ND, ND, < 0.02 | < 0.02 |
| | | | | | | 0 | Fruit[c] | 0.27, 0.22, 0.17 | 0.65 |
| | | | | | | 7 | Fruit[c] | 0.07, 0.12, 0.26 | 0.45 |
| | | | | | | 14 | Fruit[c] | 0.05, 0.08, 0.22 | 0.35 |
| | | | | | | 21 | Fruit[c] | 0.03, 0.05, 0.27 | 0.35 |
| | | | | | | 28 | Fruit[c] | < 0.02, < 0.02, 0.08 | 0.08 |
| Spain1997 *Satsuma* C001465 *ER 97 ECS741* | 35EC | Spray Interval 14 d BBCH 78, 78 | 1.05 | 3000 (0.035% ai) | 2 | 0 | Peel | 0.76, 0.74, 0.32 | 1.82 |
| | | | | | | 0 | Pulp | ND, ND, ND | < 0.02 |
| | | | | | | 7 | Peel | 0.08, 0.19, 0.46 | 0.73 |
| | | | | | | 7 | Pulp | ND, ND, ND | < 0.02 |
| | | | | | | 14 | Peel | 0.06, 0.12, 0.46 | 0.64 |
| | | | | | | 14 | Pulp | ND, ND, ND | < 0.02 |
| | | | | | | 21 | Peel | 0.06, 0.13, 0.48 | 0.67 |
| | | | | | | 21 | Pulp | ND, ND, ND | < 0.02 |
| | | | | | | 28 | Peel | 0.04, 0..07, 0.32 | 0.43 |
| | | | | | | 28 | Pulp | ND, ND, ND | < 0.02 |
| | | | | | | 0 | Fruit[c] | 0.20, 0.20, 0.09 | 0.49 |
| | | | | | | 7 | Fruit[c] | 0.03, 0.06, 0.13 | 0.22 |
| | | | | | | 14 | Fruit[c] | 0.02, 0.04, 0.13 | 0.19 |
| | | | | | | 21 | Fruit[c] | 0.02, 0.04, 0.14 | 0.20 |
| | | | | | | 28 | Fruit[c] | < 0.02, 0.03, 0.10 | 0.13 |
| Greece1997 *Climentines* C001465 *ER 97 ECS 741* | 33EC | Spray Interval 14 d BBCH 79, 81 | 1.05 | 3000 (0.035% ai) | 2 | 0 | Peel | 1.20, 0.86, 0.41 | 2.47 |
| | | | | | | 0 | Pulp | ND, ND, < 0.02 | < 0.02 |
| | | | | | | 7 | Peel | 0.14, 0.28, 0.36 | 0.78 |
| | | | | | | 7/14 | Pulp | ND, ND, < 0.02 | < 0.02 |
| | | | | | | 14 | Peel | 0.07, 0.17, 0.34 | 0.58 |
| | | | | | | 21 | Peel | 0.05, 0.12, 0.28 | 0.45 |
| | | | | | | 21 | Pulp | ND, ND, < 0.02 | < 0.02 |
| | | | | | | 28 | Peel | 0.07, 0.14, 0.42 | 0.63 |
| | | | | | | 28 | Pulp | ND, < 0.02, < 0.02 | < 0.02 |
| | | | | | | 0 | Fruit[c] | 0.51, 0.36, 0.18 | 1.05 |
| | | | | | | 7 | Fruit[c] | 0.07, 0.13, 0.16 | 0.36 |
| | | | | | | 14 | Fruit[c] | 0.03, 0.07, 0.13 | 0.23 |
| | | | | | | 21 | Fruit[c] | 0.02, 0.05, 0.10 | 0.17 |
| | | | | | | 28 | Fruit[c] | 0.03, 0.06, 0.17 | 0.26 |
| Italy1997 *Oroval* C001465 *ER 97 ECS 741* | 33EC | Spray Interval 14 days BBCH 75, 79 | 1.05 | 3000 (0.106%) | 2 | 0 | Peel | 0.87, 0.72, 0.18 | 1.77 |
| | | | | | | 0 | Pulp | < 0.02, < 0.02, ND | < 0.02 |
| | | | | | | 7 | Peel | 0.05, 0.21, 0.33 | 0.59 |
| | | | | | | 7 | Pulp | ND, ND, ND | < 0.02 |
| | | | | | | 14 | Peel | 0.02, 0.1, 0.26 | 0.38 |
| | | | | | | 14 | Pulp | ND, < 0.02, ND | < 0.02 |
| | | | | | | 21 | Peel | 0.04, 0.11, 0.34 | 0.49 |
| | | | | | | 21 | Pulp | ND, ND, ND | < 0.02 |
| | | | | | | 28 | Peel | < 0.02, 0.04, 0.15 | 0.21 |
| | | | | | | 28 | Pulp | ND, ND, < 0.02 | < 0.02 |
| | | | | | | 0 | Fruit[c] | 0.19, 0.16, 0.04 | 0.39 |
| | | | | | | 7 | Fruit[c] | < 0.02, 0.05, 0.08 | 0.13 |
| | | | | | | 14 | Fruit[c] | < 0.01, 0.04, 0.09 | 0.13 |
| | | | | | | 21 | Fruit[c] | < 0.02, 0.04, 0.10 | 0.14 |
| | | | | | | 28 | Fruit[c] | < 0.01, < 0.02, 0.05 | 0.05 |

| MANDARINS | Application | | | | | PHI | Sample | Residues of alpha, beta endosulfan, endosulfan sulfate, mg/kg | Total Residues, mg/kg |
| Country, Year of trial *Variety* Report | Form. | Method | g ai/hL | L/ha | No. | days | analysed | | |
|---|---|---|---|---|---|---|---|---|---|
| Italy1997 *Clementino comune* C001465 *ER 97 ECS741* | 33EC | Spray Interval 14 days BBCH 75, 79 | 1.05 | 3000 (0.035% ai) | 2 | 0 | Peel | 0.76, 0.64, 0.22 | 1.62 |
| | | | | | | 0 | Pulp | < 0.02, ND, ND | < 0.02 |
| | | | | | | 7 | Peel | 0.07, 0.19, 0.26 | 0.52 |
| | | | | | | 7 | Pulp | ND, ND, < 0.02 | < 0.02 |
| | | | | | | 14 | Peel | 0.03, 0.08, 0.37 | 0.48 |
| | | | | | | 14 | Pulp | ND, ND, < 0.02 | < 0.02 |
| | | | | | | 21 | Peel | 0.02, 0.07, 0.34 | 0.43 |
| | | | | | | 21 | Pulp | ND, ND, < 0.02 | < 0.02 |
| | | | | | | 28 | Peel | < 0.02,0.04, 0.24 | 0.28 |
| | | | | | | 28 | Pulp | ND, ND, < 0.02 | < 0.02 |
| | | | | | | 0 | Fruit[c] | 0.16, 0.14, 0.05 | 0.35 |
| | | | | | | 7 | Fruit[c] | 0.03, 0.06, 0.08 | 0.17 |
| | | | | | | 14 | Fruit[c] | < 0.02, 0.03, 0.10 | 0.13 |
| | | | | | | 21 | Fruit[c] | < 0.02, 0.03, 0.10 | 0.13 |
| | | | | | | 28 | Fruit[c] | < 0.02, < 0.02, 0.08 | 0.08 |
| Australia 2000 Emerald Creek *imperial* *N° 1/10/542* | EC | spray | 0.067 | 645 (10.4 g/hL) | 4 | 0 | fruit | 0.034, 0.046, 0.012 | 0.09 |
| | | | | | | 1 | fruit | 0.031, 0.076, 0.040 | 0.15 |
| | | | | | | 3 | fruit | 0.025, 0.057, 0.024 | 0.11 |
| | | | | | | 7 | fruit | 0.011,0.039, 0.033 | 0.08 |
| Australia 2000 Emerald Creek *imperial* *N° 1/10/542* | EC | spray | 0.13 | 645 | 4 | 0 | fruit | 0.037, 0.099, 0.10 | 0.24 |
| | | | | | | 3 | fruit | 0.039, 0.092, 0.048 | 0.18 |
| Australia 2000 Taylorville *kara* *N° 1/10/542* | EC | spray | 0.21 | 2000 (10.5 g/hL) | 4 | 0 | fruit | 0.064,0.072., 0.018 | 0.15 |
| | | | | | | 1 | fruit | 0.021, 0.042, 0.026 | 0.09 |
| | | | | | | 3 | fruit | 0.013, 0.033, 0.025 | 0.07 |
| | | | | | | 7 | fruit | 0.006, 0.014,0.015 | 0.04 |
| Australia 2000 Taylorville *kara* *N° 1/10/542* | EC | spray | 0.42 | 2000 | 4 | 0 | fruit | 2.90, 1.90.0.11 | 4.91 |
| | | | | | | 3 | fruit | 0.03, 0.067, 0.044 | 0.14 |

Table 35. Endosulfan residues in oranges resulting from supervised trials in Spain, Italy, Greece, and Australia.

| ORANGES | Application | | | | | PHI | Sample | Residues of alpha, beta endosulfan, endosulfan sulfate, mg/kg | Total Residues, mg/kg |
| Country, Year of trial Variety Report | Form. | Method | kg ai/ha/ applic'n | L/ha | No. | days | analysed | | |
|---|---|---|---|---|---|---|---|---|---|
| Spain1999 Navelina C016758 ER 99 ECS 758 | CS* | Spray Interval 14 d BBCH 81, 81 | 1.05 | 3000 (0.035% ) | 2 | 0 | Peel | 1.7, 0.99, 0.043 | 2.73 |
| | | | | | | 0 | Pulp | < 0.02, < 0.02, < 0.02 | < 0.02 |
| | | | | | | 0 | Fruitc | 0.59, 0.35, 0.027 | 0.97 |
| | | | | | | 7 | Peel | 1.5, 0.84, 0.059 | 2.40 |
| | | | | | | 7 | Pulp | < 0.02, < 0.02, < 0.02 | < 0.02 |
| | | | | | | 7 | Fruitc | 0.43, 0.25, 0.031 | 0.71 |
| | | | | | | 14 | Peel | 1.1, 0.65, 0.067 | 1.82 |
| | | | | | | 14 | Pulp | < 0.02, < 0.02, < 0.02 | < 0.02 |
| | | | | | | 14 | Fruitc | 0.31, 0.20, 0.034 | 0.54 |
| | | | | | | 21 | Peel | 0.73, 0.47, 0.058 | 1.26 |
| | | | | | | 21 | Pulp | < 0.02, < 0.02, < 0.02 | < 0.02 |
| | | | | | | 21 | Fruitc | 0.25, 0.16, 0.033 | 0.44 |
| | | | | | | 28 | Peel | 0.86, 0.57, 0.077 | 1.51 |
| | | | | | | 28 | Pulp | < 0.02, < 0.02, < 0.02 | < 0.02 |
| | | | | | | 28 | Fruitc | 0.27, 0.18, 0.038 | 0.49 |

| ORANGES | | Application | | | | PHI | Sample | Residues of alpha, | Total |
| Country, Year of trial Variety Report | Form. | Method | kg ai/ha/ applic'n | L/ha | No. | days | analysed | beta endosulfan, endosulfan sulfate, mg/kg | Residues, mg/kg |
|---|---|---|---|---|---|---|---|---|---|
| Italy1999 Navelina<br><br>C016758 ER 99 ECS 758 | CS* | Spray Interval<br><br>14 d BBCH 81,<br><br>83 | 1.05 | 3000 (0.035%) | 2 | 0<br>0<br><br>0<br>7<br><br>7<br>7<br>14<br>14<br>14<br>21<br>21 | Peel<br>Pulp<br><br>Fruitc<br>Peel<br><br>Pulp<br>Fruitc<br>Peel<br>Pulp<br>Fruitc<br>Peel<br>Pulp | 1.6, 0.95, 0.087<br>< 0.02, < 0.02, < 0.02<br><br>0.56, 0.34, 0.043<br>1.80, 1.3, 0.13<br><br>< 0.02, < 0.02, < 0.02<br>0.66, 0.45, 0.059<br>1.2, 0.76, 0.08<br>0.038, 0.03, 0.027<br>0.45, 0.28, 0.053<br>1.1, 0.75, 0.12<br>< 0.02, < 0.02, < 0.02 | 2.64<br>< 0.02<br><br>0.94<br>3.23<br><br>< 0.02<br>1.17<br>2.04<br>0.10<br>0.78<br>1.97<br>< 0.02 |
| | | | | | | 21<br>28<br>28<br>28 | Fruitc<br>Peel<br>Pulp<br>Fruitc | 0.41, 0.28, 0.055<br>0.95, 0.67, 0.16<br>< 0.02, < 0.02, < 0.02<br>0.35, 0.25, 0.069 | 0.74<br>1.78<br>< 0.02<br>0.67 |
| Italy1999 Navelina<br><br>C016758 ER 99 ECS758 | CS* | Spray Interval<br><br>14 d BBCH 81,<br><br>81 | 1.05 | 3000 (0.035%) | 2 | 0<br>0<br><br>0<br>7<br><br>7<br>7<br>14<br>14<br>14<br>21<br>21<br>21<br>28<br>28<br>28 | Peel<br>Pulp<br><br>Fruitc<br>Peel<br><br>Pulp<br>Fruitc<br>Peel<br>Pulp<br>Fruitc<br>Peel<br>Pulp<br>Fruitc<br>Peel<br>Pulp<br>Fruitc | 1.9, 1.2, 0.068<br>< 0.02, < 0.02, < 0.02<br><br>0.69, 0.42, 0.037<br>2.2, 1.4, 0.057<br><br>< 0.02, < 0.02, < 0.02<br>0.77, 0.49, 0.033<br>0.77, 0.49, 0.047<br>< 0.02, < 0.02, < 0.02<br>0.28, 0.19, 0.029<br>1.2, 0.82, 0.065<br>< 0.02, < 0.02, < 0.02<br>0.45, 0.30, 0.036<br>0.93, 0.64, 0.087<br>< 0.02, < 0.02, < 0.02<br>0.34, 0.24, 0.043 | 3.17<br>< 0.02<br><br>1.15<br>3.66<br><br>< 0.02<br>1.29<br>1.31<br>< 0.02<br>0.50<br>2.09<br>< 0.02<br>0.79<br>1.66<br>< 0.02<br>0.62 |
| Spain1999 Navelina<br><br>C016112 ER 99 ECS750 | CS* | Spray Interval13 d<br><br>BBCH 81, 81 | 1.05 | 3000 (0.035%) | 2 | 0<br>22 | Fruit<br>Fruit | 0.42, 0.23, 0.04<br>0.17, 0.12, 0.07 | 0.69<br>0.36 |
| Spain1999 Newhall<br><br>C016112 ER 99 ECS750 | CS* | Spray Interval 14d<br><br>BBCH 81, 83 | 1.05 | 3000 (0.035%) | 2 | 0<br>22 | Fruit<br>Fruit | 0.27, 0.14, < 0.02<br>0.31, 0.21, 0.06 | 0.41<br>0.58 |
| Spain1999 Navelina<br><br>C016112 ER 99 ECS750 | CS* | Spray Interval 14d<br><br>BBCH 81, 83 | 1.05 | 3000 (0.035%.) | 2 | 0<br>22 | Fruit<br>Fruit | 0.51, 0.26, < 0.02<br>0.46, 0.25, < 0.02 | 0.77<br>0.71 |
| Italy 2000 Navelina<br><br>C016112 ER 99 ECS750 | CS* | Spray Interval 14d<br><br>BBCH 81, 83 | 1.05 | 3000 (0.035%) | 2 | 0<br>22 | Fruit<br>Fruit | 0.51, 0.35, 0.04<br>0.34, 0.23, 0.03 | 0.90<br>0.60 |

| ORANGES | | Application | | | | PHI | Sample | Residues of alpha, | Total |
|---|---|---|---|---|---|---|---|---|---|
| Country, Year of trial Variety Report | Form. | Method | kg ai/ha/ applic'n | L/ha | No. | days | analysed | beta endosulfan, endosulfan sulfate, mg/kg | Residues, mg/kg |
| Italy 2000 Navel<br><br>C016112 ER 99 750ECS | CS* | Spray Interval14 d<br><br>BBCH 79, 81 | 1.05 | 3000 (0.035% ai) | 2 | 0<br>22 | Fruit<br>Fruit | 0.36, 0.26, 0.12<br>0.28, 0.22, 0.10 | 0.74<br>0.60 |
| Spain 1998 Navelina<br><br>C005108 ER 98 ECS 750 | CS* | Spray Interval 15d<br><br>BBCH 74, 78 | 1.05 | 3000 (0.035% ai) | 2 | 0<br>7<br><br>14<br>21<br><br>28<br>0/7<br>14/28<br>21 | Peel<br>Peel<br><br>Peel<br>Peel<br><br>Peel<br>Pulp<br>Pulp<br>Pulp | 1.4, 0.81, 0.18<br>0.90, 0.53, 0.27<br><br>0.50, 0.37, 0.35<br>0.36, 0.29, 0.39<br><br>0.36, 0.30, 0.38<br>ND, ND, ND<br>< 0.02, < 0.02 ND<br>ND, ND, ND | 2.39<br>1.68<br><br>1.22<br>1.04<br><br>1.04<br>< 0.02<br>< 0.02<br>< 0.02 |
| | | | | | | 0<br>7<br>14<br>21<br>28 | Fruitc<br>Fruitc<br>Fruitc<br>Fruitc<br>Fruitc | 0.40, 0.23, 0.06<br>0.22, 0.13, 0.07<br>0.13, 0.10, 0.09<br>0.09, 0.07, 0.09<br>0.09, 0.08, 0.10 | 0.69<br>0.42<br>0.32<br>0.25<br>0.27 |
| Spain1998 Navelina<br><br>C005108 ER 98 ECS 750 | CS* | Spray Interval<br><br>16 d BBCH 75,<br><br>78 | 1.05 | 3000 (0.035% ai) | 2 | 0<br>7<br><br>14<br>21<br><br>28<br>0<br>7/14<br>21/28<br>0<br>7<br>14<br>21<br>28 | Peel<br>Peel<br><br>Peel<br>Peel<br><br>Peel<br>Pulp<br>Pulp<br>Pulp<br>Fruitc<br>Fruitc<br>Fruitc<br>Fruitc<br>Fruitc | 0.79, 0.51, 0.07<br>0.63, 0.41, 0.17<br><br>0.50, 0.33, 0.23<br>0.53, 0.37, 0.23<br><br>0.33, 0.25, 0.21<br>< 0.02, ND, ND<br>ND, ND, ND<br>ND, ND, ND<br>0.21, 0.14, 0.03<br>0.17, 0.11, 0.05<br>0.13, 0.09, 0.07<br>0.13, 0.10, 0.06<br>0.08, 0.07, 0.06 | 1.37<br>1.21<br><br>1.06<br>1.13<br><br>0.79<br>< 0.02<br>< 0.02<br>< 0.02<br>0.38<br>0.33<br>0.29<br>0.29<br>0.21 |
| Spain1998 Navelina<br><br>C003108 ER 98 ECS 740 | 35EC | Spray Interval<br><br>14d BBCH 75<br><br>, 78 | 1.05 | 3000 (0.035% ai) | 2 | 0<br>0<br><br>23<br>23<br><br>0<br>23 | Peel<br>Pulp<br><br>Peel<br>Pulp<br><br>Fruitc<br>Fruitc | 0.85, 0.83, 0.30<br>ND, ND, ND<br><br>0.07, 0.11, 0.43<br>ND, ND, ND<br><br>0.21, 0.20, 0.08<br>0.02, 0.03, 0.11 | 1.98<br>< 0.02<br><br>0.61<br>< 0.02<br><br>0.49<br>0.16 |
| Spain1998 Navelina<br><br>C003108 ER 98 ECS 740 | 35EC | Spray Int ;14d<br><br>BBCH 75, 78 | 1.05 | 3000 (0.035% ai) | 2 | 0<br>0<br><br>23<br>23<br><br>0<br>23 | Peel<br>Pulp<br><br>Peel<br>Pulp<br><br>Fruitc<br>Fruitc | 1.30, 0.93, 0.34<br>ND, ND, ND<br><br>0.11, 0.16, 0.71<br>ND, ND, < 0.02<br><br>0.34, 0.24, 0.09<br>0.04, 0.05, 0.19 | 2.57<br>< 0.02<br><br>0.98<br>< 0.02<br><br>0.67<br>0.28 |
| Italy1998 Washington<br><br>Navel C003108 ER 98 ECS 740 | 35EC | Spray Interval<br><br>14 d BBCH 77, 81 | 1.05 | 3000 (0.035% ai) | 2 | 0<br>0<br><br>23<br>23<br>0<br><br>23 | Peel<br>Pulp<br><br>Peel<br>Pulp<br>Fruitc<br><br>Fruitc | 0.85, 0.72, 0.31<br>< 0.02, < 0.02, ND<br><br>0.07, 0.13, 0.19<br>ND, ND, ND<br>0.21, 0.18, 0.08<br><br>0.03, 0.04, 0.06 | 1.88<br>< 0.02<br><br>0.39<br>< 0.02<br>0.47<br><br>0.13 |

| ORANGES | Application | | | | | PHI | Sample | Residues of alpha, | Total |
| Country, Year of trial Variety Report | Form. | Method | kg ai/ha/ applic'n | L/ha | No. | days | analysed | beta endosulfan, endosulfan sulfate, mg/kg | Residues, mg/kg |
|---|---|---|---|---|---|---|---|---|---|
| Italy1998 Tarocco | 35EC | Spray Interval | 1.05 | 3000 (0.035% ai) | 2 | 0 | Peel | 0.78, 0.66, 0.18 | 1.62 |
| | | | | | | 0 | Pulp | < 0.02, < 0.02, < 0.02 | < 0.02 |
| C003108 ER 98 ECS 740 | | 14 d BBCH 77, | | | | 21 | Peel | 0.09, 0.20, 0.27 | 0.56 |
| | | | | | | 21 | Pulp | ND, ND, ND | < 0.02 |
| | | 81 | | | | 0 | Fruitc | 0.20, 0.17, 0.07 | 0.44 |
| | | | | | | 21 | Fruitc | 0.03, 0.04, 0.06 | 0.13 |
| Spain1997 Navel | 35EC | Spray Interval | 1.05 | 3000 (0.035% ai) | 2 | 0 | Peel | 1.1, 0.82, 0.58 | 2.50 |
| | | | | | | 0 | Pulp | ND, < 0.02, ND | < 0.02 |
| C001464 ER 97 ECS 740 | | 14 d BBCH 78, | | | | 7 | Peel | 0.42, 0.53, 0.83 | 1.78 |
| | | | | | | 7 | Pulp | ND, ND, ND | < 0.02 |
| | | 78 | | | | 14 | Peel | 0.22, 0.3, 0.64 | 1.16 |
| | | | | | | 14 | Pulp | ND, ND, ND | < 0.02 |
| | | | | | | 21 | Peel | 0.22, 0.32, 0.69 | 1.23 |
| | | | | | | 21 | Pulp | ND, ND, < 0.02 | < 0.02 |
| | | | | | | 28 | Peel | 0.12, 0.19, 0.63 | 0.94 |
| | | | | | | 28 | Pulp | ND, ND, ND | < 0.02 |
| Spain1997 Navel | 35EC | Spray Interval | 1.05 | 3000 (0.035% ai) | 2 | 0 | Fruitc | 0.26, 0.20, 0.14 | 0.60 |
| | | | | | | 7 | Fruitc | 0.08, 0.10, 0.15 | 0.32 |
| C001464 ER 97 ECS 740 | | 14 d BBCH 78, | | | | 14 | Fruitc | 0.05, 0.07, 0.14 | 0.26 |
| | | | | | | 21 | Fruitc | 0.05, 0.07, 0.14 | 0.26 |
| | | | | | | 28 | Fruitc | 0.03, 0.05, 0.13 | 0.21 |
| Spain1997 Navelina | 35EC | Spray Interval | 1.05 | 3000 (0.035% ai) | 2 | 0 | Peel | 1.1, 0.72, 0.54 | 2.36 |
| | | | | | | 0 | Pulp | ND, ND, ND | < 0.02 |
| C001464 ER 97 ECS 740 | | 14 d BBCH 78, | | | | 7 | Peel | 0.32, 0.45, 0.75 | 1.52 |
| | | | | | | 7 | Pulp | ND, ND, ND | < 0.02 |
| | | 78 | | | | 13 | Peel | 0.27, 0.36, 0.72 | 1.35 |
| | | | | | | 13 | Pulp | ND, ND, ND | < 0.02 |
| | | | | | | 21 | Peel | 0.17, 0.24, 0.57 | 0.98 |
| | | | | | | 21 | Pulp | ND, ND, ND | < 0.02 |
| | | | | | | 28 | Peel | 0.18, 0.23, 0.73 | 1.14 |
| | | | | | | 28 | Pulp | ND, ND, ND | < 0.02 |
| | | | | | | 0 | Fruitc | 0.20, 0.13, 0.10 | 0.43 |
| | | | | | | 7 | Fruitc | 0.07, 0.10, 0.16 | 0.33 |
| | | | | | | 13 | Fruitc | 0.06, 0.08, 0.16 | 0.30 |
| | | | | | | 21 | Fruitc | 0.04, 0.06, 0.13 | 0.23 |
| | | | | | | 28 | Fruitc | 0.05, 0.06, 0.17 | 0.28 |
| Greece1997 Lutsiana | 35EC | Spray Interval | 1.05 | 3000 (0.035% ai) | 2 | 0 | Peel | 0.4, 0.32, 0.16 | 0.88 |
| | | | | | | 0 | Pulp | ND, < 0.02, ND | < 0.02 |
| C001464 ER 97 ECS 740 | | 14 d BBCH 79, | | | | 7 | Peel | 0.13, 0.24, 0.30 | 0.67 |
| | | | | | | 7 | Pulp | ND, ND, ND | < 0.02 |
| Santa | | 81 | | | | 14 | Peel | 0.07, 0.12, 0.21 | 0.40 |
| | | | | | | 14 | Pulp | ND, ND, ND | < 0.02 |
| | | | | | | 21 | Peel | 0.1, 0.2, 0.32 | 0.62 |
| | | | | | | 21 | Pulp | ND, ND, ND | < 0.02 |
| | | | | | | 28 | Peel | 0.07, 0.1, 0.27 | 0.44 |
| | | | | | | 28 | Pulp | ND, ND, ND | < 0.02 |
| | | | | | | 0 | Fruitc | 0.17, 0.14, 0.11 | 0.42 |
| | | | | | | 7 | Fruitc | 0.05, 0.09, 0.11 | 0.25 |
| | | | | | | 14 | Fruitc | 0.03, 0.05, 0.09 | 0.17 |
| | | | | | | 21 | Fruitc | 0.05, 0.09, 0.14 | 0.28 |
| | | | | | | 28 | Fruitc | 0.03, 0.04, 0.11 | 0.18 |

| ORANGES | Application | | | | | PHI | Sample | Residues of alpha, | Total |
| Country, Year of trial Variety Report | Form. | Method | kg ai/ha/ applic'n | L/ha | No. | days | analysed | beta endosulfan, endosulfan sulfate, mg/kg | Residues, mg/kg |
|---|---|---|---|---|---|---|---|---|---|
| Spain1997 Navelina C001464 ER 97 ECS 740 | 35EC | Spray Interval 14 d BBCH 78, | 3.15 | 3000 (0.105% ai) | 2 | 21 21 21 | Fruit Raw juice Peel /pomace | 0.12, 0.18, 0.33 ND, ND, ND 0.32, 0.38, 0.74 | 0.62 < 0.02 1.44 |
| Spain1997 Navel Moncada, C001464 ER 97 ECS 740 | 35EC | Spray 14 d BBCH 78, | 3.15 | 3000 (0.105% ai) | 2 | 21 21 21 | Fruit Raw juice Peel /pomace | 0.09, 0.14, 0.25 ND, ND, ND 0.14, 0.19, 0.51 | 0.48 < 0.02 0.84 |
| Italy1997 Naveline C001464 ER 97 ECS 740 | 35EC | Spray Interval 14 d BBCH 81, 81 | 1.05 | 3000 (0.035% ai) | 2 | 0 0 7 7 14 14 19 19 | Peel Pulp Peel Pulp Peel Pulp Peel Pulp | 0.57, 0.47, 0.04 < 0.02, < 0.02, ND 0.3, 0.37, 0.05 ND, < 0.02, ND 0.19, 0.32, 0.10 < 0.02, < 0.02, ND 0.13, 0.28, 0.06 < 0.02, < 0.02, ND | 1.08 < 0.02 0.72 < 0.02 0.61 < 0.02 0.47 < 0.02 |
| | | | | | | 28 28 0 7 14 19 28 | Peel Pulp Fruitc Fruitc Fruitc Fruitc Fruitc | 0.13, 0.24, 0.06 ND, ND, ND 0.02, 0.18, 0.02 0.02, 0.11, 0.02 0.03, 0.10, 0.03 0.03, 0.10, 0.03 0.02, 0.08, 0.02 | 0.43 < 0.02 0.22 0.15 0.16 0.16 0.12 |
| Italy1997 Naveline C001464 ER 97 ECS 740 | 35EC | Spray Interval 14 days BBCH 81, | 3.15 | 3000 (0.105% ai) | 2 | 21 21 21 | Fruit Raw juice Peel/ pomace | 0.15, 0.26, 0.04 ND, < 0.02, < 0.02 0.46, 0.69, 0.08 | 0.45 < 0.02 1.23 |
| Italy1997 Navel C001464 ER 97 ECS 740 | 35EC | Spray Interval 14 d BBCH 76, 79 | 1.05 | 3000 (0.035% ai) | 2 | 0 0 7 7 14 14 19 19 28 28 0 7 14 19 28 | Peel Pulp Peel Pulp Peel Pulp Peel Pulp Peel Pulp Fruitc Fruitc Fruitc Fruitc Fruitc | 0.42, 0.38, 0.11 < 0.02, < 0.02, ND 0.14, 0.27, 0.13 ND, < 0.02, ND 0.11, 0.23, 0.14 ND, < 0.02, ND 0.08, 0.15, 0.17 ND, ND, ND 0.05, 0.08, 0.11 ND, ND, ND 0.11, 0.10, 0.03 0.04, 0.08, 0.04 0.03, 0.06, 0.04 0.03, 0.05, 0.05 0.02, 0.03, 0.03 | 0.91 < 0.02 0.54 < 0.02 0.48 < 0.02 0.40 < 0.02 0.24 < 0.02 0.24 0.16 0.13 0.13 0.08 |
| Australia 2000 Beverford valencia N° 1/10/544 | EC | spray | 0.21 | 2000 (10.5 g/hL) | 4 | 0 1 3 7 | Fruit Fruit Fruit Fruit | 0.035, 0.052, 0.015 0.02, 0.044, 0.019 0.01, 0.025, 0.014 0.007, 0.024, 0.023 | 0.10 0.08 0.05 0.05 |

| ORANGES | Application | | | | | PHI | Sample | Residues of alpha, | Total |
| Country, Year of trial Variety Report | Form. | Method | kg ai/ha/ applic'n | L/ha | No. | days | analysed | beta endosulfan, endosulfan sulfate, mg/kg | Residues, mg/kg |
|---|---|---|---|---|---|---|---|---|---|
| Australia 2000 Beverford valencia N° 1/10/544 | EC | spray | 0.42 | 2000 | 4 | 0<br>3 | Fruit<br>Fruit | 0.09, 0.097, 0.027<br>0.058, 0.12, 0.046 | 0.21<br>0.22 |
| Australia 2000<br><br>Taylorville valencia N° 1/10/544 | EC | spray | 0.21 | 2000 (10.5 g/hL) | 4 | 0<br>1<br><br>3<br>7 | Fruit<br>Fruit<br><br>Fruit<br>Fruit | 0.081, 0.100, 0.008<br>0.032, 0.077, 0.013<br><br>0.018, 0.047,0.013<br>0.007, 0.017, 0.009 | 0.19<br>0.12<br><br>0.08<br>0.03 |
| Australia 2000 Taylorville valencia N° 1/10/544 | EC | spray | 0.42 | 2000 | 4 | 0<br><br>+3<br>+3 | Fruit<br><br>peel<br>flesh | 0.17, 0.20, 0.029<br><br>0.10, 0.21, 0.067<br>0.011, 0.016, 0.007 | 0.40<br><br>0.38<br>0.03 |

Table 36. Endosulfan residues in apple resulting from supervised trials in Europe, South Africa, USA and Australia.

| APPLE | Application | | | | | PHI | Sample | Residues of alpha, beta | Total |
| Country, Year of trial *Variety* Report | Form. | Method | kg ai/ha/ applic'n | L/ha | No. | days | analysed | endosulfan, endosulfan sulfate, mg/kg | Residues, mg/kg |
|---|---|---|---|---|---|---|---|---|---|
| Germany1989 *Jonathan* C004071 (A49973) *PSR99/012* | 35EC | Spray | 0.40, 0.47, 0.47, 0.40 | 1140, 1330 1330 1140 (0.035 % ai) | 4 | 0<br>21<br>21<br>21<br>21<br>21 | Fruit<br>Fruit<br>Mash<br>Juice<br>Pomace<br>Wash w | | 0.42<br>0.06<br>0.02<br>0.005<br>0.075<br>0.006 |
| Germany1989 *James Greaves* C004071 (A49972) *PSR99/012* | 35EC | Spray | 0.53 | 1500 0.035%a.i | 4 | 0<br>21<br>21<br>21<br>21<br>21 | Fruit<br>Fruit<br>Mash<br>Juice<br>Pomace<br>Wash w | | 0.76<br>0.11<br>0.02<br>0.006<br>0.175<br>0.006 |
| Germany1983 *Golden Delicious* A28757 *DEU83171111* | 35 WP | Spray Interval 14 d | 0.74 | 500 (0.148%a.i | 5 | 0<br>7<br>14<br>21 | Fruit<br>Fruit<br>Fruit<br>Fruit | 2.0, 1.2, 0.08<br>0.9, 0.9, 0.07<br>0.8, 1.4, 0.09<br>0.3, 0.8, 0.1 | 3.28<br>1.87<br>2.29<br>1.20 |
| Germany1983 *Klarapfel* A28758 *DEU83171121* | 35 WP | Spray Interval 13-17 d | 0.74 | 500 (0.148 % ai) | 5 | 0<br>7<br>14<br>21 | Fruit<br>Fruit<br>Fruit<br>Fruit | 0.8, 0.8, 0.1<br>0.2, 0.3, 0.06<br>0.2, 0.4, 0.03<br>0.09, 0.2, 0.03 | 1.7<br>0.56<br>0.63<br>0.32 |
| Germany1983 *Golden Delicious* A28759 *DEU83171131* | 35 WP | Spray Interval 14-17d days | 0.74 | 500 (0.148 % ai) | 5 | 0<br>7<br>14<br>21 | Fruit<br>Fruit<br>Fruit<br>Fruit | 0.8, 0.6, 0.05<br>0.4, 0.6, 0.05<br>1.1, 1.0, 0.06<br>0.09, 0.1, 0.03 | 1.45<br>1.05<br>2.16<br>0.22 |
| Germany1983 *Golden Delicious* A28760 *DEU83170131* | 35 EC | Spray Interval 14-17d | 0.799 | 500 (0.16% ai) | 5 | 0<br>7<br>14<br>21 | Fruit<br>Fruit<br>Fruit<br>Fruit | 0.4, 0.5, 0.07<br>0.3, 0.5, 0.06<br>0.4, 0.6, 0.06<br>0.2, 0.4,0.07 | 0.97<br>0.86<br>1.06<br>0.67 |
| Germany1983 *Klarapfel* A28761 *DEU83170121* | 35 EC | Spray Interval 13- 17d | 0.799 | 500 (0.16% ai) | 5 | 0<br>7<br>14<br>21 | Fruit<br>Fruit<br>Fruit<br>Fruit | 0.8, 0.7, 0.1<br>0.2, 0.4, 0.09<br>0.09, 0.3,0.1<br>0.05, 0.3, 0.1 | 1.60<br>0.69<br>0.49<br>0.45 |
| Germany1983 *Golden Delicious* A28762 *DEU83170111* | 35 EC | Spray Interval 14 d | 0.799 | 500 (0.16 % ai) | 5 | 0<br>7<br>14<br>21 | Fruit<br>Fruit<br>Fruit<br>Fruit | 3.0, 2.2, 0.1<br>1.7, 1.8, 0.1<br>1.0, 1.5, 0.1<br>1.1, 1.2, 0.1 | 5.30<br>3.6O<br>2.60<br>2.40 |

| APPLE | Application | | | | | PHI | Sample | Residues of alpha, beta | Total |
| Country, Year of trial *Variety* Report | Form. | Method | kg ai/ha/ applic'n | L/ha | No. | days | analysed | endosulfan, endosulfan sulfate, mg/kg | Residues, mg/kg |
|---|---|---|---|---|---|---|---|---|---|
| Germany1976 *Cox Orange* A08886 *LEA 1/48/01/02-76* | 24EC* | Spray Interval 14, 21, 18, 10 d | 0.96 | 2000 (0.048 % ai) | 5 | 0 10 14 21 | Fruit Fruit Fruit Fruit | | 1.20 0.01 1.0 0.80 |
| Germany1976 *James Grieve* A08888a *LEA 2/48/01/02-76* | 24EC* | Spray Interval 11, 23, 18, 11 d | 0.96 | 2000 (0.048 % ai) | 5 | 0 10 14 21 | Fruit Fruit Fruit Fruit | | 1.10 1.00 0.70 0.30 |
| Germany1983 *Golden Delicious* A29493 *DEU83172231* | 2.82 DP | Spread Interval 14, 14, 14, 17 d | 0.705 | 25 kg Product | 5 | 0 7 14 21 | Fruit Fruit Fruit Fruit | 0.06, 0.08, 0.04 0.06, 0.1, 0.06 0.03, 0.08, 0.06 0.1, 0.09, 0.05 | 0.18 0.22 0.17 0.24 |
| Germany1983 *Golden Delicious* A29495 *DEU83172211* | 2.82 DP | Spread Interval 14 d | 0.705 | 25 kg Product | 5 | 0 7 14 21 | Fruit Fruit Fruit Fruit | 0.2, 0.1, 0.4 0.06, 0.04, 0.05 0.04, 0.07, 0.04 0.08, 0.1, 0.05 | 0.68 0.15 0.15 0.23 |
| Germany1983 *James Grieve* A29496 *DEU83172241* | 2.82 DP | Spread Interval 14 d | 0.705 | 25 kg Product | 5 | 0 7 14 21 | Fruit Fruit Fruit Fruit | 0.1, 0.1, 0.06 0.05, 0.08, 0.05 0.05, 0.1, 0.05 0.02, 0.04, 0.04 | 0.26 0.18 0.20 0.10 |
| Germany1983 *Victoria* A29494 *DEU83172221* | 2.82 DP | Spread Interval 14, 14, 13, 13 d | 0.705 | 25 kg Product | 5 | 0 7 14 21 | Fruit Fruit Fruit Fruit | 0.2, 0.1, 0.05 0.04, 0.07, 0.05 0.04, 0.07, 0.05 0.03, 0.06, 0.07 | 0.35 0.16 0.16 0.16 |
| Germany1976 *Cox Orange* A08889 *LEA 4/48/01/02-76* | 24EC* | Spray Interval 14, 21, 14, 14 d | 0.96 | 2000 (0.048 % ai) | 5 | 0 10 14 21 | Fruit Fruit Fruit Fruit | | 2.70 0.80 0.70 0.60 |
| Germany1975 *Cox Orange* A04837 *LEA 4/82/01/02-75* *(A)* | 24EC* | Spray Interval 14 d | 0.96 | 2000 (0.048 % ai) | 4 | 0 7 14 21 28 | Fruit Fruit Fruit Fruit Fruit | ND, ND, ND 0.40, 0.60, 0.05 0.20, 0.20, 0.04 0.09, 0.20, 0.03 0.20, 0.20, 0.03 | < 0.02 1.05 0.44 0.32 0.43 |
| Germany1975 *Gold Parmane* A04838 *LEA 4/82/02/02-75* *(A)* | 24EC* | Spray Interval 14 d | 0.96 | 2000 (0.048 % ai) | 4 | 0 7 14 21 28 | Fruit Fruit Fruit Fruit Fruit | 0.30, 0.30, 0.05 0.20, 0.30, 0.07 0.50, 0.70, 0.20 0.10, 0.30, 0.09 0.10, 0.20, 0.09 | 0.65 0.57 1.40 0.49 0.39 |
| Germany1975 *Freiherr von Berlepsch* A04839 *LEA 4/82/03/02-75* | 24EC* | Spray Interval 14 d | 0.96 | 2000 (0.048 % ai) | 4 | 0 7 14 21 28 | Fruit Fruit Fruit Fruit Fruit | 0.50, 0.50, ND 0.40, 0.50, 0.04 0.30, 0.50, 0.03 0.10, 0.20, 0.02 0.09, 0.10, 0.02 | 1.20 0.94 0.83 0.32 0.21 |
| Germany1975 *Jonared 1975* A05860 *LEA 1/82/01/02-75* | 24EC* | Spray Interval 13, 75 d | 0.96 | 2000 (0.048 % ai) | 3 | 0 10 20 28 | Fruit Fruit Fruit Fruit | | 0.30 0.30 0.20 0.07 |
| Germany1975 *Golden 1975* A05861 *LEA 1/82/02/02-75* | 24EC* | Spray Interval 13, 75 d | 0.96 | 2000 (0.048 % ai) | 3 | 0 10 20 28 | Fruit Fruit Fruit Fruit | | 1.50 0.90 0.40 0.20 |
| Germany1975 *Cox Orange* A05862 *LEA 2/82/01/02-75* | 24EC* | Spray Interval 14, 16, 12 d | 0.96 | 2000 (0.048 % ai) | 4 | 0 14 28 | Fruit Fruit Fruit | | 0.60 0.10 0.06 |
| Germany1974 *Golden Delicious* A02616 *1/90/02/02* | 24EC* | Spray | 0.72 | 2000 (0.036 % ai) | 1 | 0 7 14 21 28 | Fruit Fruit Fruit Fruit Fruit | | 0.58 0.27 0.19 0.21 0.18 |

| APPLE | Application | | | | | PHI | Sample | Residues of alpha, beta | Total |
| Country, Year of trial *Variety* Report | Form. | Method | kg ai/ha/ applic'n | L/ha | No. | days | analysed | endosulfan, endosulfan sulfate, mg/kg | Residues, mg/kg |
|---|---|---|---|---|---|---|---|---|---|
| Germany1974 *Golden* *Delicious* A02617 *1/90/02/04* | 24EC* | Spray Interval 30 days | 0.96 | 2000 (0.048 % ai) | 2 | 0 7 14 21 28 | Fruit Fruit Fruit Fruit Fruit | | 0.76 1.00 0.49 0.51 0.66 |
| Germany1974 *Golden* *Delicious* A02618 *1/90/02/03* | 24EC* | Spray | 0.96 | 2000 (0.048 % ai) | 1 | 0 7 14 21 28 | Fruit Fruit Fruit Fruit Fruit | | 0.71 0.39 0.26 0.36 0.36 |
| Germany1974 *Goldparmane* A02619 *4/90/01/03* | 24EC* | Spray | 0.96 | 2000 (0.048 % ai) | 1 | 0 7 14 21 28 | Fruit Fruit Fruit Fruit Fruit | 0.30, ND 0.26, ND 0.26, ND 0.12, ND 0.11, 0.17 | 0.30 0.26 0.26 0.12 0.28 |
| Germany1974 *Goldparmane* A02620 *4/90/01/04* | 24EC* | Spray Interval 31 d | 0.96 | 2000 (0.048 % ai) | 2 | 0 7 14 21 28 | Fruit Fruit Fruit Fruit Fruit | 0.57, 0.08 0.36, 0.10 0.62, ND 0.28, ND 0.45, 0.05 | 0.65 0.46 0.62 0.28 0.50 |
| Germany1974 *Cox Orange* A02621 *4/90/02/02* | 24EC* | Spray | 0.72 | 2000 (0.036 % ai) | 1 | 0 7 14 21 28 | Fruit Fruit Fruit Fruit Fruit | 0.28, ND 0.18, ND 0.15, ND 0.11, ND 0.13, ND | 0.28 0.18 0.15 0.11 0.13 |
| Germany1974 *Cox Orange* A02622 *4/90/02/03* | 24EC* | Spray | 0.96 | 2000 (0.048 % ai) | 1 | 0 7 14 21 28 | Fruit Fruit Fruit Fruit Fruit | 0.22, ND 0.29, ND 0.18, ND 0.14, ND 0.17, ND | 0.22 0.29 0.18 0.14 0.17 |
| Germany1974 *Cox Orange* 1974 A02623 *4/90/02/04* | 24EC* | Spray Interval 31 days | 0.96 | 2000 (0.048 % ai) | 2 | 0 7 14 21 28 | Fruit Fruit Fruit Fruit Fruit | 0.31, ND 0.35, ND 0.33, ND 0.25, ND 0.27, ND | 0.31 0.35 0.33 0.25 0.27 |
| Germany1974 *Goldparmane* 1974 A02624 *4/90/01/02* | 24EC* | Spray | 0.72 | 2000 (0.036 % ai) | 1 | 0 7 14 21 28 | Fruit Fruit Fruit Fruit Fruit | 0.26, ND 0.17, ND 0.16, ND 0.18, ND 0.18, ND | 0.26 0.17 0.16 0.18 0.18 |
| Germany1974 *Cox Orange* A03995 *1/90/01/02* | 24EC* | Spray | 0.72 | 2000 (0.036 % ai) | 1 | 0 7 14 21 28 | Fruit Fruit Fruit Fruit Fruit | | 0.82 0.46 0.34 0.37 0.22 |
| Germany1974 *(Goldparmane)* *1/90/01/03* | 24 EC | Spray | 0.96 | 2000 (0.048 % ai) | 1 | 0 7 14 21 28 | Fruit Fruit Fruit Fruit Fruit | | 1.23 0.96 0.61 0.27 0.31 |
| Germany1974 *Cox Orange* *1/90/01/04* A03997 | 24EC* | Spray Interval 30 d | 0.96 | 2000 (0.048 % ai) | 2 | 0 7 14 21 28 | Fruit Fruit Fruit Fruit Fruit | | 1.19 1.07 0.84 0.52 0.35 |
| Germany1974 *Golden Delicious* *2/90/01/02* A03998 | 24EC* | Spray | 0.72 | 2000 (0.048 % ai) | 1 | 0 7 14 21 28 | Fruit Fruit Fruit Fruit Fruit | | 0.60 0.30 0.15 0.15 0.04 |

| APPLE Country, Year of trial *Variety* Report | Form. | Method | kg ai/ha/ applic'n | L/ha | No. | PHI days | Sample analysed | Residues of alpha, beta endosulfan, endosulfan sulfate, mg/kg | Total Residues, mg/kg |
|---|---|---|---|---|---|---|---|---|---|
| Germany1974 *Golden Delicious* A03999 2/90/01/03 | 24EC* | Spray | 0.96 | 2000 (0.048 % ai) | 1 | 0 7 14 21 28 | Fruit Fruit Fruit Fruit Fruit | | 0.90 0.40 0.25 0.25 0.12 |
| Germany1974 *Golden delicious* 2/90/01/04 A04000 | 24EC* | Spray Interval 27 d | 0.96 | 2000 (0.048 % ai) | 2 | 0 7 14 21 28 | Fruit Fruit Fruit Fruit Fruit | | 1.70 0.90 0.50 0.30 0.12 |
| Italy 1993 Baricella, Emiglia Romagna *Cooper 4* A54359 *ER 93 ECS 705* | 35EC | (backpack) BBCH 76-77 and 81 interval sprays-55 | 0.66 | 1250 (0.0528 % ai) | 2 | 0 7 14 14 14 14 14 14 14 | Apple fruit "washed "unwashed cider mash pomace ww | 0.14, 0.15, < 0.01 0.03, 0.04, < 0.01 0.03, 0.05, 0.03 0.02, 0.03, < 0.01 0.04, 0.05, 0.02 < 0.01, < 0.01, < 0.01 < 0.01, 0.01, < 0.01 0.04, 0.07, 0.01 < 0.01, < 0.01, < 0.01 | 0.29 0.07 0.11 0.05 0.11 < 0.01 0.01 0.12 < 0.01 |
| Italy 1993 Baricella, Emiglia Romagna *Coope 4* A54359 *ER 93 ECS 705* | 35EC | (backpack) BBCH 76-77 and 81 interval sprays-55 | 1.3199 | 1250 (0.1056 % ai) | 2 | 0 7 14 14 14 14 14 14 14 | Apple fruit "washed "unwashed cider mash pomace ww | 0.44, 0.40, 0.02 0.06, 0.13, 0.02 0.03, 0.045, 0.015 0.02, 0.05, < 0.01 0.04, 0.07, 0.02 < 0.01, < 0.01, < 0.01 < 0.01, 0.01, < 0.01 0.10, 0.15, 0.04 < 0.01< 0.01, < 0.01 | 0.86 0.21 0.09 0.07 0.13 < 0.01 0.01 0.29 < 0.01 |
| Spain1999 *Red chief* C016113c *ER 99 ECS 755* | CS* | (knapsack) BBCH 79 and 85 Interval 15d | 1.05 | 1000 (0.105 % ai) | 2 | 0 3 7 14 | fruit | 0.35, 0.24, < 0.02 0.24, 0.19, < 0.02 0.28, 0.19, < 0.02 0.16, 0.13, < 0.02 | 0.59 0.43 0.47 0.29 |
| Spain1999 *Golden Smutte* C016113 *ER 99 ECS 755* | CS* | (knapsack) BBCH 79 and 85 Interval 13d | 1.05 | 1000 (0.105 % ai) | 2 | 0 3 7 14 | Fruit | 0.65, 0.44, < 0.02 0.89, 0.57, < 0.02 0.49, 0.38, < 0.02 0.21, 0.16, < 0.02 | 1.09 1.46 0.87 0.37 |
| Greece1999 *Golden* C016113 *ER 99 ECS733* | CS* | Spray (gun) BBCH 81 and 85 Interval 14d | 1.05 | 1500 (0.07 % ai) | 2 | 0 3 7 14 | Fruit | 1.20, 0.72, < 0.02 0.43, 0.27, < 0.02, 0.84, 0.78, < 0.02 0.63, 0.46, < 0.02 | 1.92 0.70 1.62 1.09 |
| Italy1999 *Mondial gala* C016113 *ER 99 ECS 755* | CS* | Spray (gun) BBCH 77 and 77 Interval 14d | 1.05 | 1500 (0.07 % ai) | 2 | 0 3 7 14 | Fruit | 0.69, 0.43, < 0.02 0.49, 0.32, < 0.02 0.64, 0.40, < 0.02 0.39, 0.23, < 0.02 | 1.12 0.81 1.04 0.62 |
| Portugal1999 *Baby Gold* *ER 99 ECS 755* | CS* | (knapsack) BBCH 78 and 81 Interval 14d | 1.05 | 1100 (0.01 % ai) | 2 | 0 3 7 14 | Fruit | 0.96, 0.57, < 0.02 1.2, 0.71, < 0.02 0.99, 0.65, < 0.02 0.22, 0.19, < 0.02 | 1.53 1.91 1.64 0.41 |
| Spain1994 *Golden Smuthe* Alfamen, A55874 *ER 94 ECS 705* | 35EC | (backpack) BBCH 78 and 81 interval sprays-51 | 0.528 | 1000 (0.0528 % ai) | 2 | 0 7 ~12 21 28 | Apple | 0.05, 0.09, < 0.01 0.03, 0.02, < 0.01 < 0.01, 0.02, < 0.01 < 0.01, 0.01, < 0.01 < 0.01, < 0.01, < 0.01 | 0.14 0.05 0.02 0.01 < 0.01 |
| Spain1994 *Golden Smuthe* A55874 *ER 94 ECS 705* | 35EC | (backpack) BBCH 78 and 81 interval sprays-51 | 1.056 | 1000 (0.1056 % ai) | 2 | 0 7 ~12 21 28 | Apple | 0.26, 0.24, 0.02 0.04, 0.08, 0.02 0.02, 0.04, 0.02 0.02, 0.03, 0.02 0.01, 0.015, 0.01 | 0.52 0.14 0.08 0.07 0.04 |

| APPLE | Application | | | | | PHI | Sample | Residues of alpha, beta | Total |
|---|---|---|---|---|---|---|---|---|---|
| Country, Year of trial *Variety* Report | Form. | Method | kg ai/ha/ applic'n | L/ha | No. | days | analysed | endosulfan, endosulfan sulfate, mg/kg | Residues, mg/kg |
| Spain1994 Zaragoza *Starkingson* A55874 *ER 94 ECS 705* | 35EC | backpack) BBCH 78 and 81 interval sprays-51 | 0.528 | 1000 (0.0528 % ai) | 2 | 0 7 12 21 28 | Apple | 0.07, 0.09, < 0.01 0.01, 0.02, < 0.01 0.02, 0.02, < 0.01 0.02, 0.03, < 0.01 < 0.01, < 0.01, < 0.01 | 0.16 0.03 0.04 0.05 < 0.01 |
| Spain1994 Zaragoza *Starkingson* A55874 *ER 94 ECS 705* | 35EC | backpack) BBCH 78 and 81 interval sprays-51 | 1.056 1.056 | 1000 (0.1056 % ai) | 2 | 0 7 12 21 28 | Apple | 0.15, 0.21, < 0.01 0.04, 0.06, 0.02 0.02, 0.045, < 0.01 0.02, 0.033, < 0.01 0.023, 0.03, < 0.01 | 0.36 0.12 0.07 0.05 0.05 |
| France1994 South *Golden Spur* A55874 *ER 94 ECS 705* | 35EC | (backpack) BBCH 76 and 80 Interval 71d | 0.528 | 990, 972 (0.0533, 0.0543 % ai) | 2 | 0 7 13 21 28 | Apple | < 0.01, < 0.01, < 0.01 0.02, 0.04, 0.01 < 0.01, < 0.01, < 0.01 < 0.01, < 0.01, < 0.01 < 0.01, < 0.01< 0.01 | < 0.01 0.07 < 0.01 < 0.01 < 0.01 |
| France1994 South *Golden Spur* A55874 *ER 94 ECS 705* | 35EC | (backpack BBCH 76-And 8( Interval 71d | 1.056 | 990, 952 (0.1067, 0.1109 % ai) | 2 | 0 7 13 21 28 | Apple | 0.34, 0.31, 0.02 0.04, 0.06, 0.02 < 0.01, < 0.01, < 0.01 0.04, 0.04, 0.05 0.03, 0.06, 0.03 | 0.67 0.12 < 0.01 0.13 0.12 |
| Italy1994 *Golden Delicious* A55874 *ER 94 ECS 705* | 35EC | (backpack) BBCH 77 And na Interval 52d | 0.7922, 0.7919 | 1500 (0.0528 % ai) | 2 | 0 7 14 21 28 | Apple | 0.33, 0.34, < 0.01 0.06, 0.095, 0.02 0.075, 0.12, 0.025 0.04, 0.07, 0.03 0.03, 0.06, 0.02 | 0.67 0.18 0.22 0.14 0.11 |
| Italy 1994 *(Golden Delicious)* *ER 94 ECS 705* | 35EC | (backpack) BBCH 77-78 and na interval sprays-52 | 1.5848, 1.5841 | 1500 (0.1056 % ai) | 2 | 0 7 14 14 14 14 | Apple fruit " "washed "unwshd must | 1.0, 0.71, 0.05 0.135, 0.225, 0.035 0.10, 0.133, 0.03 0.07, 0.12, 0.03 0.10, 0.13, 0.04 < 0.01, < 0.01, < 0.01 | 1.76 0.40 0.26 0.22 0.27 < 0.01 |
| | | | | | | 14 14 14 21 28 | puree pomace w, w Fruit Fruit | 0.02, 0.03, 0.01 0.24, 0.36, 0.09 < 0.01, < 0.01, < 0.01 0.16, 0.26, 0.05 0.075, 0.135, 0.03 | 0.06 0.69 < 0.01 0.47 0.24 |
| Italy 1994 *Imperatore* *ER 94 ECS 705* | 35EC | (backpack) BBCH 76 and 81 interval sprays-51 | 0.7922, 0.7919 | 1500 (0.0528 % ai) | 2 | 0 7 14 21 28 | Apple fruit | 0.35, 0.16, < 0.01 0.04, 0.02, < 0.01 0.01, 0.02, < 0.01 0.025, 0.04, < 0.01 < 0.01, 0.02, < 0.01 | 0.51 0.06 0.03 0.07 0.02 |
| Italy 1994 *Imperatore* *ER 94 ECS 705* | 35EC | Sprays BBCH 76-77 and 78-81 interval sprays-51 | 1.5848, 1.5841 | 1500 (0.1056 % ai) | 2 | 0 7 14 14 14 14 14 14 14 21 28 | Apple fruit " "washed "unwshd must puree pomace w, w Fruit Fruit | 0.62, 0.27, 0.01 0.11, 0.12, 0.02 0.13, 0.13, 0.01 0.055, 0.075, 0.02 0.05, 0.09, 0.045 < 0.01, < 0.01, < 0.01 < 0.01, 0.01, < 0.01 0.17, 0.26, 0.07 < 0.01, < 0.01, < 0.01 0.033, 0.046, 0.02 0.063, 0.07, 0.026 | 0.90 0.25 0.27 0.15 0.19 < 0.01 0.01 0.50 < 0.01 0.10 0.16 |
| Spain 1993 *Starking* A54359 *ER 93 ECS 705* | 35EC | (backpack) BBCH 77 and 81 interval sprays-56 | 1.0419, 1.0238 | 1973 1939 (0.0528 % ai) | 2 2 | 0 7 14 14 14 14 14 14 14 | Apple fruit Fruit "washed "unwshd mash pomace ww cider | 0.26, 0.25, < 0.01 0.05, 0.06, < 0.01 0.02, 0.03, < 0.01 0.04, 0.04, 0.01 0.03, 0.04, 0.01 < 0.01, 0.01, < 0.01 0.06, 0.10, 0.01 < 0.01, < 0.01, < 0.01 < 0.01, < 0.01, < 0.01 | 0.51 0.11 0.05 0.08 0.08 0.01 0.17 < 0.01 < 0.01 |

| APPLE | Application | | | | | PHI | Sample | Residues of alpha, beta | Total |
| Country, Year of trial *Variety* Report | Form. | Method | kg ai/ha/ applic'n | L/ha | No. | days | analysed | endosulfan, endosulfan sulfate, mg/kg | Residues, mg/kg |
|---|---|---|---|---|---|---|---|---|---|
| Spain 1993 *Starking* A54359 *ER 93 ECS 705* | 35EC | backpack) BBCH 77 and 81 Interval sprays-56 | 2.3708, 2.4428 | 2245 2313 (0.1056 % ai) | 2 | 0 | Apple | 0.50, 0.41, < 0.01 | 0.91 |
| | | | | | | 7 | fruit | 0.12, 0.16, 0.03 | 0.31 |
| | | | | | | 14 | Fruit | 0.05, 0.07, 0.02 | 0.14 |
| | | | | | | 14 | "washed | 0.03, 0.06, < 0.01 | 0.09 |
| | | | | | | 14 | "unwshd | 0.03, 0.08, 0.03 | 0.14 |
| | | | | | | 14 | cider | < 0.01, < 0.01, < 0.01 | < 0.01 |
| | | | | | | 14 | mash | 0.01, 0.02, < 0.01 | 0.03 |
| | | | | | | 14 | pomace | 0.07, 0.17, 0.025 | 0.27 |
| | | | | | | 14 | ww | < 0.01, < 0.01, < 0.01 | < 0.01 |
| France 1993 *Golden Delicious* A54359 *ER 93 ECS 705* | 35EC | (backpack) BBCH 76-77 and 81 Interval70d | 0.528 | 1000 (0.0528 % ai) | 2 | 0 | Apple | 0.05, 0.08, < 0.01 | 0.13 |
| | | | | | | 7 | fruit | 0.015, 0.03, < 0.01 | 0.05 |
| | | | | | | 13 | | 0.02, 0.04, 0.01 | 0.07 |
| | | | | | | 13 | "washed | < 0.01, 0.01, < 0.01 | 0.01 |
| | | | | | | 13 | "unwshd | 0.02, 0.03, < 0.01 | 0.05 |
| | | | | | | 13 | cider | < 0.01, < 0.01, < 0.01 | < 0.01 |
| | | | | | | 13 | mash | < 0.01, < 0.01, < 0.01 | < 0.01 |
| | | | | | | 13 | pomace | 0.03, 0.07, < 0.01 | 0.10 |
| | | | | | | 13 | ww | < 0.01,< 0.01, < 0.01 | < 0.01 |
| France South 1993 *Golden Delicious* A54359 *ER 93 ECS 705* | 35EC | (backpack BBCH 76-77 and 81 interval sprays-70 | 1.0561 | 1000 (0.1056 % ai) | 2 | 0 | Apple | 0.34, 0.29, 0.02 | 0.65 |
| | | | | | | 7 | fruit | 0.06, 0.12, 0.04 | 0.22 |
| | | | | | | 13 | | 0.03, 0.04, < 0.01 | 0.07 |
| | | | | | | 13 | "washed | 0.01, 0.03, 0.01 | 0.05 |
| | | | | | | 13 | "unwshd | 0.04, 0.07, 0.02 | 0.13 |
| | | | | | | 13 | cider | < 0.01, < 0.01, < 0.01 | < 0.01 |
| | | | | | | 13 | mash | < 0.01, 0.01, < 0.01 | 0.01 |
| | | | | | | 13 | pomace | 0.06, 0.10, 0.035 | 0.20 |
| | | | | | | 13 | ww | < 0.01, < 0.01, < 0.01 | < 0.01 |
| France 1993 *Canada Gris* A54359 *ER 93 ECS 705* | 35EC | (backpack BBCH 77-78 and 81 interval sprays-55 | 0.528 | 1000 (0.0528 % ai) | 2 | 0 | Apple | 0.09, 0.09, 0.01 | 0.19 |
| | | | | | | 7 | fruit | 0.115, 0.115, 0.01 | 0.24 |
| | | | | | | 13 | | 0.085, 0.105, 0.01 | 0.20 |
| | | | | | | 13 | "washed | 0.04, 0.06, < 0.01 | 0.10 |
| | | | | | | 13 | "unwshd | 0.105, 0.12, 0.015 | 0.24 |
| | | | | | | 13 | cider | < 0.01, < 0.01, < 0.01 | < 0.01 |
| | | | | | | 13 | mash | 0.01, 0.02, < 0.01 | 0.03 |
| | | | | | | 13 | pomace | 0.16, 0.18, 0.02 | 0.36 |
| | | | | | | 13 | ww | < 0.01, < 0.01, < 0.01 | < 0.01 |
| France 1993 St Pardon de C. Aquitaine *Canada Gris* A54359 *ER 93 ECS 705* | 35EC | (backpack) BBCH 77-78 and 81 interval sprays-55 | 1.0561 | 1000 (0.1056 % ai) | 2 | 0 | Apple | 0.21, 0.26, 0.01 | 0.48 |
| | | | | | | 7 | fruit | 0.21, 0.62, 0.03 | 0.86 |
| | | | | | | 13 | | 0.20, 0.23, 0.03 | 0.46 |
| | | | | | | 13 | "washed | 0.13, 0.13, 0.01 | 0.27 |
| | | | | | | 13 | "unwshd | 0.12, 0.18, 0.04 | 0.34 |
| | | | | | | 13 | cider | < 0.01, < 0.01, < 0.01 | < 0.01 |
| | | | | | | 13 | mash | 0.04, 0.06, < 0.01 | 0.10 |
| | | | | | | 13 | pomace | 0.58, 0.77, 0.03 | 1.38 |
| | | | | | | 13 | ww | < 0.01, < 0.01, < 0.01 | < 0.01 |
| Italy 1993 Gallo, Emiglia Romagna *Golden Delicious* A54359 *ER 93 ECS 705* | 35EC | (backpack) BBCH 76-77 and 81 interval sprays-59 | 0.7922 | 1500 (0.0528 % ai) | 2 | 0 | Apple | 0.16, 0.15, 0.01 | 0.32 |
| | | | | | | 7 | fruit | 0.06, 0.08, 0.02 | 0.16 |
| | | | | | | 14 | | 0.025, 0.03, 0.02 | 0.08 |
| | | | | | | 14 | "washed | 0.05, 0.07, 0.03 | 0.15 |
| | | | | | | 14 | "unwshd | 0.05, 0.07, 0.02 | 0.14 |
| | | | | | | 14 | cider | < 0.01, < 0.01, < 0.01 | < 0.01 |
| | | | | | | 14 | mash | < 0.01, 0.02, < 0.01 | 0.02 |
| | | | | | | 14 | pomace | 0.04, 0.07, 0.02 | 0.13 |
| | | | | | | 14 | ww | < 0.01, < 0.01, < 0.01 | < 0.01 |
| Italy 1993 Gallo, Emiglia Romagna *Golden Delicious* A54359 *ER 93 ECS 705* | 35EC | (backpack) BBCH 76-77 and 81 interval sprays-59 | 1.5841 | 1500 (0.1056 % ai) | 2 | 0 | Apple | 0.58, 0.54, 0.02 | 1.14 |
| | | | | | | 7 | fruit | 0.10.0.16, 0.025 | 0.29 |
| | | | | | | 14 | | 0.07, 0.10, 0.04 | 0.21 |
| | | | | | | 14 | "washed | 0.04, 0.06, 0.01 | 0.11 |
| | | | | | | 14 | "unwshd | 0.06, 0.08, 0.04 | 0.18 |
| | | | | | | 14 | cider | < 0.01, < 0.01, < 0.01 | < 0.01 |
| | | | | | | 14 | mash | 0.01, 0.02, < 0.01 | 0.03 |
| | | | | | | 14 | pomace | 0.13, 0.21, 0.06 | 0.40 |
| | | | | | | 14 | ww | < 0.01, < 0.01, < 0.01 | < 0.01 |

| APPLE Country, Year of trial *Variety* Report | Application Form. | Method | kg ai/ha/ applic'n | L/ha | No. | PHI days | Sample analysed | Residues of alpha, beta endosulfan, endosulfan sulfate, mg/kg | Total Residues, mg/kg |
|---|---|---|---|---|---|---|---|---|---|
| S. Africa1978 *Starking* ZA/77/03-02-02 | 24EC* | Spray | 0.24 | 2000 (0.012 % ai) | 1 | 14 | Fruit | | 0.07 |
| S. Africa1978 *Starking* ZA/77/03-02-03 | 24EC* | Spray | 0.48 | 2000 (0.024 % ai) | 1 | 14 | Fruit | | 0.05 |
| S. Africa1978 *Starking* ZA/77/03-02-04 | 24EC* | Spray | 0.72 | 2000 (0.036 % ai) | 1 | 14 | Fruit | | 0.10 |
| S. Africa1978 *Golden Delicious* ZA/77/01-04-06 | 24EC* | Spray Interval 9, 18, 22, 22, 20 d | 0.72 | 2000 (0.036 % ai) | 7 | 13 | Fruit | | 0.50 |
| S. Africa1978 *Golden Delicious* ZA/77/01-04-07 | 24EC* | Spray Interval 9, 18, 21, 22, 20 d | 0.96 | 2000 (0.048 % ai) | 7 | 13 | Fruit | | <u>0.60</u> |
| USA1972 *McIntosh* C004071 (A30341) *M-3087* | 2EC | Spray | 4.48 | | 2 | 16 | Fruit w/Peel/ core dry/Peel/cor peeled/core Wet Pomace Dry Pomace Cider | 0.045, 0.087, 0.092 0.20, 0.41, 0.26 0.47, 1.34, 0.95 ND, ND, ND 0.21, 0.42, 0.38 0.24, 1.22, 1.46 ND, ND, ND | 0.22 0.87 2.76 < 0.02 1.01 2.92 < 0.02 |
| USA1963 *Golden Delicious* A30339 R-677 | 50WP | Spray Interval 40 d | 2.46 | 4200 | 3 | 0 7 14 <u>22</u> 29 37 | | 1.66, 0.04* 1.04, 0.18 0.64, 0.14 0.64, 0.13 0.41, 0.10 0.34, 0.09 | 1.70 1.22 0.78 <u>0.77</u> 0.51 0.43 |
| USA1963 *Golden Delicious* CA A30339 R-677 | 2 EC | Spray Interval 40 d | 2.46 | 4200 | 3 | 0 7 14 <u>22</u> 29 37 | | 1.33, 0.04 0.59, 0.09 0.56, 0.16 0.42, 0.12 0.38, 0.11 0.34, 0.12 | 1.37 0.68 0.72 <u>0.54</u> 0.49 0.46 |
| USA1960 *Gravenstein* C009878 *R-385* | 50WP | Spray Interval 33 d | 3.36 | 5500 | 2 | 0 8 15 <u>21</u> | | | 0.56 <u>0.36</u> |
| USA1960 *Golden Delicious* C009878 *R-385* | 50WP | Spray Interval 33 d | 3.36 | 5500 | 2 | 0 8 15 <u>21</u> 29 | | | 2.67 1.30 0.26 <u>0.27</u> < 0.02 |
| USA 1960 *Gravenstein* C009878 *R-385* | 50WP | Spray Interval 33 d | 3.36 | 5500 | 2 | 0 8 15 <u>21</u> | | | 1.95 0.78 0.57 <u>0.16</u> |
| AUSTRALIA 2000 Bathurst *red delicious* N° 1/10/563 | EC | spray | 1.024 | 1541 | 6 | 0 7 14 21 14 14 14 14 | Fruit  w fruit juice D pomace W pomace | 1.20, 0.90, 0.029 0.25, 0.48, 0.036 0.084, 0.18, 0.022 0.085, 0.16, 0.023 0.12, 0.23, 0.034, 0.005, 0.017, < 0.005 0.72, 1.40, 0.12 0.25, 0.46, 0.046 | 2.13 0.77 0.29 0.27 0.38 0.02 2.24 0.76 |
| AUSTRALIA 2000 Bathurst *red delicious* N° 1/10/563 | EC | spray | 2.034 | 1541 | 6 | 0 7 14 21 | Fruit | 1.50, 1.30.0.065 0.35, 0.64, 0.052 0.24, 0.44, 0.039 0.20, 0.34, 0.040 | 2.87 1.04 0.72 0.58 |

| APPLE Country, Year of trial *Variety* Report | Application Form. | Method | kg ai/ha/ applic'n | L/ha | No. | PHI days | Sample analysed | Residues of alpha, beta endosulfan, endosulfan sulfate, mg/kg | Total Residues, mg/kg |
|---|---|---|---|---|---|---|---|---|---|
| AUSTRALIA 2000 Cottonvale *royal gala* N° 1/10/563 | EC | spray | 1.220 | 1833 | 6 | 14 | Fruit | 0.16, 0.22, 0.15 | 0.539 |

Table 37. Endosulfan residues in pear resulting from supervised trials in Australia.

| PEARS Country, Year of trial *Variety* Report | Application Form. | Method | kg ai/ha | L/ha | No. | PHI days | Sample analysed | Residues of alpha, beta endosulfan, Endosulfan sulfate, mg/kg | Total Residues, mg/kg |
|---|---|---|---|---|---|---|---|---|---|
| AUSTRALIA 2000 Ardmona *packham* N° 1/10/548 | EC | spray | 0.997 | 1500 | 6 | 0 7 14 21 | Fruit | 0.44, 0.45, 0.018 0.26, 0.20, 0.38 0.22, 0.37, 0.20 0.097, 0.19, 0.13 | 0.91 0.84 0.79 0.417 |
| AUSTRALIA 2000 Ardmona *packham* N° 1/10/548 | EC | spray | 1.980 | 1500 | 6 | 0 7 14 21 | Fruit | 0.86, 0.74, 0.23 0.57, 0.82, 0.30 0.48, 0.77, 0.33 0.21, 0.40, 0.24 | 1.83 1.69 1.58 0.85 |
| AUSTRALIA 2000 Paracombe *duchess* N° 1/10/548 | EC | spray | 1.330 | 2000 | 6 | 0 7 14 | Fruit | 0.74, 0.67, 0.32 0.34, 0.46, 0.35 0.062, 0.14, 0.24 | 1.73 1.15 0.44 |
| AUSTRALIA 2000 Paracombe *duchess* N° 1/10/548 | EC | spray | 2.64 | 2000 | 6 | 14 | Fruit | 0.044, 0.10, 0.23 | 0.374 |

Table 38. Endosulfan residues in cherry resulting from supervised trials in USA.

| CHERRY Country, Year of trial *Variety* Report | Application Form | Method | kg ai/ha/ applic'n | L/ha | No. | PHI days | Sample analysed | Residues of alpha, beta endosulfan, endosulfan sulfate, mg/kg | Total Residues, mg/kg |
|---|---|---|---|---|---|---|---|---|---|
| USA 1996 A57715 *B-J96R-04* US/FDA/PAM vol 1 | EC WP | Airblast spray | 3.36 3.36 | 935 935 | 1 1 | 21 21 | Sour cherry | 0.10, 0.31, 0.22 < 0.05, 0.06, < 0.05 | 0.63 0.06 |
| USA 1996 A57715 *B-J96R-04* US/FDA/PAM vol 1 | EC WP | Airblast spray | 3.36 3.36 | 944 944 | 1 1 | 21 21 | Sour cherry | 0.05, 0.17, 0.07 < 0.05, 0.06, < 0.05 | 0.29 0.06 |
| USA 1996 A57715 *B-J96R-04* US/FDA/PAM vol 1 | EC WP | Airblast spray | 3.36 3.36 | 926 926 | 1 1 | 21 21 | Sour cherry | 0.09, 0.19, 0.06 < 0.05, < 0.05, < 0.05 | 0.34 < 0.05 |
| USA 1996 A57715 *B-J96R-04* US/FDA/PAM vol 1 | EC WP | Airblast spray | 3.36 3.36 | 944 935 | 1 1 | 21 21 | Sour cherry | 0.07, 0.18, 0.12 < 0.05, 0.09, < 0.05 | 0.37 0.09 |
| USA 1996 A57715 *B-J96R-04* US/FDA/PAM vol 1 | EC WP | Airblast spray | 3.36 3.36 | 944 944 | 1 1 | 21 21 | Sour cherry | < 0.05, 0.06, 0.06 < 0.05, < 0.05, < 0.05 | 0.12 < 0.05 |
| USA 1996 A57715 *B-J96R-04* US/FDA/PAM vol 1 | EC WP | Airblast spray | 3.36 3.36 | 926 890 | 1 1 | 21 21 | Sour cherry | 0.23, 0.48, 0.14 < 0.05, < 0.05, < 0.05 | 0.85 < 0.05 |

| CHERRY | Application | | | | | PHI | Sample | Residues of alpha, beta | Total |
| Country, Year of trial *Variety* Report | Form | Method | kg ai/ha/ applic'n | L/ha | No. | days | analysed | endosulfan, endosulfan sulfate, mg/kg | Residues, mg/kg |
|---|---|---|---|---|---|---|---|---|---|
| USA 1996 A57715 *B-J96R-04* US/FDA/PAM vol 1 | EC WP | Airblast spray | 3.36 3.36 | 935 953 | 1 1 | 21 21 | Sour cherry | 0.12, 0.24, 0.17 < 0.05, < 0.05, < 0.05 | <u>0.53</u> < 0.05 |
| USA 1996 A57715 *B-J96R-04* US/FDA/PAM vol 1 | EC WP | Airblast spray | 3.36 3.36 | 981 981 | 1 1 | 21 21 | Sour cherry | 0.11, 0.31, 0.21 < 0.05, < 0.05, < 0.05 | <u>0.63</u> < 0.05 |
| USA 1996 A57715 *B-J96R-04* US/FDA/PAM vol 1 | EC WP | Airblast spray | 3.36 3.36 | 1000 990 | 1 1 | 21 21 | Sour cherry | 0.11, 0.30, 0.13 < 0.05, < 0.05, < 0.05 | <u>0.54</u> < 0.05 |
| USA 1996 A57715 *B-J96R-04* US/FDA/PAM vol 1 | EC WP | Airblast spray | 3.36 3.36 | | 1 1 | 21 21 | Sour cherry | 0.22, 0.60, 0.30 < 0.05, < 0.05, < 0.05 | <u>1.12</u> < 0.05 |
| USA 1996 A57718 *B-J96R-04* AL 6086 | EC WP | Airblast spray | 3.36 3.36 | 936 944 | 1 1 | 21 21 | Sweet cherry | 0.14, 0.26, 0.12 < 0.05, 0.08, < 0.05 | <u>0.52</u> 0.08 |
| USA 1996 A57718 *B-J96R-04* AL 6086 | EC WP | Airblast spray | 3.36 3.36 | 963 953 | 1 1 | 21 21 | Sweet cherry | 0.36, 0.64, 0.38 < 0.05, 0.07, 0.07 | <u>1.38</u> 0.14 |
| USA 1996 CA A57718 *B-J96R-04* AL 6086 | EC WP | Airblast spray | 3.36 3.36 | 936 935 | 1 1 | 21 21 | Sweet cherry | 0.09, 0.20, 0.12 < 0.05, < 0.05, < 0.05 | <u>0.41</u> < 0.05 |
| USA 1996 CA *Bing* A57718 *B-J96R-04* AL 6086 | EC WP | Airblast spray | 3.36 3.36 | 944 926 | 1 1 | 21 21 | Sweet cherry | < 0.05, 0.08, 0.06 < 0.05, 0.06, < 0.05 | <u>0.14</u> 0.06 |
| USA 1996 W *Bing* A57718 *B-J96R-04* AL 6086 | EC WP | Airblast spray | 3.36 3.36 | 935 935 | 1 1 | 21 21 | Sweet cherry | 0.12, 0.32, 0.13 0.06, 0.23, 0.05 | <u>0.57</u> 0.34 |
| USA 1996 W *Bing* A57718 *B-J96R-04* AL 6086 | EC WP | Airblast spray | 3.36 3.36 | 944 944 | 1 1 | 21 21 | Sweet cherry | 0.11, 0.25, 0.08 < 0.05, 0.10, < 0.05 | <u>0.44</u> 0.10 |
| USA 1996 O *Lambert* A57718 *B-J96R-04* AL 6086 | EC WP | Airblast spray | 3.36 3.36 | 890 898 | 1 1 | 21 21 | Sweet cherry | 0.27, 0.42, 0.23 0.05, 0.09, 0.06 | <u>0.92</u> 0.20 |
| USA 1996 O A57718 *B-J96R-04* AL 6086 | EC WP | Mist blower | 3.36 3.36 | 963 972 | 1 1 | 21 21 | Sweet cherry | < 0.05, 0.10, 0.05 0.08, 0.13, 0.10 | 0.10 0.31 |
| USA 1996 O A57718 *B-J96R-04* AL 6086 | EC WP | Mist blower | 3.36 3.36 | 963 944 | 1 1 | 21 21 | Sweet cherry | 0.06, 0.10, < 0.05 0.19, 0.37, 0.16 | 0.16 0.72 |
| USA 1996 O A57718 *B-J96R-04* AL 6086 | EC WP | Mist blower | 3.36 3.36 | 730 740 | 1 1 | 21 21 | Sweet cherry | < 0.05, 0.08, 0.06 0.22, 0.36, 0.20 | 0.14 0.78 |

Table 39. Endosulfan residues in apricots resulting from supervised trials in Australia.

| APRICOTS | Application | | | | | PHI | Sample | Residues of alpha, | Total |
| Country, Year of trial *Variety* Report | Form. | Method | kg ai/ha/ applic'n | L/ha | No. | days | analysed | beta endosulfan, endosulfan sulfate, mg/kg | Residues, mg/kg |
|---|---|---|---|---|---|---|---|---|---|
| Australia *2000* Taylorville *storey* N° 1/10/5641 | EC | spray | 1.33 | 2000 | 3 | 0 14 28 35 | Fruit | 1.70, 1.50, 0.52 0.15, 0.30, 0.36 0.11, 0.29, 0.63 0.012, 0.043, 0.21 | 3.72 1.05 0.30 0.27 |

| APRICOTS | Application | | | | | PHI | Sample | Residues of alpha, | Total |
|---|---|---|---|---|---|---|---|---|---|
| Country, Year of trial *Variety* Report | Form. | Method | kg ai/ha/ applic'n | L/ha | No. | days | analysed | beta endosulfan, endosulfan sulfate, mg/kg | Residues, mg/kg |
| Australia *2000* Taylorville *storey* N° 1/10/5641 | EC | spray | 2.64 | 2000 | 3 | 0 14 28 35 | Fruit | 5.00, 4.1, 1.1 1.00, 1.8, 1.3 0.33, 0.74, 1.1 0.16, 0.34, 0.94 | 10.20 4.10 2.17 1.44 |

Table 40. Endosulfan residues in nectarines resulting from a supervised trial in Australia.

| NECTARINE | Application | | | | | PHI | Sample | Residues of alpha, beta | Total |
|---|---|---|---|---|---|---|---|---|---|
| Country, Year of trial *Variety* Report | Form. | Method | kg ai/ha/ applic'n | L/ha | No. | days | analysed | endosulfan, endosulfan sulfate, mg/kg | Residues mg/kg |
| Australia *2000* Montacute *tasty gold* N° 1/10/536 | EC | spray | 1.33 | 2000 | 3 | 0 14 28 35 | Fruit | 1.70, 1.10, 0.031 0.26, 0.34, 0.024 0.038, 0.11, 0.029 0.099,0.27,0.059 | 2.83 0.62 0.18 0.43 |

Table 41. Endosulfan residues in peaches resulting from supervised trials in Europe, USA, and Australia.

| PEACH | Application | | | | | PHI | Sample | Residues of alpha, beta | Total |
|---|---|---|---|---|---|---|---|---|---|
| Country, Year of trial *Variety* Report | Form. | Method | kg ai/ha/ applic'n | L/ha | No. | days | analysed | endosulfan, endosulfan sulfate, mg/kg | Residues, mg/kg |
| Germany1974 *Rekord von Alfter* A06462 4/93/04/04[#] | EC* | Spray Interval: 7 days | 0.96 | 2000 (0.048 % ai) | 5 | 0 7 14 21 28 | fruit fruit fruit fruit fruit | 1.8, 0.004 0.04, 0.004 0.18, ND 0.11, 0.01 0.01, 0.005 | 1.80 0.04 0.18 0.12 0.015 |
| Germany1974 *Rekord von Alfter* A06463 4/93/04/02 | EC* | Spray Interval: 7 days | 0.72 | 2000 (0.036 % ai) | 5 | 0 7 14 21 28 | fruit fruit fruit fruit fruit | 0.90, ND 0.40, ND 0.005, ND 0.03, ND 0.01, 0.003 | 0.90 0.40 0.005 0.03 0.01 |
| Germany1974 *Rekord von Alfter* A06464 4/93/04/03 | EC* | Spray | 0.96 | 2000 (0.048 % ai) | 1 | 0 7 14 21 28 | fruit fruit fruit fruit fruit | 1.80, 0.004 0.04, 0.004 0.20, ND 0.10, 0.01 0.01, 0.005 | 1.80 0.04 0.20 0.11 0.015 |
| Germany1974 *Madame Rogliat* A06465 4/93/05/02 | EC* | Spray | 0.72 | 2000 (0.036 % ai) | 1 | 0 7 14 21 28 | fruit fruit fruit fruit fruit | 1.60, 0.005 0.30, 0.01 0.02, ND 0.05, 0.003 0.01, 0.003 | 1.61 0.31 0.02 0.05 0.01 |
| Germany1974 *Madame Rogliat* A06466 4/93/05/03 | EC* | Spray | 0.96 | 2000 (0.048 % ai) | 1 | 0 7 14 21 28 | fruit fruit fruit fruit fruit | 2.00, ND 0.40, ND 0.20, 0.003 0.10, 0.008 0.02, 0.003 | 2.00 0.40 0.203 0.11 0.02 |
| Germany1974 *Rogliat* A06467 4/93/05/04 | EC* | Spray | 0.96 | 2000 (0.048 % ai) | 1 | 0 7 14 21 | fruit fruit fruit fruit | 2.00, 0.01 0.80, 0.005 0.30, 0.02 0.10, 0.01 | 2.01 0.805 0.32 0.11 |

| PEACH | Application | | | | | PHI days | Sample analysed | Residues of alpha, beta endosulfan, endosulfan sulfate, mg/kg | Total Residues, mg/kg |
|---|---|---|---|---|---|---|---|---|---|
| Country, Year of trial *Variety* Report | Form. | Method | kg ai/ha/ applic'n | L/ha | No. | | | | |
| Germany1974 *Dixired* A06468 4/93/02/02 | EC* | Spray | 0.72 | 2000 (0.048 % ai) | 1 | 0 | fruit | 2.30, ND | 2.30 |
| | | | | | | 7 | fruit | 0.50, 0.01 | 0.51 |
| | | | | | | 14 | fruit | 0.30, 0.01 | 0.31 |
| | | | | | | 21 | fruit | 0.10, 0.01 | 0.11 |
| Germany1974 *Dixired* A06469 4/93/02/03 | EC* | Spray | 0.96 | 2000 (0.048 % ai) | 1 | 0 | fruit | 4.60, 0.01 | 4.61 |
| | | | | | | 7 | fruit | 0.20, 0.01 | 0.21 |
| | | | | | | 14 | fruit | 0.80, 0.007 | 0.81 |
| | | | | | | 21 | fruit | 0.20, 0.03 | 0.23 |
| Germany1974 *Dixired* A06470 4/93/02/04 | EC* | Spray | 0.96 | 2000 (0.048 % ai) | 1 | 0 | fruit | 4.40, 0.50 | 4.90 |
| | | | | | | 7 | fruit | 1.00, 0.20 | 1.20 |
| | | | | | | 14 | fruit | 0.70, 0.20 | 0.90 |
| | | | | | | 21 | fruit | 0.10, 0.05 | 0.15 |
| Germany1974 *Red Heaven* A06471 4/93/01/02 | EC* | Spray | 0.72 | 2000 (0.036 % ai) | 1 | 0 | fruit | 1.30, 0.003 | 1.303 |
| | | | | | | 7 | fruit | 1.40, 0.01 | 1.41 |
| | | | | | | 14 | fruit | 0.04, 0.006 | 0.05 |
| | | | | | | 21 | fruit | 0.08, 0.005 | 0.085 |
| Germany1974 *Red Heaven* A06472 4/93/01/03 | EC* | Spray | 0.96 | 2000 (0.048 % ai) | 1 | 0 | fruit | 2.40, ND | 2.40 |
| | | | | | | 7 | fruit | 0.40, 0.005 | 0.405 |
| | | | | | | 14 | fruit | -, - | |
| | | | | | | 21 | fruit | 0.08, ND | 0.08 |
| Germany1974 *Red Heaven* A06473 4/93/01/04 | EC* | Spray Interval 29 days | 0.96 | 2000 (0.048 % ai) | 2 | 0 | fruit | 2.10, 0.007 | 2.11 |
| | | | | | | 7 | fruit | 1.40, 0.06 | 1.46 |
| | | | | | | 14 | fruit | 0.70, 0.01 | 0.71 |
| | | | | | | 21 | fruit | 0.20, 0.02 | 0.22 |
| Spain 1999 *Flordastar* C017102 ER 99 ECS 754 DGM F01/97-0 | CS* | Spray , BBCH 73, 75, 76 | 0.7993 | 1500 (0.053 % ai) | 3 | 0 | fruit wo stone | 1.90, 1.00, 0.08 | 2.98 |
| | | | | | | 0 | Fruit | 1.50, 0.81, 0.06 | 2.37 |
| | | | | | | 21 | fruit wo stone | 0.25, 0.18, 0.037 | 0.467 |
| | | | | | | 21 | Fruit | 0.23, 0.17, 0.34 | 0.74 |
| Spain 1999 *Sudanell* C017102 ER 99 ECS 754 DGM F01/97-0 | CS* | handgun BBCH 75 | 0.7993 | 1500 (0.053 % ai) | 3 | 0 | fruit wo stone | 1.30, 0.78, 0.03 | 2.11 |
| | | | | | | 0 | fruit | 1.00, 0.61, 0.02 | 1.63 |
| | | | | | | 21 | fruit wo stone | 0.27, 0.20, 0.02 | 0.49 |
| | | | | | | 21 | fruit | 0.22, 0.17, < 0.02 | 0.39 |
| France (S) 1999 *Orelie* C017102 ER 99 ECS 754 DGM F01/97-0 | CS* | handgun BBCH 77 , 85, 85 | 0.7993 | 1500 (0.053 % ai) | 3 | 0 | fruit wo stone | 0.80, 0.50, < 0.02 | 1.30 |
| | | | | | | 0 | fruit | 0.77, 0.48, < 0.02 | 1.25 |
| | | | | | | 21 | fruit wo stone | 0.18, 0.18, < 0.02 | 0.36 |
| | | | | | | 21 | fruit | 0.18, 0.18, < 0.02 | 0.36 |
| Italy 1999 *Star red gold* C017102 ER 99 ECS 754 DGM F01/97-0 | CS* | Spray, BBCH 75, 75, 77 | 0.7993 | 1500 (0.053 % ai) | 3 | 0 | fruit wo stone | 0.20, 0.12, < 0.02 | 0.32 |
| | | | | | | 0 | fruit | 0.15, 0.10, < 0.02 | 0.25 |
| | | | | | | 21 | fruit wo stone | 0.09, 0.08, < 0.02 | 0.17 |
| | | | | | | 21 | fruit | 0.083 0.074, < 0.02 | 0.16 |
| Italy1999 *Federica* ER 99 ECS 754 DGM F01/97-0 | CS* | Spray, BBCH 73 75, 76 , | 0.7993 | 1500 (0.053 % ai) | 3 | 0 | fruit wo stone | 2.10, 1.20, < 0.02 | 3.30 |
| | | | | | | 0 | fruit | 1.80, 1.10, < 0.02 | 2.90 |
| | | | | | | 21 | fruit wo stone | 0.24, 0.20, < 0.02 | 0.44 |
| | | | | | | 21 | fruit | 0.22, 0.18, < 0.02 | 0.40 |
| Spain1998 *Flor Down* Benifaio, Valencia ER 98 ECS 754 DGM F01/97-0. | CS* | Spray BBCH 71, 73, 81 | 0.7993 | 1500 (0.053 % ai) | 3 | 0 | fruit wo stone | 1.00, 0.53, 0.05 | 1.58 |
| | | | | | | 7 | fruit wo stone | 0.52, 0.35, 0.04 | 0.91 |
| | | | | | | 13 | fruit wo stone | 0.27, 0.23, 0.04 | 0.54 |
| | | | | | | 20 | fruit wo stone | 0.11, 0.10, 0.02 | 0.23 |
| | | | | | | 0 | fruit | 0.84, 0.45, 0.04 | 1.33 |
| | | | | | | 7 | fruit | 0.45, 0.31, 0.04 | 0.80 |
| | | | | | | 13 | fruit | 0.24, 0.21, 0.04 | 0.49 |
| | | | | | | 20 | fruit | 0.10, 0.09, 0.02 | 0.21 |
| Spain1988 *Spind Graes* *ER 98 ECS 754* DGM F01/97-0. | CS* | Spray BBCH 71, | 0.7993 | 1500 (0.053 % ai) | 3 | 0 | fruit wo stone | 0.92, 0.48, 0.04 | 1.44 |
| | | | | | | 7 | fruit wo stone | 0.51, 0.36, 0.03 | 0.90 |
| | | | | | | 14 | fruit wo stone | 0.23, 0.16, 0.02 | 0.51 |
| | | | | | | 19 | fruit wo stone | 0.13, 0.08, < 0.02 | 0.23 |

| PEACH | Application | | | | | | | | |
|---|---|---|---|---|---|---|---|---|---|
| Country, Year of trial *Variety* Report | Form. | Method | kg ai/ha/ applic'n | L/ha | No. | PHI days | Sample analysed | Residues of alpha, beta endosulfan, endosulfan sulfate, mg/kg | Total Residues, mg/kg |
| *Spind Graes* ER 98 ECS 754 DGM F01/97-0. | | 75, 81 | | | | 0 | fruit | 0.83, 0.43, 0.04 | 1.30 |
| | | | | | | 7 | fruit | 0.47, 0.33, 0.03 | 0.83 |
| | | | | | | 14 | fruit | 0.22, 0.15, 0.02 | 0.39 |
| | | | | | | 19 | fruit | 0.12, 0.08, < 0.02 | 0.20 |
| Greece1998 *Louatel* ER 98 ECS 754 DGM F01/97-0. | CS* | Spray (gun) BBCH 71, 73, 75 | 0.7993 | 1500 (0.053 % ai) | 3 | 0 | fruit wo stone | 0.60, 0.36, < 0.02 | 0.96 |
| | | | | | | 7 | fruit wo stone | 0.74, 0.42, 0.02 | 1.18 |
| | | | | | | 14 | fruit wo stone | 0.14, 0.08, < 0.02 | 0.22 |
| | | | | | | 21 | fruit wo stone | 0.16, 0.08, < 0.02 | 0.24 |
| | | | | | | 0 | fruit | 0.51, 0.31, < 0.02 | 0.82 |
| | | | | | | 7 | fruit | 0.67, 0.38, 0.02 | 1.07 |
| | | | | | | 14 | fruit | 0.13, 0.07, < 0.02 | 0.20 |
| | | | | | | 21 | fruit | 0.15, 0.07, < 0.02 | 0.22 |
| Italy1998 *Maria Luisa* ER 98 ECS 754 DGM F01/97-0. | CS* | Mist r blower (knapsack BBCH 73, 75, 77 | 0.7993 | 1500 (0.053 % ai) | 3 | 0 | fruit wo stone | 1.20, 0.72, 0.03 | 1.95 |
| | | | | | | 7 | fruit wo stone | 0.70, 0.52, 0.03 | 1.25 |
| | | | | | | 14 | fruit wo stone | 0.32, 0.21, < 0.02 | 0.53 |
| | | | | | | 21 | fruit wo stone | 0.23, 0.20, < 0.02 | 0.43 |
| | | | | | | 0 | fruit | 1.10, 0.64, 0.03 | 1.77 |
| | | | | | | 7 | fruit | 0.63, 0.47, 0.03 | 1.13 |
| | | | | | | 14 | fruit | 0.28, 0.19, < 0.02 | 0.47 |
| | | | | | | 21 | fruit | 0.21, 0.18, < 0.02 | 0.39 |
| Italy1998 *Spring Bel* ER 98 ECS 754 DGM F01/97-0. | CS* | Mist blower (knapsack BBCH 73, 75, 78 | 0.7993 | 1500 (0.053 % ai) | 3 | 0 | fruit wo stone | 1.50, 0.90, 0.08 | 2.48 |
| | | | | | | 7 | fruit wo stone | 0.78, 0.54, 0.09 | 1.41 |
| | | | | | | 14 | fruit wo stone | 0.32, 0.27, 0.12 | 0.71 |
| | | | | | | 21 | fruit wo stone | 0.15, 0.14, 0.08 | 0.37 |
| | | | | | | 0 | fruit | 1.30, 0.81, 0.07 | 2.18 |
| | | | | | | 7 | fruit | 0.70, 0.49, 0.08 | 1.27 |
| | | | | | | 14 | fruit | 0.30, 0.25, 0.11 | 0.66 |
| | | | | | | 21 | fruit | 0.14, 0.13, 0.08 | 0.35 |
| Spain1998 *Flor Down* C002960 ER 98 ECS 742 | 35EC | Spray (gun) BBCH 71, 73, 75 | 0.800 | 1500 (0.053 % ai) | 3 | 0 | fruit wo stone | 1.40, 0.90, 0.14 | 2.44 |
| | | | | | | 20 | fruit wo stone | 0.08, 0.09, 0.04 | 0.21 |
| | | | | | | 0 | fruit | 1.20, 0.75, 0.12 | 2.07 |
| | | | | | | 20 | fruit | 0.07, 0.08, 0.04 | 0.19 |
| Spain1998 *Flor Down* C002960 ER 98 ECS 742 | | Spray (gun) BBCH 71, 73, 75 | 1.600 | 1500 (0.106 % ai) | 3 | 21 | fruit | 0.16,0.17,0.08 | 0.41 |
| Spain1998 *Spind graes* C002960 ER 98 ECS 742 | 35EC | Spray BBCH 71, 75, 81 | 0.800 | 1500 (0.053 % ai) | 3 | 0 | fruit wo stone | 0.61, 0.43, 0.10 | 1.14 |
| | | | | | | 19 | fruit wo stone | 0.04, 0.06, 0.04 | 0.14 |
| | | | | | | 0 | fruit | 1.30, 0.17, 0.11 | 1.58 |
| | | | | | | 19 | fruit | 0.08, 0.05, 0.04 | 0.17 |
| Spain1998 *Spind graes* C002960 ER 98 ECS 742 | | Spray (gun) BBCH 71, 75, 81 | 1.600 | 1500 (0.106 % ai) | 3 | 21 | fruit | 0.16,0.26,0.12 | 0.54 |
| Greece1998 *Louatel* C002960 ER 98 ECS 742 | 35EC | Spray (gun) BBCH 73, 73, 75 | 0.800 | 1500 (0.053 % ai) | 3 | 0 | fruit wo stone | 0.74, 0.42, 0.03 | 1.19 |
| | | | | | | 21 | fruit wo stone | 0.07, 0.07, 0.02 | 0.16 |
| | | | | | | 0 | fruit | 0.63, 0.36, 0.03 | 1.02 |
| | | | | | | 21 | fruit | 0.07, 0.07, 0.02 | 0.16 |
| Italy1998 *Maris Luisa* C002960 ER 98 ECS 742 | 35EC | Mist blower (knapsack BBCH 73 | 0.800 | 1500 (0.053 % ai) | 3 | 0 | fruit wo stone | 1.00, 0.68, 0.05 | 1.73 |
| | | | | | | 21 | fruit wo stone | 0.08, 0.11, 0.05 | 0.24 |
| | | | | | | 0 | fruit | 0.89, 0.60, 0.05 | 1.54 |
| | | | | | | 21 | fruit | 0.07, 0.10, 0.05 | 0.22 |
| Italy1998 *Maris Luisa* C002960 ER 98 ECS 742 | | blower (knapsack) BBCH 73, 75, 77 | 1.600 | 1500 (0.053 % ai) | 3 | 13 | fruit | 0.26,0.34,0.12 | 0.72 |
| Italy1998 *Spring Bell* C002960 ER 98 ECS 742 | 35EC | blower) (knapsack BBCH 73 , 75, 78 | 0.800 | 1500 (0.053 % ai) | 3 | 0 | fruit wo stone | 0.53, 0.36, 0.02 | 0.91 |
| | | | | | | 21 | fruit wo stone | 0.14, 0.17, 0.09 | 0.40 |
| | | | | | | 0 | fruit | 0.48, 0.33, 0.02 | 0.83 |
| | | | | | | 21 | fruit | 0.13, 0.16, 0.09 | 0.38 |

| PEACH Country, Year of trial *Variety* Report | Form. | Application Method | Application kg ai/ha/ applic'n | Application L/ha | Application No. | PHI days | Sample analysed | Residues of alpha, beta endosulfan, endosulfan sulfate, mg/kg | Total Residues, mg/kg |
|---|---|---|---|---|---|---|---|---|---|
| Spain1997 *Baby Gold 6* C001114b ER 97 ECS 742 | 35EC | Spray) (knapsack BBCH 75, 77, 89 | 0.800 | 1500 (0.053 % ai) | 3 | 0 | fruit wo stone | 0.49, 0.31, < 0.02 | 0.80 |
| | | | | | | 7 | fruit wo stone | 0.33, 0.37, 0.03 | 0.73 |
| | | | | | | 14 | fruit wo stone | 0.23, 0.33, 0.04 | 0.60 |
| | | | | | | 21 | fruit wo stone | Sample na | |
| | | | | | | 0 | fruit | 0.43, 0.27, < 0.02 | 0.70 |
| | | | | | | 7 | fruit | 0.30, 0.34, 0.03 | 0.67 |
| | | | | | | 14 | fruit | 0.21, 0.31, 0.04 | 0.56 |
| | | | | | | 21 | fruit | Sample na | |
| Spain1997 *Sudanel* C001114b *ER 97 ECS* 742 | 35EC | Spray) (knapsack BBCH 75 75, 77 | 0.800 | 1500 (0.053 % ai) | 3 | 0 | fruit wo stone | 1.20, 0.87, 0.04 | 2.11 |
| | | | | | | 7 | fruit wo stone | 0.28, 0.32, 0.03 | 0.63 |
| | | | | | | 14 | fruit wo stone | 0.15, 0.24, 0.04 | 0.43 |
| | | | | | | 21 | fruit wo stone | 0.05, 0.13, 0.03 | 0.21 |
| | | | | | | 0 | fruit | 1.05, 0.76, 0.04 | 1.85 |
| | | | | | | 7 | fruit | 0.26, 0.29, 0.03 | 0.58 |
| | | | | | | 14 | fruit | 0.14, 0.22, 0.04 | 0.40 |
| | | | | | | 21 | fruit | 0.05, 0.12, 0.03 | 0.20 |
| Italy1997 *Star Red Gold* C001114b *ER 97 ECS* 742 | 35EC | Mist blower (knapsack) BBCH 75, 75, 77 | 0.800 | 1500 (0.053 % ai) | 3 | 0 | fruit wo stone | 0.42, 0.32, < 0.02 | 0.74 |
| | | | | | | 7 | fruit wo stone | 0.14, 0.12, < 0.02 | 0.26 |
| | | | | | | 14 | fruit wo stone | 0.12, 0.16, < 0.02 | 0.28 |
| | | | | | | 21 | fruit wo stone | 0.05, 0.08, < 0.02 | 0.13 |
| | | | | | | 0 | fruit | 0.37, 0.28, < 0.02 | 0.65 |
| | | | | | | 7 | fruit | 0.13, 0.11, < 0.02 | 0.24 |
| | | | | | | 14 | fruit | 0.11, 0.15, < 0.02 | 0.26 |
| | | | | | | 21 | fruit | 0.05, 0.07, < 0.02 | 0.12 |
| Italy1997 *Lafaiette* C001114b *ER 97 ECS* 742 | 35EC | Mist blower (knapsack) BBCH 75, 75, 78 | 0.800 | 1500 (0.053 % ai) | 3 | 0 | fruit wo stone | 0.43, 0.25, < 0.02 | 0.68 |
| | | | | | | 7 | fruit wo stone | 0.22, 0.22, < 0.02 | 0.44 |
| | | | | | | 14 | fruit wo stone | 0.04, 0.05, < 0.02 | 0.09 |
| | | | | | | 21 | fruit wo stone | 0.04, 0.05, < 0.02 | 0.09 |
| | | | | | | 0 | fruit | 0.38, 0.22, < 0.02 | 0.60 |
| | | | | | | 7 | fruit | 0.21, 0.21, < 0.02 | 0.42 |
| | | | | | | 14 | fruit | 0.04, 0.05, < 0.02 | 0.09 |
| | | | | | | 21 | fruit | 0.04, 0.05, < 0.02 | 0.09 |
| Italy1997 *Maycrest* C001114b *ER 97 ECS* 742 | 35EC | Mist blower) (knapsack BBCH 73, 75, 81 | 0.800 | 1500 (0.053 % ai) | 3 | 0 | fruit wo stone | 0.75, 0.19, 0.02 | 0.96 |
| | | | | | | 7 | fruit wo stone | 0.25, 0.27, 0.02 | 0.54 |
| | | | | | | 14 | fruit wo stone | 0.04, 0.10, 0.03 | 0.17 |
| | | | | | | 21 | fruit wo stone | 0.03, 0.03, 0.02 | 0.08 |
| | | | | | | 0 | fruit | 0.75, 0.19, 0.02 | 0.96 |
| | | | | | | 7 | fruit | 0.23, 0.24, 0.02 | 0.49 |
| | | | | | | 14 | fruit | 0.04, 0.09, 0.03 | 0.16 |
| | | | | | | 21 | fruit | 0.03, 0.03, 0.02 | 0.08 |
| USA1963 *Elberta* Rockville, CA C009876 R-689 | 50WP | | 1.56+2.46 | 28618, 4114 (0.05, 0.06 % ai) | 2 | 0 | fruit | 3.70, 0.20 | 3.90 |
| | | | | | | 6 | fruit | 1.58, 0.13 | 1.71 |
| | | | | | | 14 | fruit | 1.03, 0.16 | 1.19 |
| | | | | | | 14 | fruit | 1.08, 0.18 | 1.26 |
| | | | | | | 22 | fruit | 0.43, 0.11 | 0.54 |
| | | | | | | 28 | fruit | 0.24, 0.05 | 0.29 |
| | | | | | | 35 | fruit | 0.20, 0.04 | 0.24 |
| Australia 2000 GoulburnValley *Tatura* N° 1/10/541 | EC | spray | 1.33 | 2000 | 3 | 0 | Fruit | 0.13,0.085, < 0.005 | 0.22 |
| | | | | | | 14 | | 0.39,0.53, 0.13 | 1.05 |
| | | | | | | 28 | | 0.079, 0.15, 0.069 | 0.30 |
| | | | | | | 35 | | 0.026,0.092, 0.07 | 0.19 |
| Australia 2000 GoulburnValley *Tatura* N° 1/10/541 | EC | spray | 1.98 | 1500 | 6 | 0 | Fruit | 2.90, 1.90.0.11 | 4.91 |
| | | | | | | 14 | | 0.29, 0.33, 0.05 | 0.67 |
| | | | | | | 28 | | 0.09, 0.20.0.076 | 0.37 |
| | | | | | | 35 | | 0.033, 0.11, 0.065 | 0.21 |

| PEACH | Application | | | | | PHI days | Sample analysed | Residues of alpha, beta endosulfan, endosulfan sulfate, mg/kg | Total Residues, mg/kg |
|---|---|---|---|---|---|---|---|---|---|
| Country, Year of trial *Variety* Report | Form. | Method | kg ai/ha/ applic'n | L/ha | No. | | | | |
| Australia 2000 Passchendale *Crown princess* N° 1/10/541 | EC | spray | 0.651 | 980 | 6 | 0 | Fruit | 0.81, 0.68, 0.18 | 1.73 |
| | | | | | | 14 | | 0.26, 0.35, 0.14 | 1.15 |
| | | | | | | 28 | | 0.06, 0.097, 0.053 | 0.21 |
| | | | | | | 35 | | 0.027, 0.066,0.05 | 0.14 |
| | | | | | | 0c | | 0.007, 0.007,< 0.005 | 0.02 |

# No reference provided for analytical method.

Table 42. Endosulfan residues in grapes resulting from supervised trials in Europe and USA.

| GRAPE | Application | | | | | PHI days | Sample analysed | Residues of alpha, beta endosulfan, endosulfan sulfate, mg/kg | Total Residues, mg/kg |
|---|---|---|---|---|---|---|---|---|---|
| Country, Year of trial *Variety* report | Form. | Method | kg ai/ha | L/ha | No | | | | |
| Germany 1984 *MullerThurgau* PSR99/012 | 33WP | Spray. Interval 63 days | 0.592, 1.184 | 300, 600 (0.197% ai) | 2 | 0 | Fruit | | 9.20 |
| | | | | | | 14 | Fruit | | 1.88 |
| | | | | | | 35 | Fruit | | 0.68 |
| | | | | | | 60 | Fruit | | 0.55 |
| | | | | | | 60 | Must | | 0.04 |
| | | | | | | 60 | Wine | | 0.02 |
| Germany 1984 *(MullerThurgau)* PSR99/012 | 33WP | Spray. Interval 51 days | 1.184 | 600 (0.197% ai) | 2 | 0 | Fruit | | 12.2 |
| | | | | | | 19 | Fruit | | 1.26 |
| | | | | | | 35 | Fruit | | 0.7 |
| | | | | | | 62 | Fruit | | 0.49 |
| | | | | | | 62 | Must | | 0.03 |
| | | | | | | 62 | Wine | | 0.02 |
| Germany 1984 *(MullerThurgau)* | 35MC* | Spray Interval 51 days | 1.26 | 600 (0.21% ai) | 2 | 0 | Fruit | 10.0, 5.0, 0.01 | 15.01 |
| | | | | | | 19 | Fruit | 1.30, 1.10, 0.09 | 2.49 |
| | | | | | | 35 | Fruit | 0.70, 0.60, 0.06 | 1.36 |
| | | | | | | 62 | Fruit | 0.20, 0.30, 0.09 | 0.59 |
| Germany 1984 *(MullerThurgau)* | 35MC* | Spray Spray Interval 63 days | 0.63, 1.26 | 300, 600 (0.21% ai) | 2 | 0 | Fruit | 9.00, 4.40, < 0.01 | 13.40 |
| | | | | | | 14 | Fruit | 1.10, 0.70, 0.02 | 1.82 |
| | | | | | | 35 | Fruit | 0.40, 0.40, 0.02 | 0.82 |
| | | | | | | 60 | Fruit | 0.20, 0.30, 0.10 | 0.60 |
| Spain 1999 *(Italia)* | CS* | Foliar Interval 14 days BBCH 79, | 1.05 | 800, 1000 (0.105% ai) | 2 | 0 | Fruit | 2.50, 1.30, < 0.02 | 3.80 |
| | | | | | | 7 | Fruit | 1.70, 1.00, < 002 | 2.70 |
| | | | | | | 14 | Fruit | 0.78, 0.43,< 0.02 | 1.21 |
| | | | | | | 28 | Fruit | 0.54, 0.33, < 0.02 | 0.87 |
| Spain 1999 *(Alfonso Labale)* | CS* | Spray Interval 15 days BBCH 77, | 1.05 | 800, 1000 (0.105% ai) | 2 | 0 | Fruit | 1.1, 0.62, < 0.02 | 1.72 |
| | | | | | | 7 | Fruit | 0.35, 0.21, < 0.02 | 0.56 |
| | | | | | | 14 | Fruit | 0.37, 0.25, < 0.02 | 0.62 |
| | | | | | | 28 | Fruit | 0.11, 0.077, < 0.02 | 0.19 |
| Greece 1999 *Xinimavro* ER 99 ECS 756 | CS* | Foliar Interval 13 d BBCH 79, 83 | 1.05 | 1035 (0.101% ai) | 2 | 0 | Fruit | 0.44, 0.22, < 0.02 | 0.66 |
| | | | | | | 7 | Fruit | 0.20, < 0.02, < 0.02 | 0.20 |
| | | | | | | 14 | Fruit | 0.17, 0.14, < 0.02 | 0.31 |
| | | | | | | 28 | Fruit | 0.23, 0.17, < 0.02 | 0.40 |
| Italy1999 (Trebbiano di Romagna) ER 99 ECS 756 | CS* | | 1.05 | 1000 (0.105% ai) | 2 | 0 | Fruit | 0.57, 0.29, < 0.02 | 0.86 |
| | | | | | | 7 | Fruit | 0.54, 0.33, < 0.02 | 0.87 |
| | | | | | | 15 | Fruit | 0.44, 0.30, ,0.02 | 0.76 |
| | | | | | | 28 | Fruit | 0.41, 0.28, 0.027 | 0.72 |
| Italy 1999 Vittora ER 99 ECS 756 | CS* | Foliar Interval 14 d BBCH 75, 79 BBCH 75, 79 | 1.05 | 1000 (0.105% ai) | 2 | 0 | Fruit | 4.60, 2.50, < 0.02 | 7.10 |
| | | | | | | 7 | Fruit | 2.50, 1.50, < 0.02 | 4.00 |
| | | | | | | 14 | Fruit | 1.50, 1.10, 0.02 | 2.62 |
| | | | | | | 28 | Fruit | 0.47, 0.31, < 0.02 | 0.78 |
| Italy1997 Italia ER 97 ECS 744 | 35 EC | Mist blower Interval14 d BBCH 79, 83 | 1.05 | 1000 (0.105% ai) | 2 | 0 | Fruit | 1.70, 1.10, < 0.02 | 2.80 |
| | | | | | | 7 | Fruit | 0.31, 0.49, < 0.02 | 0.80 |
| | | | | | | 14 | Fruit | 0.30, 0.43, < 0.02 | 0.73 |
| | | | | | | 22 | Fruit | 0.29, 0.43, < 0.02 | 0.72 |
| | | | | | | 28 | Fruit | 0.10, 0.17, < 0.02 | 0.27 |

| GRAPE | | Application | | | | | | | Residues of alpha, beta endosulfan, endosulfan sulfate, mg/kg | Total Residues, mg/kg |
|---|---|---|---|---|---|---|---|---|---|---|
| Country, Year of trial *Variety* report | Form. | Method | kg ai/ha | L/ha | No | PHI days | Sample analysed | | | |
| Spain 1997 Macabeo ER 97 ECS 744 | 35 EC | Foliar Spray Interval 13 days BBCH 81, 83 | 1.05 | 1000 (0.105% ai) | 2 | 0 | Fruit | | 0.59, 0.38, < 0.02 | 0.97 |
| | | | | | | 7 | Fruit | | 0.07, 0.12, 0.02 | 0.21 |
| | | | | | | 14 | Fruit | | 0.07, 0.11, 0.03 | 0.21 |
| | | | | | | 22 | Fruit | | 0.04, 0.06, < 0.02 | 0.10 |
| | | | | | | 28 | Fruit | | 0.03, 0.03, < 0.02 | 0.06 |
| Spain 1997 Bobal ER 97 ECS 744 | 35 EC | Foliar Spray Interval 13 days BBCH 81, 83 | 1.05 | 1000 (0.105% ai) | 2 | 0 | Fruit | | 0.57, 0.31, < 0.02 | 0.88 |
| | | | | | | 7 | Fruit | | 0.19, 0.25, 0.03 | 0.47 |
| | | | | | | 14 | Fruit | | 0.05 0.07, 0.02 | 0.14 |
| | | | | | | 22 | Fruit | | 0.03, 0.06, 0.02 | 0.11 |
| | | | | | | 28 | Fruit | | 0.31, ND, ND | 0.31 |
| Spain 1997 Bobal ER 97 ECS 744 | 35 EC | Foliar Spray Interval 13 days BBCH 81, 83 | 1.05 | 1000 (0.105% ai) | 2 | 0 | Fruit | | 0.31, 0.20, < 0.02 | 0.51 |
| | | | | | | 7 | Fruit | | 0.07, 0.12, < 0.02 | 0.19 |
| | | | | | | 14 | Fruit | | 0.08, 0.16, 0.02 | 0.26 |
| | | | | | | 22 | Fruit | | 0.04, 0.06, < 0.02 | 0.10 |
| | | | | | | 28 | Fruit | | < 0.02, 0.02, < 0.02 | 0.02 |
| Italy 1997 *Regina* ER 97 ECS 744 | 35 EC | Mist blower Interval 14 days BBCH 79, 83 | 1.05 | 1000 (0.105% ai) | 2 | 0 | Fruit | | 1.10, 0.57, < 0.02 | 1.67 |
| | | | | | | 7 | Fruit | | 0.08, 0.21, < 0.02 | 0.29 |
| | | | | | | 14 | Fruit | | 0.07, 0.27, 0.03 | 0.37 |
| | | | | | | 21 | Fruit | | < 0.02, 0.04, < 0.03 | 0.04 |
| | | | | | | 28 | Fruit | | < 0.02, 0.07, 0.02 | 0.09 |
| Spain 1994 *Cencibel* ER 94 ECS 730 | 35 EC | 2.408 | 0.792 | 750 (0.1056 % ai) | 3 | 0 | Fruit | | 0.16, 0.20, < 0.05 | 0.36 |
| | | | | | | 8 | Fruit | | < 0.05, 0.08, < 0.05 | 0.08 |
| | | | | | | 15 | Fruit | | < 0.05, 0.07, < 0.05 | 0.07 |
| | | | | | | 22 | Fruit | | < 0.05, < 0.05, < 0.05 | < 0.05 |
| | | | | | | 29 | Fruit | | < 0.05, < 0.05, < 0.05 | < 0.05 |
| | | | | | | 15 | Juice | | ND, < 0.05, < 0.05 | < 0.05 |
| | | | | | | 15 | Pomace | | 0.07, 0.20, 0.06 | 0.33 |
| | | | | | | 15 | Young wine | | < 0.05, < 0.05, < 0.05 | < 0.05 |
| | | | | | | 15 | Wine | | < 0.05, < 0.05, < 0.05 | < 0.05 |
| Spain 1994 *Bobal* ER 94 ECS 730 | 35 EC | | 0.792 | 750 (0.1056 % ai) | 3 | 0 | Fruit | | < 0.05, < 0.05, < 0.05 | < 0.05 |
| | | | | | | 8 | Fruit | | < 0.05, 0.10, < 0.05 | 0.10 |
| | | | | | | 15 | Fruit | | < 0.05, 0.105, < 0.05 | 0.11 |
| | | | | | | 22 | Fruit | | < 0.05, < 0.05, < 0.05 | < 0.05 |
| | | | | | | 29 | Fruit | | < 0.05, 0.05, < 0.05 | 0.05 |
| | | | | | | 15 | Juice | | < 0.05, < 0.05, < 0.05 | < 0.05 |
| | | | | | | 15 | Pomace | | < 0.05, 0.17, 0.07 | 0.24 |
| | | | | | | 15 | Young wine | | < 0.05, < 0.05, < 0.05 | < 0.05 |
| | | | | | | 15 | Wine | | < 0.05, < 0.05, < 0.05 | < 0.05 |
| Spain 1994 *Garrida* ER 94 ECS 730 | 35 EC | | 0.317 | 300 (0.1056 % ai) | 3 | 0 | Fruit | | 0.13, 0.12, < 0.05 | 0.25 |
| | | | | | | 7 | Fruit | | < 0.05, < 0.05, < 0.05 | < 0.05 |
| | | | | | | 13 | Fruit | | < 0.05, < 0.05, 0.09 | 0.09 |
| | | | | | | 20 | Fruit | | < 0.05, < 0.05, < 0.05 | < 0.05 |
| | | | | | | 13 | Juice | | < 0.05, < 0.05, < 0.05 | < 0.05 |
| | | | | | | 13 | Pomace | | < 0.05, < 0.05, < 0.05 | < 0.05 |
| | | | | | | 13 | Young wine | | < 0.05, < 0.05, < 0.05 | < 0.05 |
| | | | | | | 13 | Wine | | < 0.05, < 0.05, < 0.05 | < 0.05 |
| Italy 1994 *Sangiovese* ER 94 ECS 730 | 35 EC | | 1.267 | 1200 (0.1056 % ai) | 3 | 0 | Fruit | | 0.84, 0.685, < 0.05 | 1.53 |
| | | | | | | 7 | Fruit | | 0.91, 0.70, < 0.05 | 1.61 |
| | | | | | | 13 | Fruit | | 0.10, 0.25, < 0.05 | 0.35 |
| | | | | | | 21 | Fruit | | < 0.05, 0.12, < 0.05 | 0.12 |
| | | | | | | 28 | Fruit | | 0.05, 0.13, < 0.05 | 0.18 |
| | | | | | | 14 | Juice | | < 0.05, < 0.05, < 0.05 | < 0.05 |
| | | | | | | 14 | Pomace | | 0.13, 0.48, 0.07 | 0.68 |
| | | | | | | 14 | Young wine | | 0.065, < 0.05, < 0.05 | 0.07 |
| | | | | | | 14 | Wine | | < 0.05, < 0.05, < 0.05 | < 0.05 |

| GRAPE Country, Year of trial *Variety* report | Form. | Method | kg ai/ha | L/ha | No | PHI days | Sample analysed | Residues of alpha, beta endosulfan, endosulfan sulfate, mg/kg | Total Residues, mg/kg |
|---|---|---|---|---|---|---|---|---|---|
| Italy 1994 *Trebbiano* *TR 3T* *ER 94 ECS* 730 | 35 EC | | 1.584 | 1500 (0.1056 % ai) | 3 | 0 | Fruit | 0.80, 0.75, < 0.05 | 1.55 |
| | | | | | | 7 | Fruit | 0.26, 0.37, < 0.05 | 0.63 |
| | | | | | | 14 | Fruit | 0.20, 0.31, < 0.05 | 0.51 |
| | | | | | | 21 | Fruit | 0.11, 0.17, < 0.05 | 0.28 |
| | | | | | | 28 | Fruit | 0.12, 0.19, < 0.05 | 0.31 |
| | | | | | | 14 | Juice | < 0.05, < 0.05, < 0.05 | < 0.05 |
| | | | | | | 14 | Pomace | 0.38, 0.57, < 0.05 | 0.95 |
| | | | | | | 14 | Young wine | < 0.05, < 0.05, < 0.05 | < 0.05 |
| | | | | | | 14 | Wine | < 0.05, < 0.05, < 0.05 | < 0.05 |
| USA (FL) 1995 *Thompson Seedless* BJ-95R-0744346915 | 3EC | Airblast spray | 3.36 | 728, 637 (0.46-0.49% ai) | 2 | 4 | Fruit (a) | | 0.45 |
| | | | | | | | Fruit (b) | | 0.71 |
| | | | | | | | Juice | | < 0.05 |
| | | | | | | | Raisins | | 0.72 |
| USA1972 *Thompson seedless* A30342 *R-119B* | 50WP | Spray | 1.68 | 2800 (0.06% ai) | 2 | 7 | Wh grape | 0.75, ND* | 0.75 |
| | | | | | | | W Pomace | 0.19, ND | 0.19 |
| | | | | | | | D. Pomace | 0.58, ND | 0.58 |

Table 43. Endosulfan residues in avocados resulting from supervised trials in Australia.

| AVOCADOS Country, Year of trial *Variety* Report | Form. | Method | g ai/hL/ | L/ha | No. | PHI days | Sample analysed | Residues of alpha, beta endosulfan, Endosulfan sulfate, mg/kg | Total Residues, mg/kg |
|---|---|---|---|---|---|---|---|---|---|
| Australia 2000 Tolga *Shephard* N° 1/10/554 | EC | spray | 70 | 640 | 6 | 0 | | 0.011, 0.021, 0.040 | 0.07 |
| | | | | | | 14 | | < 0.005, < 0.005, 0.010 | 0.01 |
| | | | | | | 21 | | < 0.005, < 0.005, < 0.005 | < 0.005 |
| | | | | | | 28 | | < 0.005, < 0.005, < 0.005 | < 0.005 |
| Australia 2000 Glasshouse Mtns *Wurtz* N° 1/10/554 | EC | spray | 70 | 1000 | 5 | 0 | | 0.32, 0.29, 0.049 | 0.66 |
| | | | | | | 14 | | 0.008, 0.031, 0.026 | 0.07 |
| | | | | | | 21 | | 0.009, 0.035, 0.057 | 0.10 |
| | | | | | | 28 | | 0.015, 0.048, 0.044 | 0.11 |
| | | | | | | | co | < 0.005, 0.005, 0.007 | 0.12 |
| Australia 2000 Glasshouse Mtns *Wurtz* N° 1/10/554 | EC | spray | 140 | 1000 | 5 | 0 | | 0.55, 0.50, 0.12 | 1.17 |
| | | | | | | 14** | | 0.063, 0.20, 0.10 | 0.36 |
| | | | | | | 21 | | 0.016, 0.08, 0.066 | 0.16 |
| | | | | | | 28** | | 0.053, 0.472, 0.475 | 1.00 |
| | | | | | | 14 | peel | 0.19, 0.70, 0.35 | 1.24 |
| | | | | | | 14 | flesh | < 0.005, 0.007, 0.018 | 0.03 |
| | | | | | | 28 | peel | 0.053, 0.46, 0.44 | 0.95 |
| | | | | | | 28 | flesh | < 0.005, 0.012, 0.035 | 0.05 |
| Australia 2000 Tolga *Hass* N° 1/10/554 | EC | spray | 70 | 700 | 6 | 28 | | < 0.005, < 0.005, < 0.005 | < 0.005 |

** in the report we have ratio skin+flesh/ seed and not peel+seed /flesh

co control sample

Table 44. Endosulfan residues in custard apple resulting from supervised trials in Australia.

| CUSTARD APPLE | Application | | | | | PHI days | Sample analysed | Residues of alpha, beta endosulfan, Endosulfan sulfate, mg/kg | Total Residues, mg/kg |
|---|---|---|---|---|---|---|---|---|---|
| Country, Year of trial *Variety* Report | Form. | Method | g ai/hL/ | Water volume | No. | | | | |
| Australia 2000 Nambour *African pride* N° ½/500 | EC | spray | 70 | 8.6L/Tree | 3 | 0 7 14 28 | Fruit | 0.71 ,0.39, 0.22 0.02, 0.03, 0.05 0.01, 0.01, 0.07 ND, 0.01, 0.04 | 1.32 0.10 0.09 0.05 |
| Australia 2000 Nambour *African pride* N° ½/500 | EC | spray | 140 | 8.6L/Tree | 3 | 0 7 14 28 | Fruit | 0.84, 0.5 , 0.29 0.03, 0.08, 0.16 0.01, 0.04, 0.20 ND, 0.01, 0.06 | 1.63 0.27 0.25 0.07 |
| Australia 2000 Alstonville *Pinks mammoth* N° ½/500 | EC | spray | 70 | 8.6L/Tree | 3 | 0 7 14 28 | Fruit | 0.62, 0.35, 0.17 0.11, 0.14, 0.10 0.05, 0.07, 0.06 0.01, 0.02, 0.04 | 1.14 0.35 0.18 0.07 |
| Australia 2000 Alstonville *Pinks mammoth* | EC | spray | 140 | 8.6L/Tree | 3 | 0 7 14 28 | Fruit | 0.37, 0.62, 0.36 0.30, 0.49, 0.17 0.22, 0.32, 0.22 0.03, 0.08, 0.12 | 1.35 0.96 0.76 0.23 |

Table 45. Endosulfan residues in litchi resulting from supervised trials in Australia.

| LITCHI | Application | | | | | PHI days | Sample analysed | Residues of alpha, beta endosulfan, Endosulfan sulfate, mg/kg | Total Residues, mg/kg |
|---|---|---|---|---|---|---|---|---|---|
| Country, Year of trial *Variety* Report | Form. | Method | g ai./hl/ | L/ha | No. | | | | |
| Australia NORTH QUEENSLAND | EC | spray | 52.5 | | 4 | 0 3 7 14 | | 1.63, 1.00 ,0.69 0.51, 0.42, 0.67 0.33, 0.29, 0.66 0.11, 0.083, 0.26 | 3.32 1.60 1.28 0.45 |
| Australia NORTH QUEENSLAND | EC | spray | 104 | | 4 | 0 3 7 14 | | 2.65, 1.28 ,0.75 0.60, 0.54, 0.77 0.33, 0.32, 0.64 0.30, 0.24, 0.57 | 4.68 1.91 1.29 1.11 |
| Australia SOUTH QUEENSLAND | EC | spray | 52.5 | | 4 | 7 | | 0.24, 0.24, 0.53 | 1.01 |

Table 46. Endosulfan residues in mangoes resulting from supervised trials in Australia.

| MANGOES | Application | | | | | PHI days | Sample analysed | Residues of alpha, beta endosulfan, Endosulfan sulfate, mg/kg | Total Residues, mg/kg |
|---|---|---|---|---|---|---|---|---|---|
| Country, Year of trial *Variety* Report | Form. | Method | g ai./hl/ | L/ha | No. | | | | |
| Australia 2000 Dorroughby *Bowen* N° 1/10:537 | EC | spray | 70 | 8.00 | 2 | 0 7 14 28 | | 0.24, 0.17, 0.007 0.071, 0.067, 0.060 0.026, 0.025, 0.049 0.012, 0.015, 0.063 | 0.42 0.20 0.10 0.09 |
| Australia 2000 Tolga *Palmer* N° 1/2/500 | EC | spray | 70 | 493 | 2 | 0 7 14 28 | | 0.17, 0.15, 0.034 0.035, 0.035, 0.10 0.015, 0.013, 0.12 0.009, 0.006, 0.15 | 0.35 0.17 0.15 0.17 |

| MANGOES Country, Year of trial *Variety* Report | Form. | Application Method | g ai./hl/ | L/ha | No. | PHI days | Sample analysed | Residues of alpha, beta endosulfan, Endosulfan sulfate, mg/kg | Total Residues, mg/kg |
|---|---|---|---|---|---|---|---|---|---|
| Australia 2000 Tolga *Palmer* N° 1/2/500 | EC | spray | 140 | 667 | 2 | 0 7 14 28 | | 0.26, 0.18, 0.048 0.11, 0.11, 0.28 0.051, 0.055, 0.34 0.005, 0.005, 0.089 | 0.49 0.50 0.45 0.10 |
| Australia 2000 Wamuran *Kent* 1/10/537 | EC | spray | 70 | 400 | 2 | 28 28 | co | 0.014, 0.011, 0.20 0.008, 0.006, 0.031 | 0.22 0.05 |

Table 47. Endosulfan residues in papaya resulting from supervised trials in Australia.

| PAWPAW Country, Year of trial *Variety* Report | Form. | Application Method | g ai./hl/ | | No. | PHI days | Sample analysed | Residues of alpha, beta endosulfan, Endosulfan sulfate, mg/kg | Total Residues, mg/kg |
|---|---|---|---|---|---|---|---|---|---|
| Australia 2000 Walkamin *Ruby red* N° 1/10/539 | EC | spray | 70 | 933 | 4 | 0 7 14 21 | Fruit | 0.16, 0.12, 0.088 0.03, 0.053, 0.10 0.013, 0.024, 0.094 < 0.005, 0.005, 0.09 | 0.37 0.18 0.12 0.10 |
| Australia 2000 Walkamin *Ruby red* N° 1/10/539 | EC | spray | 140 | 1100 | 4 | 0 7 14 21 0 | Fruit | 0.37, 0.33, 0.18 0.09, 0.11, 0.13 0.025, 0.074, 0.13 < 0.005, 0.010, 0.074 0.007, 0.006, 0.01 | 0.88 0.33 0.23 0.08 0.02c |
| Australia 2000 Mareeba *Hybrid 1B* N° 1/10/539 N° 1/10/539 | EC | spray | 70 | 933 | 4 | 0 7 14 21 | Fruit | 0.11, 0076, 0.051 0.005, 0.011, 0.079 < 0.005, 0.006, 0.047 < 0.005, < 0.005, 0.045 | 0.24 0.10 0.06 0.05 |

Table 48. Endosulfan residues in persimmon resulting from supervised trials in Australia.

| PERSIMMON Country, Year of trial *Variety* Report | Form. | Application Method | g ai./hl/ | | No. | PHI days | Sample analysed | Residues of alpha, beta endosulfan, Endosulfan sulfate, mg/kg | Total Residues, mg/kg |
|---|---|---|---|---|---|---|---|---|---|
| Australia 2001 Wollongbar *Fuji* N° 1/10/545 | EC | spray | 70 | 500 | 2 | 0 7 14 28 | | 0.45, 0.37, 0.11 0.17, 0.24, 0.12 0.17, 0.23, 0.15 0.53, 0.58, 0.65 | 0.93 0.53 0.55 1.80 ** |
| Australia 2001 The Summit *Fuyu* N° 1/10/545 | EC | spray | 70 | 1719 | 2 | 0 7 14 28 | | 0.49, 0.48, 0.057 0.38, 0.42, 0.086 0.27, 0.34, 0.08 0.14, 0.21, 0.15 | 1.03 0.89 0.69 0.50 |
| Australia 2001 The Summit *Fuyu* N° 1/10/545 | EC | spray | 140 | 2131 | 2 | 0 7 14 14 28 28 | peel flesh peel flesh | 1.4, 1.2, 0.08 0.71, 0.81, 0.18 2.60, 2.80, 0.87 0.13, 0.086, 0.028 1.70, 1.90, 1.40 0.014, 0.015, 0.016 | 2.68 1.70 6.27 0.24 5.00 0.05 |

| PERSIMMON | Application | | | | | PHI | Sample | Residues of alpha, | Total |
| Country, Year of trial *Variety* Report | Form. | Method | g ai./hl/ | | No. | days | analysed | beta endosulfan, Endosulfan sulfate, mg/kg | Residues, mg/kg |
|---|---|---|---|---|---|---|---|---|---|
| Australia *2000* Glasshouse Mtns *Fuji* N° 1/10/545 | EC | spray | 70 | 680 | 2 | 28 28 | co | 0.14, 0.13, 0.45 0.17, 0.16, 0.46 | 0.72 0.79 |

** no clear comments were including in the report (sampling)

Table 49. Endosulfan residues in pineapple resulting from supervised trials in USA.

| PINEAPPLE | Application | | | | | DALA | Sample analysed | Residues of alpha, beta endosulfan, Endosulfan sulfate, mg/kg | Total Residues mg/kg |
| Country, Year of trial *Variety* report | Form. | Method | Kg ai/ha/ applic'n | L/ha | No. | | | | |
|---|---|---|---|---|---|---|---|---|---|
| USA (HI) 1996 *Smooth Cayenne* PGAAH950001P | 3EC | Spray interval 7 d | 1.68 | 1870 | 2 | 60 60 60 | Wh. Fruit Wet Pulp Juice | (0.03), 0.05, 0.03 0.07, 0.09, 0.16 (0.01), (0.01), < 0.01 | 0.11 0.32 0.02 |
| USA (HI) 1996 *Smooth Cayenne* PGAAH950001 | 3EC | Spray (interval 7 d) | 1.68 | 1870 | 2 | 60 60 60 | Wh. Fruit Wet Pulp Juice | 0.14, 0.23, 0.18 0.26, 0.33, 0.58 (0.04), 0.07, < 0.01 | 0.55 1.17 0.11 |
| USA (HI) 1996 *Smooth Cayenne* PGAAH950001P | 3EC | Spray (interval 7 d) | 1.68 | 1870 | 2 | 61 61 61 | Wh. Fruit Wet Pulp Juice | (0.04), 0.06, 0.07 0.25, 0.29, 0.55 (0.02), (0.04), < 0.01 | 0.17 1.09 0.06 |
| USA (HI) 1996 *Smooth Cayenne* PGAAH950001P | 3EC | Spray (interval 7 d) | 5.04 | 1870 | 2 | 61 61 61 | Wh. Fruit Wet Pulp Juice | 0.19, 0.26, 0.16 0.47, 0.53, 0.79 0.07, 0.10, < 0.01 | 0.61 1.79 0.17 |
| USA (HI) 1968 Oahu, HI *R-1097* | 3EC | Spray (interval 7 d) | 2.24 | 2800 | 4 | 0 7 0 7* | Wh. Fruit Wh. Fruit Bran w/o molasses Bran w molasses | | 0.25 0.20 2.56 1.75 |
| USA (HI) 1968 Oahu, HI *R-1097* | 3EC | Spray (interval 7, 7, 8 d) | 2.24 | 2800 | 4 | 0 0 21 21* | Wh. Fruit Wh. Fruit Bran w/o molasses Bran w molasses | | 0.09 0.08 1.1 1.3 |
| USA (HI) 1968 Oahu, HI *R-1097* | 3EC | Spray (interval 7, 10, 7 d) | 2.24 | 2800 | 4 | 0 0 53 53* | Wh. Fruit Wh. Fruit Bran w/o molasses Bran w molasses | | 0.13 0.22 0.93 1.1 |
| USA (HI) 1968 Oahu, HI *R-1097* | 3EC | Spray (interval 7 d) | 4.48 | 2800 | 4 | 0 7 0 7* | Wh. Fruit Wh. Fruit Bran w/o molasses Bran w molasses | | 0.72 0.62 2.8 4.1 |
| USA (HI) 1968 Oahu, HI *R-1097* | 3EC | Spray (interval 7, 7, 8 d) | 4.48 | 2800 | 4 | 0 0 21 21* | Wh. Fruit Wh. Fruit Bran w/o molasses Bran w molasses | | 0.33 0.23 5.2 2.3 |

| PINEAPPLE | Application | | | | | DALA | Sample analysed | Residues of alpha, beta endosulfan, Endosulfan sulfate, mg/kg | Total Residues mg/kg |
|---|---|---|---|---|---|---|---|---|---|
| Country, Year of trial *Variety* report | Form. | Method | Kg ai/ha/ applic'n | L/ha | No. | | | | |
| USA (HI) 1968 Oahu, HI *R-1097* | 3EC | Spray (interval 7, 10, 7 d) | 4.48 | 2800 | 4 | 0 0 53 53* | Wh. Fruit Wh. Fruit Bran w/o molasses Bran w molasses | | 0.43 0.18 2.2 1.8 |

Table 50. Endosulfan residues in head cabbage from supervised trials in Australia and USA.

| CABBAGE | Application | | | | | PHI days | Sample analysed | Residues of alpha, beta endosulfan, endosulfan sulfate, mg/kg | Total Residues, mg/kg |
|---|---|---|---|---|---|---|---|---|---|
| Country, Year of trial *Variety* Report | Form. | Method | kg ai/ha | L/ha | No. | | | | |
| Australia 2000 Darling Downs *Neptune* *N° 1/10/564* | EC | spray | 0.665 | 1000 | 3 | 0 3 7 14 | | 0.31, 0.20.0.019 0.13,0.15,0.035 0.026, 0.029,0.043 0.007,0.007,0.038 | 0.53 0.32 0.10 0.05 |
| Australia 2000 Darling Downs *neptune* *N° 1/10/564* | EC | spray | 1.32 | 1000 | 3 | 0 3 7 14 | | 0.84, 0.49,0.041 0.21, 0.24,0.062 0.09, 0.10, 0.11 0.014,0.014, 0.045 | 1.37 0.51 0.30 0.07 |
| Australia 2000 Werribee *green coronet* *N° 1/10/564* | EC | spray | 0.332 | 500 | 3 | 0 3 7 14 | | 0.035, 0.023, 0.006 0.008, 0.01, 0.008 0.006, 0.008, 0.017 < 0.005,< 0.005,< 0.005 | 0.06 0.03 0.03 < 0.005 |
| Australia 2000 Werribee *green coronet* *N° 1/10/564* | EC | spray | 0.56 | 500 | 3 | 0 3 7 14 | | 0.033, 0.019, 0.006 0.012, 0.011, 0.007 0.006, 0.012, 0.008 < 0.005,< 0.005,< 0.005 | 0.058 0.030 0.026 < 0.005 |
| USA1960 San Jose, CA *Green cabbage* R-470 | 2EC | Spray. Interval 21 d | 1.12 | 701 (0.16% ai) | 2 | 0 7 14 | Leaves Leaves Leaves | | 20.00 3.20 2.80 |
| USA1960 San Jose, CA *Green cabbage* R-470 | | | | | | 0 7 14 | Heads Heads Heads | | 0.59 ≤ 0.05 < 0.05 |
| | | | | | | 0 7 14 | Wh Heads Wh Heads Wh Heads | | 6.70 0.60 0.45 |
| USA1960 Moon Bay, CA *Red cabbage* R-470 | 2EC | Spray. Interval 7 d | 0.56,1.12 | 701 (0.16% ai) | 2 | 0 16 28 | Leaves Leaves Leaves | | 22.00 3.70 0.56 |
| | | | | | | 0 16 28 | Heads Heads Heads | | 0.52 < 0.05 < 0.05 |
| | | | | | | 0 16 28 | Wh Heads Wh Heads Wh Heads | | 6.20 1.12 0.11 |

| CABBAGE Country, Year of trial *Variety* Report | Form. | Method | kg ai/ha | L/ha | No. | PHI days | Sample analysed | Residues of alpha, beta endosulfan, endosulfan sulfate, mg/kg | Total Residues, mg/kg |
|---|---|---|---|---|---|---|---|---|---|
| USA1960 Half Moon Bay, CA *Red cabbage* R-470 | 2EC | Spray. Interval 23d | 0.56,1.12 | 701 (0.16% ai) | 2 | 0 | Leaves | | 14.00 |
| | | | | | | 12 | Leaves | | 1.84 |
| | | | | | | 0 | Heads | | 0.28 |
| | | | | | | 12 | Heads | | 0.03 |
| | | | | | | 0 | Wh Heads | | |
| | | | | | | 12 | Wh Heads | | 0.53 |
| USA 1960 R-470 | 2EC | Spray. | 1.12 | 467 (0.24 % ai) | 1 | 0 | Leaves | | 12.7 |
| | | | | | | 7 | Leaves | | 2.2 |
| | | | | | | 14 | Leaves | | 0.95 |
| | | | | | | 0 | Heads | | 0.23 |
| | | | | | | 7 | Heads | | 0.24 |
| | | | | | | 14 | Heads | | < 0.05 |
| | | | | | | 0 | Wh Heads | | 2.5 |
| | | | | | | 7 | Wh Heads | | 0.69 |
| | | | | | | 14 | Wh Heads | | 0.18 |
| USA1960 Half Moon Bay, CA *Red cabbage* R-470 | 2EC | Spray. Interval 16d | 1. 12 | 701 (0.16% ai) | 2 | 0 | Leaves | | 25.00 |
| | | | | | | 12 | Leaves | | 5.50 |
| | | | | | | 0 | Heads | | 0.33 |
| | | | | | | 12 | Heads | | 0.09 |
| | | | | | | 0 | Wh Heads | | 6.30 |
| | | | | | | 12 | Wh Heads | | 1.45 |

Table 51. Endosulfan residues in Savoy cabbage (head) from supervised trials in Germany.

| SAVOY CABBAGE Country, Year of trial *Variety* report | Form. | Method | kg ai/ha/ applic'n | L/ha | No. | PHI (Days) | Sample analysed | Residues of alpha, beta endosulfan, Endosulfan sulfate, mg/kg | Total Residues mg/kg |
|---|---|---|---|---|---|---|---|---|---|
| Germany 1983 *Gruenkopf* DEU83172611 | Dust 2.82% | Spread Interval 16, 14 days | 0.705 | | 3 | 0 | Head | 0.5, < 0.01, < 0.01 | 0.50 |
| | | | | | | 5 | Head | 0.07, < 0.01, < 0.01 | 0.07 |
| | | | | | | 10 | Head | < 0.01, < 0.01, < 0.01 | < 0.01 |
| | | | | | | 14 | Head | < 0.01, < 0.01, < 0.01 | < 0.01 |
| Germany 1983 *Vertus* DEU83172641 | Dust 2.82% | Spread Interval 14 days | 0.705 | | 3 | 0 | Head | 3.8, 1.7, < 0.01 | 5.50 |
| | | | | | | 5 | Head | 1.2, 0.7, 0.03 | 1.93 |
| | | | | | | 10 | Head | 0.07, 0.07, < 0.01 | 0.14 |
| | | | | | | 14 | Head | 0.3, 0.1, 0.02 | 0.42 |
| Germany 1983 *Wirosa* DEU83172631 | Dust 2.82% | Spread Interval 14 days | 0.705 | | 3 | 0 | Head | 0.3, 0.1, < 0.01 | 0.40 |
| | | | | | | 5 | Head | 0.03, 0.02, < 0.01 | 0.05 |
| | | | | | | 10 | Head | 0.05, 0.04, < 0.01 | 0.09 |
| | | | | | | 14 | Head | 0.08, 0.09, 0.02 | 0.19 |
| Germany1974 *Marner* PSR94/024 | 35EC | Spray | 0.35 | 1000 (0.035% ai) | 1 | 0 | Head | | 2.84 |
| | | | | | | 7 | Head | | 0.58 |
| | | | | | | 14-28 | Head | < 0.02, < 0.01, < 0.01 | < 0.01 |
| Germany1974 *Marner* PSR94/024 | 35EC | Spray | 0.35 | 1000 (0.035% ai) | 1 | 0 | Head | | 2.69 |
| | | | | | | 7 | Head | | 0.47 |
| | | | | | | 14-28 | Head | < 0.02, < 0.01, < 0.01 | < 0.01 |

| SAVOY CABBAGE | Application | | | | | PHI (Days) | Sample analysed | Residues of alpha, beta endosulfan, Endosulfan sulfate, mg/kg | Total Residues mg/kg |
|---|---|---|---|---|---|---|---|---|---|
| Country, Year of trial *Variety* report | Form. | Method | kg ai/ha/ applic'n | L/ha | No. | | | | |
| Germany 1974 *Novum* PSR94/024 | 35EC | Spray | 0.35 | 1000 (0.035% ai) | 1 | 0 | Head | | 21.46 |
| | | | | | | 7 | Head | | 2.73 |
| | | | | | | 14 | Head | | 0.63 |
| | | | | | | 21 | Head | | 0.40 |
| | | | | | | 28 | Head | | 0.13 |
| Germany 1974 Bonn *Novum* PSR94/024 | 35EC | Spray | 0.35 | 1000 (0.035% ai) | 1 | 0 | Head | | 15.25 |
| | | | | | | 7 | Head | | 2.16 |
| | | | | | | 14 | Head | | 0.40 |
| | | | | | | 21 | Head | | 0.28 |
| | | | | | | 28 | Head | | 0.11 |
| Germany 1974 *Gruenkopf* PSR94/024 | 35EC | Spray | 0.35 | 1000 (0.035% ai) | 1 | 0 | Head | | 3.73 |
| | | | | | | 7 | Head | | 0.44 |
| | | | | | | 14 | Head | < 0.008, < 0.04, < 0.05 | < 0.05 |
| | | | | | | 21 | Head | < 0.008, < 0.04, < 0.05 | < 0.05 |
| | | | | | | 28 | Head | < 0.008, < 0.04, < 0.05 | < 0.05 |
| Germany 1974 *Boeckelmanns* PSR94/024 | 35EC | Spray | 0.35 | 1000 (0.035% ai) | 1 | 0 | Head | | 3.73 |
| | | | | | | 7 | Head | | 0.07 |
| | | | | | | 14 | Head | < 0.008, < 0.04, < 0.05 | < 0.05 |
| | | | | | | 21 | Head | < 0.008, < 0.04, < 0.05 | < 0.05 |
| | | | | | | 28 | Head | < 0.008, < 0.04, < 0.05 | < 0.05 |
| Germany 1974 *Gruener* PSR94/024 | 35EC | Spray | 0.35 | 1000 (0.035% ai) | 1 | 0 | Head | | --- |
| | | | | | | 7 | Head | | 0.23 |
| | | | | | | 14 | Head | | 0.28 |
| | | | | | | 21 | Head | < 0.08, < 0.04, < 0.05 | < 0.05 |
| | | | | | | 28 | Head | < 008, < 0.04, < 0.05 | < 0.05 |
| Germany 1974 *Gruener* PSR94/024 | 35EC | Spray | 0.35 | 1000 (0.035% ai) | 1 | 0 | Head | | --- |
| | | | | | | 7 | Head | | 0.30 |
| | | | | | | 14 | Head | | 0.14 |
| | | | | | | 21 | Head | < 0.08, < 0.04, < 0.05 | < 0.05 |
| | | | | | | 28 | Head | | 0.14 |
| Germany 1974 *Gruenkopf* PSR94/024 | 35EC | Spray | 0.53 | 1000 (0.053% ai) | 1 | 0 | Head | | 7.63 |
| | | | | | | 7 | Head | | 0.06 |
| | | | | | | 14 | Head | < 0.08, < 0.04, < 0.05 | < 0.05 |
| | | | | | | 21 | Head | < 0.08, < 0.04, < 0.05 | < 0.05 |
| | | | | | | 28 | Head | < 0.08, < 0.04, < 0.05 | < 0.05 |
| Germany 1974 *Boeckelmanns* PSR94/024 | 35EC | Spray | 0.53 | 1000 (0.053% ai) | 1 | 0 | Head | | 3.83 |
| | | | | | | 7 | Head | | 0.05 |
| | | | | | | 14 | Head | < 0.008, < 0.04, < 0.05 | < 0.05 |
| | | | | | | 21 | Head | < 0.008, < 0.04, < 0.05 | < 0.05 |
| | | | | | | 28 | Head | < 0.008, < 0.04, < 0.05 | < 0.05 |
| Germany 1974 *Dr Neuers Gruener* PSR94/024 | 35EC | Spray | 0.53 | 1000 (0.053% ai) | 1 | 0 | Head | | --- |
| | | | | | | 7 | Head | | 1.20 |
| | | | | | | 14 | Head | | 0.13 |
| | | | | | | 21 | Head | < 0.08, < 0.04, < 0.05 | < 0.05 |
| | | | | | | 28 | Head | < 0.08, < 0.04, < 0.05 | < 0.05 |

| SAVOY CABBAGE | Application | | | | | PHI (Days) | Sample analysed | Residues of alpha, beta endosulfan, Endosulfan sulfate, mg/kg | Total Residues mg/kg |
|---|---|---|---|---|---|---|---|---|---|
| Country, Year of trial *Variety* report | Form. | Method | kg ai/ha/ applic'n | L/ha | No. | | | | |
| Germany 1974 *Dr Neuers Gruener* PSR94/024 | 35EC | Spray | 0.53 | 1000 (0.053% ai) | 1 | 0 | Head | | 6.90 |
| | | | | | | 7 | Head | | 1.10 |
| | | | | | | 14 | Head | < 0.08, < 0.04, < 0.05 | < 0.05 |
| | | | | | | 21 | Head | < 0.08, < 0.04, < 0.05 | < 0.05 |
| | | | | | | 28 | Head | | 0.14 |
| Germany 1976 *Vorbote* PSR94/024 | 35EC | Spray Spray Intvl 22, 21 days | 0.21 | 600 (0.035% ai) | 3 | 0 | Head | | 1.90 |
| | | | | | | 7 | Head | | 0.40 |
| | | | | | | 14 | Head | | 0.04 |
| | | | | | | 21 | Head | < 0.005, < 0.005, < 0.005 | < 0.005 |
| Germany 1976 *Iceking* PSR94/024 | 35EC | Spray Spray Intvl 18, 43 days | 0.21 | 600 (0.035% ai) | 3 | 0 | Head | | 0.20 |
| | | | | | | 7 | Head | | 0.20 |
| | | | | | | 14 | Head | | 0.20 |
| | | | | | | 21 | Head | | < 0.01 |
| Germany1976 *Boeckelmanns Westfalia* PSR94/024 | 35EC | Spray Spray Intvl 28 days | 0.21 | 600 (0.035% ai) | 3 | 0 | Head | | 0.30 |
| | | | | | | 7 | Head | | 0.10 |
| | | | | | | 14 | Head | | 0.10 |
| | | | | | | 21 | Head | | < 0.01 |
| Germany 1983 *Gruenkopf* PSR94/024 | 33WP | Spray Spray Intvl 15 days | 0.20 | 600 (0.033% ai) | 3 | 0 | Head | | 0.605 |
| | | | | | | 5 | Head | | 0.110 |
| | | | | | | 10 | Head | | 0.065 |
| | | | | | | 14 | Head | < 0.01, < 0.01, < 0.01 | < 0.01 |
| Germany 1983 *Wirosa* PSR94/024 | 33WP | Spray Spray Intvl 14 days | 0.20 | 600 (0.033% ai) | 3 | 0 | Head | | 1.71 |
| | | | | | | 5 | Head | | 1.22 |
| | | | | | | 10 | Head | | 0.09 |
| | | | | | | 14 | Head | | 0.34 |
| Germany 1983 *Vertus* PSR94/024 | 33WP | Spray Spray Intvl 14 days | 0.20 | 600 (0.033% ai) | 3 | 0 | Head | | -- |
| | | | | | | 5 | Head | | 0.10 |
| | | | | | | 10 | Head | | 0.025 |
| | | | | | | 14 | Head | | 0.035 |
| Germany 1976 *Vorbote* PSR94/024 | 35WP | Spray Spray Intvl 21, 22 days | 0.21 | 600 (0.035% ai) | 3 | 0 | Head | | 2.10 |
| | | | | | | 7 | Head | | 0.20 |
| | | | | | | 14 | Head | | 0.20 |
| | | | | | | 21 | Head | | 0.04 |
| Germany 1976 *Iceking* PSR94/024 | 35WP | Spray Spray Intvl 18, 43 days | 0.21 | 600 (0.035% ai) | 3 | 0 | Head | | 2.00 |
| | | | | | | 7 | Head | | 0.20 |
| | | | | | | 14 | Head | | 0.07 |
| | | | | | | 21 | Head | < 0.01, < 0.01, 0.01 | < 0.01 |
| Germany 1976 *Boeckelmanns* PSR94/024 | 35WP | Spray Spray Intvl 28 days | 0.21 | 600 (0.035% ai) | 3 | 0 | Head | | 1.00 |
| | | | | | | 7 | Head | | 0.10 |
| | | | | | | 14 | Head | | 0.02 |
| | | | | | | 21 | Head | | 0.04 |
| Germany *1974* *Marner* PSR94/024 | 35WP | Spray | 0.35 | 1000 (0.035% ai) | 1 | 0 | Head | | 0.88 |
| | | | | | | 7 | Head | | 0.17 |
| | | | | | | 14-28 | Head | < 0.02, < 0.01, < 0.01 | < 0.02 |
| Germany 1974 *Novurn* PSR94/024 | 35WP | Spray | 0.35 | 1000 (0.035% ai) | 1 | 0 | Head | | 17.62 |
| | | | | | | 7 | Head | | 2.92 |
| | | | | | | 14 | Head | | 1.22 |
| | | | | | | 21 | Head | | 1.10 |
| | | | | | | 28 | Head | | 0.08 |
| Germany 1974 *Gruenkopf* PSR94/024 | 35WP | Spray | 0.35 | 1000 (0.035% ai) | 1 | 0 | Head | | 4.13 |
| | | | | | | 7 | Head | | 0.07 |
| | | | | | | 14-28 | Head | < 0.008, < 0.04, < 0.05 | < 0.05 |

| SAVOY CABBAGE | Application | | | | | PHI (Days) | Sample analysed | Residues of alpha, beta endosulfan, Endosulfan sulfate, mg/kg | Total Residues mg/kg |
|---|---|---|---|---|---|---|---|---|---|
| Country, Year of trial *Variety* report | Form. | Method | kg ai/ha/ applic'n | L/ha | No. | | | | |
| Germany 1974 *Boeckelmanns* PSR94/024 | 35WP | Spray | 0.35 | 1000 (0.035% ai) | 1 | 0 | Head | | 1.03 |
| | | | | | | 7 | Head | | 0.04 |
| | | | | | | 14 | Head | < 0.008, < 0.04, < 0.05 | < 0.05 |
| | | | | | | 21 | Head | < 0.008, < 0.04, < 0.05 | < 0.05 |
| | | | | | | 28 | Head | | 0.04 |
| Germany 1974 *Dr Neuers Gruener* PSR94/024 | 35WP | Spray | 0.35 | 1000 (0.035% ai) | 1 | 0 | Head | | 3.30 |
| | | | | | | 7 | Head | | 0.13 |
| | | | | | | 14 | Head | | 0.12 |
| | | | | | | 21 | Head | < 0.08, < 0.04, < 0.05 | < 0.05 |
| | | | | | | 28 | Head | < 0.08, < 0.04, < 0.05 | < 0.05 |
| Germany 1974 *Dr Neuers Gruener* PSR94/024 | 35WP | Spray | 0.35 | 1000 (0.035% ai) | 1 | 0 | Head | | ---- |
| | | | | | | 7 | Head | | 0.70 |
| | | | | | | 14-28 | Head | < 0.08, < 0.04, < 0.05 | < 0.05 |
| Germany 1974 *Marner* PSR94/024 | 35WP | Spray | 0.53 | 1000 (0.053% ai) | 1 | 0 | Head | | 0.59 |
| | | | | | | 7 | Head | | 0.18 |
| | | | | | | 14-28 | Head | < 0.02, < 0.01, < 0.01 | < 0.02 |
| Germany 1974 *Novum* PSR94/024 | 35WP | Spray | 0.53 | 1000 (0.053% ai) | 1 | 0 | Head | | 19.46 |
| | | | | | | 7 | Head | | 3.43 |
| | | | | | | 14 | Head | | 1.17 |
| | | | | | | 21 | Head | | 0.80 |
| | | | | | | 28 | Head | | 0.09 |
| Germany 1974 *Gruenkopf* PSR94/024 | 35WP | Spray | 0.53 | 1000 (0.053% ai) | 1 | 0 | Head | | 6.03 |
| | | | | | | 7 | Head | | 0.07 |
| | | | | | | 14 | Head | < 0.008, < 0.04, < 0.05 | < 0.05 |
| | | | | | | 21 | Head | | < 0.05 |
| | | | | | | 28 | Head | < 0.008, < 0.04, < 0.05 | < 0.05 |
| Germany 1974 *Boeckelmanns* PSR94/024 | 35WP | Spray | 0.53 | 1000 (0.053% ai) | 1 | 0 | Head | | 2.13 |
| | | | | | | 7 | Head | | 0.075 |
| | | | | | | 14-28 | Head | < 0.008, < 0.04, < 0.05 | < 0.05 |
| Germany 1974 *Dr Neuers Gruener* PSR94/024 | 35WP | Spray | 0.53 | 1000 (0.053% ai) | 1 | 0 | Head | | 4.90 |
| | | | | | | 7 | Head | | 0.06 |
| | | | | | | 14 | Head | | 0.18 |
| | | | | | | 21 | Head | | < 0.05 |
| | | | | | | 28 | Head | | < 0.05 |
| Germany 1974 *Dr Neuers Gruener* PSR94/024 | 35WP | Spray | 0.53 | 1000 (0.053% ai) | 1 | 0 | Head | | 3.00 |
| | | | | | | 7 | Head | | 0.70 |
| | | | | | | 14 | Head | | 0.08 |
| | | | | | | 21 | Head | < 0.08, < 0.04, < 0.05 | < 0.05 |
| | | | | | | 28 | Head | < 0.08, < 0.04, < 0.05 | < 0.05 |

| SAVOY CABBAGE | Application | | | | | PHI (Days) | Sample analysed | Residues of alpha, beta endosulfan, Endosulfan sulfate, mg/kg | Total Residues mg/kg |
|---|---|---|---|---|---|---|---|---|---|
| Country, Year of trial *Variety* report | Form. | Method | kg ai/ha/ applic'n | L/ha | No. | | | | |
| Germany 1975 *Vorbote* LEA 2/83/01/02-75) | EC* | Spray Interval 14d | 0.216 | (0.036 % ai) | 2 | 0 5 10 14 | Head Head Head Head | | 0.50 < 0.05 < 0.05 < 0.05 |
| Germany1975 *Hammer* LEA 2/83/01/02-75) | EC* | Spray, 0.15% Interval 14d | 0.216 | 600 (0.036 % ai) | 2 | 0 7 10 14 | Head Head Head Head | | 0.90 0.60 0.40 0.20 |
| Germany1975 *King* LEA3 /83/01/02-75 | EC* | Spray, 0.15% Interval 14d | 0.216 | 600 (0.036 % ai) | 2 | 0 7 10 14 | Head Head Head Head | 0.05, ND, ND 0.007, ND, ND ND, ND,ND ND, ND, ND | 0.05 0.01 < 0.02 < 0.02 |
| Germany1975 *Boeckelmanns Westfalia* LEA 4/83/01/02-75 | EC* | Spray, 0.15% Interval 14d | 0.216 | 600 (0.036 % ai) | 2 | 0 7 10 14 | Head Head Head Head | 0.1, 0.1, ND 0.01, ND, ND 0.006, ND, ND 0.01, ND, ND | 0.20 0.01 0.01 0.01 |

Table 52. Endosulfan residues in broccoli resulting from supervised trials in USA and Australia.

| BROCCOLI | Application | | | | | PHI (Days) | Sample analysed | Residues of alpha, beta endosulfan, endosulfan sulfate, mg/kg | Total Residues mg/kg |
|---|---|---|---|---|---|---|---|---|---|
| Country, Year of trial *Variety* report | Form. | Method | Kg ai/ha/ applic'n | L/ha | No. | | | | |
| USA CA 1995/6 *BJ-95R-03* | 3EC* 50WP | Spray. interval 4 days | 1.12 1.12 | 323-330 323-326 | 3 3 | 7 7 | Broccoli Broccoli | 0.33, 0.25, 0.21 0.59, 0.57, 0.16 | 0.79 1.32 |
| USA CA1995/6 *BJ-95R-03* | 3EC 50WP | Spray. interval 4 days | 1.12 1.12 | 647-655 616-640 | 3 3 | 7 7 | Broccoli Broccoli | 0.33, 0.33, 0.08 0.14, 0.16, 0.07 | 0.74 0.37 |
| USA CA1995/6 *BJ-95R-03* | 3EC 50WP | Spray. interval 4 days | 1.12 1.12 | 640-668 649-668 | 3 3 | 7 7 | Broccoli Broccoli | 0.12, 0.16, < 0.05 0.13, 0.16, 0.07 | 0.28 0.36 |
| USA TX 1995/6 *BJ-95R-03* | 3EC 50WP | Spray. interval 4 days | 1.12 1.12 | 189-192 181-187 | 3 3 | 7 7 | Broccoli Broccoli | 0.10, 0.10, 0.36 0.20, 0.44, 0.33 | 0.56 0.97 |
| USA CA 1995/6 *BJ-95R-03* | 3EC 50WP | Spray. interval 5/4d | 1.12 1.12 | 331-337 323-324 | 3 3 | 7 7 | Broccoli Broccoli | 0.24, 0.31, 0.33 0.40, 0.41, 0.26 | 0.88 1.07 |
| USA AZ 1995/6 *BJ-95R-03* | 3EC 50WP | Spray. Spray Interval 5/ 4d | 1.12 1.12 | 96.3-99.1 96.3-98.1 | 3 3 | 7 7 | Broccoli Broccoli | 0.93, 0.72, 0.39 0.76, 0.37, 0.18 | 2.04 1.31 |
| USA CA 1995/6 *BJ-95R-03* | 3EC 50WP | Spray. Spray interval 4 days | 1.12 1.12 | | 3 3 | 7 7 | Broccoli Broccoli | 1.16, 0.94, 0.30 1.0, 0.72, 0.14 | 2.40 1.86 |
| USA OR1995/6 *BJ-95R-03* | 3EC 50WP | Spray. interval 4/7 d | 1.12 1.12 | 450-495 450-495 | 3 3 | 7 7 | Broccoli Broccoli | 0.10, 0.10, 0.06 0.27, 0.23, 0.07 | 0.26 0.57 |
| USA 1960 *R-470* | 2EC | Spray. Interval 24 d | 1.12 | 701 | 2 | 0 6 13 | Broccoli Broccoli Broccoli | | 3.1 0.22 0.16 |
| Australia 2000 Stanthorpe *Babylon* N° 1/10/594 | EC | spray | 0.533 | 802 | 3 | 0 3 7 14 | Broccoli florets | 0.55, 0.27,0.011 0.39,0.25,0.060 0.08, 0.08, 0.012 < 0.005,< 0.005,< 0.005 | 0.83 0.70 0.17 < 0.005 |
| Australia 2000 Stanthorpe *Babylon* N° 1/10/594 | EC | spray | 1.525 | 1156 | 3 | 0 3 7 14 | Broccoli florets | 1.25, 0.71, < 0.005 0.36, 0.27,0.077 0.11, 0.12, 0.055 0.08, 0.07, 0.010 | 1.96 0.71 0.29 0.16 |

| BROCCOLI | | Application | | | | PHI | Sample | Residues of alpha, | Total |
| Country, Year of trial *Variety* report | Form. | Method | Kg ai/ha/ applic'n | L/ha | No. | (Days) | analysed | beta endosulfan, endosulfan sulfate, mg/kg | Residues mg/kg |
|---|---|---|---|---|---|---|---|---|---|
| Australia 2000 Cranbourne *greenbelt* N° 1/10/594 | EC | spray | 0.465+ | 700 | 3 | 0 3 7 14 | Broccoli florets | 1.70, 0.92,< 0.005 0.51, 0.36, 0.034 0.12, 0.14, 0.025 0.08, 0.006, 0.004 | 2.62 0.89 0.29 0.09 |
| Australia 2000 Cranbourne *greenbelt* N° 1/10/594 | EC | spray | 0.924 | 700 | 3 | 0 3 7 14 | Broccoli florets | 1.91, 1.05, 0.009 0.70, 0.51, 0.047 0.27, 0.27, 0.064 0.016, 0.024, 0.020 | 2.97 1.26 0.60 0.06 |

Table 53. Endosulfan residues in Brussels sprouts from supervised trials in UK and USA.

| BRUSSELS SPROUTS | | Application | | | | PHI (Days) | Sample analysed | Residues of alpha + beta | Total Residues |
| Country, Year of trial *Variety* report | Form. | Method | kg ai/ha/ applic'n | L/ha | No. | | | endosulfan, endosulfan sulfate, mg/kg | mg/kg |
|---|---|---|---|---|---|---|---|---|---|
| UK 1976 *Onward* 2-21-01-02 | 35EC | Spray, 0.21% | 0.3 | 600 | 1 | 0 7 14 21 | Br. Sprts Br. Sprts Br. Sprts Br. Sprts | | 6.7 0.8 0.1 0.06 |
| UK 1976 *Onward* 2-21-01-03A | 35EC | Spray, 0.29% | 0.48 | 600 | 1 | 0 7 14 21 | Br. Sprts Br. Sprts Br. Sprts Br. Sprts | | 4.2 0.3 0.4 0.08 |
| USA 1965 Albion, NY *M-1575* | 2EC | Spray | 0.84 | | 14 | 0 1 3 7 10 | Br. Sprts Br. Sprts Br. Sprts Br. Sprts Br. Sprts | 0.66, < 0.05 0.88, 0.11 0.93, 0.14 0.60, 0.08 0.35, 0.10 | 0.66 0.99 1.07 0.68 0.45 |
| USA 1960 *R-470* | 2EC | Spray. Spray Interval 24 d | 0.84 | | 4 4 4 | 3 7 14 | Br. Sprts Br. Sprts Br. Sprts | | 2.41 0.88 0.94 |

Table 54. Endosulfan residues in cauliflower resulting from supervised trials in Australia, USA and Germany.

| CAULIFLOWER | | Application | | | | PHI | Sample | Residues of alpha, beta | Total |
| Country, Year of trial *Variety* Report | Form. | Method | g ai/Ha | L/ha | No. | days | analysed | endosulfan, endosulfan sulfate, mg/kg | Residues, mg/kg |
|---|---|---|---|---|---|---|---|---|---|
| Australia 2000 Medina *Galicia* N° 1/10/550 | EC | spray | 0.166 | 250 | 3 | 0 3 7 14 | Florets | 0.094, 0.066, 0.007 0.084, 0.046,0.011 0.050, 0.038,0.016 0.028,0.035,0.092 | 0.17 0.14 0.10 0.15 |
| Australia 2000 Medina *Galicia* N° 1/10/550 | EC | spray | 0.435 | 330 | 3 | 0 3 7 14 7 | Florets | 0.54, 0.35,0.014 0.054, 0.040, 0.013 0.052, 0.038, 0.019 0.008,0.007, 0.010 0.028,0.008, < 0.005 | 0.90 0.11 0.11 0.03 0.04 |

| CAULIFLOWER | Application | | | | | PHI | Sample | Residues of alpha, beta | Total |
| Country, Year of trial *Variety* Report | Form. | Method | g ai/Ha | L/ha | No. | days | analysed | endosulfan, endosulfan sulfate, mg/kg | Residues, mg/kg |
|---|---|---|---|---|---|---|---|---|---|
| Australia 2000 Werribee South *chaser* N° 1/10/550 | EC | spray | 0.315 | 474 | 3 | 0 3 7 14 | Florets | 0.042, 0.025, 0.020 0.039, 0.029, 0.006 0.010, 0.006, < 0.005 < 0.005,< 0.005,< 0.005 | 0.09 0.07 0.02 < 0.005 |
| Australia 2000 Werribee South *chaser* N° 1/10/550 | EC | spray | 0.625 | 474 | 3 | 0 3 7 14 | Florets | 0.23, 0.13, 0.005 0.11, 0.074, 0.012 0.046, 0.040, 0.008 0.006, < 0.005, < 0.005 | 0.37 0.20 0.09 0.01 |
| Germany 1983 *Erfurter Zwerg* DEU83172711 | 2.82% Dust | Spread Interval 11,7 d | 0.705 | | 3 | 0 5 10 14 | Head Head Head Head | 0.60, 0.40, 0.03 0.10, 0.20, 0.04 0.03, 0.07, 0.06 0.02, 0.02, 0.02 | 1.03 0.34 0.16 0.06 |
| Germany 1983 *Erfurter Zwerg* DEU83172721 | 2.82% Dust | Interval 13,11 d | 0.705 | | 3 | 0 5 10 14 | Head Head Head Head | 1.00, 0.80, 0.04 0.30, 0.20, 0.04 0.02, 0.03, 0.02 < 0.01, < 0.01, < 0.01 | 1.84 0.54 0.07 < 0.01 |
| Germany 1983 *Necker-Perle* DEU83172731 | 2.82% Dust | Interval 14, 12 d | 0.705 | | 3 | 0 5 10 14 | Head Head Head Head | 0.07, 0.03, < 0.01 < 0.01, < 0.01, < 0.01 < 0.01, < 0.01, < 0.01 < 0.01, < 0.01, < 0.01 | 0.10 < 0.01 < 0.01 < 0.01 |
| Germany 1983 *Celestar* DEU83172741 | 2.82% Dust | Interval 11, 14 d | 0.705 | | 3 | 0 5 10 14 | Head Head Head Head | 0.50, 0.06, < 0.01 0.09, 0.04, 0.03 0.02, 0.04, 0.02 < 0.01, < 0.01, < 0.01 | 0.56 0.16 0.08 < 0.01 |
| | | | | | | | | alpha+beta/sulfate | |
| USA 1964 *Island Queen* A48549 | 2EC | Spray Interval 8, 7, 12 d | 0.84 | 935 (0.09 % ai) | 4 | 0 4 7 11 | Heads+lvs Hds+lvs Hds+lvs Hds+lvs | 0.75, < 0.05 0.46, 0.05 0.23, 0.07 0.15, 0.09 | 0.75 0.51 0.30 0.24 |
| USA 1964 A48549 | 2EC | Spray interval 7 days | 0.84 | | 11 | 0 1 3 7 10 0 1 3 7 10 | Heads Heads Heads Heads Heads Leaves Leaves Leaves Leaves Leaves | < 0.05,< 0.05 0.41, 0.05 0.05, 0.05 0.05, 0.05 < 0.05, < 0.05 0.30, 0.08 2.58, 0.05 0.18, 0.05 0.24 , 0.07 0.34, 0.11 | < 0.05 0.46 0.10 0.10 < 0.05 0.38 2.63 0.23 0.31 0.45 |
| USA 1964 *Snowball "Y"* A48549 | 2EC | Spray | 0.75 | 100 (0.09 % ai) | 8 | 0 3 7 10 14 0 3 7 10 14 | Heads Heads Heads Heads Heads Leaves Leaves Leaves Leaves Leaves | < 0.05, < 0.05 < 0.05, < 0.05 < 0.05, < 0.05 < 0.05, < 0.05 < 0.05, < 0.05 7.36, 0.95 2.30, 1.36 2.28, 1.05 2.38, 1.60 2.05, 1.74 | < 0.05 < 0.05 ≤ 0.05 < 0.05 < 0.05 8.31 3.66 3.33 3.98 3.79 |

Table 55. Endosulfan residues in cucumbers from supervised trials in Germany, USA and Australia.

| CUCUMBER | Application | | | | | PHI | Sample | Residues of alpha, beta endosulfan, endosulfan sulfate, mg/kg | Total Residues, mg/kg |
|---|---|---|---|---|---|---|---|---|---|
| Country, Year of trial *Variety, report* | Form. | Method | kg ai/ha/ applic'n | L/ha | No. | days | analysed | | |
| Germany 1978 *Pepinex* *LEA 1/701/1-78* | 24EC* | Spray indoor interval 3,3,4 days | 0.576 | 1200 (0.048 % ai) | 4 | 0 1 2 3 | Fruit Fruit Fruit Fruit | | 1.20 0.60 0.30 0.30 |
| Germany 1978 *Sandra* *LEA 2/701/1-78* | 24EC* | Spray indoor interval 4,3,4 days | 0.576 | 1200 (0.048% ai) | 4 | 0 1 2 3 | Fruit Fruit Fruit Fruit | | 0.02 0.01 0.01 0.005 |
| Germany 1978 *Sandra* *LEA 2/701/1-78* | 24EC* | Spray indoor interval 4,3,4 days | 0.576 | 1200 (0.048 % ai) | 4 | 0 1 2 3 | Fruit Fruit Fruit Fruit | | 0.10 0.05 0.70 0.30 |
| Germany 1978 *Uniflora* *LEA 2/701/1-78* | 24EC* | Spray indoor interval 3,3,4 days | 0.576 | 1200 (0.048 % ai) | 4 | 0 1 2 3 | Fruit Fruit Fruit Fruit | | 0.06 0.006 0.20 0.20 |
| Germany 1976 *Hokus* *LEA 3/74/01/02-76* | 24EC* | Spray Interval 18 d | 0.216 | 600 (0.036% ai) | 2 | 0 7 10 14 | Fruit/peel Fruit/peel Fruit/peel Fruit/peel | | 0.04/0.1 0.02/0.1 ND/0.02 ND/0.04 |
| Germany 1975 *Bambina* *LEA 3/85/01/02-75* | 24EC* | Spray indoor | 0.96 | 2000 (0.048% ai) | 1 | 0 7 10 14 0 7 10 14 | Fruit Fruit Fruit Fruit Peel Peel Peel Peel | 0.01, ND, 0.01 ND, 0.009, 0.01 ND, ND, 0.01 ND, ND, 0.02 0.90, 0.90 ND ND, 0.03, 0.10 0.20 0.07, ND ND, ND, 0.06 | 0.02 0.02 0.01 0.02 1.80 0.13 0.27 0.06 |
| Germany 1975 *Uniflora* *LEA 2/85/ 01/02-75A* | 24EC* | Spray indoor | 0.96 | 2000 (0.048% ai) | 1 | 0 7 10 14 | Fruit/peel Fruit/peel Fruit/peel Fruit/peel | | < 0.06/0.7 0.04/0.7 0.01/0.2 < 0.06/0.09 |
| Germany 1975 *Uniflora* *LEA 4/85/01/02-75B* | 24EC* | Spray indoor | 0.96 | 2000 (0.048 % ai) | 1 | 0 7 10 14 0 7 10 14 | Fruit Fruit Fruit Fruit Peel Peel Peel Peel | ND, ND, ND ND, ND, 0.06 ND, ND, 0.08 0.02, 0.02, 0.10 1.90 1.70.0.20 0.03, 0.04, 0.10 ND, ND, 0.06 ND, ND, 0.10 | < 0.02 0.06 0.08 0.14 3.80 0.17 0.06 0.10 |
| Germany 1975 *Pepinex* *LEA 1/85/ 01/02-75* | 24EC* | Spray indoor | 0.96 | 2000 (0.048 % ai) | 1 | 0 7 10 14 | Fruit/peel Fruit/peel Fruit/.peel Fruit/peel | | 0.01/3.0 0.10/0.9 0.09/0.5 0.10/0.4 |
| USA1996 NC *Clypso* *BJ-96R-01* | 33EC* 50WP | Note: All spray intervals = 7 days Spray Spray | 1.12, 1.24, 1.12 1.12, 1.24, 1.12 | 96, 100, 92.5 98, 95.3,91.6 | 3 | 2 | Fruit Fruit | 0.15, 0.15, 0.10 0.08, 0.07, 0.07 | 0.40 0.22 |
| USA1996 SC *Poinsett 76* *BJ-96R-01* | 33EC 50WP | Spray Spray | 1.12 1.12 | 192, 195, 195 193, 194, 194 | 3 | 2 | Fruit Fruit | 0.13, 0.11, 0.12 0.08, 0.11, 0.11 | 0.36 0.30 |

| CUCUMBER | Application | | | | | PHI | Sample | Residues of alpha, | Total |
|---|---|---|---|---|---|---|---|---|---|
| Country, Year of trial Variety, report | Form. | Method | kg ai/ha/ applic'n | L/ha | No. | days | analysed | beta endosulfan, endosulfan sulfate, mg/kg | Residues, mg/kg |
| USA 1996 FL<br><br>Poinsett<br><br><br>BJ-96R-01 | 33EC<br><br><br>50WP | Spray<br><br><br>Spray | 1.12, 1.34, 1.12<br>1.12, 1.34, 1.1 | 202, 237,199<br><br><br>202, 237,199 | 3 | 2 | Fruit<br><br><br>Fruit | 0.12, 0.09, 0.11<br><br><br>0.09, 0.05, 0.09 | 0.32<br><br><br>0.23 |
| USA 1996 FL Poinsett 76 BJ-96R-01 | 33EC 50WP | Spray Spray | 1.12 1.12 | 274, 281,283 274, 276, 274 | 3 | 2 | Fruit Fruit | 0.13, 0.11, 0.07 0.07, 0.06, 0.06 | 0.31 0.19 |
| USA 1996 FL<br><br>BJ-96R-01 | 33EC<br>50WP | Spray<br>Spray | 1,1, 1.120 1.12 | 469, 464, 475<br>472, 473, 473 | 3 | 2 | Fruit<br>Fruit | 0.22, 0.18, 0.13<br>0.12, 0.09, 0.07 | 0.53<br>0.28 |
| USA 1996 MI Marketmore 76 BJ-96R-01 | 33EC 50WP | Spray Spray | 1.12 1.12 | 221, 211, 231 223, 217, 217 | 3 | 2 | Fruit Fruit | 0.17, 0.15, 0.10 0.06, 0.06, 0.06 | 0.42 0.18 |
| USA 1996 OH<br><br>Thunder BJ-96R-01 | 33EC<br><br>50WP | Spray<br><br>Spray | 1.12, 1.12,1 1.12 | 239, 236, 223<br><br>236, 236, 229 | 3 | 2 | Fruit<br><br>Fruit | 0.13, 0.09, 0.10<br><br>0.14, 0.07, 0.10 | 0.32<br><br>0.31 |
| USA 1996 WIS<br><br>Marketmore 76 BJ-96R-01 | 33EC<br><br>50WP | Spray<br><br>Spray | 1.12, 1.12,1 1.12 | 172, 183, 177.6<br>181, 183, 180 | 3 | 2 | Fruit<br><br>Fruit | 0.27, 0.20, 0.11<br><br>0.11, 0.07, 0.06 | 0.58<br><br>0.24 |
| USA 1996 OKL Straight Eight BJ-96R-01 | 33EC 50WP | Spray Spray | 1.12 1.12 | 164, 179,188 166, 179,189 | 3 | 2 | Fruit Fruit | 0.28, 0.22, 0.14 0.14, 0.08, 0.08 | 0.64 0.30 |
| USA 1996 TX BJ-96R-01 | 33EC 50WP | Spray Spray | 1.12 1.12 | 317, 305, 272 317, 305, 272 (0.35% ai)_ | 3 | 2 | Fruit Fruit | 0.10, 0.09, 0.13 0.12, 0.13, 0.11 | 0.32 0.36 |
| USA 1984 CA Spacemaster<br><br>1913(5) | 3EC | Spray. Interval: 7 d | 1.12<br>UTC<br><br>UTC | 140 (0.8% ai)_ | 6 | 0<br>3<br>0<br><br>3 | Fruit Fruit Fruit<br><br>Fruit | 0.165, 0.101, < 0.01<br>0.178, 0.122, 0.015<br>0.012, < 0.01, < 0.01<br>0.021, < 0.01, < 0.01 | 0.27<br>0.32<br>0.01<br><br>0.02 |
| USA 1984 CA 1913(5) | 50WP | Spray. Interval: 7 days | 1.12 | 140 (0.8% ai) | 6 | 0 | Fruit | 0.513, 0.229, < 0.01 | 0.74 |
| USA 1984 CA Spacemaster<br><br>1913(12) | 3EC | Spray. Interval: 7 days | 1.12<br>UTC<br><br>UTC | 140 (0.8% ai) | 6 | 0<br>3<br>0<br><br>3 | Fruit Fruit Fruit<br><br>Fruit | 0.126, 0.059, < 0.01<br>0.029, 0.022, < 0.01<br>0.0195, 0.022, 0.0388<br>0.0276, 0.023, 0.0396 | 0.19<br>0.05<br>0.08<br><br>0.09 |
| USA 1984 CA Spacemaster 1913(12) | 50WP | Spray. Interval: 7 days | 1.12 | 140 (0.8% ai) | 6 | 0<br>3 | Fruit Fruit | 0.412, 0.181, < 0.01<br>0.305, 0.183, < 0.01 | 0.59<br>0.49 |
| USA 1984 NY Victory 1913 (7) | 3EC | Spray. Interval: 7/8 days | 1.12<br>UTC<br>UTC | 327 (0.34% ai) | 5 | 0<br>3<br>0<br>3 | Fruit Fruit Fruit Fruit | 0.197, 0.23, 0.139<br>0.035, 0.038, 0.046<br>0.01, < 0.01, < 0.01<br>< 0.01, < 0.01, < 001 | 0.57<br>0.12<br>0.03<br>< 0.03 |
| USA 1984 NY Victory 1913 (7) | 50WP | | 1.12 | 327 (0.34% ai) | 5 | 0<br>3 | Fruit Fruit | 0.118, 0.096, 0.109<br>0.088, 0.080, 0.069 | 0.32<br>0.23 |

| CUCUMBER | Application | | | | | PHI | Sample | Residues of alpha, | Total |
| Country, Year of trial *Variety, report* | Form. | Method | kg ai/ha/ applic'n | L/ha | No. | days | analysed | beta endosulfan, endosulfan sulfate, mg/kg | Residues, mg/kg |
|---|---|---|---|---|---|---|---|---|---|
| Australia 2000 Darlington *coolah* N° 1/10/547 | EC | spray | 0.168 | 253 | 4 | 0 3 5 7 | Fruit Fruit Fruit Fruit | 0.065, 0.056, 0.028 0.047,0.027, 0.035 0.030, 0.017,0.029 0.051,0.031,0.035 | 0.15 0.11 0.08 0.11 |
| Australia 2000 Darlington *coolah* N° 1/10/547 | EC | spray | 0.555 | 421 | 4 | 0 3 5 7 | Fruit Fruit Fruit Fruit | 0.31, 0.20.0.054 0.12, 0.11,0.049 0.059, 0.037,0.034 0.14,0.11, 0.082 | 0.56 0.28 0.13 0.33 |
| Australia 2000 Lowood *Warmer* N° 1/10/547 | EC | spray | 0.15 | 226 | 4 | 0 3 5 7 | Fruit Fruit Fruit Fruit | 0.058, 0.037, 0.042 0.028, 0.017, 0.034 0.029, 0.017, 0.036 0.031, 0.021, 0.042 | 0.14 0.08 0.08 0.09 |
| Australia 2000 Lowood *Warmer* N° 1/10/547 | EC | spray | 0.30 | 226 | 4 | 0 3 5 7 7 | Fruit Fruit Fruit Fruit Fruit co | 0.071, 0.043, 0.050 0.041, 0.024, 0.054 0.039, 0.024, 0.050 0.044, 0.032, 0.056 0.007 < 0.005, < 0.005 | 0.17 0.12 0.11 0.13 0.007 |

Table 56. Endosulfan residues in melons from supervised trials in Australia.

| MELON | Application | | | | | PHI | Sample | Residues of alpha, | Total |
| Country, Year of trial *Variety* Report | Form. | Method | g ai./hl | L/ha | No. | days | analysed | beta endosulfan, Endosulfan sulfate, mg/kg | Residues, mg/kg |
|---|---|---|---|---|---|---|---|---|---|
| Australia 2000 Kialla West *Hiline* N° 1/10/551 | EC | spray | 66.5 | 1170 | 4 | 0 3 5 7 | Fruit | 0.39, 0.28, 0.032 0.25,0.27, 0.034 0.12, 0.11,0.037 0.089,0.097,0.045 | 0.70 0.55 0.27 0.23 |
| Australia 2000 Kialla West *Hiline* N° 1/10/551 | EC | spray | 132 | 1170 | 4 | 0 3 5 7 | Fruit | 0.89,0.90.0.04 0.26, 0.32, 0.047 0.34, 0.40.0.059 0.13, 0.16, 0.060 | 1.83 0.63 0.80 0.35 |
| Australia 2000 Fernvale *planters jumbo* N° 1/10/556 | EC | spray | 66.5 | 334 | 4 | 0 3 5 7 | Fruit | 0.30, 0.26, 0.13 0.42, 0.35, 0.21 0.23,0.30, 070 0.35, 0.36, 0.29 | 0.69 0.98 1.23 1.0 |
| Australia 2000 Fernvale *planters jumbo* N° 1/10/556 | EC | spray | 132 | 334 | 4 | 7 | Fruit | 0.10, 0.21, 0.74 0.006, 0.005, 0.019 0.008, 0.005, 0.019 0.008, 0.008, 0.019 | 1.05 0.03 0.03 0.04 |

Table 57. Endosulfan residues in melon from supervised trials in USA and Europe.

| MELON Country, Year of trial Variety, report | Form. | Method | kg ai/ha/ applic'n | L/ha | No. | PHI days | Sample analysed | Residues of alpha, beta endosulfan, Endosulfan sulfate, mg/kg | Total Residues, mg/kg |
|---|---|---|---|---|---|---|---|---|---|
| USA 1995 CA BJ-95R-05 | 35EC | Spray Interval 7 d | 1.12 | 191.6,184, 190 | 3 | 2 | Cantaloupe | 0.18, 0.27, < 0.05 | 0.45 |
| USA 1995 CA BJ-95R-05 | 50WP | Spray Interval 7d | 1.12 | 191.6,184, 190 | 3 | 2 | Cantaloupe | 0.24, 0.36, < 0.05 | 0.60 |
| USA 1995 CA BJ-95R-05 | 35EC | Spray Interval 7 d | 1.12 | 323.5, 329 326 | 3 | 2 | Cantaloupe | < 0.05, < 0.05, 0.05 | 0.05 |
| USA 1995 CA BJ-95R-05 | 50WP | Spray Interval 7 d | 1.12 | 325, 330 34.8 | 3 | 2 | Cantaloupe | 0.07, 0.11, 0.06 | 0.24 |
| USA 1995 CA BJ-95R-05 | 35EC | Spray Interval 7 d | 1.12 | | 3 | 2 | Cantaloupe | 0.16, 0.25, < 0.05 | 0.41 |
| USA 1995 CA BJ-95R-05 | 50WP | Spray Interval 7 d | 1.12 | | 3 | 2 | Cantaloupe | 0.14, 0.21, < 0.05 | 0.35 |
| USA 1995 FLA BJ-95R-05 | 35EC | Spray Interval 7 d | 1.23,1.12 1.12 | 215,204.7 202 | 3 | 2 | Cantaloupe | 0.13, 0.15, 0.06 | 0.34 |
| USA 1995 FLA BJ-95R-05 | 50WP | Spray Interval 7 d | 1.23,1.12 1.12 | 215, 204.9 202 | 3 | 2 | Cantaloupe | 0.18, 0.17, 0.05 | 0.40 |
| USA 1995 MI BJ-95R-05 | 50WP | Spray Interval 7 d | 1.12 | 190.7,195.4 187 | 3 | 2 | Cantaloupe | 0.12, 0.12, 0.06 | 0.30 |
| USA 1995 TX BJ-95R-05 | 35EC | Spray Interval 7 d | 1, 1, 1 | 188,193.5 187 | 3 | 2 | Cantaloupe | 0.10, 0.15, 0.05 | 0.30 |
| USA 1995 TX BJ-95R-05 | 50WP | Spray Interval 7 d | 1, 1, 1 | 190,186 191.6 | 3 | 2 | Cantaloupe | 0.18, 0.26, 0.05 | 0.49 |
| USA 1995 MI BJ-95R-05 | 35EC | Spray Interval 7 d | 1.12 | 190.7,190.7 186 | 3 | 2 | Cantaloupe | 0.07, 0.09, 0.06 | 0.22 |
| Spain 2000 Extra Rica Miel DR 00 EUS 131 | CS* | Sprays at 28, 14,7 d PHI BBCH 67, 72 83 | 0.530 | 600 (0.088% ai) | 3 | 0 7 7 7 | Fruit Fruit Peel Pulp | 0.04, 0.02, 0.04 0.05, 0.02, 0.05, 0.06, 0.03, < 0.02 < 0.02, < 0.02, 0.02 | 0.10 0.12 0.09 0.02 |
| Italy 2000 Calipso DR 00 EUS 131 | CS* | Sprays at 28, 14,7 d PHI BBCH 64, 76 81 | 0.530 | 800 (0.066% ai) | 3 | 0 7 7 7 | Fruit Fruit Peel Pulp | 0.10, 0.07, < 0.02 0.04, 0.04, < 0.02 0.22, 0.19, < 0.02 < 0.02, < 0.02, < 0.02 | 0.17 0.08 0.41 < 0.02 |
| Italy 2000 Proteo DR 00 EUS 131 | CS* | Sprays at 28, 14,7 d PHI BBCH 71, 84 87 | 0.530 | 800 (0.066% ai) | 3 | 0 7 7 7 | Fruit Fruit Peel Pulp | 0.26, 0.16, < 0.02 0.13, 0.07, < 0.02 0.19, 0.11, < 0.02 < 0.02, < 0.02, < 0.02 | 0.42 0.20 0.30 < 0.02 |
| Spain 1999 Piel sapo Ricamie ER 99 ECS 757 | CS* | Sprays at 28, 13,7 d PHI BBCH 73, 74 84 | 0.530 | 600 (0.088 % ai) | 3 | 0 3 7 | Fruit Fruit Fruit | 0.10, 0.05,0.03 0.07, 0.04, 0.02 0.04, 0.02, 0.02 | 0.18 0.13 0.08 |

| MELON | Application | | | | | PHI | Sample | Residues of alpha, | Total |
| Country, Year of trial *Variety, report* | Form. | Method | kg ai/ha/ applic'n | L/ha | No. | days | analysed | beta endosulfan, Endosulfan sulfate, mg/kg | Residues, mg/kg |
|---|---|---|---|---|---|---|---|---|---|
| Italy 1999 *Galia* ER 99 ECS 757 | CS* | Sprays at 28, 14,7 d PHI BBCH 66, 68 75 | 0.530 | 800 (0.088% ai) | 3 | 0 | Fruit | 0.26, 0.15, < 0.02 | 0.41 |
| | | | | | | 3 | Fruit | 0.13, 0.08, 0.04 | 0.25 |
| | | | | | | 7 | Fruit | 0.08, 0.07, 0.03 | 0.18 |
| Spain 1997 *Panal* Musk melon ER 97 ECS 745 | 35EC | Sprays at 21, 14,7 d PHI BBCH 73, 75 83 | 0.530 | 300 (0.177 % ai) | 3 | 0 | Peel | 0.13, 0.10 < 0.02 | 0.23 |
| | | | | | | 0 | Pulp | ND, ND, < 0.02 | < 0.02 |
| | | | | | | 3 | Peel | 0.02, 0.03, < 0.02 | 0.05 |
| | | | | | | 3 | Pulp | ND, ND, < 0.02 | < 0.02 |
| | | | | | | 7 | Peel | < 0.02, < 0.02, < 0.02 | < 0.02 |
| | | | | | | 7 | Pulp | ND, ND, 0.02 | 0.02 |
| | | | | | | 0 | Fruit | 0.06, 0.04, < 0.02 | 0.10 |
| | | | | | | 3/7 | Fruit | < 0.02, < 0.02, < 0.02 | < 0.02 |
| France (S) 1997 *Manta* Cantaloupe ER 97 ECS 745 | 35EC | Sprays at 21, 14,7 d PHI BBCH 71, 71 72 | 0.530 | 250 (0.212 % ai) | 3 | 0 | Peel | 0.20, 0.13, 0.05 | 0.38 |
| | | | | | | 0 | Pulp | < 0.02, < 0.02, < 0.02 | < 0.02 |
| | | | | | | 3 | Peel | 0.05, 0.09, 0.08 | 0.22 |
| | | | | | | 3 | Pulp | < 0.02, < 0.02, < 0.02 | < 0.02 |
| | | | | | | 7 | Peel | 0.02, 0.04, 0.04 | 0.10 |
| | | | | | | 7 | Pulp | ND, < 0.02, < 0.02 | < 0.02 |
| | | | | | | 0 | Fruit | 0.08, 0.05, 0.02 | 0.15 |
| | | | | | | 3 | Fruit | 0.02, 0.03, 0.03 | 0.08 |
| | | | | | | 7 | Fruit | < 0.02, 0.02, 0.02 | 0.04 |
| Portugal 2000 *Pele de Sapo* DR 00 EUS 131 | CS* | Sprays at 29, 14,7 d PHI BBCH 67, 72 83 | 0.530 | 600 (0.088% ai) | 3 | 0 | Fruit | < 0.02, < 0.02, < 0.02 | < 0.02 |
| | | | | | | 7 | Fruit | < 0.02, < 0.02, < 0.02 | < 0.02 |
| | | | | | | 7 | Peel | 0.05, 0.05, < 0.02 | 0.10 |
| | | | | | | 7 | Pulp | < 0.02, < 0.02, 0.02 | 0.02 |
| Greece 1999 *Daniel* ER 99 ECS 757 | CS* | Sprays at 29, 14,7 d PHI BBCH 73, 75 77 | 0.530 | 600 (0.088% ai) | 3 | 0 | Fruit | 0.16, 0.09, 0.04 | 0.29 |
| | | | | | | 3 | Fruit | 0.06, 0.03, 0.04 | 0.13 |
| | | | | | | 7 | Fruit | 0.05, 0.04, 0.04 | 0.13 |
| Italy 1997 *Pamir* Musk melon ER 97 ECS 745 | 35EC | Sprays at 20, 14,7 d PHI BBCH 74, 79 82 | 0.530 | 300 (0.177% ai) | 3 | 0 | Peel | 0.11, 0.07, < 0.02 | 0.18 |
| | | | | | | 0 | Pulp | < 0.02, < 0.02, 0.03 | 0.03 |
| | | | | | | 3 | Peel | < 0.02, 0.02, < 0.02 | 0.02 |
| | | | | | | 3 | Pulp | < 0.02, ND, 0.02 | 0.02 |
| | | | | | | 7 | Peel/Pulp | ND, ND, < 0.02 | < 0.02 |
| | | | | | | 0 | Fruit | 0.05, 0.03, 0.02 | 0.10 |
| | | | | | | 3 | Fruit | < 0.02, 0.02, < 0.02 | 0.02 |
| | | | | | | 7 | Fruit | ND, ND, < 0.02 | < 0.02 |
| Italy 1997 *Momo* Cantaloupe ER 97 ECS 745 | 35EC | Sprays at 21, 14,7 d PHI BBCH 72, 81 84 | 0.530 | 600 (0.088% ai) | 3 | 0 | Peel | 0.66, 0.44, 0.02 | 1.12 |
| | | | | | | 0 | Pulp | < 0.02, < 0.02, < 0.02 | < 0.02 |
| | | | | | | 3 | Peel | 0.24, 0.27, 0.02 | 0.53 |
| | | | | | | 3 | Pulp | ND, ND, ND | < 0.02 |
| | | | | | | 7 | Peel | 0.08, 0.14, < 0.02 | 0.22 |
| | | | | | | 7 | Pulp | ND, ND, < 0.02 | < 0.02 |
| | | | | | | 0 | Fruit | 0.21, 0.15, < 0.02 | 0.36 |
| | | | | | | 3 | Fruit | 0.07, 0.08, < 0.02 | 0.15 |
| | | | | | | 7 | Fruit | 0.03, 0.04, < 0.02 | 0.07 |
| Spain 1994 *Daimiel* ER 94 ECS 780 | 35EC | Sprays at 21, 14,7 d PHI BBCH 69/70, 69/70, 69/70 | 1.0561 | 400 (0.264% ai) | 3 | 0 | Peel | 0.141, 0.12, 0.095 | 0.36 |
| Spain 1994 *Rixan* Musk melon ER 94 ECS 780 | 35EC | Sprays at 21, 14,7 d PHI BBCH 70, 70 70 | 1.0561 | 300, 400, 400 (0.352% ai) (0.264% ai) | 3 | 0 | Pulp | < 0.05, < 0.05, 0.074 | 0.07 |
| | | | | | | 0 | Peel | 0.084, 0.07, 0.076 | 0.23 |
| | | | | | | 0 | Fruit | | 0.16 |
| | | | | | | 3 | Pulp | < 0.05, < 0.05, 0.074 | 0.07 |
| | | | | | | 3 | Peel | < 0.05, 0.053, 0.061 | 0.11 |
| | | | | | | 3 | Fruit | | 0.10 |
| | | | | | | 7 | Pulp | < 0.05, < 0.05, 0.081 | 0.08 |
| | | | | | | 7 | Peel | < 0.05, < 0.05, 0.052 | 0.05 |
| | | | | | | 7 | Fruit | | 0.07 |
| | | | | | | 14 | Peel | < 0.05, < 0.05, < 0.05 | < 0.05 |
| | | | | | | 21 | Peel | < 0.05, < 0.05, < 0.05 | < 0.05 |
| | | | | | | 29 | Peel | < 0.05, < 0.05, < 0.05 | < 0.05 |

| MELON Country, Year of trial Variety, report | Form. | Method | kg ai/ha/ applic'n | L/ha | No. | PHI days | Sample analysed | Residues of alpha, beta endosulfan, Endosulfan sulfate, mg/kg | Total Residues, mg/kg |
|---|---|---|---|---|---|---|---|---|---|
| Spain 1994 *Daimiel* Musk melon *ER 94 ECS 780* | 35EC | Sprays at 21, 14,7 d PHI BBCH 69/70, 69/70, 69/70 | <u>0.528</u> | 400 (0.132% ai) | 3 | 0 | Pulp | < 0.05, < 0.05, 0.076 | 0.08 |
| | | | | | | 0 | Peel | 0.102, 0.104, 0.092 | 0.30 |
| | | | | | | 0 | Fruit | | 0.19 |
| | | | | | | 3 | Pulp | < 0.05, < 0.05, 0.092 | 0.09 |
| | | | | | | 3 | Peel | < 0.05, < 0.05, 0.092 | 0.09 |
| | | | | | | 3 | Fruit | | 0.09 |
| | | | | | | 7 | Pulp | < 0.05, < 0.05, 0.091 | 0.09 |
| | | | | | | 7 | Peel | < 0.05, 0.061, 0.082 | 0.14 |
| | | | | | | 7 | Fruit | | 0.12 |
| Spain 1994 *Rixan* Musk melon *ER 94 ECS 780* | 35EC | Sprays at 21, 14,7 d PHI BBCH 70, 70 70 | 0.528 | 300, 400, 400 (0.177% ai) (0.132% ai) | 3 | 0 | Pulp | < 0.05, < 0.05, 0.095 | 0.10 |
| | | | | | | 0 | Peel | < 0.05, < 0.05, 0.05 | 0.05 |
| | | | | | | 0 | Fruit | | 0.10 |
| | | | | | | 3 | Pulp | < 0.05, < 0.05, 0.096 | 0.10 |
| | | | | | | 3 | Peel | < 0.05, < 0.05, 0.05 | 0.05 |
| | | | | | | 3 | Fruit | | 0.10 |
| | | | | | | 7 | Pulp | < 0.05, < 0.05, 0.095 | 0.10 |
| | | | | | | 7 | Peel | < 0.05, < 0.05, < 0.05 | < 0.05 |
| | | | | | | 7 | Fruit | | 0.08 |
| | | | | | | 14 | Pulp | < 0.05, < 0.05, 0.115 | 0.12 |
| | | | | | | 14 | Peel | < 0.05, < 0.05, < 0.05 | < 0.05 |
| | | | | | | 14 | Fruit | | 0.09 |
| | | | | | | 21 | Pulp | < 0.05, < 0.05, 0.09 | 0.09 |
| | | | | | | 21 | Peel | < 0.05, < 0.05, < 0.05 | < 0.05 |
| | | | | | | 21 | Fruit | | 0.07 |
| | | | | | | 29 | Fruit | | 0.06 |
| Spain 1994 *Daimiel* Musk melon *ER 94 ECS 780* | 35EC | Sprays at 21, 14,7 d PHI BBCH 69/70, 69/70, 69/70 | 0.528 | 400 (0.132% ai) | | 0 | Pulp | < 0.05, < 0.05, 0.069 | 0.07 |
| | | | | | | 0 | Peel | < 0.05, < 0.05, 0.071 | 0.07 |
| | | | | | | 0 | Fruit | | 0.07 |
| | | | | | | 3 | Pulp | 0.084, 0.064, 0.086 | 0.23 |
| | | | | | | 3 | Peel | < 0.05, < 0.05, 0.059 | 0.06 |
| | | | | | | 3 | Fruit | | 0.15 |
| | | | | | | 7 | Pulp | < 0.05, < 0.05, 0.09 | 0.09 |
| | | | | | | 7 | Peel | < 0.05, < 0.05, 0.066 | 0.07 |
| | | | | | | 7 | Fruit | | 0.08 |
| Italy 1994 *Tamaris* Musk melon *ER 94 ECS 780* | 35EC | Sprays at 21, 14,7 d PHI BBCH 64/80, 69/81, 70/82 | 0.528 | 1000 (0.053% ai) | 3 | 0 | Pulp | < 0.05, < 0.05, < 0.05 | < 0.05 |
| | | | | | | 0 | Peel | 0.432, 0.294, < 0.05 | 0.73 |
| | | | | | | 0 | Fruit | | 0.31 |
| | | | | | | 3 | Pulp | < 0.05, < 0.05, < 0.05 | < 0.05 |
| | | | | | | 3 | Peel | 0.11, 0.096, < 0.05 | 0.21 |
| | | | | | | 3 | Fruit | | 0.11 |
| | | | | | | 7 | Pulp | < 0.05, < 0.05, < 0.05 | < 0.05 |
| | | | | | | 7 | Peel | 0.112, 0.141, < 0.05 | 0.25 |
| | | | | | | 7 | Fruit | | 0.12 |
| | | | | | | 14 | Peel | < 0.05, < 0.05, < 0.05 | < 0.05 |
| | | | | | | 21 | Peel | < 0.05, < 0.05, < 0.05 | < 0.05 |
| Spain 1994 *Daimiel* *ER 94 ECS 780* | 35EC | Sprays at 21, 14,7 d PHI BBCH 69/70, 69/70, 69/70 | 1.0561 | 400 (0.264% ai) | 3 | 0 | Peel | 0.215, 0.145, 0.081 | 0.44 |
| Portugal 1999 *Branco do Ribatejo ER 99 ECS 757* | CS* | Sprays at 28, 14,7 d PHI BBCH 71, 74 84, | 0.530 | 600 (0.088% ai) | 3 | 0 | Fruit | 0.07, 0.04, 0.06 | *0.17* |
| | | | | | | 3 | Fruit | 0.06, 0.04, 0.08 | 0.18 |
| | | | | | | 7 | Fruit | 0.04, 0.03, 0.08 | 0.15 |
| Italy 1994 *Calipso* Musk melon *ER 94 ECS 780* | 35EC | Sprays at 21, 14,7 d PHI BBCH 69/75, 69/80, 69/81 | 0.528 | 1000 (0.0528 % ai) | 3 | 0 | Pulp | < 0.05, < 0.05, < 0.05 | < 0.05 |
| | | | | | | 0 | Peel | 0.469, 0.503, < 0.05 | 0.97 |
| | | | | | | 0 | Fruit | | 0.47 |
| | | | | | | 3 | Pulp | < 0.05, < 0.05, < 0.05 | < 0.05 |
| | | | | | | 3 | Peel | 0.205, 0.261, < 0.05 | 0.47 |
| | | | | | | 3 | Fruit | | 0.22 |
| | | | | | | 7 | Pulp | < 0.05, < 0.05, < 0.05 | < 0.05 |
| | | | | | | 7 | Peel | 0.162, 0.282, 0.052 | 0.50 |
| | | | | | | 7 | Fruit | | 0.19 |

| MELON Country, Year of trial Variety, report | Form. | Method | kg ai/ha/ applic'n | L/ha | No. | PHI days | Sample analysed | Residues of alpha, beta endosulfan, Endosulfan sulfate, mg/kg | Total Residues, mg/kg |
|---|---|---|---|---|---|---|---|---|---|
| Italy 1994 Calipso ER 94 ECS 780 | 35EC | Sprays at 21, 14,7 d PHI BBCH 69/75, 69/80, 69/81 | 1.0561 | 1000 (0.1056 % ai) | 3 | 0 | Peel | 0.779, 0.758, 0.079 | 1.62 |
| Spain 1992 Futuro PRS99/012 | 35EC | Spray | 0.82 | 780 (0.105% ai) | 1 | 0 3 7 15 | Fruit Fruit Fruit Fruit | | 0.81 0.28 0.23 0.11 |
| Spain 1992 Amarillo canario PRS99/012 | 35EC | Spray | 0.71 | 680 (0.105% ai) | 1 | 0 3 7 15 | Fruit Fruit Fruit Fruit | | 0.38 0.05 0.02 0.02 |
| Spain 1992 Galia PRS99/012 | 35EC | Spray (0.105%) | 0.76 | 720 (0.105% ai) | 1 | 0 3 7 15 | Fruit Fruit Fruit Fruit | | 0.97 0.63 0.50 0.22 |
| Italy 1994 Tamaris Musk melon ER 94 ECS 780 | 35EC | Sprays at 21, 14,7 d PHI BBCH 64/80, 69/81, 70/82 | 1.0561 | 1000 (0.106 % ai) | 3 | 0 0 0 3 3 3 7 7 | Pulp Peel Fruit Pulp Peel Fruit Pulp Peel | < 0.05, < 0.05, < 0.05 0.76, 0.487, < 0.05 < 0.05, < 0.05, < 0.05 0.192, 0.20 < 0.05 < 0.05, < 0.05, < 0.05 < 0.05, 0.085, < 0.05 | < 0.05 1.25 0.50 < 0.05 0.39 0.20 < 0.05 0.09 |
| | | | | | | 7 14 21 | Fruit Peel Peel | < 0.05, < 0.05, < 0.05 < 0.05, < 0.05< 0.05 | 0.08 < 0.05 < 0.05 |
| Spain 1992 Futuro PRS99/012 | 35EC | Spray (0.105%) | 0.87 | 830 (0.105 % ai) | 1 | 0 3 7 15 | Fruit Fruit Fruit Fruit | | 0.09 < 0.01 < 0.01 0.04 |

Table 58. Endosulfan residues in summer squash from supervised trials in Spain and USA.

| S. SQUASH Country, Year of trial Variety, report | Form. | Method | kg ai//ha applic'n | L/ha | No. | PHI | Sample analysed | Residues of alpha, beta endosulfan, Endosulfan sulfate, mg/kg | Total Residues mg/kg |
|---|---|---|---|---|---|---|---|---|---|
| Spain 1992 Elite PRS99/012 | 35EC | Spray | 1.09 | 1040 (0.105% ai) | 1 | 0 3 7 15 | Fruit Fruit Fruit Fruit | | 1.14 0.46 0.23 0.03 |
| Spain 1992 Senator PRS99/012 | 35EC | Spray | 1.21 | 1150 (0.105% ai) | 1 | 0 3 7 15 | Fruit Fruit Fruit Fruit | | 1.02 0.53 0.05 0.04 |
| Spain 1992 Senator PRS99/012 | 35EC | Spray | 1.37 | 1300 (0.105% ai) | 1 | 0 3 7 15 | Fruit Fruit Fruit Fruit | | 0.11 < 0.01 0.05 0.02 |
| Spain 1992 Diamante PRS99/012 | 35EC | Spray | 1.02 | 970 (0.105% ai) | 1 | 0 3 7 15 | Fruit Fruit Fruit Fruit | | 0.32 0.13 0.02 0.02 |
| USA 1996 NI Supersett BJ96R02 | 3EC 50WP | Broadcast Interval 7d Broadcast Interval 7d | 1.12,1.12 1.23 3x1.12 | 3x96.3 96.3,2x95.4 | 3 3 | 1 1 | Fruit Fruit | 0.08 < 0.05, 0.05 0.11, 0.06, 0.06 | 0.13 0.23 |

| S. SQUASH | Application | | | | | PHI | Sample analysed | Residues of alpha, beta endosulfan, Endosulfan sulfate, mg/kg | Total Residues mg/kg |
|---|---|---|---|---|---|---|---|---|---|
| Country, Year of trial *Variety, report* | Form. | Method | kg ai//ha applic'n | L/ha | No. | | | | |
| USA 1996 NC *Supersett BJ96R02* | 3EC | Broadcast Interval 7d | <u>3x1.12</u> | 2x95.4,96.3 | 3 | <u>2</u> | Fruit | 0.11, 0.05, < 0.05 | <u>0.16</u> |
| | 50WP | Broadcast Interval 7d | <u>3x1.12</u> | 2x97.2, 95.4 | | <u>2</u> | Fruit | 0.12, 0.05, < 0.05 | <u>0.17</u> |
| USA 1996 FL *Early Summer BJ96R02* | 3EC | Broadcast Interval 7d | 1.12,2x1 | 3x92.5 | 3 | 2 | Fruit | < 0.05, < 0.05, < 0.05 | <u>≤ 0.05</u> |
| | 50WP | Broadcast Interval 7d | 1.12 | 91.6,92.5, 91.6 | 3 | 2 | Fruit | 0.05, < 0.05, < 0.05 | <u>0.05</u> |
| USA 1996 MI *Lemon Drop L BJ96R02* | 3EC | Broadcast Interval 7d | <u>1.12</u> | 92.5,96,95 | 3 | <u>2</u> | Fruit | 0.09, < 0.05, < 0.05 | <u>0.09</u> |
| | 50WP | Broadcast Interval 7d | <u>1.12</u> | 92.5,96,94.4 | 3 | <u>2</u> | Fruit | 0.07, < 0.05, < 0.05 | <u>0.07</u> |
| USA 1996 CA *BJ96R02* CA | 3EC | Broadcast Interval 7d | 2x1.12,1. | 87,2x96.3 | 3 | <u>2</u> | Fruit | 0.08, < 0.05, < 0.05 | <u>0.08</u> |
| | 50WP | Broadcast Interval 7d | 2x1.12,1 | 90.8,95.4,96 | 3 | <u>2</u> | Fruit | 0.08, 0.07, < 0.05 | <u>0.15</u> |
| USA 1984 NY *Goldbar 1913 (6)* | 3EC | Spray Interval 7/8 d | 1 | 327 | 5 | 0 | Fruit | 0.142, 0.143, 0.065 | 0.35 |
| | | | | | | 3 | Fruit | 0.034, 0.021, 0.010 | 0.06 |
| | | | | | | 0 co | Fruit | < 0.01, < 0.01, 0.023 | 0.02 |
| | | | | | | 3 co | Fruit | 0.0269, 0.018, 0.022 | 0.07 |
| | 50WP | Spray Interval 7/8 d | 1 | 327 | 5 | 0 | Fruit | 0.129, 0.109, 0.057 | 0.29 |
| | | | | | | 3 | Fruit | 0.064, 0.061, 0.013 | 0.14 |
| USA 1984 CA Whittier, CA *1913 (5)* | 3EC | Spray Interval 7d | 1 | 140 | 6 | 0 | Fruit | 0.504, 0.271, 0.068 | 0.84 |
| | | | | | | 3 | Fruit | 0.098, 0.04, 0.028 | 0.17 |
| | | | | | | 0 co | Fruit | 0.02, < 0.01, < 0.01 | 0.02 |
| | | | | | | 3 co | Fruit | 0.03, < 0.01, 0.0153 | 0.05 |
| | 50WP | Spray Interval 7 d | 1 | 140 | 6 | 0 | Fruit | 1.363, 0.684, 0.068 | 2.12 |
| | | | | | | 3 | Fruit | 0.86, 0.101, 0.018 | 0.98 |
| USA 1996 CA *Black Beauty BJ96R02* | 3EC | Broadcast Interval 7d | <u>3x1.12</u> | 94,95,90 | 3 | <u>2</u> | Fruit | 0.09, 0.07, < 0.05 | <u>0.16</u> |
| | 50WP | Broadcast Interval 7d | <u>3x1.12</u> | 94.4,2x93.5 | 3 | <u>2</u> | Fruit | 0.08, 0.06, < 0.05 | <u>0.14</u> |

Table 59. Endosulfan residues in zucchini from supervised trials in Australia.

| ZUCCHINI | Application | | | | | PHI days | Sample analysed | Residues of alpha, beta endosulfan, endosulfan sulfate, mg/kg | Total Residues, mg/kg |
|---|---|---|---|---|---|---|---|---|---|
| Country, Year of trial *Variety* Report | Form. | Method | g ai/hL | L/ha | No. | | | | |
| Australia 2000 Koraleigh *Regal black N° 1/10/556* | EC | spray | 66.5 | 224 | 4 | 0 | zucchini | 0.078, 0.036, 0.046 | 0.16 |
| | | | | | | 3 | zucchini | 0.030.0.021, 0.039 | <u>0.09</u> |
| | | | | | | 5 | zucchini | 0.023, 0.012,0.032 | 0.07 |
| | | | | | | 7 | zucchini | 0.042,0.014,0.032 | 0.09 |
| Australia 2000 Koraleigh *Regal black N° 1/10/556* | EC | spray | 132 | 224 | 4 | 0 | zucchini | 0.069,0.038,0.019 | 0.13 |
| | | | | | | 3 | zucchini | 0.026, 0.011, 0.025 | 0.07 |
| | | | | | | 5 | zucchini | 0.018, 0.009,0.023 | 0.05 |
| | | | | | | 7 | zucchini | 0.026, 0.005, 0.033 | 0.06 |
| | | | | | | 7 | co | 0.007, < 0.005,< 0.005 | 0.007 |
| Australia 2000 Walkamin *Regal black N° 1/10/556* | EC | spray | 66.5 | 310 | 4 | 0 | zucchini | 0.028, 0.013, 0.038 | 0.08 |
| | | | | | | 3 | zucchini | 0.011, 0.005, 0.039 | <u>0.06</u> |
| | | | | | | 5 | zucchini | 0.011, < 0.005, 0.034 | 0.05 |
| | | | | | | 7 | zucchini | 0.005, < 0.005, 0.032 | 0.04 |
| | | | | | | 0 | co | 0.12 | |

| ZUCCHINI | | Application | | | | PHI | Sample | Residues of alpha, | Total |
| Country, Year of trial *Variety* Report | Form. | Method | g ai/hL | L/ha | No. | days | analysed | beta endosulfan, endosulfan sulfate, mg/kg | Residues, mg/kg |
|---|---|---|---|---|---|---|---|---|---|
| Australia 2000 Walkamin *Regal black* N° 1/10/556 | EC | spray | 132 | 488 | 4 | 0 | zucchini | 0.11, 0.083, 0.046 | 0.24 |
| | | | | | | 3 | zucchini | 0.006, 0.005, 0.019 | 0.03 |
| | | | | | | 5 | zucchini | 0.008, 0.005, 0.019 | 0.03 |
| | | | | | | 7 | zucchini | 0.008, 0.008, 0.019 | 0.04 |
| | | | | | | 7 | co | 0.018,, 0.005,0.095 | 0.12 |
| Australia 2000 Malanda *Gold finger* N° 1/10/556 | EC | spray | 66.5 | 266 | 4 | 0 | zucchini | 0.085, 0.035, 0.12 | 0.24 |
| | | | | | | 3 | zucchini | 0.021,0.007, 0.059 | 0.09 |
| | | | | | | 5 | zucchini | 0.013, < 0.005, 0.046 | 0.06 |
| | | | | | | 7 | zucchini | 0.012,< 0.005, 0.068 | 0.09 |
| Australia 2000 Malanda *Gold finger* N° 1/10/556 | EC | spray | 132 | 518 | 4 | 7 | zucchini | 0.016, < 0.005, 0.058 | 0.08 |
| | | | | | | 7 | co | < 0.005, < 0.005, 0.069 | 0.07c |
| Australia 2000 Wattleup *Regal black* N° 1/10/556 | EC | spray | 66.5 | 260 | 4 | 0 | zucchini | 0.17, 0.10, 0.015 | 0.28 |
| | | | | | | 3 | zucchini | 0.028,0.009, 0.012 | 0.05 |
| | | | | | | 5 | zucchini | 0.021, 0.005,0.012 | 0.04 |
| | | | | | | 7 | zucchini | 0.015,< 0.005, 0.01 | 0.03 |
| | | | | | | 7 | co | < 0.005,< 0.005,< 0.005 | < 0.005 |
| Australia 2000 Wattleup *Regal black* N° 1/10/556 | EC | spray | 132 | 260 | 4 | 7 | zucchini | 0.019, < 0.005, 0.011 | 0.03 |
| | | | | | | 7 | co | < 0.005,< 0.005,< 0.005 | < 0.005 |

Table 60. Endosulfan residues in peppers from supervised trials in Spain and USA.

| PEPPERS | | Application | | | | PHI | Sample | Residues of alpha, | Total |
| Country, Year of trial *Variety* report | Form | Method | kg ai/ha /applic'n | L/ha | No. | (Days) | analysed | +Beta endosulfan,/ Endosulfan sulfate, mg/kg | Residues, mg/kg |
|---|---|---|---|---|---|---|---|---|---|
| Spain 2000 *Dallas* R-11726 | 35EC | Spray indoor Interval 7 d BBCH 71, 73, 74 | 1.058, 1.109, 1.069 | 1002 1050 1012 (0.11% ai) | 3 | 14 | Red Pepper | 0.15,0.02 | 0.17 |
| | | | | | | 21 | Red Pepper | 0.05,0.02 | 0.07 |
| Spain 2000 *Genil* R-11726 | 35EC | Spray indoor Interval 7 d BBCH 71, 71 , 73 | 1.003, 1.062, 1.064 | 950 1006 1008 (0.11% ai) | 3 | 14 | Green Pepper | 0.37,0.03 | 0.40 |
| | | | | | | 21 | Green Pepper | 0.31,0.05 | 0.36 |
| Spain 2000 *Turia* R-11726 | 35EC | Spray indoor) Interval 8, 6 d BBCH 71, 72, 82 | 1.035, 1.073, 1.075 | 980 1016 1018 (0.11% ai) | 3 | 0 | Green Pepper | | 1.82 |
| | | | | | | 3 | Green Pepper | | 1.83 |
| | | | | | | 7 | Green Pepper | | 1.25 |
| | | | | | | 14 | Green Pepper | | 1.15 |
| | | | | | | 21 | Green Pepper | | 0.51 |
| Spain 2000 *Italic* R-11726 | 35EC | Spray indoor Spray interval 7 days BBCH 71, 73, 81 | 1.014, 1.060, 1.024 | 960 1004 970 (0.11% ai) | 3 | 0 | Green Pepper | 0.56,0.10 | 0.66 |
| | | | | | | 3 | Green Pepper | 0.34,0.09 | 0.43 |
| | | | | | | 7 | Green Pepper | 0.26,0.24 | 0.50 |
| | | | | | | 14 | Green Pepper | 0.12,0.14 | 0.26 |
| | | | | | | 21 | Green Pepper | 0.03,0.09 | 0.12 |

| PEPPERS | Application | | | | | PHI | Sample analysed | Residues of alpha, +Beta endosulfan,/ Endosulfan sulfate, mg/kg | Total Residues, mg/kg |
|---|---|---|---|---|---|---|---|---|---|
| Country, Year of trial *Variety* report | Form | Method | kg ai/ha /applic'n | L/ha | No. | (Days) | | | |
| Spain 1999 *Turia* R-11724 | 35EC | Spray Interval 7d BBCH 71, 73, 74 | 1.086, 1.096, 1.109 | 1012 1038 1050 (0.11% ai) | 3 | 14 21 | G Pepper G Pepper | 0.01,0.01 | 0.02 < 0.02 |
| Spain 1999 *Blanco* R-11724 | 35EC | Spray Interval 7d BBCH 72, 72, 73 | 1.030, 1.050, 1.065 | 976 995 1009 (0.11% ai) | 3 | 14 21 | Pepper Pepper | < 0.01,< 0.01 < 0.01,< 0.01 | < 0.01 < 0.01 |
| Spain 1999 *Estilo* R-11724 | 35EC | Spray Interval 7d BBCH 71, 69/72, 69/ 73 | 1.087, 1.090, 1.107 | 1029 1033 1048 (0.11% ai) | 3 | 0 4 7 14 21 | Green Pepper Green Pepper Green Pepper Green Pepper Green Pepper | 0.23,0.02 0.09,0.03 0.04,0.04 < 0.01,0.02 < 0.01,0.02 | 0.25 0.12 0.08 0.02 0.02 |
| Spain 1999 *La Canal* R-11724 | 35EC | Spray Interval 7d 73 | 1.048, 1.113, 1.082 | 992 1054 1025 (0.11% ai) | 3 | 0 4 7 14 21 | Green Pepper Green Pepper Green Pepper Green Pepper Green Pepper | 1.66,0.07 0.14,0.05 0.07,0.05 0.02,0.04 0.01, < 0.01 | 1.73 0.19 0.12 0.06 0.01 |
| Spain 2000 *Barbadillo* R-11725 | 35EC | Spray indoor Interval 7d BBCH 71, 71, 72 | 1.039, 1.035, 1.056 | 984 980 1000 (0.11% ai) | 3 | 14 21 | Red Pepper Red Pepper | 0.41,0.08 0.14,0.06 | 0.49 0.20 |
| Spain 2000 *Turia* R-11725 | 35EC | Spray indoor Interval 8,6 d BBCH 71, 71, 72 | 1.068, 1.044, 1.118 | 1012 988 1059 (0.11% ai) | 3 | 0 3 7 14 21 | Pepper Pepper Pepper Pepper Pepper | 0.72,0.06 0.12,0.06 0.04,0.05 0.02,0.03 < 0.01,0.03 | 0.78 0.18 0.09 0.05 0.03 |
| Spain 2000 *Mariner* R-11725 | 35EC | Spray indoor Interval 7d BBCH 72, 73, 81 | 1.076, 1.043, 1.066 | 1019 988 1010 (0.11% ai) | 3 | 14 21 | Red Pepper Red Pepper | 0.20.0.07 0.12,0.05 | 0.27 0.17 |
| Spain 2000 *Teide* R-11725 | 35EC | Spray indoor Interval 7d BBCH 71, 72, 82 | 0.850, 0.830, 0.850 | 805 786 805 (0.11% ai) | 3 | 0 3 7 14 21 | Green Pepper Green Pepper Green Pepper Green Pepper Green Pepper | 0.88,0.06 0.57,0.18 0.20.0.07 0.10.0.09 0.02,0.04 | 0.94 0.75 0.27 0.19 0.06 |
| USA 1966 Great Northern A48560 | 2EC | Interval 7 d | 1.12 | 567 (0.2% ai) | 3 | 0 2 7 | Bell Pepper Bell Pepper Bell Pepper | | 0.47* 0.05 0.02 |
| | 50WP | Interval 7 d | 1.12 | 567 (0.2% ai) | 3 | 0 2 7 | Bell Pepper Bell Pepper Bell Pepper | | 0.97 0.22 0.02 |
| | 3% Dust | Interval 7 d | 1.68 | | 3 | 0 2 7 | Bell Pepper Bell Pepper Bell Pepper | | 0.88 0.44 0.02 |
| | 2EC | Interval 7d | 1.12 | 567 (0.2% ai) | 3 | 0 2 7 | Green Pepper Green Pepper Green Pepper | | 3.15 3.30 0.65 |

Table 61. Endosulfan residues in peppers from supervised trials in Australia.

| PEPPERS Country, Year of trial *Variety* Report | Form. | Method | Kg ai/ha | L/ha | No. | PHI days | Sample analysed | Residues of alpha, beta endosulfan, Endosulfan sulfate, mg/kg | Total Residues, mg/kg |
|---|---|---|---|---|---|---|---|---|---|
| Australia 2000 Emerald Creek *Merlin* N° 1/10/559 | EC | spray | 0.735 | 500 | 3 | 0 3 7 14 | | 0.076, 0.098, 0.012 0.056, 0.087, 0.012 0.058, 0.091,0.018 0.066,0.11,0.18 | 0.19 0.16 0.17 0.36 |
| Australia 2000 Emerald Creek *Merlin* N° 1/10/559 | EC | spray | 1.47 | 500 | 3 | 0 3 7 14 | | 0.21,0.29,0.018 0.065, 0.10, 0.013 0.18, 0.27,0.048 0.065, 0.14, 0.032 | 0.52 0.18 0.50 0.24 |
| Australia 2000 Gumlu *Airies* N° 1/10/559 | EC | spray | 0.735 | 5084 | 3 | 7 14 | | 0.02, 0.053, 0.016 0.019, 0.039, 0.016 | 0.09 0.07 |
| Australia 2000 Shepparton *Target* N° 1/10/559 | EC | spray | 0.735 | 452 | 3 | 7 14 | | 0.009, 0.018, 0.010 0.006, 0.014, 0.007 | 0.04 0.03 |
| Australia 2000 Virginia *Yaspo* N° 1/10/559 | EC | spray | 0.735 | 500 | 3 | 0 3 7 14 | | 0.36, 0.39,0.13 0.054, 0.13, 0.22 0.013, 0.023, 0.039 < 0.005, < 0.005, 0.006 | 0.88 0.40 0.08 0.02 |
| Australia 2000 Virginia *Yaspo* N° 1/10/559 | EC | spray | 1.47 | 500 | 3 | 0 3 7 14 | | 0.32, 0.30, 0.061 0.037, 0.10, 0.13 < 0.005, 0.013, 0.050 0.005, 0.014, 0.015 | 0.68 0.27 0.07 0.03 |

Table 62. Endosulfan residues in tomatoes from supervised field trials in Europe, USA and Australia.

| TOMATO Country, Year of trial *Variety* report | Form. | Method | kg ai/ha/ applic'n | L/ha | No. | PHI (Days) | Sample analysed | Residues of alpha, beta endosulfan, endosulfan sulfate mg/kg | Total Residues mg/kg |
|---|---|---|---|---|---|---|---|---|---|
| Germany 1989 *Rheinglut* PSR99/012 | 35EC | Spray | 0.21 0.28 0.42 0.42 (0.035 % ai) | 600 800 1200 1200 | 4 | 0 7 7 7 7 7 7 | Fruit Fruit Washings Cooking water Cooked Fruit Puree Juice | 0.40, 0.19, < 0.01 0.01, 0.02, < 0.01 < 0.01, < 0.01, < 0.01 < 0.01, < 0.01, < 0.01 0.01, 0.02, < 0.01 < 0.01, < 0.01, < 0.01 < 0.01, < 0.01, < 0.01 | 0.59 0.03 < 0.01 < 0.01 0.03 < 0.01 < 0.01 |
| Germany 1989 *Hellfrucht* PSR99/012 | 35EC | Spray | 0.21 0.235 0.308 0.42 (0.035 % ai) | 600 675 880 1200 | 4 | 0 7 7 7 7 7 7 | Fruit Fruit Washings Cooking water Cooked Fruit Puree Juice | 0.48, 0.23, < 0.01 0.04, 0.05, < 0.01 < 0.01, < 0.01, < 0.01 < 0.01, < 0.01, < 0.01 0.05, 0.04, < 0.01 < 0.01, < 0.01, < 0.01 < 0.01, < 0.01, < 0.01 | 0.71 0.09 < 0.01 < 0.01 0.09 < 0.01 < 0.01 |

| TOMATO | Application | | | | | PHI | Sample | Residues of alpha, beta | Total |
| Country, Year of trial *Variety* report | Form. | Method | kg ai/ha/ applic'n | L/ha | No. | (Days) | analysed | endosulfan, endosulfan sulfate mg/kg | Residues mg/kg |
|---|---|---|---|---|---|---|---|---|---|
| Germany 1983 *Moneymaker* DEU8317 2911 | 2.82% Dust | Spread interval 14 d | 0.705 | 25 Kg product | 3 | 0 | Fruit | 0.20, 0.20, 0.01 | 0.41 |
| | | | | | | 5 | Fruit | 0.10, 0.10, 0.01 | 0.21 |
| | | | | | | 10 | Fruit | 0.06, 0.06, < 0.01 | 0.12 |
| | | | | | | 14 | Fruit | 0.03, 0.04, < 0.01 | 0.07 |
| Germany 1983 *Hoffmans Rentita* DEU83172921 | 2.82% Dust | Spread interval 19, 11 days | 0.705 | 25 Kg product | 3 | 0 | Fruit | 0.20, 0.10, < 0.01 | 0.30 |
| | | | | | | 5 | Fruit | 0.10, 0.10, < 0.01 | 0.20 |
| | | | | | | 10 | Fruit | 0.04, 0.06, < 0.01 | 0.10 |
| | | | | | | 14 | Fruit | 0.03, 0.06, 0.01 | 0.10 |
| Germany 1983 *Hoffmans Rentita* DEU83172941 | 2.82% Dust | Spread interval 14 days | 0.705 | 25 Kg product | 3 | 0 | Fruit | 0.40, 0.30, < 0.01 | 0.70 |
| | | | | | | 5 | Fruit | 0.10, 0.10, < 0.01 | 0.20 |
| | | | | | | 10 | Fruit | 0.06, 0.06, < 0.01 | 0.12 |
| | | | | | | 14 | Fruit | 0.06, 0.06, < 0.01 | 0.12 |
| Germany 1976 *Rheinlands Ruhm* LEA 3/67/01/02-76 | 24EC* | Spray, 0.15% Spray interval 18 days | 0.216 | 600 (0.036% ai) | 2 | 0 | Fruit | | 1.1 |
| | | | | | | 7 | Fruit | | 0.02 |
| | | | | | | 10 | Fruit | | 0.02 |
| | | | | | | 14 | Fruit | | 0.03 |
| Greece2002 *Titane* MR-510/02 C030836 | CS* | Spray. Interval 13d BBCH 87, 88 | 1.0603 | 500 (0.212% ai) | 2 | 3 | Fruit | 0.71, 0.35, ND, < 0.02(endosulfan diol) | 1.06 |
| Italy2002 *Locale di Molfetta* MR-510/02 C030836 | CS* | Spray. Interval 14d BBCH 84, 88 | 1.0603 | 500 (0.212% ai) | 2 | 3 | Fruit | 0.06, 0.04, ND, < 0.02(endosulfan diol) | 0.10 |
| Greece 2002 *Titano* 02 R 171 | 35EC | Spray. Interval 11 d BBCH 87, 88 | 0.5298 | 500 (0.1065% ai) | 2 | 0 | Fruit | 0.66, 0.34, < 0.02 | 1.00 |
| | | | | | | 2 | Fruit | 0.59, 0.40, < 0.02 | 0.99 |
| | | | | | | 3 | Fruit | 0.35, 0.32, < 0.02 | 0.67 |
| | | | | | | 7 | Fruit | 0.39, 0.35, 0.045 | 0.79 |
| Italy 2002 02 R 171 | 35EC | Spray. Interval 12d BBCH 84, 88 | 0.5298 | 500 (0.1065% ai) | 2 | 0 | Fruit | 0.095, 0.093, < 0.02 | 0.19 |
| | | | | | | 2 | Fruit | 0.017, 0.031, < 0.02 | 0.05 |
| | | | | | | 3 | Fruit | < 0.02, 0.03, < 0.02 | 0.03 |
| | | | | | | 7 | Fruit | < 0.02, 0.021, < 0.02 | 0.02 |
| Italy 2002 *PS 1296* 02 R 171 | 35EC | Spray Interval 12 d BBCH 83, 88 | 0.5298 | 500 (0.1065% ai) | 2 | 0 | Fruit | 0.087, 0.15, < 0.02 | 0.24 |
| | | | | | | 2 | Fruit | 0.089, 0.088, < 0.02 | 0.18 |
| | | | | | | 3 | Fruit | 0.048, 0.081, < 0.02 | 0.13 |
| | | | | | | 7 | Fruit | < 0.02, 0.055, < 0.02 | 0.06 |
| Spain2001 Optima 02F002 | CS* | Spray. Interval 14d BBCH 76, 86 | 0.800 | 1500 (0.0535% ai) | 2 | 3 | Single Fruit** | 0.06, 0.06, 0.01 | 0.13 |
| France2001 Felicia 02F002 | CS* | Spray. Interval 14 d BBCH 81, 81 | 0.900, 0.800 | 1688, 1500 (0.0535% ai) | 2 | 3 | Single Fruit | 0.19, 0.12, 0.01 | 0.32 |
| Greece2001 Alma 02F002 | CS* | Spray. Interval 14d BBCH 81, 87 | 0.800 | 1500 (0.0535% ai) | 2 | 3 | Single Fruit | 0.07, 0.05, 0.01 | 0.13 |

| TOMATO Country, Year of trial Variety report | Form. | Method | kg ai/ha/ applic'n | L/ha | No. | PHI (Days) | Sample analysed | Residues of alpha, beta endosulfan, endosulfan sulfate mg/kg | Total Residues mg/kg |
|---|---|---|---|---|---|---|---|---|---|
| Italy2001 Naxos 02F002 | CS* | Spray. Interval 14 d BBCH 77, 82 | 0.800 | 1500 (0.0535% ai) | 2 | 3 | Single Fruit | 0.17, 0.12, 0.01 | 0.30 |
| Spain1999 Inca ER 99 ECS 752 | CS* | Spray. Interval 13d BBCH 87, 89 | 0.530 | 300 (0.177% ai) | 2 | 0 3 | Fruit Fruit | 0.34, 0.20, ND 0.26, 0.17, ND | 0.54 0.43 |
| Greece1999 Rio Grande ER 99 ECS 752 | CS* | Spray. Interval 13d BBCH 73, 88 | 0.530 | 500 (0.106% ai) | 2 | 0 3 | Fruit Fruit | 0.13, 0.09, ND 0.13, 0.08, ND | 0.22 0.21 |
| Greece1999 Titan ER 99 ECS 752 | CS* | Spray. Interval 14d BBCH 86, 89 | 0.530 | 500 (0.106% ai) | 2 | 0 3 | Fruit Fruit | 0.16, 0.11, ND 0.09, 0.06, ND | 0.27 0.15 |
| Italy1999 PS 1296 ER 99 ECS 752 | CS* | Spray. Interval 15d BBCH 79, 88 | 0.530 | 500 (0.106% ai) | 2 | 0 3 | Fruit Fruit | 0.19, 0.11, ND 0.18, 0.10, ND | 0.30 0.28 |
| Portugal1999 H9280 F1 ER 99 ECS 752 | CS* | Spray Interval 18d BBCH 79, 83 | 0.530 | 400 (0.133% ai) | 2 | 0 3 | Fruit Fruit | 0.11, 0.07, ND 0.08, 0.05, ND | 0.18 0.13 |
| Spain 1998 Inca ER 98 ECS 752 | CS* | Spray. Interval 14d BBCH 81, 88 | 0.530 | 300 (0.177% ai) | 2 | 0 1 3 7 | Fruit Fruit Fruit Fruit | 0.22, 0.13, ND 0.17, 0.11, ND 0.12, 0.07, ND 0.10, 0.07, ND | 0.35 0.28 0.19 0.17 |
| Greece 1998 Rio Grande ER 98 ECS 752 | CS* | Spray. Interval 14d BBCH 86, 88 | 0.530 | 500 (0.177% ai) | 2 | 0 1 3 7 | Fruit Fruit Fruit Fruit | 0.08, 0.06, ND 0.06, 0.04, ND 0.13, 0.09, < 0.02 0.07, 0.05, ND | 0.14 0.10 0.22 0.12 |
| Greece 1998 Rio Grande ER 98 ECS 752 | CS* | Spray. Interval 14d BBCH 86, 88 | 0.530 | 500 (0.177% ai) | 2 | 0 1 3 7 | Fruit Fruit Fruit Fruit | 0.12, 0.07, ND 0.07, 0.06, < 0.02 0.14, 0.10, < 0.02 0.04, 0.04, ND | 0.19 0.13 0.24 0.08 |
| Italy 1998 Hypeel-244 ER 98 ECS 752 | CS* | Spray Interval 14 d BBCH 72, 85 | 0.530 | 500 (0.177% ai) | 2 | 0 1 3 7 | Fruit Fruit Fruit Fruit | 0.12, 0.07, ND 0.08, 0.05, ND 0.08, 0.05, ND 0.05, 0.02, ND | 0.19 0.13 0.13 0.07 |
| Portugal 1998 Stromboli F1 ER 98 ECS 752 | CS* | Spray. Interval 14d BBCH 81, 86 | 0.530 | 400 (0.133% ai) | 2 | 0 1 3 7 | Fruit Fruit Fruit Fruit | 0.36, 0.18, ND 0.12, 0.08, ND 0.09, 0.06, ND 0.06, 0.04, ND | 0.54 0.20 0.15 0.10 |
| Spain 1994 Red Zetor ER 94 ECS 700b | 35EC | Spray. Interval 14d GS 17, 19 | 0.264 | 350 (0.075% ai) | 2 | 0 3 7 14 20 27 | Fruit Fruit Fruit Fruit Fruit Fruit | 0.05, 0.04, < 0.01 0.03, 0.03, < 0.01 0.03, 0.04, < 0.01 < 0.01, < 0.01, < 0.01 < 0.01, < 0.01, < 0.01 < 0.01, < 0.01, < 0.01 | 0.09 0.06 0.07 < 0.01 < 0.01 < 0.01 |

| TOMATO | Application | | | | | PHI (Days) | Sample analysed | Residues of alpha, beta endosulfan, endosulfan sulfate mg/kg | Total Residues mg/kg |
|---|---|---|---|---|---|---|---|---|---|
| Country, Year of trial *Variety* *report* | Form. | Method | kg ai/ha/ applic'n | L/ha | No. | | | | |
| Spain 1994 *Red Zetor* *ER 94 ECS 700b* | 35EC | Spray Interval 14d GS 17, 19 | 0.528 | 350 (0.15% ai) | 2 | 0 | Fruit | 0.14, 0.13, < 0.01 | 0.27 |
| | | | | | | 3 | Fruit | 0.06, 0.05, < 0.01 | 0.11 |
| | | | | | | 6 | Cann. Liq. | < 0.01, < 0.01, < 0.01 | < 0.01 |
| | | | | | | 6 | Frt unwshd | 0.04, 0.04, < 0.01 | 0.08 |
| | | | | | | 6 | Frt washed | 0.04, 0.04, < 0.01 | 0.08 |
| | | | | | | 6 | Frt proc'd | 0.04, 0.04, < 0.01 | 0.08 |
| | | | | | | 6 | juice | < 0.01, < 0.01, < 0.01 | < 0.01 |
| | | | | | | 6 | Frt presv'd | 0.34, 0.24, 0.03 | 0.61 |
| | | | | | | 6 | Wash water | < 0.01, < 0.01, < 0.01 | < 0.01 |
| | | | | | | 7 | Fruit | 0.04, 0.04, < 0.01 | 0.08 |
| | | | | | | 14 | Fruit | 0.02, 0.02, < 0.01 | 0.04 |
| | | | | | | 20 | Fruit | < 0.01, < 0.01, < 0.01 | < 0.01 |
| | | | | | | 27 | Fruit | < 0.01< 0.01, < 0.01 | < 0.01 |
| Spain 1994 *Pluton* *ER 94 ECS 700b* | 35EC | Spray Interval 14d GS 17/19, 21 | 0.264 | 350 (0.075 % ai) | 2 | 0 | Fruit | 0.04, 0.04, < 0.01 | 0.08 |
| | | | | | | 3 | Fruit | < 0.01, 0.01, < 0.01 | 0.01 |
| | | | | | | 7 | Fruit | 0.01, 0.01, < 0.01 | 0.02 |
| | | | | | | 14 | Fruit | < 0.01, < 0.01, < 0.01 | < 0.01 |
| Spain 1994 *Petto 95* *ER 94 ECS 700b* | 35EC | Spray Interval 14d GS 17/19, 19 | 0.264 | 350 (0.075 % ai) | 2 | 0 | Fruit | 0.07, 0.06, < 0.01 | 0.13 |
| | | | | | | 3 | Fruit | 0.01, 0.02, < 0.01 | 0.03 |
| | | | | | | 8 | Fruit | < 0.01, 0.01, < 0.01 | < 0.01 |
| | | | | | | 14 | Fruit | < 0.01, < 0.01, < 0.01 | < 0.01 |
| | | | | | | 21 | Fruit | < 0.01, < 0.01,< 0.01 | < 0.01 |
| | | | | | | 28 | Fruit | < 0.01, < 0.01, < 0.01 | < 0.01 |
| Spain 1994 *Petto 95* *ER 94 ECS 700b* | 35EC | Spray Interval 14 days GS 17/19, 19 | 0.528 | 350 (0.15 % ai) | 2 | 0 | Fruit | 0.11, 0.06, < 0.01 | 0.17 |
| | | | | | | 3 | Fruit | 0.03, 0.04, < 0.01 | 0.07 |
| | | | | | | 8 | Fruit | 0.01, 0.02, < 0.01 | 0.03 |
| | | | | | | 14 | Fruit | < 0.01, < 0.01, < 0.01 | < 0.01 |
| | | | | | | 21 | Fruit | < 0.01, < 0.01, < 0.01 | < 0.01 |
| | | | | | | 28 | Fruit | < 0.01, < 0.01, < 0.01 | < 0.01 |
| Italy 1994 *Loni* *ER 94 ECS 700* | 35EC | Spray. Interval 14 d GS17/19, 17/19 | 0.264 | 1000 (0.0264 % ai) | 2 | 0 | Fruit | 0.01, 0.02, < 0.01 | 0.03 |
| | | | | | | 3 | Fruit | < 0.01, < 0.01, < 0.01 | < 0.01 |
| | | | | | | 7 | Fruit | < 0.01, < 0.01, < 0.01 | < 0.01 |
| | | | | | | 14 | Fruit | < 0.01, < 0.01, < 0.01 | < 0.01 |
| | | | | | | 21 | Fruit | < 0.01, < 0.01, < 0.01 | < 0.01 |
| | | | | | | 29 | Fruit | < 0.01, < 0.01, < 0.01 | < 0.01 |
| Italy 1994 *U.C. 82* *ER 94 ECS700b* | 35EC | Spray. Spray Interval 14d GS 15/17, 15/19 | 0.264 | 1200 (0.022 % ai) | 2 | 0 | Fruit | 0.03, 0.03, < 0.01 | 0.06 |
| | | | | | | 3 | Fruit | 0.027, 0.03, < 0.01 | 0.06 |
| | | | | | | 7 | Fruit | 0.025, 0.025, < 0.01 | 0.05 |
| | | | | | | 14 | Fruit | 0.01, 0.02, 0.02 | 0.05 |
| | | | | | | 21 | Fruit | < 0.01, 0.01, < 0.01 | < 0.01 |
| | | | | | | 28 | Fruit | < 0.01, < 0.01, < 0.01 | < 0.01 |
| Spain 1994 *Pluton* *ER 94 ECS 700b* | 35EC | Spray. Interval 14d GS 17/19, 21 | 0.528 | 350 (0.15 % ai) | 2 | 0 | Fruit | 0.21, 0.15, < 0.01 | 0.36 |
| | | | | | | 3 | Fruit | 0.02, 0.03, < 0.01 | 0.05 |
| | | | | | | 7 | Fruit | 0.01, 0.03, < 0.01 | 0.04 |
| | | | | | | 14 | Fruit | 0.01, 0.02, < 0.01 | 0.03 |

| TOMATO | Application | | | | | PHI | Sample | Residues of alpha, beta | Total |
| Country, Year of trial *Variety* *report* | Form. | Method | kg ai/ha/ applic'n | L/ha | No. | (Days) | analysed | endosulfan, endosulfan sulfate mg/kg | Residues mg/kg |
|---|---|---|---|---|---|---|---|---|---|
| Italy<br><br>1994<br>Emiglia<br>Romagna<br>*U.C. 82*<br>*ER 94 ECS700b* | 35EC | Spray. Spray Interval 14 days GS 15/17, 15/19 | <u>0.528</u> | 1200<br><br>(0.044<br>% ai) | 2 | 0 | Fruit | 0.17, 0.12, < 0.01 | 0.29 |
| | | | | | | 3 | Fruit | 0.04, 0.05, 0.01 | <u>0.10</u> |
| | | | | | | 7 | Cann liq | < 0.01, < 0.01, < 0.01 | < 0.01 |
| | | | | | | 7 | Frt unwshd | 0.03, 0.03, < 0.01 | 0.06 |
| | | | | | | 7 | Frt washed | 0.03, 0.03, < 0.01 | 0.06 |
| | | | | | | 7 | Frt presv'd | 0.033, 0.033, < 0.01 | 0.07 |
| | | | | | | 7 | juice | < 0.01, < 0.01, < 0.01 | < 0.01 |
| | | | | | | 7 | pomace | 0.15, 0.12, 0.02 | 0.29 |
| | | | | | | 7 | Wash water | < 0.01, < 0.01, < 0.01 | < 0.01 |
| | | | | | | 7 | Fruit | 0.03, 0.04, 0.01 | 0.08 |
| | | | | | | 14 | Fruit | 0.03, 0.03, 0.02 | 0.08 |
| | | | | | | 20 | Fruit | 0.02, 0.02, < 0.01 | 0.04 |
| | | | | | | 28 | Fruit | 0.01, 0.02, < 0.01 | 0.03 |
| Spain<br>1993<br>*Ipanema*<br>*ER 93 ECS700* | 35EC | Spray, (sprayer) hand held Interval 14d GS 17, 19 | 0.2642 | 500<br><u>(0.053% ai)</u> | 2 | 0 | Fruit | 0.10, 0.08, < 0.01 | 0.18 |
| | | | | | | <u>3</u> | Fruit | 0.03, 0.04, < 0.01 | 0.07 |
| | | | | | | 7 | Fruit | 0.02, 0.02, < 0.01 | 0.04 |
| | | | | | | 14 | Fruit | 0.01, < 0.01< 0.01 | 0.01 |
| | | | | | | 14 | Cann liq | < 0.01, < 0.01, < 0.01 | < 0.01 |
| | | | | | | 14 | Frt unwshd | < 0.01, 0.01, < 0.01 | < 0.01 |
| | | | | | | 14 | Frt washed | < 0.01, < 0.01, < 0.01 | < 0.01 |
| | | | | | | 14 | Frt presv'd | < 0.01, 0.01, < 0.01 | 0.01 |
| | | | | | | 14 | Juice | < 0.01, < 0.01, < 0.01 | < 0.01 |
| | | | | | | 14 | Paste | < 0.01, 0.05, < 0.01 | 0.05 |
| | | | | | | 14 | Pomace | 0.04, < 0.01, 0.01 | 0.04 |
| | | | | | | 14 | Wash water | < 0.01, < 0.01, < 0.01 | < 0.01 |
| Spain<br>1993<br>*Ipanema*<br>*ER 93 ECS700* | 35EC | Spray (sprayer, Hand held) Interval 14d GS 17, 19 | <u>0.528</u> | 500<br>(0.1056<br>% ai) | 2 | 0 | Fruit | 0.15, 0.10, < 0.01 | 0.25 |
| | | | | | | <u>3</u> | Fruit | 0.10, 0.09, < 0.01 | <u>0.19</u> |
| | | | | | | 7 | Fruit | 0.02, 0.025, < 0.01 | 0.05 |
| | | | | | | 14 | Fruit | 0.02, 0.02, < 0.01 | 0.04 |
| | | | | | | 14 | Cann liq | < 0.01, < 0.01, < 0.01 | < 0.01 |
| | | | | | | 14 | Frt unwshd | 0.03, 0.03, < 0.01 | 0.06 |
| | | | | | | 14 | Frt washed | 0.01, 0.02, < 0.01 | 0.03 |
| | | | | | | 14 | Frt presv'd | 0.01, 0.01, < 0.01 | 0.02 |
| | | | | | | 14 | Juice | < 0.01, < 0.01, < 0.01 | < 0.01 |
| | | | | | | 14 | Paste | < 0.01, 0.01, < 0.01 | 0.01 |
| | | | | | | 14 | Pomace | 0.08, 0.09, 0.03 | 0.20 |
| | | | | | | 14 | Wash water | < 0.01, < 0.01, < 0.01 | < 0.01 |
| Spain<br>1993<br>*Justar*<br>*ER 93 ECS700* | 35EC | Spray (sprayer, hand held) Interval 14d GS 21, 21 | 0.2642 | 500<br>(0.0528% ai) | 2 | 0 | Fruit | 0.09, 0.09, < 0.01 | 0.18 |
| | | | | | | <u>3</u> | Fruit | 0.03, 0.03, < 0.01 | 0.06 |
| | | | | | | 7 | Fruit | 0.03, 0.03, < 0.01 | 0.06 |
| | | | | | | 14 | Fruit | 0.015, 0.015, < 0.01 | 0.03 |
| | | | | | | 14 | Cann liq | < 0.01, < 0.01, < 0.01 | < 0.01 |
| | | | | | | 14 | Frt unwshd | 0.02, 0.03, < 0.01 | 0.05 |
| | | | | | | 14 | Frt washed | 0.035, 0.04, < 0.01 | 0.08 |
| | | | | | | 14 | Frt presv'd | 0.01, 0.01, < 0.01 | 0.02 |
| | | | | | | 14 | Juice | < 0.01, < 0.01, < 0.01 | < 0.01 |
| | | | | | | 14 | Paste | < 0.01, < 0.01, < 0.01 | < 0.01 |
| | | | | | | 14 | Wash water | < 0.01, < 0.01, < 0.01 | < 0.01 |
| | | | | | | 14 | Pomace | 0.07, 0.1, 0.02 | 0.19 |

| TOMATO Country, Year of trial *Variety* report | Form. | Application Method | kg ai/ha/ applic'n | L/ha | No. | PHI (Days) | Sample analysed | Residues of alpha, beta endosulfan, endosulfan sulfate mg/kg | Total Residues mg/kg |
|---|---|---|---|---|---|---|---|---|---|
| Spain 1993 *Justar* ER 93 ECS700 | 35EC | Spray (sprayer, hand held) Interval 14d GS 21, 21 | 0.528 | 500 (0.1056% ai) | 2 | 0 | Fruit | 0.24, 0.18, < 0.01 | 0.42 |
| | | | | | | <u>3</u> | Fruit | 0.09, 0.10, < 0.01 | <u>0.19</u> |
| | | | | | | 7 | Fruit | 0.04, 0.05, < 0.01 | 0.09 |
| | | | | | | 14 | Fruit | 0.03, 0.04, < 0.01 | 0.07 |
| | | | | | | 14 | Cann liq | < 0.01, < 0.01, < 0.01 | < 0.01 |
| | | | | | | 14 | Frt unwshd | 0.03, 0.03, < 0.01 | 0.06 |
| | | | | | | 14 | Frt washed | 0.02, 0.03, < 0.01 | 0.05 |
| | | | | | | 14 | Frt presv'd | 0.01, 0.02, < 0.01 | 0.03 |
| | | | | | | 14 | Juice | < 0.01, < 0.01, < 0.01 | < 0.01 |
| | | | | | | 14 | Paste | < 0.01, 0.02, < 0.01 | 0.02 |
| | | | | | | 14 | Pomace | 0.14, 0.19, 0.015 | 0.35 |
| | | | | | | 14 | Wash water | < 0.01, < 0.01, < 0.01 | < 0.01 |
| Italy 1993 Foggia *Marcoro* ER 93 ECS700 | 35EC | Spray (sprayer, hand held) Interval 14d GS 11/17, 17/19 | 0.2642 | 700 (0.037% ai) | 2 | 0 | Fruit | 0.04, 0.05, 0.03 | 0.12 |
| | | | | | | 3 | Fruit | < 0.01, < 0.01, < 0.01 | < 0.01 |
| | | | | | | 7 | Fruit | < 0.01, < 0.01, < 0.01 | < 0.01 |
| | | | | | | 14 | Fruit | < 0.01, < 0.01, < 0.01 | < 0.01 |
| | | | | | | 14 | Cann liq | < 0.01, < 0.01, < 0.01 | < 0.01 |
| | | | | | | 14 | Frt unwshd | < 0.01, < 0.01, < 0.01 | < 0.01 |
| | | | | | | 14 | Frt washed | < 0.01, < 0.01, < 0.01 | < 0.01 |
| | | | | | | 14 | Frt presv'd | < 0.01, < 0.01, < 0.01 | < 0.01 |
| | | | | | | 14 | Juice | < 0.01, < 0.01, < 0.01 | < 0.01 |
| | | | | | | 14 | Paste | < 0.01, < 0.01, < 0.01 | < 0.01 |
| | | | | | | 14 | Pomace | 0.03, 0.03, 0.01 | 0.07 |
| | | | | | | 14 | Wash water | < 0.01, < 0.01, < 0.01 | < 0.01 |
| Italy 1993 Foggia *Marcoro* ER 93 ECS700 | 35EC | Spray (sprayer, hand held) Interval 14d GS 11/17, 17/19 | <u>0.528</u> | 700 (0.0754% ai) | 2 | 0 | Fruit | 0.12, 0.08, < 0.01 | 0.20 |
| | | | | | | <u>3</u> | Fruit | 0.01, 0.02, < 0.01 | <u>0.03</u> |
| | | | | | | 7 | Fruit | < 0.01, < 0.01, < 0.01 | < 0.01 |
| | | | | | | 14 | Fruit | < 0.01, < 0.01< 0.01 | < 0.01 |
| | | | | | | 14 | Cann liq | < 0.01, < 0.01, < 0.01 | < 0.01 |
| | | | | | | 14 | Frt unwshd | < 0.01, 0.01, < 0.01 | 0.01 |
| | | | | | | 14 | Frt washed | < 0.01, 0.01, < 0.01 | 0.01 |
| | | | | | | 14 | Frt presv'd | < 0.01, < 0.01, < 0.01 | < 0.01 |
| | | | | | | 14 | Juice | < 0.01, < 0.01, < 0.01 | < 0.01 |
| | | | | | | 14 | Paste | < 0.01, < 0.01, < 0.01 | < 0.01 |
| | | | | | | 14 | Pomace | 0.05, 0.06, 0.04 | 0.15 |
| | | | | | | 14 | Wash water | < 0.01, < 0.01, < 0.01 | < 0.01 |
| Italy 1993 Emilia Romagna *V.C. 82 B* ER 93 ECS700 | 35EC | Spray (sprayer, hand held) Interval 14d GS 17/19, 19/21 | 0.2642 | 1000 (0.0264% ai) | 2 | 0 | Fruit | 0.11, 0.1, < 0.01 | 0.21 |
| | | | | | | 3 | Fruit | < 0.01, < 0.01, < 0.01 | < 0.01 |
| | | | | | | 7 | Fruit | < 0.01, < 0.01, < 0.01 | < 0.01 |
| | | | | | | 14 | Fruit | < 0.01, < 0.01< 0.01 | < 0.01 |
| | | | | | | 14 | Cann liq | < 0.01, < 0.01, < 0.01 | < 0.01 |
| | | | | | | 14 | Frt unwshd | < 0.01, 0.01, < 0.01 | 0.01 |
| | | | | | | 14 | Frt washed | < 0.01, < 0.01, < 0.01 | < 0.01 |
| | | | | | | 14 | Frt proc'c | < 0.01, 0.01, < 0.01 | 0.01 |
| | | | | | | 14 | Juice | < 0.01, < 0.01, < 0.01 | < 0.01 |
| | | | | | | 14 | Paste | < 0.01, < 0.01, < 0.01 | < 0.01 |
| | | | | | | 14 | Pomace | 0.03, 0.03, < 0.01 | 0.06 |
| | | | | | | 14 | Wash water | < 0.01, < 0.01, < 0.01 | < 0.01 |

| TOMATO Country, Year of trial Variety report | Form. | Method | kg ai/ha/ applic'n | L/ha | No. | PHI (Days) | Sample analysed | Residues of alpha, beta endosulfan, endosulfan sulfate mg/kg | Total Residues mg/kg |
|---|---|---|---|---|---|---|---|---|---|
| Italy 1994 | 35EC | Spray. Interval 14 d | 0.528 | 1000 (0.0528 | 2 | 0 | Fruit | 0.06, 0.06, < 0.01 | 0.12 |
| | | | | | | 3 | Fruit | 0.015, 0.025, < 0.01 | 0.04 |
| Loni ER 94 ECS 700b | | GS 17/19,17/19 | | % ai) | | 7 | Fruit | < 0.01, 0.01, < 0.01 | 0.01 |
| | | | | | | 14 | Fruit | < 0.01, 0.015, < 0.01 | 0.02 |
| | | | | | | 21 | Fruit | < 0.01, < 0.01, < 0.01 | < 0.01 |
| | | | | | | 29 | Fruit | < 0.01, < 0.01, < 0.01 | < 0.01 |
| Italy 1993 Emilia Romagna V.C. 82 B ER 93 ECS700 | 35EC | Spray (sprayer, Hand held) Interval 14d GS 17/19, 19/21 | 0.528 | 1000 (0.0528 % ai) | 2 | 0 | Fruit | 0.13, 0.1, < 0.01 | 0.23 |
| | | | | | | 3 | Fruit | < 0.01, 0.02, < 0.01 | 0.02 |
| | | | | | | 7 | Fruit | 0.0125, 0.015, < 0.01 | 0.03 |
| | | | | | | 14 | Fruit | 0.01, 0.01< 0.01 | 0.01 |
| | | | | | | 14 | Cann liq | < 0.01, < 0.01, < 0.01 | < 0.01 |
| | | | | | | 14 | Frt unwshd | < 0.01, 0.01, < 0.01 | 0.01 |
| | | | | | | 14 | Frt washed | 0.01, 0.01, < 0.01 | 0.02 |
| | | | | | | 14 | Frt presv'd | 0.01, 0.01, < 0.01 | 0.02 |
| | | | | | | 14 | Juice | < 0.01, < 0.01, < 0.01 | < 0.01 |
| | | | | | | 14 | Paste | < 0.01, < 0.01, < 0.01 | < 0.01 |
| | | | | | | 14 | Pomace | 0.06, 0.06, 0.015 | 0.14 |
| | | | | | | 14 | Wash water | < 0.01, < 0.01, < 0.01 | < 0.01 |
| USA 1995 CA Apex 1000 BJ-95R-09 | 3EC* 33.7% | Broadcast Spray, Interval 5d | 5.6 (5x) | 259 258 262 (2.14% ai) | 3 | 2 | Fld frt** Proc frt Puree Paste | 1.09, 1.04, < 0.05 1.09, 1.19, < 0.05 0.65, 0.71, < 0.05 1.26, 1.46, 0.06 | 2.13 2.28 1.36 2.78 |
| USA 1995 CA Apex 1000 BJ-95R-06 | 3EC 50WP | Broadcast Spray, Interval 5d | 1.12 1.12 | 2x258-262 2x258-262 | 3 3 | 2 2 | Fruit Fruit | 0.16, 0.18, < 0.05 0.14, 0.24, < 0.05 | 0.34 0.38 |
| USA1995 CA Roma BJ-95R-06 | 3EC 50WP | Broadcast Spray, Interval 4d | 1.23,1.12 1.12 | 197,189,186 189,184,187 | 3 3 | 2 2 | Fruit Fruit | 0.12, 0.13, < 0.05 0.08, 0.08, < 0.05 | 0.25 0.16 |
| USA1995 CA Rio Grande BJ-95R-06 | 3EC 50WP | Broadcast Spray, Interval 4d | 1.12 1.12 | 476,476,465 471,468,463 | 3 3 | 2 2 | Fruit Fruit | 0.20, 0.13, < 0.05 0.20, 0.25, < 0.05 | 0.33 0.45 |
| USA1995 PA Better Boy BJ-95R-06 | 3EC 50WP | Broadcast Spray, Interval 4d | 1.12 1.12 | 286,315, 305 292, 312, 307 | 3 3 | 2 2 | Fruit Fruit | 0.10, 0.14, < 0.05 0.13, 0.14, < 0.05 | 0.24 0.27 |
| USA1995 FL Heatwave BJ-95R-06 | 3EC 50WP | Broadcast Spray, Interval 4 | 1.12 1,12, 1,1 | 181, 178, 184 181, 153, 158 | 3 3 | 2 2 | Fruit Fruit | 0.10, 0.15, < 0.05 0.11, 0.16, < 0.05 | 0.25 0.27 |
| USA1995 OH Heinz 8813 BJ-95R-06 | 3EC 50WP | Broadcast Spray, Interval 5d | 1.12 1.12 | 178, 175,182 178, 175, 182 | 3 3 | 2 2 | Fruit Fruit | 0.39, 0.46, < 0.05 0.33, 0.33, < 0.05 | 0.85 0.66 |
| USA1995 CA Sureset BJ-95R-06 | 3EC 50WP | Broadcast Spray, Interval 4d | 1.12 1.12 | 321, 327, 325 322, 322,326 | 3 3 | 2 2 | Fruit Fruit | 0.20, 0.22, < 0.05 0.22, 0.23, < 0.05 | 0.42 0.45 |

| TOMATO Country, Year of trial *Variety* report | Form. | Method | Application kg ai/ha/ applic'n | L/ha | No. | PHI (Days) | Sample analysed | Residues of alpha, beta endosulfan, endosulfan sulfate mg/kg | Total Residues mg/kg |
|---|---|---|---|---|---|---|---|---|---|
| USA1995 CA | 3EC | Broadcast | <u>1.12</u> | 186, 187, 186 | 3 | <u>2</u> | Fruit | 0.11, 0.16, < 0.05 | <u>0.27</u> |
| *512* BJ-95R-06 | 50WP | Spray, Interval 5d | <u>1.12</u> | 186, 188, 187 | 3 | <u>2</u> | Fruit | 0.22, 0.25, < 0.05 | <u>0.47</u> |
| USA1995CA | 3EC | Broadcast | <u>1.12</u> | 188,189, 185 | 3 | <u>2</u> | Fruit | 0.15, 0.20, < 0.05 | <u>0.35</u> |
| *512* BJ-95R-06 | 50WP | Spray, Interval 5d | <u>1.12</u> | 188, 186, 185 | 3 | <u>2</u> | Fruit | 0.10, 0.17, < 0.05 | <u>0.27</u> |
| USA1995 CA | 3EC | Broadcast | 1.12,1.12 0.88 | 375, 374, 302 | 3 | <u>2</u> | Fruit | 0.33, 0.40, < 0.05 | <u>0.73</u> |
| *Sunny* BJ-95R-06 | 50WP | Spray, Interva4d | 1.12,1.12 0.88 | 375, 376, 287 | 3 | <u>2</u> | Fruit | 0.37, 0.46, < 0.05 | <u>0.83</u> |
| USA1995 FL *Agroset* | 3EC | Broadcast Spray, | <u>1.12</u> | 785,785,785 | 3 | <u>2</u> | Fruit | < 0.05, < 0.05, < 0.05 | <u>≤ 0.05</u> |
| BJ-95R-06 | 50WP | Interval 4d | <u>1.12</u> | 785,785,785 | 3 | <u>2</u> | Fruit | < 0.05, < 0.05, < 0.05 | <u>≤ 0.05</u> |
| USA1995 FL | 3EC | Broadcast | <u>1.12</u> | 235.6,234, 235.6 | 3 | <u>2</u> | Fruit | < 0.05, < 0.05, < 0.05 | <u>≤ 0.05</u> |
| *Apex 1000* BJ-95R-06 | 50WP | Spray, Interval 4d | <u>1.12</u> | 236, 234, 234.6 | 3 | <u>2</u> | Fruit | < 0.05, < 0.05, < 0.05 | <u>≤ 0.05</u> |
| USA 1984 *Sunny* PSR99/012 PSR99/012 | 3EC | Spray at 0, 7, 14, 47, 54 days Pre-harvest before harvest | 1.12 | 587 | 5 | 1 hr 3 7 14 21 | Fruit Fruit Fruit Fruit Fruit | 0.049, 0.033, < 0.01 0.020, 0.024, < 0.01 < 0.01, < 0.01, < 0.01 0.011, 0.017, < 0.01 0.014, 0.023, < 0.01 | 0.08 0.04 < 0.01 0.03 0.04 |
| USA 1984 FL *Sunny* PSR99/012 | 3EC | Spray at 0, 7, 14, 47, 54 days Pre-harvest | 1.12 | 587 | 5 | 1 hr <u>3</u> 7 14 21 | Fruit Fruit Fruit Fruit Fruit | < 0.01, < 0.01, < 0.01 0.010, 0.016, < 0.010 < 0.01, < 0.01, < 0.01 < 0.01, < 0.01, < 0.01 < 0.01, 0.023, < 0.01 | < 0.01 <u>0.03</u> < 0.01 < 0.01 0.02 |
| USA 1984 FL *Sunny* PSR99/012 | 50WP | Spray at 0, 7, 14, 47, 54 days Pre-harvest | 1.12 | 587 | 5 | 1 hr <u>3</u> 7 14 21 | Fruit Fruit Fruit Fruit Fruit | 0.061, 0.024, < 0.01 0.034, 0.039, < 0.01 0.015, 0.027, < 0.01 < 0.01, < 0.01, < 0.01 < 0.01, < 0.01, < 0.01 | 0.09 <u>0.07</u> 0.04 < 0.01 < 0.01 |
| USA 1984 FL *Sunny* PSR99/012 | 3EC | Spray at 0, 7, 15, 52, 59 d Pre-harvest Seeds and peel Puree 10/11% Puree 16% | 1.12 solids | 587 | 5 | 0.5 hr 0.5 hr | Fruit* Fruit* Ch fr Sds, pl Puree P slds P slds Fruit co Ch fr co Sds, pl co Puree co P slds co P slds co | 0.049, 0.058, < 0.01 0.047, 0.081, < 0.01 0.082, 0.081, < 0.01 0.571, 0.546, 0.047 0.018, 0.019, < 0.01 < 0.01, 0.025, < 0.01 < 0.01, 0.033, < 0.01 < 0.01, < 0.01, < 0.01 < 0.01, < 0.01, < 0.01 0.0172, 0.01, < 0.01 < 0.01, < 0.01, < 0.01 < 0.01, < 0.01, < 0.01 < 0.01, < 0.01, < 0.01 | 0.11 0.13 0.16 1.16 0.04 0.03 0.03 < 0.01 < 0.01 0.03 < 0.01 < 0.01 < 0.01 |

| TOMATO Country, Year of trial *Variety* report | Form. | Method | kg ai/ha/ applic'n | L/ha | No. | PHI (Days) | Sample analysed | Residues of alpha, beta endosulfan, endosulfan sulfate mg/kg | Total Residues mg/kg |
|---|---|---|---|---|---|---|---|---|---|
| USA 1984 FL  *Sunny*  *PSR99/012* | 50WP | Spray Sprays at 0, 7, 14, 47, 54 days Pre-harvest | 1.12 | 587 | 5 | 1 hr | Fruit | 0.074, 0.036, < 0.01 | 0.11 |
|  |  |  |  |  |  | 3 | Fruit | 0.015, 0.014, < 0.01 | 0.03 |
|  |  |  |  |  |  | 7 | Fruit | < 0.01, 0.02, < 0.01 | 0.02 |
|  |  |  |  |  |  | 14 | Fruit | < 0.01, 0.015, < 0.01 | 0.02 |
|  |  |  |  |  |  | 21 | Fruit | < 0.01, < 0.01, < 0.01 | < 0.01 |
| USA 1984FL *Sunny* *PSR99/012* | 3EC | Spray at 0, 7, 15, 52, 59 d Pre-harvest | 1.12 | 587 | 5 | 0 | Fruit* | 0.049, 0.058, < 0.01 | 0.11 |
|  |  |  |  |  |  | 0 | Fruit* | 0.047, 0.081, < 0.01 | 0.13 |
|  |  |  |  |  |  |  | Dry P | 1.354, 1.245, 0.111 | 2.71 |
|  |  |  |  |  |  |  | Fruit co | < 0.01, < 0.01, < 0.01 | < 0.01 |
|  |  |  |  |  |  |  | Dry P co | 0.084, 0.053, 0.031 | 0.17 |
| USA 1984 CA *na* *PSR99/012* | 50WP | Spray at 0, 7, 13, 52, 59 d Pre-harvest | 1.12 | 374 | 5 | 0 | Fruit | 0.304, 0.249, < 0.01 | 0.55 |
|  |  |  |  |  |  | 3 | Fruit | 0.117, 0.176, < 0.01 | 0.29 |
|  |  |  |  |  |  | 7 | Fruit | 0.039, 0.092, < 0.01 | 0.13 |
|  |  |  |  |  |  | 14 | Fruit | 0.017, 0.076, < 0.01 | 0.09 |
|  |  |  |  |  |  | 21 | Fruit | 0.011, 0.097, 0.013 | 0.12 |
|  | 3EC | Spray at 0, 7, 13, 52, 59 d Pre-harvest | 1.12 | 374 | 5 | 0 | Fruit | 0.201, 0.222, < 0.01 | 0.42 |
|  |  |  |  |  |  | 3 | Fruit | 0.111, 0.161, 0.011 | 0.28 |
|  |  |  |  |  |  | 7 | Fruit | 0.016, 0.049, < 0.01 | 0.07 |
|  |  |  |  |  |  | 14 | Fruit | 0.018, 0.049, 0.008 | 0.08 |
|  |  |  |  |  |  | 21 | Fruit | 0.011, 0.032, < 0.01 | 0.04 |
| USA 1984 CA *na* *PSR99/012* | 3EC | Spray at 0, 7, 14, 66, 73 d Pre-harvest | 1.12 | 374 (0.3% ai) | 5 | 0 | Fruit | 0.045, 0.052, < 0.01 | 0.10 |
|  |  |  |  |  |  | 0 | Fruit | 0.073, 0.076, < 0.01 | 0.15 |
|  |  |  |  |  |  |  | Fruit co | < 0.01, < 0.01, < 0.01 | < 0.01 |
|  |  |  |  |  |  |  | Fruit co | < 0.01, < 0.01, < 0.01 | < 0.01 |
| USA 1984 CA *na* *PSR99/012* | 3EC | Spray at 0, 7, 14, 66, 73 d Pre-harvest | 1.12 | 374 | 5 | 0 | Fruit | 0.045, 0.052, < 0.01 | 0.10 |
|  |  |  |  |  |  | 0 | Fruit | 0.073, 0.076, < 0.01 | 0.15 |
|  |  |  |  |  |  |  | Wh Pk | < 0.01, < 0.01, < 0.01 | < 0.01 |
|  |  |  |  |  |  |  | Juice | 0.023, 0.021, < 0.01 | 0.04 |
|  |  |  |  |  |  |  | Paste | 0.026, 0.036, 0.026 | 0.09 |
|  |  |  |  |  |  |  | Ckd S,P | 1.830, 1.975, 0.138 | 3.94 |
|  |  |  |  |  |  |  | Dry SP | 1.492, 2.827, 0.244 | 4.56 |
|  |  |  |  |  |  |  | Fruit co | < 0.01, < 0.01, < 0.01 | < 0.01 |
|  |  |  |  |  |  |  | Fruit co | < 0.01, < 0.01, < 0.01 | < 0.01 |
| USA 1984 CA *Better Boy* *03R/1919 (4)* | 3EC | Spray Interval 7 days | 1.12 | 140 (0.8% ai) | 5 | 0 | Fruit | 0.133 0.146, 0.037 | 0.32 |
|  |  |  |  |  |  | 3 | Fruit | 0.065, 0.112, 0.032 | 0.21 |
|  |  |  |  |  |  | 7 | Fruit | 0.011, 0.037, 0.019 | 0.07 |
|  |  |  |  |  |  | 14 | Fruit | < 0.01, < 0.01, < 0.01 | < 0.01 |
|  |  |  |  |  |  | 21 | Fruit | < 0.01, < 0.01, < 0.01 | < 0.01 |
| USA 1984 CA *Better Boy* *03R/1919 (4)* | 50WP | Spray Interval 7 days | 1.12 | 140 (0.8% ai) | 5 | 0 | Fruit | 0.115, 0.105, 0.017 | 0.24 |
|  |  |  |  |  |  | 3 | Fruit | 0.021, 0.060, 0.023 | 0.10 |
|  |  |  |  |  |  | 7 | Fruit | 0.019, 0.050, 0.019 | 0.09 |
|  |  |  |  |  |  | 14 | Fruit | < 0.01, < 0.01, < 0.01 | < 0.01 |
|  |  |  |  |  |  | 21 | Fruit | < 0.01, < 0.01, < 0.01 | < 0.01 |
| USA 1984NU *XP27P2* *E184-USA-* *03R/1913 (1)* | 3EC | Spray Interval 7d | 1.12 | 776 (0.145% ai) | 5 | 0 | Fruit | 0.149, 0.077, < 0.01 | 0.23 |
|  |  |  |  |  |  | 3 | Fruit | 0.016, 0.022, < 0.01 | 0.04 |
|  |  |  |  |  |  | 7 | Fruit | 0.01, 0.017, 0.014 | 0.04 |
|  |  |  |  |  |  | 14 | Fruit | 0.010, 0.016, 0.021 | 0.05 |
|  |  |  |  |  |  | 21 | Fruit | < 0.01, 0.21, 0.011 | 0.22 |
| USA 1984 NI *XP27P2* *E184-USA-* *03R/1913 (1)* | 50WP | Spray Interval 7d | 1.12 | 776 (0.145% ai) | 5 | 0 | Fruit | 0.091, 0.033, < 0.01 | 0.12 |
|  |  |  |  |  |  | 3 | Fruit | 0.040, 0.033, < 0.01 | 0.07 |
|  |  |  |  |  |  | 7 | Fruit | 0.091, 0.083, 0.079 | 0.25 |
|  |  |  |  |  |  | 14 | Fruit | 0.028, 0.035, 0.020 | 0.08 |
|  |  |  |  |  |  | 21 | Fruit | 0.010, 0.016, 0.012 | 0.04 |

| TOMATO | Application | | | | | PHI | Sample | Residues of alpha, beta | Total |
| Country, Year of trial *Variety report* | Form. | Method | kg ai/ha/ applic'n | L/ha | No. | (Days) | analysed | endosulfan, endosulfan sulfate mg/kg | Residues mg/kg |
|---|---|---|---|---|---|---|---|---|---|
| USA 1984 MI *Pik Red* | 3EC | Spray Interval 20, 11, 9 days | 1.12 | 580 (0.193% ai) | 4 | 0 | Fruit | 0.159, 0.247, 0.016 | 0.42 |
| | | | | | | <u>3</u> | Fruit | 0.143, 0.16, 0.026 | <u>0.33</u> |
| | | | | | | 7 | Fruit | 0.045, 0.063, 0.024 | 0.13 |
| *E184-USA-03R/1913 (2)* | | | | | | 14 | Fruit | 0.046, 0.086, 0.048 | 0.18 |
| | | | | | | 21 | Fruit | 0.033, 0.055, 0.03 | 0.12 |
| USA 1984 MI *Pik Red* | 50WP | Spray Interval 20, 11, 9 days | <u>??</u> | 62 | 4 | 0 | Fruit | 0.153, 0.215, 0.017 | 0.38 |
| | | | | | | 3 | Fruit | 0.135, 0.188, 0.05 | 0.37 |
| | | | | | | 7 | Fruit | 0.067, 0.093, 0.027 | 0.19 |
| *E184-USA-03R/1913 (2)* | | | | | | 14 | Fruit | 0.039, 0.069, 0.06 | 0.17 |
| | | | | | | 21 | Fruit | 0.03, 0.045, 0.028 | 0.10 |
| Australia 2000 Walkamin *Zola N° 1/10/552* | EC | spray | 0.735 | 493 | 3 | 0 | | 0.037, 0.045, 0.007 | 0.09 |
| | | | | | | 3 | | 0.013,0.037, 0.006 | <u>0.06</u> |
| | | | | | | 7 | | 0.005, 0.021,0.009 | 0.04 |
| | | | | | | 14 | | < 0.005,0.008,0.007 | 0.02 |
| Australia 2000 Walkamin *Zola N° 1/10/552* | EC | spray | 1.47 | 552 | 3 | 0 | | 0.054, 0.064, 0.006 | 0.12 |
| | | | | | | 3 | | 0.025, 0.074, 0.010 | 0.11 |
| | | | | | | 7 | | 0.007, 0.028,0.010 | 0.05 |
| | | | | | | 14 | | 0.006, 0.033, 0.014 | 0.05 |
| Australia2000 Caffey *Thunder N° 1/10/552* | EC | spray | 0.735 | 673 | 3 | 0 | | 0.044, 0.032, 0.007 | 0.08 |
| | | | | | | 3 | | < 0.005, < 0.005, < 0.005 | <u>≤ 0.005</u> |
| | | | | | | 7 | | < 0.005, < 0.005, < 0.005 | < 0.005 |
| | | | | | | 14 | | < 0.005, < 0.005, < 0.005 | < 0.005 |
| Australia 2000 Goulburn Valley *Granades N° 1/10/552* | EC | spray | 0.735 | 452 | 3 | 0 | | 0.027, 0.026, 0.006 | 0.06 |
| | | | | | | 3 | | 0.025,0.035, 0.009 | <u>0.07</u> |
| | | | | | | 7 | | 0.007, 0.013,0.007 | 0.03 |
| | | | | | | 14 | | 0.009,0.016,0.013 | 0.04 |
| Australia 2000 Goulburn Valley *Granades N° 1/10/552* | EC | spray | 1.47 | 452 | 3 | 0 | | 0.072, 0.061, 0.012 | 0.15 |
| | | | | | | 3 | | 0.020, 0.033, 0.009 | 0.06 |
| | | | | | | 7 | | 0.016, 0.032,0.018 | 0.07 |
| | | | | | | 14 | | 0.014, 0.025, 0.018 | 0.06 |
| Australia 2000 Mancini *Early nema N° 1/10/552* | EC | spray | 0.735 | 421 | 3 | 0 | | 0.037, 0.044, < 0.005 | 0.08 |
| | | | | | | 3 | | 0.032,0.053, 0.009 | <u>0.09</u> |
| | | | | | | 7 | | 0.030, 0.052,0.008 | 0.09 |
| | | | | | | 14 | | 0.009,0.011, < 0.005 | 0.02 |

** endosulfan diol

Table 63. Endosulfan residues in tomatoes from supervised trials indoor in Europe.

| TOMATO (Indoor) | Application | | | | | PHI | Sample | Residues of alpha, beta | Total |
| Country, Year of trial *Variety, report* | Form. | Method | kg ai/ha/ applic'n | L/ha | No. | (Days) | analysed | endosulfan, Endosulfan sulfate mg/kg | Residue, mg/kg |
|---|---|---|---|---|---|---|---|---|---|
| Germany 1975 *Hellfrucht* LEA 4/84/01/02-75B | 24EC* | Spray, 0.2% | 0.72 | 1500 (0.048% ai) | 1 | 0 | Fruit | 0.60, 0.40, ND | 1.00 |
| | | | | | | 7 | Fruit | 0.09, 0.09, ND | 0.18 |
| | | | | | | 10 | Fruit | 0.10, 0.10, ND | 0.20 |
| | | | | | | 14 | Fruit | 0.02, 0.03, ND | 0.05 |

| TOMATO (Indoor) Country, Year of trial *Variety, report* | Form. | Method | kg ai/ha/ applic'n | L/ha | No. | PHI (Days) | Sample analysed | Residues of alpha, beta endosulfan, Endosulfan sulfate mg/kg | Total Residue, mg/kg |
|---|---|---|---|---|---|---|---|---|---|
| Application spanning Form–No |  |  |  |  |  |  |  |  |  |
| Germany 1975 *Hildaris* LEA3/84/01/02-75 | 24EC* | Spray, 0.2% | 0.72 | 1500 (0.048 % ai) | 1 | 0 | Fruit | 0.09, 0.08, ND | 0.17 |
|  |  |  |  |  |  | 7 | Fruit | 0.02, 0.07, ND | 0.09 |
|  |  |  |  |  |  | 10 | Fruit | 0.02, 0.05, ND | 0.07 |
|  |  |  |  |  |  | 14 | Fruit | 0.01, 0.03, ND | 0.04 |
| Germany 1975 *Hellfrucht* LEA 2/84/01/02-75A | 24EC* | Spray, 0.2% | 0.96 | 2000 (0.048% ai) | 1 | 0 | Fruit |  | 0.30 |
|  |  |  |  |  |  | 7 | Fruit |  | 0.05 |
|  |  |  |  |  |  | 10 | Fruit |  | < 0.05 |
|  |  |  |  |  |  | 14 | Fruit |  | < 0.05 |
| Germany 1975 *Refa* LEA1/84/01/02-75 | 24EC* | Spray, 0.2% | 0.96 | 2000 (0.048 % ai) | 1 | 0 | Fruit |  | 0.80 |
|  |  |  |  |  |  | 7 | Fruit |  | 0.10 |
|  |  |  |  |  |  | 10 | Fruit |  | 0.10 |
|  |  |  |  |  |  | 14 | Fruit |  | 0.10 |
| Spain 2001 *Optima* 01 R 642 | 35EC | Spray interval 14 d BBCH 76, 86 | 0.800 | 1500 (0.053% ai) | 2 | 0 | Fruit | 0.03, 0.03, < 0.02 | 0.06 |
|  |  |  |  |  |  | 3 | Fruit | 0.04, 0.03, < 0.02 | <u>0.07</u> |
| France2001 *Felicia* 01 R 642 | 35EC | Spray interval 14 d BBCH 81, 81 | 0.885, 0.800 | 1660 1500 (0.053% ai) | 2 | 0 | Fruit | 0.06, 0.05, < 0.02 | 0.11 |
|  |  |  |  |  |  | 3 | Fruit | 0.09, 0.07, < 0.02 | <u>0.16</u> |
| Greece 2001 *Alma* 01 R 642 | 35EC | Spray indoor interval 14 d BBCH 81, 87 | 0.800 | 1500 (0.053 % ai) | 2 | 0 | Fruit | 0.13, 0.08, < 0.02 | 0.21 |
|  |  |  |  |  |  | <u>3</u> | Fruit | 0.13, 0.08, < 0.02 | <u>0.21</u> |
| Italy 2001 *Naxos* 01 R 642 | 35EC | Spray indoor interval 14 d BBCH 77, 82 | 0.800 | 1500 (0.053 % ai) | 2 | 0 | Fruit | 0.18, 0.12, < 0.02 | 0.30 |
|  |  |  |  |  |  | <u>3</u> | Fruit | 0.08, 0.09, < 0.02 | <u>0.17</u> |
| Spain 2001 *Optima* 01 R 641 | CS* | Spray indoor interval 14 d BBCH 76, 86 | 0.800 | 1500 (0.053% ai) | 2 | 0 | Fruit | 0.08, 0.05, < 0.02 | 0.13 |
|  |  |  |  |  |  | 3 | Fruit | < 0.02, 0.03, < 0.02 | 0.03 |
|  |  |  |  |  |  | 7 | Fruit | 0.07, 0.04, < 0.02 | <u>0.11</u> |
| France 2001 *Felicia* 01 R 641 | CS* | Spray indoor interval 14 d BBCH 81, 81 | 0.900 0.800 | 1688, 1500 (0.053% ai) | 2 | 0 | Fruit | 0.14, 0.08, < 0.02 | 0.22 |
|  |  |  |  |  |  | 3 | Fruit | 0.22, 0.13, < 0.02 | <u>0.35</u> |
|  |  |  |  |  |  | 7 | Fruit | 0.17, 0.10, < 0.02 | 0.27 |
| Greece 2001 *Alma* 01 R 641 | CS* | Spray indoor interval 14 d BBCH 81, 87 | 0.800 | 1500 (0.053% ai) | 2 | 0 | Fruit | 0.12, 0.07, < 0.02 | 0.19 |
|  |  |  |  |  |  | 3 | Fruit | 0.14, 0.07, < 0.02 | <u>0.21</u> |
|  |  |  |  |  |  | 7 | Fruit | 0.12, 0.06, < 0.02 | 0.18 |
| Italy 2001 *Naxos* 01 R 641 | CS* | Spray indoor interval 14 d BBCH 77, 82 | 0.800 | 1500 (0.053% ai) | 2 | 0 | Fruit | 0.15, 0.09, < 0.02 | 0.24 |
|  |  |  |  |  |  | 3 | Fruit | 0.27, 0.14, < 0.02 | <u>0.41</u> |
|  |  |  |  |  |  | 7 | Fruit | 0.22, 0.12, < 0.02 | 0.34 |
| Spain 1998 *Genaro* ER 98 ECS 753 | CS* | Spray indoor interval 14 d BBCH 72, 74 | 0.7986 0.8865 | 1500 1665 (0.053% ai) | 2 | 0 | Fruit | 0.20, 0.10, ND | 0.30 |
|  |  |  |  |  |  | 1 | Fruit | 0.18, 0.09, ND | 0.27 |
|  |  |  |  |  |  | 3 | Fruit | 0.15, 0.08, ND | <u>0.23</u> |
|  |  |  |  |  |  | 7 | Fruit | 0.14, 0.09, ND | 0.23 |
| Greece 1998 *Arleta* ER 98 ECS 753 | CS* | Spray indoor interval 14 d BBCH 81, 85 | 0.7986 | 1500 (0.053% ai) | 2 | 0 | Fruit | 0.19, 0.11, ND | 0.30 |
|  |  |  |  |  |  | 1 | Fruit | 0.12, 0.07, ND | 0.19 |
|  |  |  |  |  |  | 3 | Fruit | 0.11, 0.06, ND | 0.17 |
|  |  |  |  |  |  | 7 | Fruit | 0.13, 0.07, ND | <u>0.20</u> |
| Greece 1998 *Arleta* ER 98 ECS 753 | CS* | Spray indoor interval 14 d BBCH 87, 87 | 0.7986 | 1500 (0.053 % ai) | 2 | 0 | Fruit | 0.15, 0.09, ND | 0.24 |
|  |  |  |  |  |  | 1 | Fruit | 0.19, 0.12, ND | 0.31 |
|  |  |  |  |  |  | 3 | Fruit | 0.15, 0.09, ND | <u>0.24</u> |
|  |  |  |  |  |  | 7 | Fruit | 0.06, 0.04, ND | 0.10 |
| Italy 1998 *Vermone* ER 98 ECS 753 | CS* | Spray indoor interval 14 d BBCH 75, 77 | 0.7986 | 1500 (0.053 % ai) | 2 | 0 | Fruit | 0.32, 0.17, ND | 0.49 |
|  |  |  |  |  |  | 1 | Fruit | 0.45, 0.24, ND | 0.69 |
|  |  |  |  |  |  | 3 | Fruit | 0.44, 0.21, ND | <u>0.65</u> |
|  |  |  |  |  |  | 7 | Fruit | 0.27, 0.14, < 0.02 | 0.41 |

| TOMATO (Indoor) | Application | | | | | PHI (Days) | Sample analysed | Residues of alpha, beta endosulfan, Endosulfan sulfate mg/kg | Total Residue, mg/kg |
|---|---|---|---|---|---|---|---|---|---|
| Country, Year of trial *Variety, report* | Form. | Method | kg ai/ha/ applic'n | L/ha | No. | | | | |
| Portugal 1998 *Zapata* ER 98 ECS 753 | CS* | Spray indoor interval 14 d BBCH 73, 79 | 0.7986 | 1500 (0.053 % ai) | 2 | 0 | Fruit | 0.19, 0.11, ND | 0.30 |
| | | | | | | 1 | Fruit | 0.16, 0.10, ND | 0.26 |
| | | | | | | 3 | Fruit | 0.17, 0.11, ND | 0.28 |
| | | | | | | 7 | Fruit | 0.07, 0.04, ND | 0.11 |
| Spain 1994 Andalucia *Presto* ER 94 ECS 701 | 35EC | Spray indoor (motorised knapsack) interval 14d GS 22, 23 | 1.074, 0.809 | 2033 1533 (0.053 % ai) | 2 | 0 | Fruit | 0.12, 0.09, < 0.01 | 0.21 |
| | | | | | | 3 | Fruit | 0.04, 0.06, < 0.01 | 0.10 |
| | | | | | | 7 | Fruit | 0.04, 0.04, 0.02 | 0.10 |
| | | | | | | 14 | Fruit | 0.01, 0.02, 0.02 | 0.05 |
| | | | | | | 21 | Fruit | < 0.01, < 0.01, 0.01 | < 0.01 |
| | | | | | | 29 | Fruit | < 0.01, < 0.01, < 0.01 | < 0.01 |
| Spain 1994 Andalucia *Presto* ER 94 ECS 701 | 35EC | Spray indoor (motorised knapsack) interval 14d GS 22, 23 | 1.919, 1.655 | 1817 1567 (0.1056 % ai) | 2 | 0 | Fruit | 0.17, 0.14, < 0.01 | 0.31 |
| | | | | | | 3 | Fruit | 0.11, 0.15, 0.03 | 0.29 |
| | | | | | | 7 | Fruit | 0.09, 0.10.0.04 | 0.23 |
| | | | | | | 14 | Fruit | 0.04, 0.07, 0.04 | 0.15 |
| | | | | | | 21 | Fruit | 0.04, 0.06, 0.03 | 0.13 |
| | | | | | | 29 | Fruit | < 0.01, 0.02, 0.02 | 0.04 |
| Spain 1994 *Caruso* ER 94 ECS 701 | 35EC | Spray indoor (motorised knapsack) interval 14d GS 22, 23 | 0.616, 0.720 | 1167 1364 (0.053% ai) | 2 | 0 | Fruit | 0.06, 0.07, < 0.01 | 0.13 |
| | | | | | | 3 | Fruit | 0.02, 0.03, < 0.01 | 0.05 |
| | | | | | | 7 | Fruit | 0.01, 0.02, < 0.01 | 0.03 |
| | | | | | | 14 | Fruit | < 0.01, 0.02, 0.01 | 0.03 |
| | | | | | | 21 | Fruit | < 0.01, < 0.01, < 0.01 | < 0.01 |
| | | | | | | 29 | Fruit | < 0.01, < 0.01, < 0.01 | < 0.01 |
| Spain 1994 *Caruso* ER 94 ECS 701 | 35EC | Spray indoor (motorised knapsack) interval 14d GS 22, 23 | 1.168, 1.121 | 1106 1061 0.1056% ai) | 2 | 0 | Fruit | 0.08, 0.08, < 0.01 | 0.16 |
| | | | | | | 3 | Fruit | 0.07, 0.10, 0.04 | 0.21 |
| | | | | | | 7 | Fruit | 0.03, 0.06, 0.04 | 0.13 |
| | | | | | | 14 | Fruit | 0.02, 0.03, 0.02 | 0.07 |
| | | | | | | 21 | Fruit | < 0.01, 0.02, 0.01 | 0.03 |
| | | | | | | 29 | Fruit | < 0.01, < 0.01, 0.01 | 0.01 |
| Italy 1994 *Vemone* ER 94 ECS 701 A | 35EC | Spray indoor (motorised knapsack) interval 14d GS 11/17, 11/21 | 0.898, 0.898 | 1700 1700 (0.053% ai) | 2 | 0 | Fruit | 0.19, 0.18, < 0.01 | 0.37 |
| | | | | | | 3 | Fruit | 0.14, 0.12, 0.01 | 0.27 |
| | | | | | | 7 | Fruit | 0.05, 0.07, 0.02 | 0.14 |
| | | | | | | 14 | Fruit | 0.01, 0.03, 0.01 | 0.05 |
| | | | | | | 21 | Fruit | < 0.01, < 0.01, < 0.01 | < 0.01 |
| | | | | | | 28 | Fruit | < 0.01, < 0.01, < 0.01 | < 0.01 |
| Italy 1994 *Vemone* ER 94 ECS 701 | 35EC | Spray indoor (motorised knapsack) interval 14d GS 11/17, 11/21 | 1.795, 1.795 | 1700 1700 0.1056% ai) | 2 | 0 | Fruit | 0.45, 0.4, < 0.01 | 0.85 |
| | | | | | | 3 | Fruit | 0.35, 0.35, 0.02 | 0.72 |
| | | | | | | 7 | Fruit | 0.15, 0.31, 0.02 | 0.48 |
| | | | | | | 14 | Fruit | 0.06, 0.11, 0.04 | 0.21 |
| | | | | | | 21 | Fruit | 0.01, 0.04, 0.02 | 0.07 |
| | | | | | | 28 | Fruit | 0.01, 0.02, 0.02 | 0.05 |
| Italy 1994 *San Marzano (Italdor)* ER 94 ECS 701 | 35EC | Spray indoor (knapsack) interval 14d GS 15/17, 15/21 | 1.056, 1.056 | 2000 2000 (0.053% ai) | 2 | 0 | Fruit | 0.2, 0.1, < 0.01 | 0.30 |
| | | | | | | 3 | Fruit | 0.06, 0.05, < 0.01 | 0.11 |
| | | | | | | 7 | Fruit | 0.03, 0.043, 0.01 | 0.08 |
| | | | | | | 14 | Fruit | 0.033, 0.057, 0.033 | 0.12 |
| | | | | | | 21 | Fruit | 0.01, 0.03, 0.02 | 0.06 |
| | | | | | | 27 | Fruit | < 0.01, < 0.01, 0.01 | < 0.01 |
| Italy 1994 *San Marzano (Italdor)* ER 94 ECS 701 | 35EC | Spray indoor (knapsack) interval 14d GS 15/17, 15/21 | 2.112, 2.112 | 2000 2000 0.1056% ai) | 2 | 0 | Fruit | 0.45, 0.25, 0.02 | 0.72 |
| | | | | | | 3 | Fruit | 0.35, 0.23, 0.02 | 0.60 |
| | | | | | | 7 | Fruit | 0.053, 0.06, 0.02 | 0.13 |
| | | | | | | 14 | Fruit | 0.1, 0.113, 0.04 | 0.25 |
| | | | | | | 21 | Fruit | 0.03, 0.04, 0.04 | 0.11 |
| | | | | | | 27 | Fruit | 0.02, 0.02,  0.02 | 0.06 |
| Spain 1993 Andalucia *Prieto* ER 93 ECS 701 | 35EC | Spray indoor (motorised knapsack) interval 14d | 0.5376 | 1018 (0.053 % ai) | 2 | 0 | Fruit | 0.13, 0.06, < 0.01 | 0.19 |
| | | | | | | 3 | Fruit | 0.04, 0.05, < 0.01 | 0.09 |
| | | | | | | 7 | Fruit | 0.01, 0.03, < 0.01 | 0.04 |
| | | | | | | 14 | Fruit | < 0.01, 0.02, < 0.01 | 0.02 |

| TOMATO (Indoor) | Application | | | | | PHI (Days) | Sample analysed | Residues of alpha, beta endosulfan, Endosulfan sulfate mg/kg | Total Residue, mg/kg |
|---|---|---|---|---|---|---|---|---|---|
| Country, Year of trial *Variety, report* | Form. | Method | kg ai/ha/ applic'n | L/ha | No. | | | | |
| Spain 1993 Andalucia *Prieto* ER 93 ECS 701 | 35EC | Spray indoor (motorised knapsack) interval 14d | 1.0752 | 1018 (0.1056 % ai) | 2 | 0 <u>3</u> 7 14 | Fruit Fruit Fruit Fruit | 0.25, 0.12, < 0.01 0.06, 0.13, < 0.01 0.03, 0.08, 0.02 0.02, 0.05, 0.02 | 0.37 <u>0.19</u> 0.13 0.09 |
| Italy 1993 Ampulia *Majorca* ER 93 ECS 701 | 35EC | Spray indoor (motorised knapsack) interval 14d GS 11/19, | 0.8975 | 1700 (0.053 % ai) | 2 | 0 <u>3</u> 7 14 | Fruit Fruit Fruit Fruit | 0.18, 0.12, < 0.01 0.03, 0.04, < 0.01 0.11, 0.175, 0.03 0.02, 0.03, 0.015 | 0.30 0.07 <u>0.32</u> 0.07 |
| Italy 1993 Ampulia *Majorca* ER 93 ECS 701 | 35EC | Spray indoor (motorised knapsack) interval 14d GS 11/19, | 1.7954 | 1700 (0.1056 % ai) | 2 | 0 <u>3</u> 7 14 | Fruit Fruit Fruit Fruit | 0.48, 0.31, < 0.01 0.13, 0.22, 0.023 0.025, 0.035, 0.01 0.025, 0.045, 0.02 | 0.79 <u>0.37</u> 0.07 0.09 |

Table 64. Endosulfan residues in eggplant from supervised trials in Australia.

| EGGPLANT | Application | | | | | PHI days | Sample analysed | Residues of alpha, beta endosulfan, Endosulfan sulfate, mg/kg | Total Residues, mg/kg |
|---|---|---|---|---|---|---|---|---|---|
| Country, Year of trial *Variety* Report | Form. | Method | kg ai/ha/ applic'n | L/ha | No. | | | | |
| Australia *2001* Koraleigh *Grace* 1/10/540& | EC | Spray | 0.73 | 224 | 3 | 0 3 7 14 | Fruit | 0.034,0.030,< 0.005 < 0.005,0.007< 0.005 < 0.005,< 0.005,< 0.005 < 0.005,< 0.005,< 0.005 | 0.07 0.01 <u>< 0.005</u> < 0.005 |
| Australia *2001* Glasshouse Mtns *Venus* 1/10/540& | EC | Spray | 1.46 | 334 | 3 | 0 3 7 14 | Fruit | 0.058, 0.081, 0.016 | 0.15 |
| Australia *2001* Glasshouse Mtns *Venus* 1/10/540& | EC | Spray | 0.73 | 334 | 3 | 0 3 7 14 | Fruit | 0.33, 0.21, 0.26 0.043, 0.062, 0.031 0.014, 0.029, 0.012 | 0.57 0.14 <u>0.06</u> |
| Australia *2001* Shepparton *Black pearl* 1/10/540& | EC | Spray | 0.73 | 452 | 3 | 0 3 7 14 | Fruit | 0.015,0.011, 0.006 < 0.005,< 0.005< 0.005 < 0.005,< 0.005,< 0.005 < 0.005,< 0.005,< 0.005 | 0.03 < 0.005 <u>< 0.005</u> < 0.005 |
| Australia *2001* Gumlu *Black pearl* 1/10/540& | EC | Spray | 0.73 | 352 | 3 | 0 3 6 14 | Fruit | 0.014,0.014,< 0.005 < 0.005<,0.005< 0.005 < 0.005,< 0.005,< 0.005 0.006, < 0.005,< 0.005 | 0.03 < 0.005 < 0.005 <u>0.006</u> |

Table 65. Endosulfan residues in sweet corn from supervised trials in Australia.

| SWEET CORN | Application | | | | | PHI | Sample | Residues of alpha, beta | Total |
|---|---|---|---|---|---|---|---|---|---|
| Country, Year of trial *Variety* Report | Form. | Method | kg ai/ha/ applic'n | L/ha | No. | days | analysed | endosulfan, Endosulfan sulfate, mg/kg | Residues, mg/kg |
| Australia 2000 Mulgowie *Golden sweet 1/10/560* | EC | Spray | 0.73 | 226 | 3 | 0<br>3<br>7<br>14 | Fruit | < 0.005<,0.005< 0.005<br>< 0.005,< 0.005,< 0.005<br>< 0.005,< 0.005,< 0.005<br>< 0.005,< 0.005,< 0.005 | < 0.005<br>< 0.005<br>< 0.005<br>< 0.005 |
| Australia 2000 Mulgowie *Golden sweet 1/10/560* | EC | Spray | 1.46 | 226 | 3 | 0<br>3<br>7 | Fruit | < 0.005<,0.005< 0.005 | < 0.005 |
| Australia 2000 Warragal *Honey sweet 1/10/560* | EC | Spray | 0.73 | 347 | 3 | 0<br>3<br>7<br>14 | Fruit | < 0.005<,0.005< 0.005<br>< 0.005,< 0.005,< 0.005<br>< 0.005,< 0.005,< 0.005<br>< 0.005,< 0.005,< 0.005 | < 0.005<br>< 0.005<br>< 0.005<br>< 0.005 |
| Australia 2000 Warragal *Honey sweet 1/10/560* | EC | Spray | 1.46 | 347 | 3 | 7 | Fruit | < 0.005,< 0.005,< 0.005 | < 0.005 |
| Australia 2000 Koraleigh *Golden sweet 1/10/560* | EC | Spray | 0.73 | 205 | 3 | 0<br>3<br>7<br>14 | Fruit | < 0.005<,0.005< 0.005<br>< 0.005,< 0.005,< 0.005<br>< 0.005,< 0.005,< 0.005<br>< 0.005,< 0.005,< 0.005 | < 0.005<br>< 0.005<br>< 0.005<br>< 0.005 |
| Australia 2000 Koraleigh *Golden sweet 1/10/560* | EC | Spray | 0.73 | 224 | 3 | 7 | Fruit | < 0.005,< 0.005,< 0.005 | < 0.005 |

Table 66. Endosulfan residues in beans from supervised trials in Germany, USA and Australia.

| PHASEOLUS BEANS | Application | | | | | PHI (Days) | Sample analysed | Residues of alpha, beta endosulfan, Endosulfan sulfate, mg/kg | Total Residues mg/kg |
|---|---|---|---|---|---|---|---|---|---|
| Country, Year of trial *Variety, report* | Form. | Method | kg ai /ha/ applic'n | L/ha | No. | | | | |
| Germany 1983 *Filetty* DEU83172811 | 2.82% Dust | Spread Interval 14, 21 d | <u>0.705</u> | | 3 | 0<br>5<br><br>11<br>14 | Bean<br>Bean<br><br>Bean<br>Bean | 0.10, 0.07, ND<br>0.02, 0.04, 0.01<br><br>ND, 0.02, 0.02<br>ND, 0.01, 0.01 | 0.17<br>0.06<br><br>0.04<br>0.02 |
| Germany 1983 *Marona* DEU83172821 | 2.82 % Dust | Spread Interval 15d | <u>0.705</u> | | 3 | 0<br><br>5<br><br>11 | Bean<br><br>Bean<br><br>Bean | 0.20 0.10 0.03<br><br>0.05, 0.05, 0.09<br><br>0.02, 0.01, 0.09 | 0.33<br><br>0.19<br><br>0.12 |
| Germany 1983 *Sotexa* DEU83172831 | 2.82 % Dust | Spread interval 14, 12, d | <u>0.705</u> | | 3 | 0<br><br>5<br><br>10<br>14 | Bean<br><br>Bean<br><br>Bean<br>Bean | 0.10 0.07, 0.02<br><br>ND, ND, 0.05<br><br>ND, ND, 0.02<br>ND, ND, 0.03 | 0.19<br><br>0.05<br><br>0.02<br>0.03 |

| PHASEOLUS BEANS | Application | | | | | PHI (Days) | Sample analysed | Residues of alpha, beta endosulfan, Endosulfan sulfate, mg/kg | Total Residues mg/kg |
|---|---|---|---|---|---|---|---|---|---|
| Country, Year of trial *Variety, report* | Form. | Method | kg ai /ha/ applic'n | L/ha | No. | | | | |
| Germany 1983 *Dublette* DEU83172841 | 2.82% Dust | Spread Interval 14 d | <u>0.705</u> | | 3 | 0<br>5<br>10<br>14 | Bean<br>Bean<br>Bean<br>Bean | 0.70, 0.40, 0.02<br>0.06, 0.10, 0.03<br>0.01, 0.03, 0.03<br>ND, 0.02, 0.03 | 1.12<br>0.19<br>0.07<br>0.07 |
| Germany 1974 *Bravo* 1/74/01/02 | 3% Dust | Spread | 0.9 | | 1 | 0<br>7<br>14<br>21<br>28<br>35 | Bean<br>Bean<br>Bean<br>Bean<br>Bean<br>Bean | | 46.6<br>0.64<br>0.59<br>0.22<br>0.08<br>< 0.05 |
| Germany 1974 *Favorit* 2/74/01/02 | 3% Dust | Spread | 0.9 | | 1 | 0<br>7<br>14<br>21<br>28 | Bean<br>Bean<br>Bean<br>Bean<br>Bean | | 21.8<br>2.3<br>0.2<br>0.1<br>0.04 |
| Germany 1974 Hattersheim *Favorit* 4/74/01/02 | 3% Dust | Spread | 0.9 | | 1 | 0<br>7<br>14<br>21<br>28 | Bean<br>Bean<br>Bean<br>Bean<br>Bean | 2.90, 0.2<br>0.04, 0.05<br>0.02, 0.02<br>0.03, 0.04<br>ND, 0.02 | 3.10<br>0.09<br>0.04<br>0.07<br>0.02 |
| Germany 1974 Hattersheim *Sotexa* 4/74/02/02 | 3% Dust | Spread | 0.9 | | 1 | 0<br>7<br>14<br>21<br>28 | Bean<br>Bean<br>Bean<br>Bean<br>Bean | 6.20 1.0<br>0.04, 0.06<br>0.02, 0.02<br>0.01, 0.02<br>ND, 0.02 | 7.20<br>0.10<br>0.04<br>0.03<br>0.02 |
| Germany 1974 *Kaskade* 3/74/02/02 | 3% Dust | Spread | 0.9 | | 1 | 0<br>7<br>14<br>21<br>28 | Bean<br>Bean<br>Bean<br>Bean<br>Bean | 0.05, 0.04, 0.02<br>ND, ND, 0.02<br>ND, ND, 0.08<br>ND, ND, 0.009<br>ND, ND, 0.002 | 0.11<br>0.02<br>0.08<br>0.01<br>0.01 |
| Germany 1974 *Kaskade* 3/74/01/02 | 3% Dust | Spread | 0.9 | | 1 | 0<br>7<br>14<br>21<br>28 | Bean<br>Bean<br>Bean<br>Bean<br>Bean | 0.05, 0.01, 0.01<br>0.02, ND, 0.01<br>ND, ND, 0.004<br>ND, ND, 0.002<br>ND, ND, ND | 0.07<br>0.03<br>0.01<br>0.01<br>< 0.01 |
| USA 1965 NY Lima beans M-1610 | 50WP | Spray Interval 37, 7 d | 0.56 | | 3 | 0<br>1<br>4<br>7<br>11<br>20<br>32<br>11<br><br>20<br><br>32<br>11<br>20<br>32 | Beans, pods<br>Beans, pods<br>Beans, pods<br>Beans, pods<br>Beans, pods<br>Beans, pods<br>Beans, pods<br>Shelled beans<br>Shelled beans<br>Shelled beans<br>Pods<br>Pods<br>Pods | 1.48, 0.38<br>0.68, 0.14<br>0.35, 0.13<br>0.15, 0.08<br>0.48, 0.35<br>0.22, 0.14<br>0.09, 0.10<br>< 0.05, < 0.05<br><br>< 0.05, 0.05<br><br>< 0.05, < 0.05<br>0.38, 0.20<br>0.18, 0.12<br>0.23, 0.20 | 1.86<br>0.82<br><u>0.48</u><br>0.23<br>0.83<br>0.36<br>0.19<br>< 0.05<br><br>0.05<br><br>< 0.05<br>0.58<br>0.30<br>0.43 |

| PHASEOLUS BEANS | Application | | | | | PHI (Days) | Sample analysed | Residues of alpha, beta endosulfan, Endosulfan sulfate, mg/kg | Total Residues mg/kg |
|---|---|---|---|---|---|---|---|---|---|
| Country, Year of trial *Variety, report* | Form. | Method | kg ai /ha/ applic'n | L/ha | No. | | | | |
| USA 1965 NY Lima beans M-1610 | 50WP | Spray Interval 37, 7 d | 0.56 | | 3 | 0 | Beans, pods | 0.77, 0.10 | 0.87 |
| | | | | | | 1 | Beans, pods | 0.18, 0.10 | 0.28 |
| | | | | | | 4 | Beans, pods | 0.19, 0.12 | <u>0.31</u> |
| | | | | | | 7 | Beans, pods | 0.11, 0.06 | 0.17 |
| | | | | | | 11 | Beans, pods | 0.20, 0.14 | 0.34 |
| | | | | | | 20 | Beans, pods | 0.06, 0.08 | 0.14 |
| | | | | | | 32 | Beans, pods | < 0.05, 0.05 | 0.05 |
| | | | | | | 11 | Shelled beans | < 0.05, < 0.05 | < 0.05 |
| | | | | | | 20 | Shelled beans | < 0.05, 0.05 | 0.05 |
| | | | | | | 32 | Shelled beans | < 0.05, < 0.05 | < 0.05 |
| | | | | | | 11 | Pods | 0.21, 0.15 | 0.36 |
| | | | | | | 20 | Pods | 0.09, 0.12 | 0.21 |
| | | | | | | 32 | Pods | 0.06, 0.07 | 0.13 |
| USA 1965 NY Snap beans M-1592 | 50WP | Spray Interval 2, 8 d | 0.56 | | 3 | 0 | Snap beans | 2.30 ND | 2.30 |
| | | | | | | 3 | Snap beans | 0.25, ND | <u>0.25</u> |
| | | | | | | 5 | Snap beans | 0.07, 0.05 | 0.12 |
| | | | | | | 7 | Snap beans | 0.10, 0.05 | 0.15 |
| | | | | | | 10 | Snap beans | 0.09, 0.06 | 0.15 |
| | | | | | | 20 | Snap beans | < 0.05, 0.09 | 0.09 |
| | | | | | | 28 | Snap beans | < 0.05, 0.05 | 0.05 |
| | | | | | | 35 | Snap beans | < 0.05, 0.12 | 0.12 |
| | | | | | | 40 | Snap beans | < 0.05, 0.06 | 0.06 |
| USA 1965 NY M-1592 | 50WP | Spray Interval 2, 8 days | <u>1.12</u> | | 3 | 0 | Snap beans | 2.88, ND | 2.88 |
| | | | | | | 3 | Snap beans | 0.59, 0.05 | <u>0.64</u> |
| | | | | | | 5 | Snap beans | 0.18, 0.11 | 0.29 |
| | | | | | | 7 | Snap beans | 0.18, 0.12 | 0.30 |
| | | | | | | 10 | Snap beans | 0.20, 0.18 | 0.38 |
| | | | | | | 20 | Snap beans | 0.06, 0.14 | 0.20 |
| | | | | | | 28 | Snap beans | < 0.05, 0.09 | 0.09 |
| | | | | | | 35 | Snap beans | 0.05, 0.20 | 0.25 |
| | | | | | | 40 | Snap beans | < 0.05, 0.11 | 0.11 |
| Australia 2000 Glen Allyn *festina* *1/10/538* | EC | SPRAY | 0.73 | 426 | 3 | 7 | Beans | 0.032, 0.006, 0.11 | 0.148 |
| | | | | | | 10 | | 0.015, 0.005,0.062 | 0.082 |
| | | | | | | 14 | | 0.014,0.008, 0.028 | 0.050 |
| Australia 2000 Glen Allyn *festina* *1/10/538* | EC | SPRAY | 1.46 | 500 | 3 | 7 | Beans | 0.18, 0.12, 0.58 | 0.88 |
| | | | | | | 10 | | 0.035, 0.019,019 | 0.24 |
| | | | | | | 14 | | 0.034,0.022, 0.11 | 0.167 |
| | | | | | | 7c | | 0.11, 0.01, 0.007 | 0.127 |
| Australia 2000 Don *montano* *1/10/538* | EC | SPRAY | 0.73 | 533 | 3 | 0 | Beans | 0.30, 0.23, 0.055 | 0.58 |
| | | | | | | 3 | | 0.081, 0.066, 0.09 | 0.237 |
| | | | | | | 7 | | < 0.005,< 0.005,< 0.005 | < 0.005 |
| | | | | | | 14 | | < 0.005,< 0.005,< 0.005 | < 0.005 |
| Australia 2000 Don *montano* *1/10/538* | EC | SPRAY | 1.46 | 533 | 3 | 7 | Beans | 0.050, 0.033, 0.11 | 0.193 |
| Australia 2000 Goulburn Valley *dwarf* *1/10/538* | EC | SPRAY | 0.73 | 452 | 3 | 0 | Beans | 0.15, 0.10, 0.039 | 0.29 |
| | | | | | | 3 | | 0.055,0.048,0.035 | 0.143 |
| | | | | | | 7 | | 0.022, 0.021,0.049 | 0.092 |
| | | | | | | 14 | | 0.006,0.006,0.025 | 0.037 |

Table 67. Endosulfan residues in peas from supervised trials in Australia.

| PEAS Country, Year of trial *Variety* Report | Form. | Method | kg ai/ha/ applic'n | L/ha | No. | PHI days | Sample analysed | Residues of alpha, beta endosulfan, Endosulfan sulfate, mg/kg | Total Residues, mg/kg |
|---|---|---|---|---|---|---|---|---|---|
| Australia 2000 Lockyer Valley *epic* *1/10/557* | EC | SPRAY | 0.73 | 500 | 3 | 0 3 7 14 | Peas in pod | 0.42, 0.48, 0.095 0.062, 0.15,0.10 0.009,0.026, 0.047 0.005, 0.010, 0.022 | 0.995 0.312 0.082 0.037 |
| Australia 2000 Lockyer Valley *epic* *1/10/557* | EC | SPRAY | 1.46 | 500 | 3 | 0 3 7 14 | Peas in pod | 1.50,1.40, 0.20 0.21, 0.45,0.28 0.06,0.11, 0.24 0.013, 0.037, 0.089 | 3.10 0.94 0.41 0.139 |
| Australia 2000 Don *Small sieve freezer* *1/10/557* | EC | SPRAY | 0.73 | 533 | 3 | 0 3 7 14 | Peas in pod | 0.46, 0.55, 0.05 0.068, 0.12, 0.17 0.015, 0.022, 0.087 0.006, 0.006, 0.018 | 1.06 0.358 0.124 0.03 |
| AUSTRALIA 2000 Don *Small sieve freezer* *1/10/557* | EC | SPRAY | 1.46 | 533 | 3 | 7 | Peas in pod | 0.033, 0.10, 0.20 | 0.333 |
| AUSTRALIA 2000 Werribee *melbourne market* *1/10/557* | EC | SPRAY | 0.73 | 500 | 3 | 0 3 7 28 | Peas in pod | 1.00, 0.81, 0.15 0.17,0.20.0.33 0.067, 0.099, 0.20 < 0.005, 0.005, 0.008 | 1.96 0.70 0.366 0.018 |

Table 68. Endosulfan residues in soybeans from supervised trials in Australia and Brazil.

| SOYBEAN Country, Year of trial *Variety* | Form. | Method | kg ai/ha/ applic'n | L/ha | No. | PHI (Days) | Sample analysed | Residues of alpha, beta endosulfan, Endosulfan sulfate, mg/kg | Total Residues, mg/kg |
|---|---|---|---|---|---|---|---|---|---|
| Australia 1981 *Forrest* PSR99/010 | 35EC | Spray (3.675% ai) | 0.74 | 20 | 1 | 21 28 | Seeds Seeds | < 0.02, < 0.02, < 0.02 < 0.02, < 0.02, 0.02 | < 0.02 0.02 |
| Australia 1981 *Forrest* A30088 | 35EC | Spray (3.675% ai) | 1.47 | 40 | 1 | 21 28 | Seeds Seeds | < 0.02, < 0.02, < 0.02 0.02, < 0.02, 0.13 | < 0.02 0.15 |
| Australia 1981 *Forrest* | 250ULV | Spray (24.0% ai) | 0.72 | 3 | 1 | 21 28 | Seeds Seeds | | 0.015 0.02 |
| Australia 1981 *Forrest* | 250ULV | Spray (24.0% ai) | 1.44 | 6 | 1 | 21 28 | Seeds Seeds | | 0.02 0.02 |
| Brazil 1974 *Santa Rosa* | 35EC | Spray (0.105% ai) | 0.42 | 400 | 3 | 62 | Seeds | ND, ND, 0.20 | 0.20 |
| Brazil 1974 *Santa Rosa* (A01812) | 35EC | Spray (0.105% ai) | 0.42 | 400 | 4 | 13 | Seeds | 0.03, 0.04, 0.10 | 0.17 |
| Brazil 1975 *Davies* A07560) | 35EC | Spray (0.131% ai) | 0.53 | 400 | 1 | 13 | Seeds | ND, 0.20, 0.01 | 0.21 |
| Brazil 1977 *IAC-3* (A13732) | 35EC | Spray (0.131% ai) | 0.53 | 700 | 1 | 103 | Seeds | 0.05, 0.04 | 0.09 |

| SOYBEAN Country, Year of trial Variety | Form. | Method | kg ai/ha/ applic'n | L/ha | No. | PHI (Days) | Sample analysed | Residues of alpha, beta endosulfan, Endosulfan sulfate, mg/kg | Total Residues, mg/kg |
|---|---|---|---|---|---|---|---|---|---|
| Brazil 1977 *IAC-3* (A13733) | 35EC | Spray (0.075% ai) | 0.53 | 700 | 2 | 66 | Seeds | 0.08, 0.25 | 0.33 |
| Brazil 1977 *IAC-3* (A13735) | 35EC | Spray (0.075% ai) | 0.53 | 700 | 3 | 36 | Seeds | 0.12, 0.33 | 0.45 |
| Brazil 1977 *IAC-3* (A13734) | 35EC | Spray (0.075% ai) | 0.53 | 700 | 1 | 66 | Seeds | 0.08, 0.25 | 0.33 |
| Brazil 1977 *IAC-3* (A13730) | 35EC | Spray (0.075% ai) | 0.53 | 700 | 1 | 36 | Seeds | 0.11, 0.04 | 0.15 |
| Brazil 1977 *IAC-3* (A13731) | 35EC | Spray (0.075% ai) | 0.53 | 700 | 2 | 36 | Seeds | 0.13, 0.29 | 0.42 |
| Brazil 1977 *Santa Rosa* (A13738) | 35EC | Spray (0.075% ai) | 0.53 | 700 | 1 | 22 | Seeds | 0.05, 0.04 | 0.09 |
| Brazil 1977 *Santa Rosa* (A13736) | 35EC | Spray (0.075% ai) | 0.53 | 700 | 2 | 22 | Seeds | 0.10, 0.15 | 0.25 |
| Brazil 1977 *Santa Rosa* (A13737) | 35EC | Spray (0.075% ai) | 0.53 | 700 | 3 | 22 | Seeds | 0.12, 0.19 | 0.31 |
| Brazil 1978 *Santa Rosa* (A16115) | 35EC | Spray (0.075% ai) | 0.53 | 700 | 1 | 90 | Seeds | ND, ND, ND | < 0.01 |
| Brazil 1978 *Santa Rosa* (A16114) | 35EC | Spray (0.075% ai) | 0.53 | 700 | 2 | 62 | Seeds | | 0.50 |
| Brazil 1978 *Santa Rosa* (A16111) | 35EC | Spray (0.075% ai) | 0.53 | 700 | 3 | 31 | Seeds | | 0.40 |
| Brazil 1978 *Santa Rosa* (A16113) | 35EC | Spray (0.075% ai) | 0.53 | 700 | 2 | 31 | Seeds | | 0.10 |
| Brazil 1978 *Santa Rosa* (A16116) | 35EC | Spray (0.075% ai) | 0.53 | 700 | 1 | 62 | Seeds | | 0.20 |
| Brazil 1978 *Santa Rosa* (A16112) | 35EC | Spray (0.075% ai) | 0.53 | 700 | 2 | 31 | Seeds | | 0.30 |
| Brazil 1978 *Santa Rosa* (A16124) | 35EC | Spray (0.075% ai) | 0.53 | 700 | 1 | 90 | Seeds | | 0.05 |
| Brazil 1978 *Santa Rosa* (A16121) | 35EC | Spray (0.075% ai) | 0.53 | 700 | 2 | 61 | Seeds | | 0.20 |
| Brazil 1978 *Santa Rosa* (A16118) | 35EC | Spray (0.075% ai) | 0.53 | 700 | 3 | 29 | Seeds | | 0.30 |
| Brazil 1978 *Santa Rosa* (A16120) | 35EC | Spray (0.075% ai) | 0.53 | 700 | 2 | 29 | Seeds | | 0.20 |
| Brazil 1978 *Santa Rosa 16117* | 35EC | Spray (0.075% ai) | 0.53 | 700 | 1 | 31 | Seeds | | 0.10 |
| Brazil 1978 *Santa Rosa* (A16123) | 35EC | Spray (0.075% ai) | 0.53 | 700 | 1 | 61 | Seeds | | 0.20 |

| SOYBEAN | Application | | | | | PHI | Sample | Residues of alpha, | Total |
| Country, Year of trial *Variety* | Form. | Method | kg ai/ha/ applic'n | L/ha | No. | (Days) | analysed | beta endosulfan, Endosulfan sulfate, mg/kg | Residues, mg/kg |
|---|---|---|---|---|---|---|---|---|---|
| Brazil 1978 *Santa Rosa* (A16119) | 35EC | Spray (0.075% ai) | <u>0.53</u> | 700 | 2 | <u>29</u> | Seeds | | <u>0.20</u> |
| Brazil 1978 *Santa Rosa* (A16122) | 35EC | Spray (0.075% ai) | <u>0.53</u> | 700 | 1 | <u>29</u> | Seeds | | <u>0.08</u> |
| Brazil 1979 *Santa Rosa* (A17983) | 35EC | Spray (0.075% ai) | 0.53 | 700 | 1 | 101 | Seeds | ND, 0.02, 0.08 | 0.10 |
| Brazil 1979 *Santa Rosa* (A17982) | 35EC | Spray (0.075% ai) | 0.53 | 700 | 2 | 71 | Seeds | ND, 0.04, 0.3 | 0.34 |
| Brazil 1979 *Santa Rosa* (A17979, 17978) | 35EC | Spray (0.075% ai) | 0.53 | 700 | 3 | 41 41 41 | Seeds Crude oil Press cake | 0.02, 0.02, 0.3 0.1, 0.7, 0.6 ND, ND, ND | 0.34 1.40 < 0.02 |
| Brazil 1979 *Santa Rosa* (A17981) | 35EC | Spray (0.075% ai) | 0.53 | 700 | 2 | 41 | Seeds | 0.02, 0.06, 0.20 | 0.28 |
| Brazil 1979 *Santa Rosa* (A17984) | 35EC | Spray (0.075% ai) | 0.53 | 700 | 1 | 71 | Seeds | ND, ND, 0.30 | 0.34 |
| Brazil 1979 *Santa Rosa* (A17980) | 35EC | Spray (0.075% ai) | 0.53 | 700 | 2 | 41 | Seeds | 0.01, 0.05, 0.50 | 0.56 |
| Brazil 1979 *Santa Rosa* (A17985) | 35EC | Spray (0.075% ai) | 0.53 | 700 | 1 | 41 | Seeds | 0.02, 0.05, 0.20 | 0.27 |
| Brazil 1979 *Santa Rosa* (A17993) | 35EC | Spray (0.075% ai) | 0.53 | 700 | 1 | 91 | Seeds | ND, ND, ND | < 0.02 |
| Brazil 1979 *Santa Rosa* (A17990) | 35EC | Spray (0.075% ai) | 0.53 | 700 | 2 | 62 | Seeds | ND, 0.05, 0.20 | 0.25 |
| Brazil 1979 *Santa Rosa* (A17986) | 35EC | Spray (0.075% ai) | <u>0.53</u> | 700 | 3 | <u>30</u> | Seeds | 0.10, 0.10, 0.40 | <u>0.60</u> |
| Brazil 1979 *Santa Rosa* (A17987) | 35EC | Spray (0.075% ai) | 0.53 | 700 | 3 | 30 30 | Crude oil Press cake | 0.10, 0.30, 0.30 ND, ND, ND | 0.70 < 0.02 |
| Brazil 1979 *Santa Rosa* (A17989) | 35EC | Spray (0.075% ai) | <u>0.53</u> | 700 | 2 | <u>30</u> | Seeds | ND, ND, ND | <u>< 0.02</u> |
| Brazil 1979 *Santa Rosa* (A17992) | 35EC | Spray (0.075% ai) | 0.53 | 700 | 1 | 62 | Seeds | ND, ND, 0.10 | 0.10 |
| Brazil 1979 *Santa Rosa* (A17988) | 35EC | Spray (0.075% ai) | <u>0.53</u> | 700 | 2 | <u>30</u> | Seeds | ND, ND, 0.20 | <u>0.20</u> |
| Brazil 1979 *Santa Rosa* (A17991) | 35EC | Spray (0.075% ai) | <u>0.53</u> | 700 | 1 | <u>30</u> | Seeds | ND, ND, 0.05 | <u>0.05</u> |
| Brazil 1978 *Parana* (A16110) | 250 ULV | Spray | 0.50 (25.% ai | 2 | 2 | 32 | Seeds Crude oil Press cake | | 0.30 1.30 0.03 |

Table 69. Endosulfan residues in beetroot resulting from supervised trials in Australia.

| BEETROOT | Application | | | | | PHI | Sample | Residues of alpha, | Total |
| Country, Year of trial *Variety* Report | Form. | Method | kg ai/ha | L/ha | No. | days | analysed | beta endosulfan, Endosulfan sulfate, mg/kg | Residues, mg/kg |
|---|---|---|---|---|---|---|---|---|---|
| Australia 2000 Lockyer Valley *Detroit short top* *N° 1/8/534* | EC | spray | 0.735 | 1000 | 4 | 0 7 14 21 | Root | 0.18,0.11,0.10 0.10, 0.11,011 0.062, 0.063, 0.075 0.080, 0.080, 0.090 | 0.39 0.32 0.20 0.25 |
| Australia 2000 Lockyer Valley *Detroit short top* *N° 1/8/534* | EC | spray | 1.47 | 1000 | 5 | 0 7 14 21 | Root | 0.38, 0.27, 0.13 0.25, 0.22, 0.16 0.20, 0.20, 0.20 0.16, 0.15, 0.15 | 0.78 0.63 0.60 0.46 |

Table 70. Endosulfan residues in carrot resulting from supervised trials in Australia.

| CARROT | Application | | | | | PHI | Sample | Residues of alpha, | Total |
| Country, Year of trial *Variety* Report | Form. | Method | kg ai/ha | L/ha | No. | days | analysed | beta endosulfan, Endosulfan sulfate, mg/kg | Residues, mg/kg |
|---|---|---|---|---|---|---|---|---|---|
| Australia 2000 Virginia *Ricardo* *N° 1/10/565* | EC | spray | 0.735 | 400 | 3 | 0 7 14 21 | Root | 0.016, 0.030, 0.028 0.017, 0.034, 0.025 0.020, 0.021, 0.019 0.034, 0.054, 0.046 | 0.07 0.08 0.06 <u>0.13</u> |
| Australia 2000 Virginia *Ricardo* *N° 1/10/565* | EC | spray | 1.47 | 400 | 3 | 0 7 14 21 14co | Root  co | 0.040, 0.066, 0.046 0.066, 0.12, 0.062 0.044, 0.080, 0.062 0.066, 0.13, 0.082 *0.011, < 0.005, 0.007* | 0.15 0.25 0.19 0.28 *0.02* |
| Australia 2000 Virginia *ricarto* *N° 1/10/565* | EC | spray | 0.735 | 400 | 3 | 14 21 14co | Root  co | 0.019, 0.043, 0.033 0.018, 0.023, 0.019 *0.013, < 0.005, 0.005* | <u>0.10</u> 0.06 *0.02* |
| Australia 2000 SILVAN VIC *flakie* *N° 1/10/565* | EC | spray | 0.735 | 387 | 3 | 0 7 14 21 | Root | < 0.005,< 0.005,< 0.005 < 0.005,< 0.005,< 0.005 < 0.005,< 0.005,< 0.005 < 0.005,< 0.005,< 0.005 | < 0.005 < 0.005 <u>≤ 0.005</u> < 0.005 |
| Australia 2000 Silvan *flakie* *N° 1/10/565* | EC | spray | 1.47 | 387 | 3 | 0 7 14 21 | Root | < 0.005,< 0.005,< 0.005 < 0.005,< 0.005,0.005 < 0.005,< 0.005,< 0.005 0.005,< 0.005,0.006 | < 0.005 0.01 < 0.005 0.01 |
| Australia 2000 Medina *ivar* *N° 1/10/565* | EC | spray | 0.735 | 250 | 3 | 14 21 | Root | 0.013, 0.012, 0.012 0.019, 0.019, 0.016 | <u>0.04</u> 0.05 |

Table 71. Endosulfan residues in potato from supervised trials in Europe, USA and Australia.

| POTATO Country, Year of trial Variety Report | Form. | Method | kg ai/ha | L/ha | No. | PHI days | Sample analysed | Residues of alpha, beta endosulfan, Endosulfan sulfate, mg/kg | Total Residues, mg/kg |
|---|---|---|---|---|---|---|---|---|---|
| Germany 1983 Nicola PSR96/058 | 28DP 2.82% ai. | Spread interval 14d | 0.705 | 25 kg formulation by ha | 2 | 0<br>5<br>10<br>14 | Tuber<br>Tuber<br>Tuber<br>Tuber | | 0.015<br>0.015<br>0.015<br>0.015 |
| Germany 1983 Grata PSR96/058 | 28DP 2.82% ai. | Spread interval 12, 14 d | 0.705 | 25 kg formulation by ha | 3 | 0<br>6<br>10<br>14 | Tuber<br>Tuber<br>Tuber<br>Tuber | | 0.015<br>0.015<br>0.015<br>0.015 |
| Germany 1983 Grata PSR96/058 | 28DP 2.82% ai. | Spread interval 14d | 0.705 | 25 kg formulation by ha | 2 | 0<br>5<br>11<br>14 | Tuber<br>Tuber<br>Tuber<br>Tuber | | 0.015<br>0.015<br>0.015<br>0.015 |
| Germany 1976 Frigga PSR96/058 | 35WP | Spread interval 20d 20 days | 0.210 | 600 (0.035 % ai) | 2 | 13<br>19<br>23<br>28 | Tuber<br>Tuber<br>Tuber<br>Tuber | | 0.015<br>0.015<br>0.015<br>0.015 |
| Germany 1976 Erstling PSR96/058 | 35WP | Spread interval 13d | 0.210 | 600 (0.035 % ai) | 2 | 0<br>13<br>20<br>28 | Tuber<br>Tuber<br>Tuber<br>Tuber | | 0.015<br>0.015<br>0.015<br>0.015 |
| Germany 1976 Marion PSR96/058 | 35WP | Spread interval 8d | 0.210 | 600 (0.035 % ai) | 2 | 20<br>24<br>28<br>20<br>24<br>28 | Peel<br>Peel<br>Peel<br>Tuber<br>(w/o Peel)<br>Tuber | | 0.015<br>0.015<br>0.015<br>0.015<br>0.015<br>0.015 |
| Germany 1977 Saskia 2/713/01/02-77A | 24EC* | Spray interval 20 d | 0.216 | 600 (0.036 % ai) | 2 | 0<br>7<br>14<br>21 | Tuber<br>Tuber<br>Tuber<br>Tuber | | 0.01<br>ND<br>ND<br>ND |
| Germany 1977 Holl. Erstlinge 4/713/01/02-77A | 24EC* | Spray interval 21d | 0.216 | 600 (0.036 % ai) | 2 | 0<br>7<br>14<br>21 | Tuber<br>Tuber<br>Tuber<br>Tuber | | ND<br>ND<br>ND<br>ND |
| Germany 1977 Holl. Erstlinge 1/713/01/02-77 | 24EC* | Spray interval 21d | 0.216 | 600 (0.036 % ai) | 2 | 0<br>7<br>14<br>21 | Tuber<br>Tuber<br>Tuber<br>Tuber | | ND<br>ND<br>ND<br>ND |
| Germany 1977 Marion 3/713/01/02-77A | 24EC* | Spray interval 21 d | 0.216 | 600 (0.036 % ai) | 2 | 0<br>7<br>14<br>21 | Tuber<br>Tuber<br>Tuber<br>Tuber | | ND<br>ND<br>ND<br>ND |
| Spain 1994 Quebec ER 94 ECS 770 | 35EC | Spray interval 14 d BBCH 59/61, 65/71 | 0.528 | 300 (0.176% ai) | 2 | 0<br>7<br>13<br>21<br>27 | Tuber<br>Tuber<br>Tuber<br>Tuber<br>Tuber | < 0.01, < 0.01, < 0.01<br>< 0.01, < 0.01, < 0.01<br>< 0.01, < 0.01, < 0.01<br>< 0.01, < 0.01, < 0.01<br>< 0.01, < 0.01, < 0.01 | < 0.01<br>< 0.01<br>< 0.01<br>< 0.01<br>< 0.01 |
| Spain 1994 Spunta ER 94 ECS 770 | 35EC | Spray interval 14 d BBCH 35, 395 | 0.528 | 300 (0.176% ai) | 2 | 0<br>7<br>14<br>21<br>28 | Tuber<br>Tuber<br>Tuber<br>Tuber<br>Tuber | < 0.01, < 0.01, < 0.01<br>< 0.01, < 0.01, < 0.01<br>< 0.01, < 0.01, < 0.01<br>< 0.01, < 0.01, < 0.01<br>< 0.01, < 0.01, < 0.01 | < 0.01<br>< 0.01<br>< 0.01<br>< 0.01<br>< 0.01 |
| France (S) 1994 Spunta ER 94 ECS 770 | 35EC | Spray interval 14 d BBCH 31/35, 39/41 | 0.528 | 200 (0.264% ai) | 2 | 0<br>7<br>15<br>21<br>29 | Tuber<br>Tuber<br>Tuber<br>Tuber<br>Tuber | < 0.01, < 0.01, < 0.01<br>< 0.01, < 0.01, < 0.01<br>< 0.01, < 0.01, < 0.01<br>< 0.01, < 0.01, < 0.01<br>< 0.01, < 0.01, < 0.01 | < 0.01<br>< 0.01<br>< 0.01<br>< 0.01<br>< 0.01 |

| POTATO Country, Year of trial *Variety* Report | Form. | Method | kg ai/ha | L/ha | No. | PHI days | Sample analysed | Residues of alpha, beta endosulfan, Endosulfan sulfate, mg/kg | Total Residues, mg/kg |
|---|---|---|---|---|---|---|---|---|---|
| Germany 1976 *Hollers* PSR96/058 | 35WP | Spread interval 28d | 0.210 | 600 (0.035 % ai) | 2 | 0<br>13<br>20<br>28 | Tuber<br>Tuber<br>Tuber<br>Tuber | | 0.015<br>0.015<br>0.015<br>0.015 |
| Italy 1994 *Sigunde* ER 94 ECS 770 | 35EC | Spray interval 14 d BBCH 41/51, 69/75 | 0.528 | 900, 1000 (0.0528% ai) | 2 | 0<br>7<br>14<br>21 | Tuber<br>Tuber<br>Tube<br>Tuber | < 0.01, < 0.01, < 0.01<br>< 0.01, < 0.01, < 0.01<br>< 0.01, < 0.01, < 0.01<br>< 0.01, < 0.01, < 0.01 | < 0.01<br>< 0.01<br>< 0.01<br>< 0.01 |
| Italy 1994 *Liseta* ER 94 ECS 770 | 35EC | Spray interval 14 d BBCH 59/61, 61/71 | 0.528 | 300 (0.176% ai) | 2 | 0<br>7<br>14<br>21<br>28 | Tuber<br>Tuber<br>Tuber<br>Tuber<br>Tuber | < 0.01, < 0.01, < 0.01<br>< 0.01, < 0.01, < 0.01<br>< 0.01, < 0.01, < 0.01<br>< 0.01, < 0.01, < 0.01<br>< 0.01, < 0.01, < 0.01 | < 0.01<br>< 0.01<br>< 0.01<br>< 0.01<br>< 0.01 |
| Spain 1994 *Quebec* ER94ECS770 | 35EC | Spray interval 14 d | 0.528 | 300 (0.176% ai) | 2 | 0<br>7<br>13<br>21<br>27 | Tuber<br>Tuber<br>Tuber<br>Tuber<br>Tuber | < 0.01, < 0.01,< 0.01<br>< 0.01, < 0.01, < 0.01<br>< 0.01, < 0.01, < 0.01<br>< 0.01, < 0.01, < 0.01<br>< 0.01, < 0.01, < 0.01 | < 0.01<br>< 0.01<br>< 0.01<br>< 0.01<br>< 0.01 |
| Spain 1994 *Spunta* ER94ECS770 | 35EC | Spray interval 14 d | 0.528 | 300 (0.176% ai) | 2 | 0<br>7<br>14<br>21<br>27 | Tuber<br>Tuber<br>Tuber<br>Tuber<br>Tuber | < 0.01, < 0.01,< 0.01<br>< 0.01, < 0.01, < 0.01<br>< 0.01, < 0.01, < 0.01<br>< 0.01, < 0.01, < 0.01<br>< 0.01, < 0.01, < 0.01 | < 0.01<br>< 0.01<br>< 0.01<br>< 0.01<br>< 0.01 |
| France 1994 *Nicola* ER94ECS770 | 35EC | Spray interval 14 d | 0.528 | 200 (0.264% ai) | 2 | 0<br>7<br>15<br>22<br>29 | Tuber<br>Tuber<br>Tuber<br>Tuber<br>Tuber | < 0.01, < 0.01,< 0.01<br>< 0.01, < 0.01, < 0.01<br>< 0.01, < 0.01, < 0.01<br>< 0.01, < 0.01, < 0.01<br>< 0.01, < 0.01, < 0.01 | < 0.01<br>< 0.01<br>< 0.01<br>< 0.01<br>< 0.01 |
| Italy 1994 *Sigunde* ER94ECS770 | 35EC | Spray interval 14 d | 0.528 | 900, 1000 (0.0528% ai) | 2 | 0<br>7<br>14<br>21 | Tuber<br>Tuber<br>Tuber<br>Tuber | < 0.01, < 0.01,< 0.01<br>< 0.01, < 0.01, < 0.01<br>< 0.01, < 0.01, < 0.01<br>< 0.01, < 0.01, < 0.01 | < 0.01<br>< 0.01<br>< 0.01<br>< 0.01 |
| Italy 1994 *Liseta* Maggiore ER94ECS770 | 35EC | Spray interval 14 d | 0.528 | 300 (0.176% ai) | 2 | 0<br>7<br>14<br>21<br>28 | Tuber<br>Tuber<br>Tuber<br>Tuber<br>Tuber | < 0.01, < 0.01,< 0.01<br>< 0.01, < 0.01, < 0.01<br>< 0.01, < 0.01, < 0.01<br>< 0.01, < 0.01, < 0.01<br>< 0.01, < 0.01, < 0.01 | < 0.01<br>< 0.01<br>< 0.01<br>< 0.01<br>< 0.01 |
| USA (WA) 1995 *R1612B/U022* | 3EC* 33.7% | Broadcast spray Interval 4d | 5;6** | 190 | 3 | 1 | Potato<br>Potato<br>Flakes<br>Chips<br>Wet peel | < 0.05, < 0.05, < 0.05<br>< 0.05, < 0.05, < 0.05<br>< 0.05, < 0.05, < 0.05<br>< 0.05, < 0.05, < 0.05<br>< 0.05, < 0.05, < 0.05 | < 0.05<br>< 0.05<br>< 0.05<br>< 0.05<br>< 0.05 |
| Australia 2000 Torquay *Sequoia* N° 1/10/562 | EC | spray | 0.735 | 350 | 3 | 0<br>7<br>14<br>21 | Tuber<br>Tuber<br>Tuber<br>Tuber | 0.005, 0.005, < 0.005<br>< 0.005,< 0.005, < 0.005<br>< 0.005,< 0.005, < 0.005<br>< 0.005,< 0.005, 0.005 | 0.01<br>< 0.005<br>< 0.005<br>0.005 |
| Australia 2000 Torquay *Sequoia* N° 1/10/562 | EC | spray | 1.47 | 350 | 3 | 0<br>7<br>14<br>21 | Tuber<br>Tuber<br>Tuber<br>Tuber | < 0.005,< 0.005, < 0.005<br>< 0.005,< 0.005, < 0.005<br>< 0.005,< 0.005, < 0.005<br>< 0.005,< 0.005, < 0.005 | < 0.005<br>< 0.005<br>< 0.005<br>< 0.005 |

| POTATO Country, Year of trial *Variety* Report | Form. | Method | kg ai/ha | L/ha | No. | PHI days | Sample analysed | Residues of alpha, beta endosulfan, Endosulfan sulfate, mg/kg | Total Residues, mg/kg |
|---|---|---|---|---|---|---|---|---|---|
| Australia 2000 Lockyer Valley *Sebago* N° 1/10/562 | EC | spray | 0.735 | 500 | 3 | 0 | Tuber | < 0.005,< 0.005,0.008 | 0.008 |
| | | | | | | 7 | Tuber | < 0.005,< 0.005,0.007 | 0.007 |
| | | | | | | 14 | Tuber | < 0.005,< 0.005, 0.007 | 0.007 |
| | | | | | | 21 | Tuber | < 0.005,< 0.005,0.006 | 0.006 |
| Australia 2000 Lockyer Valley *Sebago* N° 1/10/562 | EC | spray | 1.47 | 500 | 3 | 0 | Tuber | < 0.005,< 0.005,0.007 | 0.007 |
| | | | | | | 7 | Tuber | < 0.005,< 0.005,0.006 | 0.006 |
| | | | | | | 14 | Tuber | < 0.005,< 0.005,0.008 | 0.008 |
| | | | | | | 21 | Tuber | < 0.005,< 0.005, 0.008 | 0.008 |
| Australia 2000 Medina *Delaware* N° 1/10/565 | EC | spray | 0.735 | 250 | 3 | 0 | Tuber | 0.26, 0.15,< 0.005 | 0.41 |
| | | | | | | 7 | Tuber | < 0.005,< 0.005,< 0.005 | < 0.005 |
| | | | | | | 14 | Tuber | < 0.005,< 0.005,< 0.005 | < 0.005 |
| | | | | | | 21 | Tuber | < 0.005,< 0.005,< 0.005 | < 0.005 |
| Australia 2000 Virginia *Collabeen* N° 1/10/565 | EC | spray | 1.47 | 300 | 3 | 14 | Tuber | < 0.005,< 0.005,< 0.005 | < 0.005 |

Table 72. Endosulfan residues in sweet potato from supervised trials in USA and Australia.

| SWEET POTATO Country, Year of trial *Variety* | Form. | Method | kg ai/ha/ applic'n | L/ha | No. | PHI Days | Sample analysed | Residues of alpha, beta endosulfan, Endosulfan sulfate, mg/kg | Total Residues mg/kg |
|---|---|---|---|---|---|---|---|---|---|
| USA 1996 NC *BJ-96R-05* | 3EC* | Foliar spray | 1, 34 | 10.1, 96.3 | 2 | 1 | Sw.Potato | < 0.05, < 0.05, < 0.05 | < 0.05 |
| | 50WP | | 1, 34 | 91.6,90.8 | 2 | 1 | Sw.Potato | < 0.05, < 0.05, < 0.05 | < 0.05 |
| | 3EC | 1 soil incorp, | 3x1.12 | 93.5, 91.6,94.4 | 3 | 1 | Sw.Potato | < 0.05, < 0.05, < 0.05 | < 0.05 |
| | 50WP | + 2 fol. spray | 3x1.12 | 96, 95, 95 | 3 | 1 | Sw.Potato | < 0.05, < 0.05, < 0.05 | < 0.05 |
| USA 1996 NC *BJ-96R-05* | 3EC | Foliar spray | 1, 34 | 92.5,94. | 2 | 1 | Sw.Potato | < 0.05, < 0.05, < 0.05 | < 0.05 |
| | 50WP | | 1, 34 | 90.8,93.5 | 2 | 1 | Sw.Potato | < 0.05, < 0.05, < 0.05 | < 0.05 |
| | 3EC | 1 soil incorp, | 3x1.12 | 93.5,92.5,94.4 | 3 | 1 | Sw.Potato | < 0.05, < 0.05, < 0.05 | < 0.05 |
| | 50WP | + 2 fol. spray | 3x1.12 | 94.4,91.6,93.5 | 3 | 1 | Sw.Potato | < 0.05, < 0.05, < 0.05 | < 0.05 |
| USA 1996 SC *BJ-96R-05* | 3EC | Foliar spray | 1, 34 | 98, 102.9 | 2 | 1 | Sw.Potato | < 0.05, < 0.05, < 0.05 | < 0.05 |
| | 50WP | | 1, .34 | 100, 103.8 | 2 | 1 | Sw.Potato | < 0.05, < 0.05, < 0.05 | < 0.05 |
| | 3EC | 1 soil incorp, | 3x1.12 | 99,99,102 | 3 | 1 | Sw.Potato | < 0.05, < 0.05, < 0.05 | < 0.05 |
| | 50WP | + 2 fol. spray | 3x1.12 | 97, 99, 102 | 3 | 1 | Sw.Potato | < 0.05, < 0.05, < 0.05 | < 0.05 |
| USA *Escambia Co.,* 1996 FL *BJ-96R-05* | 3EC | Foliar spray | 1, 34 | 98,98 | 2 | 1 | Sw.Potato | < 0.05, < 0.05, < 0.05 | < 0.05 |
| | 50WP | | 1, .34 | .93.5,98 | 2 | 1 | Sw.Potato | < 0.05, < 0.05, < 0.05 | < 0.05 |
| | 3EC | 1 soil incorp, | 3x1.12 | 115,96.3, 98 | 3 | 1 | Sw.Potato | < 0.05, < 0.05, < 0.05 | < 0.05 |
| | 50WP | + 2 fol. spray | 3x1.12 | 80.4, 95, 97 | 3 | 1 | Sw.Potato | < 0.05, < 0.05, < 0.05 | < 0.05 |
| USA *Seminole Co., FL* 1996 FL *BJ-96R-05* | 3EC | Foliar spray | 1, 34 | 91.6,96.3 | 2 | 1 | Sw potato | < 0.05, < 0.05, < 0.05 | < 0.05 |
| | 50WP | | 1, .34 | 95.4,93.5 | 2 | 1 | Sw potato | < 0.05, < 0.05, < 0.05 | < 0.05 |
| | 3EC | 1 soil incorp, | 3x1.12 | 84, 95,88 | 3 | 1 | Sw potato | < 0.05, < 0.05, < 0.05 | < 0.05 |
| | 50WP | + 2 fol. spray | 3x1.12 | 88, 98, 94.4 | 3 | 1 | Sw potato | < 0.05, < 0.05, < 0.05 | < 0.05 |

| SWEET POTATO | Application | | | | | PHI Days | Sample analysed | Residues of alpha, beta endosulfan, Endosulfan sulfate, mg/kg | Total Residues mg/kg |
|---|---|---|---|---|---|---|---|---|---|
| Country, Year of trial Variety | Form. | Method | kg ai/ha/ applic'n | L/ha | No. | | | | |
| USA | 3EC | Foliar spray | 1, 34 | 93.5,92.5 | 2 | 1 | Sw potato | < 0.05, < 0.05, < 0.05 | < 0.05 |
| Greenville Co., | 50WP | | 1, .34 | 93.5, 92.5 | 2 | 1 | Sw potato | < 0.05, < 0.05, < 0.05 | < 0.05 |
| MS | 3EC | 1 soil incorp, | 3x1.12 | 100,93.5,92.5 | 3 | 1 | Sw potato | < 0.05, < 0.05, < 0.05 | ≤ 0.05 |
| 1996 MS BJ-96R-05 | 50WP | + 2 fol. spray | 3x1.12 | 100,93.5,92.5 | 3 | 1 | Sw potato | < 0.05, < 0.05, < 0.05 | ≤ 0.05 |
| USA | 3EC | Foliar spray | 1, 34 | 91.6,93.5 | 2 | 1 | Sw potato | < 0.05, < 0.05, < 0.05 | < 0.05 |
| Pattison Co., TX | 50WP | | 1, .34 | 93.5, 94.4 | 2 | 1 | Sw potato | < 0.05, < 0.05, < 0.05 | < 0.05 |
| 1996 TX | 3EC | 1 soil incorp, | 3x1.12 | 94.4, 93.5,93.5 | 3 | 1 | Sw potato | < 0.05, < 0.05, < 0.05 | ≤ 0.05 |
| BJ-96R-05 | 50WP | + 2 fol. spray | 3x1.12 | 94.4,94.4, 94.4 | 3 | 1 | Sw potato | < 0.05, < 0.05, < 0.05 | ≤ 0.05 |
| USA | 3EC | Foliar spray | 1, 34 | 94.4, 94.4 | 2 | 1 | Sw potato | < 0.05, < 0.05, < 0.05 | < 0.05 |
| Fresno Co., CA | 50WP | | 1, .34 | 94.4, 95.4 | 2 | 1 | Sw potato | < 0.05, < 0.05, < 0.05 | < 0.05 |
| 1996 CA | 3EC | 1 soil incorp, | 3x1.12 | 3x93.5 | 3 | 1 | Sw potato | < 0.05, < 0.05, < 0.05 | ≤ 0.05 |
| | 50WP | + 2 fol. spray | 3x1.12 | 2x93.5,95.4 | 3 | 1 | Sw potato | < 0.05, < 0.05, < 0.05 | ≤ 0.05 |
| Australia | EC | spray | 0.735 | 300 | 3 | 0 | Sw potato | < 0.005,0.005,< 0.005 | 0.005 |
| 2000 | | | | | | 7 | Sw potato | < 0.005,< 0.005, < 0.005 | < 0.005 |
| Southedge | | | | | | 14 | Sw potato | < 0.005,< 0.005, < 0.005 | ≤ 0.005 |
| sentinal N° 1/8/503 | | | | | | 21 | Sw potato | < 0.005,< 0.005, < 0.005 | < 0.005 |
| Australia | EC | spray | 1.47 | 500 | 3 | 0 | Sw potato | 0.005,< 0.005,< 0.005 | 0.005 |
| 2000 | | | | | | 7 | Sw potato | 0.007,0.008,< 0.005 | 0.015 |
| Southedge | | | | | | 14 | Sw potato | 0.007, 0.006,< 0.005 | 0.013 |
| sentinal N° 1/8/503 | | | | | | 21 | Sw potato | < 0.005,< 0.005,0.0065 | 0.007 |

Table 73. Endosulfan residues in sugar beet from supervised trials in Italy.

| SUGARBEET | Application | | | | | PHI (Days) | Sample analysed | Residues of alpha, beta endosulfan, Endosulfan sulfate, mg/kg | Total Residues, mg/kg |
|---|---|---|---|---|---|---|---|---|---|
| Country, Year of trial Variety | Form. | Method | kg ai/ha/ applic'n | L/ha | No. | | | | |
| Italy 1998 | 35EC | Spray | 0.630 | 600 | 2 | 0 | Root | < 0.02, < 0.02, < 0.02 | < 0.02 |
| Dorotea ER 98 ECS746 | | Interval 30 d BBCH39 | | (0.105% ai) | | 27 | Root | ND, ND, < 0.02 | < 0.02 |
| Italy 1998 | 35EC | Spray | 0.630 | 600 | 2 | 0 | Root | 0.05, 0.03, < 0.02 | 0.08 |
| Adige ER 98 ECS 746 | | Interval 33d BBCH39 | | (0.105% ai) | | 26 | Root | ND, < 0.02, 0.03 | 0.03 |
| Italy 1998 | 35EC | Spray | 0.630 | 300, 600 | 2 | 0 | Root | 0.02, < 0.02, < 0.02 | 0.02 |
| Alesia ER 98 ECS 746 | | Interval 32d BBCH38/39 | | (0.105% ai) | | 28 | Root | ND, < 0.02, 0.03 | 0.03 |
| Italy 1998 | 35EC | Spray | 0.630 | 300, | 2 | 0 | Root | < 0.02, < 0.02, < 0.02 | < 0.02 |
| Monodoro ER 98 ECS 746 | | Interval 32 d BBCH 38/39 | | 600 (0.10% ai)5 | | 28 | Root | ND, < 0.02, < 0.02 | < 0.02 |
| Italy1997 | 35EC | Spray | 0.630 | 600 | | 0 | Root | 0.02 ,< 0.02, < 0.02 | 0.02 |
| Bianca | | Interval 20 d | | (0.105 | | 14 | Root | ND, ND, < 0.02, | < 0.02 |
| ER 97 ECS 746 | | BBCH 39 | | % ai) | | 28 | Root | ND, ND, < 0.02 | < 0.02 |
| Italy1997 | 35EC | Spray | 0.630 | 600 | 2 | 0 | Root | 0.02, < 0.02, 0.03 | 0.05 |
| Adige | | Interval 20 d | | (0.105 | | 14 | Root | < 0.02, < 0.02, 0.03 | 0.03 |
| ER 97 ECS 746 | | BBCH 39 | | % ai) | | 28 | Root | ND, ND, < 0.02 | < 0.02 |
| Italy1997 | 35EC | Spray | 0.630 | 600 | 2 | 0 | Root | < 0.02, < 0.02, < 0.02 | < 0.02 |
| Monodoro | | Interval 20 d | | (0.105 | | 14 | Root | ND, < 0.02, < 0.02 | < 0.02 |
| ER 97 ECS 746 | | BBCH 39 | | % ai) | | 28 | Root | ND, ND, 0.02 | 0.02 |
| Italy1997 | 35EC | Spray | 0.630 | 600 | 2 | 0 | Root | < 0.02, < 0.02, 0.02 | 0.02 |
| Formula | | Interval 20 d | | (0.105 | | 14 | Root | ND, < 0.02, 0.04 | 0.04 |
| ER 97 ECS 746 | | BBCH 39,49 | | % ai) | | 28 | Root | ND, ND, 0.03 | 0.03 |

| SUGARBEET | Application | | | | | PHI | Sample | Residues of alpha, | Total |
| Country, Year of trial *Variety* | Form. | Method | kg ai/ha/ applic'n | L/ha | No. | (Days) | analysed | beta endosulfan, Endosulfan sulfate, mg/kg | Residues, mg/kg |
|---|---|---|---|---|---|---|---|---|---|
| Italy1997 *Rizor* *ER 97 ECS 746* | 35EC | Spray Interval 20 d BBCH 39,49 | 0.630 | 600 (0.105 % ai) | 2 | 0 14 28 | Root Root Root | < 0.02, ND, < 0.02 ND, ND, 0.02 ND, ND, < 0.02 | < 0.02 0.02 < 0.02 |

Table 74. Endosulfan residues in celery from supervised trials in USA and Australia.

| CELERY | Application | | | | | PHI | Sample | Residues of alpha, | Total |
| Country, Year of trial *Variety, report* | Form | Method | Rate kg ai/ha/ applic'n | Water volume L/ha | No. | Days | analysed | beta endosulfan, Endosulfan sulfate, mg/kg | Residues mg/kg |
|---|---|---|---|---|---|---|---|---|---|
| USA 1997/8 MI *BJ-97R-02* | 3EC* 50WP | (backpack). (backpack). | 1.12 1.12 | 271,280.5 (0.414% ai) | 1 1 | 14 14 | Celery Celery | 0.83, 0.37, 0.26 0.83, 0.31, 0.27 | 1.46 1.41 |
| USA1997/8 CA *Pacifica* *BJ-97R-02* | 3EC* 50WP | Spray (backpack). | 1 1 | 19 20 (0.6% ai) | 1 1 | 14 14 | Celery Celery | 1.09, 0.79, 0.46, 1.28, 1.44, 1.00 | 2.34 3.72 |
| USA 1997/8 CA *TA Special #1* *BJ-97R-02* | 3EC* 50WP | Spray (backpack). | 1 1 | 30 30 (0.414% ai) | 1 1 | 14 14 | Celery Celery | 0.35, 0.24, 0.15 0.17, 0.29, 0.09 | 0.74 0.55 |
| USA1997/8 CA *Conquistador* *BJ-97R-02* | 3EC* 50WP | Spray (backpack). | 1 1 | 31 30 (0.414% ai) | 1 1 | 14 14 | Celery Celery | 0.32, 0.26, 0.14 0.38, 0.73, 0.17 | 0.72 1.28 |
| USA 1997/8 CA *Conquistador* *BJ-97R-02* | 3EC* 50WP | Spray . (backpack). | 1.12 1.12 | 271 290 (0.414% ai) | 1 1 | 14 14 | Celery Celery | 1.20, 0.83, 0.40 1.56, 1.52, 0.39 | 2.43 3.47 |
| USA 1995 CA *Matador* *U022/R141C* | 3EC* 50WP | Spray (directed) | 1.12 1.12 | 649 648 (0.17% ai) | 1 1 | 4 4 | Celery Celery | 1.19, 1.15, 0.13 1.41, 1.37, 0.15 | 2.47 2.93 |
| USA 1995 CA *T&A Special #1* *U022/R141C* | 3EC* 50WP | Spray (directed) | 1.12 1.12 | 635 646 (0.174% ai) | 1 1 | 4 4 | Celery Celery | 1.61, 1.19, 0.34 1.36, 1.47, 0.24 | 3.14 3.07 |
| USA 1995 CA *TA Special #1* *U022/R141C* | 3EC* 50WP | Spray (directed) | 1.12 1.12 | 650 663 (0.17% ai) | 1 1 | 4 4 | Celery Celery | 0.87, 1.49, 0.24 1.61, 2.0, 0.14 | 2.60 3.75 |
| USA 1995 MI *Florida 683K* *U022/R141C* | 3EC* 50WP | Spray broadcast | 1.12 1.12 | 220.7 220.7 (0.51% ai) | 1 1 | 4 4 | Celery Celery | 0.77, 0.75, 0.23 1.24, 1.14, 0.23 | 1.75 2.61 |
| USA 1995 FL *1622* *U022/R141C* | 3EC* 50WP | Spray broadcast | 1.12 1.12 | 416 416 (0.27% ai) | 1 1 | 4 4 | Celery Celery | 0.46, 0.29, 0.25 0.63, 0.47, 0.26 | 1.00 1.36 |
| USA 1995 CA *Conquistador* *U022/R141C* | 3EC* 50WP | Spray broadcast | 1.12 1.12 | 909 942 (0.12% ai) | 1 1 | 4 4 | Celery Celery | 1.89, 1.77, 0.39 2.32, 2.14, 0.55 | 4.05 5.01 |
| USA 1965 CA *F-M 96* *R-90* | 2EC | Spray | 1.12 | | 3 | 0 7 14 21 28 | Celery Celery Celery Celery Celery | | 17.8-17.9 2.0-2.6 0.83-1.09 0.53-0.58 0.37-0.39 |
| USA 1965 CA *F-M 96* *R-908* | 2EC | Spray (2 qt/40 gal) | 1.12 | | 3 | 0 7 14 21 | Celery Celery Celery Celery | | 11.2-13.9 1.72-1.77 1.08-1.27 0.79-0.96 |
| USA 1965 CA *F-M 96* *R-908* | 2EC | Spray (2 qt/40 gal) | 1.12 | | 3 | 0 7 14 21 28 | Celery Celery Celery Celery Celery | | 13.5-17.7 3.3-3.5 0.85-1.17 0.60-0.72 0.47-0.51 |

| CELERY | Application | | | | | PHI | Sample | Residues of alpha, beta endosulfan, Endosulfan sulfate, mg/kg | Total Residues mg/kg |
|---|---|---|---|---|---|---|---|---|---|
| Country, Year of trial *Variety, report* | Form | Method | Rate kg ai/ha/ applic'n | Water volume L/ha | No. | Days | analysed | | |
| Australia 2000 Toowoomba *American stringless* *N° 1/10/535* | EC | spray | 0.232 | 350 | 3 | 0 | Celery | 2.5, 1.5, 0.22 | 4.20 |
| | | | | | | 3 | Celery | 0.58, 0.39,0.25 | 1.20 |
| | | | | | | 7 | Celery | 0.31, 0.16, 0.12 | 0.59 |
| | | | | | | 10 | Celery | 0.56, 0.29,0.26 | 1.10 |
| Australia 2000 Toowoomba *American stringless* *N° 1/10/535* | EC | spray | 0.707 | 575 | 3 | 7 | Celery | 1.1, 0.63, 0.48 | 2.20 |
| | | | | | | 10 | Celery | 1.6, 0.85,0.58 | 3.0 |
| | | | | | | 7 | Celery co | 0.054,0.024,0.045 | 0.12 |
| | | | | | | 10 | Celery co | 0.064,0.029,.058 | 0.15 |
| Australia 2000 Cranbourne *Summit* *N° 1/10/535* | EC | spray | 0.465 | 700 | | 0 | Celery | 0.18, 0.12, 0.053 | 0.36 |
| | | | | | | 3 | Celery | 0.16, 0.13,0.068 | 0.36 |
| | | | | | | 7 | Celery | 0.12, 0.076, 0.062 | 0.26 |
| | | | | | | 10 | Celery | 0.14, 0.09, 0.062 | 0.29 |
| Australia 2000 Cranbourne *Summit* *N° 1/10/535* | EC | spray | 0.865 | 700 | | 7 | Celery | 0.71, 0.43, 0.15 | 1.30 |
| | | | | | | 10 | Celery | 0.34, 0.20.0.081 | 0.62 |
| | | | | | | 7 | Celery co | < 0.005,< 0.005,< 0.005 | 0.015 |
| | | | | | | 10 | Celery co | 0.006,0.013,< 0.005 | 0.019 |

Table 75. Endosulfan residues in rhubarb from supervised trials in Australia.

| RHUBARB | Application | | | | | PHI | Sample | Residues of alpha, beta endosulfan, Endosulfan sulfate, mg/kg | Total Residues, mg/kg |
|---|---|---|---|---|---|---|---|---|---|
| Country, Year of trial *Variety* Report | Form. | Method | Rate g ai/hL/ | Water volume L/ha | No. | days | analysed | | |
| Australia 2000 Hamton *Sydney crimson* *N° 1/10/558* | EC | spray | 70 | 185 | 2 | 0 | rhubarb | 0.015,0.29,0.26 | 0.57 |
| | | | | | | 3 | rhubarb | 0.016,0.0612,0.013 | 0.09 |
| | | | | | | 7 | rhubarb | 0.012,0.036,0.011 | 0.059 |
| | | | | | | 10 | rhubarb | 0.017,0.042,0.02 | 0.079 |
| Australia 2000 Hamton *Sydney crimson* *N° 1/10/558* | EC | spray | 140 | 185 | 2 | 7 | rhubarb | 0.026,0.10.0.039 | 0.16 |
| | | | | | | 10 | rhubarb | 0.015, 0.06,0.028 | 0.10 |
| | | | | | | 7 | rhubarb co | < 0.005,< 0.005,< 0.005 | <.0.005 |
| | | | | | | 10 | rhubarb co | < 0.005,< 0.005,< 0.005 | < 0.015 |
| Australia 2000 Mt Tambourine *Big red* *N° 1/10/558* | EC | spray | 70 | 334 | 2 | 0 | rhubarb | 2.10,1.50.0.094 | 3.70 |
| | | | | | | 3 | rhubarb | 0.41,0.33,0.065 | 0.80 |
| | | | | | | 7 | rhubarb | 0.098,0.17,0.072 | 0.34 |
| | | | | | | 10 | rhubarb | 0.056,0.087,0.053 | 0.20 |
| Australia 2000 Mt Tambourine *Big red* *N° 1/10/558* | EC | spray | 140 | 334 | 2 | 7 | rhubarb | 0.19,0.034,0.20 | 0.73 |
| | | | | | | 10 | rhubarb | 0.063, 0.12,0.063 | 0.25 |
| | | | | | | 7 | rhubarb co | < 0.005,< 0.005,< 0.005 | < 0.005 |
| | | | | | | 10 | rhubarb co | < 0.005,< 0.005,< 0.005 | < 0.005 |

Table 76. Endosulfan residues in hazelnuts from supervised trials in Italy.

| HAZELNUT | Application | | | | | PHI | Sample analysed | Residues of alpha, beta endosulfan, Endosulfan sulfate, mg/kg | Total Residues, mg/kg |
|---|---|---|---|---|---|---|---|---|---|
| Country, Year of trial *Variety* | Form. | Method | Rate kg ai/ha/ applic'n | Water volume L/ha | No. | days | | | |
| Italy 1999 *Gentile Tonda Romana 99335/11- FPHN* | 33EC | Spray (backpack) Interval 14d BBCH 81, 81 | 0.796 | 992 (0.08 % ai) | 2 | 0 | Nutmeat | ND, ND, ND | < 0.02 |
| | | | | | | 7 | Nutmeat | ND, ND, ND | < 0.02 |
| | | | | | | 14 | Nutmeat | ND, ND, ND | < 0.02 |
| | | | | | | 21 | Nutmeat | ND, ND, ND | ≤ 0.02 |
| | | | | | | 28 | Nutmeat | ND, ND, ND | ≤ 0.02 |
| Italy 1999 *Gentile Tonda Romana 99335/11- FPHN* | 33EC | Spray (backpack) | 0.796 | 991 (0.08 % ai) | 2 | 0 | Nutmeat | ND, ND, ND | < 0.02 |
| | | | | | | 7 | Nutmeat | ND, ND, ND | < 0.02 |
| | | | | | | 14 | Nutmeat | ND, ND, ND | < 0.02 |
| | | | | | | 21 | Nutmeat | ND, ND, ND | ≤ 0.02 |
| | | | | | | 28 | Nutmeat | ND, ND, ND | ≤ 0.02 |
| Italy 1998 *Gentile Tonda Romana 98059/11- FPHN* | 33EC | | 0.288 | 1000 (0.028 % ai) | 2 | 0 | Nutmeat | ND, ND, ND | < 0.02 |
| | | | | | | 7 | Nutmeat | ND, ND, ND | < 0.02 |
| | | | | | | 14 | Nutmeat | ND, ND, ND | ≤ 0.02 |
| | | | | | | 21 | Nutmeat | ND, ND, ND- | < 0.02 |
| | | | | | | 28 | Nutmeat | ND, ND, ND | < 0.02 |
| Italy 1998 *Gentile Tonda Romana 98059/11- FPHN* | 33EC | | 0.2688 | 1000 (0.028 % ai) | 2 | 0 | Nutmeat | ND, ND, ND | < 0.02 |
| | | | | | | 7 | Nutmeat | ND, ND, ND | < 0.02 |
| | | | | | | 14 | Nutmeat | ND, ND, ND | ≤ 0.02 |
| | | | | | | 21 | Nutmeat | ND, ND, ND | < 0.02 |
| | | | | | | 28 | Nutmeat | ND, ND, ND | < 0.02 |

Table 77. Endosulfan residues in macadamia from supervised trials in Australia.

| MACADAMIA | Application | | | | | PHI | Sample analysed | Residues of alpha, beta endosulfan, Endosulfan sulfate, mg/kg | Total Residues, mg/kg |
|---|---|---|---|---|---|---|---|---|---|
| Country, Year of trial *Variety* Report | Form. | Method | g ai/hL | L/ha | No. | days | | | |
| Australia *2001* Dorroughby *334* N° 1/10/532 | EC | spray | 52.5 | 1400 | 3 | 0 | | < 0.005,< 0.005,< 0.005 | < 0.005 |
| | | | | | | 1 | | < 0.005,< 0.005,< 0.005 | < 0.005 |
| | | | | | | 2 | | < 0.005,< 0.005,< 0.005 | ≤ 0.005 |
| | | | | | | 4 | | < 0.005,< 0.005,< 0.005 | < 0.005 |
| Australia *2001* Dorroughby *334* N° 1/10/532 | EC | spray | 105 | 1400 | 3 | 0 | | < 0.005,< 0.005,< 0.005 | < 0.005 |
| | | | | | | 1 | | < 0.005,< 0.005,< 0.005 | < 0.005 |
| | | | | | | 2 | | < 0.005,< 0.005,< 0.005 | < 0.005 |
| | | | | | | 4 | | < 0.005,< 0.005,< 0.005 | < 0.005 |
| Australia *2001* Tolga *344* N° 1/10/532 | EC | spray | 52.5 | 1250 | 3 | 0 | | < 0.005,< 0.005,< 0.005 | < 0.005 |
| | | | | | | 1 | | < 0.005,< 0.005,< 0.005 | < 0.005 |
| | | | | | | 2 | | < 0.005,< 0.005,< 0.005 | ≤ 0.005 |
| | | | | | | 4 | | < 0.005,< 0.005,< 0.005 | < 0.005 |
| Australia *2001* Glasshouse Mtns *344/711* N° 1/10/532 | EC | spray | 52.5 | 1400 | 3 | 0 | | < 0.005,< 0.005,< 0.005 | < 0.005 |
| | | | | | | 1 | | < 0.005,< 0.005,< 0.005 | < 0.005 |
| | | | | | | 2 | | < 0.005,< 0.005,< 0.005 | ≤ 0.005 |
| | | | | | | 4 | | < 0.005,< 0.005,< 0.005 | < 0.005 |

Table 78. Endosulfan residues in cotton from supervised trials in Spain and Greece.

| COTTON | Application | | | | | PHI | Sample | Residues of alpha, | Total |
|---|---|---|---|---|---|---|---|---|---|
| Country, Year of trial *Variety*, report | Form. | Method | kg ai/ha / applic'n | L/ha | No. | days | analysed | beta endosulfan, Endosulfan sulfate, mg/kg | Residues mg/kg |
| Spain 2002 tabladilla 02 R 172 | CS | spray | 0.840 | 550 0.15% | 3 | 0<br><br>20 | Bolls<br><br>seeds | 0.65, 0.56, 0.02<br><br>0.07, 0.09, 0.08 | 1.23<br><br>0.24 |
| Spain 2002 *Roca* 02 R 172 | CS | spray | 0.840 | 550 0.15% | 3 | 0<br><br>20 | Bolls<br><br>seeds | 0.53, 0.46, < 0.02<br><br>< 0.02, < 0.02, < 0.02 | 0.99<br><br>≤ 0.02 |
| Greece 2002 *Sig 125* 02 R 172 | CS | spray | 0.840 | 600 0.14% | 3 | 0<br><br>28 | Bolls<br><br>seeds | 0.93, 0.77, 0.09<br><br>< 0.02, 0.03, < 0.02 | 1.79<br><br>0.03 |
| Greece 2002 *Sig 125* 02 R 172 | CS | spray | 0.840 | 600 0.14% | 3 | 0<br><br>28 | Bolls<br><br>seeds | 0.34, 0.35, 0.13<br><br>< 0.02, < 0.02, < 0.02 | 0.82<br><br>< 0.02 |
| Spain 2002 *tabladilla* 02 R 170 | 33 EC | spray | 0.84 | 550 0.15% | 3 | 0<br><br>20 | Bolls<br><br>seeds | 0.17, 0.29, 0.04<br><br>< 0.02, 0.02, 0.04 | 0.50<br><br>0.06 |
| Spain 2002 *Roca* 02 R 170 | 33 EC | spray | 0.84 | 550 0.15% | 3 | 0<br><br>20 | Bolls<br><br>seeds | 0.50, 0.52, 0.04<br><br>< 0.02, < 0.02, < 0.02 | 1.06<br><br>≤ 0.02 |
| Greece 2002 *Sig 125* 02 R 170 | 33 EC | spray | 0.84 | 600 0.14% | 3 | 0<br><br>28 | Bolls<br><br>seeds | 0.94, 1.30, 0.28<br><br>< 0.02, < 0.02, 0.04 | 2.52<br><br>0.04 |
| Greece 2002 *Sig 125* 02 R 170 | 33 EC | spray | 0.84 | 600 0.14% | 3 | 0<br><br>28 | Bolls<br><br>seeds | 0.12, 0.19, 0.11<br><br>< 0.02, < 0.02, < 0.02 | 0.42<br><br>< 0.02 |
| Greece 2002 *Midas* 02 R 170 | 33 EC | spray | 0.84 | 600 0.14% | 3 | 0<br><br>28 | Bolls<br><br>seeds | 1.40, 1.80, 0.21<br><br>< 0.02, < 0.02, < 0.02 | 3.41<br><br>< 0.02 |
| Spain 2001 *Sonia* 02 R 170 | CS | spray | 0.84 | 1000 0.084% | 3 | 0<br>7<br>14<br>21 | Bolls<br>Bolls<br>Bolls<br>Seeds | 0.46, 0.36, 0.15<br>0.10, 0.08, 0.04<br>0.05, 0.04, 0.03<br>< 0.02, < 0.02, < 0.02 | 0.97<br>0.22<br>0.12<br>≤ 0.02 |
| Spain 2001 *Sonia* 01 R 170 | 33 EC | spray | 0.84 | 1000 0.084% | 3 | 0<br>7<br>14<br><br>14 | Bolls<br>Bolls<br>Bolls<br><br>Seeds | 0.18, 0.15, < 0.02<br>< 0.02, 0.03, 0.06<br>< 0.02, < 0.02, < 0.02<br><br>< 0.02, < 0.02, < 0.02 | 0.33<br>0.11<br>< 0.02<br><br>< 0.02 |
| Spain 2001 *Sonia* 01 R 170 | CS | spray | 0.84 | 1000 0.084% | 3 | 0<br>7<br>14<br>14 | Bolls<br>Bolls<br>Bolls<br>Seeds | 0.67, 0.40, 0.02<br>0.34, 0.16, 0.07<br>0.09, 0.07, 0.03<br>< 0.02, < 0.02, < 0.02 | 1.09<br>0.57<br>0.19<br>< 0.02 |
| Spain 2001 01 R 170 | 33 EC | spray | 0.84 | 1000 0.084% | 3 | 0<br>7<br>14<br>14 | Bolls<br>Bolls<br>Bolls<br>Seeds | 1.10, 0.69, 0.16<br>0.05, 0.03, 0.17<br>0.04, 0.12, 0.27<br>< 0.02, < 0.02, < 0.02 | 1.95<br>0.25<br>0.43<br>< 0.02 |
| Spain 2001 *bravada* 01 R 170 | CS | spray | 0.84 | 1000 0.084% | 3 | 0<br>7<br>14<br>14 | Bolls<br>Bolls<br>Bolls<br>Seeds | 0.89, 0.52, 0.40<br>0.33, 0.36, 0.17<br>0.25, 0.26, 0.17<br>< 0.02, < 0.02, < 0.02 | 1.81<br>0.86<br>0.68<br>< 0.02 |

| COTTON Country, Year of trial *Variety,* report | Application Form. | Method | kg ai/ha / applic'n | L/ha | No. | PHI days | Sample analysed | Residues of alpha, beta endosulfan, Endosulfan sulfate, mg/kg | Total Residues mg/kg |
|---|---|---|---|---|---|---|---|---|---|
| Spain 2001 01 R 170 | CS | spray | 0.49 | 1000 0.05% | 3 | 0 7 14 14 | Bolls Bolls Bolls Seeds | 0.54, 0.28, 0.04 0.08, 0.07, 0.04 0.10, 0.10, 0.06 < 0.02, < 0.02, < 0.02 | 0.86 0.19 0.26 < 0.02 |
| Greece 2001 01 R 170 | 33 EC | spray | 0.84 | 600 0.14% | 3 | 0 7 14 14 | Bolls Bolls Bolls Seeds | 0.14, 0.10, < 0.02 < 0.02, 0.06, 0.05 < 0.02, 0.02, < 0.02 < 0.02, < 0.02, < 0.02 | 0.26 0.13 0.02 < 0.02 |
| Greece 2001 *SG 125* 01 R 170 | CS | spray | 0.84 | 600 0.14% | 3 | 0 7 14 14 | Bolls Bolls Bolls Seeds | < 0.02, < 0.02, < 0.02 0.15, 0.18, 0.06 0.10, 0.11, 0.02 0.04, 0.04, 0.03 | < 0.02 0.39 0.23 0.11 |
| Greece 2001 *SG 125* 01 R 170 | CS | spray | 0.49 | 600 0.08% | 3 | 0 7 14 14 | Bolls Bolls Bolls Seeds | 0.05, 0.03, < 0.02 1.40, 0.75,< 0.02 1.00, 0.49, < 0.02 0.07, 0.06, < 0.02 | 0.08 2.15 1.49 0.13 |
| Greece 2001 *SG 125* 01 R 170 | CS | spray | 0.84 | 600 0.14% | 3 | 0 7 14 14 | Bolls Bolls Bolls Seeds | 0.12, 0.11, < 0.02 2.20, 1.20, < 0.02 2.3, 1.10, 0.04 0.29, 0.20, < 0.02 | 0.23 3.40 3.44 0.49 |
| Australia 1974*Delta pine* PSR 99/011 | 35 EC | spray | 0.74 | 10-20 3.7-7.4% | 13 | 44 | Seeds | | ≤ 0.02 |
| Australia 1974*Delta pine* PSR 99/011 | 35 EC | spray | 0.74 | 11 6.73% | 15 | 25 | Seeds | | 0.035 |
| Spain 1992 *Crema 111* PSR 99/011 | 35EC | spray | 0.63 | 600 0.105% | 1 | 0 3 7 15 | Seeds Seeds Seeds Seeds | | 2.99 0.78 0.27 0.05 |
| Spain 1992 *Stoneville 506* PSR 99/011 | 35EC | spray | 0.63 | 600 0.105% | 1 | 0 3 7 15 | Seeds Seeds Seeds Seeds | | 2.96 0.35 0.30 0.05 |
| Spain 1992 *Crema 111* PSR 99/011 | 35EC | spray | 1.00 | 956 0.105% | 1 | 0 3 7 15 | Seeds Seeds Seeds Seeds | | 0.91 0.20 0.17 0.02 |
| Spain 1992 *Cocker 310* PSR 99/011 | 35EC | spray | 1.00 | 956 0.105% | 1 | 0 3 7 14 | Seeds Seeds Seeds Seeds | | 0.86 0.22 0.22 0.25 |
| Spain 1992 *Stoneville* PSR 99/011 | 35EC | spray | 1.00 | 950 0.105% | 1 | 0 3 7 14 | Seeds Seeds Seeds Seeds | | 0.79 0.62 0.25 0.03 |
| Spain 1992 *Crema 111* PSR 99/011 | 35EC | spray | 1.00 | 950 0.105% | 1 | 0 3 7 14 | Seeds Seeds Seeds Seeds | | 0.68 0.10 0.10 0.12 |
| Spain 1992 *Max 9* PSR 99/011 | 35EC | spray | 1.11 | 1058 0.105% | 1 | 0 3 7 14 | Seeds Seeds Seeds Seeds | | 1.39 0.24 0.11 0.07 |
| Spain 1992 *Cocker 310* PSR 99/011 | 35EC | spray | 111 | 1058 0.105% | 1 | 0 3 7 14 | Seeds Seeds Seeds Seeds | | 1.83 0.40 0.11 0.10 |

Table 79. Endosulfan residues in cocoa from supervised trials in Brazil, Ghana and Ivory Coast.

| COCOA Country, Year of trial Variety | Form | Method | kg ai/ha / applic'n | L/ha | No. | PHI days | Sample analysed | Residues of alpha, beta endosulfan, Endosulfan sulfate, mg/kg | Total Residues mg/kg |
|---|---|---|---|---|---|---|---|---|---|
| Brazil1982 Forasteiro (A25749) | 35EC | Foliar Spray Interval 21 d | 0.35 | 120 (0.292%) | 2 | 30 45 | Fruit Fruit | < 0.01, < 0.01, < 0.01 < 0.01, < 0.01, < 0.01 | < 0.01 < 0.01 |
| Brazil1982 Forasteiro | 35EC | Foliar Spray Interval 21 d | 0.70 | 120 (0.583%) | 2 | 30 45 | Fruit Fruit | < 0.01, < 0.01, < 0.01 < 0.01, < 0.01, < 0.01 | < 0.01 < 0.01 |
| Brazil1982 Forasteiro (A25748) | 35EC | Foliar Spray Interval 21 d | 0.70 | 120 (0.583%) | 2 | 30 45 | Fruit Fruit | < 0.01, < 0.01, < 0.01 < 0.01, < 0.01, < 0.01 | < 0.01 < 0.01 |
| Brazil1982 Forasteiro (A25747) | 35EC | Foliar Spray Interval 21 d | 0.35 | 120 (0.292%) | 2 | 30 45 | Fruit Fruit | < 0.01, < 0.01, < 0.01 < 0.01, < 0.01, -- | < 0.01 < 0.01 |
| Ghana Ahensan 2002 Several hybrids AF1-FPCC 20013056/ C032568 | CS* | Spray, BBCH 67-81 and 66-84 (Spray #1: 42 Days before harvest Spray #2: 13 day PHI) | 0.243, 0.273 | 58, 66 (1.25, 1.24 % v/v) | 2 | 0 7 13/20 13 13 20 27 27 27 | Beans Beans Beans Ferm.beans Dry beans Ferm.beans Beans Ferm.beans Dry beans | 0.01, < 0.01, ND < 0.01, < 0.01, ND ND, ND, ND < 0.01, ND, ND ND, ND, ND < 0.01, < 0.01, ND ND, ND, ND < 0.01, < 0.01, ND < 0.01, ND, ND | 0.01 ≤ 0.01 < 0.01 < 0.01 < 0.01 < 0.01 ≤ 0.01 < 0.01 < 0.01 |
| Ghana Ahensan 2002 Several hybrids 20013056/ AF1-FPCC | 500EC | Spray, BBCH 67-81 and 66-85 (42 d and 13 d before harvest | 0.256, 0.263 | 62, 63 58, 66 (0.78, 0.79%) | 2 | 0/7/13 13/20 13 20/27 20/27 27 | Beans Ferm.beans Dry beans Beans Dry beans Ferm.beans | ND, ND, ND ND, ND, ND ND, ND, ND ND, ND, ND ND, ND, < 0.01 ND, ND, < 0.01 | ≤ 0.01 < 0.01 < 0.01 < 0.01 < 0.01 < 0.01 |
| Ghana Fumso 2002 Amazon 20013056/ AF1-FPCC | CS* | Spray, BBCH 67-77 and 84 (Spray 1: 41 Days before harvest Spray 2: 14 day PHI) | 0.26, 0.26 | 62, 62 (1.25% v/v) | 2 | 0 7 14 14 14 20/27 20 20 27 27 | Beans Beans Beans Ferm.beans Dry beans Beans Ferm.beans Dry beans Ferm.beans Dry beans | < 0.01, ND, ND 0.02, 0.01, ND 0.01, < 0.01, ND < 0.01, < 0.01, ND 0.01,< 0.01, ND ND, ND, ND < 0.01, ND, ND ND, ND, < 0.01 < 0.01, < 0.01, ND < 0.01, ND, < 0.01 | < 0.01 0.03 0.01 < 0.01 0.01 < 0.01 < 0.01 < 0.01 < 0.01 < 0.01 |
| Ghana Fumso 2002 Amazon 20013056/ AF1-FPCC | 500EC | Spray BBCH 67-77 and 84 ( 41 d and 14 d before harvest) | 0.259, 0.25 | 62, 60 (0.79% v/v) | 2 | 0 7/14 14/20: 14 20/27 20/27 27 | Beans Beans Ferm.beans Dry beans Beans Dry beans Ferm.beans | ND, ND, ND ND, ND, ND ND, ND, ND ND, ND, ND ND, ND, ND ND, ND, < 0.01 ND, ND, ND | < 0.01 ≤ 0.01 < 0.01 < 0.01 < 0.01 < 0.01 < 0.01 |
| Ghana Tafo 2002 Hybrids 20013056/ AF1-FPCC | CS* | Spray BBCH 67-84 and 66-82 ( 40 d and 13 d before harvest) | 0.246, 0. 256 | 59, 61 (1.24, 1.25% v/v) | 2 | 0 6 13/27 13 13 20 20/27 20/27 | Beans Beans Beans Ferm.beans Dry beans Beans Ferm.beans Dry beans | 0.02, < 0.01, ND < 0.01, < 0.01, ND ND, ND, ND ND, ND, < 0.01 ND, ND, < 0.01 < 0.01, ND, ND < 0.01, ND, < 0.01 ND, ND, < 0.01 | 0.02 ≤ 0.01 < 0.01 < 0.01 < 0.01 < 0.01 < 0.01 < 0.01 |

| COCOA Country, Year of trial *Variety* | Form | Method | kg ai/ha / applic'n | L/ha | No. | PHI days | Sample analysed | Residues of alpha, beta endosulfan, Endosulfan sulfate, mg/kg | Total Residues mg/kg |
|---|---|---|---|---|---|---|---|---|---|
| Ghana Tafo  2002 Hybrids 20013056/ AF1-FPCC | 500EC | Spray BBCH 67-84  and 66-86 ( 40 d and 13 d before harvest) | 0. 254 0.255, | 61, 61 (0.79, 0.78% | | 0 6/13  13 13/20 20/27 20/27 27 | Beans Beans  Ferm.beans Dry beans Beans Ferm.beans Dry beans | < 0.01, < 0.01, ND ND, ND, ND  ND, ND, ND ND, ND, < 0.01 ND, ND, ND ND, ND, < 0.01 ND, ND, 0.02 | < 0.01 ≤ 0.01  < 0.01 < 0.01 ≤ 0.01 < 0.01 0.02 |
| Ghana Bososo  2002 Several hybrids  20013056/ AF1-FPCC | CS* | Spray BBCH 67-76  and 66-85 (42 d and 14d before harvest) | 0.250, 0. 243 | 60, 58 (1.24, 1.25% v/v) | 2 | 0 7  14/21 14/21  14/21 27 27 27 | Beans Beans  Beans Ferm.beans  Dry beans Beans Ferm.beans Dry beans | < 0.01, < 0.01, ND < 0.01, ND, ND  ND, ND, ND < 0.01, ND, < 0.01  ND, ND, < 0.01 ND, ND, ND < 0.01, ND, < 0.01 ND, ND, < 0.01 | < 0.01 ≤ 0.01  < 0.01 < 0.01  < 0.01 < 0.01 < 0.01 < 0.01 |
| Ghana Bososo  2002 *Several hybrids*  20013056/ *AF1-FPCC* | 500EC | Spray BBCH 67-76  and 66-85 (42 d and 14d before harvest) | 0.247, 0. 272 | 59, 65 (0.79%, v/v) | 2 | 0/7/14 14/21  14/21 21/27  27 27 | Beans Ferm.beans  Dry beans Beans  Ferm.beans Dry beans | ND, ND, ND ND, ND, < 0.01  ND, ND, < 0.01 ND, ND, ND  ND, ND, 0.01 ND, ND, 0.01 | ≤ 0.01 < 0.01  < 0.01 ≤ 0.01  0.01 0.01 |
| Ivory Coast Yakassé Mé  2001 *Criollo*  *Forasteiro* 20013056/ *IC1-FPCC* | CS* | Spray BBCH 59-73  and 73-89 (41 d and 14d before harvest) | 0.256, 0.262 | 60, 62 (1.24, 1.25% v/v) | 2 | 0 0  2 7  14 14 14 | Beans Ferm.bean  Beans Beans  Beans Ferm.bean Dry beans | < 0.01, < 0.01, ND ND, ND, < 0.01  0.02, < 0.01, ND 0.02, 0.01, ND  < 0.01, ND, < 0.01 < 0.01, < 0.01, < 0.01 ND, ND, < 0.01 | < 0.01 < 0.01  0.02 0.03  < 0.01 < 0.01 < 0.01 |
| Ivory Coast Yakassé Mé  2001 *Criollo /*  *Forasteiro* 20013056/ *IC1-FPCC* | 500EC | Spray BBCH 59-73  and 73-89 (41 d and 14d before harvest) | 0.279, 0.27 | 63, 63 (0.84, 0.83% v/v) | 2 | 0 0  2 7/14  14 14 | Beans Ferm.bean  Beans Beans  Ferm.bean Dry beans | < 0.01, < 0.01, < 0.01 ND, ND, ND  ND, < 0.01, ND ND, ND, ND  ND, ND, < 0.01 ND, ND, < 0.01 | < 0.01 < 0.01  ≤ 0.01 < 0.01  < 0.01 < 0.01 |
| Ivory Coast  Akoupe 2001 *Frances*  20013056/ *IC1-FPCC* | CS* | Spray  BBCH 61-73 and 76-89 (39 d and 14d before harvest) | 0.25, 0.25 | 58, 64  (1.26% v/v) | 2 | 0/14  0 3 7  14 14 | Beans  Ferm.bean Beans Beans  Ferm.bean Dry beans | ND, ND, ND  ND, ND, ND 0.03, 0.02, ND 0.03, 0.03, ND  < 0.01, < 0.01, ND ND, < 0.01, < 0.01 | < 0.01  < 0.01 0.05 0.06  < 0.01 < 0.01 |
| Ivory Coast  Akoupe 2001 *Frances*  20013056/ *IC1-FPCC* | 500EC | Spray  BBCH 61-73 and 76-89 (39 d and 14d before harvest) | 0.26, 0.26 | 58, 61  (0.83% v/v) | 2 | 0  0 3/14 7  14 14 | Beans  Ferm.bean Beans Beans  Ferm.bean Dry beans | < 0.01, < 0.01, ND  ND, < 0.01, < 0.01 ND, ND, ND < 0.01, < 0.01, ND  ND, ND, < 0.01 ND, ND, < 0.01 | < 0.01  < 0.01 ≤ 0.01 < 0.01  < 0.01 < 0.01 |

| COCOA Country, Year of trial *Variety* | Form | Method | kg ai/ha / applic'n | L/ha | No. | PHI days | Sample analysed | Residues of alpha, beta endosulfan, Endosulfan sulfate, mg/kg | Total Residues mg/kg |
|---|---|---|---|---|---|---|---|---|---|
| Ivory Coast Monteso 2001 *Ghana* 20013056/ IC1-FPCC | CS* | Spray, BBCH 61-74 and 73-89 (40 d and 14d before harvest) | 0.24, 0.26 | 56, 61 (1.27% v/v) | 2 | 0 | Beans | 0.01, 0.01, ND | 0.02 |
| | | | | | | 0 | Ferm.bean | ND, ND, ND | < 0.01 |
| | | | | | | 3 | Beans | 0.02, 0.01, ND | 0.03 |
| | | | | | | 7 | Beans | < 0.01, < 0.01, ND | < 0.01 |
| | | | | | | 14 | Beans | ND, ND, ND | < 0.01 |
| | | | | | | 14 | Ferm.bean | < 0.01, < 0.01, < 0.01 | < 0.01 |
| | | | | | | 14 | Dry beans | ND, ND, ND | < 0.01 |
| Ivory Coast Monteso 2001 *Ghana* 20013056/ IC1-FPCC | 500EC | Spray BBCH 61-74 and 73-89 (40 d and 14d before harvest) | 0.25, 0.28 | 57, 63 (0.83% v/v) | 2 | 0 | Beans | ND, < 0.01, ND | < 0.01 |
| | | | | | | 0/14 | Ferm.bean | ND, ND, ND | < 0.01 |
| | | | | | | 3 | Beans | ND, < 0.01, ND | ≤ 0.01 |
| | | | | | | 7/14 | Beans | ND, ND, ND | < 0.01 |
| | | | | | | 14 | Dry beans | ND, ND, ND | < 0.01 |
| Ivory Coast Akoupe 2001 *Ghana* 20013056/ IC1-FPCC | CS* | Spray, BBCH 61-74 and 76-89 (40 d and 14d before harvest) | 0.24, 0.25 | 57, 58 (1.26% v/v) | 2 | 0 | Beans | ND, ND, < 0.01 | < 0.01 |
| | | | | | | 0 | Ferm.bean | ND, ND, < 0.01 | < 0.01 |
| | | | | | | 3/14 | Beans | ND, ND, ND | < 0.01 |
| | | | | | | 7 | Beans | 0.01, < 0.01, ND | 0.01 |
| | | | | | | 14 | Ferm.bean | ND, ND, ND | < 0.01 |
| | | | | | | 14 | Dry beans | ND, ND, ND | < 0.01 |
| Ivory Coast Akoupe 2001 Ghana 20013056/ IC1-FPCC | 500EC | Spray BBCH 61-74 and 76-89 (40 d and 14d before harvest) | 0.25, 0.26 | 57, 59 (0.83% v/v) | 2 | 0 | Beans | ND, < 0.01, ND | < 0.01 |
| | | | | | | 0/14 | Ferm.bean | ND, ND, ND | < 0.01 |
| | | | | | | 3/14 | Beans | ND, ND, ND | ≤ 0.01 |
| | | | | | | 7 | Beans | ND, < 0.01, ND | <.0.01 |
| | | | | | | 14 | Dry beans | ND, ND, ND | < 0.01 |
| Ivory Coast 1983 *PSR94/034* (A28024) | 50EC | Foliar Spray Interval 21 d | 0.25 | 40 (0.625%) | 2 | 28 | Seed | < 0.01, < 0.01, < 0.01 | < 0.01 |
| Ivory Coast 1983 Bingerville A28025) | 50EC | Foliar Spray Interval 21 d | 0.25 | 40 (0.625%) | 2 | 10 | Seed | < 0.01, < 0.01, 0.01 | 0.01 |
| Ivory Coast 1983 Bingerville (A28026) | 50EC | Foliar Spray | 0.25 | 40 (0.625%) | 2 | 2 | Seed | < 0.01, 0.05, 0.03 | 0.08 |

Table 80. Endosulfan residues in coffee from supervised trials in Brazil, Columbia, Guatemala and Mexico.

| COFFEE Country, Year of trial *Variety, report* | Form | Method | kg ai/ha/ applic'n | L/ha | No. | PHI days | Sample analysed | Residues of alpha, beta endosulfan, Endosulfan sulfate, mg/kg | Total Residues mg/kg |
|---|---|---|---|---|---|---|---|---|---|
| Brazil 1994 *Catuai* AA930040 | 35EC | Spray, ~ 90, 60, 30 days before harvest. | 0.70 | 500 (0.14% ai) | 3 | 32 | Green bean | < 0.01, < 0.01, 0.01 | 0.01 |
| Brazil 1994 *Catuai* AA930040 | 35EC | | 0.70 | (0.14% ai) | 3 | 33 | Green bean | < 001, 0.02, 0.02 / 0.02, 0.04, 0.05 mean | 0.04 / 0.11 / 0.08 |
| Brazil 1994 *Catuai* AA930040 | 35EC | | 0.70 | (0.14% ai) | 3 | 33 | Green bean | 0.02, 0.04, 0.03 / 0.01, 0.02, 0.01 mean | 0.09 / 0.04 / 0.06 |

| COFFEE Country, Year of trial *Variety, report* | Form. | Method | Application kg ai/ha/ applic'n | Application L/ha | Application No. | PHI days | Sample analysed | Residues of alpha, beta endosulfan, Endosulfan sulfate, mg/kg | Total Residues mg/kg |
|---|---|---|---|---|---|---|---|---|---|
| Brazil 1994 *Catuai* *AA930040* | 35EC | Spray Interval 30 d | 2.10 | (0.42% ai) | 3 | 33 | Green bean Ground roast coffee Instant coffee | 0.04, 0.08, 0.04 < 0.01, < 0.01, < 0.01 | 0.16 < 0.01 < 0.01 |
| Colombia1994 *Colombia* *AA930040* | 35EC | | 0.70 | (0.14% ai) | 3 | <u>30</u> | Green bean | < 0.01,< 0.01,< 0.01 | <u>< 0.01</u> |
| Colombia1994 *Colombia* *AA930040* | 35EC | | 0.70 | (0.14% ai) | 3 | <u>30</u> | Green bean | < 0.01,< 0.01,< 0.01 | <u>< 0.01</u> |
| Colombia1994 *Colombia* *AA930040* | 35EC | | 0.70 | (0.14% ai) | 3 | <u>30</u> | Green bean | < 0.01,< 0.01,< 0.01 | <u>< 0.01</u> |
| Colombia1994 *AA930040* | 35EC | Spray Interval 30 d | 2.10 | (0.42% ai) | 3 | 33 | Green bean roast coffee Inst. coffee | < 0.01,< 0.01,< 0.01 < 0.01,< 0.01,< 0.01 < 0.01,< 0.01,< 0.01 | < 0.01 < 0.01 < 0.01 |
| Guatemala1994 *Catimor* *AA930040* | 35EC | | 0.70 | (0.14% ai) | 3 | <u>32</u> | Green bean | 0.02,0.01,0.04 0.04,0.02,0.04 <u>mean</u> | 0.07 0.10 <u>0.09</u> |
| Guatemala 1994 *Caturra* *AA930040* | 35EC | | 0.70 | (0.14% ai) | 3 | <u>32</u> | Green bean | < 0.01,< 0.01,0.02 < 0.01,< 0.01,0.03 <u>mean</u> | 0.02 0.03 <u>0.03</u> |
| Mexico 1994 Veracruz *Catimor* *AA930040* | 35EC | | 0.70 | (0.14% ai) | 3 | <u>31</u> | Green bean | < 0.01,< 0.01,0.02 < 0.01,< 0.01,0.02 <u>mean</u> | 0.02 0.02 <u>0.02</u> |
| Mexico 1994 Veracruz *Tipica* *AA930040* | 35EC | | 0.70 | (0.14% ai) | 3 | <u>31</u> | Green bean | < 0.01,< 0.01,0.02 < 0.01,< 0.01,0.02 <u>mean</u> | 0.02 0.02 <u>0.02</u> |

Table 81. Endosulfan residues in tea from supervised trials in India.

| TEA Country, Year of trial *Variety* | Form. | Method | Application kg ai/ha / applic'n | Application L/ha | Application No. | PHI days | Sample analysed | Total Residues, mg/kg |
|---|---|---|---|---|---|---|---|---|
| India 1971 Cinchona (elev. 3000ft Above MSL) *(A31719)* *PSR94/028* | 35EC | Spray | 0.88 | 350 (0.250 % ai) | 3 | 1 7 15  1 7 15 1 7 15 1 | Dried green tea Dried green tea Dried green tea  Proc. Black tea Proc. Black tea Proc. Black tea Tea infusion from green tea Tea infusion from green tea Tea infusion from green tea Tea infusion from black tea | 6.0-18.2 2.1-4.8 0.7-1.2  8.4-29.6 16.3-35.0 2.4-11.4 0.013 0.016 0.006-007 0.014 |
| India 1971 Akkamalai (elev. 5000ft Above MSL) *(A31719)* *PSR94/028* | 35EC | Spray | 0.44 | 350 (0.125 % ai) | 3 | 1 7 15  1 7 15 1 1 1 | Dried green tea Dried green tea Dried green tea  Proc. Black tea Proc. Black tea Proc. Black tea Tea infusion from green tea Tea infusion from green tea Tea infusion from black tea | 6.2-37.5 16.2-24.1 2.5-4.0  15.0-36.4 4.0-12.7 2.7-3.3 0.027 0.041 0.086 |

| TEA Country, Year of trial *Variety* | Form. | Method | Application kg ai/ha / applic'n | L/ha | No. | PHI days | Sample analysed | Total Residues, mg/kg |
|---|---|---|---|---|---|---|---|---|
| India 1971 Akkamalai (elev. 5000ft Above MSL) *(A31719)* *PSR94/028* | 35EC | Spray | 0.88 | 350 (0.250 % ai) | 3 | 1 7 15  1 7 15 1 1 1 | Dried green tea Dried green tea Dried green tea  Proc. Black tea    Tea infusion from green tea Tea infusion from green tea Tea infusion from black tea | 14.4-49.7 3.9-13.6 1.9-5.3  31.1-84.0 6.8-14.8 3.2-9.9 0.101 0.062 0.107 |
| India 1970 Tocklai Exptl Sta *(A31718)* *PSR94/028* | 35EC | Spray | 0.88 | 100 (0.875 % ai) | 3 | 1 2 4 7  1 2 4 7 | Dry tea Dry tea Dry tea Dry tea  Tea infusion Tea infusion Tea infusion Tea infusion | 19.7-25.6 15.4-18.1 4.9-8.4 2.3-4.2  0.028-0.030 0.014-0.017 0.003-0.007 0.001-0.002 |
| India 1970 Tocklai Exptl Sta *(A31718)* *PSR94/028* | 35EC | Spray | 1.75 | 100 (1.750 % ai) | 3 | 1 2 4 7  1 2 4 7 | Dry tea Dry tea Dry tea Dry tea  Tea infusion Tea infusion Tea infusion Tea infusion | 93-108 22.9-42.7 6.3-9.0 2.1-2.3  0.097-0.158 0.026-0.032 0.008-0.016 0.001-0.002 |
| India 1971 Cinchona (elev. 3000ft Above MSL) *(A31719)* *PSR94/028* | 35EC | Spray | 0.44 | 350 (0.125% ai) | 3 | 1 7 15 1 7 15 1 7 15 1 | Dried green tea Dried green tea Dried green tea Proc. Black tea Proc. Black tea Proc. Black tea Tea infusion from green tea Tea infusion from green tea Tea infusion from green tea Tea infusion from black tea | 2.2-4.2 1.1-5.0 0.7-1.2 7.8-15.6 4.5-16.1 0.8-1.6 0.016 0.006 0.003 0.043 |

Table 82. Endosulfan residues in sugar beet leaves and head from supervised trials in Italy.

| SUGARBEET Country, Year of trial *Variety* | Form. | Method | Application kg ai/ha/ applic'n | L/ha | No. | PHI (Days) | Sample analysed | Residues of alpha, beta endosulfan, Endosulfan sulfate, mg/kg | Total Residues, mg/kg |
|---|---|---|---|---|---|---|---|---|---|
| Italy 1998  *Dorotea* ER 98 ECS746 | | Spray sprayer, hand) Interval 30 d BBCH39 | 0.630 | 600 (0.105% ai) | 2 | 0 27 | Lves+ hd Lves+ hd | 3.40, 2.50, 0.29 0.02, 0.04, 0.30 | 6.19 0.36 |
| Italy 1998  *Adige* ER 98 ECS 746 | | Spray sprayer, hand) Interval 33d BBCH39 | 0.630 | 600 (0.105% ai) | 2 | 0 26 | Lves+ hd Lves+ hd | 4.10, 2.70, 0.25 0.11, 0.14, 0.57 | 7.05 0.82 |
| Italy 1998  *Alesia* ER 98 ECS 746 | | Spray sprayer, hand) Interval 32d BBCH38/39 | 0.630 | 300, 600  (0.105% ai) | 2 | 0 28 | Lves+ hd Lves+ hd | 3.60, 2.00, 0.32 < 0.02, < 0.02, 0.31 | 5.92 0.31 |

| Italy 1998 Monodoro ER 98 ECS 746 | | Spray sprayer, hand) Interval 32 d BBCH 38/39 | 0.630 | 300, 600 (0.10% ai)5 | 2 | 0 28 | Lves+ hd Lves+ hd | 2.90 1.70 0.41 < 0.02, < 0.02, 0.28 | 5.01 0.28 |
|---|---|---|---|---|---|---|---|---|---|
| Italy 1997 Adige ER 97 ECS 746 | 35EC | Spray sprayer, hand) Interval 20 d BBCH 39 | 0.630 | 600 (0.105 % ai) | 2 | 0 14 28 | Leaves+ hd Leaves+ hd Leaves+ hd | 4.40 2.80 1.00 0.13, 0.19, 0.49 0.04, 0.09, 0.49 | 8.20 0.81 0.62 |
| Italy 1997 Monodoro ER 97 ECS 746 | 35EC | Spray sprayer, hand) Interval 20 d BBCH 39 | 0.630 | 600 (0.105 % ai) | 2 | 0 14 28 | Leaves+ hd Leaves+ hd Leaves+ hd | 4.40 2.80 0.54 0.04, 0.08, 0.57 0.02, 0.05, 0.35 | 7.74 0.69 0.42 |
| Italy 1997 Formula Romagna ER 97 ECS 746 | 35EC | Spray sprayer, hand) Interval 20 d BBCH 39,49 | 0.630 | 600 (0.105 % ai) | 2 | 0 14 28 | Leaves+ hd Leaves+ hd Leaves+ hd | 2.60 1.70 0.20 0.05, 0.11, 0.59 < 0.02, 0.03, 0.11 | 4.50 0.75 0.14 |
| Italy 1997 Rizor ER 97 ECS 746 | 35EC | Spray sprayer, hand) Interval 20 d BBCH 39,49 | 0.630 | 600 (0.105 % ai) | 2 | 0 14 28 | Leaves+ hd Leaves+ hd Leaves+ hd | 4.60 2.90 0.44 0.07, 0.14, 0.78 0.02, 0.04, 0.17 | 7.94 0.99 0.23 |
| Italy 1997 Bianca ER 97 ECS 746 | 35EC | Spray sprayer, hand) Interval 20 d BBCH 39 | 0.630 | 600 (0.105 % ai) | 2 | 0 14 28 | Leaves+hd Leaves+hd Leaves+hd | 3.40 2.30 0.39 0.13, 0.21, 0.53 0.03, 0.03, 0.40 | 6.09 0.87 0.46 |

Table 83. Endosulfan residues in forage and vines beans from supervised trials in USA.

| VINE BEANS | Application | | | | | PHI | Sample analysed | Residues of alpha + beta endosulfan, Endosulfan sulfate, mg/kg | Total Residues, mg/kg |
|---|---|---|---|---|---|---|---|---|---|
| Country, Year of trial Variety | Form. | Method | kg ai/ha/ applic'n | L/ha | No. | (Days) | | | |
| USA1965 NY Lima beans M-1610 | 50W | Interval 37, 7 d | 0.56 | | 3 | 11 20 32 | Bean forage Bean forage Bean forage | 2.11, 2.09 0.97, 1.75 1.56, 3.86 | 4.20 2.72 5.42 |
| USA 1965 NY Snap beans M-1592 | 50WP | Spray Interval 2, 8 d | 0.56 | | 3 | 0 10 20 28 35 40 | Vines Vines Vines Vines Vines Vines | 50, 2.3 2.27, 1.65 0.96, 3.15 0.35, 0.81 0.95, 0.76 0.29, 0.41 | 52.3 3.92 4.11 1.16 1.71 0.70 |
| USA 1965 NY M-1592 | 50WP | Spray Interval 2, 8 days | 1.12 | | 3 | 0 10 20 23 35 40 | Vines Vines Vines Vines Vines Vines | 87.5, ND 5.4, 4.31 1.08, 3.95 0.68, 2.31 0.63, 1.54 0.61, 1.53 | 87.5 9.71 5.03 2.99 2.17 2.14 |
| USA 1965 NY Lima beans M-1610 | 50WP | Spray Interval 37, 7 d | 0.56 | | 3 | 0 11 20 32 | Bean forage Bean forage Bean forage Bean forage | 40.3, 1.62 5.80, 4.22 3.99, 3.70 4.72, 9.16 | 41.92 10.02 7.69 13.88 |

Table 84. Endosulfan residues in pea hay from supervised trials in Australia.

| PEA HAY | Application | | | | | PHI | Sample | Residues of alpha, | Total |
| Country, Year of trial *Variety* Report | Form. | Method | kg ai/ha/ applic'n | L/ha | No. | days | analysed | beta endosulfan, Endosulfan sulfate, mg/kg | Residues, mg/kg |
|---|---|---|---|---|---|---|---|---|---|
| Australia 2000 Lockyer Valley *epic* *1/10/557* | EC | SPRAY | 0.73 | 500 | 3 | 7 14 | | 0.42, 1.60, 1.105 0.30, 0.81, 1.10 | 3.12 2.21 |
| Australia 2000 Lockyer Valley *epic* *1/10/557* | EC | SPRAY | 1.46 | 500 | 3 | 7 14 7c | | 0.81,350, 3.80 0.33, 1.10, 2.40 0.006,0.005, < 0.005 | 8.11 3.83 0.011 |
| Australia 2000 Werribee *melbourne market* | EC | SPRAY | 0.73 | 500 | 3 | 28 | | 0.015, 0.041, 0.066 | 0.122 |

Table 85. Endosulfan residues in cocoa shell from supervised trials in Ghana and Ivory Coast.

| COCOA | Application | | | | | PHI | Sample | Residues of alpha, | Total |
| Country, Year of trial *Variety, report* | Form. | Method | kg ai/ha / applic'n | L/ha | No. | days | analysed | beta endosulfan, Endosulfan sulfate, mg/kg | Residues mg/kg |
|---|---|---|---|---|---|---|---|---|---|
| Ghana2002 Ahensan *Several hybrids* 20013056/ C032568 | CS* | Spray, BBCH 67-81 and 66-84 | 0.243, 0.273 | 58, 66 (1.25, 1.24 % v/v) | 2 | 0 7 13 20 27 | Shell Shell Shell Shell Shell | 0.13, 0.07, < 0.01 0.09, 0.04, ND 0.15, 0.08, < 0.01 0.12, 0.06, < 0.01 0.07, 0.03, ND | 0.20 0.13 0.23 0.18 0.10 |
| Ghana2002 Ahensan *Several hybrids* 20013056/ | 500EC | Spray, BBCH 67-81 and 66-85 | 0.256, 0.263 | 62, 63 58, 66 (0.78, 0.79%) | 2 | 0 7 13 20 | Shell Shell Shell Shell | 0.06, 0.06, < 0.01 ND, ND, < 0.01 < 0.01, ND, < 0.01 < 0.01, ND, < 0.01 | 0.12 < 0.01 < 0.01 < 0.01 |
| Ghana2002 Fumso *Amazon* 20013056/ | CS* | Spray, BBCH 67-77 and 84 | 0.26, 0.26 | 62, 62 (1.25% v/v) | 2 | 0 7 14 20 27 | Shell Shell Shell Shell Shell | 0.09, 0.05, < 0.01 0.24, 0.13, < 0.01 0.11, 0.08, < 0.01 0.07, 0.03, < 0.01 0.10, 0.06, < 0.01 | 0.14 0.37 0.19 0.10 0.16 |
| Ghana2002 Fumso *Amazon* 20013056/ | 500EC | Spray , BBCH 67-77 and 84 | 0.259, 0.25 | 62, 60 (0.79% v/v) | 2 | 0 7 14 20 | Shell Shell Shell Shell | 0.04, 0.05, < 0.01 ND, ND, ND ND, < 0.01, ND ND, < 0.01, ND | 0.09 < 0.01 < 0.01 < 0.01 |
| Ghana2002 Tafo *Hybrids* 20013056/ | CS* | Spray, BBCH 67-84 and 66-82 | 0.246, 0. 256 | 59, 61 (1.24, 1.25% v/v) | 2 | 0 6 13 20 | Shell Shell Shell Shell | 0.26, 0.15, < 0.01 0.19, 0.10, < 0.01 0.25, 0.12, 0.01 0.09, 0.04, < 0.01 | 0.41 0.29 0.38 0.13 |
| Ghana2002 Tafo *Hybrids* 20013056/ | 500EC | Spray, BBCH 67-84 And 66-86 | 0.255, 0. 254 | 61, 61 (0.79, 0.78% v/v) | 2 | 0 6 13 20 | Shell Shell Shell Shell | 0.09, 0.08, ND ND, 0.01, ND ND, < 0.01, ND ND, < 0.01, ND | 0.17 0.01 < 0.01 < 0;01 |
| Ghana2002 Bososo *Several hybrids* 20013056/ | CS* | Spray, BBCH 67-76 and 66-85 | 0.250, 0. 243 | 60, 58 (1.24, 1.25% v/v) | 2 | 0 7 14 21 | Shell Shell Shell Shell | 0.15, 0.07, ND 0.19, 0.10, < 0.01 0.06, 0.03, < 0.01 0.05, 0.03, < 0.01 | 0.22 0.29 0.09 0.08 |
| Ghana Bososo 2002 | 500EC | Spray, BBCH 67-76 and 66-85 | 0.247, 0. 272 | 59, 65 (0.79%, v/v) | 2 | 0 7 14 | Shell Shell Shell | 0.05, 0.06, ND ND, ND, ND ND, ND, < 0.01 | 0.11 < 0.01 < 0.01 |

| COCOA | Application | | | | | PHI | Sample | Residues of alpha, | Total |
| Country, Year of trial *Variety, report* | Form. | Method | kg ai/ha / applic'n | L/ha | No. | days | analysed | beta endosulfan, Endosulfan sulfate, mg/kg | Residues mg/kg |
|---|---|---|---|---|---|---|---|---|---|
| Ivory Coast Yakassé Mé 2001 *Criollo /* | CS* | Spray, BBCH59-73 and73-89 | 0.256, 0.262 | 60, 62 (1.24, 1.25% v/v) | 2 | 0 | Shell | 0.06, 0.03, ND | 0.09 |
| | | | | | | 2 | Shell | 0.05, 0.03, ND | 0.08 |
| | | | | | | 7 | Shell | 0.03, 0.01, ND | 0.04 |
| | | | | | | 14 | Shell | 0.04, 0.01, < 0.01 | 0.05 |
| Ivory Coast Yakassé Mé 2001 | 500EC | Spray, BBCH 59-73 and 73-89 | 0.279, 0.27 | 63, 63 (0.84, 0.83% v/v) | 2 | 0 | Shell | 0.07, 0.07, 0.02 | 0.16 |
| | | | | | | 2 | Shell | ND, < 0.01, ND | < 0.01 |
| | | | | | | 7 | Shell | ND, < 0.01, < 0.01 | < 0.01 |
| Ivory Coast Akoupe 2001 *Frances* | CS* | Spray, BBCH 61-73 and 76-89 | 2x0.25, | 58, 64 (1.26% v/v) | 2 | 0 | Shell | 0.11, 0.07, ND | 0.18 |
| | | | | | | 3 | Shell | 0.07, 0.04, ND | 0.11 |
| | | | | | | 7 | Shell | 0.07, 0.05, < 0.01 | 0.12 |
| | | | | | | 14 | Shell | 0.02, < 0.01, ND | 0.02 |
| Ivory Coast Akoupe 2001 *Frances* | 500EC | Spray, BBCH 61-73 and 76-89 | 2x0.26 | 58, 61 (0.83% v/v) | 2 | 0 | Shell | 0.02, 0.02, ND | 0.04 |
| | | | | | | 3 | Shell | ND, < 0.01, ND | < 0.01 |
| | | | | | | 7 | Shell | ND, < 0.01, ND | < 0.01 |
| | | | | | | 14 | Shell | ND, ND, ND | < 0.01 |
| Ivory Coast Monteso 2001 *Ghana* | CS* | Spray, BBCH 61-74 and 73-89 | 0.24, 0.26 | 56, 61 (1.27% v/v) | 2 | 0 | Shell | 0.04, 0.02, ND | 0.06 |
| | | | | | | 3 | Shell | 0.07, 0.04, ND | 0.11 |
| | | | | | | 7 | Shell | 0.04, 0.02, ND | 0.06 |
| | | | | | | 14 | Shell | 0.01, < 0.01, ND | 0.01 |
| Ivory Coast Monteso | 500EC | Spray , BBCH 61-74 | 0.25, 0.28 | 57, 63 (0.83% v/v) | 2 | 0 | Shell | 0.01, 0.01, ND | 0.02 |
| | | | | | | 3 | Shell | ND, < 0.01, ND | < 0.01 |
| Ivory Coast Akoupe | CS* | Spray, BBCH 61-74 | 0.24, 0.25 | 57, 58 (1.26% v/v) | 2 | 0 | Shell | 0.04, 0.02, ND | 0.06 |
| | | | | | | 3 | Shell | < 0.01, < 0.01, ND | < 0.01 |
| | | | | | | 7 | Shell | 0.02, < 0.01, ND | 0.02 |
| Ivory Coast Akoupe | 500EC | Spray , BBCH 61-74 | 0.25, 0.26 | 57, 59 (0.83% v/v) | 2 | 0 | Shell | 0.01, 0.01, ND | 0.02 |
| | | | | | | 3 | Shell | ND, ND, ND | < 0.01 |

Table 86. Endosulfan residues in cotton lint from supervised trials in Europe.

| COTTON | Application | | | | | PHI | Sample | Residues of alpha, | Total |
| Country, Year of trial *Variety*, report | Form. | Method | kg ai/ha / applic'n | L/ha | No. | days | analysed | beta endosulfan, Endosulfan sulfate, mg/kg | Residues mg/kg |
|---|---|---|---|---|---|---|---|---|---|
| Spain 2002 tabladilla | CS | spray | 0.840 | 550 0.15% | 3 | 0 | Bolls | 0.65, 0.56, 0.02 | 1.23 |
| | | | | | | 20 | Lint | 0.37, 0.43, 0.52 | 1.32 |
| Spain 2002 *Roca 02 R 172* | CS | spray | 0.840 | 550 0.15% | 3 | 0 | Bolls | 0.53, 0.46, < 0.02 | 0.99 |
| | | | | | | 20 | Lint | 0.09, 0.10, 0.07 | 0.26 |
| Greece 2002 *Sig 12502R172* | CS | spray | 0.840 | 600 0.14% | 3 | 0 | Bolls | 0.93, 0.77, 0.09 | 1.79 |
| | | | | | | 28 | Lint | 0.16, 0.23 0.17 | 0.56 |
| Greece 2002 *Sig 125 2R172* | CS | spray | 0.840 | 600 0.14% | 3 | 0 | Bolls | 0.34, 0.35, 0.13 | 0.82 |
| | | | | | | 28 | Lint | 0.02, 0.02, 0.02 | 0.06 |
| Spain 2002 *tabladilla 170* | 33 EC | spray | 0.84 | 550 0.15% | 3 | 0 | Bolls | 0.17, 0.29, 0.04 | 0.50 |
| | | | | | | 20 | Lint | 0.06, 0.13, 0.44 | 0.63 |
| Spain 2002 *Roca 02R170* | 33 EC | spray | 0.84 | 550 0.15% | 3 | 0 | Bolls | 0.50, 0.52, 0.04 | 1.06 |
| | | | | | | 20 | Lint | 0.02, 0.10, 0.16 | 0.28 |
| Greece 2002 *Sig 12502R170* | 33 EC | spray | 0.84 | 600 0.14% | 3 | 0 | Bolls | 0.94, 1.30, 0.28 | 2.52 |
| | | | | | | 28 | Lint | 0.02, 0.09 0.23 | 0.34 |
| Greece 2002 *Sig 12502R170* | 33 EC | spray | 0.84 | 600 0.14% | 3 | 0 | Bolls | 0.12, 0.19, 0.11 | 0.42 |
| | | | | | | 28 | Lint | < 0.02,< 0.02, 0.03 | 0.03 |
| Greece 2002 *midas02R170* | 33 EC | spray | 0.84 | 600 0.14% | 3 | 0 | Bolls | 1.40, 1.80, 0.21 | 3.41 |
| | | | | | | 28 | Lint | 0.10, 0.35, 0.45 | 0.90 |
| Spain 2001 *Sonia 02R170* | CS | spray | 0.84 | 1000 0.084% | 3 | 0 | Bolls | 0.46, 0.36, 0.15 | 0.97 |
| | | | | | | 21 | Lint | 0.15, 0.11, 0.16 | 0.42 |
| Spain 2001 *sonia01R170* | 33 EC | spray | 0.84 | 1000 0.084% | 3 | 0 | Bolls | 0.18, 0.15, < 0.02 | 0.33 |
| | | | | | | 14 | Lint | < 0.02, 0.04, 0.09 | 0.13 |
| Spain 2001 *sonia01R170* | CS | spray | 0.84 | 1000 0.084% | 3 | 0 | Bolls | 0.67, 0.40, 0.02 | 1.09 |
| | | | | | | 14 | Lint | 0.03, 0.02, 0.05 | 0.10 |

| COTTON | Application | | | | | PHI | Sample | Residues of alpha, | Total |
|--------|------|--------|-------------|-------|-----|------|----------|---------------------------|----------|
| Country, Year of trial *Variety,* report | Form. | Method | kg ai/ha / applic'n | L/ha | No. | days | analysed | beta endosulfan, Endosulfan sulfate, mg/kg | Residues mg/kg |
| Spain 2001 01 R 170 | 33 EC | spray | 0.84 | 1000 0.084% | 3 | 0 14 | Bolls Lint | 1.10, 0.69, 0.16 < 0.02, < 0.02, 0.07 | 1.95 0.07 |
| Spain 2001 *bravada 170* | CS | spray | 0.84 | 1000 0.084% | 3 | 0 14 | Bolls Lint | 0.89, 0.52, 0.40 0.16, 0.34, 0.36 | 1.81 0.86 |
| Spain 2001 01 R 170 | CS | spray | 0.49 | 1000 0.05% | 3 | 0 14 | Bolls Lint | 0.54, 0.28, 0.04 0.16, 0.33, 0.32 | 0.86 0.81 |
| Greece 2001 *SG 1251R170* | CS | spray | 0.84 | 600 0.14% | 3 | 0 14 | Bolls Lint | < 0.02, < 0.02, < 0.02 0.08, 0.09, 0.04 | < 0.02 0.21 |
| Greece 2001 *SG 125 01R170* | CS | spray | 0.84 | 600 0.14% | 3 | 0 14 | Bolls Lint | 0.12, 0.11, < 0.02 1.90, 1.10, 0.05 | 0.23 3.05 |
| Greece 2001 *SG 12501R170* | CS | spray | 0.49 | 600 0.08% | 3 | 0 14 | Bolls Lint | 0.05, 0.03, < 0.02 0.95, 0.53, 0.03 | 0.08 1.51 |
| Greece 2001 01 R 170 | 33 EC | spray | 0.84 | 600 0.14% | 3 | 0 14 | Bolls Lint | 0.14, 0.10, < 0.02 <.0.02, 0.06, 0.08 | 0.24 0.14 |

# FATE OF RESIDUES IN STORAGE AND PROCESSING

## *In storage*

No data for endosulfan in storage was submitted.

## *In processing*

The meeting received information on the fate of endosulfan residues during processing for potato, tomato, citrus, apples, peach, grape, pineapple, soybean, coffee and tea.

A study was provided on the fate of endosulfan to hydrolysis conditions likely during commercial food processing. Maurer (2002) investigated the hydrolysis of [6, 7, 8, 9-U-$^{14}$C]-endosulfan under conditions representing food processing operations. The treatment was carried out at two incubation rates: 0.1 mg/L and 1.0 mg/L. Each experiment was conducted using replicate samples. Residues were analysed using radio-HPLC and radio-TLC. α -endosulfan and β endosulfan were the main components following pasteurisation at both treatment levels; small amounts of endosulfan-diol were also found.

In the brewing, baking and boiling simulation the hydrolysis product endosulfan diol was the major single compound at both incubation levels. The sum of α and β endosulfan represented nearly half of the applied radioactivity. Furthermore one degradation product at the 0.1 mg/L and three degradation products at the 1.0 mg/L level were observed representing less than 3.4% of applied radioactivity each.

The sterilisation process resulted in a complete degradation of endosulfan. Endosulfan diol was the major degradation product amounting to approximately 75% of the radioactivity applied. With one exception, the compounds represent more polar compounds than endosulfan diol. None of the other reference standards used in HPLC or TLC investigations in this study corresponded to one of the resulting radio-peaks in the chromatograms. However, each of these non-identified components represented only 0.5 to 8.1% of applied radioactivity (mean values).

Table 87. Recovery data after the simulation of processing

| Process | rate | α -Endosulfan | | β Endosulfan | | Endosulfan diol | | Identified | | Sum of N.I. | |
|---|---|---|---|---|---|---|---|---|---|---|---|
| | mg/L | %applied | mg/L | %applied | mg/L | %applied | mg/L | %applied | mg/L | %applied | |
| Pasteurisation | 0.1 | 68.29 | 0.07 | 29.15 | 0.03 | | | 97.44 | 0.1 | | |
| pH 4, 90 °C | 1 | 64.95 | 0.64 | 26.35 | 0.259 | 3.98 | 0.039 | 95.28 | 0.938 | 5.8 | |
| Baking, boiling | 0.1 | 34.28 | 0.036 | 13.04 | 0.014 | 41.97 | 0.044 | 89.29 | 0.094 | 2.9 | |
| pH 5, 100 °C | 1 | 29.92 | 0.298 | 11.87 | 0.118 | 49.27 | 0.491 | 91.05 | 0.907 | 6.4 | |
| Sterilisation | 0.1 | | | | | 71.68 | 0.075 | 71.68 | 0.075 | 23.0 | |
| pH 6, 120 °C | 1 | | | | | 75.72 | 0.749 | 75.72 | 0.749 | 24.3 | |

Potatoes were processed into potato flakes and chips (Brady, 1997) and wet peel (J. Englar). Residues in the raw material were below the LOQ even treated at five times the rate. Therefore no transfer factor can be estimated.

Seven studies were performed on processed tomato. Tomato fruits were treated twice at two field sites in Greece and Italy in 2002 at the rate of 1.06 kg ai/ha. and a PHI of 3 days (Erbel and Ertz, 003). Tomatoes were washed by moving them slowly in water. A reduction of residue concentrations occurs in the washing water, peeling water, juice, peeled fruits, canned peeled and canned unpeeled fruits. A small reduction of residue concentration may occur for washed fruit. Residues concentrated in peel (transfer factors were 12.1 and 16.7) and wet pomace (transfer factors were 11.5 and 5.6). Most residues were located on the exterior of the tomatoes.

In the second study tomato fruits were treated twice in two field sites in Spain and Italy in 1994, at 6 days before harvest, at growth stage 17/19, at a rate of 0.528 kg/ha (Sonder et al., 1996b). Wash water and canning liquid contained no residues above the LOQ. There were no residues in the juice. Residues were found only in pomace at 0.29 and 0.61 mg/kg. Transfer factors were as shown in the table below. Apparent residues in untreated samples were < 0.01 mg/kg.

In the third study tomato fruits were treated twice in two field sites in Spain and Italy in 1993., 14 days before harvest, at growth stage 17/19, at a rate of 0.528 kg/ha (Sonder et al., 1996a). Tomatoes were processed into paste, juice and into canned unpeeled tomatoes. Fruit contained total residues of 0.07 mg/kg at the sites in Spain, and at or about 0.02 mg/kg at the sites in Italy. The paste, made from tomatoes from the Spanish sites, contained residues at or about 0.02 mg/kg, tomato juice < 0.02 mg/kg, preserved tomatoes (canned and unpeeled) 0.025 and 0.035 mg/kg, and pomace 0.20 and 0.35 mg/kg. Residues in processed fractions for tomatoes from the two sites in Italy were less than the LOQ except for the pomace (0.15 mg/kg and 0.14 mg/kg). Apparent residues in untreated samples were below 0.01 mg/kg.

In the fourth study tomato fruits were treated three times, the last 2 days before harvest at a rate of 5.6 kg ai/ha at a field site in Fresno, California in 1995 (Brady, 1997b). Samples were processed into paste, puree and juice. In the treated samples total residues in the RAC were 2.28 mg/kg (the range was 1.99 to 2.69 mg/kg). In the puree total residues were 1.36 (the range was 1.31 to 1.39 mg/kg) and in the paste 2.78 mg/kg (the range was 1.59 to 3.49 mg/kg). The residue transfer factor for puree was 0.6 and for paste was 1.2. Apparent residues in untreated samples were < 0.05 mg/kg.

In the fifth study tomato fruits were treated four times at the rate of 0.2 to 0.4 kg ai/ha at a field site in Germany 1989 (Huth, 1999a). The tomatoes were harvested 7 days after the last application. Raw tomato fruit samples from the two sites, containing residues of 0.035 and 0.095 mg/kg were processed into cooked fruits, puree and juice. Cooked fruits contained about the same residue concentration as the raw samples, but puree and juice in both studies were free of residues.

In the sixth study (Huth, 1999a.) tomatoes were sprayed 5 times at a rate of 1.12 kg ai/ha/ application at 59, 52, 15, 7 and 0 days before harvest. (Florida 1984). Tomato samples from study A32878 were processed to seeds and peel, puree, puree 10–11% dry matter solids and puree 16% dry matter solids. Results were corrected for residues found in UTCs. Raw tomatoes contained endosulfan residues of 0.12 mg/kg. The transfer factor for seeds and peel was 9.4 and for dry pomace was 22. The transfer factor for puree was 0.33. Apparent residues in untreated samples were < 0.01 mg/kg, except for one sample of seeds and peel, where alpha endosulfan was found at 0.017 mg/kg, and for dry pomace, where analytes were found up to 0.084 mg/kg. In the seventh study tomatoes were sprayed 5 times at a rate of 1.12 kg ai/ha/application at 73, 66, 14, 7 and 0 days before harvest. Sprays were made at 364 L (Huth, 1999a) Tomato samples were processed to juice, paste, cooked skins and peel, and dried skins and peel.

Table 88. Processing of Tomatoes to Tomato Juice, Pomace and Canned Tomatoes.

| Processed Fraction | Residues of α, β Endosulfan, Endosulfan sulfate and endosulfan diol(mg/kg) | Total Endosulfan residues (mg/kg) | processing factors for α, β Endosulfan and Endosulfan diol | processing Factor total |
|---|---|---|---|---|
| Balance study for Trial R 10572 (Greece 2002) | | | | |
| Tomato RAC | 0.71, 0.35, ND,< 0.02 | 1.06 | | |
| Washed Tomatoes | 0.44, 0.26, ND,< 0.02 | 0.70 | 0.62, 0.74 | 0.66 |
| Washing Water | 0.03, 0.02, 0.0007, 0.001 | 0.06 | 0.04, 0.06, 0.05 | 0.05 |
| Peeled Tomatoes | < 0.02, < 0.02, ND, ND | < 0.02 | | < 0.02 |
| Peel | 11.4, 6.51, 0.03, 0.18 | 18.12 | 16.1, 18.6, 9.0 | 16.7 |
| Peeling Water | 0.05, 0.03, 0.0001, 0.002 | 0.08 | 0.07, 0.09, 0.1 | 0.06 |
| Tomato Raw Juice | 0.12, 0.09, ND, ND | 0.21 | 0.17, 0.26, - | 0.20 |
| Tomato Juice pasteurised | 0.12, 0.09, ND, ND | 0.21 | 0.17, 0.26, - | 0.20 |
| Wet Tomato Pomace | 3.65, 2.36, < 0.02,0.06 | 6.07 | 5.14, 6.74, 3.0 | 5.72 |
| Canned Peeled Tomatoes (pasteurised) | 0.05, 0.03, ND, ND | 0.08 | 0.07, 0.09, - | 0.07 |
| Canned Unpeeled Tomatoes (sterilised) | 0.32, 0.15, ND, < 0.02 | 0.47 | 0.45, 0.43, - | 0.44 |
| Follow Up study for Trial R 10572 (Greece 2002) | | | | |
| Tomato Juice pasteurised | 0.18, 0.10, ND, < 0.02 | 0.28 | 0.25, 0.29 | 0.27 |
| Canned Peeled Tomatoes | 0.05, 0.04, ND, 0.02 | 0.11 | 0.07, 0.11, 1.0, | 0.10 |
| Balance study for Trial R 10573 (Italy 2002) | | | | |
| Tomato RAC | 0.06, 0.04, ND | 0.10 | | |
| Washed Tomatoes | 0.06, 0.04, ND | 0.10 | 1, | 1 |
| Washing Water | 0.005, 0.004, 0.00014 | 0.009 | 0.08, 0.1 | 0.09 |
| Peeled Tomatoes | ND, ND, ND | < 0.02 | -, - | < 0.02 |
| Peel | 0.62, 0.53, 0.06 | 1.21 | 10.3, 13. 3 | 12.1 |
| Peeling Water | 0.01, 0.007, 0.0005 | 0.02 | 0.17, 0.2 | 0.2 |
| Tomato Raw Juice | < 0.02, < 0.02, ND | < 0.02 | | < 0.20 |
| Tomato Juice pasteurised | < 0.02, < 0.02, ND | < 0.02 | | < 0.02 |
| Wet Tomato Pomace | 0.63, 0.45, 0.07 | 1.15 | 10.5, 11.3 | 11.5 |
| Canned Peeled Tomatoes (pasteurised) | < 0.02, ND, ND | < 0.02 | | < 0.02 |
| Canned Unpeeled Tomatoes (sterilised) | 0.03, 0.02, < 0.02 | 0.05 | 0.5, 0.5 | 0.5 |
| Follow Up study for Trial R 10573 (Italy 2002) | | | | |
| Tomato Juice (pasteurised) | < 0.02, < 0.02, ND | < 0.02 | | < 0.20 |
| Canned Peeled Tomatoes | < 0.02, < 0.02, ND | < 0.02 | | < 0.20 |

Table 89. Processing of Tomatoes to Tomato Juice, Pomace and Canned Tomatoes.

| Processed Fraction | Residues of α, β Endosulfan and Endosulfan sulfate (mg/kg) | Total Endosulfan residues (mg/kg) | Transfer Factor | Residues of α, β Endosulfan and Endosulfan sulfate (mg/kg) | Total Endosulfan residues (mg/kg) | processing Factor |
|---|---|---|---|---|---|---|
| | Spain, ESP 000103 (1994) | | | Italy, ITA000203 | | |
| Unwashed Fruit | 0.04, 0.04, < 0.01 | 0.08 | | 0.03, 0.03, < 0.01 | 0.06 | |
| Washed Fruit | 0.04, 0.04, < 0.01 | 0.08 | 1 | 0.03, 0.03, < 0.01 | 0.06 | 1 |
| Preserved Fruit (unpeeled) | 0.04, 0.04, < 0.01 | 0.08 | 1 | 0.033, 0.033, < 0.01 | 0.07 | 1.1 |
| Canning Liquid | < 0.01, < 0.01, < 0.01 | < 0.01 | | < 0.01, < 0.01, < 0.01 | < 0.01 | < 0.17 |
| Juice | < 0.01, < 0.01, < 0.01 | < 0.01 | < 0.12 | < 0.01, < 0.01, < 0.01 | < 0.01 | < 0.17 |
| Pomace | 0.34, 0.24, 0.03 | 0.61 | 7.6 | 0.15, 0.12, 0.02 | 0.29 | 4.5 |

| Processed Fraction | Residues of α, β Endosulfan and Endosulfan sulfate (mg/kg) | Total Endosulfan residues (mg/kg) | Transfer Factor | Residues of α, β Endosulfan and Endosulfan sulfate (mg/kg) | Total Endosulfan residues (mg/kg) | processing Factor |
|---|---|---|---|---|---|---|
| Wash Water | < 0.01, < 0.01, < 0.01 | < 0.01 | | < 0.01, < 0.01, < 0.01 | < 0.01 | |
| | Spain 1, ESP000203 | | | Spain-2, ESP000303 | | |
| Unwashed Fruit | 0.03, 0.03, < 0.01 | 0.06 | | 0.03, 0.03, < 0.01 | 0.06 | |
| Washed Fruit | 0.01, 0.02, < 0.01 | 0.03 | 0.50 | 0.02, 0.03, < 0.01 | 0.05 | 0.84 |
| Preserved Fruit (unpeeled) | 0.01, 0.01, < 0.01 | 0.02 | 0.33 | 0.01, 0.02, < 0.01 | 0.03 | 0.50 |
| Canning Liquid | < 0.01, < 0.01, < 0.01 | < 0.01 | < 0.16 | < 0.01, < 0.01, < 0.01 | < 0.01 | < 0.16 |
| Juice | < 0.01, < 0.01, < 0.01 | < 0.01 | < 0.16 | < 0.01, < 0.01, < 0.01 | < 0.01 | < 0.16 |
| Paste | < 0.01, 0.01, < 0.01 | 0.01 | 0.16 | < 0.01, 0.02, < 0.01 | 0.02 | 0.33 |
| Pomace | 0.08, 0.09, 0.03 | 0.20 | 3.30. | 0.14, 0.19, 0.015 | 0.35 | 5.80. |
| Wash Water | < 0.01, < 0.01, < 0.01 | < 0.01 | < 0.16 | < 0.01, < 0.01, < 0.01 | < 0.01 | < 0.16 |
| | Italy-1, ESP000103 | | | Italy-2, ESP000203 | | |
| Unwashed Fruit | < 0.01, 0.01, < 0.01 | 0.01 | | < 0.01, 0.01, < 0.01 | 0.01 | |
| Washed Fruit | < 0.01, 0.01, < 0.01 | 0.01 | | 0.01, 0.01, < 0.01 | 0.02 | |
| Preserved Fruit (unpeeled) | < 0.01, < 0.01, < 0.01 | < 0.01 | | 0.01, 0.01, < 0.01 | 0.02 | |
| Canning Liquid | < 0.01, 0.01, < 0.01 | 0.01 | | < 0.01, < 0.01, < 0.01 | < 0.01 | |
| Juice | < 0.01, < 0.01, < 0.01 | < 0.01 | | < 0.01, < 0.01, < 0.01 | < 0.01 | |
| Paste | < 0.01, < 0.01, < 0.01 | < 0.01 | | < 0.01, < 0.01, < 0.01 | < 0.01 | |
| Pomace | 0.05, 0.06, 0.04 | 0.15 | | 0.06, 0.06, 0.015 | 0.14 | |
| Wash Water | < 0.01, < 0.01, < 0.01 | < 0.01 | | < 0.01, < 0.01, < 0.01 | < 0.01 | |

Table 90. Processing of Tomatoes to Puree and Paste in USA

| Processed Fraction | Residues of α, β Endosulfan and Endosulfan sulfate (mg/kg) | Total Endosulfan residues (mg/kg) | Transfer Factor (Total Endosulfan) |
|---|---|---|---|
| Fruit | 1.29, 1.40, < 0.05 A<br>1.04 1.12, < 0.05 B<br>0.93, 1.06, < 0.05 C | 2.69<br>2.16<br>1.98 | |
| Puree | 0.66, 0.72, < 0.05 A<br>0.67, 0.72, < 0.05 B<br>0.63, 0.68, < 0.05 C | 1.38<br>1.39<br>1.31 | 0.51<br>0.64<br>0.66 |
| Paste | 0.74, 0.85, < 0.05 A<br>1.60, 1.83, 0.06 B<br>1.44, 1.76, 0.05 C | 1.59<br>3.49<br>3.25 | 0.59<br>1.62<br>1.63 |

Table 91. Processing of Tomatoes to Cooked Fruit, Puree and Juice.

| Study No., Location | Sample analysed | Residues of α, β Endosulfan and Endosulfan sulfate (mg/kg) | Total Residues (mg/kg) | Transfer Factor |
|---|---|---|---|---|
| A49970 DEU89170721 Bonheim | Fruit | 0.01, 0.02, < 0.01 | 0.03 | |
| | Washings | < 0.01, < 0.01, < 0.01 | < 0.01 | |
| | Cooking water | < 0.01, < 0.01, < 0.01 | < 0.01 | |
| | Cooked Fruit | 0.01, 0.02, 0.01 | 0.04 | |
| | Puree | < 0.01, < 0.01, < 0.01 | < 0.01 | |
| | Juice | < 0.01, < 0.01, < 0.01 | < 0.01 | |

Table 92. Processing of Tomatoes to Cooked Fruit, Puree and Juice.

| Study No., Location | Sample analysed | Residues of α, β Endosulfan and Endosulfan sulfate (mg/kg) | Total Residues (mg/kg) | Transfer Factor |
|---|---|---|---|---|
| A49971 DEU89170741 Kestlerbach | Fruit | 0.04, 0.05, < 0.01 | 0.09 | |
| | Washings | < 0.01, < 0.01, < 0.01 | < 0.01 | |
| | Cooking water | < 0.01, < 0.01, < 0.01 | < 0.01 | |
| | Cooked Fruit | 0.05, 0.04, < 0.01 | 0.09 | 1.0 |
| | Puree | < 0.01, < 0.01, < 0.01 | < 0.01 | < 0.01 |
| | Juice | < 0.01, < 0.01, < 0.01 | < 0.01 | < 0.01 |

Table 93. Processing of Tomatoes to Puree and Pomace

| Study No. Location | Sample analysed | Residues of α, β Endosulfan and Endosulfan sulfate(mg/kg) | Total Residues, mg/kg | Control sample | Transfer Factor |
|---|---|---|---|---|---|
| A32878 | Fruit* | 0.049, 0.058, < 0.01 | | | |
| | | 0.047, 0.081, < 0.01 | 0.12 | < 0.01, < 0.01, < 0.01 | |
| | Chopped fruit | 0.082, 0.081, < 0.01 | 0.16 | < 0.01, < 0.01, < 0.01 | 1.4 |
| A32879 | Seeds, peel | 0.571, 0.546, 0.047 | 1.15 | 0.017, 0.01, < 0.01 | 10 |
| | Puree | 0.018, 0.019, < 0.01 | 0.04 | < 0.01, < 0.01, < 0.01 | 0.33 |
| | Puree 10/11% solids | < 0.01, 0.025, < 0.01 | 0.03 | < 0.01, < 0.01, < 0.01 | 0.27 |
| A32879 | Puree 16% solids | < 0.01, 0.033, < 0.01 | 0.03 | < 0.01, < 0.01, < 0.01 | 0.27 |
| A32881 | Dry Pomace | 1.354, 1.245, 0.111 | 2.71 | 0.084, 0.053, 0.031 | 24.6 |

* Duplicate analyses of same sample

Table 94. Processing of Tomatoes to Tomato Juice, Paste and Cooked Skins and Peel.

| Study No. in C004071 | Sample analysed | Residues of α, β Endosulfan and Endosulfan sulfate (mg/kg) | Total Residues (mg/kg) | Transfer Factor |
|---|---|---|---|---|
| A32880 | Fruit* | 0.045, 0.052, < 0.01 | 0.09 | |
| | Whole Pack (peeled) | < 0.01, < 0.01, < 0.01 | < 0.01 | < 0.10 |
| | Juice | 0.023, 0.021, < 0.01 | 0.044 | 0.49 |
| | Paste | 0.026, 0.036, 0.026 | 0.09 | 1 |
| | Cooked skins and peel | 1.830, 1.975, 0.138 | 3.94 | 43.7 |
| | Dry skins and peel | 1.492, 2.827, 0.244 | 4.56 | 50.2 |

Soybean plants were treated with 3 applications from 0.50 to 0.53 kg ai/ha and harvested between 30 and 43 days in three trials in Brazil (Huth, 1999). The transfer factors were 4.1 and 4.3 for crude oil.

A second study (Dorr and Krebs, 1982) from Brazil, where EC and ULV formulations were applied at the rate of 0.17 to 0.5 kg ai/ha, a transfer factor of 1.67 for crude oil was found.

A third study (Fox, 1979) was conducted by fortification with a mixture of alpha, beta and endosulfan sulfate at 0.06, 0.42 and 0.36 mg/kg respectively of untreated soybeans. After steaming at 120°C during 60 minutes, 60% and 28% of the dose of alpha and beta endosulfan were recovered. For refining the fortification was 0.1, 0.3 and 0.3 for 0.7 and 1.4 mg/kg of alpha, beta and sulfate endosulfan and 0.1, 0.7 and 0.6 in a second experiment.

The report (Their, 1979) describes experiments in which untreated soybean flour was spiked with (0.02 mg/kg α-endosulfan + 0.02 mg/kg β-endosulfan + 0.3 mg/kg endosulfan sulfate) in one experiment, and 0.35 mg/kg α/β-endosulfan mixture in a second experiment. The flour was baked for 2 hours at 200°C. Residues were reduced to 7 to 22% of the initially spiked amount.

Table 95. Processing of Soybean seeds to Crude oil and Press cake.

| Processed Fraction | Residues of α, β Endosulfan and Endosulfan sulfate (mg/kg) | Total residues (mg/kg) | Transfer Factor | Residues of α, β Endosulfan and Endosulfan sulfate (mg/kg) | Total residues (mg/kg) | Transfer Factor | Residues of α, β Endosulfan and Endosulfan sulfate (mg/kg) | Total residues (mg/kg) | Transfer Factor |
|---|---|---|---|---|---|---|---|---|---|
|  | A17979 (Hugh) | | | A17987I | | | A16110 (Hugh) | | |
| seeds | 0.02, 0.02, 0.3 | 0.34 | | | 0.60 | | | 0.30 | |
| crude oil | 0.1, 0.7, 0.6 | 1.40 | 4.1 | 0.1, 0.3, 0.3 | 0.70 | 1.16 | | 1.30 | 4.33 |
| press cake | (0.01), (0.01), < 0.01 | (0.02 | 0.06 | (0.01), (0.01), < 0.01 | (0.02 | 0.05 | (0.01), (0.01), < 0.01 | (0.02) | 0.07 |

Table 96. Processing of Soybean seeds to Crude and Refined Oil and Baking.

| Process | Crushed grain | Bran | Crude oil | Refined oil |
|---|---|---|---|---|
| Residues of α, β Endosulfan and Endosulfan sulfate(mg/kg) | < 0.01,< 0.01,< 0.01 to 0.09 | < 0.01,< 0.01,< 0.01 to 0.2 | < 0.01,< 0.01, 0.02 to 0.2 | < 0.01,< 0.01,< 0.01 to 0. 2 |
| process | | Baking Before | Baking after | |
| Residues of α, β Endosulfan and Endosulfan sulfate(mg/kg) | | 0.02,0.02,0,.3 | 0.004,0;02,0.05 | |
| Residues of α, β Endosulfan and Endosulfan sulfate(mg/kg) | | 0.21,0.14,< 0.01 | 0.004,0;01<,0.01 | |

Cocoa was treated twice with CS and EC formulations in Ghana in 2002 (Balluth, 2003). The rates of applications ranged from 0.493 to 0.520 kg ai/ha. Samples were taken at 13/14, 20/21 and 27 days after the last application. Cocoa pods were separated and the beans and flesh were wrapped in untreated banana leaves, placed in wooden boxes, wrapped in plastic and sealed. After this period the fermentation was stopped. The beans were cleaned manually and frozen. Fermented beans were dried using natural sunlight in the same box used for fermentation. Residues in eight trials were mostly ND or less than the LOQ (< 0.01 mg/kg) in beans, fermented beans and dry beans. No transfer factors could be calculated.

In the second study by Balluth in 2002, cocoa was treated twice with CS and EC formulations in the Ivory Coast in 2001. The application rates were equivalent, the seasonal total ranging from 0.50 to 0.518 kg ai/ha. The spray interval was 25/27 days, and samples were taken 14 days after the last application. Residues in eight trials were mostly ND or less than the LOQ (< 0.01 mg/kg) in beans, fermented beans and dry beans. No transfer factors could be calculated.

Gomez (1996) studied the magnitude endosulfan residues in coffee and processed fractions. Coffee in two plots in 1994 (Brazil and one in Colombia) were sprayed three times at 30 day intervals at a 3× rate of 2.1 kg ai/ha/application, in a water volume of 500 L/ha. Green coffee beans were harvested 33 days after the last application.

The green coffee beans were roasted at 350–430 °F for 6 minutes, cooled and ground. No residue of any analyte was found in the processed fractions, roast coffee and instant coffee.

Table 97. Processing of Green Coffee to Roasted Coffee.

| Processed Fraction | Residues of α, β Endosulfan and Endosulfan sulfate (mg/kg) | Total residues (mg/kg) | Transfer Factor | Residues of α, β Endosulfan and Endosulfan sulfate (mg/kg) | Total residues (mg/kg) | Transfer Factor |
|---|---|---|---|---|---|---|
|  | Brazil | | | Colombia | | |
| Green beans | 0.04, 0.08, 0.04 | 0.16 | | < 0.01, < 0.01, < 0.01 | < 0.01 | |
| Roast coffee | < 0.01, < 0.01, < 0.01 | < 0.01 | < 0.06 | < 0.01, < 0.01, < 0.01 | < 0.01 | |
| Instant coffee | < 0.01, < 0.01, < 0.01 | < 0.01 | < 0.06 | < 0.01, < 0.01, < 0.01 | < 0.01 | |

## RESIDUES IN ANIMAL COMMODITIES

The Meeting received two lactating dairy cow feeding studies and one lactating goat study which provided information indicating that residues in animal tissues and milk were likely through exposure to residues in the animal's diet.

### *Direct animal treatment*

No study provided

### *Farm animal feeding study*

#### *Lactating cows*

A dairy study (Keller and Bowman, 1959) was conducted with four groups of lactating Holstein cows using $^{14}$C-endosulfan for 30 days. Targeted treatment rates were 0, 0.3, 3 and 30 ppm based on the diet. All animals exhibited normal appearance and behaviour. Food consumption and milk production were within normal limits. Residue levels at the end of the dosing period were proportional to dose in all tissues indicating absence of bioaccumulation. The highest residues were measured in the liver. Analysis in blood showed a gradual rise reaching a plateau at 21 days. In the recovery period of 14 days the residue levels declined significantly, though in most cases not below the detection limit.

Table 98. Residues (µg/g) in cows after 30 days application of $^{14}$C-endosulfan.

| Dietary Dose | 0.3 ppm | | 3.0 ppm | | 30.0 ppm | |
|---|---|---|---|---|---|---|
| Recovery period | none | 14 days | none | 14 days | none | 14 days |
| Liver | 0.3 | 0.1 | 2.5 | 1.1 | 25 | 16 |
| Kidneys | 0.05 | 0.05 | 0.4 | 0.1 | 6 | 1 |
| Omental fat** | 0.07 | < 0.02 | 0.7 | 0.1 | 7 | 0.1 |

**Limit of detection in omental fat: 0.02 µg/g

In a second study, groups of four lactating cows (animals weighing 520–680 kg) were dosed daily via corn oil in the diet with endosulfan 0, 4, 12 and 30 ppm in the diet for 28 consecutive days (Peatman *et al.*, 1999). Milk was collected daily and frozen for residue analysis. All cows were sacrificed on day 29 of dosing (day 28 for control animals) and samples of muscle, liver, kidney and composite fat were taken. Samples were analysed by the method of FDA Pesticide Analytical Manual. The LOQs for α-endosulfan, β-endosulfan and endosulfan sulfate in milk, liver, kidney and muscle were each 0.01 mg/kg. In fat the LOQ for each analyte was 0.05 mg/kg.

Whole milk samples were analysed for all treated cows for days -1, 1, 4, 7, 10, 13, 16, 19, 22, 25 and 28 of dosing (day 9 also analysed for group 3). A residue plateau in milk was established in all three groups, between day 10 and day 13 of dosing (mean value).

For all samples analysed, the residue values for α-endosulfan and β-endosulfan were mostly below the respective LOQs (< 0.01 mg/kg). The maximum residues for α-endosulfan and β-endosulfan in tissues were 0.01, 0.02 mg/kg (liver) and 0.002, 0.08 mg/kg (fat). Residues of endosulfan sulfate were significant and accounted for the major portion of any residue measured.

Samples of whole milk, cream and skim milk were analysed from the 12 ppm dose group day 9 samples to provide some indication on the distribution of residues between milk and milk fat. Results are shown below.

Table 99. Endosulfan residues in milk and animal tissues.

| Substrate | Nominal dose level (ppm diet) | Number of days dosing | Mean Residues of α- β- endosulfan, and endosulfan sulfate | Transfer factor for Endosulfan Sulfate |
|---|---|---|---|---|
| Whole milk | 4 | 10-28 | < 0.01, < 0.01, 0.07 | 0.018 |
| | 12 | 10-28 | ND, < 0.01, 0.27 | 0.02 |
| | 30 | 10-28 | ND, < 0.01, 0.62 | 0.02 |
| Muscle | 4 | 28 | ND, < 0.01, 0.04 | 0.01 |
| | 12 | 28 | < 0.01, < 0.01, 0.21 | 0.02 |
| | 30 | 28 | < 0.01, < 0.01, 0.76 | 0.025 |
| Liver | 4 | 28 | < 0.01, < 0.01, 0.71 | 0.18 |
| | 12 | 28 | < 0.01, < 0.01, 2.0 | 0.17 |
| | 30 | 28 | ND, < 0.01, 3.2 | 0.11 |
| Kidney | 4 | 28 | < 0.01, < 0.01, 0.07 | 0.02 |
| | 12 | 28 | < 0.01, < 0.01, 0.31 | 0.03 |
| | 30 | 28 | ND, < 0.01, 0.67 | 0.02 |
| Fat | 4 | 28 | ND, < 0.05, 1.4 | 0.35 |
| | 12 | 28 | ND, < 0.05, 4.7 | 0.39 |
| | 30 | 28 | < 0.05, 0.06, 9.9 | 0.33 |

Table 100 Endosulfan sulfate residues in milk, skim milk and cream (mg/kg).

| Days dosing | Nominal dose level (ppm diet) | Whole milk | Skim milk | Cream |
|---|---|---|---|---|
| 9 | 12 | 0.23 | 0.17(0.12,0.26;0.13) | 1.0 (0.81,0.89,1.4) |

This limited data provides a transfer factor (mean) of 4.3 from whole milk to cream.

Table 101. Mean endosulfan sulfate in whole milk in mg/kg.

| day | 4 ppm in the diet | 12 ppm | 30ppm |
|---|---|---|---|
| 1 | < 0.01 | 0.02 | 0.04 |
| 4 | 0.07 | 0.26 | 0.53 |
| 7 | 0.06 | 0.23 | 0.50 |
| 9 | | 0.23 | |
| 10 | 0.07 | 0.27 | 0.56 |
| 13 | 0.07 | 0.27 | 0.61 |
| 16 | 0.07 | 0.24 | 0.62 |
| 19 | 0.07 | 0.26 | 0.66 |
| 22 | 0.06 | 0.24 | 0.64 |
| 25 | 0.07 | 0.34 | 0.56 |
| 28 | 0.06 | 0.28 | 0.66 |

*Depuration*

Data from the animals depurated for up to 21 days after cessation of dosing showed that residues fell significantly once dosing stopped.

Table 102. Residue levels in tissues after 21 days depuration, mean residues of as α-endosulfan, β-endosulfan and endosulfan sulfate.

| Days depuration | Milk | Muscle | Liver | Kidney | Fat |
|---|---|---|---|---|---|
| -2 | < 0.01, < 0.01, 0.94 | | | | |
| 0 | | < 0.01, < 0.01, 0.76 | ND, < 0.02, 3.2 | ND < 0.01, 0.67 | < 0.05, 0.07, 9.9 |
| 1 | < 0.01, < 0.01, 0.69 | | | | |
| 4 | < 0.01, < 0.01, 0.18 | | | | |
| 7 | < 0.01, < 0.01, 0.09 | ND, ND, 0.06 | ND, ND, 0.76 | ND, ND, 0.10 | ND, ND, 5.1 |
| 10 | < 0.01, < 0.01, 0.15 | | | | |

Mean data from animals in top dose group (28 days dosing).

In Australia a feeding study was conducted by Mawhinney (2001). Rations containing endosulfan were fed for twelve days, then replaced with clean feed and the rate of depletion of endosulfan sulfate residues in the fat of the trial animals was measured for a period of 28 days. Samples of subcutaneous fat were collected at days 1, 7, 14 and 21 post-treatment by means of biopsy. All trial animals were slaughtered at day 28 when samples of subcutaneous fat, perirenal fat, liver and kidney were collected from each carcass.

Table 103. Feeding diet and doses.

| Treatment Group | Ration | Composition of Daily Endosulfan Dose (mg) | | | |
|---|---|---|---|---|---|
| | | α-endosulfan | β-endosulfan | Endosulfan Sulfate | Total |
| 1 | Feedlot ration | 0.1 | 0.3 | 0.6 | 1 |
| 2 | Lucerne hay based diet containing incurred residues of endosulfan. | 42 | 40 | 180 | 262 |
| 3 | Lucerne hay based diet | 2 | 6 | 12 | 20 |
| 4 | Lucerne hay based diet | 5 | 15 | 30 | 50 |
| 5 | Lucerne hay based diet | 5 | 15 | ---- | 20 |

The depletion rates calculated from data from treatment groups 2, 3, 4 and 5 were consistent with an average half-life of 7.2 days and a 95% confidence limit of 6.6 - 8.0 days. For treatment group 1, it was much longer at 25.4 days with a confidence limit of 18.8 - 39.1 days.

Group 1 animals were fed the lowest concentration of total endosulfan in their rations and none of the residual fat concentrations of endosulfan sulfate exceeded 0.04 mg/kg at any time in the trial.

From the trial data, feed contaminated at around 0.7 mg/kg, fed for around 12 days, would be expected to give rise to residues of endosulfan sulfate around 0.2 mg/kg in the fat of cattle.

Similarly, concentrations of total endosulfan in animal feed at 0.03 mg/kg could be expected to give rise to residues of endosulfan sulfate at around 0.01 mg/kg in the fat of cattle if fed for 12 days.

It was not possible, in this study, to confirm that the plateau concentration had been reached, in each case, by day 12, when the dosed rations were withdrawn. However, comparison of the bio magnification in these animals with that in the associated DAN.092 trial (these animals had been exposed for 30–35 days) strongly suggests it had been reached.

Under the conditions of this study, the bio magnification factor for total endosulfan, when it passes from fodder and is stored as endosulfan sulfate in the fatty tissues of cattle, was around 0.3.

There was no significant difference in the concentrations of endosulfan sulfate in the fat collected from the two subcutaneous sites, but on average, the subcutaneous fat residual concentrations were 1.6 times higher than those in the corresponding perirenal fats. On average the concentrations of endosulfan sulfate in subcutaneous fats were some four times higher than the concentrations in the corresponding liver tissues and some ten times higher than the concentrations in the corresponding kidney tissues.

*Lactating goats*

The distribution of endosulfan (technical grade) was investigated by Indranignsih *et al.*, (1993) in lactating goats following repeated oral administration. Twelve adult lactating feral goats (25 to 40 kg body weight), each with one kid, were dosed orally with 1 mg/kg of non-labelled endosulfan for a period of 28 days using gelatine capsules. The applied dose of 1 mg/kg body weight corresponded to 29 ppm in the diet. Feed and water were given *ad libitum*. Groups of 3 animals were sacrificed for tissue collection 1, 8, 15, and 21 days after the last treatment. Milk and venous blood samples were taken from each animal before being killed.

The highest residues were detected in organs and tissues of goats, which were slaughtered 24 hours after the 28-day feeding with endosulfan. These residues are presented in the Table 104.

Table 104. Endosulfan residues in organs/tissues/milk of goats 24 hours after daily dosing with technical grade Endosulfan at 1 mg/kg body weight for a period of 28 days ( in mg/kg).

| Organs/ tissues | Alpha-Endosulfan | Beta–Endosulfan | Endosulfan sulfate | Total Endosulfan residues ** | Clearance half-life (d) |
|---|---|---|---|---|---|
| Liver | 0.010 | 0.021 | 0.097 | 0.128 | 3,1 |
| Kidney | 0.220 | 0.059 | 0.012 | 0.291 | -Not recorded |
| Fat | 0.015 | 0.002 | 0.040 | 0.057 | 1.4 |
| Muscle | 0.033 | 0.009 | < 0.001 | 0.043 | 1.1 |
| Milk | | | | 0.020 | |

** Total Endosulfan residues = sum of $\alpha$- and $\beta$-endosulfan and endosulfan sulfate. Values < 0.001 were taken as 0.001.

The residues in all organs and tissues decreased significantly (< 0.01 mg/kg) until the next sampling point (day 8), however with one remarkable exemption: The residues in the excretion organs, the kidneys, increased from 0.29 (day 1) to 0.47 mg/kg (day 8) and decreased again to approximately 0.2 mg/kg (day 15 of the 28-day feeding). The residues in milk became undetectable after one week. At day 21 residues in all tissues were non detectable.

The clearance of the residues is relatively rapid with half-lives in the range of 1–3 days.


## RESIDUES IN FOOD IN COMMERCE OR AT CONSUMPTION


### Monitoring data

Systematic monitoring of residues of endosulfan has been carried out for several years. Typical of these programs is the "Monitoring of Pesticide Residues in Products of Plant Origin in the European Union, Norway, Iceland (and Liechtenstein)".

### Monitoring in Europe in 1999, 2000 and 2001

The results for endosulfan are given in Tables 105–7.

Table 105. Results from the EU coordinated monitoring program for endosulfan residues.

| Year | N° samples | N° samples Without residues | N° samples With residues Below MRL | % | N° samples With residues Above MRL | % | Residues maximum |
|---|---|---|---|---|---|---|---|
| 2000 | 3318 | 3277 | 41 | 11.24 | 0 | 0 | 0.36 (head cabbage, EC-MRL:1.0) |
| 2001 | 8478 | 8125 | 326 | 3.85 | 27 | 0.32 | 3.50 (lettuce, EC-MRL:1.0/0..05) |

Table 106. Results from the EU coordinated monitoring program for pesticide residues for some pesticide analysed for in cauliflower, peppers, wheat grains, and melons.

| Pesticide As example | N° samples | N° samples Without residues | N° samples With residues Below MRL | % | N° samples With residues Above MRL | % | Residues maximum |
|---|---|---|---|---|---|---|---|
| Endosulfan | 4071 | 3387 | 678 | 16.7 | 6 | 0.15 | 1.5 (peppers, EC-MRL:1.0) |

Some pesticides found most often with residues at or below the MRL (national or EC—MRL) and pesticides exceeding the MRLs (national or EC—MRLs) for different commodities in 1997, 1998 and 1999 are given in Table 107.

Table 107 Comparison of pesticides found most often and pesticides exceeding MRLs (national or EC-MRLs) analysed on different commodities in 1997, 1998 and 1999.

| | Program 1997 | | Program 1998 | | Program 1999 | |
|---|---|---|---|---|---|---|
| | Mandarin, pear, banana, bean, potato | | Orange, peach, ,carrot, spinach | | Cauliflower, pepper, wheat, melon | |
| | % samples With residues Below MRL | % samples With residues Above MRL | % samples With residues Below MRL | % samples With residues Above MRL | % samples With residues Below MRL | % samples With residues Above MRL |
| Deltamethrin | na | na | 0.38 | 0 | 0.58 | 0 |
| Diazinon | 0.55 | 0 | 1.1 | 0.10 | 0.34 | 0.02 |
| Endosulfan | 1.3 | 0 | 2.0 | 0.02 | 16.7 | 0.15 |
| Iprodione | 1.3 | 0.13 | 4.0 | 0.30 | 1.1 | 0 |

## Monitoring in Australia 2003-2004 and 2004-2005

Endosulfan was included in the Australian National Residue Survey program in 2003-2004 and 2004-2005 (Hamilton NRS, 2004 NRS 2005).

Table 108. Survey program in Australia for endosulfan.

| Commodity | Limit of reporting, mg/kg | Australian MRL, mg/kg | Number of analyses 2003-2004 | Number of analyses 2004-2005 | Number of residues |
|---|---|---|---|---|---|
| Apple | 0.05 | 2 | 214 | 221 | 0 |
| Barley | 0.02 | 0.2 | 280 | 73 | 0 |
| Buffalo fat | 0.02 | 0.2 | 10 | 10 | 0 |
| Camel fat | 0.02 | 0.2 | 10 | 10 | 0 |
| Canola | 0.02 | 1 | 57 | 19 | 0 |
| Cattle fat | 0.02 | 0.2 | 610 | 1096 | 0 |
| Deer fat | 0.02 | 0.2 | 25 | 26 | 0 |
| Field pea | 0.02 | 1 | 42 | 9 | 0 |
| Game pig fat | 0.02 | 0.2 | 66 | 75 | 0 |
| Goat fat | 0.02 | 0.2 | 97 | 99 | 0 |
| Honey | 0.02 | not set | 13 | | 0 |
| Horse fat | 0.02 | 0.2 | 10 | 19 | 0 |
| Kangaroo fat | 0.02 | 0.2 | 77 | 75 | 0 |
| Lupin | 0.02 | 1 | 51 | 21 | 0 |
| Macadamia nut | 0.05 | 2 | 120 | 120 | 0 |
| Oats | 0.02 | 0.2 | 32 | 17 | 0 |
| Onion | 0.05 | 0.2 | 136 | 101 | 0 |
| Ostrich fat | 0.02 | 0.2 (not set)5 | 24 | 28 | 0 |
| Pear | 0.05 | 2 | 71 | 71 | 0 |
| Pecan nuts | 0.05 | 2 | 30 | | 0 |
| Pig fat | 0.02 | 0.2 | 96 | 299 | 0 |
| Sheep fat | 0.02 | 0.2 | 753 | 725 | 0 |

| Commodity | Limit of reporting, mg/kg | Australian MRL, mg/kg | Number of analyses 2003-2004 | Number of analyses 2004-2005 | Number of residues |
|---|---|---|---|---|---|
| Sorghum | 0.02 | 0.2 | 72 | 31 | 0 |
| Wheat | 0.02 | 0.2 | 729 | 181 | 0 |
| Wheat bran | 0.02 | 0.2 | 33 | 10 | 0 |
| Wheat flour | 0.02 | 0.2 | 33 | 10 | 0 |

## NATIONAL MAXIMUM RESIDUES LIMITS

Table 109. EU MRLs of endosulfan.

| COMMODITY | EU MRL | NEW EU MRL[#] | FR | B | NL | GE | IT | UK |
|---|---|---|---|---|---|---|---|---|
| Citrus fruit | 1.0 | 0.05* | 1.00 | | 0.50 | | 1.00 | |
| Tree nuts | 0.1 | 0.1* | 0.10 | 0.10 | 0.10 | 0.10 | 0.10 | 0.10 |
| Pome fruit | 1.0 | 0.3 | 1.00 | | 0.30 | | 1.00 | |
| Stone fruit (peaches) | 1.0 | 0.05* | 0.50 | | 0.50 | | 1.00 | |
| Table and wine grapes | 1.0 | 0.5 | | | 0.50 | | 1.00 | |
| Strawberries | | | 1.00 | | | | 1.00 | |
| Raspberries | 1.0 | 0.05* | 0.50 | | | | 1.00 | |
| Blackberries | | | 1.00 | | | | 1.00 | |
| Red blackcurrants | | | 1.00 | | | | 1.00 | |
| Other berries | 0.05 | 0.05* | | 0.05 | | | | |
| Wild berries and fruit | 0.05 | 0.05* | | | | 0.05 | | 0.05 |
| Kiwis | 1.0 | 0.05* | | 1.0 | | 1.00 | 1.00 | 1.00 |
| Olives | 1.0 | 0.05* | | | | | | |
| Beetroot | 0.2 | 0.05* | 0.20 | 0.20 | | | 0.20 | 0.20 |
| Carrots | 0.2 | 0.05* | 0.20 | 0.20 | | 0.05 | 0.20 | 0.20 |
| Celeriac | 0.2 | 0.05* | 0.20 | 0.20 | | 0.05 | 0.20 | 0.20 |
| Radishes | 0.2 | 0.05* | 0.20 | 0.20 | | 0.05 | 0.20 | 0.05 |
| Swedes | 0.2 | 0.05* | 0.20 | 0.20 | | 0.05 | 0.20 | 0.20 |
| Turnips | 0.2 | 0.05* | 0.20 | 0.20 | | 0.05 | 0.20 | 0.20 |
| Onions | 1.0 | 0.05* | | | | 1.00 | 1.00 | 1.00 |
| Peppers | 1.0 | 1.0 | | | 1 | 1.00 | 1.00 | |
| Tomatoes | 1.0 | 0.5 | | | 0.50 | 1.00 | | |
| Cucurbits edible peel | 1.0 | 0.05* | 0.5 | | 0.30 | 1.00 | 1.00 | |
| Cucurbits inedible peel | 1.0 | 0.05* | 0.5 | | | 1.00 | 1.00 | |
| Sweet corn | 0.05 | 0.05* | | | | 0.05 | | |
| Flowering brassica | 1.0 | 0.05* | 1.00 | | | 0.05 | | |
| Head brassica | 1.0 | 0.05* | | | | 0.05 | 1.00 | |
| Leafy brassica | 1.0 | 0.05* | | | | 0.05 | | |
| Kohlrabi | 0.05 | 0.05* | | 0.05 | | 0.05 | | |
| Lettuce and similar | 1.0 | 0.05* | 1.00 | | | 0.05 | 1.00 | |
| Spinach an similar | 1.0 | 0.05* | | | | | 1.00 | |
| Legume vegetables | 1.0 | 0.05* | 0.50 | | | | 1.00 | |
| Asparagus | | | 1.00 | 0.05 | | 0.05 | | |
| Cardoons | 1.0 | 0.05* | | | | | | |
| Celery | 1.0 | 0.05* | | | | | 1.00 | |
| Globe artichokes | 1.0 | 0.05* | 1.00 | | | | 1.00 | |
| Leeks | 1.0 | 0.05 | | | | | 1.00 | |

| COMMODITY | EU MRL | NEW EU MRL[#] | FR | B | NL | GE | IT | UK |
|---|---|---|---|---|---|---|---|---|
| Cultivated mushrooms | 1.0 | 0.05 | | | | | | |
| Cotton seed | 0.3 | 5 | | | 0.3 | | | |
| Other oilseeds | 0.1 | 0.1* | | | 0.5soya) | | | |
| Potatoes | | 0.05* | | 0.10 | 0.05 | 0.05 | 0.20 | |
| Tea | 30.0 | 30 | | | 30 | | | |
| Cereals | 0.1 | 0.05* | | | | | | |
| Animal fats | 0.1 | | | | 0.1* | | | |
| Milk | 0.004 | | | | 0.004 | | | |

# Directive 29 June 2006

## Table 110. MRLs in USA, Canada, Japan, Mexico, South Africa and Codex.

| Commodity | CAC | Canada | Japan | Mexico | South Africa | USA | USA |
|---|---|---|---|---|---|---|---|
| Alfalfa forage (green) | 1.0 | | 0.50 | 0.3 | | 0.3 | |
| Alfalfa, hay | | | | | | 1.0 | |
| Almond hulls | | | | | | 1.0 | 1.0 |
| Almonds | | | | | | 0.2 | 0.2N |
| Apples | | 2.0 | 0.50 | 2.0 | 0.5 | 2.0 | 2.0 |
| Apricots | | 0.5 | 0.50 | 2.0 | 0.5 | 2.0 | 2.0 |
| Artichokes | | 1.0 | 0.50 | | 2.0 | | |
| Barley | | | 0.50 | 0.1 | | 0.1 | 0.1 (N) |
| Barley straw | | | | | | 0.2 | 0.2 (N) |
| Broccoli | 0.5 | 2.0 | 0.50 | 2.0 | | 2.0 | 2.0 |
| Bean | 0.5 | 2.0 | 0.50 | 2.0 | 1.0 | 2.0 | 2.0 |
| Blueberries | | | | | | 0.1 | 0.1 (N) |
| Brussels sprouts | | 2.0 | 0.50 | 2.0 | | 2.0 | 2.0 |
| Cabbage, Savoy | 2.0 | 2.0 | 0.50 | 2.0 | | 2.0 | 2.0 |
| Cabbages, Head | 1.0 | 2.0 | 0.50 | | | 2.0 | 2.0 |
| Cocoa beans | 0.1 | | 0.50 | | | | |
| Carrot | 0.2 | | 0.50 | | | 0.2 | 0.2 |
| Cattle fat | | | | | | 0.2 | 0.2 |
| Cattle liver | | | | | | | 0.2 |
| Cattle meat | | | | | | 0.2 | 0.2 |
| Cauliflower | 0.5 | 1.0 | 0.50 | 2.0 | | 2.0 | 2.0 |
| Celery | 2.0 | 1.0 | 0.50 | 2.0 | | 2.0 | 2.0 |
| Cherries | 1.0 | 2.0 | 0.50 | | 0.5 | 2.0 | 2.0 |
| Citrus | | | | | 1.0 | | |
| Clover | 1.0 | | 0.50 | | | | |
| Coffee beans | 0.1 | | 0.50 | 0.04 | 0.5 | | |
| Common bean (pods and/or immature seeds) | 0.5 | | 0.50 | | | | |
| Corn | | | | 2?0 | | 0.2 | |
| Cotton seed | 1.0 | | 0.50 | 1.0 | 0.2 | 1.0 | 1.0 |
| Cotton seed oil, Crude | 0.5 | | 0.50 | | | | |
| Cucumber | 0.5 | 2.0 | 0.50 | 2.0 | 0.5 | 2.0 | 2.0 |
| Eggplant | | 2.0 | 0.50 | 2.0 | | 2.0 | 2.0 |
| Fruits (except as otherwise listed) | | | 0.50 | | | | |
| Garden pea (young pods) | 0.5 | 0.5 | 0.50 | 2.0 | 0.5 | | |
| Grapes | 1.0 | 1.0 | 0.50 | 2.0 | 0.5 | 2.0 | 2.0 |

| Commodity | CAC | Canada | Japan | Mexico | South Africa | USA | USA |
|---|---|---|---|---|---|---|---|
| Hogs, meat | | | | | | | 0.2 |
| Hops | | | | 20 | | | |
| Kale | 1.0 | | 0.50 | | | 2.0 | 2.0 |
| Lettuce, Head | 1.0 | 2.0 | 0.50 | 2.0 | | 2.0 | |
| Lettuce, Leaf | 1.0 | | 0.50 | | | | 2.0 |
| Maize | 0.1 | | 0.50 | 2.0 | | | 2.0 |
| Meat (from mammals other than marine mammals) | | | 0.50 | | | | |
| Melons, except watermelon | 0.5 | 1.0 | 0.50 | 2.0 | | 2.0 | 2.0 |
| Milks | 0.004 | | 0.50 | | | 0.5 | 0.5 |
| Onion, Bulb | 0.2 | | 0.50 | | | | |
| Oranges, Sweet, Sour | 0.5 | | 0.50 | | | | |
| Peach | 1.0 | 2.0 | 0.50 | 2.0 | 0.5 | 2.0 | 2.0 |
| Pear | | 2.0 | 0.50 | 2.0 | 0.5 | | 2.0 |
| Pecan | | | | | | 0.20 | 2.0 |
| Pepper | | 1.0 | 0.50 | 2.0 | 1.00 | 2.0 | 2.0 |
| Pineapple | 2.0 | | 0.50 | 2.0 | 0.05 | 2.0 | 2.0 |
| Plums (including prunes) | 1.0 | 2.0 | 0.50 | 2.0 | 0.5 | 2.0 | 2.0 |
| Pome fruits | 1.0 | 2.0 | 0.50 | | | | |
| Potato | 0.2 | | 0.5 | 0.20 | 0.05 | 0.2 | 0.2 (N) * |
| Pumpkins | | 2.0 | 0.50 | 2.0 | | 2.0 | 2.0 |
| Rape seed | 0.5 | | 0.50 | | 1.0 | 0.2 | |
| Rice | 0.1 | | 0.50 | | | | |
| Sorghum | | | | | | | |
| Soya bean (dry) | 1.0 | | 0.50 | | | | |
| Spinach | 2.0 | 2.0 | 0.50 | | 2.0 | | 2.0 |
| Squash, Summer | 0.5 | 1.0 | 0.50 | 2.0 | 2.0 | | 2.0 |
| Sugar beet | 0.1 | | 0.50 | | 0.1 | | |
| Sugar beet leaves or tops | 1.0 | | 0.50 | | | | |
| Sugar cane | | | | 0.5 | 0.1 | 0.5 | |
| Sunflower seed | 1.0 | | 0.50 | 0.2 | 0.1 | 2.0 | |
| Strawberries | | 1.0 | 0.50 | 2.0 | | 2.0 | |
| Sweet potato | 0.2 | | 0.50 | | | 0.2 | |
| Tea, Green, Black | 30.0 | | 0.5 | | | | |
| Tobacco | | | | | 2.0 | | |
| Tomato | 0.5 | 1.0 | 0.50 | 2.0 | 0.5 | 2.0 | |
| Trefoil | 1.0 | | 0.50 | | | | |
| Vegetables | | | 0.50 | | | | |
| Wheat grain | 0.2 | | 0.50 | 0.1 | 0.5 | 0.1 | |
| Wheat straw | | | | 0.1 | | 0.2 | |

## APPRAISAL

Endosulfan was listed in the periodic re-evaluation programme at the 36th Session of the CCPR for periodic review by the 2006 JMPR. The toxicology of endosulfan was reviewed within the periodic review program by the 1998 JMPR.

The Meeting received extensive information on the metabolism and environmental fate, methods of analysis, stability of residues in storage, registered use patterns, residue supervised trials data, farm animal feeding studies and the fate of residues during processing.

### Animal metabolism

The Meeting received animal metabolism studies with endosulfan in rats, dairy cows, lactating sheep and laying hens.

Initial metabolism of endosulfan in rats involves either sulfoxidation to endosulfan sulphate, a fat-soluble metabolite, followed by desulfation to the diol, or direct hydrolysis to the diol followed by oxidation to the ether, the hydroxy ether, the dihydroxy ether, and to the main metabolite in urine and faeces, the lactone. A number of unidentified polar metabolites are probably the conjugates of known metabolites. The majority of an oral dose was excreted in the faeces (70–90%) and urine (9–20%) as polar metabolites. Highest radioactivity concentrations were observed in liver and kidney followed by fat. Repeated administration of radiolabeled endosulfan or a 2 year feeding study in rats did not show a bioaccumulation of residues in fatty tissues.

Dairy cows were dosed with [$^{14}$C]-endosulfan at a dose rate equivalent to 22 ppm in the diet for five consecutive days, equivalent to 0.64 mg/kg bw/day. Radioactivity was detected in all edible tissues and milk at between 0.05 and 3.57 mg/kg parent equivalent. The parent compound alpha and beta isomers were detected in tissues from 2 to 15%. The major metabolite identified in all tissues, including fat and milk, was endosulfan sulphate (12–89%) with endosulfan lactone being found in kidney and liver tissue, indicating that the endosulfan is readily cleaved following dosing to a dairy cow. Metabolites other than endosulfan sulphate reported in liver and kidney tissue were produced as a result of enzymatic and acid hydrolysis of polar material that predominated in these two tissues.

Following a single dose of [$^{14}$C]-endosulfan (methylene labelling) to two lactating East Friesian sheep at a dose rate equivalent to 0.3 ppm in the diet, approximately 90% of the administered $^{14}$C-material was excreted in the urine and faeces. Endosulfan diol and endosulfan hydroxyether, but not parent, were found in urine while endosulfan was the major component of the residue in faeces. 1–2% of the radioactivity was found in milk collected 0–17 days after administration. The main metabolite was endosulfan sulphate with the highest concentration of 0.15 mg/kg (0–24 hours after dosing) and was clearly concentrated in cream. At sacrifice, 40 days after dosing, the radioactivity level was less than 0.02 mg eq/kg in most of the organs and tissues, with exemption of liver having a peak level of 0.03 mg eq/kg.

Laying hens were dosed with [$^{14}$C]-endosulfan at a dose rate equivalent to 10 ppm in the diet for 12 consecutive days; the radioactivity was detected in all edible tissues at a level ranging between 0.013 and 0.974 mg/kg parent equivalent. The major metabolite identified in all tissues (excluding egg white) was endosulfan sulphate (36–65%), with a small percentage of unchanged α- and β-endosulfan also seen, plus the products of hydrolysis and oxidation namely endosulfan diol and endosulfan lactone.

In summary, the primary residues found in animal tissues were the parent, endosulfan, both alpha and beta isomers, and to a larger extent, endosulfan sulphate. The metabolism studies are consistent with the view that the parent is converted to the sulphate *in situ* and the sulphate is more likely to be measured in tissues than the parent compound. While liver appears to be the target organ for metabolism of endosulfan, the above residue components are clearly present in significant amounts in fat. The high presence of these metabolites in fat is consistent with endosulfan being a fat-soluble pesticide. However, endosulfan and endosulfan sulphate do not bioaccumulate in organisms due to the extensive metabolism with enzymatic hydrolysis of endosulfan and endosulfan sulphate forming more polar metabolites.

### Plant metabolism

The meeting received plant metabolism studies with endosulfan on tomato, cucumber, apple, sugar beet and soybean.

Young tomato plants were treated three times with $^{14}$C-labelled endosulfan at intervals of 7 days, each time at an application rate of 635 g ai/ha. 90% of the total radioactive residues were extracted from tomato fruit with acetone/water and shown to consist of the parent isomers, α- and β-endosulfan and the metabolite endosulfan sulphate. In leaves, trace amounts of free and considerable amounts of conjugated endosulfan diol were also observed.

A young <u>apple</u> tree was treated with [14]C-labelled formulated endosulfan at a rate which corresponded to 1.5 kg ai/ha. 90% of the total radioactive residues could be extracted from apples with acetone/water. These residues consisted almost exclusively of the parent isomers α- and β-endosulfan and to a very low extent the metabolite endosulfan sulphate. In leaves, endosulfan sulphate occurred as a major metabolite accounting for approximately 50% of the total radioactive residues. Only traces of endosulfan diol could be detected. The portion of non-extractable residues increased up to approximately 10% at day 21 after treatment.

<u>Cucumber</u> plants were treated three times with [14]C -labelled endosulfan at intervals of 7 days, each time at a nominal application rate of 530 g ai/ha. The total radioactive residues in the leaves decreased from 185 mg/kg to 52 mg/kg parent equivalent 0 to 14 days after the last treatment. The corresponding levels in the fruit decreased only from 0.23 to 0.18 mg/kg eq. After 14 days and the third treatment with endosulfan, the major components α- and β-endosulfan and endosulfan sulphate contributed approximately 50% of the total radioactive residues. Several smaller components did not exceed 0.05 mg/kg eq each.

<u>Sugar beet</u> plants were treated twice at 630 g ai/ha each and harvested 21 days later. In roots, 93.4% of TRR were extractable leaving 6.6% of TRR non-extractable. The organo-soluble radioactivity in roots consisted mainly of endosulfan sulphate (59.6% of TRR) followed by α- and β-endosulfan. In sugar beet leaves, more than 93% of TRR were extractable. In total, 51.9% of the TRR were identified in the leaves. A further 32.7% of the TRR was characterised as polar radioactivity. α--endosulfan, β-endosulfan and endosulfan sulphate were the major residue components in all plant parts.

<u>Soybean</u> plants were treated twice at 530 g ai/ha each. Applications were made at forage stage 61 days before harvest and hay stage 38 days before harvest. In forage just after the first treatment 98.5% of TRR were extractable with 75.4% on the plant surface. In hay 87% of the TRR were extractable and in beans at harvest 94.5%. In beans and hay the major metabolite was endosulfan sulphate with respectively 78.4 and 51.2% of the TRR. β-endosulfan and α-endosulfan were detected at 5 and 1.5% of the TRR for these two parts of the plant, respectively.

The metabolism of endosulfan in plants was characterised by decreasing levels of α-endosulfan and increasing levels of β-endosulfan and the subsequent formation of endosulfan sulphate which is the major metabolite.

### Environmental fate in soil

The aerobic degradation of endosulfan in soil starts with the modification of the 7-membered dioxothiepin ring. Oxidation results in the formation of the main metabolite endosulfan sulphate. The microbially induced hydrolysis of endosulfan and of endosulfan sulphate leads to ring opening of the 7-membered ring and formation of endosulfan diol. The endosulfan diol is then condensed to endosulfan ether (minor pathway) or oxidised to endosulfan hydroxy carboxylic acid and its condensation product endosulfan lactone. The chlorinated bicyclic carbon skeleton was shown to be completely degraded by considerable formation of labelled carbon dioxide in the soil metabolism study with ring labelled endosulfan sulphate.

The half lives in the laboratory were in the range of 12–39 days for α-endosulfan, 58–264 days for β-endosulfan and about 150 days for endosulfan sulphate. It should be noted here that the former laboratory degradation studies lack in the microbial activity due to the small soil samples employed and the long incubation period without re-fertilisation of the soil microbes. Therefore, degradation studies in the field are a more realistic approach. In the field, the degradation half life is shortened to 7–21 days under Southern European summer conditions. However at colder fall and winter temperatures, the half life increased to 75–93 days. It appears that the alpha isomer degrades faster (with a half life of 6–11 days) than the beta isomer (with a half life of 19–36 days) in the field.

The main soil metabolite endosulfan sulphate is more persistent than isomers of the parent, and degrades in the field with a half life of approximately 75–161 days depending on the study conditions. Other metabolites only appear at a low level in soil and are deemed not to be relevant.

A multi-year study showed only a slight increase in soil residue levels, from the first year, to form a relatively constant plateau level in subsequent years, even in Northern Europe with cold to moderate temperatures. There does not, therefore, appear to be significant long-term accumulation of endosulfan and its sulphate in soil. Furthermore, the plateau level decreased following termination of the application.

In a rotational crop study endosulfan residues taken up by root and leafy vegetable crops, sown immediately after soil treatment at a 6× exaggerated application rate, were generally lower than the corresponding residues in soil. The highest residues were 0.2 mg/kg in the leaf and 0.3 mg/kg in the tuber of carrots, being the critical crop at the application rate of 6.6 kg ai/ha. It should be noted that there were some varying residue levels reported applying to non-mature plants at the earlier samplings. Therefore, a significant reduction in the absolute residue level in rotational crops may be expected under normal circumstances, such as; when a 1× rate is used, when there is interception of the spray by the plants reducing the proportion reaching the (non-target) soil, and when partial degradation of the pesticide in soil could occur during the interval between application of endosulfan and planting of the rotational crop.

The Meeting concluded that the presence of endosulfan residues in succeeding crops from foliar application is unlikely to be significant.

### Methods of residue analysis

Methods of analysis of residues of endosulfan in plants and animal products used GC/ECD.

The methods for plant material have been validated on a wide range of crops and processed products. The principle of most methods involves a solvent extraction step followed by different matrix dependant clean up steps such as GPC, Florisil or silica gel column chromatography. The final determination is carried out by GC mostly with ECD. For enforcement purposes of plant material the method derived from the Dutch multi-residue method MRM-1 is suitable. The limit of quantification (LOQ) is typically about 0.02 mg/kg for α-endosulfan, β-endosulfan and endosulfan sulphate.

For the analysis of animal matrices, after extraction with an appropriate solvent and partition in acetonitrile, α-endosulfan, β-endosulfan and endosulfan sulphate were determined after purification by GC/ECD. The LOQ is typically about 0.025 mg/kg for α-endosulfan, β-endosulfan and endosulfan sulphate.

### Stability of residues in stored analytical samples

The storage stability of endosulfan and its important metabolites was tested in plant materials and animal tissues and products. The results of all the studies indicate that the compounds are stable in frozen storage in the tested plant commodities for 18 to 24 months and in animal commodities for at least one year.

### Definition of the residue

Based on the results of various plant and animal metabolism studies, endosulfan (α- and β- isomer) and its main metabolite endosulfan sulphate are the relevant residue components.

Results from metabolism studies on the distribution ratio of residues between muscle and fat show that the residues are fat soluble which is confirmed by the log $P_{OW}$ of 4.6-4.7 for α-endosulfan and 4.3-4.8 for β-endosulfan and 3.8 for endosulfan sulphate. Endosulfan residues are considered as fat soluble.

The Meeting concluded that the residue definition for enforcement and dietary intake purposes in plant and animal commodities is the sum of α- and β- isomer and its main metabolite endosulfan sulphate.

### Results of supervised residue trials

#### Citrus fruits

Endosulfan is registered for foliar application to citrus fruits in Angola, Australia, Central America, Chile, Morocco, Mozambique, Saudi Arabia and South Africa. The GAP in Australia for citrus fruits is 10.5 g ai/hL with a PHI of 3 days. Endosulfan residues from supervised trials conducted in Australia according to the GAP were: 0.03, 0.16 and 0.19 mg/kg for lemons; 0.07 and 0.11 mg/kg for mandarins; and 0.05 and 0.08 mg/kg for oranges.

The Meeting considered seven supervised trials insufficient to estimate a maximum residue level for citrus fruit and withdraw the previous recommendation for oranges, sweet, sour (0.5 mg/kg).

#### Pome fruits

Endosulfan is registered in apples in Australia, Canada, Central America, Chile, China, Japan, Namibia, Saudi Arabia, South Africa, the USA and Zimbabwe. Results of supervised trials in Australia were reported, but those trials were not conducted according to the GAP of Australia (66.5 g ai/hL and a PHI of 28 days).

Endosulfan residues from five trials in the USA according to the US GAP (3.36 kg ai/ha/year, three applications at 66.5 g ai/hL with a PHI 21 days) were 0.16, 0.27, 0.36, 0.54 and 0.77 mg/kg.

Endosulfan residue from one trial in South Africa following GAP (1.18 kg ai /ha and a PHI of 14 days) were 0.60 mg/kg. The Meeting considered that the residues were from the same population and thus could be combined. Endosulfan residues in trials that matched GAP in ranked order were: 0.16, 0.27, 0.36, 0.54, 0.60 and 0.77 mg/kg.

Endosulfan is registered in pear in Australia, Canada, Central America, Chile, Cyprus, Greece, Japan, South Africa and the USA. Results of four supervised trials in Australia were reported, but those trials were not conducted according to the GAP (66.5 g ai/hL and a PHI of 28 days).

The Meeting considered there were insufficient trials to recommend a maximum residue level for pome fruits. The previous recommendation of 1 mg/kg for pome fruit was withdrawn.

#### Cherries

Supervised trials on sweet and sour cherries were performed in the USA according to GAP (3.36 kg ai/ha/year, 2 × 260 g ai/hL with a PHI of 21 days; 350 EC formulation).

Endosulfan residues obtained in sour cherry trials were as follows:

EC formulation (airblast spray): 0.12, 0.29, 0.34, 0.37, 0.53, 0.54, 0.63 (2), 0.85, 1.1 mg/kg

WP formulation (airblast spray): < 0.05 (7), 0.06 (2), 0.09 mg/kg.

Endosulfan residues obtained in sweet cherry trials were as follows:

EC formulation (airblast spray): 0.14, 0.41, 0.44, 0.52, 0.57, 0.92, 1.4 mg/kg

EC formulation (mist blower): 0.1, 0.14, 0.16 mg/kg

WP formulation (airblast spray): < 0.05, 0.06, 0.08, 0.1, 0.14, 0.20, 0.34 mg/kg and

WP formulation (mist blower): 0.31, 0.72, 0.78 mg/kg.

The residues obtained using WP and EC formulations from airblast and mist blower sprayers do not represent the same population. As a result only residues obtained from the application of the EC formulation with an airblast sprayer were considered. The results for the trials on sour and sweet cherries were combined, resulting in endosulfan residues in ranked order were: 0.12, 0.14, 0.29, 0.34, 0.37, 0.41, 0.44, 0.52, 0.53, 0.54, 0.57, 0.63 (2), 0.85, 0.92, 1.1, and 1.4 mg/kg.

The Meeting recommended a maximum residue level for cherries of 2 mg/kg to replace the previous recommendation of 1 mg/kg, an HR value of 1.4 mg/kg and an STMR value of 0.53 mg/kg.

### Apricot, nectarine and peach

The Meeting received results of supervised trials on apricot, nectarine and peach conducted in Australia, but there is no GAP for these commodities. For peach, supervised trials were also reported from Europe (no GAP available) and the USA. The US trials were not conducted according to the GAP (3.36 kg ai /ha/year, 2 × 340 g ai/hL with a PHI of 21 or 2 × 66.5 g ai/hL and a PHI of 30 days).

The Meeting considered there were insufficient trials to recommend a maximum residue level for apricot, nectarine, or peach. The previous recommendation for peach of 1 mg/kg was withdrawn.

### Plums (including prunes)

Neither residue data nor information on GAP for the use of endosulfan in plums was submitted.

The Meeting recommended withdrawal of the previous recommendation of 1 mg/kg for plums (including prunes).

### Grapes

Endosulfan is registered for use on grapes in Canada, Central America, Chile, Croatia, Japan, Namibia, South Africa, Turkey and the USA.

Endosulfan residues from one trial in the USA conducted according to the GAP (3.36 kg ai/ha/year, 3 × 70 g ai/hL and a PHI of 7 days) was 0.75 mg/kg.

The Meeting considered one supervised residue trial insufficient to estimate a maximum residue level for grapes. The previous recommendation for grapes of 1 mg/kg was recommended for withdrawal.

### Pineapple

The Meeting received results of supervised trials conducted on pineapple in the USA. As these trials were not conducted according to the US GAP (2.5 kg ai/ha, 3.36 kg ai/ha/year, with a PHI of 7 days), the Meeting could not consider them for the estimation of a maximum residue limit for pineapple.

The Meeting recommended withdrawal of the previous recommendation of 2 mg/kg (Po).

### Other tropical fruits (avocado, custard apple, litchi, mango, papaya, persimmon)

The Meeting received results of supervised trials conducted in Australia on avocado, custard apple, litchi, mango, pawpaw (papaya) and persimmon.

In Australia, the GAP specifies an application rate of 70 g ai/hL for avocado, custard apple, mango and persimmon and an application rate of 52.5 g ai/hL for litchi and papaya. The PHI is 7 days, except for avocado (14 days).

Endosulfan residues obtained from the trials in Australia according to the corresponding GAPs were 0.01 and 0.11 mg/kg for avocado; 0.1 and 0.35 mg/kg for custard apple; 1.0 and 1.3

mg/kg for litchi; 0.17 and 0.20 mg/kg for mango; 0.1 and 0.18 mg/kg for papaya; and 0.55 and 0.89 mg/kg for persimmons.

The Meeting decided to combine endosulfan residues for avocado, custard apple, mango and papaya for mutual support, the residues in ranked order were: 0.01, 0.1 (2), 0.11, 0.17, 0.18, 0.20 and 0.35 mg/kg.

The Meeting estimated a maximum residue level of 0.5 mg/kg, an HR value of 0.35 mg/kg and an STMR value of 0.14 mg/kg for avocado, custard apple, mango and papaya.

The Meeting decided to combine endosulfan residues for litchi and persimmon for mutual support, with the residues being, in ranked order, 0.55, 0.89, 1.0, and 1.3 mg/kg.

The Meeting estimated a maximum residue level of 2 mg/kg, an HR value of 1.3 mg/kg and an STMR value of 0.95 mg/kg for litchi and persimmon.

## Onion, bulb

Neither residue data nor information on GAP on the use of endosulfan in onions was submitted.

The Meeting recommended withdrawal of the previous recommendation of 0.2 mg/kg for onion, bulb.

## Cabbages, head

Endosulfan is registered for foliar application on cabbage in Australia, Canada, Central America, Chile, Japan, New Zealand, Turkey and the USA.

Endosulfan residues in head cabbages from two trials in the USA according to that countries GAP (1.27 kg ai /ha with a PHI of 7 days) were 0.05 and 0.24 mg/kg, and from two trials in Australia according to its GAP (0.735 kg ai /ha with a PHI of 7 days) 0.026 and 0.1 mg/kg. The Meeting considered four supervised trials insufficient to estimate a maximum residue level for cabbage.

The Meeting recommended withdrawal of the previous recommendations of 1 mg/kg for head cabbage and of 2 mg/kg for Savoy cabbage.

## Brussels sprouts

Endosulfan is registered for foliar application to Brussels sprouts in Canada, Central America, Namibia, South Africa and the USA.

Two trials from the USA were done according to the Canadian GAP (0.7 kg ai /ha with a PHI of 7 days), resulting in endosulfan residues of 0.68 and 0.94 mg/kg.

The Meeting considered two supervised trials insufficient to recommend a maximum residue level for Brussels sprouts.

## Broccoli

Endosulfan is registered for foliar application to broccoli in Australia, Canada, Central America and in the USA.

Endosulfan residues from 17 trials from the USA, according to the GAP (1.27 kg ai /ha with a PHI of 7 days) were 0.22, 0.26, 0.28, 0.36, 0.37, 0.56, 0.57, 0.74, 0.79, 0.88, 0.97, 1.07, 1.31, 1.32, 1.86, 2.04, and 2.40 mg/kg.

Endosulfan residues from three trials in Australia, conforming to that countries GAP (0.735 kg ai /ha with a PHI of 7 days) were 0.17, 0.29 and 0.60 mg/kg.

The Meeting considered the trials to all be from similar populations and decided to combine the data for the purpose of maximum residue level recommendation. Endosulfan residues in ranked order were (n = 20): 0.17, 0.22, 0.26, 0.28, 0.29, 0.36, 0.37, 0.56, 0.57, <u>0.60</u>, <u>0.74</u>, 0.79, 0.88, 0.97, 1.07, 1.31, 1.32, 1.86, 2.04, and 2.4 mg/kg.

The Meeting estimated a maximum residue level for broccoli of 3 mg/kg to replace the previous recommendation of 0.5 mg/kg, an HR value of 2.4 mg/kg and an STMR value of 0.67 mg/kg.

*Cauliflower*

Endosulfan is registered for foliar application to cauliflower in Australia, Canada, Central America, Japan, New Zealand and the USA.

Endosulfan residues in cauliflower from two US trials conforming to the Canadian GAP (0.875 kg ai/ha with a PHI of 7 days) were < 0.05 and 0.1 mg/kg. Residues from one Australian trial at GAP (0.735 kg ai/ha with a PHI of 7 days) was 0.09 mg/kg. The Meeting considered three supervised trials insufficient to recommend a maximum residue level for cauliflower.

The Meeting recommended withdrawal of the previous recommendations of 0.5 mg/kg for cauliflower.

*Cucumber*

The Meeting received results of supervised trials on cucumbers conducted in Europe, Australia and the USA. No GAP was available for cucumber in Europe. The GAP in Australia specifies an application concentration of 70 g ai/hL and a PHI of 3 days. The GAP in the USA specifies an application rate of 1.27 kg ai/ha (3.36 kg ai/ha/year) and a PHI of 2 days.

Endosulfan residues from 20 trials in the USA at the GAP, in ranked order, were 0.18, 0.19, 0.22, 0.23, 0.24, 0.28, 0.30 (2), <u>0.31</u> (2), 0.32 (3), 0.36 (2), 0.40, 0.42, 0.53, 0.58 and 0.64 mg/kg. Endosulfan residues from two trials on cucumbers in Australia at the GAP were 0.09 and 0.11 mg/kg. The combined residues in cucumber were 0.09, 0.11, 0.18, 0.19, 0.22, 0.23, 0.24, 0.28, 0.30 (2), <u>0.31</u> (2), 0.32 (3), 0.36 (2), 0.40, 0.42, 0.53, 0.58 and 0.64 mg/kg.

The Meeting estimated a maximum residue level of 1 mg/kg to replace the previous recommendation of 0.5 mg/kg, an STMR value of 0.31 mg/kg and an HR value of 0.64 mg/kg.

*Melons, except watermelon*

The Meeting received results of supervised trials on melons conducted in Europe, Australia and the USA. No GAP was available for melons in Europe. The GAP in Australia specifies an application concentration of 70 g ai/hL and a PHI of 3 days. The GAP in the USA specifies an application rate of 1.27 kg ai/ha (3.36 kg ai/ha/year) and a PHI of 2 days.

Endosulfan residues from 12 trials in the USA at the GAP were 0.05, 0.22, 0.24, 0.30 (2), 0.34, 0.35, 0.40, 0.41, 0.45, 0.49 and 0.60 mg/kg in the whole fruit. Endosulfan residues from two trials on melons in Australia at the GAP were 0.55 and 1.2 mg/kg in the whole fruit. The combined residue data, in rank order, were: 0.05, 0.22, 0.24, 0.30 (2), 0.34, <u>0.35</u>, <u>0.40</u>, 0.41, 0.45, 0.49, 0.55, 0.60 and 1.2 mg/kg in the whole fruit.

No pulp samples were analyzed in the US and Australian trials. The Meeting decided to use results for pulp and whole fruit reported for trials in Southern Europe, obtaining pulp to whole fruit ratios of < 0.1, < 0.13, 0.17, < 0.25 (2), < 0.29, < 0.50, <1.0 (2) with a median of < 0.25.

Based on the whole fruit data, the Meeting estimated a maximum residue level of 2 mg/kg to replace the previous recommendation of 0.5 mg/kg. Based on the melon pulp vs. whole fruit residue ratio, the Meeting estimated an STMR value of 0.09 mg/kg and an HR value of 0.3 mg/kg for melon pulp.

*Squash, summer*

The Meeting received results of supervised trials on summer squash conducted in Spain and the USA and on zucchini in Australia.

Endosulfan residues from 12 trials in the USA at the GAP, in ranked order, were < 0.05, 0.05, 0.07, 0.08, 0.09, 0.13, 0.14, 0.15, 0.16 (2), 0.17 and 0.23 mg/kg. For zucchini, residues from four Australian trials, at the GAP, were 0.05, 0.06, and 0.09 (2) mg/kg. The combined summer squash and zucchini residues were < 0.05, 0.05, 0.05, 0.06, 0.07, 0.08, 0.09 (3), 0.13, 0.14, 0.15, 0.16 (2), 0.17 and 0.23 mg/kg.

The Meeting estimated a maximum residue level for summer squash of 0.5 mg/kg which confirms the previous recommendation, an STMR of 0.09 mg/kg and an HR of 0.23 mg/kg.

*Peppers*

Endosulfan is registered for use as a foliar spray on peppers in Australia, Canada, Cyprus, Greece, and the USA. The Meeting received results of supervised trials on peppers conducted in the USA, Australia and Spain. No GAP was available from Spain and the GAP from Greece did not specify a PHI.

Endosulfan residues in two trials from the USA according to Canadian GAP (1.125 kg ai/ha with a PHI of 2 days) were 0.05 and 0.22 mg/kg. Endosulfan residues in two Australian trials at the GAP (66.5 g ai/hL with a PHI of 3 days) were 0.36 and 0.40 mg/kg.

The Meeting considered four trials insufficient to recommend a maximum residue limit for peppers.

*Tomato*

Endosulfan is registered for use as a foliar spray on tomatoes in Angola, Australia, Canada, Central America, Chile, Cyprus, Ecuador, Greece, Japan, Morocco, Mozambique, Namibia, New Zealand, South Africa, Spain, the USA, Venezuela and Zimbabwe. The Meeting received results of supervised trials on tomatoes conducted in the USA, Australia, Germany, Greece, Italy, Portugal and Spain.

In the USA, endosulfan is registered for the use on tomatoes at 1.27 kg ai/ha (3.36 kg ai/ha/year) with a PHI of 2 days. In field trials in the USA that matched the GAP, the endosulfan residues were 0.03, 0.04, < 0.05, < 0.05, 0.07, 0.1, 0.16, 0.21, 0.22, 0.24, 0.25, 0.25, 0.25, 0.27, 0.27, 0.27, 0.27, 0.28, 0.29, 0.33, 0.33, 0.34, 0.35, 0.38, 0.42, 0.45, 0.45, 0.47, 0.66, 0.73, 0.83 and 0.85 mg/kg (n = 32).

In Australia, endosulfan residues from field trials after application of 0.15 – 0.17 kg ai/hL and a PHI of 3 days were < 0.005, 0.06, 0.07 and 0.09 mg/kg but did not match the GAP (66.5 g ai/hL).

Endosulfan residues conducted outdoor in Southern Europe according to the GAP of Spain (0.53 kg ai/ha and a PHI of 3 days) were: 0.03, 0.03, 0.04, 0.05, 0.07, 0.07, 0.1, 0.11, 0.13, 0.13, 0.13, 0.15, 0.15, 0.19, 0.19, 0.19, 0.21, 0.22, 0.24, 0.28, 0.43 and 0.79 mg/kg (n = 22).

Endosulfan residues conducted indoor in Southern Europe according to the GAP of Spain were: 0.05, 0.07, 0.09, 0.1, 0.11, 0.12, 0.16, 0.17, 0.19, 0.20, 0.21, 0.21, 0.21, 0.23, 0.24, 0.27, 0.28, 0.29, 0.32, 0.35, 0.37, 0.41, 0.60, 0.65 and 0.72 mg/kg (n = 25).

The Meeting agreed to combine the results of the trials in the USA and Southern Europe, resulting in endosulfan residues of (in ranked order):

0.03(3), 0.04, 0.04, < 0.05, < 0.05, 0.05, 0.05, 0.07 (4), 0.09, 0.1(3), 0.11, 0.11, 0.12, 0.13(3), 0.15, 0.15, 0.16, 0.16, 0.17, 0.19(4), 0.20, 0.21(5), 0.22, 0.22, 0.23, 0.24(3), 0.25(3), 0.27(4), 0.28(3), 0.29, 0.29, 0.32, 0.33, 0.33, 0.34, 0.35, 0.35, 0.37, 0.38, 0.41, 0.42, 0.43, 0.45, 0.45, 0.47, 0.60, 0.65, 0.66, 0.72, 0.73, 0.79, 0.83 and 0.85 mg/kg (n = 79).

The Meeting estimated a maximum residue level for tomatoes of 1 mg/kg to replace the previous recommendation of 0.5 mg/kg, an HR value of 0.85 mg/kg and an STMR value of 0.22 mg/kg.

*Eggplant*

The Meeting received results of supervised trials on eggplant from Australia. Four trials were conducted according to the GAP of Australia (0.735 kg ai/ha with a PHI of 7 days), resulting in endosulfan residues of < 0.005, < 0.005, 0.006 and 0.06 mg/kg.

The Meeting estimated a maximum residue level for eggplant of 0.1 mg/kg, an HR value of 0.06 mg/kg and an STMR value of 0.006 mg/kg.

*Sweet corn*

The Meeting received results of supervised trials on sweet corn from Australia. As no GAP was available for Australia, the Meeting was not able to recommend a maximum residue limit for sweet corn.

*Lettuce and kale*

Endosulfan is registered for use on lettuce and kale from the USA. No residue data for lettuce and kale were submitted.

The Meeting recommended withdrawal of the previous recommendations of 1 mg/kg for kale, lettuce, head and lettuce, leaf.

*Spinach*

Endosulfan is registered for use on spinach in the USA. No residue data for spinach were submitted.

The Meeting recommended withdrawal of the previous recommendations of 2 mg/kg for spinach.

*Beans*

Endosulfan is registered for use as a foliar spray on beans in Angola, Canada, Central America, Chile, Japan, Peru, Myanmar, Namibia, South Africa, the USA and Zimbabwe. The Meeting received results of supervised trials on beans from Germany, Australia and the USA.

Only one trial in the USA conformed to the US GAP (3.36 kg ai/ha/year with a PHI of 3 days). In this trial the total endosulfan residue was 0.64 mg/kg.

The Meeting considered four supervised trials insufficient to estimate a maximum residue level for beans. The previous recommendations for broad bean (green pods and immature seeds) and common bean (pods and/or immature seeds) of 0.5 mg/kg were withdrawn.

*Peas*

The Meeting received results of supervised trials on peas from Australia. No GAP was available for Australia, therefore the Meeting was not able to recommend a maximum residue limit for peas.

The Meeting recommended withdrawal of the previous recommendation for garden pea (young pods) of 0.5 mg/kg.

*Soybean (dry)*

Endosulfan is registered for use on soybean in Australia, Brazil, Central America, Chile, Iran and Zimbabwe.

The Meeting received results of supervised trials on soybeans from Australia and Brazil. The trials in Australia were not conducted according to the GAP of Australia (350 g ai/ha with a PHI of 1 day).

Eighteen trials conducted in Brazil conformed to the Brazilian GAP (0.525 kg ai/ha and a PHI of 30 days). Endosulfan residues obtained in these trials were < 0.02, 0.05, 0.08, 0.09, 0.1 (2), 0.15, 0.20 (3), 0.25, 0.30 (2), 0.31, 0.40, 0.42, 0.45 and 0.60 mg/kg.

The Meeting estimated a maximum residue level for soybeans of 1 mg/kg which confirms the previous recommendation, and an STMR value of 0.2 mg/kg.

*Carrot and beetroot*

The Meeting received results of supervised trials on carrot and beetroot from Australia. The GAP of Australia for carrot and beetroot specifies a maximum application rate of 0.735 kg ai/ha and a PHI of 14 days.

For carrot, four trials were conducted according to the GAP, with endosulfan residues being < 0.005, 0.04, 0.1 and 0.13 mg/kg. Only one trial on beetroot conformed to the GAP, resulting in 0.25 mg/kg of endosulfan.

The Meeting considered five supervised trials insufficient to estimate a maximum residue level for carrot and beetroot. The Meeting recommended withdrawal of the previous recommendation for carrot of 0.2 mg/kg.

*Potato and sweet potato*

Endosulfan is registered for use as a foliar spray on sweet potatoes in Australia, Japan and the USA and on potatoes in Australia, Canada, Central America, Chile, Iran, Japan, New Zealand, Peru, Turkey, the USA and Zimbabwe.

The Meeting received results of supervised trials on potatoes in Australia, Europe (no GAP), and the USA. Three trials in Australia were conducted according to the GAP of Australia (0.735 kg ai /ha with a PHI of 14 days), resulting in endosulfan residues of < 0.005, 0.005 and 0.007 mg/kg. In a single trial reported in the USA, endosulfan residues < 0.05 mg/kg occurred at the rate of 5.56 kg ai/ha with 3 applications and a PHI of 1 day (US GAP: 3.36 kg ai/ha/year and a PHI of 1 day). Endosulfan residues in all trials on potatoes were below the LOQ of 0.05 mg/kg (even for two Australian trials conducted at a double application rate as compared to the GAP in Australia).

The Meeting received results of supervised trials on sweet potato in Australia and the USA. One trial in Australia was at the GAP (0.735 kg ai/ha with a PHI of 14 days) and the endosulfan residue was < 0.005 mg/kg. Sixteen trials in the USA were conducted according to the US GAP (3.36 kg ai/ha/year with a PHI of 1 day), resulting in endosulfan residues < 0.05 mg/kg.

The Meeting decided to use the results of supervised trials on sweet potato to support the recommendation for potato. The Meeting estimated a maximum residue level for potato and sweet potato of 0.05* mg/kg, an HR value of 0.05 and an STMR value of 0.05 mg/kg. The Meeting decided to withdraw the previous recommendations for potato and sweet potato of 0.2 mg/kg.

*Sugar beet*

Endosulfan is registered for foliar application to sugar beet in Canada, Chile and Japan. The Meeting received results of supervised trials on sugar beet in Italy. No GAP was available for Europe; therefore the Meeting was not able to recommend a maximum residue limit for sugar beet.

The Meeting recommended withdrawal of the previous recommendation of 0.1 mg/kg for sugar beet.

*Celery*

Endosulfan is registered for foliar application to celery in Canada, Central America, Australia and in the USA. The Meeting received results of supervised trials on celery from Australia and the USA.

Two trials in Australia were conducted according to Australian GAP (66.5 g ai/hL with a PHI of 7 days). Endosulfan residues were 0.29 and 1.1 mg/kg.

Twelve trials in the USA were conducted according to the GAP of the USA (1.12 kg ai/ha/year with a PHI of 4 days). Endosulfan residues were 1.0, 1.4, 1.8, 2.5, 2.6 (2), 2.9, 3.1 (2), 3.8, 4.1 and 5.0 mg/kg.

The Meeting considered the trials to all be from similar populations and decided to combine the residue data obtained from the US and Australian trials. Endosulfan residues in ranked order were: 0.29, 1.0, 1.1, 1.4, 1.8, 2.5, <u>2.6</u> (2), 2.9, 3.1 (2), 3.8, 4.1 and 5.0 mg/kg.

The Meeting estimated a maximum residue level for celery of 7 mg/kg to replace the previous recommendation of 2 mg/kg, an HR value of 5.0 mg/kg and an STMR value of 2.6 mg/kg.

*Rhubarb*

The Meeting received results of supervised trials on rhubarb in Australia. Endosulfan is not registered for use as a foliar spray on rhubarb in Australia and therefore the Meeting was not able to recommend a maximum residue limit for rhubarb.

*Hazelnuts and macadamia nuts*

Endosulfan is registered for foliar spraying on hazelnuts in Poland, Spain and Turkey and on macadamia nuts in Australia.

For hazelnuts, the Meeting received results of supervised trials from Italy. Two of the trials were performed according to the GAP of Spain (105 g ai/hL with a PHI of 30 days). Endosulfan residues were < 0.02 mg/kg.

For macadamia nuts, results of four supervised trials from Australia were reported. Three of the trials were conducted according to the GAP of Australia (70 g ai/hL with a PHI of 2 days) and one trial at 50% above the GAP. Endosulfan residues were < 0.005 mg/kg in all four trials.

The Meeting decided to use the results for hazelnuts and macadamia nuts for mutual support and estimated a maximum residue level for hazelnuts and macadamia nuts of 0.02(*) mg/kg, an HR of 0 mg/kg and a STMR of 0 mg/kg .

*Cotton seed*

Endosulfan is registered for use on cotton in Angola, Australia, Benin, Brazil, Burkina, Central America, China, Cyprus, Ecuador, Ethiopia, Greece, India, Iran, Ivory Coast, Madagascar, Mali, Morocco, Mozambique, Myanmar, Namibia, Pakistan, Peru, South Africa, Spain, Sudan, Thailand, Togo, Turkey, the USA, Venezuela and Zimbabwe. The Meeting received results of supervised trials on cotton conducted from Australia, Greece and Spain.

One trial in Australia, conducted according to Australian GAP (0.735 kg ai/ha with a PHI of 56 days), had a residue < 0.02 mg/kg.

Seven trials in Southern Europe with a CS formulation were according to the GAP of Southern Europe (0.84 kg ai/ha with a PHI of 21 days), resulting in endosulfan residues of < 0.02 (5) and 0.24 mg/kg. Three trials with an EC formulation in Southern Europe according the same GAP resulted in endosulfan residues of < 0.02 (3) and 0.06 mg/kg.

The Meeting considered the trials to all be from similar populations and decided to combine the residue data obtained from Australia and Southern Europe. Combined endosulfan residues, in ranked order, were: < 0.02 (9), 0.06 and 0.24 mg/kg.

The Meeting estimated a maximum residue level for cotton seed of 0.3 mg/kg to replace the previous recommendation of 1 mg/kg, and a STMR of 0.02 mg/kg.

*Rape seed*

Endosulfan is registered for use on oil seed in Australia. No residue data for rape seed were submitted.

The Meeting recommended withdrawal of the previous recommendation of 0.5 mg/kg for rape seed.

*Sunflower seed*

Endosulfan is registered for use on oil seed in Australia. No residue data for sunflower seed were submitted.

The Meeting recommended withdrawal of the previous recommendation of 1 mg/kg for sunflower seed.

*Maize*

Endosulfan is registered for use on cereals in Australia. No residue data for maize were submitted.

The Meeting recommended withdrawal of the previous recommendation of 0.1 mg/kg for maize.

*Rice*

Neither residue data nor information on GAP of the use of endosulfan in rice were submitted.

The Meeting recommended withdrawal of the previous recommendation of 0.1 mg/kg for rice.

*Wheat*

Endosulfan is registered for use on cereals (barley, oats, rye, wheat) in Australia and the USA. No residue data for cereals were submitted.

The Meeting recommended withdrawal of the previous recommendation of 0.2 mg/kg for wheat.

*Cocoa beans*

Endosulfan is registered for use on cocoa in Brazil, Cameroon, Ivory Coast, Malaysia and Nigeria. The Meeting received results of supervised trials on cocoa from Brazil, Ghana and the Ivory Coast.

The trials in Brazil were conducted at application rates below the rate specified in the Brazilian GAP (87.5 g ai/hL with a PHI of 30 days).

Eight trials from Ghana and one trial from the Ivory Coast were conducted according to the GAP of Cameroon (0.26 kg ai/ha, with a PHI of 28 day), resulting in endosulfan residues in beans of < 0.01 (9) mg/kg.

Nine trials from the Ivory Coast conformed to the GAP of the Ivory Coast (0.250 g ai/ha, with the PHI not specified). The highest residues from these trials were selected for consideration: < 0.01 (5), 0.01, 0.03 (2), 0.06, and 0.08 mg/kg.

Endosulfan residues obtained in the trials from Ghana and the Ivory Coast in ranked order were: < 0.01 (14), 0.01, 0.03 (2), 0.06 and 0.08 mg/kg.

The Meeting estimated a maximum residue level for cocoa beans of 0.2 mg/kg to replace the previous recommendation of 0.1 mg/kg, and an STMR value of 0.01 mg/kg.

*Coffee beans*

Endosulfan is registered for use on coffee in Brazil, Cameroon, Central America, Cuba, Ecuador, Namibia, Peru, South Africa, Sudan, Thailand and Zimbabwe.

Three trials from Colombia, two trials from Mexico, two trials from Guatemala and three trials from Brazil were conducted according to the GAP of Cuba (0.613 kg ai/ha with a PHI of 30 days). Endosulfan residues in ranked order were: < 0.01 (3), 0.01, 0.02 (2), 0.03, 0.06, 0.08 and 0.09 mg/kg.

The Meeting estimated a maximum residue level for coffee beans of 0.2 mg/kg to replace the previous recommendation of 0.1 mg/kg, and a STMR of 0.02 mg/kg.

*Tea*

Endosulfan is registered for use on tea in China, Japan and Malaysia. The Meeting received results of supervised trials from India, which could not be matched against provided GAPs from China, Japan or Malaysia. The Meeting was not able to recommend a maximum residue limit for tea.

The Meeting recommended withdrawal of the previous recommendation of 30 mg/kg for tea, green and black.

## Fate of residues during processing

The hydrolysis of $^{14}$C-endosulfan under conditions representing food processing operations was investigated. Following pasteurisation, baking, boiling and sterilisation simulation, $\alpha$-endosulfan, $\beta$-endosulfan and the hydrolysis product endosulfan-diol were the main components found.

The effect of processing on the level of residues of endosulfan has been studied in oranges, apples, peaches, grapes, pineapples, tomatoes, potatoes, soybeans, coffee beans, cacao beans and tea.

The processing factors (PF) shown below were calculated from the total residues for the commodities for which MRLs, STMRs and HRs were estimated. The mean PF was calculated from three values, otherwise the median PF was calculated.

| RAC | Processed product | No. | PF | Mean/median PF |
|-----|------------------|-----|-----|----------------|
| Tomatoes | juice | 10 | < 0.1, < 0.12, < 0.16, < 0.16, < 0.17, < 0.20, < 0.20, 0.20, 0.27, 0.49 | < 0.185 |
| | paste | 5 | 0.16, 0.33, 0.59, 1.0, 1.62, 1.63 | 0.59 |
| | puree | 5 | < 0.1, 0.33, 0.51, 0.64, 0.66 | 0.51 |
| | fruit, peeled and canned | 4 | 0.075, 0.1, < 0.20, < 0.20 | 0.15 |
| | fruit, unpeeled and canned | 6 | 0.33, 0.44, 0.50, 0.50, 1.0,1.1 | 0.50 |
| Soybeans | crude oil | 3 | 1.17, 4.1, 4.33 | 3.2 |
| Coffee beans | ground roast coffee | 1 | < 0.063 | < 0.063 |
| | instant coffee | 1 | < 0.063 | < 0.063 |

Tomatoes were processed into juice, paste, puree, peeled canned fruit and unpeeled canned fruit with processing factors of < 0.185, 0.59, 0.51, 0.15 and 0.50, respectively. Based on the STMR value of 0.22 mg/kg for tomato, the STMR-Ps were 0.04 mg/kg, 0.13 mg/kg, 0.11 mg/kg, 0.03 mg/kg,

0.11 mg/kg, for residues in tomato juice, paste, puree, peeled canned fruit and unpeeled canned fruit, respectively.

Soya beans were processed into crude oil with a processing factor of 3.2. Based on the STMR value of 0.2 mg/kg for soya beans, the STMR-P was 0.64 mg/kg for soybean crude oil.

The Meeting recommended a maximum residue limit of 2 mg/kg for soybean crude oil, based on the highest residue of 0.6 mg/kg for soya beans and the processing factor of 3.2.

Coffee beans were processed into roasted coffee and instant coffee with a processing factor of < 0.063 for both. Based on the STMR value of 0.02 mg/kg for coffee beans, the STMR-Ps were 0.0013 mg/kg for roasted coffee and instant coffee.

For cotton seed, no processing studies were submitted. The previous recommendation of 0.5 mg/kg for cotton seed oil, crude, was recommended for withdrawal.

## Farm animal dietary burden

The Meeting estimated the dietary burden of endosulfan residues in livestock (farm animals) on the basis of the livestock diets listed in Appendix IX of the *FAO Manual* (FAO 2002).

The maximum dietary burden calculations include the highest residues (HR) and STMR-P values which are used for the estimation of maximum residue levels in animal commodities such as milk, eggs, meat and offal. The STMR dietary burden calculations for livestock allow an estimate of the median residues in milk, eggs, meat and offal that can be used in the chronic dietary assessments and in this case STMR and STMR-P values for feeds are used.

The percentage dry matter (DM) is taken as 100% where highest residues and STMR values are expressed on a dry weight basis.

*Calculation of the dietary burden for maximum residue estimation*

| Commodity | Group | Residue (mg/kg) | % DM | highest residue or STMR | Diet content (%) | | | Residue (mg/kg) | | Contribution |
|---|---|---|---|---|---|---|---|---|---|---|
| | | | | | Beef cattle | Dairy cows | Poultry | Beef cattle | Dairy cows | Poultry |
| Cotton seed | SO | 0.24 | 88 | HR | 10 | 25 | NU | 0.027 | 0.068 | NU |
| Soya bean | VD | 0.6 | 89 | HR | 15 | 15 | 20 | 0.1 | 0.1 | 0.135 |
| Potato | VR | 0.05 | 20 | HR | 75 | 40 | NU | 0.19 | 0.1 | NU |
| Total | | | | | 100 | 80 | 20 | 0.32 | 0.27 | 0.13 |

The calculated highest dietary burdens for beef cattle, dairy cattle and poultry are 0.32, 0.27 and 0.13 ppm, respectively.

*Calculation of the dietary burden for STMR estimation*

| Commodity | Group | Residue (mg/kg) | % DM | highest residue or STMR | Diet content (%) | | | Residue (mg/kg) | | Contribution |
|---|---|---|---|---|---|---|---|---|---|---|
| | | | | | Beef cattle | Dairy cows | Poultry | Beef cattle | Dairy cows | Poultry |
| Cotton seed | SO | 0.02 | 88 | STMR | 10 | 25 | NU | 0.002 | 0.006 | NU |
| Soya bean | VD | 0.2 | 89 | STMR | 15 | 15 | 20 | 0.034 | 0.034 | 0.045 |
| Potato | VR | 0.05 | 20 | STMR | 75 | 40 | NU | 0.188 | 0.1 | NU |
| Total | | | | | 100 | 80 | 20 | 0.22 | 0.14 | 0.04 |

The STMR dietary burdens for beef cattle, dairy cattle and poultry are 0.22, 0.14 and 0.04 ppm, respectively.

*Animal commodity maximum residue levels*

The livestock dietary burdens used for the estimation of the maximum residue levels for animal commodities are 0.32 ppm for beef cattle, 0.27 ppm for dairy cattle and 0.13 ppm for poultry. The livestock dietary burdens used for the STMR estimation for dietary risk assessment are 0.22 ppm for beef cattle, 0.14 ppm for dairy cattle and 0.04 ppm for poultry.

For poultry, the maximum dietary burden is estimated as 0.13 ppm. As a poultry feeding study was not provided, the poultry metabolism study is used to estimate maximum residue levels for eggs and poultry tissues. In the poultry metabolism study, hens were orally dosed for 12 days at levels of $^{14}$C endosulfan ranging 10 to 12 ppm. Scaling the TRR in eggs and poultry tissues for a maximum dietary burden of 0.13 ppm, residues in eggs, poultry muscle/fat and liver are 0.011 mg/kg, 0.013 mg/kg and 0.006 mg/kg respectively. The validated method of analysis for poultry tissues and eggs was conducted at concentrations of 0.025 mg/kg for each component of the residue definition, and as residues in poultry tissues and eggs are expected to be less than the validated LOQ of all of the components of the residue definition, the Meeting recommended maximum residue levels of 0.03* mg/kg for eggs, poultry meat and poultry edible offal. The STMR and HR values for eggs, poultry meat and poultry edible offal were 0.025 mg/kg.

For cattle, the maximum dietary burden for beef cattle and dairy cows is 0.32 and 0.27 ppm, respectively. The dietary burden for beef cattle will determine the estimates for meat, fat and edible offal while the dietary burden for dairy cows will determine the estimate for milk.

The maximum dietary burden of 0.32 ppm is below the lowest dose level in the cattle feeding study of 4 ppm. The target tissue for endosulfan residues in animal tissues is fat. The variation in residues in fat with dose level is significant and it is noted that at the 4 ppm dose level residues in composite fat were 1.2, 1.4 and 1.7 mg/kg. Using the highest residue of 1.7 mg/kg and scaling to 0.32 ppm, leads to an estimated residue of 0.14 mg/kg in fat. The Meeting noted that as the samples in the feeding study were composited fats and not from individual fat depots and as residues in meat producing animals are likely to be higher than milk producing animals, the Meeting recommended a maximum residue level of 0.2 mg/kg in meat on a fat basis. The previous recommendation of 0.1 mg/kg (fat) for meat (from mammals other than marine mammals) was withdrawn.

Similarly, scaling for residues in liver and kidney against the highest residues in the dose group leads to estimates of 0.078 mg/kg for liver and 0.006 mg/kg in kidney. On the basis of the estimates, the Meeting recommended maximum residue levels of 0.1 mg/kg for liver and 0.03* mg/kg for kidney.

For milk, residues in whole milk following dosing at 4 ppm ranged from 0.05 mg/kg to 0.08 mg/kg. Scaling for a dietary burden of 0.27 ppm, leads to an estimate of 0.005 mg/kg endosulfan in whole milk. Endosulfan is defined as fat-soluble, and residues in cream following dosing at 12 ppm ranged 0.81–1.42 mg/kg. Based on a dietary burden of 0.27 ppm for a dairy animal, residues in cream result in an estimate of 0.032 mg/kg. The Meeting recommended maximum residue levels of 0.1 mg/kg for milk fat and of 0.01 mg/kg for whole milk. The previous recommendation of 0.004 mg/kg F was withdrawn.

For dietary risk assessment, the STMR values are 0.09 mg/kg for meat/fat, 0.0039 mg/kg for muscle, 0.003 mg/kg for milk, 0.034 mg/kg for cream or milk fat, 0.054 mg/kg for liver and 0.004 mg/kg for kidney. The estimated HR values were 0.14 mg/kg for meat/fat, 0.0056 mg/kg for muscle, 0.078 mg/kg for liver and 0.006 for kidney.

## RECOMMDENDATIONS

On the basis of the data from supervised trials, the Meeting concluded that the residue concentrations listed below are suitable for establishing MRLs and for assessing IEDIs and IESTIs.

Definition of the residue (for compliance with the MRL and for estimation of the dietary intake):

*Sum of alpha endosulfan, beta endosulfan and endosulfan sulfate.* This definition applies to plant and animal commodities.

The residue is fat soluble.

| CCN | Commodity | MRL, mg/kg | | STMR or STMR-P, mg/kg | HR or HR/P mg/kg |
|-----|-----------|------------|---|----------|-----------|
| | | New | Previous | | |
| FI 0326 | Avocado | 0.5 | | 0.14 | 0.35 |
| VP 0522 | Broad bean (green pods and immature seeds) | W | 0.5 | | |
| VB 0400 | Broccoli | 3 | 0.5 | 0.67 | 2.4 |
| VB 0403 | Cabbage, Savoy | W | 2 | | |
| VB 0041 | Cabbages, Head | W | 1 | | |
| SB 0715 | Cacao beans | 0.2 | 0.1 | 0.01 | |
| VR 0577 | Carrot | W | 0.2 | | |
| VB 0404 | Cauliflower | W | 0.5 | - | - |
| VS 0624 | Celery | 7 | 2 | 2.6 | 5.0 |
| FS 0013 | Cherries | 2 | 1 | 0.53 | 1.4 |
| SB 0716 | Coffee beans | 0.2 | 0.1 | 0.02 | |
| | Coffee beans, roasted | | | 0.0013 | |
| | Coffee, instant | | | 0.0013 | |
| VP 0526 | Common bean (pods and/or immature beans) | W | 0.5 | | |
| SO 0691 | Cotton seed | 0.3 | 1 | 0.02 | |
| OC 0691 | Cotton seed oil, Crude | W | 0.5 | | |
| VC 0424 | Cucumber | 1 | 0.5 | 0.31 | 0.64 |
| FI 0322 | Custard apple | 0.5 | | 0.14 | 0.35 |
| PE 0112 | Eggs | 0.03* | | 0.025 | 0.025 |
| VO 0440 | Eggplant | 0.1 | | 0.006 | 0.06 |
| VP 0528 | Garden pea (young pods) | W | 0.5 | | |
| FB 0269 | Grapes | W | 1 | | |
| TN 0666 | Hazelnuts | 0.02* | | 0 | 0 |
| VL 0480 | Kale | W | 1 | | |
| MO 0098 | Kidney of cattle, goats, pigs and sheep | 0.03* | | 0.004 | 0.006 |
| VL 0482 | Lettuce, Head | W | 1 | | |
| VL 0483 | Lettuce, Leaf | W | 1 | | |
| FI 0343 | Litchi | 2 | | 0.95 | 1.3 |
| MO 0098 | Liver of cattle, goats, pigs and sheep | 0.1 | | 0.054 | 0.078 |
| TN 0669 | Macadamia nuts | 0.02* | | 0 | 0 |
| GC0645 | Maize | W | 0.1 | | |
| MM 0095 | Meat (from mammals other than marine mammals) | 0.2 (fat) | 0.1 (fat) | fat 0.09 muscle 0.0039 | fat 0.14 muscle 0.0056 |
| FI 0345 | Mango | 0.5 | | 0.14 | 0.35 |
| VC 0046 | Melons, except watermelon | 2 | 0.5 | 0.09 | 0.3 |
| ML 0106 | Milks | 0.01 | 0.004 F | 0.003 | |
| FM 0183 | Milk fats | 0.1 | | 0.034 | |

| CCN | Commodity | MRL, mg/kg | | STMR or STMR-P, mg/kg | HR or HR/P mg/kg |
|---|---|---|---|---|---|
| | | New | Previous | | |
| VA 0385 | Onion, Bulb | W | 0.2 | | |
| FC 0004 | Oranges, Sweet, Sour | W | 0.5 | | |
| FI 0350 | Papaya | 0.5 | | 0.14 | 0.35 |
| FS 0247 | Peach | W | 1 | | |
| FI 0352 | Persimmon | 2 | | 0.95 | 1.3 |
| FI 0353 | Pineapple | W | 2 Po | | |
| FS 0014 | Plums (including prunes) | W | 1 | | |
| FP 0009 | Pome fruits | W | 1 | | |
| VR 0589 | Potato | 0.05* | 0.2 | 0.05 | 0.05 |
| PM 0110 | Poultry meat | 0.03* | | 0.025 | 0.025 |
| PO 0111 | Poultry, edible offal of | 0.03* | | 0.025 | 0.025 |
| SO 0495 | Rape seed | W | 0.5 | | |
| GC 0649 | Rice | W | 0.1 | | |
| VD 0541 | Soya bean (dry) | 1 | 1 | 0.2 | |
| OC 0541 | Soya bean oil, crude | 2 | | 0.64 | |
| VL 0502 | Spinach | W | 2 | | |
| VC 0431 | Squash, Summer | 0.5 | 0.5 | 0.09 | 0.23 |
| VR 0596 | Sugar beet | W | 0.1 | | |
| SO 0702 | Sunflower seed | W | 1 | | |
| VR 0508 | Sweet potato | 0.05* | 0.2 | 0.05 | 0.05 |
| DT 1114 | Tea, Green, Black | W | 30 | | |
| VO 0448 | Tomato | 1 | 0.5 | 0.22 | 0.85 |
| JF 0448 | Tomato juice | | | 0.04 | |
| | Tomato paste | | | 0.13 | |
| | Tomato puree | | | 0.11 | |
| | Tomato canned fruit, unpeeled | | | 0.11 | |
| | Tomato canned fruit, peeled | | | 0.03 | |
| GC 0654 | Wheat | W | 0.2 | | |

## DIETARY RISK ASSESSMENT

### Long-term intake

The evaluation of endosulfan resulted in recommendations for MRLs and STMR values for raw and processed commodities. Where data on consumption were available for the listed food commodities, dietary intakes were calculated for the 13 GEMS/Food Consumption Cluster Diets. The results are shown in Annex 3 of the 2006 JMPR Report.

The International Estimated Daily Intakes (IEDI) of endosulfan, based on estimated STMRs were 3–20% of the maximum ADI (0.006 mg/kg bw). The Meeting concluded that the long-term intake of residues of endosulfan from uses that have been considered by the JMPR is unlikely to present a public health concern.

### Short-term intake

The International Estimated Short Term Intake (IESTI) of endosulfan calculated for the commodities for which residue levels were estimated. The results are shown in Annex 4 of the 2006 JMPR Report.

The IESTI of endosulfan calculated on the basis of the recommendations made by the JMPR represented for children 0–390% and for the general population 0–210% of the ARfD (0.02 mg/kg bw). The IESTI for broccoli for children was 390% and for the general population 210% of the ARfD, for celery 270% for children and 120% for the general population, 120% for cherries for children, and 110% for tomato for children.

The Meeting concluded that the short-term intake of residues of endosulfan resulting from the uses that have been considered by the JMPR, except the uses on broccoli, celery, cherries and tomatoes, is unlikely to present a public health concern.

The Meeting noted that no residue data relating to an alternative GAP were submitted. The information provided to the JMPR precludes an estimate that the dietary intake would be below the ARfD for consumption of broccoli, celery, cherries and tomatoes by children and broccoli and celery for the general population.

The meeting noted that the ARfD of endosulfan was established in 1998. Since then improvements in the toxicological assessment have been made, including the introduction of compound specific assessment factors. Consequently, it is recommended that the ARfD of endosulfan be reassessed at a future meeting for possible refinements.

## REFERENCES

### Author, Date, Title, Institute, Report references Document Number

Albrecht and Rexer, K. 1982. Endosulfan substance, production grade (physical form) unpublished report WIR 0072 (39) A24344

Albrecht and Rexer, K. 1981. Endosulfan pure active ingredient (solubility in organic solvents) unpublished report WIR 0004 (17)A21252

Albrecht, Kappes 1974. Endosulfan Pure (Melting Point) unpublished report WIR 0004 (26) A09432

Albrecht, Kappes and Frensch 1974. Endosulfan pure, (Density) unpublished report WIR 0004 (06) A09427

Albrecht, Rexer K. 1982. Endosulfan substance, production grade unpublished report WIR0072(37) A24343

Albrecht, Rexer K. 1982. Endosulfan substance, production grade unpublished report WIR0072(40) A24345

Balluff, M. 2001. Field Soil Dissipation of AE F002671 (Endosulfan) Following a Single Application to Bare (pre-emergence) Cotton Plots at 1 Location in Greece, 2000 Arbeitsgemeinschaft GAB Biotechnologie GmbH and IFU Umweltanalytik GmbH unpublished report20003033/GR1-FS C018180

Bowman, James S. 1959. Subacute Feeding - Dairy Cows, preliminary report unpublished report A14205

Buerkle, W.L. 1995. Metabolism in cucumber (Cucumis sativus) following three treatments with the 14C-labelled test substance at 7-day intervals and a nominal rate of 530 g ai/ha each, Endosulfan. Code: Hoe 002671 00 ZE97 0005 unpublished report CM93/039 A56011

Buerkle, W.L., Wuerz, S., Mueller, A. 1989. Metabolism in tomato plants after three applications at a rate of 635 g/ha Endosulfan-14C Hoe 002671 unpublished report CM031/87 A44894

Christ and Kellner 1968. Investigations with endosulfan-14C in mice unpublished report A53842

Craine, E.M. 1986. A Dermal Absorption Study in Rats with 14C-Endosulfan unpublished report A35730

Craine, Elliott M. 1988. A Dermal Absorption Study in Rats with 14C-Endosulfan with Extended Test Duration unpublished report       A39677

Czarnecki, J. J. and Mayasich, J. M. 1992. Terrestrial Field Dissipation of Endosulfan Applied to Cropped and Bare ground Plots in California published report Report of Bio/dynamics, Inc. NJ. and Research for Hire, CA, USA A51819

Davies, D.J. 2002. Comparative Study on the Penetration through Human and Rat Skin unpublished report C021864

Dorough, H.W., Huhtanen, K., Marshall, T.C. and Bryant, H.E. 1978, published report Pest. Biochem. Physiol. Vol. 8. 241-252. 1978 A14276 CO 21864

Gildemeister, H. 1985. Hoe 002671-14-C, Aerobic Aquatic Metabolism with the Insecticide Endosulfan unpublished report Report (B)106/85 of Hoechst AG, Analytisches Laboratorium A31182

Gildemeister, H. and Jordan, J. 1984. Aerobic Soil Metabolism of the Insecticide Hoe 002671 (Endosulfan) unpublished report (B)176/84 of Hoechst AG, Analytisches Laboratorium    A29680

Goerlitz, G. and Kloeckner, Ch. 1982. Hydrolysis of Hoe 02671 (endosulfan) unpublished report (B) 90(82) of Hoechst AG, Analytical Laboratory A31069

Goerlitz, G. and Rutz, U. 1988. Abiotic hydrolysis of the two isomers Hoe 052618 (a-Endosulfan), Hoe 052619 (b-Endosulfan) as a function of pH unpublished report CP014/88 of Hoechst AG Analytisches Laboratorium A40003

Goerlitz, G., Rutz, U. 1989. Abiotic hydrolysis of the two isomers Hoe 052618 (a-Endosulfan) and Hoe 052619 (b-Endosulfan) as a function of pH unpublished report CPO14/88 A40003

Gorbach 1965. Investigations on Thiodan in the Metabolism of Milk Sheep unpublished report A14209

Gorbach, S.C., Christ, O.E., Kellner, H.M., Kloss, G. and Boerner, E. 1968 Metabolism of Endosulfan in Milk Sheep published report J. Agr. Food Chem. Vol 6. page 950. 1968 A14216

Gorlitz, G. 1990. Hoe 002671, water solubility in the non-neutral range unpublished report OE90/175 A45268

Gorlitz, G. and Eyrich, U. 1986. Fat-Solubility (Hoe 002671, Hoe 052618, Hoe 052619) unpublished report CP076/85 A32583

Hacker, L. A. 1989. Endosulfan (Thiodan 3EC), Field Dissipation Study of Terrestrial Uses on Tomatoes in Georgia/USA unpublished report of Landis Associates, Inc. GA, USA A42193

Hammel, K. 2004. Kinetic Evaluation of the Dissipation of Endosulfan and its Metabolites Endosulfan Sulfate, Endosulfan Diol, Endosulfan Hydroxy Carboxylic Acid in Aerobic Water-Sediment Test Systems unpublished report MEF-318/03of Bayer CropScience C042131

Hardy, I. 2001. Endosulfan : Field Dissipation Study in Spain unpublished report26644 of Aventis CropScience UK Ltd. C015651

Huth, G. 1996. Endosulfan active substance, Emulsifiable concentrate 352 g/L. Additional information on physical and chemical properties of active substance and plant protection product unpublished report AL 20/79 A55218

Indranignsih, McSweeney C.S., Ladds P.W. 1993. Residues of endosulfan in the tissues of lactating goats. published report Australian Vet. Journal. Vol. 70. pages 59-72 A51447

Jonas, W. 2002. Degradation of [14C]Endosulfan in two Aerobic Water/Sediment Systems unpublished report NA 019404 of Natec Inst. C022921

Kappes, A. 1974a. Endosulfan, physical state, colour unpublished report A02525

Kappes, A. 1974b. Endosulfan, odour unpublished report WIR 0004 (14)A09429

Keller, John G. 1959. Subacute Feeding Study – Dairy Cows. (Supplement to Report dated March 20, 1959) unpublished report    A14206

Kellner and Eckert. 1983. Hoe 02671-14C Pharmacokinetics and residue determinations after oral and intravenous administration to rats unpublished report A49475

Klais, O. and Rexer, K. 1995. Endosulfan substance, technical. (Code Hoe 002671 00 ZD97 0003). Determination of the Oxidising Properties unpublished report A53949

Krebs, B., Eickhoff, H., Raquet, H., Their, W. 1986. Quantitation of residues in vegetable crops following uptake from contaminated soil. unpublished report DEU 85172641, LEA/R86/013 A53399

Lachmann, G. Siegemund, B. 1987. Hoe 002671-(5a,9a-14-C). Dermal Absorption of 14C-Endosulfan in Rhesus Monkeys unpublished report A36685

Leah, J.M. and Reynolds, C.M.M. 1995. Endosulfan-Distribution, elimination and the nature of the metabolite residues in the milk and edible tissues of a lactating cow unpublished report TOX/95/142-3    A57041

Leist, K.-H.; Mayer, D. 1987. Endosulfan – Active ingredient Technical (Code: Hoe 002671 0I ZD97 0003), 30-Day Feeding rat Study in Adult Male Wistar unpublished report A37112

Mester, T. C. 1990. Final Report Endosulfan (LX165-03), Terrestrial/Runoff Study on Cotton in California with Furrow Irrigation unpublished report Hoechst AG, Analytisches Laboratorium and Landis International GA USA A42997

Mislankar S and Tull P. 2003. Metabolism of [14C]-endosulfan in soybeans, unpublished report of Bayer CropScience, RTP, NC, USA edition no. M-24126501-1, unpublished report  601 BJ B004326

Needham, D. 1998. 1st Amendment to Report No. TOX/97142-4 (Agredoc No. A59694), Report Title: Endosufan - [C14]: Distibution, metabolism and excretion in the rat following a single oral administration of 1 mg or 6mg/kg bodyweight unpublished report A67544

Needham, D. 2001.Endosulfan-[14C]: Rat-Analysis of polar metabolites following a single oral dose of 6 mg/kg bodyweight. unpublished report TOX 97098A C010989

Needham, D. and Gutierrez Giulianotti, L. 1997. Endosufan - [C14]: Distibution, metabolism and excretion in the rat following a single oral administration of 1 mg or 6mg/kg bodyweight unpublished report TOX/97/142-4 A59694

Needham, D., Creedy, C.L., Hennings, P.A. 1998. Endosulfan-[14C]: Toxicokinetics in the rat following repeated daily oral administration of 1 mg/kg bodyweight for up to 28 days. unpublished report TOX 97099       A67138

Noctor, J.C. and John, S.A. 1995. 14C)-Endosulfan: Rates of penetration through human and rat skin determined using an in vitro system unpublished report A54103

Paarlar 1988. Photochemical Degradability of alpha and beta Endosulfan and Endosulfan sulfate in Air unpublished report A39963

Peatman, M.H., Reynolds, C.M., Bright, J.H.M. and Godfrey, T.L. 1999. Residues of a-Endosulfan, b-Endosulfan and Endosulfan sulphate in milk and edible cattle tissues following 28 days feeding to lactating cows. unpublished report RESID/98/8       C003624

Rexer, K. and Albrecht. 1982. Endosulfan substance, production grade (odour) unpublished report WIR 0072 (14) A24333

Rexer, K. and Albrecht 1982b. Endosulfan substance, production grade (flash point) unpublished report WIR 0072 (13) A24332

Rexer, K. and Albrecht 1982c. Endosulfan substance, production grade (explosiveness) unpublished report WIR 0072 (422) A24347

Rexer, K. and Maier 1990. Endosulfan substance, technical (colour) unpublished report WIR 0072 (11) A43649

Reynolds, C.M.M. 1995. Endosulfan-Distribution, elimination and the nature of the metabolite residues in the eggs and edible tissues of the laying hen unpublished report TOX/95/142-2 A56354

Rochling, Rexer, K.and Maier. 1990. Endosulfan Substance, Pure (Boiling Point) unpublished report WIR 0004 (29)A43118

Rochling, and Rexer, K. 1990. Endosulfan substance pure, Chemical stability unpublished report WIR 0004 (31C) A43119

Sarafin, K.and Asshauer, J. 1987a. Hoe 052618  and Hoe 052619 (a-Endosulfan) and  b-Endosulfan)-Partition Coefficient Octanol/Water unpublished report (B)124/87      A36576

Sarafin, K.and Asshauer, J. 1987b. Hoe 052618  and Hoe 052619 (a-Endosulfan) and  b-Endosulfan)-Solubility in water unpublished report (B)154/87 A36704

Sarafin, R. 1986. 1H-NMR-Spectrum (beta Endosulfan) unpublished Report(B) 2/86 A32480

Sarafin, R. 1987. Hoe 002671 (Endosulfan), Hoe 052618 (a-Endosulfan), and Hoe 052619 (b-Endosulfan)-Vapour Pressures unpublished report(B)153/87 A36734

Sarafin, R. 1985a. 1H-NMR-Spectrum (alpha Endosulfan) unpublished  Report(B) 266/85 A32477

Sarafin, R. 1985b. Infrared (IR)-Absorption-Spectrum (alpha Endosulfan) unpublished report(B)267/85      A32479

Sarafin, R. 1985c. Infrared (IR)-Absorption-Spectrum (beta Endosulfan) unpublished report(B)269/85      A32582

Sarafin, R. and Winterscheidt, G. 1985. Mass-Spectrum (alpha Endosulfan) unpublished report (B)241/85   A32478

Sarafin, R. and Winterscheidt, G. 1985b. Mass-Spectrum (beta Endosulfan) unpublished report (B)240/85   A32481

Schnoeder, F. [14C]AE F051327: Soil Metabolism and Degradation unpublished report 1651-1490-019 of Covance Laboratories GmbH C019647

Schnoeder, F. Amendment to the Final Report [14C]AE F051327: Soil Metabolism and Degradation unpublished report 1651-1490-019 of Covance Laboratories GmbH C020629

Schwab, W. 1995. Metabolism in apples (Malus sylvestris va. Domestica) following single treatment of a young tree with C14-labelled test substance Endosulfan. Code: Hoe 002671 00 ZE97 0005 unpublished report CM93/040 A53662

Selzer, J. 2001. Metabolism in Sugar Beets (Beta Vulgaris) Treated Twice with Endosulfan at an application Rate of 630 g a.s./ha Each unpublished report CM99/131 C015970

Smeykal, H. 2001. Endosulfan Substance Pure (Melting Point) unpublished report 20010418,01 C013620

Smeykal, H. 2001. Endosulfan Substance, Pure (Boiling Point) unpublished report 20010418,01 C013620

Stumpf, K. Comments Regarding the Dutch Hazard Assessment of Endosulfan Concerning the Bioavailability in Water/Sediment Systems. unpublished report OE90/252 of Hoechst AG, Produktentwicklung GB-C, Oekologie A44231

Stumpf, K. and Schink, C. 1988. Hoe 002671-14C: Photodegradation of alpha-Endosulfan (Hoe 052618) and beta-Endosulfan (Hoe 052619) in Water unpublished report CB074/87      A37588

Stumpf, K., Dambach, P. and Lenz O. Hoe 002671, Hoe 052618, Hoe 052619, Metabolism of 14C-labelled Endosulfan in Five Soils under Aerobic Conditions unpublished report CB88/037 of AgrEvo GmbH, Umweltforschung      A53618

Stumpf, K., Gildemeister, H., Dambach, P. and von Fleischbein, I. 1986. Hoe 002671-14C, Aerobic Metabolism of Endosulfan in Soil and the Influence of Increased Microbial Biomass at 28°C unpublished report CB017/86 of Hoechst AG, Analytisches Laboratorium   A39429

Tiirmaa, H., Krebs and Sochor, H. 1993. Degradation of endosulfan in soil after applications of Thiodan 50 WP over several seasons on an apple orchard unpublished report of 98/144 of Hoechst AG, Geschaeftsbereich Landwirtschaft, Produktentwicklung Oekologie II and Hoechst Holland N.V. Agro Chemie      A53771

Weller, O. 1990a. Hoe 002671, dissociation constant (pK value) unpublished report OE90-183 A45269

Weller, O. 1990b. Henry constants of Hoe 052618 (alpha-Endosulfan) and Hoe 052619 (beta-Endosulfan) unpublished report OE90/180    A43544

Wink, O. 1985a. UV-VIS-Spectrum (alpha Endosulfan) unpublished report(B)108/85    A31065

Wink, O. 1985b. UV-VIS-Spectrum (beta Endosulfan) unpublished report(B)109/85 A31063

## Methods of analysis

Diot, R. and Kieken, J.-L. 2004. Storage Stability of Residues of Endosulfan Lactone and Endosulfan Diol in Sugar Beet Leaves During Deep Freeze Storage for up to 24 Months unpublished report01-130 C040163

Garner M.A. and Snowdon P.J. 1995. Validation of analytical method, crops, gas chromatography unpublished report RESID/95/28 A55596

Haines, B.K. 2001. Independent Laboratory Validation for the Determination of Residues of Deltamethrin in Lettuce, Oranges, Milk and Fat, and Endosulfan in Lettuce and Oranges Using Method DGM F01/97-1 unpublished report XEN00-31 B003259

Hong Li. 1999. Determination of endosulfan alpha, beta, sulphate, lactone and diol residues in wheat grain, forage and straw, and sugar beet roots and tops unpublished report 01-130    C040163

Huff D.K., Winkler D.A. 1997. Validation of the analytical method  in animal tissues, egg (white and yolk) and dairy matrices based upon FDA pesticide analytical manual, volume I multi-residue methodology for the determination of Endosulfan (alpha, beta and sulfate) unpublished report95-0061 A57847

Idstein H., Junker H. and Becker D. 1995. Determination of Hoe 002671 (Endosulfan) and Hoe 051327 (endosulfan sulpfhate) in potatoes by gas chromatography (modified DFG S19 unpublished report  AL 003/95-0method). A55564

Martens R. 1999. Enforcement method and validation for water by GC Deltamethrin, endosulfan Code: AE F032640, AE F002671 unpublished report EMF11/99-0 C005528

Martens R. 1998a. Analytical method and validation for the determination of residues of endosulfan and deltamethrin by GC Deltamethrin, endosulfan Code: AE F032640 and AE F002671 unpublished report DGM F01/97-0  C000413

Martens R. 1998b. Analytical method and validation for the determination of residues of endosulfan and deltamethrin by GC, Amendment 1. unpublished report DGM F01/97-0 C001652

Martens R. 1998c. Validation of analytical method DGM F01/97-0 for residues of endosulfan and deltamethrin in cucumber, orange, melon and tomato Deltamethrin, endosulfan Code: AE F032640, AE F002671 unpublished report CR97/027 C001152

Martens R. 2000a. Data generation and enforcement method for residues on plant material by GC Deltamethrin, Endosulfan Code: AE F032640, AE F002671 unpublished report DGM F01/97-1 C007949

Martens R. 2000b. Validation of analytical method DGM F01/97-0 for dry crops (grain) Deltamethrin Endosulfan unpublished report CR99/025 C006935

Seefeld F. 1990. Validation report. Analysis of endosulfan residues in soil unpublished  Report Oec11/90    C008891

Werner H.-J., Klante G. and Merz H.D. 1987. Residue determination in soil, water, urine and plant-material of the active ingredient and Endosulfan-sulfate as well as in soil, water, urine and plant-material of Endosulfandiol and endosulfan-lactone unpublished report AL 60/86 A34558

Winkler, D.A. 1997. Freezer Storage Stability of Endosulfan (a, b and Sulfate) On Crop Raw Agricultural Commodities and Processed Commodities unpublished report 95-0072 A57831

Winkler, D.A. 1998a. Freezer Storage Stability of Endosulfan (a, b and Sulfate) On Animal Tissue and Dairy Matrices unpublished report96-0046 A67512

Winkler, D.A. 1998b. Freezer Storage Stability of Endosulfan (a, b and Sulfate) On Crop Raw Agricultural Commodities and Processed Commodities-Amendment No. 1 to Final Report unpublished report95-0072 A67528

Wrede A.2002. Analytical method and validation of endosulfan in cotton by GC-MSD Code: AE F002671 unpublished report AM 02/03   C022533

**Residue studies**

Anon    1995    Report on plant protection residue trial (Solanum tuberosum) Code: Hoe 002671 unpublished report ER94ECS770/BBA        A55993

Anon    1977a    Pflanzenschutzmittel-Rueckstaende unpublished report LEA2/48/01/02-76    A08888

Anon    1977b    Pflanzenschutzmittel-Rueckstaende unpublished report LEA2/713/01/02-77A A12270

Anon    1977c    Pflanzenschutzmittel-Rueckstaende unpublished report LEA3/713/01/02-77A A12271

Anon    1977d    Pflanzenschutzmittel-Rueckstaende unpublished report LEA4/713/01/02-77A A12272

Anon    1977e    Pflanzenschutzmittel-Rueckstaende unpublished report LEA1/713/01/02-77    A12767

Balluff M. 1999. Determination of residues of endosulfan in hazelnuts following two applications of AE F002761 00 EC33 B331 under field conditions at 2 locations in Italy, 1998 Code: AE F002671 00 EC33 B331 unpublished report98059/I1-FPHN    C004070

Balluff M. 2000. Determination of residues of endosulfan in hazelnuts following two applications of AE F002671 00 EC33 B329 under field conditions at 2 locations in Italy, 1999 unpublished report99335/I1-FPHN C008366

Balluff M. 2000A. Field Residue Study for the Determination of Residues of Endosulfan after the Maximum Number of Applications with Arasulfan 35EC in Green or Red Pepper at 4 different Site in Spain, 1999 unpublished report99172/S1-FPGP    C044590

Balluff M. 2000B. Field Residue Study for the Determination of Residues of Endosulfan after the Maximum Number of Applications with Arasulfan 35EC in Green or Red Pepper at 4 different Sites (Greenhouse) in Spain, 1999 unpublished report 99172/S1-FPGP        C044592

Balluff M. 2000 C. Field Residue Study for the Determination of Residues of Endosulfan after the Maximum Number of Applications with Arasulfan 35EC in Green or Red Pepper at 4 different Sites (Greenhouse) in Spain, 2000 unpublished report 99172/S2-FPGP        C044588

Balluff M. 2002. Field residue study for the determination of residues of endosulfan in cacao after the maximum number of applications with Thiodan 26 CS (AE F002671 00 CS26 E218) and Thiodan 500 EC (AE F002671 00 EC43 A508) in Ivory Coast, Western Africa, 2001 / 2002 unpublished report 20013056/IC1-FPCC  C020887

Balluff M. 2003. Field residue study for the determination of residues of endosulfan in cacao after the maximum number of applications with Thiodan 26 CS (AE F002671 00 CS26 E218) and Thiodan 500 EC (AE F002671 00 EC434 A508) under field conditions in Western Africa, 2002 unpublished report 20013056/AF1-FPCC C032568

Bodnaruk K. 2001. Field trials to determine the level of endosulfan in carrots at harvest following 3 applications to the crops unpublished report 1/8/565   AH04007

Bodnaruk K. 2001. Field trials to determine the level of endosulfan in beetrot at harvest following 4 applications to the crops unpublished report 1/8/534   AH04007

Bodnaruk K. 2001. Field trials to determine the level of endosulfan in potatoes at harvest following 3 applications to the crops unpublished report 1/7/562       AH04007

Bodnaruk K. 2001. Field trials to determine the level of endosulfan in sweet potatoes at harvest following 3 applications to the crops unpublished report 1/8/533 AH04007

Bodnaruk K. 2001. Field trials to determine the level of endosulfan in rhubarb at harvest following 2 applications to the crops unpublished report 1/10/558       AH04007

Bodnaruk K. 2001. Field trials to determine the level of endosulfan in celery at harvest following 3 applications to the crops unpublished report 1/10/535 AH04007

Bodnaruk K. 2001. Field trials to determine the level of endosulfan in apricots at harvest following 3 applications to the crops unpublished report 1/9/561       AH04007

Bodnaruk K. 2001. Field trials to determine the level of endosulfan in peaches at harvest following 3 applications to the crops unpublished report1/9/541       AH04007

Bodnaruk K. 2001. Field trials to determine the level of endosulfan in nectarines at harvest following 3 applications to the crops unpublished report1/9/536       AH04007

Bodnaruk K. 2001. Field trials to determine the level of endosulfan in peas at harvest following 3 applications to the crops unpublished report1/10/557 AH04007

Bodnaruk K. 2001. Field trials to determine the level of endosulfan in beans at harvest following 3 applications to the crops unpublished report1/10/538 AH04007

Bodnaruk K. 2001. Field trials to determine the level of endosulfan in tomatoes at harvest following 3 applications to the crops unpublished report1/10/552       AH04007

Bodnaruk K. 2001. Field trials to determine the level of endosulfan in egg plants at harvest following 3 applications to the crops unpublished report1/10/540    AH04007

Bodnaruk K. 2001. Field trials to determine the level of endosulfan in sweet corn at harvest following 3 applications to the crops unpublished report1/10/580    AH04007

Bodnaruk K. 2001. Field trials to determine the level of endosulfan in capsicums at harvest following 3 applications to the crops unpublished report1/10/559    AH04007

Bodnaruk K. 2001. Field trials to determine the level of endosulfan in pears at harvest following 6 applications to the crops unpublished report1/8/553 AH04007

Bodnaruk K. 2001. Field trials to determine the level of endosulfan in apples at harvest following 6 applications to the crops unpublished report1/8/563 AH04007

Bodnaruk K. 2001. Field trials to determine the level of endosulfan in macadamia at harvest following 3 applications to the crops unpublished report1/10/532    AH04007

Bodnaruk K. 2001. Field trials to determine the level of endosulfan in rockmelons at harvest following 4 applications to the crops unpublished report1/10/551    AH04007

Bodnaruk K. 2001. Field trials to determine the level of endosulfan in zucchini at harvest following 4 applications to the crops unpublished report1/10/556    AH04007

Bodnaruk K. 2001. Field trials to determine the level of endosulfan in cucumbers at harvest following 4 applications to the crops unpublished report1/10/547    AH04007

Bodnaruk K. 2001. Field trials to determine the level of endosulfan in lemons at harvest following 4 applications to the crops unpublished report1/10/543 AH04007

Bodnaruk K. 2001. Field trials to determine the level of endosulfan in mandarins at harvest following 4 applications to the crops unpublished report1/10/542    AH04007

Bodnaruk K. 2001. Field trials to determine the level of endosulfan in oranges at harvest following 4 applications to the crops unpublished report1/10/544 AH04007

Bodnaruk K. 2001. Field trials to determine the level of endosulfan in cabbage at harvest following 3 applications to the crops unpublished report1/5/564 AH04007

Bodnaruk K. 2001. Field trials to determine the level of endosulfan in broccoli at harvest following 3 applications to the crops unpublished report1/5/594 AH04007

Bodnaruk K. 2001. Field trials to determine the level of endosulfan in cauliflower at harvest following 3 applications to the crops unpublished report1/5/550    AH04007

Bodnaruk K. 2001. Field trials to determine the level of endosulfan in pawpaws at harvest following 4 applications to the crops unpublished report1/10/539    AH04007

Bodnaruk K. 2001. Field trials to determine the level of endosulfan in persimmons at harvest following 2 applications to the crops unpublished report1/10/545    AH04007

Bodnaruk K. 2001. Field trials to determine the level of endosulfan in avocadoes at harvest following 6 applications to the crops unpublished report1/10/554    AH04007

Bodnaruk K. 2001. Field trials to determine the level of endosulfan in mangoes at harvest following 4 applications to the crops unpublished report1/10/537    AH04007

Bodnaruk K. 2001. supervised residue trials for endosulfan in custard apple unpublished report1/2/500 AH04007

Brady, S.S. 1997a. Magnitude of residues in or on cucumbers resulting from three applications of Phaser(R) EC or Phaser(R) WP insecticide USA 1995 unpublished report BJ95R05    A55818

Brady S.S. 1997b. Magnitude of Endosulfan residues in or on tomatoes resulting from three applications of Phaser(R) EC or Phaser WP insecticide USA, 1995 Endosulfan unpublished report BJ-95R-06  A55822

Brady S.S. 1997c. Magnitude of residues in or on celery resulting from a single application of Phaser(R) EC or Phaser WP insecticide, USA, 1995 Endosulfan  unpublished report BJ-95R-04   A55830

Brady S.S. 1997d. Magnitude of Endosulfan residues in or on potatoes and processed potato commodities resulting from three applications of Phaser(R) EC insecticide at an exaggerated rate, USA, 1995 Endosulfan unpublished report BJ95R08 A57705

Brady S.S. 1997e. Magnitude of Endosulfan residues in or on summer squash resulting from three applications of Phaser(R) insecticide, USA, 1996 unpublished report BJ96R02 A57714

Brady S.S. 1997f. Magnitude of Endosulfan residues in or on tomatoes and processed tomato commodities resulting from three applications of Phaser(R) EC insecticide at an exaggerated rate, USA, 1995 Endosulfan unpublished report BJ95R09 A57707

Brady S.S. 1997g. Magnitude of residues in or on grapes and processed grape commodities resulting from two applications of Phaser(R) EC insecticide Endosulfan unpublished report BJ95R07 A55834

Brady S.S. 1997h. Magnitude of Endosulfan residues in or on broccoli resulting from three applications of Phaser(R) EC or Phaser WP insecticide, USA, 1995 Endosulfan unpublished report BJ95R03   A55827

Brady S.S. 1997i.  Magnitude of residues in or on cucumbers resulting from three applications of Phaser(R) insecticide Endosulfan unpublished report BJ96R01 A57716

Brady S.S. 1997j.  Magnitude of Endosulfan residues in or on sweet cherries resulting from one application of Phaser(R) insecticide, USA, 1996 Endosulfan unpublished report BJ96R03    A57718

Brady S.S. 1997k. Magnitude of Endosulfan residues in or on sweet potatoes resulting from two to three applications of Phaser(R) insecticide Endosulfan unpublished report BJ96R05       A57719

Brady S.S. 1997l.  Magnitude of Endosulfan residues in or on sour cherries resulting from one application of Phaser(R) insecticide, USA, 1996 Endosulfan unpublished report BJ96R04     A57715

Brady S.S. 1998.  Magnitude of endosulfan residues in or on celery resulting from a single application of Phaser(R) insecticide, USA, 1997 unpublished report BJ97R002 C000900

Buxton R.W. 1961. 15-Thiodan (Residues on cole crops) unpublished report R-470 A48547

Erbel A., Ertz K.  2003. Processing of tomatoes treated with endosulfan CS (330 g/L) into tomato juice and canned tomatoes unpublished report MR-510/02 C030836

Fabian   1974. Pflanzenschutzmittel-Rueckstaende unpublished report2/74/01/02 A05393

Gomez C.A. 1996. Magnitude of the residue of endosulfan in coffee raw agricultural commodities and processed fractions unpublished report AA930040       A55745

Gorbach S. 1970.  Determination of Endosulfan (alpha- und beta-Isomer) and Endosulfan-Sulfate in Dry Tea and Tea Infusion Prepared Thereof. unpublished report(B)49/70    A53967

Hees M., Idstein H., Junker H. 1996. Determination of residues of Hoe 002671 to establish a maximum residue level following 2 applications in tomatoes under greenhouse conditions Endosulfan emulsifiable concentrate 352 g/L unpublished report ER93ECS701 Code: Hoe 002671 00 EC33 B324    A54361

Helgers A. 1999a. Decline of residues in sugar beet, Italy 1997 Endosulfan emulsifiable concentrate 352 g/L unpublished report ER97ECS746 Code: AE F002671 00 EC33 B331    A67554

Helgers A. 1999b. Residues at harvest in sugar beet Italy 1998 Endosulfan emulsifiable concentrate 352 g/L Code: AE F002671 00 EC33 B331 unpublished report ER98ECS746 C000486

Hinstridge P. 1960. 15 – Thiodan (Residue on apples) unpublished report R-385  C009878

Hinstridge P.A. 1963a. 15-Thiodan and Thiodan-sulfate Residues on apples unpublished report R-677 A30339

Hinstridge P.A. 1963b. 15-Thiodan and thiodan sulfate residues in peaches unpublished report R-689 C009876

Hinstridge P.A. 1965a. Residues on peppers 15-Thiodan and Thiodan-sulfate unpublished report R-1001 A48560

Hinstridge P.A. 1965b. 15-Thiodan and Thiodan-sulfate Residues on celery unpublished report R-908 A30347

Hinstridge P.A. 1968. Thiodan and Thiodan-sulfate (Residues on fresh pineapple and pineapple bran) unpublished report R-1097    A26128

Hinstridge P.A. 1972. 15-Thiodan  (Residues in or on grapes and grape pomace) unpublished report R-1198 A30342

Hoppe T. 1974a      Pflanzenschutzmittel-Rueckstaende unpublished report4/74/01/02    A05394

Hoppe T. 1974b      Pflanzenschutzmittel-Rueckstaende unpublished report4/74/02/02    A05395

Hoppe T. 1974c      Pflanzenschutzmittel-Rueckstaende unpublished report4/93/04/04    A06462

Hoppe T. 1974d      Pflanzenschutzmittel-Rueckstaende unpublished report4/93/04/02    A06463

Hoppe T. 1974e      Pflanzenschutzmittel-Rueckstaende unpublished report4/93/04/03    A06464

Hoppe T. 1974f      Pflanzenschutzmittel-Rueckstaende unpublished report4/93/05/02    A06465

Hoppe T. 1974g      Pflanzenschutzmittel-Rueckstaende unpublished report4/93/05/03    A06466

Hoppe T. 1974h      Pflanzenschutzmittel-Rueckstaende unpublished report4/93/05/04    A06467

Hoppe T. 1974i      Pflanzenschutzmittel-Rueckstaende unpublished report4/93/02/02    A06468

Hoppe T. 1974j   Pflanzenschutzmittel-Rueckstaende unpublished report4/93/02/03   A06469

Hoppe T. 1974k   Pflanzenschutzmittel-Rueckstaende unpublished report4/93/02/04   A06470

Hoppe T. 1974l   Pflanzenschutzmittel-Rueckstaende unpublished report4/93/01/02   A06471

Hoppe T. 1974m   Pflanzenschutzmittel-Rueckstaende unpublished report4/93/01/03   A06472

Hoppe T. 1974n   Pflanzenschutzmittel-Rueckstaende unpublished report4/93/01/04   A06473

Hoppe T. 1974o   Pflanzenschutzmittel-Rueckstaende unpublished report4/90/02/02   A02621

Hoppe T. 1974p   Pflanzenschutzmittel-Rueckstaende unpublished report4/90/02/03   A02622

Hoppe T. 1974q   Pflanzenschutzmittel-Rueckstaende unpublished report4/90/02/04   A02623

Hoppe T. 1974r   Pflanzenschutzmittel-Rueckstaende unpublished report4/90/01/02   A02624

Hoppe T. 1974s   Pflanzenschutzmittel-Rueckstaende unpublished report4/90/01/03   A02619

Hoppe T. 1975a   Pflanzenschutzmittel-Rueckstaende unpublished report3/74/02/02   A05398

Hoppe T. 1975b   Pflanzenschutzmittel-Rueckstaende unpublished report3/74/01/02   A05399

Hoppe T. 1975c   Pflanzenschutzmittel-Rueckstaende unpublished report LEA4/84/01/02-75B A04698

Hoppe T. 1975d   Pflanzenschutzmittel-Rueckstaende unpublished report LEA3/84/01/02-75 A05870

Hoppe T. 1975e   Pflanzenschutzmittel-Rueckstaende unpublished report LEA3/85/01/02-75 A04697

Hoppe T. 1975f   Pflanzenschutzmittel-Rueckstaende unpublished report LEA4/85/01/02-75B A04699

Hoppe T. 1975g   Pflanzenschutzmittel-Rueckstaende unpublished report LEA4/82/01/02-75A A04837

Hoppe T. 1975h   Pflanzenschutzmittel-Rueckstaende unpublished report LEA4/82/02/02-75A A04838

Hoppe T. 1975i   Pflanzenschutzmittel-Rueckstaende unpublished report LEA4/82/03/02-75A A04839

Hoppe T. 1976a   Pflanzenschutzmittel-Rueckstaende unpublished report LEA3/83/01/02-75 A05866

Hoppe T. 1976b   Pflanzenschutzmittel-Rueckstaende unpublished report LEA4/83/01/02-75 A05867

Hoppe T. 1977c   Pflanzenschutzmittel-Rueckstaende unpublished report LEA3/67/01/02-76 A08914

Hoppe T. 1977d   Pflanzenschutzmittel-Rueckstaende unpublished report LEA3/74/01/02-76 A08916

Hoppe T. 1977e   Residues of Plant Protection Chemicals unpublished report2-21-01-02 A10195

Hoppe T. 1977f   Residues of Plant Protection Chemicals unpublished report2-21-01-03A A13597

Hoppe T. 1978g   Residues of plant protection chemicals unpublished report ZA/77/03-02-02 A15532

Hoppe T. 1978h   Residues of plant protection chemicals unpublished report ZA/77/03-02-03 A15534

Hoppe T. 1978i   Residues of plant protection chemicals unpublished report ZA/77/03-02-04 A15536

Hoppe T. 1978j   Residues of plant protection chemicals unpublished report ZA/77/01-04-06 A15538

Hoppe T. 1978k   Residues of plant protection chemicals unpublished report ZA/77/01-04-07 A15540

Hoppe T. 1978l   Pflanzenschutzmittel-Rueckstaende unpublished report LEA1/701/1-78 A16217

Hoppe T. 1978m   Pflanzenschutzmittel-Rueckstaende unpublished report LEA2/701/1-78 A16219

Hoppe T. 1978n.   Pflanzenschutzmittel-Rueckstaende unpublished report LEA3/701/1-78 A16221

Hoppe T. 1978o   Pflanzenschutzmittel-Rueckstaende. (Gewaechshausgurke) unpublished report LEA4/701/1-78 A16223

Huth, G. 1999a. Collection of residues data summary from supervised trials and processing studies in Pomefruit, Grapes, Tomatoes and Melon. unpublished report PRS99/012   C004071

Huth, G. 1999b. Residues data summary from supervised trials in Cotton unpublished report PSR99/011   C004073

Huth, G. 1999c. Residues data summary from supervised trials in Soybeans unpublished report PSR99/010 C004072

Huth, G. and Wurm, W. 1976. Pflanzenschutzmittel-Rueckstaende unpublished report LEA1/57/01/02-76   A57141

Idstein H., Junker H., Klein E.H-J.   1996a   Determination of residues of Hoe 002671 to establish a maximum residue level following 2 applications in potatoes Endosulfan  emulsifiable concentrate 352 g/L Code: Hoe 002671 00 EC33 B325 unpublished report ER94ECS770   A55214

Idstein H., Junker H., Klein E.H.-J.    1996b    Residue trials in apples to establish a maximum residue level Determination of active substances and the metabolite decline following 2 applications in apples and processing to apple puree and apple juice Endosulfan emulsifiable concentrate 352 g/L unpublished report ER94ECS705 A55874

Idstein H., Junker H., Sonder K.-H.    1996    Determination of residues of Hoe 002671 to establish a maximum residue level following 2 applications in tomatoes under greenhouse conditions Endosulfan emulsifiable concentrate 352 g/L Code: Hoe 002671 00 EC33 B325 unpublished report ER94ECS701      A54360

Jackson M.C. 1997. Residues in pineapple processing fractions of Thiodan 3EC unpublished report PGAH950001P A59907

Klein E. H.-J. 1999d. Residues at harvest in oranges European Union (southern zone) 1998. Endosulfan emulsifiable concentrate (EC) 32.9% w/w (=352 g/L) Code: AE F002671 00 EC33 B331 unpublished report ER98ECS740    C003108

Klein E.H.-J. 1998a. Residue trials on grapes to confirm a maximum residue level.  Determination of active substance and metabolite decline following two (2) applications. European Union (southern zone) 1997 Endosulfan emulsifiable concentrate 352 g/L Code: AE F002671 00 EC unpublished report ER97ECS744 C001151

Klein E.H.-J. 1998b. Residue trials on peaches to establish a maximum residue level.  Determination of active substance and metabolite decline following 3 applications. European Union (southern zone) 1997 Endosulfan emulsifiable concentrate 352 g/L Code: AE F002671 00 EC33 unpublished report ER97ECS742 C001114

Klein E.H.-J. 1999a. Decline of residues in mandarin tree, European union, southern zone 1997 Endosulfan emulsifiable concentrate (EC) 32.9% w/w (= 352 g/L) Code: AE F002671 00 EC33 B331 unpublished report ER97ECS741    C001465

Klein E.H.-J. 1999b. Decline of residues in orange fruits and transfer of residues into juice European Union (southern zone) 1997 Endosulfan emulsifiable concentrate (EC) 32.9% w/w (= 352 g/L) Code: AE F002671 00 EC33 B331 unpublished report ER97ECS740      C001464

Klein E.H.-J. 1999c. Residues at harvest in peaches European Union (southern zone) 1998. Endosulfan emulsifiable concentrate (EC) 32.9% w/w (= 352 g/L) Code: AE F002671 00 EC33 B331 unpublished report ER98ECS742    C002960

Klein E.H.-J. 1999d. Residues at harvest in oranges European Union (southern zone) 1998 Endosulfan emulsifiable concentrate (EC) 32.9% w/w (=352 g/L) Code: AE F002671 00 EC33 B331 unpublished report ER98ECS740    C003108

Klein E.H.-J. 1999e. Residues at harvest in mandarins European Union (southern zone) 1998 Endosulfan emulsifiable concentrate (EC) 32.9% w/w (=352 g/L) Code: AE F002671 00 EC33 B331 unpublished report ER98ECS741    C003107

Klein E.H.-J. 2003a. Residues at harvest in cotton European Union (Southern zone) 2002. Endosulfan, AE F002671 suspension of microcapsules (CS) 25.78% w/w (= 330 g/L) Code: AE F002671 00 CS26 E218 unpublished report02R172    C029816

Klein E.H.-J. 2003b. Residues at harvest in cotton European Union (Southern zone) 2002. Endosulfan, AE F002671 emulsifiable concentrate (EC) 32.9% w/w (= 352 g/L) Code: AE F002671 00 EC33 B333 unpublished report02R170    C029815

Klein E.H.-J. 2003c. Decline of residues in tomatoes and following rotational crop Europen Union (Southern zone) 2002 Endosulfan, AE F002671 emulsifiable concentrate (EC) 32.9% w/w (= 352 g/L) Code: AE F002671 00 EC33 B333 unpublished report02R171    C032692

Klein E.H.-J., Buerstell H. 1997. Residue trials in vine to establish a Maximum Residue Level Determination of active substance and the metabolite decline following 3 applications in vine and processed fractions European Union, (southern zone), 1994. Endosulfan EC  (emulsifiable concen unpublished report ER94ECS730    A55225

Krebs, B. 1994a.    Endosulfan Residues data summary from supervised trials in Savoy cabbage unpublished report PRS94/024 A53963

Krebs, B. 1994b.    Endosulfan Residues data summary from supervised trials in Cacao unpublished report PSR94/034    A53973

Martens R. 1999.    Transfer of residues of endosulfan and its metabolite into processed commodities of peaches European Union (southern zone) 1998 Endosulfan emulsifiable concentrate 352 g/L Code: AE F002671 00 EC33 B331 unpublished report CR98/010    C003180

Rose, S  1984a    Analytical report. Analysis of Endosulfan (Hoe 002671) in tomato. unpublished report EI84-USA-03R/1913(4)    A30309

Rose, S  1984b    Analytical report. Analysis of Endosulfan (Hoe 002671) in tomato. unpublished report EI84-USA-03R/1913(1)    A32875

Rose, S 1984c   Analytical report. Analysis of Endosulfan (Hoe 002671) in cucumber and cantaloupe unpublished report1913(7)   A32884

Rose, S 1984d   Analytical report. Analysis of Endosulfan (Hoe 002671) in squash unpublished report EI84-USA-02R/1913(6)   A30310

Rose, S 1984e   Analytical report. Analysis of Endosulfan (Hoe 002671) in squash, cucumber and cantaloupe unpublished report EI84-USA-02R/1913(5)   A32882

Rose, S 1985   Analytical report. Analysis of Endosulfan (Hoe 002671 in cucumber unpublished report1913(12)   A32883

Ryan Keith, Rose Sam, Novak Roger 1984. Analytical report  Analysis in tomato of Endosulfan Code: Hoe 002671 unpublished report1913(2)   A32876

Sing N. 2000.   Analytical report of analysis in litchis of endosulfan longan association unpublished report LAA-3A AH04007

Sonder K.-H. 1996. Residues to establish a maximum residue level following 3 applications A54358 in musk melons (Cucumis melo) under field conditions European Union (southern zone), 1994 of Hoe 002671 and its metabolites Endosulfan emulsifiable concentrate 352 g/L unpublished report ER94ECS780

Sonder K.-H. 2002a. Residues at harvest in tomatoes (indoor) European Union (Southern zone) 2001 Endosulfan emulsifiable concentrate (EC) 32.9% w/w (= 352 g/L) C020750 Code: AE F002671 00 EC33 C703 unpublished report01R642

Sonder K.-H.   2002b   Residues at harvest in tomatoes (protected) European Union (Southern zone) 2001 Endosulfan suspension of micro capsules (CS) 25.78% w/w (= 330 g/L) Code: AE F002671 00 CS26 E218 unpublished report01R641 C020749

Sonder K.-H., Idstein H., Junker H.   1996a   Determination of residues of Hoe 002671 to establish a maximum residue level following 2 applications in tomatoes for industrial use under field conditions Endosulfan  emulsifiable concentrate 352 g/L  Code: Hoe 02671 00 EC33 B324 unpublished report ER93ECS700 A54363

Sonder K.-H., Idstein H., Junker H.   1996b   Determination of residues of Hoe 002671 to establish a maximum residue level following 2 applications in tomatoes for industrial use under field conditions Endosulfan emulsifiable concentrate 352 g/L Code: Hoe 002671 00 EC33 B325 unpublished report ER94ECS700 A54362

Sonder K.-H., Idstein H., Junker H.   1996c   Determination of residues of Hoe 002671 to establish a maximum residue level following 2 applications in apples Endosulfan emulsifiable concentrate 352 g/L  Code: Hoe 002671 00 EC33 B324 unpublished report ER93ECS705 A54359

Specht   1974a   Pflanzenschutzmittel-Rueckstaende unpublished report1/90/02/02   A02616

Specht   1974b   Pflanzenschutzmittel-Rueckstaende unpublished report1/90/02/04   A02617

Specht   1974c   Pflanzenschutzmittel-Rueckstaende unpublished report1/90/02/03   A02618

Specht   1974d   Pflanzenschutzmittel-Rueckstaende unpublished report1/90/01/04B A04293

Specht   1975   Pflanzenschutzmittel-Rueckstaende unpublished report LEA2/82/01/02-75A   A05862

Specht   1976a   Pflanzenschutzmittel-Rueckstaende unpublished report LEA1/83/01/02-75   A05863

Specht   1976b   Pflanzenschutzmittel-Rueckstaende unpublished report LEA1/85/01/02-75   A06339

Specht   1976c   Pflanzenschutzmittel-Rueckstaende unpublished report LEA1/82/01/02-75   A05860

Specht   1976d   Pflanzenschutzmittel-Rueckstaende unpublished report LEA1/82/02/02-75   A05861

Specht   1977a   Pflanzenschutzmittel-Rueckstaende unpublished report LEA1/48/01/02-76   A08886

Specht   1977b   Pflanzenschutzmittel-Rueckstaende unpublished report LEA4/48/01/02-76   A08889

Stanovick Richard P. 1965a. Determination of residues in or on cauliflower Thiodan I, II and sulfate unpublished report M-1541 A48549

Stanovick Richard P. 1965b. Determination of Thiodan I, II and sulfate Residues in or on brussels sprouts unpublished report M-1575 A48555

Stanovick Richard P. 1965c. Determination of Thiodan I, II and sulfate residues in or on snap beans (beans and forage) unpublished report M-1592   A48556

Stanovick Richard P.   1965d   Determination of Thiodan I, II and sulfate residues in or on lima beans  (beans and pods, shelled beans, pods and forage) unpublished report M-1610   A48557

Thier, A.W. 1974a. Pflanzenschutzmittel-Rueckstaende unpublished report1/74/01/02   A05392

Thier, A.W. 1974b. Pflanzenschutzmittel-Rueckstaende unpublished report1/90/01/02   A03995

Thier, A.W. 1974c. Pflanzenschutzmittel-Rueckstaende unpublished report1/90/01/03    A03996

Thier, A.W. 1974d. Pflanzenschutzmittel-Rueckstaende unpublished report1/90/01/04    A03997

Thier, A.W. 1974e. Pflanzenschutzmittel-Rueckstaende unpublished report2/90/01/02    A03998

Thier, A.W. 1974f. Pflanzenschutzmittel-Rueckstaende unpublished report2/90/01/03    A03999

Thier, A.W. 1974g. Pflanzenschutzmittel-Rueckstaende unpublished report2/90/01/04    A04000

Thier, A.W. 1975a. Pflanzenschutzmittel-Rueckstaende unpublished report LEA2/83/01/02-75A A05865

Thier, A.W. 1975b. Pflanzenschutzmittel-Rueckstaende unpublished report LEA2/84/01/02-75A A05868

Thier, A.W. 1975c. Pflanzenschutzmittel-Rueckstaende unpublished report A05871

Thier, A.W. 1976. Pflanzenschutzmittel-Rueckstaende unpublished report LEA1/84/01/02-75 A05869

Thier, A.W. 1984a. Berichtsbogen fuer Rueckstandsuntersuchungen unpublished report DEU83I72831 A28389

Thier, A.W. 1984b. Berichtsbogen fuer Rueckstandsuntersuchungen unpublished report DEU83I72841 A28390

Thier, A.W. 1984c. Berichtsbogen fuer Rueckstandsuntersuchungen unpublished report DEU83I72811 A28387

Thier, A.W. 1984d. Berichtsbogen fuer Rueckstandsuntersuchungen unpublished report DEU83I72821 A28388

Thier, A.W. 1984e. Berichtsbogen fuer Rueckstandsuntersuchungen unpublished report  DEU83I72711 A28593

Thier, A.W. 1984f. Berichtsbogen fuer Rueckstandsuntersuchungen unpublished report DEU83I72721 A28594

Thier, A.W. 1984g. Berichtsbogen fuer Rueckstandsuntersuchungen unpublished report DEU83I72731 A28595

Thier, A.W. 1984h. Berichtsbogen fuer Rueckstandsuntersuchungen unpublished report DEU83I72741 A28596

Thier, A.W. 1984i. Berichtsbogen fuer Rueckstandsuntersuchungen unpublished report DEU83I71111 A28757

Thier, A.W. 1984j. Berichtsbogen fuer Rueckstandsuntersuchungen unpublished report DEU83I71121 A28758

Thier, A.W. 1984k. Berichtsbogen fuer Rueckstandsuntersuchungen unpublished report DEU83I71131 A28759

Thier, A.W. 1984l. Berichtsbogen fuer Rueckstandsuntersuchungen unpublished report DEU83I70121 A28761

Thier, A.W. 1984m. Berichtsbogen fuer Rueckstandsuntersuchungen unpublished report DEU83I72211 A29495

Thier, A.W. 1984n. Berichtsbogen fuer Rueckstandsuntersuchungen unpublished report DEU83I72241 A29496

Thier, A.W. 1984o. Berichtsbogen fuer Rueckstandsuntersuchungen unpublished report DEU83I70131 A28760

Thier, A.W. 1984p. Berichtsbogen fuer Rueckstandsuntersuchungen unpublished report DEU83I70111 A28762

Thier, A.W. 1984q. Berichtsbogen fuer Rueckstandsuntersuchungen unpublished report DEU83I72231 A29493

Thier, A.W. 1984r. Berichtsbogen fuer Rueckstandsuntersuchungen unpublished report DEU83I72221 A29494

Thier, A.W. 1984s. Berichtsbogen fuer Rueckstandsuntersuchungen unpublished report DEU83I72611 A28919

Thier, A.W. 1984t. Berichtsbogen fuer Rueckstandsuntersuchungen unpublished report DEU83I72641 A28921

Thier, A.W. 1984u. Berichtsbogen fuer Rueckstandsuntersuchungen unpublished report DEU83I72631 A28920

Thier, A.W. 1984v. Berichtsbogen fuer Rueckstandsuntersuchungen unpublished report DEU83I72911 A28922

Thier, A.W. 1984w. Berichtsbogen fuer Rueckstandsuntersuchungen unpublished report DEU83I72921 A28923

Thier, A.W. 1984x. Berichtsbogen fuer Rueckstandsuntersuchungen unpublished report DEU83I72941 A28924

Thier, A.W. 1985a. Berichtsbogen fuer Rueckstandsuntersuchungen unpublished report DEU84I70941 A30909

Thier, A.W. 1985b. Berichtsbogen fuer Rueckstandsuntersuchungen unpublished report DEU84I70926 A30910

Welcker H. 1998. Decline of residues in melon; European Union southern zone 1997. Endosulfan emulsifiable concentrate 32.9% w/w (= 352 g/L) Code: AE F002671 00 EC33 B331 unpublished report ER97ECS745 C001310

Welcker H. 2001a. Residues at harvest in melon European Union southern zone (2000) Endosulfan suspension of microcapsules (CS) 25.78% w/w = 330 g/L Code: AE F002671 00 CS26 E215 unpublished report DR00EUS131 C015857

Welcker H. 2001b. Residues at harvest in mandarin European Union southern zone (1999) Endosulfan suspension of microcapsules (CS), 25.78% w/w (= 330 g/L) Code: AE F002671 00 CS26 E206  unpublished report ER99ECS751 C016672

Welcker H. 2001c. Decline of residues in apple European Union southern zone (1999) Endosulfan suspension of microcapsules (CS), 25.78% w/w (= 330 g/L) Code: AE F002671 00 CS26 E206 unpublished report ER99ECS755 C016113

Welcker H. 2001d. Decline of residues in grape vine European Union southern zone (1999) Endosulfan suspension of microcapsules (CS), 25.78% w/w (= 330 g/L) Code: AE F002671 00 CS26 E206 unpublished report ER99ECS756 C016114

Welcker H. 2001e. Residues at harvest in oranges European Union (southern zone) 1999 Endosulfan, AE F002671 suspension of microcapsules (CS) 25.78% w/w (= 330 g/L) Code: AE F002671 00 CS26 E206 unpublished report ER99ECS750    C016112

Welcker H. 2002a. Residues at harvest in peaches European Union (Southern zone) 1999 Endosulfan suspension of microcapsules (CS) 25.78% w/w (=330 g/L) Code: AE F002671 00 CS26 E206 unpublished report ER99ECS754 C017102

Welcker H. 2002b. Decline of residues in orange European Union (Southern zone) 1999. Endosulfan suspension of microcapsules (CS) 25.78% w/w (=330 g/L) Code: AE F002671 00 CS26 E206 unpublished report ER99ECS758 C016758

Welcker H. 2002c. Decline of residues in cotton European Union (Southern zone) 2001 Endosulfan suspension of micro-capsules (CS) 25.78% w/w (= 330 gL) and emulsifiable concentrate (EC) 32.9% w/w (= 352 g/L) Code: AE F002671 00 CS26 E218, AE F002671 00 EC33 C702 unpublished report01R170 C022557

Welcker H., Martens R. 1999b. Decline of residues in protected tomatoes European Union (southern zone) 1998 Endosulfan suspension of microcapsules (CS) 25.78% w/w (= 330 g/L) Code: AE F002671 00 CS26 E105 unpublished report ER98ECS753    C004455

Welcker H., Martens R.    1999a    Decline of residues in field tomatoes European Union (southern zone) 1998 Endosulfan suspension of microcapsules (CS) 25.78% w/w (= 330 g/L) Code: AE F002671 00 CS26 E105 unpublished report ER98ECS752    C004416

Welcker H., Martens R.    1999c    Decline of residues in orange tree European Union (southern zone) 1998 Endosulfan suspension of microcapsules (CS) 25.78% w/w (= 330 g/L) Code: AE F002671 00 CS26 E105 unpublished report ER98ECS750    C005108

Welcker H., Martens R.    1999d    Decline of residues in mandarin European Union (southern zone) 1998 Endosulfan suspension of microcapsules (CS) 25.78% w/w (= 330 g/L) Code: AE F002671 00 CS26 E105 unpublished report ER98ECS751    C005296

Welcker H., Martens R.    1999e    Decline of residues in peaches European Union (southern zone) 1998 Endosulfan suspension of microcapsules (CS) 25.78% w/w (= 330 g/L) Code: AE F002671 00 CS26 E105 unpublished report ER98ECS754    C004586

Welcker H., Martens R.    2001a    Decline of residues in melon European Union (southern zone) 1999 Endosulfan suspension of microcapsules 25.78% w/w (= 300 g/L) Code: AE F002671 00 CS26 E206 unpublished report ER99ECS757 C009844

Welcker H., Martens R.    2001b    Residues at harvest in field tomatoes European Union (southern zone) 1999 Endosulfan suspension of micocapsules 25.78% w/w (= 300 g/L) Code: AE F002671 00 CS26 E206 unpublished report ER99ECS752 C009843

Wrede A. 2002. Residues at harvest in tomato (single fruit) 2001 Endosulfan Code: AE F002671 unpublished report 02F002 C022705

## Fate of Residues in Storage and Processing

Balluff, M. 1999. Residue Analysis of Juice, Oil and Pellets Gained After an Industrial Processing of Oranges Treated with Endosulfan in Spain,1998  npublished report 98096/S1-FIOR    C004190

Balluff, M. 2002. Field residue study for the determination of residues of endosulfan in  cacao after the maximum number of applications with Thiodan 26 CS (AE F002671 00 CS26 E218) and Thiodan 500 EC (AE F002671 00 EC43 A508) in Ivory Coast, Western Africa, 2001 / 2002 unpublished report 20013056/IC1-FPCC  C020887

Balluff, M. 2003. Field residue study for the determination of residues of endosulfan in cacao after the maximum number of applications with Thiodan 26 CS (AE F002671 00 CS26 E218) and Thiodan 500 EC (AE F002671 00 EC434 A508) under field conditions in Western Africa, 2002 unpublished report20013056/AF1-FPCC    C032568

Brady S.S. 1997a. Magnitude of Endosulfan residues in or on potatoes and processed potato commodities resulting from three applications of Phaser(R) EC insecticide at an exaggerated rate, USA, 1995 Endosulfan unpublished report BJ95R08 A57705

Brady S.S. 1997b. Magnitude of Endosulfan residues in or on tomatoes and processed tomato commodities resulting from three applications of Phaser(R) EC insecticide at an exaggerated rate, USA, 1995 Endosulfan unpublished report BJ95R09 A57707

Brady S.S. 1997c. Magnitude of residues in or on grapes and processed grape commodities resulting from two applications of Phaser(R) EC insecticide Endosulfan unpublished report BJ95R07    A55834

Erbel A. and Ertz K. 2003. Processing of tomatoes treated with endosulfan CS (330 g/L) into tomato juice and canned tomatoes unpublished report MR-510/02    C030836

Gomez C.A. 1996. Magnitude of the residue of endosulfan in coffee raw agricultural commodities and processed fractions unpublished report AA930040          A55745

Huth, G. 1999a. Collection of residues data summary from supervised trials and processing studies in Pomefruit, Grapes, Tomatoes and Melon. unpublished report PRS 99/012 (A49970, A49971)        C004071

Huth, G. 1999b. Residues data summary from supervised trials in Soybeans unpublished report PSR99/010 C004072

Idstein H., Junker H., and Klein E.H.-J. 1996. Residue trials in apples to establish a maximum residue level Determination of active substances and the metabolite decline following 2 applications in apples and processing to apple puree and apple juice Endosulfan emulsifiable concentrate 352 g/L unpublished report ER94ECS705          A55874

Jackson M.C. 1997. Residues in pineapple processing fractions of Thiodan 3EC unpublished report  PGAH950001P A59907

Klein E.H.-J. 1999. Decline of residues in orange fruits and transfer of residues into juice European Union (southern zone) 1997 Endosulfan emulsifiable concentrate (EC) 32.9% w/w (= 352 g/L) Code: AE F002671 00 EC33 B331 unpublished report ER97ECS740 C001464

Klein E.H.-J. and Buerstell H. 1997. Residue trials in vine to establish a Maximum Residue Level Determination of active substance and the metabolite decline following 3 applications in vine and processed fractions European Union, (southern zone), 1994 Endosulfan EC  (emulsifiable concentrate) unpublished report ER94ECS730          A55225

Krebs, B. 1994. Endosulfan. Residues data summary from supervised trials and processing studies in tea unpublished report PSR94/028 A53967

Martens R. 1999. Transfer of residues of endosulfan and its metabolite into processed commodities of peaches European Union (southern zone) 1998 Endosulfan emulsifiable concentrate 352 g/L Code: AE F002671 00 EC33 B331 unpublished report CR98/010    C003180

Maurer, M. 2002. Endosulfan: Investigation of the Nature of the Potential Residue in the products of Industrial Pocessing or Household Preparation unpublished  report CP01/022        C018814

Sonder K.-H., Idstein H. and Junker H. 1996a. Determination of residues of Hoe 002671 to establish a maximum residue level following 2 applications in tomatoes for industrial use under field conditions Endosulfan emulsifiable concentrate 352 g/L Code: Hoe 002671 00 EC33 B324 unpublished report ER93ECS700          A54363

Sonder K.-H., Idstein H. and Junker H. 1996b. Determination of residues of Hoe 002671 to establish a maximum residue level following 2 applications in tomatoes for industrial use under field conditions Endosulfan emulsifiable concentrate 352 g/L Code: Hoe 002671 00 EC33 B325 unpublished report ER94ECS700          A54362

Sonder K.-H., Idstein H. and Junker H. 1996c. Determination of residues of Hoe 002671 to establish a maximum residue level following 2 applications in apples Endosulfan emulsifiable concentrate 352 g/L Code: Hoe 002671 00 EC33 B324 unpublished report ER93ECS705          A54359

**Information and Data from Farm Animal Feeding and External Animal Treatment Studies.**

Peatman, M.H., Reynolds, C.M., Bright, J.H.M. and Godfrey, T.L. Mawhinney H. 1999. Residues of alpha-Endosulfan, beta-Endosulfan and Endosulfan sulphate and metabolites in milk and edible cattle tissues following 28 days feeding to lactating cows. unpublished report characteristics of endosulfan depletion in the fat of cattle following ingestion of contamined feed unpublished report RESID/99/8 or DAQ 109 C003624

**Chapter 8. Residues in Food in Commerce and at Consumption**

ano EU   2001. Monitoring of Pesticide Residues in Products of Plant Origin in the European Union, Norway and Iceland. 1999. June 2001 published report

ano EU   2002. Monitoring of Pesticide Residues in Products of Plant Origin in the European Union, Norway and Iceland. 2000. April 2002 published report

ano EU   2003. Monitoring of Pesticide Residues in Products of Plant Origin in the European Union, Norway, Iceland and Liechtenstein, 2001 March 2003 published report

ano  2005 Australian Pesticides & Veterinary Medicines Authority. 2005. Maximum Residue Limits: the MRL standard - maximum residue limits in food and animal feedstuff. The 20th Australian Total Diet Survey. Food Standards Australia New Zealand. www.foodstandards.gov.au  published report

Simpson P and Hamilton D. 2006. Endosulfan Queensland department of primary industries and fisheries submission for JMPR 2006 published report DPI& F

Simpson P and Hamilton D. 2005. Endosulfan Queensland department of primary industries and fisheries submission for JMPR 2005 published report DPI & F

# FENAMIPHOS (85)

*First draft prepared by Dr. Yukiko Yamada, National Food Research Institute, Tsukuba, Japan.*

## EXPLANATION

Fenamiphos is a systemic organophosphorus nematicide which is registered for use in more than 60 countries. It was first reviewed by the JMPR in 1974 with subsequent residue evaluations in 1977, 1978 and 1980. The compound was evaluated for residues under the periodic review program in 1999 where the Meeting confirmed the existing residue definition, "sum of fenamiphos, its sulfoxide and sulfone, expressed as fenamiphos". The 1999 JMPR recommended 21 new maximum residue levels including six confirmed maximum residue levels while withdrawing nine previously recommended maximum residue levels.

The compound was evaluated toxicologically in 1974, 1985, 1987, 1997 (periodic review) and 2002. The 1997 JMPR allocated an ADI of 0-0.0008 mg/kg bw and concluded that the available data did not permit the establishment of an ARfD different from the ADI. The 2002 JMPR allocated a new ARfD of 0.003 mg/kg bw.

The International Estimated Daily Intakes for the five GEMS/Food regional diets calculated by the 1999 JMPR were in the range of 3–14% of the ADI. The International Estimated Short-term Intakes calculated by the 2002 JMPR for peppers, pineapple and tomato for the general population exceeded the ARfD while those for carrot, grapes, peppers, pineapple, tomato and watermelon exceeded the ARfD for children. The 2003 JMPR refined the IESTI using the variability factor of 3. The refined IESTIs calculated still exceeded the ARfD for tomato for general population and for grapes, sweet peppers, pineapple, tomato and watermelon for children.

The 33rd Session of the Codex Committee on Pesticide Residues (CCPR) in 2001 advanced all the proposed draft MRLs to Step 5 and decided not to advance these MRLs beyond Step 7 until intake concerns were resolved. The 34th and 35th CCPR returned all draft MRLs to Step 6 due to intake concerns. The 36th CCPR decided to return the MRLs for peppers, tomato and watermelon to Step 6 because of acute intake concerns and to advance the remaining MRLs to Step 8, which was subsequently adopted by the Codex Alimentarius Commission in 2004.

The 37th CCPR in 2005 noted that acute intake concerns existed for peppers, tomato and watermelon and decided to return these MRLs to Step 6. It also decided to delete the existing Codex MRLs for carrot, grapes and pineapple. Since this was the third time that the MRLs for peppers, tomato and watermelon were returned to Step 6 due to intake concerns, the CCPR decided to request JMPR to review GAPs that might result in lower MRL recommendations and add fenamiphos to the Priority List for review of GAPs for MRL proposal.

The 38th CCPR in 2006 decided to return the MRLs for peppers, tomato and watermelon to Step 6 noting the acute intake concerns identified by JMPR for these commodities, and agreed to request that JMPR consider using alternative GAPs to recommend lower MRLs for these commodities. The current Meeting received GAP information and new residue trial data on these commodities.

In addition, the Meeting received new trial data on eggplant and melons. Although there was no acute intake concern related to the MRL for melon, the previous data on melons was reviewed because it formed the basis of the existing MRL for watermelon. The data on eggplant was not reviewed because it is outside of the task entrusted to the current Meeting by the CCPR.

## USE PATTERN

Fenamiphos may be applied pre-planting, at planting, in established crops, or in seedbeds and nurseries. For effective control, fenamiphos formulations should be incorporated into the soil in the zones of root growth, as these are exposed to nematodes. Fenamiphos, fenamiphos sulfoxide and fenamiphos sulfone all exhibit nematicidal activity, thereby providing prolonged activity.

Due to acute intake concerns related to residues in some commodities, GAP in Europe has been modified since 2003 to minimize residues in these commodities.

GAP information for countries in Europe, Near East, Central and South America and Australia was provided for the formulations of capsule suspension, emulsifiable concentrate, fine granule and granule. Labels in Australia, Greece, Italy, the Netherlands and Spain were provided. The information relevant to the use for eggplant, melons, peppers, tomato and watermelon is summarized in Table 1.

Information on use for eggplant, melons, peppers, tomato and watermelon contained in the 1999 JMPR Evaluation was also extracted in Table 1.

Table 1. Registered uses of fenamiphos.

| Crop | Country | Form. | Application | | | | PHI, days | Notes |
|------|---------|-------|-------------|--|--|--|------|-------|
| | | | Timing | Method | Rate, kg ai/ha | No | | |
| Vegetables | Greece | 10GR | Before planting | Incorporation | 6-8 4 8 g/hole | 1 | | c |
| | | 400EC | | In irrigation water or spraying | 4-8 6-8 4 | 1 | 60 | d |
| Cucurbits | Australia | 400EC | Pre-planting or pre-transplanting | Soil treatment | 9.6 | 1 | - | |
| Melons | Argentina | 240CS | | In irrigation water, dripping | 3.2-4.0 | 1 | 90 | |
| Melons | Belize | 15GR | | Spreading/incorp. | 2.6-5.1 | 1 | 60 | |
| Melons | Brazil | 400EC | At sowing | In irrigation water | 4 | 1 | - | |
| Melons/ Watermelon | Colombia | 10GR | | Spreading | 0.5-0.8 g ai/plant | | 90 | † |
| Melons | Costa Rica | 10GR 15GR | | Spreading Spreading | 0.2-2.5 2.6-5.1 | 1 1 | 60 60 | |
| Melons/ Watermelon | Dominican Rep. | 10GR 12GR 15GR | | Spreading/incorp. | 2.5-5.0 0.72-1.2 2.5-5.1 | 1 | 60 | † |
| Melons | El Salvador | 10GR 15GR | | Spreading/incorp. Spreading/incorp. | 2.5-5.0 2.6-5.1 | 1 1 | 60 60 | |
| Watermelon | El Salvador | 4GR | | Spreading/incorp. | 0.6-1.2 | 1 | 60 | |
| Melons | Guatemala | 10GR 10GR 15GR | | Spreading/incorp. Spreading Spreading/incorp. | 2.5-5.0 5 2.6-5.1 | 1 1 1 | 60 60 60 | |
| Watermelon | Guatemala | 4GR | | Spreading/incorp. | 0.6-1.2 | 1 | 60 | |
| Melons | Honduras | 10GR 15GR | | Spreading/incorp. Spreading/incorp. | 2.5-5.0 2.6-5.1 | 1 1 | 60 60 | |
| Melons | Italy | 5GR | | Soil incorporation | 9.6-12 | 1 | 60 | |
| Melons/ watermelon | Italy | 240CS | From transplanting to abt 10 d later | In irrigation water | 10 | 1 | 60 | |
| Melons | Jordan | 10GR | | Spreading/incorp. | 6 | 1 | 90 | |
| Melons | Lebanon | 400EC | | High-vol. spraying followed by irrigation | 4-8 (1.3-2.7 kg ai/hl) | 1 | 90 | |
| Melons | Morocco | 10GR | | Spreading/incorp. | | 1 | 90 | |
| Melons | Libya | 10GR | | Spreading | | 1 | 90 | |
| Melons | Nicaragua | 15GR | | Spreading/incorp. | 2.6-5.1 | 1 | 60 | |
| Melons/ Watermelon | Panama | 10GR 15GR | | Spreading/incorp. Spreading/incorp. | 2.5-5.0 2.6-5.1 | 1 1 | 60 60 | |
| Melons | Spain | 10GR | Pre-planting or pre-sowing | Spreading/incorp. | 5-10 | 1 | 60 | |

| Crop | Country | Form. | Application | | | | PHI, days | Notes |
|------|---------|-------|-------------|--|--|--|-----------|-------|
| | | | Timing | Method | Rate, kg ai/ha | No | | |
| Melons/ Watermelon | Spain | 400EC | | Spraying | 4.8-10 | 2 | 60 | a |
| | | 240CS | Before start of flowering | Soil treatment | 4.8-9.6 | 2 | 60 | b |
| Peppers, Sweet | Argentina | 240CS | | In irrigation water, dripping | 3.2-4.0 | 1 | 90 | |
| Peppers | Dominican Rep. | 12GR | | Spreading/incorp. | 0.72-1.2 | 1 | 60 | † |
| Peppers, Sweet | El Salvador | 4GR | | Spreading/incorp. | 0.6-1.2 | 1 | 60 | |
| Peppers, Sweet | Guatemala | 4GR | | Spreading/incorp. | 0.6-1.2 | 1 | 60 | |
| Peppers | Honduras | 12GR | | Spreading/incorp. | 0.72-1.2 | 1 | 60 | † |
| Peppers, Sweet | Iraq | 400EC | | Spraying/incorp. | 8 | 1 | - | |
| Peppers, Sweet | Italy | 240CS | From transplanting to abt 10 d later | In irrigation water | 10 | 1 | 60 | |
| Peppers, Sweet | Libya | 10GR | | Spreading | | 1 | 90 | |
| Peppers | Spain | 10GR | Pre-planting or pre-sowing | Spreading/incorp. | 5-10 | 1 | 60 | |
| | | 400EC | | Spraying | 4.8-10 | 2 | 60 | a |
| | | 240CS | Before start of flowering | Soil treatment | 4.8-9.6 | 2 | 60 | b |
| Tomato | Algeria | 10GR | | Put down | 3 | 1 | 90 | |
| Tomato | Angola | 10GR | | Spreading/incorp. | | 1 | - | |
| Tomato | Argentina | 240CS | | In irrigation water, dripping | 3.2-4.0 | 1 | 90 | |
| Tomato | Australia | 400EC | Pre-planting or pre-transplanting | Spraying | 9.6 | 1 | - | |
| | | 10GR | Within 7 d of planting | Sprinkling | 11 | 1 | - | |
| | | 5GR | Within 7 d of planting | Sprinkling over soil | 13 | 1 | - | Not Tasmania |
| Tomato | Belize | 15GR | | Spreading/incorp. | 2.6-5.1 | 1 | 60 | |
| Tomato | Brazil | 10GR | At planting | Incorporation | 3-4 | 1 | 90 | |
| Tomato | Chile | 400EC | To planted or sown crop | Spraying/incorp. | 4-10 | 1 | 45 | |
| | | | | Drench | 2.8-4.8 | 1 | 45 | |
| Tomato | Colombia | 10GR | Before or after sowing | Spreading | 15 | 1 | 60 | |
| | | | | Spreading | 20 | 1 | 60 | |
| | | | At transplanting | Spreading | 8.2 | 1 | 60 | |
| | | | | Spreading | 16 | 1 | 60 | |
| Tomato | Costa Rica | 10GR | | Spreading/incorp. | 2.5-5.0 | 1 | 60 | |
| | | 15GR | | Spreading/incorp. | 2.1-2.6 | 1 | 60 | |
| Tomato | Dominican Rep. | 400EC | | Drench | 2.8-3.4 | 1 | 60 | |
| Tomato | Ecuador | 10GR | | Spreading | 2-4 | 1 | 60 | |
| | | 15GR | | Spreading | 3-6 | 1 | 60 | |
| Tomato | El Salvador | 10GR | | Spreading/incorp. | 2.5-5.0 | 1 | 60 | |
| | | 15GR | | Spreading/incorp. | 2.6-5.1 | 1 | 60 | |
| | | 4GR | | Spreading/incorp. | 0.6-1.2 | 1 | 60 | |
| Tomato | Guatemala | 400EC | | Spraying | 2.8-3.4 | 1 | 60 | |
| | | 10GR | | Spreading | 5 | 1 | 60 | |
| | | 10GR | | Spreading/incorp. | 2.5-5.0 | 1 | 60 | |
| | | 15GR | | Spreading/incorp. | 2.6-5.1 | 1 | 60 | |
| | | 4GR | | Spreading/incorp. | 0.6-1.2 | 1 | 60 | |
| Tomato | Honduras | 10GR | | Spreading/incorp. | 2.5-5.0 | 1 | 60 | |
| | | 15GR | | Spreading/incorp. | 2.6-5.1 | 1 | 60 | |
| Tomato | Iraq | 400EC | | Spraying/incorp. | 8 | 1 | - | |

| Crop | Country | Form. | Application | | | | PHI, days | Notes |
|------|---------|-------|------|--------|----------|-----|------|-------|
| | | | Timing | Method | Rate, kg ai/ha | No | | |
| Tomato | Italy | 5GR | | Soil incorporation | 9.6-12 | 1 | 60 | |
| | | 240CS | From transplanting to abt 10 d later | In irrigation water | 10 | 1 | 60 | |
| Tomato | Jordan | 10GR | | Spreading/incorp. | 6 | 1 | 90 | |
| Tomato | Lebanon | 400EC | | High-vol. spraying followed by irrigation | 4-8 (1.3-2.7 kg ai/hl) | 1 | 90 | |
| Tomato | Libya | 10GR | | Spreading | 4.5 | 1 | 90 FI 60 GH | |
| Tomato | Malawi | 400EC 10GR | | Spraying/incorp. Spreading/incorp. | | 1 1 | - - | |
| Tomato | Morocco | 400EC 10GR | | In irrigation water, dripping Spreading/incorp. | 4.8 (0.08-0.12 kg ai/hl) 5 | 1 1 | 60 90 | |
| Tomato | Mozambique | 400EC 10GR | | Spraying/incorp. Spreading/incorp. | | 1 1 | 42 - | |
| Tomato | Namibia | 400EC 10GR | | Spraying/incorp. Spreading/incorp. | | 1 1 | - - | |
| Tomato | Nicaragua | 15GR | | Spreading/incorp. | 2.5-5.0 | 1 | 60 | |
| Tomato | Panama | 10GR 15GR | | Spreading/incorp. Spreading/incorp. | 2.5-5.0 2.6-5.1 | 1 1 | 60 60 | |
| Tomato | Paraguay | 400EC | | In irrigation water | 3.2-4.0 | 1 | - | |
| Tomato | Peru | 10GR | | Spreading | 1.3-2.5 | 1 | 90 | |
| Tomato | Portugal | 10FG | 1st, 1-2 w pre-planting 2nd, 30 d after 1st | Spreading | 3.4 (1st) 3.0 (2nd) | 2 | - | |
| Tomato | Saudi Arabia | 10GR 10GR | | Spreading Spreading | 10-15 1.5-3.0 | 1 1 | - - | |
| Tomato | S. Africa | 240CS 400EC 10GR | Pre-planting | Spraying Spraying/incorp. Spreading/incorp. | 9.8 | 1 1 1 | - - - | |
| Tomato | Spain | 10GR 400EC 240CS | Pre-planting or pre-sowing Before start of flowering | Spreading/incorp. Spraying Soil treatment | 5-10 4.8-10 4.8-9.6 | 1 2 2 | 60 60 60 | a b |
| Tomato | Tunisia | 10GR | | Spreading/incorp. | 10 | 1 | 20 | |
| Tomato | Turkey | 400EC 10GR | | Drench/incorp. Spreading in furrow | 10 10 | 1 1 | 90 - | |
| Tomato | Uruguay | 240CS | | Drench | 1.6-9.6 | 1-2 | - | |
| Tomato | Zambia | 400EC 10GR | | Spraying/incorp. Spreading followed by irrigation | | 1 1 | - - | |
| Tomato | Zimbabwe | 400EC 10GR | | Spraying/incorp. Spreading/incorp. | | 1 1 | 42 - | |

† Included in the 1999 JMPR Evaluation.

a. Up to two applications per season can be given by dividing the dose in half if the duration of the crop makes it possible to observe the safety period. The first application should be before sowing or transplanting or immediately afterwards and the second during the rooting period of the crop before the start of flowering.

b. First application, just before transplanting or sowing or immediately afterwards; and second application, during the rooting period before the start of flowering.

c. To the entire surface with incorporation before sowing or planting, 6-8 kg ai/ha; to the sowing or planting drills, 4 kg ai/ha; to the planting holes, 8 g ai/hole.

d. Vegetables include tomato, eggplant, peppers, melons, watermelon, cabbage and cauliflower. With water on transplantation, 4-8 kg ai/ha; before sowing or planting by spraying the entire soil surface with simultaneous incorporation, 6-8 kg ai/ha; spraying the seed or planting drills or with a dropwise irrigation system, 1 kg ai/ha.

## RESIDUES RESULTING FROM SUPERVISED TRIALS

Results of new trials on melons, watermelon, eggplant, sweet peppers and tomato conducted in 1998-1999 were provided. The reports contained necessary information such as trial locations, varieties, methods and timing of application, sampling and analysis including recovery information. The residues in control plots were all below the LOQ and therefore not recorded in the following tables. Residue data are not adjusted for recovery. When residues were not detected they are shown as below the LOQ. Residues, application rates and spray concentrations have generally been rounded to two significant figures or, for residues near the LOQ, to one significant figure.

The summary of trials on these crops is also extracted from the 1999 JMPR Evaluation and reproduced in the following tables.

Data according to the maximum GAP is double-underlined. Data from trials conducted under conditions within GAP and used for the estimation of maximum residue level is single-underlined.

### Melons

Fenamiphos is widely registered for use on melon and watermelons. In Europe the GAP had recently been changed as shown in Table 1 in order to minimize the resulting fenamiphos residues.

Six new trials were conducted in 1998 and 1999 with fenamiphos and imidacloprid (CS 246) on melon in greenhouse: one in France, four in Italy and one in Spain (Deissler & Ohs, 1999d; Anderson & Elke, 2000b). The combination product (CS 246) was applied once to the plant at a rate of 10 kg fenamiphos/ha (0.252 kg imidacloprid/ha) via a drip irrigation system. Water rates were 7000 – 100000 L/ha, depending on the given situation in the greenhouses in which the tests were performed.

The relevant trials were extracted from the 1999 JMPR Evaluation. Among those trials, there are nine trials conducted (Heinemann & Ohs, 1997a, Blass 1998a) with fenamiphos alone (CS 240) and with the same application rate and technique as the new trials with the CS formulation.

Trial data are summarized in Table 2. Except where otherwise described in a footnote, procedural recovery of analysis in each study was acceptable.

Table 2. Residues of fenamiphos in melons from trials in Australia, France, Guatemala, Italy, Mexico and Spain.

| MELONS Location Year variety | G or F | Application | | | PHI | Sample | Residues mg/kg | Reference |
|---|---|---|---|---|---|---|---|---|
| | | Form. | kg ai/ha | No | | | | |
| Newly submitted trial data | | | | | | | | |
| Italy (Fondi) 1998 Proteo | G | 240CS (246CS) | 10 in irrigation water (before flowering) | 1 | 59 | Fruit | < 0.02 | RA-2033/98 1178-98 |
| Italy (Ravenna) 1999 Drake | G | 240CS (246CS) | 10 in irrigation water (at 1st flowering) | 1 | 30 | Fruit | 0.03 | RA-2058/99 0378−99 |
| | | | | | | Pulp | 0.03 | |
| | | | | | | Peel | 0.03 | |
| | | | | | | Whole fruit | 0.03 | |
| | | | | | 60 | Fruit | 0.04 | |
| | | | | | 70 | Fruit | < 0.02 | |
| | | | | | 90 | Fruit | < 0.02 | |
| | | | | | | Pulp | < 0.02 | |
| | | | | | | Peel | < 0.02 | |
| | | | | | | Whole fruit | < 0.02 | |
| Spain (corvera) 1999 Olmedo-Pieldesapo | G | 240CS (246CS) | 10 in irrigation water (at 1st flowering) | 1 | 32 | Fruit | 0.06 | RA-2058/99 0379-99 |
| | | | | | | Pulp | 0.06 | |
| | | | | | | Peel | 0.05 | |
| | | | | | | Whole fruit | 0.06 | |
| | | | | | 62 | Fruit | 0.04 | |
| | | | | | 69 | Fruit | 0.03 | |

| MELONS Location Year variety | G or F | Application | | | PHI | Sample | Residues mg/kg | Reference |
|---|---|---|---|---|---|---|---|---|
| | | Form. | kg ai/ha | No | | | | |
| | | | | | 89 | Fruit | < 0.02 | |
| | | | | | | Pulp | < 0.02 | |
| | | | | | | Peel | < 0.02 | |
| | | | | | | Whole fruit | < 0.02 | |
| France (Montbartier) 1999 Figaro | G | 240CS (246CS) | 10 in irrigation water (7th flower open) | 1 | 30 | Fruit | 0.56 | RA-2058/99 0380-99 |
| | | | | | | Pulp | 0.38 | |
| | | | | | | Peel | 0.75 | |
| | | | | | | Whole fruit | 0.52 | |
| | | | | | 61 | Fruit | 0.08 | |
| | | | | | 70 | Fruit | 0.05 | |
| | | | | | 90 | Fruit | 0.04 | |
| | | | | | | Pulp | 0.02 | |
| | | | | | | Peel | 0.04 | |
| | | | | | | Whole fruit | 0.03 | |
| Italy (Badia Polesine) 1999 Baggio | G | 240CS (246CS) | 10 in irrigation water (1st flower open) | 1 | 28 | Fruit | 0.05 | RA-2058/99 0485-99 |
| | | | | | | Pulp | 0.02 | |
| | | | | | | Peel | 0.04 | |
| | | | | | | Whole fruit | 0.03 | |
| | | | | | 59 | Fruit | < 0.02 | |
| | | | | | 69 | Fruit | < 0.02 | |
| | | | | | 88 | Fruit | < 0.02 | |
| | | | | | | Pulp | < 0.02 | |
| | | | | | | Peel | < 0.02 | |
| | | | | | | Whole fruit | < 0.02 | |
| Italy (Gavello) 1999 Harper | G | 240CS (246CS) | 10 in irrigation water (1st flower open) | 1 | 30 | Fruit | < 0.02 | RA-2058/99 0487-99 |
| | | | | | | Pulp | < 0.02 | |
| | | | | | | Peel | < 0.02 | |
| | | | | | | Whole fruit | < 0.02 | |
| | | | | | 59 | Fruit | < 0.02 | |
| | | | | | 70 | Fruit | < 0.02 | |
| | | | | | 88 | Fruit | < 0.02 | |
| | | | | | | Pulp | < 0.02 | |
| | | | | | | Peel | < 0.02 | |
| | | | | | | Whole fruit | < 0.02 | |
| Trial data contained in the 1999 JMPR Monograph | | | | | | | | |
| Italy (Latina) 1996 Proteo | G | 240CS | 10 | 1 | 50 | Whole fruit | 0.041, 0.042 | 0281-96 |
| | | | | | | Pulp | < 0.02, < 0.02 | |
| | | | | | 60 | Whole fruit | < 0.02, < 0.02 | |
| | | | | | | Pulp | < 0.02, < 0.02 | |
| | | | | | 90 | Whole fruit | < 0.02 | |
| | | | | | | Pulp | < 0.02 | |
| Italy (Ravenna) 1996 Drake | G | 240CS | 10 | 1 | 50 | Whole fruit | 0.034, 0.036 | 0376-96 |
| | | | | | | Pulp | 0.022, 0.021 | |
| | | | | | 60 | Whole fruit | 0.02, 0.021 | |
| | | | | | | Pulp | < 0.02, < 0.02 | |
| | | | | | 90 | Whole fruit | < 0.02 | |
| | | | | | | Pulp | < 0.02 | |
| Italy (Latina) 1996 Mambo | G | 240CS | 10 | 1 | 50 | Whole fruit | < 0.02, < 0.02 | 0377-96 |
| | | | | | | Pulp | < 0.02, < 0.02 | |
| | | | | | 60 | Whole fruit | < 0.02, < 0.02 | |
| | | | | | | Pulp | < 0.02, < 0.02 | |
| | | | | | 90 | Whole fruit | < 0.02 | |
| | | | | | | Pulp | < 0.02 | |
| Italy (Verona) 1996 Golden Star | G | 240CS | 10 | 1 | 50 | Whole fruit | 0.046, 0.044 | 0378-96 |
| | | | | | | Pulp | 0.021, 0.025 | |
| | | | | | 59 | Whole fruit | 0.025, 0.028 | |
| | | | | | | Pulp | < 0.02, < 0.02 | |
| | | | | | 80 | Whole fruit | < 0.02, < 0.02 | |
| | | | | | | Pulp | < 0.02, < 0.02 | |
| | | | | | 90 | Whole fruit | < 0.02 | |
| | | | | | | Pulp | < 0.02 | |

| MELONS Location Year variety | G or F | Application Form. | Application kg ai/ha | Application No | PHI | Sample | Residues mg/kg | Reference |
|---|---|---|---|---|---|---|---|---|
| Italy (Verona Sth) 1997 Super market | G | 240CS | 10 | 1 | 50 | Whole fruit Pulp | 0.03 0.03 | 0045-97 |
| | | | | | 60 | Whole fruit Pulp | < 0.02 < 0.02 | |
| | | | | | 70 | Whole fruit Pulp | < 0.02 < 0.02 | |
| Italy (Verona Sth) 1997 Super market | G | 240CS | 10 | 1 | 50 | Whole fruit Pulp | < 0.02 < 0.02 | 0554-97 |
| | | | | | 61 | Whole fruit Pulp | < 0.02 < 0.02 | |
| | | | | | 70 | Whole fruit Pulp | < 0.02 < 0.02 | |
| Italy (Ravenna) 1997 Drake | G | 240CS | 10 | 1 | 60 | Whole fruit Pulp | < 0.02 < 0.02 | 0555-97 |
| | | | | | 70 | Whole fruit Pulp | < 0.02 < 0.02 | |
| Italy (Ravenna) 1997 Crido | G | 240CS | 10 | 1 | 50 | Whole fruit Pulp | < 0.02 < 0.02 | 0557-97 |
| | | | | | 61 | Whole fruit Pulp | < 0.02 < 0.02 | |
| | | | | | 67 | Whole fruit Pulp | < 0.02 < 0.02 | |
| Italy (Ravenna Nth) 1997 | G | 240CS | 10 | 1 | 50 | Whole fruit Pulp | 0.03 0.03 | 0801-97 |
| | | | | | 60 | Whole fruit Pulp | < 0.02 < 0.02 | |
| | | | | | 70 | Whole fruit Pulp | < 0.02 < 0.02 | |
| Australia (QLD) 1971 Hales best | F | 400EC | 8.9 by boom spray and incorporation with rotary hoe | 1 | 112 | mature fruit | < 0.01 | 33/71a[3] |
| Australia (QLD) 1971 Hales best | F | 400EC | 8.9 by boom spray and incorporation with rotary hoe | 1 | 77 | mature fruit | < 0.01 | 33/71b[4] |
| Brazil (Sao Paulo) 1995 Valenciano | F | 400EC | 4 spraying | 1 | 90 | Whole fruit | < 0.02 | BRA-2009-96-A[5] |
| Brazil (Sao Paulo) 1995 Valenciano | F | 400EC | 8 spraying | 1 | 90 | Whole fruit | < 0.02 | BRA-2009-96-B[5] |
| Guatemala (Zacapa) 1987 Mayan sweet | F | 10GR | 10 spreading/incorp. | 1 | 85 | fruit | < 0.05 | GUA-36-87-A[1] |
| Guatemala (Zacapa) 1987 Mayan sweet | F | 10GR | 10 spreading/incorp. | 1 | 71 | fruit | < 0.05 | GUA-36-87-B[1] |
| Guatemala (Zacapa) 1987 Mayan sweet | F | 10GR | 4 spreading/incorp. | 1 | 85 | fruit | < 0.05 | GUA-36-87-C[1] |
| Guatemala (Zacapa) 1987 Mayan sweet | F | 10GR | 4 spreading/incorp. | 1 | 71 | fruit | < 0.05 | GUA-36-87-D[1] |
| Mexico (Durango) 1983 Sierra gold | | 15GR | 3 | 1 | 64 | pulp peel whole fruit | < 0.01 < 0.01 < 0.01 | 96784 |
| Mexico (Durango) 1983 Sierra gold | | 15GR | 3 | 1 | 64 | pulp peel whole fruit | < 0.01 < 0.01 < 0.01 | 96784 |

| MELONS Location Year variety | G or F | Application | | | PHI | Sample | Residues mg/kg | Reference |
|---|---|---|---|---|---|---|---|---|
| | | Form. | kg ai/ha | No | | | | |
| Mexico (Coahuila) 1983 Imperial 45 | | 15GR | 3 | 1 | 62 | pulp peel whole fruit | < 0.01 < 0.01 < 0.01 | 96784 |
| Mexico (Coahuila) 1983 Imperial 45 | | 15GR | 3 | 1 | 63 | pulp peel whole fruit | < 0.01 0.02 < 0.01 | 96784 |
| Italy (Borgo Piave) 1989 Charantes | F | 5GR | 15 spreading | 1 | 85 100 105 | pulp peel whole fruit pulp peel whole fruit pulp peel whole fruit | < 0.02 < 0.02 < 0.02 < 0.02 < 0.02 < 0.02 < 0.02 < 0.02 < 0.02 | 0064-89 [2] |

1. Applied at sowing or 14 days before sowing. Recovery, 68% at 0.05 mg/kg.

2. Applied 18 days before planting.

3. Applied 35 days before sowing.

4. Applied 4 days before sowing.

5. Applied at planting.

## Watermelon

Four new trials were conducted in 1998 and 1999 with fenamiphos and imidacloprid (CS 246) on watermelon in greenhouse: three in Italy and one in Spain (Deissler & Ohs, 1999d; Anderson & Elke, 2000b). The combination product (CS246) was applied once to the plant at a rate of 10 kg fenamiphos/ha (0.252 kg imidacloprid/ha) via a drip irrigation system. Water rates were 7000–100,000 L/ha, depending on the given situation in the greenhouses in which the tests were performed.

The relevant trials were extracted from the 1999 JMPR Evaluation and the trial data is summarized in Table 3. The procedural recovery of analysis in each study was acceptable.

Table 3. Residues of fenamiphos in watermelon from trials in Italy and Spain.

| WATERMELON Location Year Variety | G or F | Application | | | PHI | Sample | Residues mg/kg | Reference |
|---|---|---|---|---|---|---|---|---|
| | | Form. | kg ai/ha | No | | | | |
| Newly submitted trial data | | | | | | | | |
| Italy (Trinitapoli) 1998 Crimson sweet | G | 240CS (246CS) | 10 in irrigation water (before flowering) | 1 | 60 | Fruit | 0.02 | RA-2033/98 1179-98 |
| Italy (Ravenna) 1999 Trophi | G | 240CS (246CS) | 10 in irrigation water (5th flower open) | 1 | 30 60 70 90 | Fruit Pulp Peel Whole fruit Fruit Fruit Fruit Pulp Peel Whole fruit | < 0.02 0.04 0.02 0.03 0.04 < 0.02 < 0.02 < 0.02 < 0.02 < 0.02 | RA-2058/99 0381-99 |

| WATERMELON Location Year Variety | G or F | Application Form. | kg ai/ha | No | PHI | Sample | Residues mg/kg | Reference |
|---|---|---|---|---|---|---|---|---|
| Spain (La Hoya Elche) 1999 Reina de Corazones | G | 240CS (246CS) | 10 in irrigation water (3rd flower initial visible) | 1 | 42 | Fruit | 0.15 | RA-2058/99 0382-99 |
| | | | | | | Pulp | 0.19 | |
| | | | | | | Peel | 0.17 | |
| | | | | | | Whole fruit | 0.18 | |
| | | | | | 59 | Fruit | 0.16 | |
| | | | | | 70 | Fruit | 0.16 | |
| | | | | | 87 | Fruit | 0.11 | |
| | | | | | | Pulp | 0.14 | |
| | | | | | | Peel | 0.08 | |
| | | | | | | Whole fruit | 0.11 | |
| Italy (Vittoria) 1999 Crimson sweet | G | 240CS (246CS) | 10 in irrigation water (before flowering) | 1 | 60 | Fruit | < 0.02 | RA-2058/99 0383-99 |
| | | | | | 70 | Fruit | < 0.02 | |
| | | | | | 90 | Fruit | < 0.02 | |
| | | | | | | Pulp | < 0.02 | |
| | | | | | | Peel | < 0.02 | |
| | | | | | | Whole fruit | < 0.02 | |
| Trial data contained in the 1999 JMPR Monograph | | | | | | | | |
| Italy (Latina) 1988 Crimson Sweet | F | 5GR | 10 spreading | 1 | 99 | fruit | < 0.02, < 0.02 | 0197-88[1] |
| | | | | | 109 | | < 0.02, < 0.02 | |
| Italy (Borgo Piave) 1989 Crimson Sweet | F | 5GR | 10 spreading | 1 | 85 | Pulp | < 0.02, < 0.02 | 0062-89[2] |
| | | | | | | Peel | < 0.02, < 0.02 | |
| | | | | | | Whole fruit | < 0.02, < 0.02 | |
| | | | | | 100 | Pulp | < 0.02, < 0.02 | |
| | | | | | | Peel | < 0.02, < 0.02 | |
| | | | | | | Whole fruit | < 0.02, < 0.02 | |
| | | | | | 105 | Pulp | < 0.02, < 0.02 | |
| | | | | | | Peel | < 0.02, < 0.02 | |
| | | | | | | Whole fruit | < 0.02, < 0.02 | |

1. Results in summary form. Applied 20 days before planting.

2. Results in summary form. Applied 2 days before planting.

*Peppers, sweet*

Fenamiphos is widely registered for use on sweet pepper. In Europe the GAP has recently been changed to minimize the residues, as shown in Table 1.

A total of ten new trials were conducted with fenamiphos and imidacloprid in greenhouses in southern Europe in 1998 and 1999: one in France, two in Greece, four in Italy, one in Portugal and two in Spain (Deissler & Ohs, 1999b; Anderson & Elke, 2000c). Fenamiphos and imidacloprid (CS 246) was applied once to the plant at a rate of 10 kg fenamiphos/ha (0.252 kg imidacloprid/ha) via a drip irrigation system. Water rates were 7000 – 24000 L/ha, depending on the given situation in the greenhouses in which the tests were performed.

The relevant trials were extracted from the 1999 JMPR Evaluation. Among those trials, there are nine trials conducted in Italy (Heinemann & Ohs, 1997c, Blass, 1998a) using fenamiphos CS240 with the same application rate and technique as the new trials in Italy, which are summarized in Table 4. Except where otherwise described, procedural recovery of analysis in each trial was acceptable.

Table 4. Residues of fenamiphos in peppers from trials in France, Greece, Italy, Portugal and Spain.

| PEPPERS Location Year Variety | G or F | Application Form. | kg ai/ha | No | PHI | Sample | Residues mg/kg | Reference |
|---|---|---|---|---|---|---|---|---|
| Newly submitted trial data | | | | | | | | |
| Italy (Pozzo Ribaudo) 1998 Lux (sweet) | G | 240CS (246CS) | 10 in irrigation water (17 d after flowing) | 1 | 60 | fruit | < 0.02 | RA-2034/98 1180-98 |

| PEPPERS Location Year Variety | G or F | Application Form. | Application kg ai/ha | Application No | PHI | Sample | Residues mg/kg | Reference |
|---|---|---|---|---|---|---|---|---|
| Italy (Santa Croce di Camerina) 1999  Lux (sweet) | G | 240CS (246CS) | 10 in irrigation water | 1 | 60 | fruit | < 0.02 | RA-2034/98 1403-98 |
| Italy (Zapponeta) 1999 Marconi R. (sweet) | G | 240CS (246CS) | 10 in irrigation water (before flowering) | 1 | 60 70 90 | fruit | < 0.02 < 0.02 < 0.02 | RA-2059/99 0351-99 |
| Spain (Arenys de Munt) 1999 Plana (sweet) | G | 240CS (246CS) | 10 in irrigation water (before flowering) | 1 | 46 60 70 90 | fruit | 0.04 0.02 < 0.02 < 0.02 | RA-2059/99 0352-99 |
| Italy (Vittoria) 1999 Cosmos | G | 240CS (246CS) | 10 in irrigation water (4th fruit reaching typical size) | 1 | 30 60 70 90 | fruit | 0.07 0.04 0.02 < 0.02 | RA-2059/99 0367-99 |
| France (St.Remy) 1999 Mariner (sweet) | G | 240CS (246CS) | 10 in irrigation water (3rd flower) | 1 | 30 60 70 90 | fruit | 0.97 0.44 0.28 0.17 | RA-2059/99 0368-99 [a] |
| Portugal (Silveira) 1999 Lido (sweet) | G | 240CS (246CS) | 10 in irrigation water (1st flower bud visible) | 1 | 30 62 70 91 | fruit | 0.03 < 0.02 < 0.02 < 0.02 | RA-2059/99 0369-99 |
| Spain (Arenys de Munt) 1999 Italiano (sweet) | G | 240CS (246CS) | 10 in irrigation water (before flowering) | 1 | 46 60 70 90 | fruit | 0.09 < 0.02 < 0.02 < 0.02 | RA-2059/99 0370-99 |
| Greece (Crete) 1999 Drango (sweet) | G | 240CS (246CS) | 10 in irrigation water (7th flower bud visible) | 1 | 33 62 71 91 | fruit | < 0.02 < 0.02 < 0.02 < 0.02 | RA-2059/99 0480-99 |
| Greece (Crete) 1999 Drango (sweet) | G | 240CS (246CS) | 10 in irrigation water (3rd flower open) | 1 | 33 62 71 91 | fruit | < 0.02 < 0.02 < 0.02 < 0.02 | RA-2059/99 0481-99 |
| Trial data contained in the 1999 JMPR Monograph | | | | | | | | |
| Italy (Latina) 1996 Sonar | G | 240CS | 10 | 1 | 30 60 90 | | 0.119, 0.108 < 0.02, < 0.02 < 0.02 | 0277-96 |
| Italy (Ragusa) 1996 Lux | G | 240CS | 10 | 1 | 31 60 90 | | 0.041, 0.033 < 0.02, < 0.02 < 0.02 (green) < 0.02 (yellow) | 0393-96 |
| Spain (Almeria) 1996 Anibal | G | 240CS | 10 | 1 | 31 60 90 | | 0.067, 0.071 0.02, 0.02 < 0.02 | 0394-96 |
| Spain (Almeria) 1996 Drago | G | 240CS | 10 | 1 | 31 60 90 | | 0.187, 0.177 0.07, 0.064 < 0.02 | 0395-96 |
| Italy (Latina) 1997 Gordo | G | 240CS | 10 | 1 | 30 60 90 | | < 0.02, < 0.02 < 0.02, < 0.02 < 0.02 | 0099-97 |
| Italy (Ragusa) 1996 Soldi | G | 240CS | 10 | 1 | 30 60 90 | | 0.091, 0.088 0.110, 0.096 < 0.02 | 0558-97 |
| Spain (Almeria) 1996 Roldan | G | 240CS | 10 | 1 | 30 60 90 | | < 0.02, < 0.02 < 0.02, < 0.02 < 0.02 | 0559-97 |
| Portugal (Lissabon) 1997 Sonar | G | 240CS | 10 | 1 | 30 60 90 | | 0.306, 0.287 0.080[8], 0.065 < 0.02 | 0560-97 |

| PEPPERS Location Year Variety | G or F | Application Form. | kg ai/ha | No | PHI | Sample | Residues mg/kg | Reference |
|---|---|---|---|---|---|---|---|---|
| Spain (Murcia) 1985 Gedeon | G | 400EC | 10 drip irrigation | 1 | 1 15 29 56 85 | | < 0.05, < 0.05 0.18, 0.18 0.28, 0.37 0.19, 0.20 0.25, 0.26 | 5207-84 [3] |
| Spain (Alicante) 1986 Gedeon | G | 400EC | 5 drip irrigation | 2 | 90 118 153 | | 0.13, 0.17 0.05, 0.06 0.05, 0.05 | 5203-86 [4] |
| Spain (Alicante) 1987 Lamuyo | F | 400EC | 10 drench spray | 1 | 50 | | 0.08, 0.08 | 5217-87 [6] |
| Spain (Semillas Llad) 1987 Hungaro | F | 400EC | 10 spraying/incorp. | 1 | 75 | | 0.05, 0.06 | 5218-87 [7] |
| Italy (Lonigo) 1980 | F | 5GR | 10 | 1 | 84 | | 0.05 | 5205-80 [1] |
| Italy (Lonigo) 1980 | F | 5GR | 10 | 1 | 84 | | 0.05 | 5206-80 [1] |
| Spain (San Javier) 1985 Gedeon | G | 10GR | 10 spreading/incorp. followed by irrigation | 1 | 1 15 29 56 84 | | < 0.05, < 0.05 < 0.05, < 0.05 0.05, 0.1 0.09, 0.1 0.31, 0.35 | 5208-84 [2] |
| Italy (Sabaudia) 1987 Eldor | F | 5GR | 14.4 spreading | 1 | 65 81 | | < 0.02, < 0.02 < 0.02, < 0.02 | 5224-87 [5] |
| Spain (Alicante) 1987 Lamuyo | F | 10GR | 10 in-furrow with slight incorp. | 1 | 63 | | < 0.05, < 0.05 | 5235-87 |
| Spain (Semillas Llad) 1987 Hungaro | F | 10GR | 10 spreading/incorp. | 1 | 75 | | 0.06, 0.06 | 5236-87 |

a. The calculation for the application via the irrigation system was done for the whole plot, but the application itself was only performed in the middle of the plot. Therefore a higher amount of fenamiphos per plant was applied, which could lead to higher residues.

1. Recoveries not indicated. Treatment applied 20 days before planting.

2. Applied at the fruiting stage.

3. Applied at the fruiting stage.

4. 1st at planting and 2nd 14 days after planting.

5. Application 11 days after planting. Irrigation every 8-10 days by sprinkler.

6. Application 13 days after planting (8-leaf stage).

7. Application at sowing.

8. In the 1999 JMPR Evaluation, the value was shown as 0.80. It was corrected to 0.080 in the current Evaluation as the value in the original study report was 0.080 mg/kg.

*Tomato*

Fenamiphos is registered very widely for use on tomato. In Europe the GAP had recently been changed as shown in Table 1 in order to minimize the resulting fenamiphos residues.

A total of seven new trials were conducted with fenamiphos & imidacloprid (CS 246) on tomato in southern Europe in greenhouse: one in France, one in Greece, three in Italy, one in Portugal and one in Spain (Deissler & Ohs, 1999a; Anderson & Elke, 2000a). Fenamiphos and imidacloprid (CS 246) was applied once to the plant at a rate of 10 kg fenamiphos/ha (0.252 kg imidacloprid/ha) via a drip irrigation system. Water rates were 9000–24000 L/ha, depending on the given situation in the greenhouses in which the tests were performed.

The relevant trials were extracted from the 1999 JMPR Evaluation. Among those trials, there are eight trials conducted in Italy, Portugal and Spain (Heinemann & Ohs, 1997b, Blass, 1998b) using fenamiphos CS240 alone with the same application rate and technique as in the new trials in Italy.

Trial data are summarized in Table 5. Except where otherwise described, procedural recovery of analysis in each trial was acceptable.

Table 5. Residues of fenamiphos in tomato from trials in Australia, Brazil, Greece, Italy, Portugal, South Africa and Spain.

| TOMATO Location Year Variety | G or F | Application | | | PHI | Sample | Residues mg/kg | Reference |
|---|---|---|---|---|---|---|---|---|
| | | Form. | kg ai/ha | No | | | | |
| Newly submitted trial data | | | | | | | | |
| Italy (Pozzo Bollente) 1998 Felicie | G | 240CS (246CS) | 10 in irrigation water (more than a month after the 1st flower) | 1 | 60 | Fruit | < 0.02 | RA-2035/98 1181-98 |
| Spain (Arenys de Munt) 1999 Bond | G | 240CS (246CS) | 10 in irrigation water (more than a month after the 1st flower) | 1 | 29 60 70 90 | fruit | 0.12 < 0.02 < 0.02 < 0.02 | RA-2051/99 0373-99 |
| Italy (S. Croce di Camerina) 1999 Felicia | G | 240CS (246CS) | 10 in irrigation water (8 d after the 1st flower) | 1 | 30 60 70 90 | fruit | 0.03 < 0.02 < 0.02 < 0.02 | RA-2051/99 0375-99 |
| Portugal (Santarem (Vale di cavalos)) 1999 Indalo | G | 240CS (246CS) | 10 in irrigation water (14 d after the 1st flower) | 1 | 30 60 71 90 | fruit | 0.07 < 0.02 < 0.02 < 0.02 | RA-2051/99 0376-99 |
| Italy (Zapponeta) 1999 Camone | G | 240CS (246CS) | 10 in irrigation water (before flowering) | 1 | 30 60 70 90 | fruit | < 0.02 < 0.02 < 0.02 < 0.02 | RA-2051/99 0377-99 |
| France (Noves) 1999 Isabella Grefle | G | 240CS (246CS) | 10 in irrigation water (abt 1 month after the 1st flower) | 1 | 30 60 70 90 | fruit | < 0.02 0.04 0.03 < 0.02 | RA-2051/99 0483-99 |
| Greece (Thessaloniki) 1999 9070 | G | 240CS (246CS) | 10 in irrigation water (abt 1 month after the 1st flower) | 1 | 46 60 70 90 | fruit | 0.10 0.22 0.14 0.03 | RA-2059/99 0370-99 |
| Trial data contained in the 1999 JMPR Monograph | | | | | | | | |
| Italy (Latina) 1996 Sidonia | G | 240CS | 10 | 1 | 30 60 90 | | 0.381, 0.368 0.02, 0.02 < 0.02 | 0278-96 [14] |
| Italy (Sicily) 1996 Sidonia | G | 240CS | 10 | 1 | 30 60 90 | | 0.081, 0.070 < 0.02, < 0.02 < 0.02 | 0384-96 [14] |
| Portugal (Lissabon) 1996 Indalo | G | 240CS | 10 | 1 | 50 61 70 | | < 0.02, < 0.02 < 0.02, < 0.02 < 0.02 | 0385-96 [14] |
| Spain (Almeria) 1996 Garbo | G | 240CS | 10 | 1 | 32 60 90 | | 0.066, 0.070 0.092, 0.080 0.042 | 0386-96 [14] |
| Italy (Latina) 1996 Arletta | G | 240CS | 10 | 1 | 30 60 90 | | < 0.02, < 0.02 < 0.02, < 0.02 < 0.02 | 0046-97 [15] |
| Italy (Ragusa) 1997 Cencara | G | 240CS | 10 | 1 | 30 60 90 | | 0.079, 0.081 < 0.02, < 0.02 < 0.02 | 0561-97 [15] |

| TOMATO Location Year Variety | G or F | Application | | | PHI | Sample | Residues mg/kg | Reference |
|---|---|---|---|---|---|---|---|---|
| | | Form. | kg ai/ha | No | | | | |
| Spain (Barcelona) 1997 Alboran | G | 240CS | 9.4 | 1 | 31 60 90 | | 0.205, 0.167 0.142, 0.144 0.032 | 0562-97 [15] |
| Portugal (Lissabon) 1997 Indalo | G | 240CS | 10 | 1 | 30 60 90 | | < 0.02, < 0.02 < 0.02, < 0.02 < 0.02 | 0563-97 [15] |
| Australia (QLD) 1971 Grosse lisse | F | 400EC | 8.9 by boom to soil; rotary hoed in | 1 | 81 | | < 0.05 | 11/71a [2] |
| Australia (QLD) 1971 Grosse lisse | F | 400EC | 11.2 by boom to soil; rotary hoed in | 1 | 81 | | < 0.05 | 11/71b [2] |
| Australia (QLD) 1971 Grosse lisse | F | 400EC | 13.4 by boom to soil; rotary hoed in | 1 | 81 | | < 0.05 | 11/71c [2] |
| Australia (QLD) 1971 Grosse lisse | F | 400EC | 8.7 by boom spray | 1 | 127 161 | | 0.15 < 0.05 | 12/71a [3] |
| Australia (QLD) 1971 Grosse lisse | F | 400EC | 8.7 by boom spray | 1 | 127 161 | | < 0.05 < 0.05 | 12/71b [3] |
| South Africa 1976 | F | 400EC | 10 | 1 | 58 73 88 | | < 0.05 < 0.05 < 0.05 | 311/880/P 163 [6] |
| South Africa 1984 Hibberdene | F | 400EC | 1 g ai/m, 30 cm band | 1 | 86 99 112 | | 0.17 0.14 0.06 | 311/88694 /B40 [7] |
| South Africa 1984 Hibberdene | F | 400EC | 1 g ai/m, 30 cm band + 0.5 g/m 40cm band | 1 + 1 | 35 48 61 | | 0.15 0.07 0.11 | 311/88694 /B40 [7] |
| South Africa 1984 Hibberdene | F | 400EC | 1 g ai/m, 30 cm band + 0.5 g/m 40cm band | 1 + 1 * | 0 7 13 20 33 | | < 0.05 0.10 < 0.05 < 0.05 < 0.05 | 311/88694 /B40 [7] |
| Spain (Alicante) 1984 Restino | G | 400EC | 10 drip irrigation | 1 | 1 15 29 62 | | < 0.05, < 0.05 0.33, 0.41 0.33, 0.37 0.19, 0.27 | 5206-84 [8] |
| Spain (San Javier) 1986 Carmelo | G | 400EC | 5 | 2 | 55 84 112 | | 0.1, 0.13 < 0.01, 0.01 < 0.01, < 0.01 | 5202-86 [9] |
| Spain (Moreno) 1986 A-7 | F | 400EC | 10 | 1 | 60 | | < 0.02, < 0.02 | 0077-88 [10] |
| Spain (El Masnou) 1988 Carmelo | F | 400EC | 10 | 1 | 66 | | < 0.02, < 0.02 | 0078-88 [13] |
| South Africa 1984 Hibberdene | F | 10GR+ 400EC | 0.5 g/m 30cm or 40cm band | 1 + 1 | 35 48 61 | | 0.36 0.25 0.16 | 311/88694 /B40 [7] |
| Australia (QLD) 1971 Grosse lisse | F | 5GR | 11.2 scattered by hand to soil; left on surface | 1 | 78 | | < 0.05 | 11/71d [1] |
| Brazil (Agrocica) 1985 | F | 10GR | 2 spreading | 1 | 124 | | < 0.1 | BRA-78900-85A [4] |
| Brazil (Sao Paulo) 1989 Santa Cruz-Okada | F | 10GR | 5 spreading | 1 | 70 94 | | < 0.1 < 0.1 | BRA-LYPES89-1-A [5] |

| TOMATO Location Year Variety | G or F | Application | | | PHI | Sample | Residues mg/kg | Reference |
|---|---|---|---|---|---|---|---|---|
| | | Form. | kg ai/ha | No | | | | |
| Brazil (Sao Paulo) 1989 Santa Cruz-Okada | F | 10GR | 10 spreading | 1 | 94 | | < 0.1 | BRA-LYPES89-1-B [5] |
| South Africa 1976 | F | 10GR | 10 | 1 | 58 73 88 | | < 0.05 < 0.05 < 0.05 | 311/880/P 163 [6] |
| South Africa 1984 Hibberdene | F | 10GR | 1 g ai/m, 30 cm band | 1 | 86 99 112 | | 0.30 0.14 0.10 | 311/88694 /B40 [7] |
| Spain (St. Boi de Llobregat) 1988 A-7 | F | 10GR | 10 | 1 | 60 | | < 0.02, < 0.02 | 0075-88 [11] |
| Spain (Sr. Jordana) 1988 Carmelo | F | 10GR | 10 | 1 | 66 | | < 0.02, < 0.02 | 0076-88 [12] |
| South Africa 1984 Hibberdene | F | 250EW | 1 g ai/m, 30 cm band | 1 | 86 99 112 | | 0.12 0.09 0.06 | 311/88694 /B40 [7] |

\* Second treatment applied 1 week before harvest.

1. Applied 3 days after transplanting.

2. Applied 1 day before transplanting.

3. Applied 21days before transplanting.

4. Applied at planting. No recovery data given.

5. Applied at transplanting.

6. No field data given.

7. No field data given. All results corrected for recovery.

8. Applied at fruit development stage.

9. Applied 28 and 42 days after planting; last application at fruit development.

10. Applied 31 days after transplanting; flowering stage.

11. Applied 30 days after transplanting; flowering stage.

12. Applied 6 days after transplanting.

13. Applied 6 days after planting

14. Applied 6 to 49 days after planting.

15. Applied 21 to 87 days after planting.

## APPRAISAL

Fenamiphos is a systemic organophosphorus nematicide which is registered for use in more than sixty countries. It was reviewed by the JMPR on several occasions including the periodic review of toxicological data in 1997 and residue data in 1999. The current ADI is set at 0–0.0008 mg/kg bw and the ARfD at 0.003 mg/kg bw. The residue definition was "sum of fenamiphos and its sulfoxide and sulfone, expressed as fenamiphos."

Due to acute intake concerns, the MRL for peppers, tomato and watermelon were returned to Step 6 at the 34th, 35th, 36th, 37th and 38th Sessions of the Codex Committee on Pesticide Residues. The 37th Session of the CCPR in 2005 requested JMPR to review GAPs that might result in lower MRL recommendations and added fenamiphos in the Priority List. The 38th CCPR in 2006 reiterated the request to JMPR to consider alternative GAPs to determine whether lower MRLs for these commodities could be recommended.

The current Meeting received GAP information and new residue trial data on these commodities. In addition, it received new trial data on melons and egg plant. Although there was no acute intake concern associated with the MRL for melon, the data on melons were reviewed because the existing Codex MRL for watermelon was based melons trial data. The data on egg plant was not reviewed as this was outside the purview of the current Meeting.

### Results of supervised residue trials

The Meeting received results of new trials on melons, watermelon, sweet peppers and tomato conducted in 1998–1999. Information from trials on these crops submitted to the 1999 JMPR was also used in the current evaluation. A number of trials with a CS formulation were reported by the 1999 JMPR but not used for estimating maximum residue levels because the GAP had been pending. As the GAP has since been approved, these trials were also included for consideration.

Labels from Australia, Greece, Italy, Portugal and Spain were made available to the Meeting.

The existing Codex MRL for watermelon, as recommended by the 1999 JMPR, was based on residues in melons from the supervised trials in accordance with GAP. Consequently, the Meeting again reviewed the results of supervised trials on melons although no short-term intake concern had previously been identified for melons.

### Melons

The Meeting received the results of six new melon trials conducted in France, Italy and Spain using a CS formulation applied in a glasshouse. Nine indoor trials in Italy with a CS formulation; two field trials from Australia and two field trials from Brazil with an EC formulation; and four field trials from Guatemala, one field trial from Italy and four trials from Mexico with GR formulations were reviewed by the 1999 JMPR.

Table 6. The current relevant GAP information for melons.

| Crop | Country | Form. | Application | | | | | Notes |
|------|---------|-------|-------------|---|---|---|---|-------|
| | | | Timing | Method | Rate, kg ai/ha | No | PHI, days | |
| Melons/ watermelon | Italy | 240CS | From transplanting to about 10 d later | In irrigation water | 10 | 1 | 60 | |
| Melons/ watermelon | Spain | 240CS | Before start of flowering | Soil treatment | 4.8-9.6 | 2 | 60 | a |
| Cucurbits | Australia | 400EC | Pre-planting or pre-transplanting | Soil treatment | 9.6 | 1 | - | |
| Melons | Brazil | 400EC | At sowing | In irrigation water | 4 | 1 | - | |
| Melons | Guatemala | 10GR | | Spreading/incorp. | 2.5-5.0 | 1 | 60 | |
| | | 10GR | | Spreading | 5 | 1 | 60 | |
| | | 15GR | | Spreading/incorp. | 2.6-5.1 | 1 | 60 | |
| Melons | Italy | 5GR | | Soil incorporation | 9.6-12 | 1 | 60 | |

a. First application, just before transplanting or sowing or immediately afterwards; and second application, during the rooting period before the start of flowering.

Six new indoor trials with a CS formulation were conducted in 1998 and 1999 and evaluated against the GAP of the countries where the trials were conducted. In the case of one French trial, Italian GAP was used. Five trials were in accordance with the maximum GAP and residues in fruit in rank order were: < 0.02 (3) and 0.04 (2) mg/kg.

Nine indoor trials were conducted with a CS formulation in Italy in 1996 and 1997. Although no information was available on the timing of application, the application rate, number of application and PHI were in accordance with the GAP of Italy. Residues in fruit at or around the PHI of 60 days were: < 0.02 mg/kg (7), 0.02 and 0.03 mg/kg.

The residues reported in pulp of samples taken 60 days after application were all below the LOQ of 0.02 mg/kg. However, pulp was not analyzed in the two trials that showed the residue level of

0.04 mg/kg in whole fruit taken 60 days after application. Regardless of compliance with GAP, where whole fruit analysis resulted in a residue concentration range of 0.03–0.05 mg/kg, residue levels in pulp were < 0.02– 0.03 mg/kg.

GAP in Argentina for a CS formulation allows one application at a rate of 3.2–4.0 kg ai/ha, with a PHI of 90 days but no trials matched this use pattern.

Two field trials from Australia with an EC formulation were conducted within GAP and residues in fruit were < 0.01 mg/kg (2). One field trial from Brazil with an EC formulation was conducted in accordance with Brazilian GAP and residues in fruit were < 0.02 mg/kg. Even with double rate, residues were still below the LOQ. No data were available for residues in pulp but they were expected to be around the same level as or lower than those in whole fruit.

There is no reported GAP for an EC formulation that would lead to lower residues.

Among trials using GR formulations were, one field trial from Guatemala, conducted in accordance with the current GAP of Guatemala, and four Mexican trials also within the GAP of Guatemala. Residues in fruit were: < 0.01 (4) and < 0.05 mg/kg. Residues in pulp were reported for the Mexican trials and were below the LOQ of 0.01 mg/kg.

Using the GAP of Costa Rica for the GR formulation (0.2–2.5 kg ai/ha, one application, with a 60 day PHI) would exclude trial results from Guatemala.

*Watermelon*

The Meeting also received the results of four new trials conducted on watermelon in Italy and Spain using a CS formulation in a greenhouse. The 1999 JMPR reviewed two field trials conducted in Italy with a GR formulation.

Among four new trials, two trials were in accordance with GAP and residues were: < 0.02 and 0.02 mg/kg.

In two Italian trials with the GR formulation, samples were not taken and analyzed 60 days after the application and therefore not used for estimating an HR.

The data were insufficient for estimating an HR for watermelon.

As the GAP for melons and watermelon is identical in many countries including Australia, Italy and Spain, the Meeting agreed to use data on melons for short-term intake estimation for watermelon as in the 1999 JMPR. As residue concentrations in pulp are not available for all trials following respective GAP, and the residue concentrations in whole fruit and pulp do not differ significantly due to the systemic nature of fenamiphos, the Meeting decided to use the residue concentrations in whole fruit to estimate an HR.

Since the residues arising from the use of these three formulations did not differ significantly, the Meeting concluded that it was not appropriate to disregard the results of trials with certain formulation(s) when estimating an HR. Based on the combined residues (< 0.01 (6), < 0.02 (11), 0.02, 0.03, 0.04 (2) and < 0.05 mg/kg), the Meeting estimated an HR of 0.04 mg/kg and noted that this HR would result in IESTI for general population being 120% and that for children being 310% of the ARfD (0.003 mg/kg bw).

The Meeting concluded that none of the residue data relating to available GAP suggests a lower maximum residue level than the current proposal.

However, as the residues in two new trials were 0.04 mg/kg, the Meeting decided to recommend a new maximum residue level of 0.05 mg/kg for fenamiphos in watermelon to replace the current proposal of 0.05* mg/kg. The Meeting estimated an STMR of 0.02 mg/kg and an HR of 0.04 mg/kg.

The Meeting noted that these recommendations are also valid for melons, except watermelon.

*Pepper, sweet*

The Meeting received the results of 10 new trials conducted from France, Greece, Italy, Portugal and Spain using a CS formulation in glasshouse. Four indoor trials in Italy, one indoor trial in Portugal and three indoor trials in Spain with a CS formulation; two field and two indoor trials in Spain with an EC formulation; and three field trials from Italy and one indoor and two field trials from Spain with a GR formulation were reviewed by the 1999 JMPR.

The relevant current GAP information is as follows:

Table 7. The current relevant GAP information for peppers.

| Crop | Country | Form. | Application | | | | | Notes |
|------|---------|-------|-------------|--------|-------------|-----|------------|-------|
| | | | Timing | Method | Rate, kg ai/ha | No | PHI, days | |
| Peppers, Sweet | Italy | 240CS | From transplanting to about 10 d later | In irrigation water | 10 | 1 | 60 | |
| Peppers, Sweet | Spain | 10GR | Pre-planting or pre-sowing | Spreading/incorp. | 5-10 | 1 | 60 | |
| | | 400EC | | Spraying | 4.8-10 | 2 | 60 | a |
| | | 240CS | Before start of flowering | Soil treatment | 4.8-9.6 | 2 | 60 | b |

a. Up to two applications per season can be given, by dividing the dose in half if the duration of the crop makes it possible to observe the safety period. The first application should be before sowing or transplanting or immediately afterwards and the second during the rooting period of the crop before the start of flowering.

b. First application, just before transplanting or sowing or immediately afterwards; and second application, during the rooting period before the start of flowering.

Ten new trials were evaluated against the GAP of Italy/Spain. Four trials were conducted in compliance with GAP and residues found were: < 0.02 (3) and 0.02 mg/kg. In two other trials in Italy and two trials from Greece, although application took place after the specified timing residues were below the LOQ of 0.02 mg/kg.

Although no information was available on the timing of application, the application rate, number of applications and PHI of trials conducted with the CS formulation in 1996 and 1997 were in accordance with GAP in Italy. Residues at or around the PHI of 60 days were: < 0.02 (4), 0.02, 0.07, 0.08 and 0.11 mg/kg.

GAP in Argentina for a CS formulation allows one application at the rate of 3.2–4.0 kg ai/ha, with a PHI of 90 days but no trials matched this condition.

Three trials from Spain with an EC formulation were conducted in accordance with Spanish GAP and residues found were: 0.06, 0.08 and 0.26 mg/kg.

There is no reported GAP for the EC formulation that would lead to lower residues.

Three trials in Spain with GR formulation were conducted in accordance with GAP in Spain and residues were: < 0.05, 0.06 and 0.35 mg/kg.

The GAP in some countries allowed lower application rates for GR formulations (single application up to 1.2 kg ai/ha), however, no trials matched such GAP.

Since the residues arising from the use of these three formulations were not significantly different the Meeting concluded that it is not appropriate to disregard results from trials with certain formulation(s) when estimating HR. Based on combined residues (< 0.02 (7), 0.02 (2), < 0.05, 0.06 (2), 0.07, 0.08 (2), 0.11, 0.26 and 0.35 mg/kg) the Meeting estimated an HR of 0.35 mg/kg, the same value as that estimated by the 1999 JMPR. The IESTI for the general population is 100% and that for children is 110% of the ARfD.

The Meeting concluded that none of the residue data relating to available GAP suggests a lower maximum residue level to replace the current proposal of 0.5 mg/kg for fenamiphos in peppers.

*Tomato*

The Meeting received the results of seven new trials conducted in France, Greece, Italy, Portugal and Spain using a CS formulation in glasshouses. The 1999 JMPR reviewed four trials from Italy, two trials from Portugal and two trials from Spain with CS a formulation in greenhouse; five field trials from Australia, four field trials from South Africa and two field and two indoor trials from Spain with an EC formulation; one field trial from Australia, four field trials from Brazil, two field trials from South Africa and two field trials from Spain with GR formulations; one field trial from South Africa with both GR and EC formulations; and one field trial from South Africa with an EW formulation.

The current relevant country GAPs are shown below.

Table 8. The current relevant GAP information for tomatoes.

| Crop | Country | Form. | Application | | | | | Notes |
|------|---------|-------|-------------|--|--|--|--|-------|
| | | | Timing | Method | Rate, kg ai/ha | No | PHI, days | |
| Tomato | Australia | 400EC | Pre-planting or pre-transplanting | Spraying | 9.6 | 1 | - | |
| | | 10GR | Within 7 days of planting | Sprinkling | 11 | 1 | - | |
| | | 5GR | Within 7 days of planting | Sprinkling over soil | 13 | 1 | - | Not Tasmania |
| Tomato | Brazil | 10GR | At planting | Incorporation | 3-4 | 1 | 90 | |
| Tomato | Italy | 240CS | From transplanting to about 10 days later | In irrigation water | 10 | 1 | 60 | |
| Tomato | S. Africa | 240CS 400EC 10GR | Pre-planting | Spraying Spraying/incorp. Spreading/incorp. | 9.8 | 1 1 1 | - - - | |
| Tomato | Spain | 10GR | Pre-planting or pre-sowing | Spreading/incorp. | 5-10 | 1 | 60 | |
| | | 400EC | | Spraying | 4.8-10 | 2 | 60 | a |
| | | 240CS | Before start of flowering | Soil treatment | 4.8-9.6 | 2 | 60 | b |

a. Up to two applications per season can be given, by dividing the dose in half if the duration of the crop makes it possible to observe the safety period. The first application should be before sowing or transplanting or immediately afterwards and the second during the rooting period of the crop before the start of flowering.

b. First application, just before transplanting or sowing or immediately afterwards; and second application, during the rooting period before the start of flowering.

Seven new trials were evaluated against GAP in Italy/Spain. One trial was conducted in compliance with GAP and residues were < 0.02 mg/kg. In four other trials from Italy, where application took place later than the specified timing, residues of samples taken 60 days after application were below the LOQ of 0.02 mg/kg.

Although no information was available on the timing of application, application rate, number of applications and PHI, trials with a CS formulation conducted in 1996 and 1997 were in accordance with Italian GAP. Residues at or around the PHI of 60 days were: < 0.02 (5), 0.02, 0.09 and 0.14 mg/kg.

The GAP in Argentina for a CS formulation allows one application at a rate of 3.2–4.0 kg ai/ha, with a PHI of 90 days, however no trials matched this use pattern.

Four trials from Australia, with an EC formulation, were in compliance with Australian GAP and residues found were < 0.05 (3) and 0.15 mg/kg. In four trials from Spain matching Spanish GAP, residues found were < 0.02 (2), 0.13 and 0.27 mg/kg. As no field data was provided fro the trials from South Africa, data from those trials were not used in this evaluation.

The GAP in some countries for the EC formulation allow lower application rates, i.e., single applications at 2.8–3.4 or 3.2–4.0 kg ai/ha. However, no submitted trials matched such a GAP.

One trial from Australia, with a GR formulation, was in compliance with Australian GAP, had residues found of < 0.05 mg/kg. One trial from Brazil, in compliance with Brazilian GAP, had residues found of < 0.1 mg/kg. As no field data was provided for the trials from South Africa, data from those trials were not used in this evaluation.

Two trials from Spain were conducted in accordance with Spanish GAP with residue found of < 0.02 (2) mg/kg.

There is no reported GAP for GR formulations that would lead to lower residues.

Since the residues arising from the use of these three formulations were not significantly different, the Meeting concluded that it was not appropriate to disregard the results of trials with certain formulation(s) when estimating HR. Based on combined residues (< 0.02 (10), 0.02, < 0.05 (4), 0.09, < 0.1, 0.13, 0.14, 0.15 and 0.27 mg/kg), the Meeting estimated an HR of 0.27 mg/kg, similar to that estimated by the 1999 JMPR (0.30 mg/kg). The IESTI for general population is 100% and that for children is 280% of the ARfD.

The Meeting concluded that none of the residue data relating to available GAPs suggests a lower maximum residue level to replace the current proposal of 0.5 mg/kg for fenamiphos on tomato.

## RECOMMENDATIONS

On the basis of the data from supervised trials on melons, except watermelon, and on watermelon, the Meeting concluded that the residue concentrations below are suitable for establishing MRLs and for assessing dietary intakes.

Definition of the residue: *Sum of fenamiphos and its sulfoxide and sulfone, expressed as fenamiphos*

| Commodity | | Recommended MRL mg/kg | | STMR/ STMR-P mg/kg | HR/HR-P mg/kg |
|---|---|---|---|---|---|
| CCN | Name | New | Previous | | |
| VC 0046 | Melons, except watermelon | 0.05 | 0.05* | 0.02 | 0.04 |
| VC 0432 | Watermelon | 0.05 | 0.05* | 0.02 | 0.04 |

The Meeting made no recommendations to amend the MRLs for peppers and tomato.

## DIETARY RISK ASSESSMENT

### Long-term intake

As the ADI had not been modified since the International Estimated Dietary Intakes of fenamiphos were calculated last time in 1999 (3–14% of the maximum ADI of 0.0008 kg/kg for five regional diets); and the STMRs estimated by the current Meeting for melons, except watermelon and watermelon were identical to those estimated in 1999, IEDI calculation was not conducted by the current Meeting.

### Short-term intake

The International Estimated Short-Term Intakes (IESTIs) of fenamiphos by general population and by children were calculated for melons, except watermelon; and watermelon for which HRs were estimated. The ARfD is 0.003 mg/kg and the calculated IESTIs for children up to 6 years of age range from 90 to 310% and those for general population from 40 to 120% of the ARfD. The information provided to the JMPR precludes an estimate that the short-term dietary intake would be below the ARfD for consumption of watermelon by children and by the general population.

# REFERENCES

**Author, Date, Title, Institute, Report Reference, Document No.**

Anderson, C.; Elke, K. 2000. Determination of Residues of Imidacloprid and Fenamiphos on Melon and Watermelon after Application of NTN 33893 & SRA 3886 246 CS in the Greenhouse in Italy, Spain and France, Bayer AG, report no.: RA-2058/99, Date: 2000-07-20, (Amendment no. 1,Date: 2000-08-15, Amendment no. 2, Date: 2000-10-05), unpublished RA-2058/99

Anderson, C.; Elke, K. 2000a. Determination of Residues of Imidacloprid and Fenamiphos after Application of NTN 33893 & SRA 3886 246 CS on Aubergine and Tomato in the Greenhouse in Italy, Spain, Portugal, France and Greece, Bayer AG, report no.: RA-2051/99, Date: 2000-07-17, (including Amendment no, 1, date: 2000-08-15 and Amendment no. 2, date: 2000-08- 23), unpublished RA-2051/99

Anderson, C.; Elke, K. 2000c. Determination of Residues of Imidacloprid and Fenamiphos after Application of NTN 33893 & SRA 3886 246 CS on Pepper in the Greenhouse in Italy, Spain, Portugal, France and Greece, Bayer AG, report no.: RA-2059/99, Date: 2000-07-28 (including Amendment no, 1,Date: 2000-08-23 and Amendment no. 2, Date: 2000-10-05), unpublished RA-2059/99

Deissler, A. Ohs, P. 1999. Determination of residues of NTN 33893 & SRA 3886 246 CS in/on melon and watermelon in the greenhouse in Italy, Bayer AG, report nos. RA-2033/98, M-016124-01-1, Date: 1999-06-30, unpublished RA-2033/98

Deissler, A. Ohs, P. 1999 Determination of residues of NTN 33893 & SRA 3886 246 CS in/on tomato and aubergine in the greenhouse in Italy, Bayer AG, report no. RA-2035/98, Date: 1999-06-23, unpublished RA-2035/98

Deissler, A. Ohs, P. 1999b. Determination of residues of NTN 33893 & SRA 3886 246 CS in/on pepper in the greenhouse in Italy, Bayer AG, report no. RA-2034/98, Date: 1999-06-24, unpublished RA-2034/98

FAO & WHO 2000. Pesticide residues in food – 1999, Evaluation 1999 Part 1-Residues (FAO Plant Production and Protection Paper 157), pp.235-372

# FAO TECHNICAL PAPERS

## FAO PLANT PRODUCTION AND PROTECTION PAPERS

1      Horticulture: a select bibliography, 1976 (E)
2      Cotton specialists and research institutions in selected countries, 1976 (E)
3      Food legumes: distribution, adaptability and biology of yield, 1977 (E F S)
4      Soybean production in the tropics, 1977 (C E F S)
4 Rev.1 Soybean production in the tropics (first revision), 1982 (E)
5      Les systèmes pastoraux sahéliens, 1977 (F)
6      Pest resistance to pesticides and crop loss assessment – Vol. 1, 1977 (E F S)
6/2    Pest resistance to pesticides and crop loss assessment – Vol. 2, 1979 (E F S)
6/3    Pest resistance to pesticides and crop loss assessment – Vol. 3, 1981 (E F S)
7      Rodent pest biology and control – Bibliography 1970-74, 1977 (E)
8      Tropical pasture seed production, 1979 (E F** S**)
9      Food legume crops: improvement and production, 1977 (E)
10     Pesticide residues in food, 1977 – Report, 1978 (E F S)
10 Rev. Pesticide residues in food 1977 – Report, 1978 (E)
10 Sup. Pesticide residues in food 1977 – Evaluations, 1978 (E)
11     Pesticide residues in food 1965-78 – Index and summary, 1978 (E F S)
12     Crop calendars, 1978 (E/F/S)
13     The use of FAO specifications for plant protection products, 1979 (E F S)
14     Guidelines for integrated control of rice insect pests, 1979 (Ar C E F S)
15     Pesticide residues in food 1978 – Report, 1979 (E F S)
15 Sup. Pesticide residues in food 1978 – Evaluations, 1979 (E)
16     Rodenticides: analyses, specifications, formulations, 1979 (E F S)
17     Agrometeorological crop monitoring and forecasting, 1979 (C E F S)
18     Guidelines for integrated control of maize pests, 1979 (C E)
19     Elements of integrated control of sorghum pests, 1979 (E F S)
20     Pesticide residues in food 1979 – Report, 1980 (E F S)
20 Sup. Pesticide residues in food 1979 – Evaluations, 1980 (E)
21     Recommended methods for measurement of pest resistance to pesticides, 1980 (E F)
22     China: multiple cropping and related crop production technology, 1980 (E)
23     China: development of olive production, 1980 (E)
24/1   Improvement and production of maize, sorghum and millet – Vol. 1. General principles, 1980 (E F)
24/2   Improvement and production of maize, sorghum and millet – Vol. 2. Breeding, agronomy and seed production, 1980 (E F)
25     Prosopis tamarugo: fodder tree for arid zones, 1981 (E F S)

26     Pesticide residues in food 1980 – Report, 1981 (E F S)
26 Sup. Pesticide residues in food 1980 – Evaluations, 1981 (E)
27     Small-scale cash crop farming in South Asia, 1981 (E)
28     Second expert consultation on environmental criteria for registration of pesticides, 1981 (E F S)
29     Sesame: status and improvement, 1981 (E)
30     Palm tissue culture, 1981 (C E)
31     An eco-climatic classification of intertropical Africa, 1981 (E)
32     Weeds in tropical crops: selected abstracts, 1981 (E)
32 Sup.1 Weeds in tropical crops: review of abstracts, 1982 (E)
33     Plant collecting and herbarium development, 1981 (E)
34     Improvement of nutritional quality of food crops, 1981 (C E)
35     Date production and protection, 1982 (Ar E)
36     El cultivo y la utilización del tarwi – Lupinus mutabilis Sweet, 1982 (S)
37     Pesticide residues in food 1981 – Report, 1982 (E F S)
38     Winged bean production in the tropics, 1982 (E)
39     Seeds, 1982 (E/F/S)
40     Rodent control in agriculture, 1982 (Ar C E F S)
41     Rice development and rainfed rice production, 1982 (E)
42     Pesticide residues in food 1981 – Evaluations, 1982 (E)
43     Manual on mushroom cultivation, 1983 (E F)
44     Improving weed management, 1984 (E F S)
45     Pocket computers in agrometeorology, 1983 (E)
46     Pesticide residues in food 1982 – Report, 1983 (E F S)
47     The sago palm, 1983 (E F)
48     Guidelines for integrated control of cotton pests, 1983 (Ar E F S)
49     Pesticide residues in food 1982 – Evaluations, 1983 (E)
50     International plant quarantine treatment manual, 1983 (C E)
51     Handbook on jute, 1983 (E)
52     The palmyrah palm: potential and perspectives, 1983 (E)
53/1   Selected medicinal plants, 1983 (E)
54     Manual of fumigation for insect control, 1984 (C E F S)
55     Breeding for durable disease and pest resistance, 1984 (C E)
56     Pesticide residues in food 1983 – Report, 1984 (E F S)
57     Coconut, tree of life, 1984 (E S)
58     Economic guidelines for crop pest control, 1984 (E F S)
59     Micropropagation of selected rootcrops, palms, citrus and ornamental species, 1984 (E)
60     Minimum requirements for receiving and maintaining tissue culture propagating material, 1985 (E F S)
61     Pesticide residues in food 1983 – Evaluations, 1985 (E)

62 Pesticide residues in food 1984 – Report, 1985 (E F S)

63 Manual of pest control for food security reserve grain stocks, 1985 (C E)

64 Contribution à l'écologie des aphides africains, 1985 (F)

65 Amélioration de la culture irriguée du riz des petits fermiers, 1985 (F)

66 Sesame and safflower: status and potentials, 1985 (E)

67 Pesticide residues in food 1984 – Evaluations, 1985 (E)

68 Pesticide residus in food 1985 – Report, 1986 (E F S)

69 Breeding for horizontal resistance to wheat diseases, 1986 (E)

70 Breeding for durable resistance in perennial crops, 1986 (E)

71 Technical guideline on seed potato micropropagation and multiplication, 1986 (E)

72/1 Pesticide residues in food 1985 – Evaluations – Part I: Residues, 1986 (E)

72/2 Pesticide residues in food 1985 – Evaluations – Part II: Toxicology, 1986 (E)

73 Early agrometeorological crop yield assessment, 1986 (E F S)

74 Ecology and control of perennial weeds in Latin America, 1986 (E S)

75 Technical guidelines for field variety trials, 1993 (E F S)

76 Guidelines for seed exchange and plant introduction in tropical crops, 1986 (E)

77 Pesticide residues in food 1986 – Report, 1986 (E F S)

78 Pesticide residues in food 1986 – Evaluations – Part I: Residues, 1986 (E)

78/2 Pesticide residues in food 1986 – Evaluations – Part II: Toxicology, 1987 (E)

79 Tissue culture of selected tropical fruit plants, 1987 (E)

80 Improved weed management in the Near East, 1987 (E)

81 Weed science and weed control in Southeast Asia, 1987 (E)

82 Hybrid seed production of selected cereal, oil and vegetable crops, 1987 (E)

83 Litchi cultivation, 1989 (E S)

84 Pesticide residues in food 1987 – Report, 1987 (E F S)

85 Manual on the development and use of FAO specifications for plant protection products, 1987 (E** F S)

86/1 Pesticide residues in food 1987 – Evaluations – Part I: Residues, 1988 (E)

86/2 Pesticide residues in food 1987 – Evaluations – Part II: Toxicology, 1988 (E)

87 Root and tuber crops, plantains and bananas in developing countries – challenges and opportunities, 1988 (E)

88 Jessenia and Oenocarpus: neotropical oil palms worthy of domestication, 1988 (E S)

89 Vegetable production under arid and semi-arid conditions in tropical Africa, 1988 (E F)

90 Protected cultivation in the Mediterranean climate, 1990 (E F S)

91 Pastures and cattle under coconuts, 1988 (E S)

92 Pesticide residues in food 1988 – Report, 1988 (E F S)

93/1 Pesticide residues in food 1988 – Evaluations – Part I: Residues, 1988 (E)

93/2 Pesticide residues in food 1988 – Evaluations – Part II: Toxicology, 1989 (E)

94 Utilization of genetic resources: suitable approaches, agronomical evaluation and use, 1989 (E)

95 Rodent pests and their control in the Near East, 1989 (E)

96 Striga – Improved management in Africa, 1989 (E)

97/1 Fodders for the Near East: alfalfa, 1989 (Ar E)

97/2 Fodders for the Near East: annual medic pastures, 1989 (Ar E F)

98 An annotated bibliography on rodent research in Latin America 1960-1985, 1989 (E)

99 Pesticide residues in food 1989 – Report, 1989 (E F S)

100 Pesticide residues in food 1989 – Evaluations – Part I: Residues, 1990 (E)

100/2 Pesticide residues in food 1989 – Evaluations – Part II: Toxicology, 1990 (E)

101 Soilless culture for horticultural crop production, 1990 (E)

102 Pesticide residues in food 1990 – Report, 1990 (E F S)

103/1 Pesticide residues in food 1990 – Evaluations – Part I: Residues, 1990 (E)

104 Major weeds of the Near East, 1991 (E)

105 Fundamentos teórico-prácticos del cultivo de tejidos vegetales, 1990 (S)

106 Technical guidelines for mushroom growing in the tropics, 1990 (E)

107 Gynandropsis gynandra (L.) Briq. – a tropical leafy vegetable – its cultivation and utilization, 1991 (E)

108 Carambola cultivation, 1993 (E S)

109 Soil solarization, 1991 (E)

110 Potato production and consumption in developing countries, 1991 (E)

111 Pesticide residues in food 1991 – Report, 1991 (E)

112 Cocoa pest and disease management in Southeast Asia and Australasia, 1992 (E)

113/1 Pesticide residues in food 1991 – Evaluations – Part I: Residues, 1991 (E)

114 Integrated pest management for protected vegetable cultivation in the Near East, 1992 (E)

115 Olive pests and their control in the Near East, 1992 (E)

116 Pesticide residues in food 1992 – Report, 1993 (E F S)

117 Quality declared seed, 1993 (E F S)

118 Pesticide residues in food 1992 – Evaluations – Part I: Residues, 1993 (E)

119 Quarantine for seed, 1993 (E)

120 Weed management for developing countries, 1993 (E S)

120/1 Weed management for developing countries, Addendum 1, 2004 (E F S)

121 Rambutan cultivation, 1993 (E)

122 Pesticide residues in food 1993 – Report, 1993 (E F S)

123 Rodent pest management in eastern Africa, 1994 (E)

124 Pesticide residues in food 1993 – Evaluations – Part I: Residues, 1994 (E)

125 Plant quarantine: theory and practice, 1994 (Ar)

126 Tropical root and tuber crops – Production, perspectives and future prospects, 1994 (E)

127 Pesticide residues in food 1994 – Report, 1994 (E)

128 Manual on the development and use of FAO specifications for plant protection products – Fourth edition, 1995 (E F S)

129 Mangosteen cultivation, 1995 (E)

130 Post-harvest deterioration of cassava – A biotechnology perspective, 1995 (E)

131/1 Pesticide residues in food 1994 – Evaluations – Part I: Residues, Volume 1, 1995 (E)

131/2 Pesticide residues in food 1994 – Evaluations – Part I: Residues, Volume 2, 1995 (E)

132 Agro-ecology, cultivation and uses of cactus pear, 1995 (E)

133 Pesticide residues in food 1995 – Report, 1996 (E)

134 (Number not assigned)

135 Citrus pest problems and their control in the Near East, 1996 (E)

136 El pepino dulce y su cultivo, 1996 (S)

137 Pesticide residues in food 1995 – Evaluations – Part I: Residues, 1996 (E)

138 Sunn pests and their control in the Near East, 1996 (E)

139 Weed management in rice, 1996 (E)

140 Pesticide residues in food 1996 – Report, 1997 (E)

141 Cotton pests and their control in the Near East, 1997 (E)

142 Pesticide residues in food 1996 – Evaluations – Part I Residues, 1997 (E)

143 Management of the whitefly-virus complex, 1997 (E)

144 Plant nematode problems and their control in the Near East region, 1997 (E)

145 Pesticide residues in food 1997 – Report, 1998 (E)

146 Pesticide residues in food 1997 – Evaluations – Part I: Residues, 1998 (E)

147 Soil solarization and integrated management of soilborne pests, 1998 (E)

148 Pesticide residues in food 1998 – Report, 1999 (E)

149 Manual on the development and use of FAO specifications for plant protection products – Fifth edition, including the new procedure, 1999 (E)

150 Restoring farmers' seed systems in disaster situations, 1999 (E)

151 Seed policy and programmes for sub-Saharan Africa, 1999 (E F)

152/1 Pesticide residues in food 1998 – Evaluations – Part I: Residues, Volume 1, 1999 (E)

152/2 Pesticide residues in food 1998 – Evaluations – Part I: Residues, Volume 2, 1999 (E)

153 Pesticide residues in food 1999 – Report, 1999 (E)

154 Greenhouses and shelter structures for tropical regions, 1999 (E)

155 Vegetable seedling production manual, 1999 (E)

156 Date palm cultivation, 1999 (E)

156 Rev.1 Date palm cultivation, 2002 (E)

157 Pesticide residues in food 1999 – Evaluations – Part I: Residues, 2000 (E)

158 Ornamental plant propagation in the tropics, 2000 (E)

159 Seed policy and programmes in the Near East and North Africa, 2000

160 Seed policy and programmes for Asia and the Pacific, 2000 (E)

161 Silage making in the tropics with particular emphasis on smallholders, 2000 (E S)

162 Grassland resource assessment for pastoral systems, 2001, (E)

163 Pesticide residues in food 2000 – Report, 2001 (E)

164 Seed policy and programmes in Latin America and the Caribbean, 2001 (E S)

165 Pesticide residues in food 2000 – Evaluations – Part I, 2001 (E)

166 Global report on validated alternatives to the use of methyl bromide for soil fumigation, 2001 (E)

167 Pesticide residues in food 2001 – Report, 2001 (E)

168 Seed policy and programmes for the Central and Eastern European countries, Commonwealth of Independent States and other countries in transition, 2001 (E)

169 Cactus (Opuntia spp.) as forage, 2003 (E S)

170 Submission and evaluation of pesticide residues data for the estimation of maximum residue levels in food and feed, 2002 (E)

171 Pesticide residues in food 2001 – Evaluations – Part I, 2002 (E)

172 Pesticide residues in food, 2002 – Report, 2002 (E)

173 Manual on development and use of FAO and WHO specifications for pesticides, 2002 (E S)

174 Genotype x environment interaction – Challenges and opportunities for plant breeding and cultivar recommendations, 2002 (E)

175/1 Pesticide residues in food 2002 – Evaluations – Part 1: Residues – Volume 1 (E)

175/2 Pesticide residues in food 2002 – Evaluations – Part 1: Residues – Volume 2 (E)

176 Pesticide residues in food 2003 – Report, 2004 (E)

177 Pesticide residues in food 2003 – Evaluations – Part 1: Residues, 2004 (E)

178 Pesticide residues in food 2004 – Report, 2004 (E)

179 Triticale improvement and production, 2004 (E)

180 Seed multiplication by resource-limited farmers - Proceedings of the Latin American workshop, 2004 (E)

181 Towards effective and sustainable seed-relief activities, 2004 (E)

182/1 Pesticide residues in food 2004 – Evaluations – Part 1: Residues, Volume 1 (E)

182/2 Pesticide residues in food 2004 – Evaluations – Part 1: Residues, Volume 2 (E)

183 Pesticide residues in food 2005 – Report, 2005 (E)

184/1 Pesticide residues in food 2005 – Evaluations – Part 1: Residues, Volume 1 (E)

184/2 Pesticide residues in food 2005 – Evaluations – Part 1: Residues, Volume 2 (E)

185 Quality declared seed system, 2006 (E F S)

186 Calendario de cultivos – América Latina y el Caribe, 2006 (S)

187 Pesticide residues in food 2006 – Report, 2006 (E)

188 Weedy rices – origin, biology, ecology and control, 2006 (E S)\

189/1 Pesticide residues in food 2006 – Evaluations – Part 1: Residues, Volume 1 (E)

189/2 Pesticide residues in food 2006 – Evaluations – Part 1: Residues, Volume 2 (E)

Availability: May 2007

| | | | |
|---|---|---|---|
| Ar | – Arabic | Multil | – Multilingual |
| C | – Chinese | * | Out of print |
| E | – English | ** | In preparation |
| F | – French | | |
| P | – Portuguese | | |
| S | – Spanish | | |

*The FAO Technical Papers are available through the authorized FAO Sales Agents or directly from Sales and Marketing Group, FAO, Viale delle Terme di Caracalla, 00153 Rome, Italy.*

| | | | |
|---|---|---|---|
| Ar | – Arabic | Multil | – Multilingual |
| C | – Chinese | * | Out of print |
| E | – English | ** | In preparation |
| F | – French | | |
| P | – Portuguese | | |
| S | – Spanish | | |